ENVIRONMENTAL SCIENCE

FOR A CHANGING WORLD

CANADIAN EDITION

ENVIRONMENTAL SCIENCE

FOR A CHANGING WORLD

CANADIAN EDITION

MARNIE BRANFIREUN
Western University

SUSAN KARR
Carson-Newman University

JENEEN INTERLANDI
Science Writer

ANNE HOUTMAN
Rochester Institute of Technology

With contributions by Alison McCook and Karen Ing

W. H. FREEMAN   SCIENTIFIC AMERICAN

Publisher: Susan Winslow
Senior Acquisitions Editor: Jerry Correa
Developmental Editor: Andrea Gawrylewski
Project Manager: Karen Misler
Associate Director of Marketing: Debbie Clare
Media and Supplements Editor: Betsye Mullaney
Editorial Assistant: Jane Taylor
Art Direction, Cover, Design, and Illustrations: MGMT. design
Photo Editor and Researcher: Stephanie Heimann
Copy-editing: First Folio Resource Group Inc.
Art Manager: Matthew McAdams
Director of Production: Ellen Cash
Printing and Binding: RR Donnelley

ISBN-13: 978-1-4641-5420-1
ISBN-10: 1-4641-5420-1

Printed in the United States of America

First printing

W. H. Freeman and Company
41 Madison Avenue
New York, NY 10010
Houndmills, Basingstoke RG21 6XS, England
www.whfreeman.com

BRIEF CONTENTS

UNIT 1
FOUNDATIONS AND TOOLS OF THE TRADE

UNIT 2
ECOLOGY, PATTERNS, AND PROCESSES

UNIT 3
EARTH'S RESOURCES, CURRENT CHALLENGES, AND SUSTAINABLE OPTIONS

DETAILED CONTENTS

UNIT 1
FOUNDATIONS AND TOOLS OF THE TRADE

UNIT 3
EARTH'S RESOURCES, CURRENT CHALLENGES, AND SUSTAINABLE OPTIONS

UNIT 4
ENERGY: A WICKED PROBLEM WITH MANY CONSEQUENCES

CARSON

UNIT 5
TOWARD A SUSTAINABLE FUTURE

AUTHORS

MARNIE BRANFIREUN, MSc, is an Instructor in the Centre for Environment and Sustainability at Western University in London, Ontario. She teaches in both the Undergraduate Environmental Science Program and the Masters in Environment and Sustainability Program. Her BSc (University of Manitoba) focused on aquatic ecology, during which she worked for the Department of Fisheries and Oceans monitoring benthic organisms in Manitoba rivers and at the Experimental Lakes Area. Her MSc (McGill University) was on mercury cycling and plant decomposition in boreal peatlands. As an ecologist, she has been engaged in environmental education, conservation, restoration, and monitoring for over 20 years.

SUSAN KARR, MS, is an Instructor in the biology department of Carson-Newman University in Jefferson City, Tennessee, and has been teaching for more than 15 years. She has served on campus and community environmental sustainability groups and helps produce an annual "State of the Environment" report on the environmental health of her county. In addition to teaching non-majors courses in environmental science and human biology, she teaches an upper-level course in animal behaviour where she and her students train dogs from the local animal shelter in a program that improves the animals' chances of adoption. She received degrees in animal behaviour and forestry from the University of Georgia.

JENEEN INTERLANDI, MA, MS, is a science writer who contributes to *Scientific American* and *The New York Times Magazine*. Previously, she spent four years as a staff writer for *Newsweek*, where she covered health, science, and the environment. In 2009, she received a Kaiser Foundation fellowship for global health reporting and travelled to Europe and Asia to cover outbreaks of drug-resistant tuberculosis. Jeneen has worked as a researcher at both Harvard Medical School and Lamont Doherty Earth Observatory. She was a 2013 Nieman Fellow. Jeneen holds Master's degrees in environmental science and journalism, both from Columbia University in New York.

ANNE HOUTMAN, DPhil, is Dean of the School of Natural Sciences, Mathematics and Engineering, and Professor of Biological Sciences at California State University Bakersfield. Her research interests are in the behavioural ecology of birds. She is strongly committed to evidence-based, experiential education and has been an active participant in the national dialogue on science education—how best to teach science to future scientists and future science "consumers"—for almost 20 years. Anne received her doctorate in zoology from the University of Oxford and conducted postdoctoral research at the University of Toronto.

Dear Reader,

Having worked in environmental education for over 20 years for both universities and public organizations, I am driven to help students understand the environmental issues that face us all. Despite the seriousness of such problems as climate change, engaging the minds and imagination of students has become an increasingly daunting task in the age of instant digital and social media. This challenge is even greater in introductory courses and courses for non-majors, where science may not be the students' primary focus.

This textbook takes a unique approach that sets core environmental science content within storylines that capture student imagination. These stories have drama, suspense, and a human dimension that gives the science a real-life context, making it easier to relate to, understand, and remember. Vivid, high-quality infographics provide detail and clarity without slowing down the story. This makes environmental science more relevant and meaningful to students, translating into a greater respect for the natural world, and for science. Students not only build knowledge, but also a greater willingness to act on that knowledge. I am very excited to bring the innovative *Environmental Science for a Changing World* to a Canadian audience.

This text has been organized to facilitate use of content to suit the particular priorities of the instructor or course, by providing flexibility to focus on sub-topics. For example there are entire chapters on several energy sources, rather than one on energy in general. The text also focuses on building core competencies for the non-major: environmental literacy, science literacy, and information literacy. End-of-chapter and online exercises provide further opportunities to develop these competencies, as well as critical thinking skills.

Environmental Literacy: The scientific, social, political, and economic facets of environmental issues are examined, with the focus on the scientific concepts and drivers underlying the issues. Material is presented in a balanced way, particularly for controversial topics. Sustainable solutions are presented.

Science Literacy: Each chapter includes experimental evidence and graphical data representation, and describes the day-to-day work of scientists, giving students many opportunities to evaluate evidence and understand the process of science.

Information Literacy: Students must be able to both find information and assess its quality. We explain how to effectively search for and find scientific information, and how to critically analyze that information.

Environmental Science for a Changing World conveys the excitement of a rapidly changing, critical area of study that is relevant to everyone. I am fortunate to be part of a terrific, dynamic production team that I thank for developing the strong, engaging content upon which this Canadian edition was based. Each contributor shares my desire to promote environmental citizenship amongst students by fostering their ability to understand, evaluate, and articulate environmental issues. My most sincere hope is that these students find this to be a transformative experience that allows them to create positive change in the world.

Sincerely,

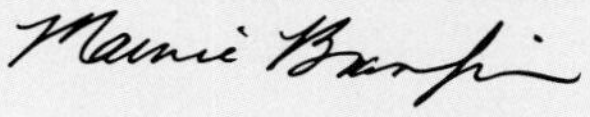

Marnie Branfireun

ENGAGING STORIES YOUR STUDENTS WILL READ

ENGAGING STORIES ILLUMINATE PRESSING ENVIRONMENTAL ISSUES

SCIENCE IS TAUGHT IN CONTEXT

It was his mother's death in the fall of 1984, from a particularly aggressive form of breast cancer, that drove Paul Cox back to the Samoan rainforest. Cox had first visited the South Pacific island in 1973, through an undergraduate research program with Brigham Young University, where he was majoring in botany. Since then, the Utah native had earned a Ph.D. from Harvard and made a career studying plant physiology in the United States.

↑ Paul Cox and family in Samoa, 1986

◉WHERE IS SAMOA?

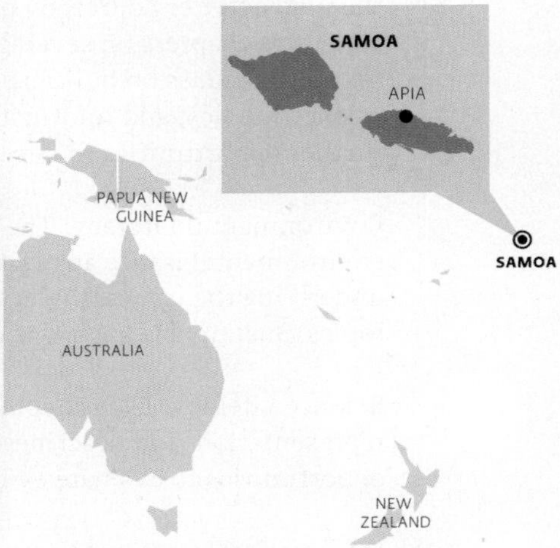
SAMOA
APIA
PAPUA NEW GUINEA
SAMOA
AUSTRALIA
NEW ZEALAND

Why did Cox suspect that Samoa might hold the key to a cancer cure? Tropical regions like Samoa—warm, lush, close to the equator—contain the greatest concentration and variety of plant and animal life forms on Earth. This variety is called **biodiversity**. Countries of this region

Cancer Institute, 'If there is even a 1% chance of finding something, it's worth taking a look,'" he recalls. "They said 'We think there is like a 3% chance.' So I went." Six months after his mother's funeral, with his wife and four young children in tow, Cox returned to Samoa.

Biodiversity benefits humans and other species.

Why did Cox suspect that Samoa might hold the key to a cancer cure? Tropical regions like Samoa—warm, lush, close to the equator—contain the greatest concentration and variety of plant and animal life forms on Earth. This variety is called **biodiversity**. Countries of this region have both high **species diversity** and high **genetic diversity**. They also usually have high **ecological diversity**, a wide variety of communities and ecosystems with

---

biodiversity The variety of life on Earth; it includes species, genetic, and ecological diversity.
species diversity The variety of species, including how many are present (richness) and their abundance relative to each other (evenness).
genetic diversity The heritable variation among individuals of a single population or within the species as a whole.
ecological diversity The variety within an ecosystem's structure, including many communities, habitats, niches, and trophic levels.

SHORT CHAPTERS, TARGETED LEARNING

CHAPTERS ARE 16-20 PAGES LONG

↑ Sea bream fish farm, Shikoku, Japan

grow freshwater fish like tilapia. Canada, too, is trying its hand at aquaculture, which is projected to grow over time. Already, it's become the fourth largest producer of farmed salmon. Sadly, cod aquaculture has been less successful, as they are difficult to raise and easily substituted by other forms of white-fleshed fish, creating less of a market. But most fish farms are family or village operations—just big enough to provide a few dozen people with a steady food supply. Zohar wanted a fish farm that could feed a modern city, or even a country. He also wanted one that could raise large marine species. "With marine aquaculture, we are talking about fish that one, we are running out of, and two, are the most beneficial to human health," he says.

But there was a huge barrier to realizing this vision: most commercially important marine fish—like sea bream, sea bass, and tuna—would not reproduce in captivity. Some scientists had suggested recreating spawning grounds—specific areas where fish come every year to deposit their eggs and sperm—in captive breeding sites. But that seemed wildly impractical to Zohar. "Most of these fish travel hundreds to thousands of miles to reach

their spawning grounds," he says. "They move through a wide range of temperatures, water depths, and salinities, and nobody had any clue which of these variables was the key to getting them to reproduce, let alone how to recreate all of that in a finite space." So instead he tracked several species across the ocean and measured their hormonal

ocean fish. Before long, elaborate net pen operations began dotting coastlines around the world. This simple technology has enabled fish farms to out-produce traditional fishers. In 2009, aquaculture crossed the threshold of providing more than half of all seafood consumed worldwide, with Asia leading the way, and China alone producing some 63% of all farmed marine species. But, it has also led to a range of environmental problems including depletion of the populations of smaller fish harvested as food for aquaculture species, excess nutrient release into coastal waters, and ecosystem damage from the aquaculture

total allowable catch (TAC) The maximum amount (weight or numbers of fish or shellfish) of a particular species that can be harvested per year or fishing season in a given area; meant to prevent overfishing.
sustainable fishery A fishery that ensures that fish stocks are maintained at healthy levels, the ecosystem is fully functional, and fishing activity does not threaten biological diversity.
aquaculture Fish farming; the rearing of aquatic species in tanks, ponds, or ocean net pens.

total allowable catch (TAC) The maximum amount (weight or numbers of fish or shellfish) of a particular species that can be harvested per year or fishing season in a given area; meant to prevent overfishing.
sustainable fishery A fishery that ensures that fish stocks are maintained at healthy levels, the ecosystem is fully functional, and fishing activity does not threaten biological diversity.
aquaculture Fish farming; the rearing of aquatic species in tanks, ponds, or ocean net pens.

A RUNNING MARGIN GLOSSARY MEANS STUDENTS ENCOUNTER IMPORTANT TERMS IN THE CONTEXT OF THE STORY

SKILLS STUDENTS WILL USE IN THE FUTURE

SCIENCE LITERACY: UNDERSTANDING ENVIRONMENTAL ISSUES AND DATA-BASED PROBLEMS BUILD CRITICAL THINKING AND QUANTITATIVE SKILLS

ANALYZING THE SCIENCE

The following graph shows the costs of extracting oil in different parts of the world.

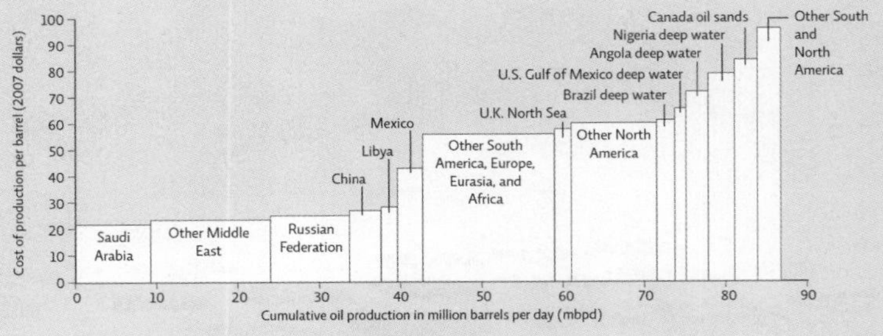

Cost of production per barrel (2007 dollars)
0
10
20
30
40
50
60
70
80
90
100
Saudi Arabia
Other Middle East
Russian Federation
China
Libya
Mexico
Other South America, Europe, Eurasia, and Africa
U.K. North Sea
Other North America
Brazil deep water
U.S. Gulf of Mexico deep water
Angola deep water
Nigeria deep water
Canada oil sands
Other South and North America
Cumulative oil production in million barrels per day (mbpd)

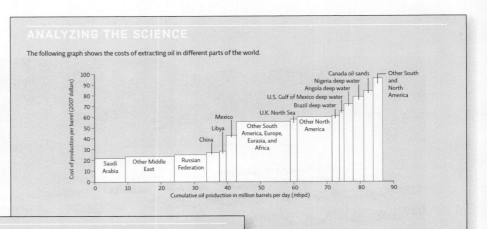

EVALUATING NEW INFORMATION

Oil from the Canadian oil sands is transported via pipelines. The proposed Northern Gateway pipeline would carry oil from Alberta, through British Columbia to the port city of Kitimat for access to Asian markets. This route takes the pipeline through rural and urban areas as well as First Nations' lands and faces opposition by those concerned about environmental risks of the pipeline and those opposed to extracting oil sands oil in general. Proponents point to the economic benefits that come with the pipeline.

Go to the following websites and read the articles about this issue: www.pipeupagainstenbridge.ca/ and www.northerngateway.ca/.

Evaluate the websites and work with the information to answer the following questions:

1. Are these reliable information sources? Do they have a clear and transparent agenda?
   a. Who runs these websites? Do the organizations' credentials make them reliable or unreliable? Explain.
   b. Do you suspect any bias on the part of these organizations? What might that bias be and how could you verify their major claims?
2. Briefly summarize what you believe are the most important pros and cons of the Northern Gateway pipeline. Based on your reading, do you support this pipeline project? Justify your answer.
3. What else would you like to know about this project? Do an Internet search to find more information. List at least two more sources, including their urls. In addition, answer Questions 1a and 1b for each source. Based on the new information you have gathered, has your opinion about the project changed? Why or why not?

INFORMATION LITERACY: WEB-BASED PROBLEMS HELP STUDENTS DECIPHER NEW INFORMATION

CRITICAL THINKING: FURTHER CASE STUDIES AND ANALYSIS EXERCISES BUILD CONNECTIONS

MAKING CONNECTIONS

FRACKING IN CANADA

Background: Western Canada has the biggest natural gas shale fracking operations in North America, with several major sites in British Columbia and Alberta supported in part by government financial incentives such as reduced royalties and assistance in road building and pipeline costs. The regulation of shale gas production on private land is largely controlled by each provincial or territorial government. In early 2011, Quebec Environment Minister Pierre Arcand announced a halt to all shale gas development in that province while the process and its impacts on the nearby St. Lawrence River are more thoroughly studied.

Case: You are a resident of Quebec, living in a community near a proposed natural gas drilling site. Shale gas development could bring a big economic boost to your area but the environmental concerns about fracking have alarmed some community members, who oppose it. You have been asked to research what is known about the environmental impacts of fracking and present the information at a public consultation meeting, so members of council can make an informed decision about what is right for the community.

1. Be sure to include the following in your explanation:
   a. A description of the fracking process and where it is currently being used on a regional and local scale
   b. Possibilities for economic improvement and job creation related to the project.
   c. General environmental concerns related to fracking.
2. Prior to your move to Quebec, you lived in Alberta and became familiar with the much publicized case of Jessica Ernst and her contaminated well water, contamination that occurred only after drilling began near her home. You decide to present a detailed case study about Jessica Ernst's problems. Be sure to include the following in your presentation:
   a. A detailed report of Ernst's case, including her credentials, the particular problems she faced, and the testing record of her well water
   b. Responses of oil companies to the people affected by fracking
   c. Responses from Alberta Environment and Water and other government agencies to both the oil companies and the people affected by fracking
3. In your summation to the council, provide your opinion of fracking and whether or not it should be permitted in your community.
   a. What is your recommendation to the council? Why?
   b. What other types of information might the council members want before they make a decision? Where would you suggest they go to find the answers to their questions?

SOLUTIONS TO ENVIRONMENTAL PROBLEMS INSPIRE STUDENTS

ENVIRONMENTAL SOLUTIONS ARE DISCUSSED IN EVERY CHAPTER

BIOMIMICRY—USING NATURE'S MODELS AND STRATEGIES TO SOLVE HUMAN PROBLEMS SUSTAINABLY—IS A FREQUENT THEME IN THE CHAPTERS

their livelihoods; but U.S. ranchers, like the ones who owned Horse Creek, will also suffer.

Scientists around the world have spent decades trying to prevent or even reverse desertification, to little avail. These days, most experts tend to agree that beyond a certain point, recovering grasslands that have swirled into deserts is impossible. "Most of our efforts to reverse desertification have failed dismally," says Richard Teague, an ecologist and rancher in West Texas, where some of the most rapid desertification is taking place. "There's been one notable exception and that's halfway around the world."

Taking our cues from nature, we can learn to use rangelands sustainably.

As the sun rises over Zimbabwe (formerly Rhodesia) in southern Africa, herds of antelope and zebra traverse a patchwork of temperate and tropical grasslands, feeding steadily on reedy stalks and short, fat shrubs; elephants and wildebeest splash around a precious watering hole, well fed and content. The animals may not realize it, but they have stumbled upon the African Center for Holistic Management (ACHM), 6500 acres of thriving rangeland in the heart of an otherwise parched and ailing prairie. [INFOGRAPHIC 12.5]

Perhaps nowhere else on Earth is such an oasis more urgently needed. Because the region is too arid to support
ng provides the only livelihood for most
ng there; about 75% of all land is used to
erds of cattle, goat, and sheep, and even
enough. With population, and thus
ouths in need of food, rising steadily,
wded more and more livestock onto lands
and drier by the day. Already stressed
e, that land is crumbling quickly into
desert. And as viable pastures become increasingly difficult to find, neighboring tribes have descended further into violent conflict—sometimes killing each other over a few stalks of grass.

Taking our cues from nature, we can learn to use rangelands sustainably.

The ACHM was established in 1992 by Allan Savory, a Rhodesian-born scientist-turned-rancher. Before then,

Infographic 12.5 | **SOUTHERN AFRICA MAP**

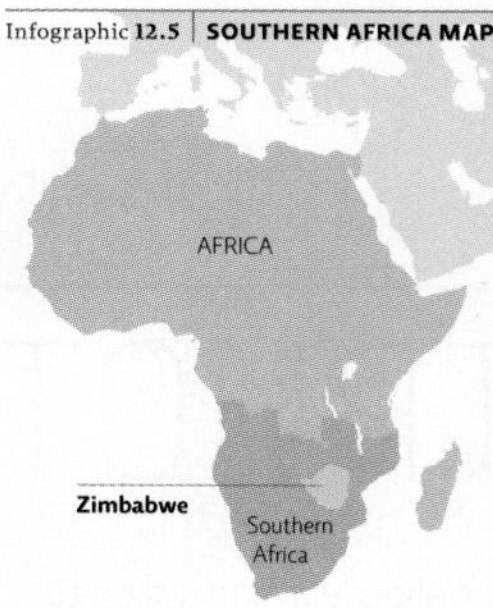

AFRICA
Zimbabwe
Southern Africa

↑ Biologist Allan Savory, squatting beside a patch of dry grasses in the desert where he teaches holistic land management.

d and to seek counsel from Savory, who is
dited for the dramatic recovery.

me by his expertise in a circuitous way. He
1950s working as a research biologist and game
the British Colonial Service. At the time, the
vernment was culling thousands of wild herds
t to create more land for farming. "Zimbabwe

BRING IT HOME

➲ PERSONAL CHOICES THAT HELP

Understanding the factors that influence how populations change can help us manage species that are either becoming invasive in nature or are facing extinction. How people view species and their connection to our world has a large impact on how management plays out.

Individual Steps

→ Learn more about wolves at the International Wolf Center (www.wolf.org).

→ Use the Internet and books on wildlife to research what your area might have been like prior to human settlement. Which species have been extirpated; which ones have been introduced? How have wildlife populations changed as a result of human action?

→ See if you can recognize distribution patterns in the wild. Do some flowers or trees grow in clumps or patches? Can you find species that appear to have a random or uniform distribution?

Group Action

→ Explore organizations that support predator preservation, such as Defenders of Wildlife and Keystone Conservation, for suggestions on how you can help educate others about the importance of predators.

→ Join a local, regional, or national group that works to monitor, protect, and restore wildlife habitats, such as the Nature Conservancy of Canada's Conservation Volunteers.

Policy Change

→ Caribou *(Rangifer tarandus)* is an iconic Canadian species that is currently listed as threatened under Canada's Species at Risk Act. In Banff National Park, falling caribou numbers are linked to rising wolf populations brought on by high elk populations. Reintroduction of caribou has been proposed using captive-bred animals. Research the issue and write a letter to Parks Canada that reflects your position on this proposal.

BRING IT HOME OFFERS WAYS THAT STUDENTS CAN ADDRESS ENVIRONMENTAL ISSUES ON THE INDIVIDUAL, GROUP, AND POLITICAL LEVEL

RESOURCES TARGET THE MOST CHALLENGING CONCEPTS AND SKILLS IN THE COURSE

CLASSROOM ACTIVITIES, ANIMATIONS, TUTORIALS, AND ASSESSMENT MATERIALS ARE ALL BUILT AROUND THE CONCEPTS AND SKILLS THAT ARE MOST DIFFICULT FOR STUDENTS TO MASTER

INSTRUCTOR RESOURCES

EnviroPortal An online learning space to help instructors administer their courses by combining our fully customizable eBook with a robust set of instructor resources. These include quizzes, news feeds, videos, tutorials, interactive infographics, the Test Bank, and homework management tools.

LearningCurve Activities that use a game-like interface to guide students through a series of questions tailored to their individual level of understanding.

Interactive eBook A complete online version of the textbook fully integrated with links right where you need them. You can even personalize the eBook just as you would a printed textbook, with highlighting, book-marking, and notes.

Story Abstracts The abstracts offer a brief story synopsis, providing interesting details relevant to the chapter and to the online resources not found in the book.

Chapter Topic Overviews Chapter overviews list specific lesson outcomes for each Guiding Question; page numbers in the chapter where each Guiding Question is covered and answered; and a list of the key terms associated with each Guiding Question. Topic overviews make it easy for instructors to plan their lecture and transition to the book.

Videos Videos from an array of trusted sources bring the stories of the book to life and make the material meaningful to students. Each video includes assessment questions to gauge student understanding.

Clicker Questions Designed as interactive in-class exercises, these questions reinforce core concepts and uncover misconceptions.

Test Bank
A collection of over a thousand questions, organized by chapter and Guiding Question, presented in a sortable, searchable platform. The Test Bank features multiple-choice and short-answer questions, and uses infographics and graphs from the book.

Optimized Art (JPEGs and layered PowerPoint slides) Infographics are optimized for projection in large lecture halls and split apart for effective presentation.

Layered or Active PowerPoint Slides
PowerPoint slides for select figures deconstruct key concepts, sequences, and processes in a step-by-step format, allowing instructors to present complex ideas in clear, manageable parts.

Lecture Outlines for PowerPoint Adjunct professors and instructors who are new to the discipline will appreciate these detailed companion lectures, perfect for walking students through the key ideas in each chapter. These rich, prebuilt lectures make it easy for instructors to transition to the book.

Team-Based Learning Activities Developed by author Susan Karr, these classroom activi-ties use proven active-learning techniques to engage students in the material and inspire critical thinking. These activities are intended for an instructor who is interested in taking an active learning approach to the course.

Instructor Resources DVD Combines a variety of Instructor Resources—Optimized Art files, Lecture Outlines for PowerPoint, Test Bank questions, and more—in one convenient package.

Course Management System e-packs available for Blackboard, WebCT, and other course management platforms.

ORGANIZED BY GUIDING QUESTIONS

STUDENT AND INSTRUCTOR RESOURCES ARE ARRANGED TO SUPPORT THE LEARNING GOALS OF EACH CHAPTER

SUPPORT ENVIRONMENTAL, SCIENCE, AND INFORMATION LITERACY

SUPPLEMENTARY MATERIALS EXPLORE TIMELY ENVIRONMENTAL ISSUES, DEVELOP CRUCIAL SCIENCE LITERACY SKILLS SUCH AS DATA ANALYSIS AND GRAPH INTERPRETATION, AND PROVIDE PRACTICE IN EVALUATING SOURCES OF INFORMATION

STUDENT RESOURCES

EnviroPortal An online learning space combining a fully customizable eBook and student resources. Students can access a variety of study tools, including quizzes, flashcards, animated interactive infographics, a lecture art notebook, and tutorials.

LearningCurve This set of formative assessment activities uses a game-like interface to guide students through a series of questions tailored to their individual level of understanding. A personalized study plan is generated based upon their quiz results. LearningCurve is available to students in the EnviroPortal.

Tutorials Our tutorials give students the opportunity to explore how science is done by analyzing data and drawing conclusions. Students build critical thinking skills with a variety of media tools, including:

- **Evaluating Sources of Information** These tutorials encourage students to examine real-world environmental issues and think critically about the opposing sides.
- **Graphing** Graphing tutorials let students build and analyze graphs, using their critical thinking skills to predict trends, identify bias, and make cause-and-effect connections.

Video Case Studies Videos from an array of trusted sources bring the stories of the book to life and allow students to apply their environmental, scientific, and information literacy skills. Each video includes questions that engage students in the critical thinking process.

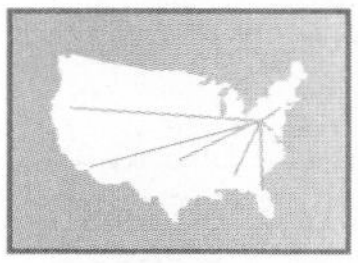

Interactive Infographics/Animations All infographics in the text include an animated interactive tutorial or an infographic activity.

Online Study Guide Organized by the chapter's Guiding Questions, the study guide elaborates on each infographic in the book and provides questions to promote students' critical thinking.

Interactive eBook A complete online version of the textbook, fully integrated with links right where you need them. You can even personalize the eBook just as you would a printed textbook, with highlighting, book-marking, and notes.

Key Term Flashcards Students can drill and learn the most important terms in each chapter using interactive flashcards.

Lecture Art Notebook The infographics for each chapter are available as PDF files that students can download and print before lectures.

Free Book Companion Website Features most student resources in an online format.

ACKNOWLEDGEMENTS

From Marnie Branfireun...

Working on the first Canadian edition of *Environmental Science for a Changing World* has been a new and exciting experience, and one that I have greatly enjoyed. This has been a truly collaborative effort, with contributions from many talented individuals. I extend my huge appreciation to the team which I have had the great good fortune to join in this endeavour. My own contributions to this text are built upon the strong foundation of those who have gone before me, and whose continued efforts have made this such a strong work. My co-author Susan Karr is a highly skilled and devoted environmental educator. Her extensive work on the U.S. edition has made this new edition possible, and working with her has been one of the most enjoyable parts of this project. I was extremely grateful to have her experience and knowledge to draw upon. I also thank co-author Anne Houtman for her work on the U.S. edition, and Jeneen Interlandi, Alison McCook, and Melinda Wenner Moyer, the gifted science writers who have made this work such a pleasure to read by creating the narrative structure unique to this text. I thank W. H. Freeman Senior Acquisitions Editor Jerry Correa for inviting me to join this project, and for his personal support. Karen Misler has been a brilliant Project Manager with an amazing ability to see what needs to be done, and keep all the contributors on track. I truly appreciate her patience and assistance throughout this process. Developmental Editor Andrea Gawrylewski is an excellent editor with an insight into content that has kept the work clear and focused. The copyediting team of Arleane Ralph and Debbie Smith from First Folio Resource Group Inc. has done a wonderful job of ensuring that this edition works for a Canadian audience, and that through all the various edits, the storyline maintains continuity and sense.

James Dauray, Kelly Cartwright, Shamili Sandiford, JodyLee Estrada Duek, and Michelle Cawthorn made excellent contributions to the Bring it Home feature and end-of-chapter materials. Thanks also go to the graphics team at MGMT. design for their skills, vision, and patience, and to Ellen Cash who expertly coordinated the production of the book.

Finally I would like to thank the many people I have worked with over the last 25 years in the areas of environmental education, outreach, conservation, and research, whose friendship and collegiality have made this often demanding and challenging work a pleasure. In particular I thank my students, who keep me thinking about how to improve how I teach this important material; especially those students who amaze me with their passion and hard work. I also thank my peers and mentors, past and present, from the University of Manitoba, McGill University, the Experimental Lakes Area, the Freshwater Institute, the Royal Botanical Gardens, the Hamilton Naturalists Club, Toronto and Region Conservation, and Western University. In all these places I have found support and inspiration, witnessed incredible dedication, and have been driven to persist in the face of adversity. To those wonderful people, and to my friends and family, my warmest thanks.

From Susan Karr...

It is amazing what you can accomplish when you work with talented and highly skilled people. I want to thank W. H. Freeman Acquisitions Editor Jerry Correa for his vision for this edition, and Developmental Editor Andrea Gawrylewski for her insights and outstanding editorial skills in crafting each chapter. I also want to thank Jeneen Interlandi, Alison McCook, and Melinda Wenner Moyer, the talented writers who have made these chapters such a pleasure to read; and our fabulously detail-oriented Project Manager, Karen Misler. A special thanks goes to Western University professor Marnie Branfireun for her expertise and tremendous attention to detail in bringing the Canadian focus to these chapters. I also owe a debt of gratitude to my environmental science students over the years for their questions, interests, demands, and passion for learning that have always challenged and inspired me. Finally, I want to thank my husband, Steve, for supporting me in so many different ways it is impossible to count them all.

From Jeneen Interlandi...

Each of these chapters is a story—of scientists and everyday people, often doing extraordinary things. It has been my great pleasure to tell those stories here. For that, I thank each and every one of my sources. Their time and patience are what made this book possible.

I would also like to thank Susan, with whom it has been an honour to work, and the entire W. H. Freeman team for their tireless efforts.

REVIEWERS

We would like to extend our deep appreciation to the following instructors who reviewed, tested, and advised on the book manuscript at various stages of development.

CHAPTER REVIEWERS

Matthew Abbott, *Des Moines Area Community College*
David Aborn, *University of Tennessee at Chattanooga*
Michael Adams, *Pasco-Hernando Community College*
Shamim Ahsan, *Metropolitan State College of Denver*
John Anderson, *Georgia Perimeter College*
Deniz Ballero, *Georgia Perimeter College*
Marcin Baranowski, *Passaic County Community College*
Brad Basehore, *Harrisburg Area Community College*
Sean Beckmann, *Rockford College*
David Berg, *Miami University*
Joe Beuchel, *Triton College*
Aaron Binns, *Florida State University*
Karen Blair, *Pennsylvania State University*
Barbara Blonder, *Flagler College*
Steve Blumenshine, *California State University, Fresno*
Ralph Bonati, *Pima Community College*
Polly Bouker, *Georgia Perimeter College*
Richard Bowden, *Allegheny College*
Jennifer Boyd, *University of Tennessee at Chattanooga*
Scott Brame, *Clemson University*
Allison Breedveld, *Shasta College*
Mary Brown, *Lansing Community College*
Brett Burkett, *Collin College*
Alan Cady, *Miami University*
Elena Cainas, *Broward College*
Deborah Carr, *Texas Tech University*
Mary Kay Cassani, *Florida Gulf Coast University*
Niccole Cerveny, *Mesa Community College*
Lu Anne Clark, *Lansing Community College*
Jennifer Cole, *Northeastern University*
Eric Compas, *University of Wisconsin, Whitewater*
Jason Crean, *Saint Xavier University & Moraine Valley Community College*
Michael Dann, *Pennsylvania State University*
James Dauray, *College of Lake County*
Michael Denniston, *Georgia Perimeter College*
Robert Dill, *Bergen Community College*
Craig Dilley, *Des Moines Area Community College*
JodyLee Estrada Duek, *Pima Community College*
Don Duke, *Florida Gulf Coast University*
James Dunn, *Grand Valley State University*
Kathy Evans, *Reading Area Community College*
Brad Fiero, *Pima Community College*
Linda Fitzhugh, *Gulf Coast Community College*
Steven Forman, *University of Illinois at Chicago*
Michael Golden, *Grossmont College*
Sherri Graves, *Sacramento City College*
Michelle Groves, *Oakton Community College*
Myra Carmen Hall, *Georgia Perimeter College*
Sally Harms, *Wayne State College*
Stephanie Hart, *Lansing Community College*
Wendy Hartman, *Palm Beach State College*
Alan Harvey, *Georgia Southern University*
Keith Hench, *Kirkwood Community College*
Robert Hollister, *Grand Valley State University*
Tara Holmberg, *Northwestern Connecticut Community College*
Jodee Hunt, *Grand Valley State University*
Meshagae Hunte-Brown, *Drexel University*
Kristin Jacobson, *Illinois Central College*
Jason Janke, *Metropolitan State College of Denver*
David Jeffrey, *Georgia Perimeter College*
Thomas Jurik, *Iowa State University*
Charles Kaminski, *Middlesex Community College*
Michael Kaplan, *College of Lake County*
John Keller, *College of Southern Nevada*
Myung-Hoon Kim, *Georgia Perimeter College*
Elroy Klaviter, *Lansing Community College*
Paul Klerks, *University of Louisiana at Lafayette*
Janet Kotash, *Moraine Valley Community College*
Jean Kowal, *University of Wisconsin, Whitewater*
John Krolak, *Georgia Perimeter College*
James Kubicki, *Pennsylvania State University*
Diane LaCole, *Georgia Perimeter College*
Katherine LaCommare, *Lansing Community College*
Andrew Lapinski, *Reading Area Community College*
Jennifer Latimer, *Indiana State University*
Stephen Lewis, *California State University, Fresno*
Eric Lovely, *Arkansas Tech University*
Marvin Lowery, *Lone Star College System*
Steve Luzkow, *Lansing Community College*
Steve Mackie, *Pima Community College*
Nilo Marin, *Broward College*
Eric Maurer, *University of Cincinnati*
Costa Mazidji, *Collin College*
DeWayne McAllister, *Johnson County Community College*
Vicki Medland, *University of Wisconsin, Green Bay*
Alberto Mestas-Nunez, *Texas A&M University, Corpus Christi*
Chris Migliaccio, *Miami Dade College*
Jessica Miles, *Palm Beach State College*
Dale Miller, *University of Colorado, Boulder*
Kiran Misra, *Edinboro University of Pennsylvania*
Scott Mittman, *Essex County College*
Edward Mondor, *Georgia Southern University*
Zia Nisani, *Antelope Valley College*
Ken Nolte, *Shasta College*
Kathleen Nuckolls, *University of Kansas*
Segun Ogunjemiyo, *California State University*
Bruce Olszewski, *San José State University*
Jeff Onsted, *Florida International University*
Nancy Ostiguy, *Pennsylvania State University*
Daniel Pavuk, *Bowling Green State University*
Barry Perlmutter, *College of Southern Nevada*
Craig Phelps, *Rutgers, The State University of New Jersey*
Neal Phillip, *Bronx Community College*
Keith Putirka, *California State University, Fresno*
Bob Remedi, *College of Lake County*
Erin Rempala, *San Diego City College*
Angel Rodriquez, *Broward College*
Dennis Ruez, *University of Illinois at Springfield*
Robert Ruliffson, *Minneapolis Community and Technical College*
Melanie Sadeghpour, *Des Moines Area Community College*
Seema Sah, *Florida International University*
Jay Sah, *Florida International University*
Shamili Sandiford, *College of DuPage*
Waweise Schmidt, *Palm Beach State College*
Jeffery Schneider, *State University of New York at Oswego*
William Shockner, *Community College of Baltimore County*
David Serrano, *Broward College*
Patricia Smith, *Valencia College*
Dale Splinter, *University of Wisconsin, Whitewater*
Jacob Spuck, *Florida State University*
Craig Steele, *Edinboro University*
Rich Stevens, *Monroe Community College*
Michelle Pulich Stewart, *Mesa Community College*
John Sulik, *Florida State University*
Donald Thieme, *Valdosta State University*
Jamey Thompson, *Hudson Valley Community College*
Susanna Tong, *University of Cincinnati*
Michelle Tremblay, *Georgia Southern University*
Karen Troncalli, *Georgia Perimeter College*
Caryl Waggett, *Allegheny College*
Meredith Wagner, *Lansing Community College*
Deena Wassenberg, *University of Minnesota*
Kelly Watson, *Eastern Kentucky University*
Jennifer Willing, *College of Lake County*
Danielle Wirth, *Des Moines Area Community College*
Janet Wolkenstein, *Hudson Valley Community College*
Douglas Zook, *Boston University*

FOCUS GROUP PARTICIPANTS

John Anderson, *Georgia Perimeter College*
Teri Balser, *University of Wisconsin, Madison*
Elena Cainas, *Broward College*
Mary Kay Cassani, *Florida Gulf Coast University*
Kelly Cartwright, *College of Lake County*
Michelle Cawthorn, *Georgia Southern University*
Mark Coykendall, *College of Lake County*
JodyLee Estrada Duek, *Pima Community College*
Jason Janke, *Metropolitan State College of Denver*
Janet Kotash, *Moraine Valley Community College*
Jean Kowal, *University of Wisconsin, Whitewater*
Nilo Marin, *Broward College*
Edward Mondor, *Georgia Southern University*
Brian Mooney, *Johnson & Wales University, North Carolina*
Barry Perlmutter, *College of Southern Nevada*
Matthew Rowe, *Sam Houston State University*
Shamili Sandiford, *College of DuPage*
Ryan Tainsh, *Johnson & Wales University*
Michelle Tremblay, *Georgia Southern University*
Kelly Watson, *Eastern Kentucky University*

COMPARATIVE REVIEWERS

Buffany DeBoer, *Wayne State College*
Dani DuCharme, *Waubonsee Community College*
James Eames, *Loyola University Chicago*
Bob East, *Washington & Jefferson College*
Matthew Eick, *Virginia Polytechnic Institute and State University*
Kevin Glaeske, *Wisconsin Lutheran College*
Rachel Goodman, *Hamdpen-Sydney College*
Melissa Hobbs, *Williams Baptist College*
David Hoferer, *Judson University*
Paul Klerks, *University of Louisiana at Lafayette*
Troy Ladine, *East Texas Baptist University*
Jennifer Latimer, *Indiana State University*
Kurt Leuschner, *College of the Desert*
Quent Lupton, *Craven Community College*
Jay Mager, *Ohio Northern University*
Steven Manis, *Mississippi Gulf Coast Community College*
Nancy Mann, *Cuesta College*
Heidi Marcum, *Baylor University*
John McCarty, *University of Nebraska at Omaha*
Chris Poulsen, *University of Michigan*
Mary Puglia, *Central Arizona College*
Michael Tarrant, *University of Georgia*
Melissa Terlecki, *Cabrini College*
Jody Terrell, *Texas Woman's University*

CLASS TEST PARTICIPANTS

Mary Kay Cassani, *Florida Gulf Coast College*
Ron Cisar, *Iowa Western Community College*
Reggie Cobb, *Nash Community College*
Randi Darling, *Westfield State University*
JodyLee Estrada Duek, *Pima Community College*
Catherine Hurlbut, *Florida State College at Jacksonville*
James Hutcherson, *Blue Ridge Community College*
Janet Kotash, *Moraine Valley*
Offiong Mkpong, *Palm Beach State College*
Edward Mondor, *Georgia Southern University*
Anthony Overton, *East Carolina University*
Shamilli Sandiford, *College of DuPage*
Keith Summerville, *Drake University*

CASE STUDY SURVEY PARTICIPANTS

Matthew Abbott, *Des Moines Area Community College*
Daphne Babcock, *Collin College*
Nancy Bain, *Ohio University*
Paul Baldauf, *Nova Southeastern University*
Brad Basehore, *Harrisburg Area Community College*
William Berry, *University of California, Berkeley*
Ralph Bonati, *Pima Community College*
Heather Bradley, *Truckee Meadows Community College*
Allison Breedveld, *Shasta College*
Robert Bruck, *North Carolina State University*
Nancy Butler, *Kutztown University*
Daniel Capuano, *Hudson Valley Community College*
Lu Anne Clark, *Lansing Community College*
Jennifer Cole, *Northeastern University*
David Corey, *Midlands Technical College*
Michael Denniston, *Georgia Perimeter College*
Robert Dill, *Bergen Community College*
Karen Gaines, *Eastern Illinois University*
Floyd Hayes, *Pacific Union College*
Robert Hollister, *Grand Valley State University*
Susan Hitchins, *Itasca Community College*
Janet Kotash, *Moraine Valley Community College*
Paul Kramer, *Farmingdale State College*
Subbu Krishna, *Collin College*
Andrew Lapinski, *Reading Area Community College*
Eric Maurer, *University of Cincinnati*
DeWayne McAllister, *Johnson County Community College*
Patricia Menchaca, *Mount San Jacinto Community College*
Myles Miller, *Philadelphia University*
Tamera Minnick, *Mesa State College*
Brian Mooney, *Johnson & Wales University*
Jean Pan, *University of Akron*
Frank Phillips, *McNeese State University*
Janet Puhalla, *Missouri State University*
Bob Remedi, *College of Lake County*
Jennifer Richter, *University of New Mexico*
Virginia Rivers, *Truckee Meadows Community College*
Shamili Sandiford, *College of DuPage*
Debora Scheidemantel, *Pima Community College*
Erik Scully, *Towson University*
Yosef Shapiro, *Broward College*
Rich Stevens, *Monroe Community College*
Peter Strom, *Rutgers University*
Jamey Thompson, *Hudson Valley Community College*
Jonah Triebwasser, *Marist and Vassar Colleges*
Thomas Wilson, *University of Arizona*
Janet Wolkenstein, *Hudson Valley Community College*

NEW IN THIS EDITION

CHAPTER	NEW STORY OR CANADIAN EXAMPLE:
1 Environmental Literacy	
2 Science Literacy and the Process of Science	
3 Information Literacy	NEW STORY: Canada's BPA regulation
4 Human Populations	
5 Ecological Economics and Consumption	NEW STORY: Profile of Mountain Equipment Co-op—a Canadian company
6 Ecosystems and Nutrient Cycling	
7 Population Ecology	NEW EXAMPLE: Wolf tracking data from Canada
8 Community Ecology	NEW STORY: Restoring the Acadian forest
9 Biodiversity	
10 Evolution and Extinction	
11 Forests	
12 Grasslands	
13 Marine Ecosystems	
14 Fisheries and Aquaculture	NEW EXAMPLE: Canadian aquaculture research
15 Freshwater Resources	
16 Water Pollution	NEW STORY: Getting the Great Lakes back to health
17 Solid Waste	NEW EXAMPLE: Canadian waste management, regulations, and policy
18 Agriculture	
19 Coal	
20 Oil and Natural Gas	NEW STORY: Oil Sands in Canada
21 Air Pollution	
22 Climate Change	NEW EXAMPLE: Canadian species' migration
23 Nuclear Power	NEW EXAMPLE: Canadian nuclear disposal policy, and nuclear development
24 Sun, Wind, and Water Energy	
25 Biofuels	
26 Urbanization and Sustainable Communities	NEW EXAMPLE: Urbanization and green buildings in Toronto

ENVIRONMENTAL SCIENCE

FOR A CHANGING WORLD

CANADIAN EDITION

CHAPTER 1 ENVIRONMENTAL LITERACY

ON THE ROAD TO COLLAPSE

What lessons can we learn from a vanished Viking society?

CORE MESSAGE

Humans are a part of the natural world and are dependent on a healthy, functioning planet. We put pressure on the planet in a variety of ways, but our choices can help us move toward sustainability.

GUIDING QUESTIONS

After reading this chapter, you should be able to answer the following questions:

→ What constitutes the "environment" and what fields of study collaborate under the umbrella of environmental science?

→ What are some of the environmental dilemmas that humans face and why are many of these considered "wicked problems"?

→ What challenges does humanity face in dealing with environmental issues and how can environmental literacy help us make more informed decisions?

→ What does it mean to be sustainable and what are the characteristics of a sustainable ecosystem?

→ What can human societies and individuals do to encourage sustainably?

The remains of Hvalsey, a Viking settlement church, in southern Greenland.

Although not much of a tourist destination, Greenland offers some spectacular sights—colossal ice sheets, a lively seascape, rare and precious wildlife (whales, seals, polar bears, eagles). But on his umpteenth trip to the island, Thomas McGovern was not interested in any of that. What he wanted to see was the garbage—specifically, the ancient, fossilized garbage that Viking settlers had left behind some seven centuries ago.

McGovern, an archaeologist at the City University of New York, had been on countless expeditions to Greenland over the past 40 years. Digging through layers of peat and permafrost, he and his team had unearthed a museum's worth of artifacts that, when pieced together, told the story of the Greenland Vikings. But as thorough as their expeditions had been, that story was still maddeningly incomplete.

Here's what they knew so far: A thousand or so years ago, an infamous Viking by the name of Erik the Red led a small group of followers across the ocean from Norway, to a vast expanse of snow and ice that he had dubbed Greenland. Most of Greenland was not green. In fact, it was a forbidding place marked by harsh winds and sparse vegetation. But tucked between two fjords along the southwestern coast, protected from the elements by jagged, imposing cliffs, the Vikings found a string of verdant meadows, brimming with wildflowers. They quickly set up camp here, and proceeded to build a society similar to the one they had left behind in Norway. They farmed, hunted, and raised livestock. They also built barns and churches as elaborate as the ones back home. They established an economy and a legal system, traded goods with mainland Europe, and at their peak, reached a population of 5000 (a large number in those days).

WHERE IS THE VIKING SETTLEMENT IN GREENLAND?

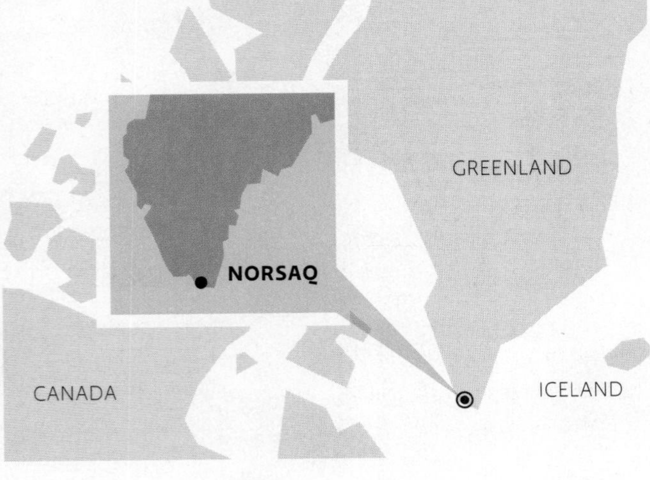
GREENLAND
NORSAQ
CANADA
ICELAND

And then, after 450 years of prosperity, they disappeared—seemingly into thin air—leaving little more than the beautiful, tragic ruins of a handful of barns and churches in their wake.

The how and why of this vanishing act remained a tantalizing mystery, one that has drawn hundreds of scientists—McGovern among them—to Greenland each summer. Recently, some of McGovern's colleagues had begun to suspect that disturbances in the natural environment—a cooling climate, loss of soil, problems with the food supply—may have been the deciding factors.

While other researchers probed ice sheets and soil deposits in search of clues, McGovern stuck to the garbage heaps, or *middens*, as Vikings called them. Every farmstead had one, and every generation of the farmstead's owners threw their waste into it. The result was an archaeological treasure trove: fine-grain details about what people ate, how they dressed, and the kinds of objects they filled their homes with. It gave McGovern and his team a clear picture of how they lived.

If they dug deep enough, McGovern thought, it might also explain how they died.

Environmental science is all encompassing.

For a modern developed society like Canada, it can be difficult to imagine a time and place when the natural world held such sway over our fate. Our food comes from a grocery store, our water from a tap; even our air is artificially heated and cooled to our liking. These days, it seems more logical to consider societal conflict, or even

environment The biological and physical surroundings in which any given living organism exists.
environmental science An interdisciplinary field of research that draws on the natural and social sciences and the humanities in order to understand the natural world and our relationship to it.

Infographic 1.1 | ENVIRONMENTAL SCIENCE IS HIGHLY INTERDISCIPLINARY

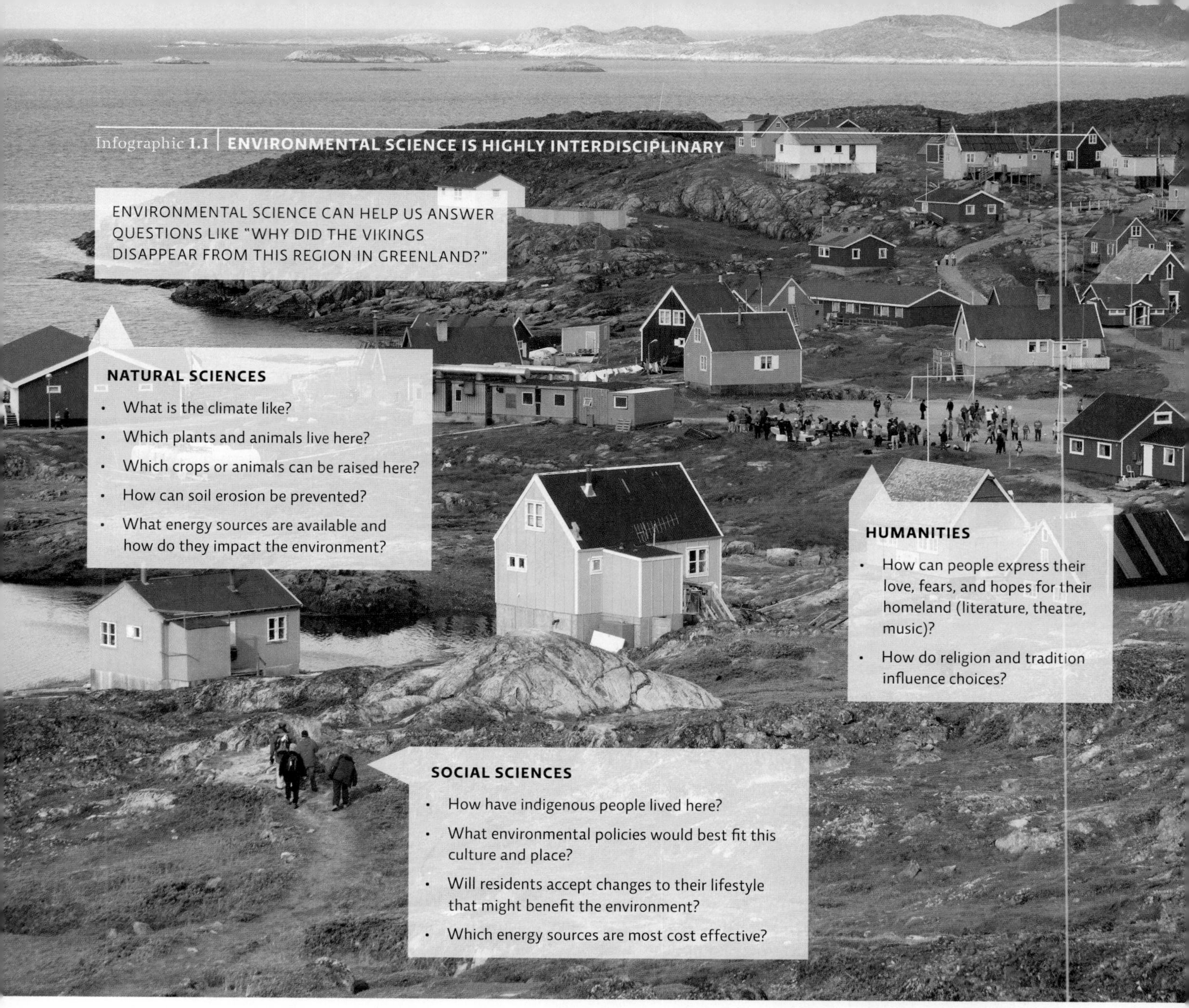
ENVIRONMENTAL SCIENCE CAN HELP US ANSWER QUESTIONS LIKE "WHY DID THE VIKINGS DISAPPEAR FROM THIS REGION IN GREENLAND?"
NATURAL SCIENCES
• What is the climate like?
• Which plants and animals live here?
• Which crops or animals can be raised here?
• How can soil erosion be prevented?
• What energy sources are available and how do they impact the environment?
HUMANITIES
• How can people express their love, fears, and hopes for their homeland (literature, theatre, music)?
• How do religion and tradition influence choices?
SOCIAL SCIENCES
• How have indigenous people lived here?
• What environmental policies would best fit this culture and place?
• Will residents accept changes to their lifestyle that might benefit the environment?
• Which energy sources are most cost effective?

↑ Environmental science studies the natural world and how humans interact with and impact it. We must look to the natural and social sciences as well as to the humanities to help us understand our world and effectively address environmental issues and environmental questions such as, "Why did the Vikings disappear from this region in Greenland and how do humans live now in such a harsh environment?"

collapse, through the lens of politics or economics. But, as we will see time and again throughout this book, the natural environment—and how we interact with it—plays a leading role in the sagas that shape human history; this is as true today as it was in the time of the Vikings.

Environment is a broad term that describes the surroundings or conditions (including living and nonliving components) in which any given organism exists. **Environmental science**—a field of research that is used to understand the natural world and our relationship to it—is extremely interdisciplinary. It relies on a range of natural and applied sciences (such as ecology, geology, chemistry, and engineering) to unlock the mysteries of the natural world, and to look at the role and impact of humans in the world. It also draws on social sciences (such as anthropology, psychology, and economics) and the humanities (such as art, literature, and music) to understand the ways that humans interact with, and thus impact, the ecosystems around them. [INFOGRAPHIC 1.1]

Infographic **1.2** | **DIFFERENT APPROACHES TO SCIENCE HAVE DIFFERENT GOALS AND OUTCOMES**

↓ Environmental science is used to systematically collect and analyze data to draw conclusions and use these conclusions to propose reasonable courses of action.

EMPIRICAL SCIENCE IS USED TO INVESTIGATE THE NATURAL WORLD

↑ Through observation, glaciologists study and record the rate of glacier melt in Greenland; it is increasing dramatically in some places.

IN APPLIED SCIENCE, KNOWLEDGE IS USED TO ADDRESS PROBLEMS OR NEEDS

↑ Engineers harness the power of water by diverting glacial meltwater to produce hydroelectric power. This power can be converted to hydrogen fuel and can provide energy to remote areas.

Environmental science is an **empirical science**: it scientifically investigates the natural world through systematic observation and experimentation. It is also an **applied science**: we use its findings to inform our actions and, in the best cases, to bring about positive change. [INFOGRAPHIC 1.2]

The ability to understand environmental problems is referred to as **environmental literacy**. Such literacy is crucial to helping us become better stewards of Earth. Environmental problems can be extremely complicated and tend to have multiple causes, each one difficult to address. We must also understand that because of their complexity, any given response to an environmental problem involves significant **trade-offs** and no one response is likely to present the ultimate solution. Scientists refer to such problems as "wicked problems." In confronting them, we must consider not only their environmental but also their economic and social causes and consequences. Scientists refer to this trifecta as the **triple bottom line**. [INFOGRAPHIC 1.3]

In his book *Collapse*, University of California at Los Angeles biologist Jared Diamond details how wicked problems can lead to a society's ultimate demise. He identifies five factors in particular that determine whether any given society will succeed or fail: natural climate change; failure to properly respond to environmental changes; self-inflicted environmental damage; hostile neighbours; and loss of friendly neighbours. According to Diamond, the relative impact of each factor varies by society.

The situation of the Greenland Vikings was a rare case. It turns out that, like a perfect storm, all five of these factors conspired together.

The Greenland Vikings' demise was caused by natural events and human choices.

Greenland's interior is covered by vast ice sheets that stretch toward the horizon—3000 metres thick and more than 250 000 years old. To residents of the hard land, these ice sheets are not good for much—they create harsh winds and brutal cold—but to climate scientists, they're a treasure trove. As snow falls, it absorbs various particles from the atmosphere and lands on the ice sheets. As time passes, the snow and particles compact into ice,

empirical science A scientific approach that investigates the natural world through systematic observation and experimentation.
applied science Research whose findings are used to help solve practical problems.
environmental literacy A basic understanding of how ecosystems function and of the impact of our choices on the environment.
trade-offs The imperfect and sometimes problematic responses that we must at times choose between when addressing complex problems.
triple bottom line The combination of the environmental, social, and economic impacts of our choices.

Infographic **1.3** | **WICKED PROBLEMS**

↓ Wicked problems are difficult to address because, in many cases, each stakeholder hopes for a different solution. Solutions that address wicked problems usually involve trade-offs, so there is no clear "winner." One example of a wicked problem is climate change. There are many causes of the current climate change we are experiencing, both natural and anthropogenic (caused by human actions), and the effects of climate change will be varied for different species and people depending on where they live and their ability to adapt to the changes.

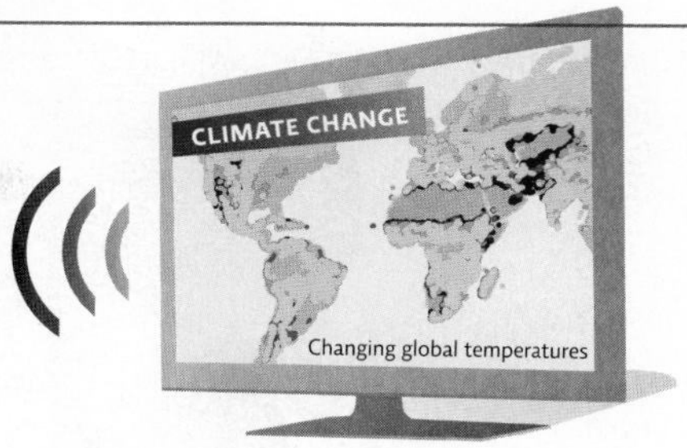
CLIMATE CHANGE
Changing global temperatures

SOME OF THE CAUSES:

- Burning fossil fuels
- Deforestation
- Methane from agriculture
- Affluence and consumption

SOME OF THE CONSEQUENCES:

- Sea level rise
- Habitat loss and species endangerment
- Spread of tropical disease
- Agriculture: worse in some areas, better in others

POSSIBLE ACTIONS (AND POTENTIAL TRADE-OFFS) INCLUDE:

- Alternative energy sources (less pollution but can be costly)
- Irrigation (increases crop yields but can cause water shortages and soil problems)
- Reforestation projects (lessen CO_2 in atmosphere and increase habitats but may take land needed for agriculture or other uses)
- Protecting flood-prone areas with dams, dikes, and diversion channels (may protect cities and farms but may fragment aquatic habitats and isolate species' populations)

freezing in time perfect samples of the atmosphere as it existed when that snow first fell. By analyzing those ice-trapped particles—dust, gases, chemicals, even the water molecules themselves—scientists can get a pretty good idea of what was happening to the climate at any given time. "It's like perfectly preserved slices of atmosphere from the past," says Lisa Barlow, a geologist and climate researcher at the University of Colorado at Boulder. "It gives us additional clues as to what was going on."

To uncover those clues, a team of scientists and engineers picked an accessible segment of ice sheet, not far from the Viking settlements, drilled from the surface all the way down to the bedrock below, and extracted a 12-centimetre-wide, 3000-metre-long cylinder of ice, which they then divided and dispersed among a handful of labs across the globe, including Jim White's light stable isotope lab—also at the University of Colorado at Boulder. Analysis of thin sequential segments of ice showed that when the Vikings first arrived in Greenland the temperature was anomalously higher than average for the last 1000 years—atmospheric temperature can be deduced from the amount of oxygen-18 or Deuterium (heavy Hydrogen) present in the sample. By the time the Vikings had vanished, temperatures had lowered so much that scientists qualified the period as a mini ice-age, overlain on seasonal changes. "It's no wonder they didn't make it," says Barlow. "With lower temperatures, livestock would have starved for lack of hay over the long winter, and self-inflicted environmental damage made the situation worse."

Indeed, in addition to natural climate change (Diamond's first factor), the Vikings also suffered from self-inflicted environmental damage (Diamond's second factor).

In addition to that single ice core, scientists have analyzed hundreds of mud cores taken from lake beds around the Viking settlements. These mud samples—which contain large amounts of soil that was blown into the lakes during Viking times—indicate that soil erosion had become a significant problem long before the region descended into a mini ice age. "This wasn't a climate problem," says Bent Fredskild, a Danish scientist who extracted and studied many of the mud cores. "This was self-inflicted. It happened the same way that soil erosion

ECLIPSE Ice Drill

↑ Scientist from the National Snow and Ice Data Center at the University of Colorado at Boulder working with an ice core drill.

happens today—they overgrazed the land, and once it was denuded, there was nothing to anchor the soil in place. So the wind carried it away."

Overgrazing wasn't their only mistake. The Greenland Vikings also used grassland to insulate their houses against the cold of winter; typical insulation consisted of 1.8-metre-long slabs of turf, and a typical home took about 4 hectares of grassland to insulate. On top of that, they chopped down the forests, harvesting enough timber to not only provide fuel and build houses but also to make the innumerable wooden objects to which they had become accustomed back in Norway.

Greenland's ecosystem was far too fragile to endure such pressure, especially as the settlement grew from a few hundred to a few thousand. The short, cool growing season meant that plants developed slowly, which in turn meant that the land could not recover quickly enough from the various assaults to protect the soil.

sustainable development Development that meets present needs without compromising the ability of future generations to do the same.

As climate cooling and overharvesting conspired to destroy pasturelands, summer hay yields shrank. When scientists counted the fossilized remains of insects that lived in the fields and haylofts of Viking Greenland, they found that their numbers fell dramatically in the settlement's final years. "The falloff in insects tells us that hay production dwindled to the point of crisis," says Fredskild. Without hay, livestock could not survive the ever-colder, ever-longer winters. And without livestock, the Vikings themselves went hungry. As scientists soon discovered, they needn't have.

Responding to environmental problems and working with neighbours help a society cope with changes.

Back in his Manhattan lab, McGovern sorts through hundreds of animal bones collected from various Greenland middens. By examining the bones and making careful note of which layers they were retrieved from, McGovern can tell what the people ate and how their diets changed over time. "This is a pretty typical set of remains for these people from this region and time period," he says, leaning over a shiny metal tray of neatly arranged bone fragments. Some are the bones of cattle imported from Europe. Others are the remains of sheep and goats; still others of local wildlife such as caribou. Conspicuously absent, McGovern says, are fish of any kind.

"If we look at a comparable pile of bones from [Norwegian settlers of] the same time period, from Iceland, we see something very different," McGovern explains. "We have fish bones and bird bones and little fragments of whale bones. Most of it, in fact, is fish—including a lot of cod." It turns out that while the Greenland Vikings were guilty of Diamond's third factor, failure to respond to the natural environment, their Icelandic cousins were not.

Like their cousins in Greenland, the Vikings who settled Iceland at about the same time were initially fooled into thinking that their newly-discovered land could sustain their cow-farming, wood-dependent ways: plants and animals looked similar to those back in Norway, and grasslands seemed lush and abundant. They cleared about 80% of Iceland's forests and allowed their cows, sheep, and goats to chew the region's grasslands down to nothing before finally noticing how profound the differences between Iceland and Norway actually were: growing seasons in Iceland were shorter, both soil and vegetation were much more fragile, and because the land could not rebound quickly, cow farming was unsustainable.

But once they saw that their old ways would not work in this new country, the Icelandic Vikings made changes. Not only did they switch from beef to fish, they also began conserving their wood and abandoned the highlands, where soil was especially fragile. And, as a result of these and other adaptations, they survived and prospered. The Icelanders responded to the limitations of their natural environment in a way that allowed them to meet present needs without compromising the ability of future generations to do the same—an approach known today as **sustainable development**.

Some of the most telling clues to the mystery of the Greenland Viking's demise come not from the Viking colonies but from another group of people who lived nearby: the Inuit. The Inuit arrived in the Arctic centuries before the Vikings. They were expert hunters of ringed seal—an exceedingly difficult-to-catch but very abundant food source. They knew how to heat and light their homes with seal blubber (instead of firewood). And they loved to fish.

Fishing is not nearly as labour-intensive as raising cattle, and in the lakes and fjords of Greenland, it provides an easy, reliable source of protein. A comparison of Inuit and Viking middens shows that even as the Greenland Vikings were scraping off every last bit of meat and marrow from their cattle bones, the Inuit had more food than they could eat.

The Vikings might have learned from their Inuit neighbours; by adapting some of their customs, they might have survived the Little Ice Age and gone on to prosper as the Icelandic Vikings did. But excavations show that virtually no Inuit artifacts made their way into Viking settlements. And according to written records, the Norse detested the Inuit who, on at least one occasion, attacked the Greenland colony; they called the Inuit *skraelings*, which is Norse for "wretches," considered them inferior, and refused to seek their friendship or their counsel. In addition to these hostile neighbours (Diamond's fourth factor), the Greenland Vikings also suffered a loss of friendly neighbours (Diamond's fifth factor).

As the productivity of the Viking colonies declined, so did visits from European ships. As time wore on, it became apparent that the Greenland Vikings could expect very little in the way of trade; royal and private ships that had visited every year came less and less often. After a while, they did not come at all. For the Greenland Vikings, who depended on the Europeans for iron, timber, and other essential supplies, this loss proved devastating. Among other things, it meant that, as the weather grew colder, and food supplies dwindled, they had no one to turn to for help.

Humans are an environmental force that impacts Earth's ecosystems.

When it comes to the environment, modern societies are not as different from the Vikings as one might assume. Vikings chose livestock and farming methods that were ill-suited to Greenland's climate and natural environment. We too use farming practices that strip away topsoil and diminish the soil's fertility. We have overharvested our forests, and in so doing have triggered a cascade of environmental consequences: loss of vital habitat and biodiversity, soil erosion, and water pollution. We have overfished and overhunted and have allowed invasive species to devastate some of our most valuable ecosystems.

In part, these problems stem from a disconnect in our understanding of the relationship between our actions and their environmental consequences. For example, unless they live nearby, many people in Canada do not realize that vast areas of forest and wetland are flooded to produce hydroelectricity, the main source of power in many provinces. Flooding destroys thousands of hectares of habitat and diverts thousands of kilometres of streams and rivers. These hydrologic alterations, and the building of large dams, allow the creation of large reservoirs to supply 'megaprojects' in northern Ontario and Quebec. We are slow to make the connection between this extensive flooding and mercury contamination of fish.

> In part, these problems stem from a disconnect in our understanding of the relationship between our actions and their environmental consequences.

We also face a suite of new problems that did not trouble the Vikings. Chief among them is population growth; as we will discuss in subsequent chapters, global population is poised to top 9 billion come 2050. The sheer volume of people will strain Earth's resources like never before. This is relevant because every environment has a **carrying capacity**—the population size that an area can support indefinitely—but some of our actions are decreasing carrying capacity, even as our population swells (see Chapter 4 for more on human populations and carrying capacity). In addition, we generate more pollution than the Vikings did, and much of what we generate is more toxic.

Environmental scientists evaluate the impact any population has on its environment—due to the resources it takes and the waste it produces—by calculating its **ecological footprint** (see Chapter 5). Some analysts feel we have already surpassed the carrying capacity of Earth; our collective footprint already surpasses what Earth can support over a long period of time.

Another serious consequence of larger populations, increasing affluence, and more sophisticated technology is **anthropogenic** climate change. While the Vikings had to contend with periodic warming and cooling periods that were part of the natural climate cycle, the vast majority of scientists today conclude that modern humans are faced with rapidly warming temperatures caused largely by our own use of greenhouse gas-emitting fossil fuels.

Scientists have coined the term "ecocide"—wilful destruction of the natural environment—to describe this constellation of forces, which, as Diamond writes, have "come to overshadow nuclear war and emerging diseases as the biggest threat to global civilization."

We have something else in common with the Vikings of Greenland: our attitudes frequently prevent us from responding effectively to environmental changes. According to Diamond, the Vikings were stymied by their own sense of superiority. They considered themselves masters of their surroundings and so they did not notice signs that they were causing irreparable harm to their environment, nor did they bother to learn the ways of their Inuit neighbours who had managed to survive in the same region for centuries before their arrival.

MANY ENVIRONMENTAL PROBLEMS CAN BE TRACED TO THREE UNDERLYING CAUSES

↑ Population Size: There are more than 7 billion people on the planet. Though some have more impact than others, our sheer numbers contribute to many environmental problems.

↑ Resource Use: We use resources faster than they are replaced, and convert matter into forms that don't readily decompose.

↑ Pollution: We generate pollution that compromises our own health and that of ecosystems.

Infographic **1.4** | **OUR ATTITUDES AFFECT HOW WE RESPOND TO PROBLEMS**

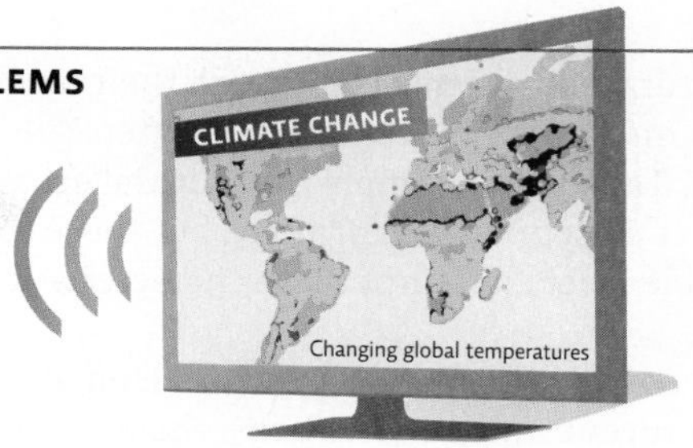
CLIMATE CHANGE
Changing global temperatures

↓ Attitudes affect how we respond to problems; some attitudes cause problems because they prevent us from responding to problems.

TECHNOLOGICAL FIX
Human ingenuity & science will be able to solve problems we run into (someone else will fix it).
EVALUATE AND RESPOND
Move forward based on the best available evidence (we have options).
GLOOM AND DOOM
There is nothing we can do (give up).
ROSY OPTIMISM
Don't worry—it will all work out (ignore the problem).
FRONTIER
Resources are here for us to use and we'll find others when needed (nature will provide the solutions).

Our own attitudes can also help or hurt our ability to respond to changes. These attitudes tend to fall into one of a few groups, and range from pessimistic to more optimistic to pragmatic. [INFOGRAPHIC 1.4]

The United Nations' Millennium Ecosystem Assessment looks at how environmental problems affect humans and makes recommendations about addressing those problems. According to its 2005 assessment, a consensus report of more than 1500 scientists, human actions are so straining the environment that the ability of the planet's ecosystems to sustain future generations is gravely imperiled. But there is hope: if we act now, the report's authors write, we can still reverse much of the damage. Some of the best lessons about how we can do this come from the natural environment itself.

Human societies can become more sustainable.

Unlike their counterparts in Greenland, the Icelandic Vikings responded to their environment and adopted more **sustainable** practices. They didn't have to look far for a model of sustainability: natural ecosystems are sustainable. This means they use resources—namely, energy and matter—in a way that ensures that those resources continue to be available.

To survive, organisms need a constant, dependable source of energy. But as energy passes from one part of an ecosystem to the next, the usable amount declines; therefore, new inputs of energy are always needed. A sustainable ecosystem is one that makes the most of **renewable energy**—energy that comes from an infinitely available or easily replenished source. For almost all natural ecosystems, that energy source is the sun. Photosynthetic organisms such as plants trap solar energy and convert it to a form that they can readily use, or that can be passed up the food chain to other organisms.

Unlike energy, matter (anything that has mass and takes up space) can be recycled and reused indefinitely; the key is not using it faster than it is recycled. Naturally sustainable ecosystems waste nothing; they recycle matter so that the waste from one organism ultimately becomes a resource for another. They also keep populations in check so that the resources are not overused and there is enough food, water, and shelter for all.

Lastly, sustainable ecosystems depend on local **biodiversity** (the variety of species present) to perform many of the jobs just mentioned; different species have different ways of trapping and using energy and matter, the net result of which boosts productivity and efficiency (see Chapter 8). And predators, parasites, and competitors serve as natural population checks.

---

carrying capacity The population size that a particular environment can support indefinitely.
ecological footprint The land area needed to provide the resources and assimilate the waste of a person or population.
anthropogenic Caused by or related to human action.
sustainable A method of using resources in such a way that we can continue to use them indefinitely.
renewable energy Energy that comes from an infinitely available or easily replenished source.
biodiversity The variety of species on Earth.

Thus, natural ecosystems live within their means and each organism contributes to the ecosytem's overall function. This is not to imply that natural ecosystems are perfect places of total harmony, but those that are sustainable meet all four of these characteristics.

Human ecosystems are another story. Humans tend to rely on **nonrenewable resources**—those whose supply is finite or is not replenished in a timely fashion. The most obvious example of this is our reliance on fossil fuels such as coal, natural gas, and petroleum, culled from deep within Earth, to power our society. Fossil fuels are replenished only over vast geologic time—far too slowly to keep pace with our rampant consumption of them. On top of that, we have a hard time keeping our population under control, despite (or maybe because of) all the advances of modern technology. We also generate volumes of waste, much of it toxic, and have yet to fully master the art of recycling.

Increasingly, however, we humans are looking to nature to help us learn how to change our ways. In Janine Benyus' book *Biomimicry* (the term used to describe the strategy of mimicking nature), Benyus explains that biomimicry involves using nature as a *model*, *mentor*, and *measure* for our own systems. Emulating nature (nature as model) gives us an example of *what* to do; it can also teach us how to do it (nature as mentor) and the level of response that is appropriate (nature as measure). As we'll see in later chapters, scientists are using biomimicry to design more sustainable methods of growing crops and livestock for human consumption, and of trapping and using energy. [INFOGRAPHIC 1.5]

Despite such efforts, many critics say that modern global societies are not acting nearly as quickly as they could or should. Once again, the charge echoes those archaeologists have levelled at the Vikings of Greenland. To understand what prevents us from changing our ways even in the face of brewing calamity, it helps to understand why the Greenland Vikings failed to do the same.

Humanity faces some challenges in dealing with environmental issues.

Experts agree that if the Greenland Vikings had simply switched from cows to fish they might have avoided their own demise. Such a switch would have saved at least some of the pastureland, not to mention the tremendous amount of time and labour it took to raise cattle in such an ill-suited environment. But cows were a status symbol in Europe, beef was a coveted delicacy, and for reasons that still elude researchers, the Greenland Vikings had a

Infographic 1.5 | **SUSTAINABLE ECOSYSTEMS CAN BE A USEFUL MODEL FOR HUMAN SOCIETIES**

SUSTAINABLE ECOSYSTEMS:	WE CAN MIMIC THIS BY:
RELY ON RENEWABLE ENERGY	**USING SUSTAINABLE ENERGY SOURCES** We can move away from nonrenewable fuels such as fossil fuels by turning to sustainable energy sources such as solar, wind, geothermal, and biomass (harvested at sustainable rates).
USE MATTER SUSTAINABLY	**USING MATTER CONSERVATIVELY AND SUSTAINABLY** We can reduce our waste by recovering, reusing, and recycling matter that we do use; we would also benefit from minimizing the toxins we create or release into the environment that degrade our natural resources.
HAVE POPULATION CONTROL	**GETTING HUMAN POPULATION GROWTH UNDER CONTROL** While predation controls many natural populations, there are many ways to reduce human birth rates without increasing the death rate through war or disease.
DEPEND ON LOCAL BIODIVERSITY TO MEET THE FIRST THREE REQUIREMENTS	**DEPENDING ON LOCAL HUMAN CONTRIBUTIONS AND BIODIVERSITY** Protecting biodiversity will help us achieve the three above goals; we can emulate nature by using a variety of local energy sources, building materials, and crops, and by exploring the many ideas and innovations that come from a diverse human community.

↑ A replica of the first church built in Brattahlid, Greenland—a Viking settlement founded by Erik the Red more than 1000 years ago.

cultural taboo against fish. It was a taboo they clung to, even as they starved to death.

"It's clear that the Viking's decisions made them especially vulnerable to the climatic and environmental changes that descended upon them," says McGovern. "When you're building wooden cathedrals in a land without trees, when you create a society reliant on imports in a situation where it's difficult to travel back and forth to the homeland, when you absolutely refuse to collaborate with your neighbours in such a harsh, unforgiving environment, you are setting yourself up for trouble." Ultimately, he says, the decisions this society made played as much of a role in its demise as the actual environmental changes did.

nonrenewable resources Resources whose supply is finite or not replenished in a timely fashion.
social traps Decisions by individuals or groups that seem good at the time and produce a short-term benefit, but that hurt society in the long run.
tragedy of the commons The tendency of an individual to abuse commonly held resources in order to maximize his or her own personal interest.
time delay Actions that produce a benefit today set into motion events that cause problems later on.
sliding reinforcer Actions that are beneficial at first but that change conditions such that their benefit declines over time.

The problem, McGovern says, is not that the Greenland Vikings made so many mistakes in their early days, but that they were unwilling to change later on. "Their conservatism and rigidity, which we can see in many different aspects, seems to have kept them on the same path, maybe even prodded them to try even harder—build bigger churches, etc.—instead of trying to adapt."

Decisions by individuals or groups that seem good at the time and produce a short-term benefit, but that hurt society in the long run, are called **social traps**. The **tragedy of the commons** is a social trap, first described by ecologist Garrett Hardin in 1968, that often emerges when many people are using a commonly held resource such as water or public land. Each person will act in a way to maximize his or her own benefit but as everyone does this, the resource becomes overused or damaged. Herders might put more animals on a common pasture because they are driven by the idea that "if I don't use it, someone else will." We do the same thing today as we overharvest forests and oceans or release toxins into the air and water. Other social traps include the **time delay** and **sliding reinforcer** traps—actions that, like the tragedy of the commons, have a negative effect later on. [INFOGRAPHIC 1.6]

Infographic **1.6** | **SOCIAL TRAPS**

↓ Social traps are decisions that seem good at the time and produce a short-term benefit but that hurt society (usually in the long run).

TRAGEDY OF THE COMMONS ↓ When resources aren't "owned" by anyone (they are commonly held), individuals who try to maximize their own benefit end up harming the resource itself.

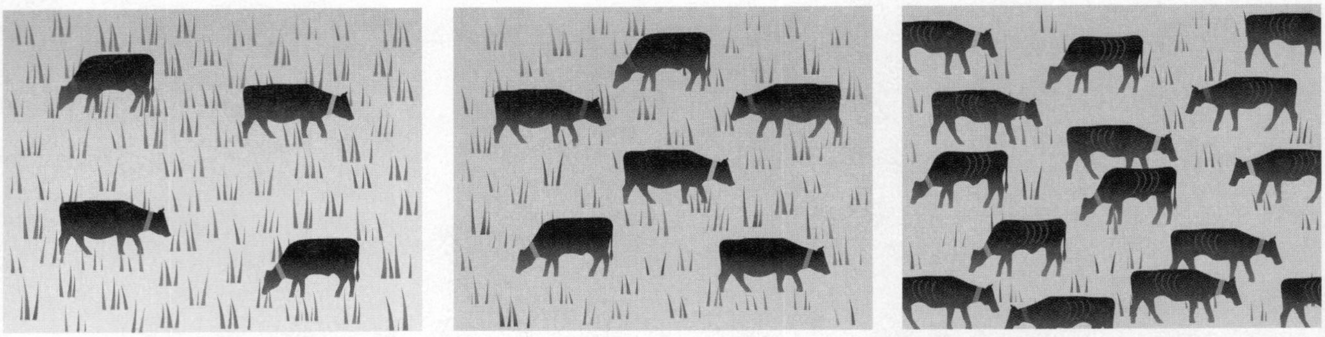

This common resource (pasture) can support four animals sustainably. Each of the four farmers benefits equally.

If one farmer adds two more cows, the commons becomes degraded. All four farmers share in the degradation (less milk produced per cow), but the farmer with three cows gets all the benefit from adding the cows.

The other farmers also must add cows to return to the same level of production as before. **Over time, the commons degrades even more and at some point will no longer support any animals.**

TIME DELAY ↓ Actions that produce a benefit today set into motion events that cause problems later on.

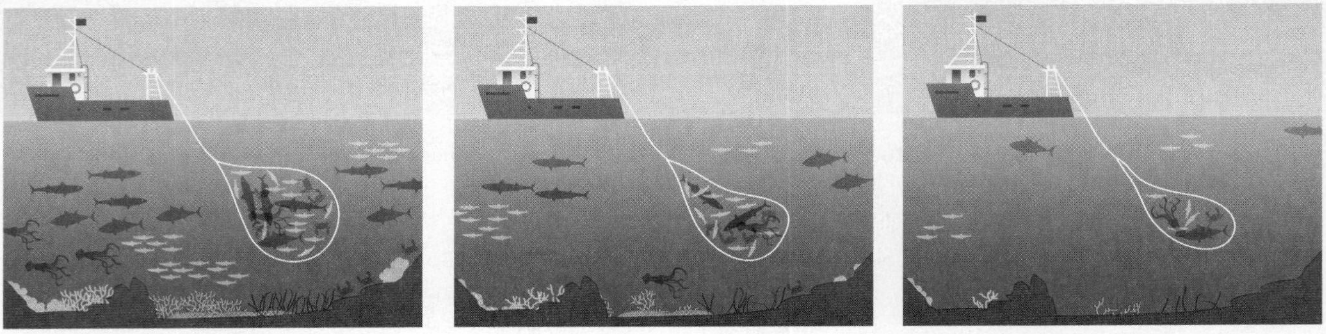

Modern fishing techniques that use giant nets can harvest large numbers of fish. Fishers try to get the most fish possible in the short term.

If more fish are taken over a number of years than are replaced naturally through fish reproduction, the population decreases. Other populations also decline as the fish they depend on are depleted.

After a decade or so, overfishing may so deplete the population that fishers cannot catch enough to meet needs. **The effects of decisions about how many fish to harvest are not felt until later.**

SLIDING REINFORCER ↓ Actions that are beneficial at first may change conditions such that their benefit declines over time.

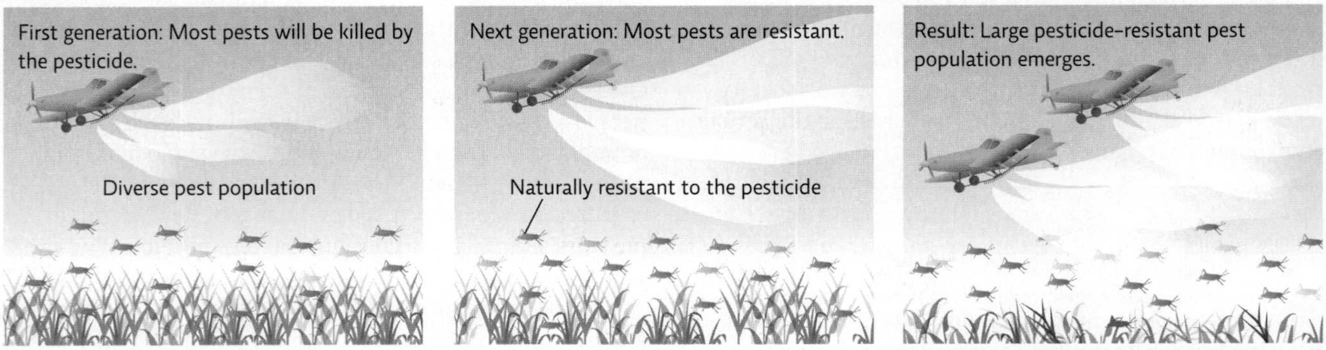
First generation: Most pests will be killed by the pesticide.
Diverse pest population
Next generation: Most pests are resistant.
Naturally resistant to the pesticide
Result: Large pesticide-resistant pest population emerges.

Pesticide application can reduce pest numbers on a crop, but a few pests might survive.

The surviving pests reproduce. The pesticide is only helpful if more is applied or a more toxic pesticide is used. This can be harmful to other organisms and to humans.

In addition to becoming resistant to the pesticide, the pest population may even become bigger if the pest's predators are also killed. **What was helpful at first is no longer helpful and is even harmful over time.**

Infographic **1.7** | **WEALTH INEQUALITY**

↓ As there was in the Greenland Viking colony, there is disparity in wealth worldwide. The World Bank estimates that about 40% of the world's population lives on less than $2 a day; almost half of the 2.2 billion children worldwide live in poverty. A small percentage of our population (perhaps 20%) controls 80% of all resources. Actions by the wealthy today that degrade the environment degrade it for all sooner or later.

Percentage of the population living on less than $2 per day
Under 2%
2-5%
6-20%
21-40%
41-60%
61-80%
Over 80%
No data

Education is our best hope for avoiding such traps. When people are aware of the consequences of their decisions, they are more likely to examine the trade-offs to determine whether long-term costs are worth the short-term gains and, hopefully, to make different choices when necessary. In Canada, environmental laws, such as the Canadian Environmental Protection Act and the Species at Risk Act, have gone a long way toward protecting our natural environment from various social traps, but enforcing those laws is a constant challenge.

Social traps are not the only challenge societies face. Another obstacle to sustainable growth is wealth inequality. In the Greenland Viking colony, wealth at first insulated the people in power from the environmental problems and they didn't feel the strain of the decline until it was too late. Today, wealthier nations are less affected by resource availability, while 2 billion or more people lack adequate resources to meet their needs. In fact, just 20% of the population controls roughly 80% of all the world's resources. On one hand, the affluent minority (of which Canada is a part) uses more than its share. Deep pockets allow us to exploit resources for wants, not just needs, and to exploit them all over the world, so that we can spare our own natural environments at the expense of someone else's. For example, our demand for coffee and other luxury food and beverage products drives deforestation in Central and South America where the tree flourishes. Because we are far away from the trees, the environmental fall-out is easy to ignore at the moment, but as it did for the Vikings, it may reach a point where it affects everyone.

On the other hand, the underprivileged also exploit the environment in an unsustainable way. With limited access to external resources, they are often forced to overexploit their immediate surroundings just to survive. The above example applies here, too. For an impoverished landowner in Costa Rica or Brazil, chopping down trees may be the only way she can feed her family—to her, harvesting the forest is more valuable than preserving it.

There are societal costs, as well, to this inequality. A growing gap between the haves and have-nots exacerbates tension and strife all over the world. In fact, fighting over resources has long been one of the contributing factors to societal decline and collapse. [INFOGRAPHIC 1.7]

Conflicting worldviews are another challenge to sustainable living. Because our **worldviews**—the windows through which we view our world and existence—are influenced by cultural, religious, and personal experiences, they vary across countries and geographic regions, even within a society. People's worldviews determine their **environmental ethic**, or how they interact with their natural environment; they also impact how they

worldviews The window through which one views one's world and existence.
environmental ethic The personal philosophy that influences how a person interacts with his or her natural environment and thus affects how one responds to environmental problems.

Infographic **1.8** | **WORLDVIEWS AND ENVIRONMENTAL ETHICS**

↓ People's environmental worldviews describe how they see themselves in relation to the world around them. Their worldview influences their environmental ethic, which in turn influences how they interact with the natural world. We present three common worldviews here (there are others). Which one most closely aligns with your own view?

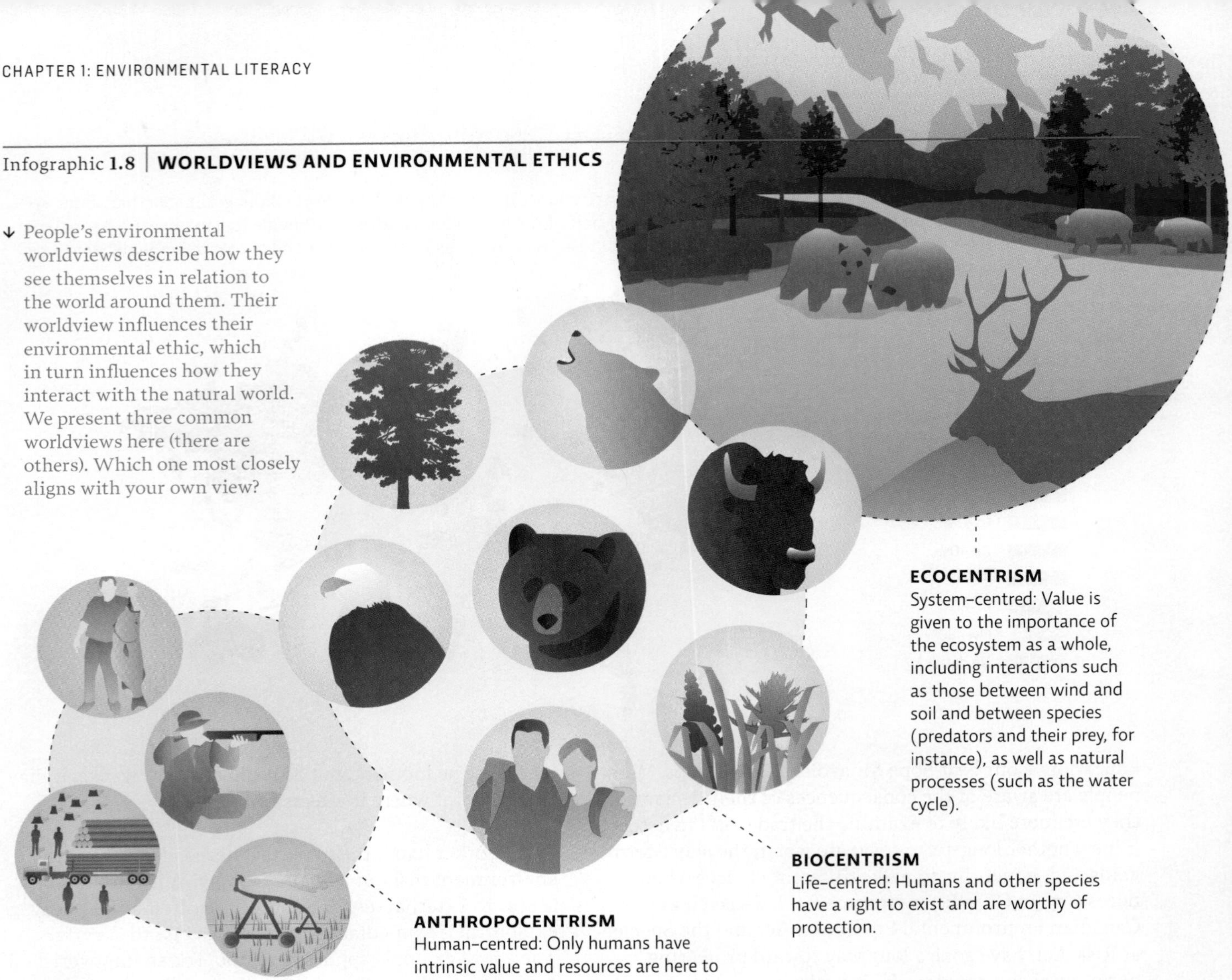
ECOCENTRISM
System-centred: Value is given to the importance of the ecosystem as a whole, including interactions such as those between wind and soil and between species (predators and their prey, for instance), as well as natural processes (such as the water cycle).
BIOCENTRISM
Life-centred: Humans and other species have a right to exist and are worthy of protection.
ANTHROPOCENTRISM
Human-centred: Only humans have intrinsic value and resources are here to meet human needs and wants.

respond to environmental problems. When different people or groups, with different worldviews, approach environmental problems, they are bound to draw very different conclusions about how best to proceed.

The Vikings had an **anthropocentric worldview**—one where only human lives and interests are important. Other species meanwhile, were seen as having **instrumental value**—meaning the Vikings valued them only for what they could get out of them. Forests were nothing more than a source of timber; grasslands a source of home insulation and a feeding ground for cattle.

A **biocentric** worldview values all life. From a biocentric standpoint, every organism has an inherent right to exist, regardless of its benefit (or harm) to humans; each organism has **intrinsic value**. This worldview would lead us to be mindful of our choices and avoid actions that indiscriminately harm other organisms or put entire species in danger of extinction. An **ecocentric worldview** values the ecosystem as an intact whole, including all of the ecosystem's organisms and the nonliving processes that occur within the ecosystem. Considering the same forests and grassland from an ecocentric worldview, the Vikings might have decided to protect both, not just for the resources they could harvest but to protect the complex processes that can produce those resources only when they remain intact. [INFOGRAPHIC 1.8]

anthropocentric worldview A human-centred view that assigns intrinsic value only to humans.
instrumental value The value or worth of an object, organism, or species is based on its usefulness to humans.
biocentric worldview A life-centred approach that views all life as having intrinsic value, regardless of its usefulness to humans.
intrinsic value The value or worth of an object, organism, or species is based on its mere existence.
ecocentric worldview A system-centred view that values intact ecosystems, not just the individual parts.

An ecocentric worldview values all living creatures and nonliving processes of an ecosystem, from animals and wildflowers to nutrient cycling in the soil and water flow from snowmelt. Sassalb mountain range, Switzerland.

Infographic **1.9** | **CANADIAN ENVIRONMENTAL HISTORY**

↓ The predominant worldview of a society shapes its actions and the choices made by its citizens at any given point in time. Throughout history, notable individuals with different worldviews have greatly influenced the evolution of our relationship with the natural world. The gradual shift from unrestrained use to conservation to sustainability that was seen in Canada and some other developed countries has now become an international movement.

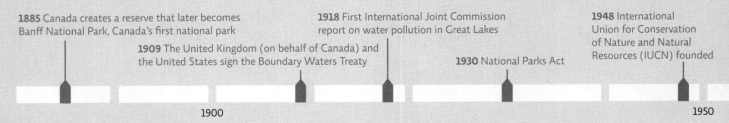

1885 Canada creates a reserve that later becomes Banff National Park, Canada's first national park

1909 The United Kingdom (on behalf of Canada) and the United States sign the Boundary Waters Treaty

1918 First International Joint Commission report on water pollution in Great Lakes

1930 National Parks Act

1948 International Union for Conservation of Nature and Natural Resources (IUCN) founded

1900

1950

EXPANSION

With the technological advancements of the Industrial Revolution in the 1800s, efficiency of resource extraction and manufacturing of goods from raw products expanded.

RESOURCE MANAGEMENT

In the late 1800s and early 1900s, the first pieces of restrictive environmental legislation were passed, creating national parks, forests, and reserves to limit resource use.

Maurice Strong is a legend in Canadian environmental policy development. While his early background was in energy and finance business, he went on to become Secretary General of both the 1972 United Nations Conference on the Human Environment and the 1992 Rio Environmental Summit. He was also the first Executive Director of the United Nations Environment Programme (UNEP).

Elizabeth May is probably Canada's best-known environmental activist, and the first leader of Canada's Green Party. Formerly Executive Director of the Sierra Club of Canada, she became the first elected Green Party Member of Parliament in May 2011.

Each of these worldviews may be found among citizens of Canada today, but we can also trace a historical and gradual change in the way Canada has viewed the natural world. In fact, we can see these worldviews expressed in the ethical positions of many prominent environmental scientists, politicians, and citizens whose actions are associated with landmark events in environmental history. [INFOGRAPHIC 1.9]

Back in Greenland, in the silt-covered ruins of a Viking farmhouse, archaeologists found the bones of a hunting dog and a newborn calf. The knife marks covering both indicate that the animals were butchered and eaten. "It shows how desperate they were," says McGovern. "They would not have eaten a baby calf, or a hunting dog, unless they were starving." By then, it was probably far too late for the Greenland Vikings to adapt in any meaningful way. Their tale had already been written—in ice cores and mud cores and in hundreds of years' worth of midden heaps.

Our own story is being written now, in much the same way. But unlike our forebears, we still have enough time to shape our own narrative. Will it be dug up, a thousand years hence, from the ruins of what we leave behind? Or will it be passed down by the voices of our successors, who continue to thrive long after we are gone? Ultimately, the answer is up to us.◉

Select references in this chapter:
Benyus, J. 1997. *Biomimicry*. New York: William Morrow and Company.
Diamond, J. 2005. *Collapse*. New York: Viking Press.
McGovern, T. H. 1980. *Human Ecology*, 8: 245–275.

- **1962** *Silent Spring* published
- **1971** Department of the Environment created
- **1972** United Nations Environment Programme is created
- **1972** Canada and United States sign the first Great Lakes Water Quality Agreement
- **1987** Montreal Protocol developed to curb CFC production
- **1988** Canada adopts Environmental Protection Act (CEPA)
- **1988** United Nations creates Intergovernmental Panel on Climate Change
- **1991** Canada–United States Air Quality Accord to curb acid rain
- **1997** Kyoto Protocol to reduce global greenhouse gas emissions
- **2002** Canadian Species at Risk Act (SARA) passed
- **2005** First Millennium Ecosystem Assessment
- **2011** 191 countries ratify the Kyoto Protocol to reduce greenhouse gas emissions, but the United States does not ratify it, and Canada withdraws from it

2000

POLLUTION CONCERNS

In the 1960s, concerns over Great Lakes pollution stimulated research and policy development. The International Joint Commission advised Canada and the United States to reduce phosphorus discharges to the Great Lakes to control eutrophication, which was causing massive algal blooms and dead zones in Lake Erie.

INTERNATIONAL EFFORTS

Today, improving the environment requires international cooperation and coordinated efforts. International groups such as the Intergovernmental Panel on Climate Change work to understand climate change; many nonprofit organizations address environmental issues, not just in their own region or nation, but around the world.

Rachel Carson's 1962 book, *Silent Spring*, opened a new dialogue on the dangers of toxins in our environment. She provided evidence that these chemicals could have unexpected effects on wildlife and that we too were vulnerable to these poisons (see Chapter 3).

Wangari Maathai (1940 - 2011) began the Greenbelt Movement in Kenya in 1977 to reforest severely damaged areas of her country. The movement has spread to more than 15 African nations, and to the United States and Europe, and has planted more than 30 million trees (see Chapter 11).

BRING IT HOME

➔ PERSONAL CHOICES THAT HELP

The concept of sustainability unites three main goals: environmental health, economic profitability, and social and economic equity. All sorts of people, philosophies, policies, and practices contribute to these goals; concepts of sustainable living apply on every level, from the individual to the society as a whole. In other words, every one of us participates. Throughout this book, you will have the opportunity to learn about personal actions that can help address environmental issues but a good starting place is to learn about your own environment and the place you call home.

Individual Steps

→ Discover your local environment. What parks or natural areas are close by? Does your campus have any natural areas? Visit one and spend a little time just observing nature and your own reactions. Write down your thoughts or share you experiences with a friend.

→ Are there restaurants, grocery stores, or other retail venues accessible through public transportation or within walking distance of your campus or home? Try walking or riding a bus to these businesses instead of driving to others farther away for a week or two. Is this a reasonable option for you? Why or why not?

Group Action

→ Discover your own interests. There is a group for every interest—from outdoor recreation, wildlife viewing and preservation, and environmental education, to transportation and air-quality issues. Get involved with organizations working to improve environmental issues or address social change and human rights. One person can make a difference but a group of people can cause a sea change.

Policy Change

→ Discover what's happening in your community. Read the newspapers and monitor blogs covering environmental and quality-of-life issues. Alert your local, regional, and national representatives about the issues you care about, and vote for government officials who support the causes you support.

UNDERSTANDING THE ISSUE

CHECK YOUR UNDERSTANDING

1. Environmental science is:
a. the physical surroundings or conditions in which any given organism exists or operates.
b. an interdisciplinary field of research that seeks to understand the natural world and our relationship to it.
c. the ability to understand environmental problems.
d. a problem that is extremely complicated and tends to have multiple causes, each of which can be difficult to address.

2. According to the United Nations' Millennium Ecosystem Assessment:
a. human actions are straining the planet's ecosystems, imperilling the ability to sustain future generations.
b. sustainable development now ensures that the planet's ecosystems will sustain future generations in perpetuity.
c. human actions have irreversibly damaged the planet's ecosystems; they will not sustain future generations.
d. developing technological fixes to reverse the damage caused to the planet's ecosystems is the only way to assure sustaining future generations.

3. Which of the following demonstrates a lack of environmental literacy?
a. A cooling climate caused summer hay yields to shrink, which lowered livestock survival over the longer, colder winters.
b. The Greenland Viking middens show no evidence of the Vikings eating locally available fish—instead they continued to eat beef, scraping off every last bit of meat and marrow from the bones.
c. The Inuit were expert hunters of ringed seal, which were an abundant food source in the Arctic.
d. The Icelandic Vikings switched to eating fish when they realized that grazing cattle were not suited to Iceland's shorter growing season.

4. Fishing while allowing sufficient numbers to be left behind for the fish population to regenerate reflects:
a. gloom and doom thinking.
b. a frontier attitude.
c. sustainable resource use.
d. tragedy of the commons.

5. Which of the following is NOT an example of a social trap?
a. Sliding reinforcer trap
b. Time delay trap
c. The tragedy of the commons
d. Education trap

6. Jack is opposed to the selective killing of deer that have vastly overpopulated the local forest. Jill argues that the ecosystem will be destroyed if some are not removed. While Jack is _______ as he sees deer as having intrinsic value, Jill is _______ as she values not just the species but the ecosystem processes as well.
a. biocentric; ecocentric
b. ecocentric; biocentric
c. anthropocentric; ecocentric
d. biocentric; anthropocentric

WORK WITH IDEAS

1. Why are environmental problems considered "wicked problems"? What factors make addressing these wicked problems a challenge?

2. According to Jared Diamond, what are the five factors that determine whether any given society will succeed or fail? How did these factors play out for the Vikings in Greenland versus those in Iceland, and what were the consequences for each society? Provide evidence to support your conclusions.

3. In what ways is contemporary society both similar to and different from the Greenland Viking society? What can we learn from that society's mistakes to ensure that we do not suffer the same fate?

4. How can human societies become more sustainable? What are the challenges to achieving sustainability?

ANALYZING THE SCIENCE

The following data show the number of different types of livestock grazed in the world and in Nigeria from 1961–2008. Use these graphs and this table to answer the next 5 questions.

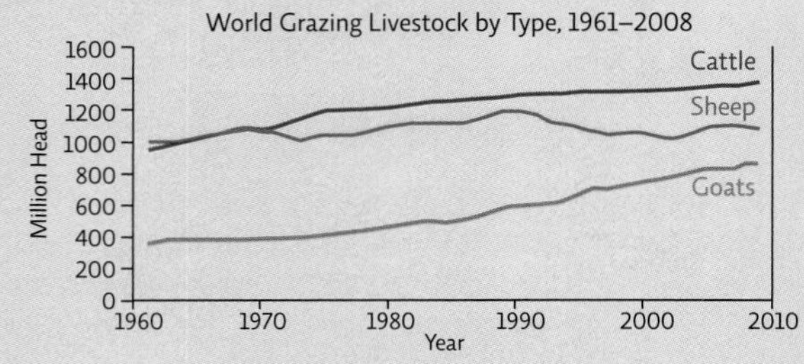
World Grazing Livestock by Type, 1961–2008
Million Head
Year
Cattle
Sheep
Goats
1600
1400
1200
1000
800
600
400
200
0
1960
1970
1980
1990
2000
2010

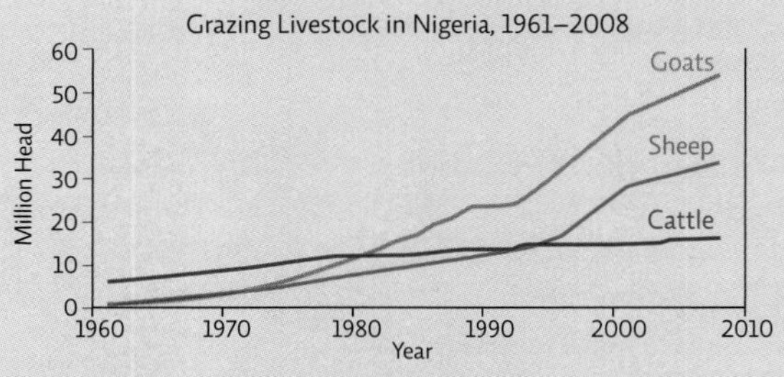
Grazing Livestock in Nigeria, 1961–2008
Million Head
Year
Goats
Sheep
Cattle
60
50
40
30
20
10
0
1960
1970
1980
1990
2000
2010

Grazing Livestock Population (in millions)

Type of livestock	World in 1961	World in 2008	% change	Nigeria in 1961	Nigeria in 2008	% change
Cattle	942	1372		6	16.3	
Goats	349	864		0.6	53.8	
Sheep	994	1086		1	33.9	

INTERPRETATION

1. How many different livestock types are included in the graphs and what are the trends in the numbers?

2. Which animals constituted the bulk of grazing livestock around 2008 on a worldwide basis? In Nigeria? Was this true in 1961 (the first year for which data are reported)? Use the data to explain your responses.

3. Using the data from the table, calculate the percent change for each type of animal. According to your calculations which animal changed the most in each case (i.e., worldwide and in Nigeria)?

Hint: percent change = $\frac{(\text{2nd value} - \text{1st value})}{\text{1st value}} \times 100$

ADVANCE YOUR THINKING

4. Unlike cattle, goats are very flexible in what they can eat, but their sharp hooves also pulverize soil more easily. The data show that the growth in goat populations is particularly dramatic in a developing country like Nigeria. What might explain this pattern? And what might be some potential consequences?

5. How might the increase in the size of the goat herd be a potential social trap? What are some ways that we could avoid this trap?

EVALUATING NEW INFORMATION

The Lorax, a children's book by Dr. Seuss, tells the story of the Lorax, a fictional character who speaks for the trees against the greedy Once-ler who represents industry. Written in 1971, *The Lorax* was banned in parts of the United States for being an allegorical political commentary. Today the book is used for educating children about environmental concerns (see http://www.seussville.com/loraxproject/). Even so, some people consider the book inappropriate for young children due to its "doom and gloom" environmentalism.

The book *The Truax*, by Terri Birkett, involves a forest industry representative offering a logging-friendly perspective to an anthropomorphic tree, known as the Guardbark. This story was criticized for containing skewed arguments, and in particular a nonchalant attitude toward endangered species. Hundreds of thousands of copies of this book have been read by elementary school students across North America.

Read both books. You can find *The Lorax* at your local public library and *The Truax* can be downloaded as a pdf from http://woodfloors.org/truax.pdf.

Evaluate the stories and work with the information to answer the following questions:

1. What are the credentials of the author of each book? In each case, do the person's credentials make him or her reliable/unreliable as a storyteller? Explain.

2. Connect each story to the key concepts in the chapter:
   a. What are the underlying attitudes and worldviews of each story?
   b. Does each story reflect social traps, and if so in what way?
   c. How might each story contribute to environmental literacy? Explain.
   d. What does each story have to say about sustainability? Explain.

3. What supporting evidence can you find for the main message in each story? In the story itself? From doing some research?

4. What is your response to each story? What do you agree and disagree with in each case? Explain.

MAKING CONNECTIONS

THE ARCTIC AND THE FUTURE OF A GLOBAL COMMONS

Background: According to a 2008 United States Geological Survey report, the Arctic has 90 billion barrels of oil—13% of the undiscovered oil in the world. Now, evidence suggests that warming in the Arctic has already resulted in earlier break-up of ice in the spring and thinner ice year round. Computer models further suggest that by 2080 Arctic sea ice will completely disappear during the summer months. Ironically, the disappearance of ice will make exploration of oil resources in the Arctic more feasible, while the burning of oil and other fossil fuels have been blamed for the loss of ice in the first place.

As temperatures increase and sea ice declines, many arctic species will find it difficult to survive. Meanwhile, the need to build a massive infrastructure for oil drilling will impact ecologically intact areas. And a single oil spill can have serious impacts, especially given that there is no effective method for containing and cleaning up an oil spill in ice conditions.

Although no country owns the North Pole, there is growing concern among the Inuit that new measures are needed to preserve and sustainably use the Arctic's resources. Though not opposed to development, the Inuit want to ensure that their communities benefit without compromising the environment.

Case: The Inuit have selected you to help them develop a vision for the future of the Arctic. Based on your research and analysis, write a position paper on how Arctic resources should be utilized and how this region should be governed to ensure sustainability and consideration of the "triple-bottom line."

In your report include the following:

a. An assessment of the importance of the Arctic region as a global commons and the challenges of governing it as such.
b. An analysis of how the situation in the Arctic presents a "wicked problem."
c. A discussion of how we can prevent "ecocide" in the Arctic—specific strategies to minimize self-inflicted environmental damage, appropriately respond to ongoing environmental changes, and ensure global cooperation to sustainably utilize resources in the Arctic.
d. The potential challenges to developing and implementing a sustainable Arctic policy and how these challenges could be addressed.

Set-up for launch of the high-altitude research balloon for ozone testing. McMurdo Station, Antarctica.

SCIENCE AND THE SKY

Solving the mystery of disappearing ozone

CORE MESSAGE

Depletion of the stratospheric ozone caused by synthetic chemicals allows more dangerous solar radiation to reach Earth's surface, threatening the health and well-being of many plants and animals, including humans. Researchers employed the scientific process to collect physical evidence, analyze it, and report their findings to the scientific community. This helped us understand what was causing the depletion, and guided wise policy decisions at the national and international level to address this tragedy of the commons.

GUIDING QUESTIONS

After reading this chapter you should be able to answer the following questions:

→ What kinds of questions are under the purview of science and why is science limited in this way? Why do we say science is a "process" and that conclusions are always open to further study?

→ How are scientific hypotheses generated and tested? What are the two main types of scientific studies and why do we need both types?

→ Why is ozone in the stratosphere beneficial to life on Earth and how is it being depleted? What is the evidence that ozone depletion might be harmful to the health of humans or other organisms?

→ How did scientists use the process of science to help us understand what was happening with the ozone layer and what unknowns still exist?

→ How did scientists, policy-makers, and world leaders take the science about ozone depletion and turn it into policy?

There's a point of no return halfway into the 9-hour flight from New Zealand to the Antarctic. Once that 4.5-hour mark passes, if something goes wrong with the plane, there's nowhere to stop in the Southern Ocean for repairs.

Susan Solomon is very familiar with that trip's all-important midway point. During her very first flight to the southernmost continent, the pilot told the passengers that their plane was not working properly: the front ski at the nose of their C-130 was frozen and couldn't be lowered into position, so landing on the packed snow at the Antarctic research station would be impossible. They had to turn around.

It was August, 1986—late winter in the Antarctic—and the atmospheric chemist was on her way to the southern continent to investigate a mystery: why was the stratospheric ozone above the South Pole disappearing? Suddenly, the remoteness of where she was going hit home.

"I remember, as we were flying back to New Zealand, thinking, 'Wow, I really am going to the Antarctic,'" Solomon says.

The next night, the atmospheric scientist and her team from various research institutions, including the National Oceanic and Atmospheric Administration (NOAA), managed to make it to Antarctica, landing at McMurdo Station. On her very first excursion to the Antarctic,

◉WHERE IS McMURDO STATION?

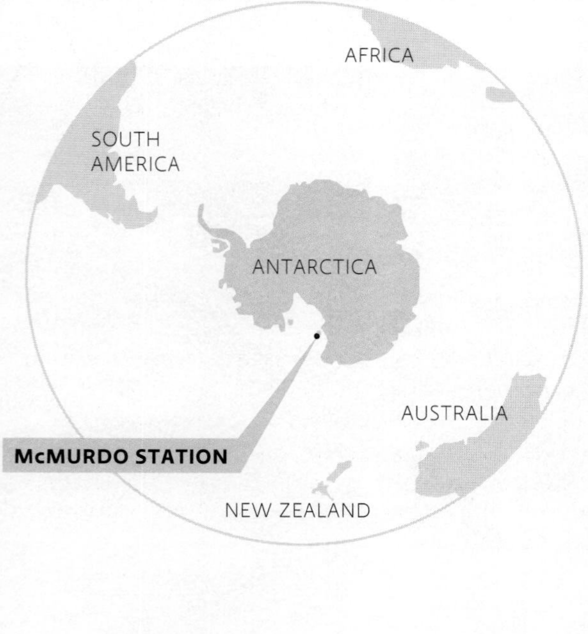
AFRICA
SOUTH AMERICA
ANTARCTICA
AUSTRALIA
McMURDO STATION
NEW ZEALAND

Solomon and her colleagues collected the initial data that would eventually grab the world's attention and settle a long-standing, hard-fought scientific debate that was taking place on an international stage.

Science gives us tools to observe the natural world.

Solomon and the team had decided to fly to the other end of the world after reading a scientific paper published the year before, in which Joe Farman of the British Antarctic Survey and his colleagues showed that, since the late 1970s, the ozone layer had thinned by about a third during the Antarctic spring.

The British team had collected nearly three decades of data in Antarctica, starting in 1957 with on-the-ground instruments. Like all good scientists, Farman and his team depended on **observations** (information detected with the senses or with equipment that extends our senses) of the natural world. His team collected data on the atmosphere's composition (*lower than normal ozone levels*) and then used these observations to draw conclusions or make **inferences**—explanations of what else might be true or what might have caused the observed phenomenon.

observations Information detected with the senses—or with equipment that extends our senses.
inferences Conclusions we draw based on observations.
atmosphere Blanket of gases that surrounds Earth and other planets.
troposphere Region of the atmosphere that starts at ground level and extends upward about 11 kilometres.
stratosphere Region of the atmosphere that starts at the top of the troposphere and extends up to about 50 kilometres; contains the ozone layer.
ozone Molecule with 3 oxygen atoms that absorbs UV radiation in the stratosphere.

Farman's group also connected their results to studies by other researchers that had shown higher concentrations of an important compound: chlorofluorocarbons (CFCs), which in turn produce atmospheric chlorine (Cl). The concentrations of these chemicals seemed to increase at a rate matching the disappearance of ozone. The observation of a decrease in ozone did not come from just a few readings, but represented data collected at two different sites over more than a dozen seasons. *Replication* within a study (multiple test subjects or measurements) and between studies (independent tests that collect the same data, preferably conducted by other researchers) increase the reliability of the data. When replicates produce similar results it is less likely that the original data was "a fluke" or an unusual response. Farman's research exemplified this hallmark of good science—in this case multiple data points at two different testing sites. His team's results have subsequently been confirmed by other researchers.

↑ Top: The McMurdo Station, the largest human settlement in the Antarctic. Bottom: Susan Solomon at the McMurdo Station in 1987.

Farman's team inferred that the ozone depletion in the Antarctic was somehow connected to the increased presence of chlorine compounds in the atmosphere.

It was a serious proposition: without ozone, the world as we know it would not exist. Ozone is a key element of the **atmosphere**, the blanket of gases surrounding our planet that is made up of discernable layers, which differ in temperature, density, and gas composition.

The lowest level, the **troposphere**, extends about 11 kilometres up. This level is familiar to us—it is the air we breathe and where our weather occurs. The next level in the atmospheric blanket, the **stratosphere**, rises about 50 kilometres above Earth's surface. The stratosphere is much less dense than the troposphere but contains a "layer" of **ozone** (abbreviated as O_3, because it contains 3 oxygen atoms), a region where most of the atmosphere's ozone is found.

Infographic **2.1** | **THE ATMOSPHERE AND UV RADIATION**

↓ The atmosphere is composed of vertical layers that differ in temperature and chemical composition. There is not much ozone in the atmosphere, but most of what ozone there is occurs in a "layer" in the stratosphere. Ozone is important because it prevents some UV-radiation from reaching Earth's surface.

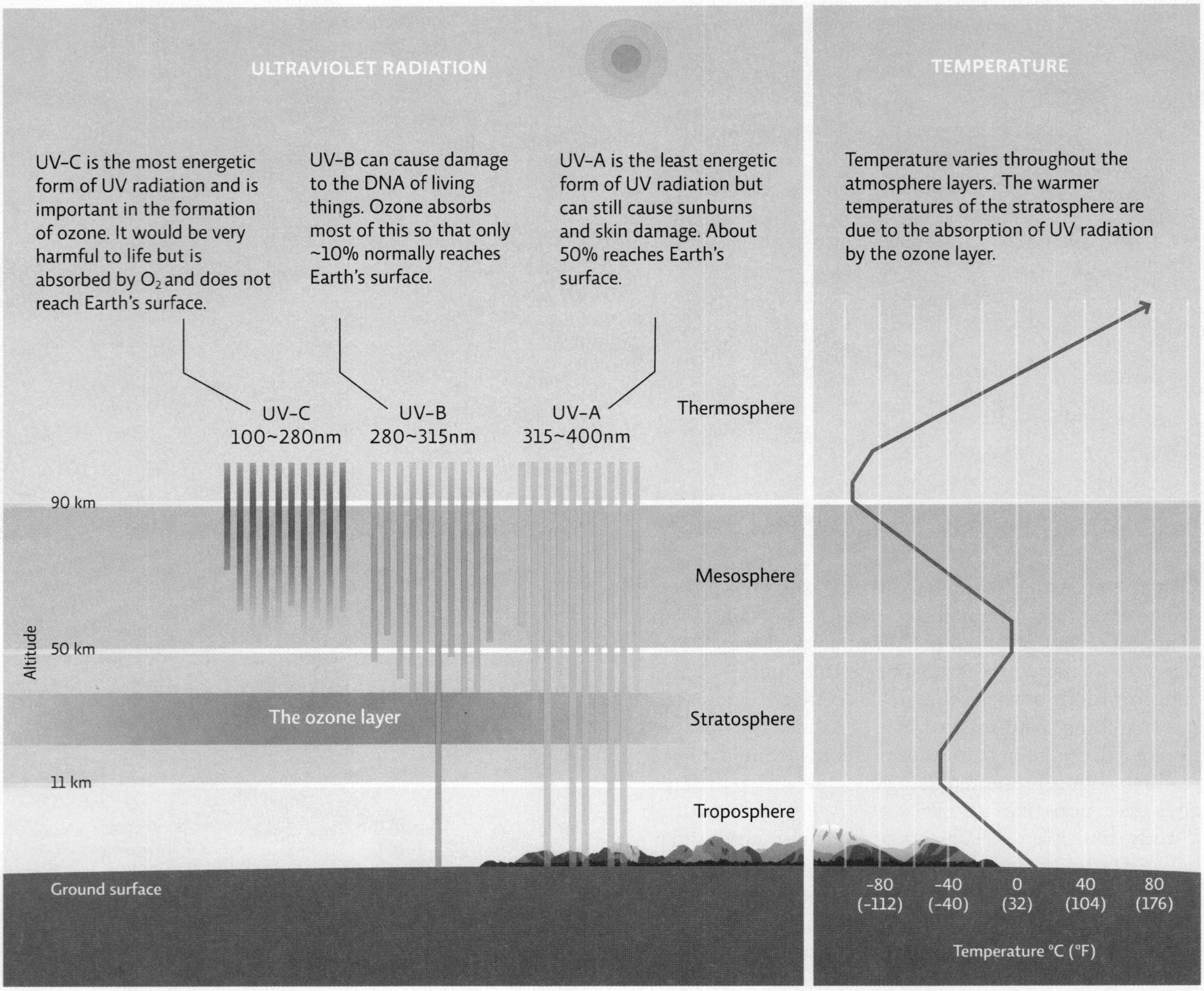
ULTRAVIOLET RADIATION
TEMPERATURE
UV-C is the most energetic form of UV radiation and is important in the formation of ozone. It would be very harmful to life but is absorbed by O_2 and does not reach Earth's surface.
UV-B can cause damage to the DNA of living things. Ozone absorbs most of this so that only ~10% normally reaches Earth's surface.
UV-A is the least energetic form of UV radiation but can still cause sunburns and skin damage. About 50% reaches Earth's surface.
Temperature varies throughout the atmosphere layers. The warmer temperatures of the stratosphere are due to the absorption of UV radiation by the ozone layer.
UV-C 100~280nm
UV-B 280~315nm
UV-A 315~400nm
Thermosphere
Mesosphere
Stratosphere
Troposphere
90 km
50 km
11 km
Altitude
The ozone layer
Ground surface
-80 (-112)
-40 (-40)
0 (32)
40 (104)
80 (176)
Temperature °C (°F)

Without ozone, the world as we know it would not exist.

Solar radiation enters the atmosphere every day, including three forms of **ultraviolet (UV) radiation**: UV-A, UV-B, and UV-C. Ozone molecules in the stratosphere absorb much of the UV-B, a vital service since UV-B can damage cells and biological molecules like DNA. In humans, exposure to UV-B radiation increases the risk of cataracts, skin damage, and cancer. Fortunately, ozone prevents most of the UV-B from reaching Earth's surface, where it can harm organisms. This stratospheric (good) ozone should not be confused with ground-level (bad) ozone found in the troposphere. Ground-level ozone is a component of smog and is harmful to living things (see Chapter 21). [INFOGRAPHIC 2.1]

Susan Solomon became a scientist because she was curious about the natural world around her. **Science** is both a body of knowledge (facts and explanations) and the process used to get that knowledge. Understanding the process is more important than the "facts," since facts may change as more information is collected through the scientific process. This process is a powerful tool that allows us to gather evidence to test our ideas, *and* to evaluate the quality of that evidence.

Science, however, is limited to asking questions about the *natural* world—not all questions are open to science. Scientific investigation, in both the natural and social sciences, is based on data gathered through **empirical evidence**, or observations. Only physical phenomena that can be objectively observed—meaning data that could be collected by *anyone* in the same place, using the same equipment, etc.—are fair game for science. Scientists can gather empirical evidence about the environment and living things using a wide variety of tools, including natural tools, such as their eyes, ears, and other senses. Phenomena that are not objectively observable (What is my dog thinking? Do ghosts exist?) and ethical or religious questions (Is economic growth more important than environmental health? What is the meaning of life?) cannot be empirically studied, and therefore are not under the purview of science.

Arriving at McMurdo Station, Solomon knew she had to apply the scientific process to understand why the ozone layer was thinning in the Antarctic. And she already had a culprit in mind.

The scientific view of CFCs did not change overnight.

Throughout Solomon's childhood, she was surrounded by items that contained the synthetic molecules known as CFCs. The compounds were first developed in the 1930s as a commercial coolant, to replace more toxic ammonia and sulphur dioxides, and were therefore included in refrigerators, air conditioners, and other household and industrial items. By the time Solomon reached graduate school at the University of California, Berkeley, in the late 1970s, CFCs were being used in everything from hairspray to foam containers for fast food.

CFCs contain atoms of carbon, fluorine, and chlorine. For example, CFC_{12}, a common refrigerant, has the molecular formula CCl_2Fl_2—one carbon atom (C) is bound to two chlorine (Cl) and two fluorine (Fl) atoms. By tweaking the balance of these atoms, chemists synthesized CFCs that they believed were stable, nonflammable, and harmless to people and the environment—qualities that led to their broad acceptance.

Over time, these items emitted CFC molecules into the air; in 1971, British scientist James Lovelock detected CFCs in the atmosphere over England. Since CFCs were considered safe, the news was not a cause for concern. But it caught the attention of scientist Sherwood Rowland of the University of California, Irvine, who wondered what would happen to this completely synthetic substance in the atmosphere. He and his colleague Mario Molina set out to look for answers.

In 1974, they proposed that CFCs were not entirely harmless in the atmosphere. The compounds were specifically designed by chemists to be stable, and Molina and Rowland realized that CFCs would stay aloft for a remarkably long time in the atmosphere, residing there and accumulating for 100 years or more. And once in the stratosphere, the molecules would be exposed to UV light so intense that it would break them apart. The process would release solitary chlorine atoms, which previous research had shown could chemically react with—and destroy—ozone, aligning with Farman's observations from Antarctica. The two chemists calculated that at the rate CFCs were being produced in 1972, the chemicals could destroy 6% of the ozone layer. And manufacturers were making more CFCs every year.

But the scientific view of CFCs didn't change overnight--because all conclusions in science are considered open to revision (because our understanding of a concept or process will change as scientists learn more), more evidence was needed to overturn the prevailing conclusion that CFCs were safe.

The scientific method systematically rules out explanations.

Farman's data was compelling, but would it hold up to scrutiny? Less than a year after his research was published in 1985, in the journal *Nature*, NASA scientists published their own report, verifying what Farman had observed. This corroboration strengthened Farman's conclusions. That ozone was depleting much faster than earlier projections set off alarm bells. These data provided a **correlation** between the presence of CFCs and ozone depletion—both occurred together. It suggested that CFCs may be related to ozone decline, but did not establish a

---

ultraviolet (UV) radiation Short-wavelength electromagnetic energy emitted by the Sun.

science A body of knowledge (facts and explanations) about the natural world, and the process used to get that knowledge.

empirical evidence Information gathered via observation of physical phenomena.

correlation Two things occur together—but it doesn't necessarily mean that one caused the other.

cause-and-effect relationship. The two trends could occur together by coincidence, or something else entirely could be causing both to occur.

To get a better picture of what was going on, scientists needed to further apply the **scientific method**, in which they would work logically and systematically to design studies specific to the question being asked. The fact that the ozone layer was thinning above the Antarctic triggered great debate among scientists, resulting in more than one possible explanation, known as **hypotheses**, for what was occurring. Some researchers thought that Antarctic air was mixing and lifting lower, low-ozone air into the stratosphere, changing the concentration. Other researchers believed that solar activity was creating nitrogen oxides (NO_x) which could be destroying ozone, as the amount of NO_x fluctuated with sunlight in the Antarctic. But Susan Solomon had a different idea.

Solomon was a young scientist when she visited Antarctica in 1986 and 1987, the beginning of her long relationship with the southern, icy world. While trying to understand why ozone was disappearing over the region, Solomon kept thinking about temperature. October is spring in Antarctica, and scientists knew that cold spring winds would swirl around, producing a cyclone of air in the atmosphere (a polar vortex), keeping cold air in place over the poles and leading to the formation of polar clouds in the stratosphere. Solomon proposed the hypothesis that cloud particles in the polar stratospheric clouds were providing surfaces for the reactions that would free chlorine molecules (Cl_2) from CFCs. In sunlight, the chlorine molecules would then break up into chlorine atoms. These isolated chlorine atoms destroyed ozone—particularly in the Antarctic spring, when sunlight streamed in.

Scientific hypotheses must be **testable** (generate **predictions** about what we could objectively observe if we conducted the test). In turn, these predictions must be **falsifiable**, meaning that it would be possible to produce evidence that shows the prediction is wrong. (Predictions based on untestable ideas—such as, "reincarnation exists" —are not falsifiable and therefore are not considered suitable for science.) In this way, if that falsifying evidence does not appear, it may be reasonable to conclude that the evidence supports the prediction.

Notice we do not claim that the hypothesis is *proven*, only that it is supported (or confirmed). This is a hallmark of the tentative nature of science. "Proven" suggests we have the final answer; science, however, is open ended, and no matter how much evidence accumulates, there are always new questions to ask and new studies to conduct that could alter our conclusions. But this does not mean

cause-and-effect relationship An association between two variables that identifies one (the effect) occurring as a result of or in response to the other (the cause).
scientific method Procedure scientists use to empirically test a hypothesis.
hypothesis A possible explanation for what we have observed that is based on some previous knowledge.
testable A possible explanation that generates predictions for which empirical evidence can be collected to verify or refute the hypothesis.
prediction A statement that identifies what is expected to happen in a given situation.
falsifiable An idea or a prediction that can be proved wrong by evidence.

Infographic 2.2 | **CERTAINTY IN SCIENCE**

↓ There are degrees of certainty in science—we know some ideas are better than others. The more evidence we have in support of an idea, especially when the evidence comes from different lines of inquiry, the more certain we are that we are on the right track. But since all scientific information is open to further evaluation, we do not expect or require "absolute" proof.

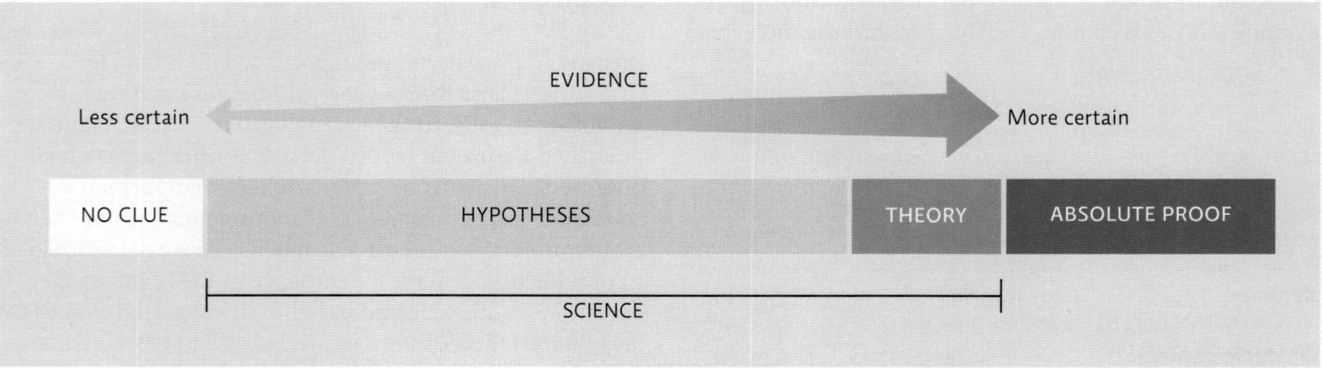
EVIDENCE
Less certain
More certain
NO CLUE
HYPOTHESES
THEORY
ABSOLUTE PROOF
SCIENCE

Infographic 2.3 | **SCIENTIFIC PROCESS**

↓ Scientists work from previous knowledge and observation to ask new questions and pose possible explanations (hypotheses) for what they observe. They then design a study to gather evidence to test predictions made from their hypotheses. The scientific method is not really a linear sequence. Rather, it is more of a cycle that scientists move through in whatever order of "steps" best suits their needs.

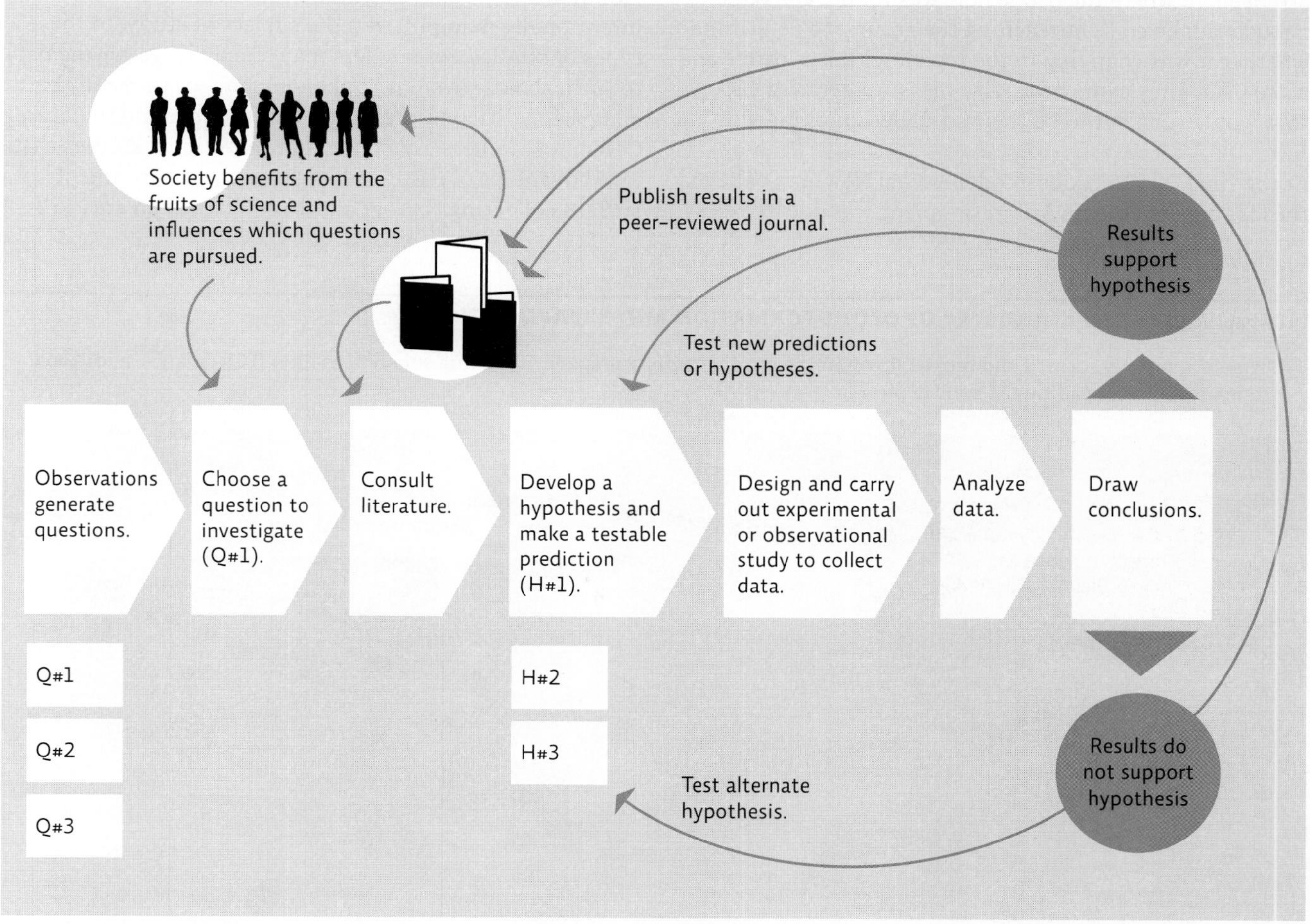
Society benefits from the fruits of science and influences which questions are pursued.
Publish results in a peer-reviewed journal.
Results support hypothesis
Test new predictions or hypotheses.
Observations generate questions.
Choose a question to investigate (Q#1).
Consult literature.
Develop a hypothesis and make a testable prediction (H#1).
Design and carry out experimental or observational study to collect data.
Analyze data.
Draw conclusions.
Q#1
Q#2
Q#3
H#2
H#3
Test alternate hypothesis.
Results do not support hypothesis

that all possible explanations are equally valid. This is precisely the reason why hypotheses must be tested again and again and in different ways. As evidence mounts in support of a hypothesis, the probability that it is wrong lessens and it becomes unreasonable to reject the hypothesis in favour of another, less supported explanation. But, in keeping with the tentative nature of science, even the best-supported hypotheses will be revised or even abandoned if new data strongly supports a new conclusion. [INFOGRAPHIC 2.2]

Once a study is conducted and data gathered, the data are evaluated to determine whether they confirm or fail to confirm the hypothesis, but the scientific process doesn't end there. If a hypothesis is rejected, alternative hypotheses can be tested. If a hypothesis is confirmed, the researchers should repeat the study to validate the data. They can also generate new predictions that test the same hypothesis from different angles. As evidence mounts from replicate studies and from multiple predictions, we become more confident in our data and conclusions. [INFOGRAPHIC 2.3]

Different types of studies amass a body of evidence.

Solomon's hypothesis generated the prediction that the stratosphere would contain high levels of chlorine monoxide, or ClO. If polar clouds and sunlight were causing chlorine to react with ozone, then the atmosphere should contain many molecules of chlorine bound to individual oxygen atoms. This prediction was falsifiable, since Solomon might not find high levels of ClO.

On both of her trips to the Antarctic, her team raised balloons into the air which measured the composition of the atmosphere where the ozone hole was found, and simultaneously analyzed light reaching the ground to determine whether it was changing in the region with less ozone and more ClO. They came back with measurements of ClO that would turn out to be their so-called smoking gun.

At the time, NASA ozone-modeller Paul Newman believed the loss of ozone in the Antarctic spring was due to excess solar activity. But when the ClO measurements from Solomon's team streamed out of a fax machine in NASA's Goddard Research Center, he knew the evidence supported Solomon's polar cloud hypothesis. [INFOGRAPHIC 2.4]

Solomon's experiment is an example of an **observational study**—collecting data in the real world without intentionally manipulating the subject of study. In these types of studies, researchers may simply be gathering data to learn about a system or phenomenon, or they may be comparing different groups or conditions found in nature. Often, researchers can conduct observational studies that take advantage of natural changes in the environment such as collecting "before and after" data in an area to

Infographic **2.4** | **THE CHEMISTRY OF OZONE FORMATION AND BREAKDOWN**

↓ Ozone is naturally formed and broken down in the stratosphere, but ozone-depleting substances like CFCs catalyze additional ozone breakdown, leading to ozone depletion.

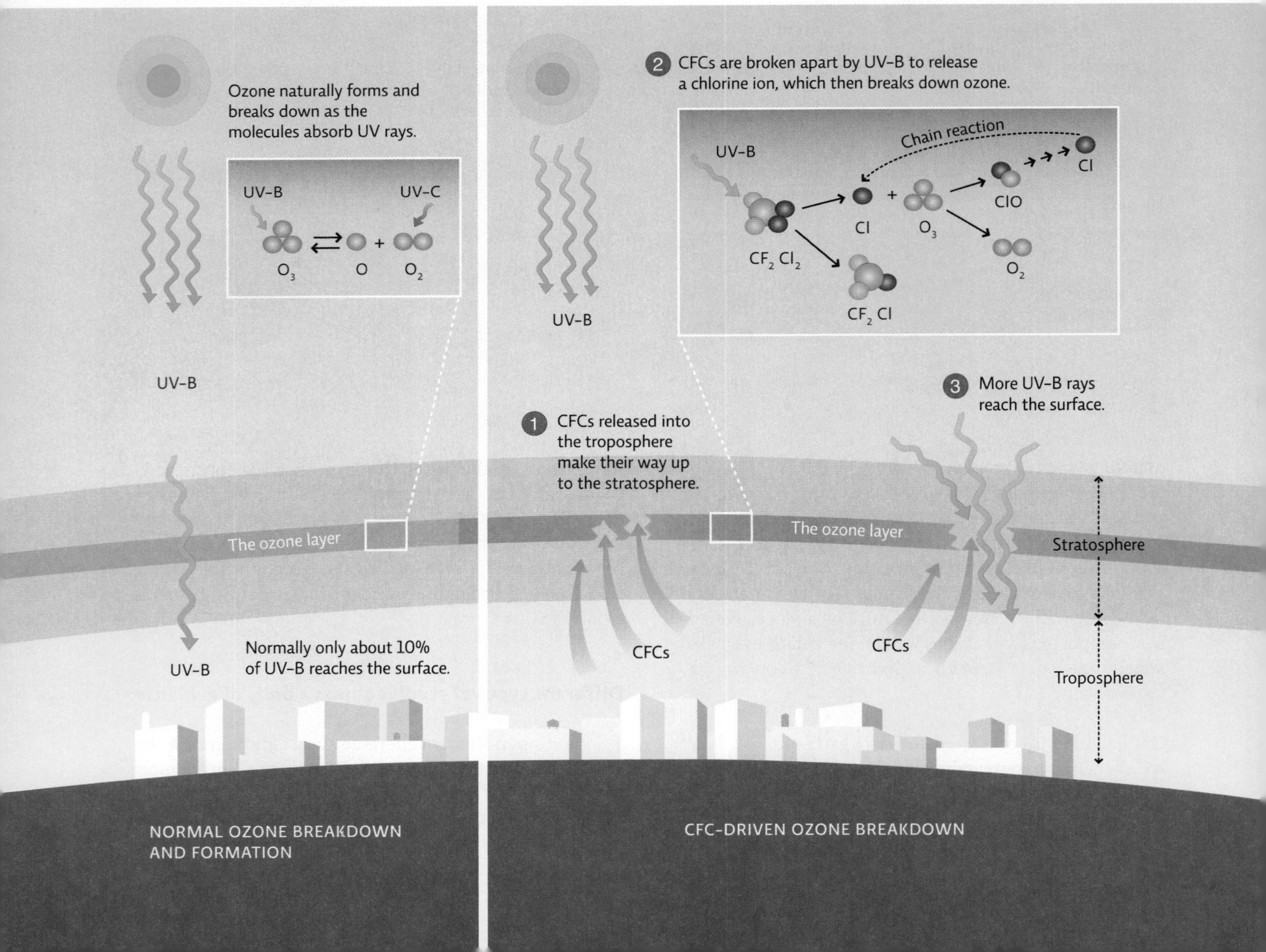

Ozone naturally forms and breaks down as the molecules absorb UV rays.
UV-B
UV-C
O3
O
O2
UV-B
The ozone layer
UV-B
Normally only about 10% of UV-B reaches the surface.
NORMAL OZONE BREAKDOWN AND FORMATION
2 CFCs are broken apart by UV-B to release a chlorine ion, which then breaks down ozone.
UV-B
Chain reaction
Cl
ClO
Cl
O3
CF2 Cl2
CF2 Cl
O2
UV-B
1 CFCs released into the troposphere make their way up to the stratosphere.
3 More UV-B rays reach the surface.
The ozone layer
Stratosphere
CFCs
CFCs
Troposphere
CFC-DRIVEN OZONE BREAKDOWN

↑ Researchers at the Finnish Ultraviolet International Research Center in the Arctic Circle research the effects of increased UV-B radiation on plant growth and biology. UV-B radiation at high latitudes damages living organisms, and can alter their DNA (genetic material).

examine the effect of a natural perturbation such as a flood or volcanic eruption, or in this case comparing areas with different levels of ozone depletion. These types of opportunities represent valuable "natural experiments" that can allow us to test cause-and-effect hypotheses, just as a controlled **experimental study**, in which the researcher intentionally manipulates the conditions of the experiment in a lab or field setting, allows us to do. For instance, much research is underway on the effects on living organisms of ozone depletion and elevated UV exposures. Previous studies have shown that UV exposure can increase the incidence of skin cancer. A reasonable question would be: is ozone depletion causing more skin cancer? This could be used to generate an experimental hypothesis such as: lower ozone levels will lead to more cases of skin cancer. This hypothesis generates many predictions; some would be tested with observational studies and others experimentally.

In 2002, Chilean researchers Jaime Abarca and Claudio Casiccia investigated this hypothesis by evaluating skin cancer incidence in residents of Punta Arenas, Chile, a region exposed to higher UV-B due to its proximity to the ozone hole. Because of the ozone hole, global air circulation patterns have actually shifted; the southern hemisphere jet stream now circulates at a latitude closer to the South Pole, taking the ozone-depleted air over the more populated regions of the Southern Hemisphere like southern Chile. They predicted that more cases of skin cancer would be seen there in years when ozone depletion was high. To test this they calculated skin cancer rates of the population between 1987 and 1993, and compared those to rates from the same population between 1994 and 2000. The results of their study showed that non-melanoma skin cancer rates were significantly higher in times when ozone depletion was higher. This evidence supports their hypothesis and correlates living close to the Antarctic ozone hole with one's chance of developing skin cancer. This observational study does not manipulate any variable—no people were exposed to higher or lower levels of UV-B or intentionally relocated to areas of higher or lower ozone depletion. The researchers simply collected data that was available in the population and

observational study Research that gathers data in a real-world setting without intentionally manipulating any variable.

experimental study Research that manipulates a variable in a test group and compares the response to that of a control group that was not exposed to the same variable.

then evaluated it to see if the two variables—ozone depletion and skin cancer incidence—were correlated.

We can't ethically manipulate people to further test this hypothesis but we can conduct an experimental study on cells or model organisms such as mice. Australian researcher Scott Menzies and his colleagues did just that in the early 1990s by testing the prediction that mice exposed to high UV-B would develop more skin cancer than mice exposed to normal levels of UV-B. This experiment compared two groups—the **control group** of mice was exposed to normal levels of UV-B and the **test group** was exposed to the same amount of UV-B received in areas where ozone depletion has been observed. The two groups are identical in every way except for the test variable (in this case, the amount of UV-B radiation exposure). The inclusion of a control group is key because it allows researchers to attribute any differences seen between the two groups to the single test variable that was altered. If researchers only looked at the test group, they would have no way to determine whether the incidence of skin cancer was higher than normal.

In an experimental study, we have both an **independent variable** and a **dependent variable**. We manipulate the independent variable (in this case, the amount of UV-B radiation), and measure the dependent variable (the incidence of skin cancer) to see if it is affected. In other words, if development of skin cancer is *dependent* on UV-B radiation, then we should see skin cancer incidence change as UV-B exposure changes. Scientists often represent their data on a graph, on which the x-axis (horizontal axis) displays the independent variable and the y-axis (vertical axis) shows the response (dependent variable). Menzies' data showed a clear difference between his control group (none developed skin cancer) and his test group (100% developed skin cancer). It is rare to see such an absolute difference and, as with any study, we would like to see these results replicated, but a well-designed and well-conducted study increases our confidence that the results are valid. [INFOGRAPHIC 2.5]

Both observational and experimental studies gather data systematically to produce scientifically valid evidence. In contrast, anecdotal accounts (individual occurrences or observations) represent data that was not systematically collected or tested and cannot be compared to any control (many uncontrolled variables exist). While anecdotes are not considered acceptable scientific evidence, a claim based on anecdotal accounts could be tested to see if a correlation or cause-and-effect relationship exists.

Solomon and her team reported their evidence to the international scientific community by publishing their results in *Nature* and the *Journal of Geophysical Research* in 1987. These are **peer-reviewed** journals—meaning that before results are published, they are reviewed by a group of outside experts. Studies that are not well-designed or well-conducted are not accepted for publication. Therefore, peer-reviewed published research represents high-quality scholarship in the field. In this case, the reviewers already knew that CFCs were present in the stratosphere, and lab studies showed they were capable of destroying ozone. Now, Solomon and her team were presenting evidence that, at the time of year when ozone was dropping, the Antarctic stratosphere contained high levels of ClO, demonstrating that free chlorine atoms were reacting with ozone.

The evidence was amassing that CFCs were contributing to ozone depletion, but that didn't mean the other hypotheses would be immediately abandoned. Research on these alternative hypotheses continued but scientists were not finding evidence to support the other hypotheses' predictions. If lower, ozone-poor air was lifting and mixing with the stratosphere, then researchers should observe gases moving upward in the atmosphere; instead, they saw the opposite: air seemed to be flowing downward. If solar activity was creating nitrogen oxides (NO_x) that were destroying ozone, as NASA modeller Paul Newman had once believed, then scientists should have observed increases in NO_x in the South Pole. But when they took measurements, they found that NO_x levels were actually decreasing. This observation provided further support for Solomon's hypothesis, which predicted that NO_x levels should be low, not high—because otherwise NO_x would combine with ClO molecules and prevent them from interacting with ozone.

The multiple lines of evidence are so compelling that they have elevated the "CFC hypothesis" to the status of **theory**—a widely accepted explanation that has been extensively and rigorously tested. (But just like well-supported hypotheses, we do not claim that a theory is *proven*. Even well-substantiated theories are always open to further study.) This differs from the casual meaning of

control group The group in an experimental study that the test group's results are compared to; ideally, the control group will differ from the test group in only one way.

test group The group in an experimental study that is manipulated somehow such that it differs from the control group in only one way.

independent variable The variable in an experiment that the researcher manipulates or changes to see if it produces an effect.

dependent variable The variable in an experiment that is evaluated to see if it changes due to the conditions of the experiment.

peer-reviewed Researchers submit a report of their work to a group of outside experts who evaluate the study's design and results of the study to determine whether it is of high-enough quality to publish.

theory A widely accepted explanation of a natural phenomenon that has been extensively and rigorously tested scientifically.

Infographic **2.5** | **HOW DO SCIENTISTS COLLECT EVIDENCE TO ANSWER QUESTIONS?**

← Scientists collect evidence to test ideas. Experimental studies are used when the test subjects can be intentionally manipulated while observational studies allow a look at entire ecosystems or other complex systems.

BACKGROUND KNOWLEDGE
UV light can cause skin cancer.

QUESTION
Is ozone depletion causing more skin cancer?

HYPOTHESIS
Lower ozone levels will lead to more cases of skin cancer.

Scientifically test the hypothesis.

OBSERVATIONAL STUDY (Abarca, *et al.*)
Prediction: The incidence of skin cancer in an area where ozone depletion is occurring will be higher than it was in times of less ozone depletion.
Procedure: Collect data on skin cancer incidence.

RESULTS

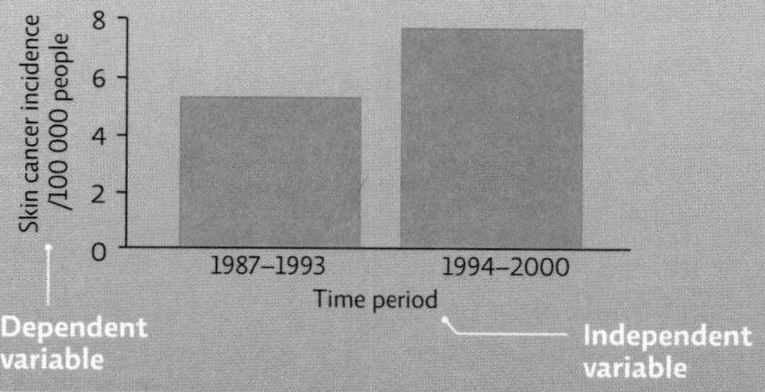
Skin cancer incidence /100 000 people
0
2
4
6
8
1987–1993
1994–2000
Time period
Dependent variable
Independent variable

EXPERIMENTERS' CONCLUSION: Living close to the ozone hole increases one's chance of developing skin cancer.

What else could have caused this increase in skin cancer?

EXPERIMENTAL STUDY (Menzies, *et al.*)
Prediction: Mice exposed to UV-B levels comparable to those seen in areas where ozone depletion is observed will develop more skin cancer than mice exposed to normal levels of UV-B.
Procedure: Expose mice to different UV-B levels and monitor them for skin cancer.

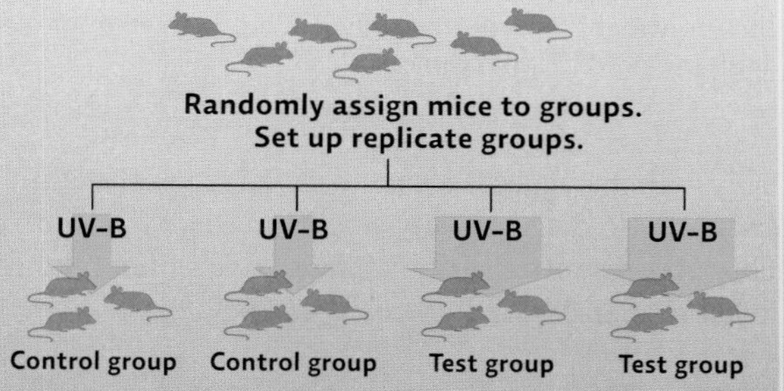
Randomly assign mice to groups.
Set up replicate groups.
UV-B
UV-B
UV-B
UV-B
Control group
Control group
Test group
Test group

RESULTS

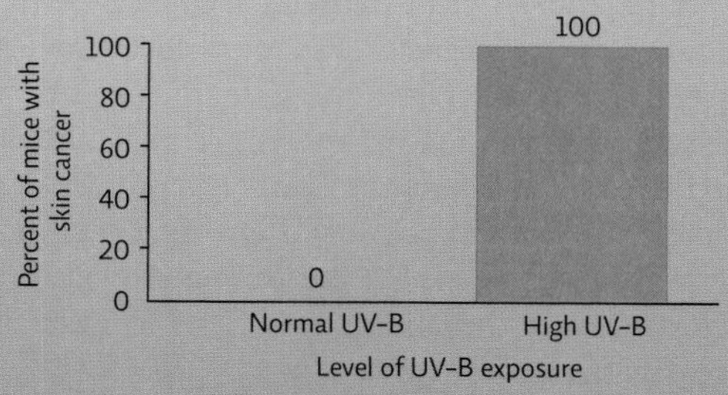
Percent of mice with skin cancer
0
20
40
60
80
100
0
100
Normal UV-B
High UV-B
Level of UV-B exposure

EXPERIMENTERS' CONCLUSION: Exposure to UV-B at levels seen in ozone-depleted areas caused skin cancer in these mice.

Is this evidence that this would occur in other species?

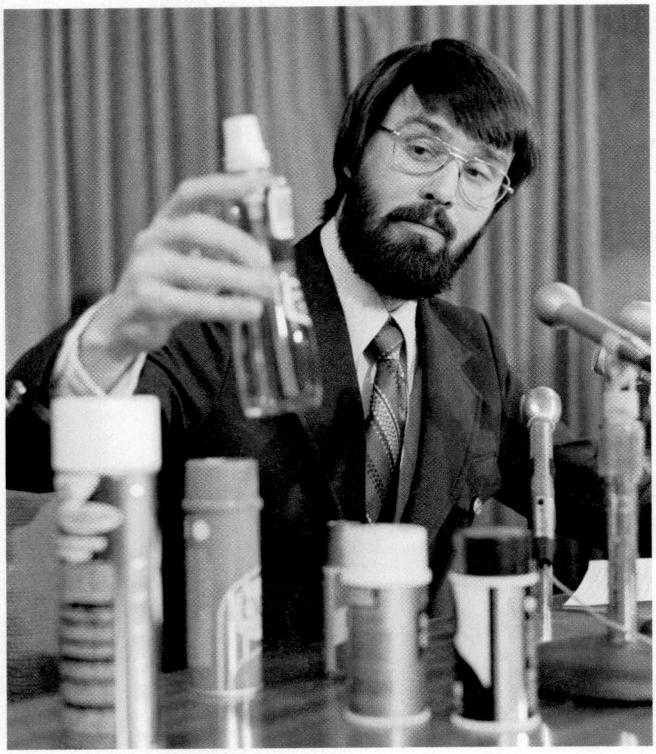

↑ In 1975, U.S. Representative Perry Bullard of Michigan introduced legislation to outlaw aerosol cans, 12 years before the Montreal Protocol was signed. This is an example of applying precautionary principles.

theory, which suggests a speculative idea without much substance. To discount any scientific theory as "just a theory" represents a serious flaw in one's understanding of what a scientific theory really is.

As mentioned earlier, in science there are *degrees of certainty*; we know some things better than others. The more evidence we have in support of an idea, especially from different types of experiments, the more certain we are that we are on the right track. These degrees of certainty are expressed mathematically in terms of probabilities using **statistics**. A mathematical analysis of the data is done to determine the probability that the occurrence of a phenomenon (in this case the experiment's result) is a random event, rather than being caused by the variable being investigated.

This type of analysis of the data allows us to quantitatively assign a level of certainty to our conclusions. In statistics, this probability is expressed as a *p-value* that represents the likelihood our conclusions are wrong. Scientists generally require a high probability (at least 95%) that their conclusions are correct ($p \leq 0.05$ represents a level of certainty of 95%). This means that there is only a 5% chance we have incorrectly accepted or rejected our hypothesis. The more evidence that accumulates in favour of a particular conclusion, the more certain we are that it is likely to be correct and can be used to inform our decisions. Just because we don't know exactly what will happen if we continue to release CFCs into the atmosphere (including how fast the ozone layer will deplete, or when it could jeopardize human health) doesn't mean we don't know enough to take action. [For more on statistics and experimental design, see Appendix 3.]

Indeed, the loss of ozone was starting to have the effects that the studies in Chile and Australia were showing. Other observational studies revealed that UV-B levels were increasing in high- and midlatitude regions. Research by NASA scientists showed that the amount of UV-B reaching the ground in the midlatitudes in the mid-1990s had increased over 1979 levels. Other research showed that UV-B levels were 45% higher than normal in the spring of 1990 at the southern tip of Argentina.

In the meantime, skin cancer incidence has increased in human populations (as have other skin disorders and eye problems such as cataracts) and other organisms are also affected by extra UV-B exposure. Photosynthesis rates in marine organisms such as the tiny phytoplankton of the Antarctic sea are lower than normal in the spring because of increased UV-B exposure—a troublesome fact since phytoplankton forms the base of the Antarctic food chain. Decreases in photosynthesis are also troubling because they could lead to lower agricultural productivity and reduce the value of these areas as "carbon sinks"—areas where carbon is stored, keeping it out of the atmosphere (see Chapter 22 on climate change).

The international community got together to meet the problem head on.

Even while scientists were still trying to understand why the ozone layer was depleting, they knew they had to do something about it. In 1985, a group of experts from around the world met in Vienna to discuss ways to research and solve the problem. This set the stage for the international community to come together in Montréal in 1987, to produce a plan to deal with the problem of ozone depletion. It would involve sacrifices—notably, phasing out dangerous chemicals such as CFCs—and only two dozen governments signed on to the **Montreal Protocol** that September. By 2009, the protocol (treaty) was eventually ratified by all 196 countries in the world when East Timor signed on.

Infographic **2.6** | **THE MONTREAL PROTOCOL AND ITS AMENDMENTS HAVE BEEN EFFECTIVE**

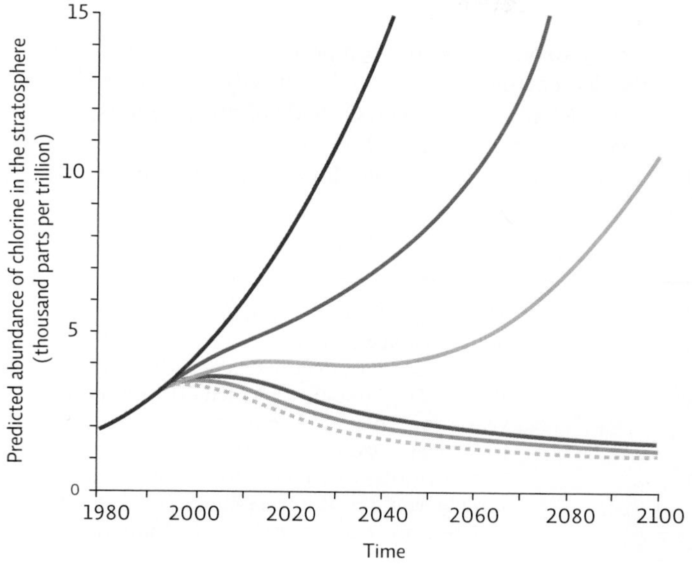
15
10
5
0
Predicted abundance of chlorine in the stratosphere (thousand parts per trillion)
1980
2000
2020
2040
2060
2080
2100
Time
No protocol
Montreal Protocol 1987
London Amendment 1990
Copenhagen Amendment 1992
Beijing Amendment 1999
Zero emissions

← Actual and projected change over time for total global emissions of ozone-depleting substances (ODS) with and without the Montreal Protocol and its amendments. Adjustments to the phase-out schedule of various ODSs in the form of amendments represent the success of adaptive management in dealing with complex environmental issues.

The Montreal Protocol, administered by the United Nations, outlined a series of deadlines over the next decade for cutting back production of CFCs. Governments would have to put in place their own plans for achieving a desired outcome, or **policy**, for reducing CFCs.

Policy has been described as *translating our values into action*. Science provides information we can use to make informed decisions to protect our health, safety, or environment. Society then decides what "ought" to be addressed (do we address ozone depletion?) and hopefully uses the best scientific information available to set reasonable policies that take into account the ecological, economic, social, and political issues at stake.

Concerns raised by researchers in the 1970s led to a ban on CFCs in most aerosol products in 1978 in several countries, including Canada. As evidence mounted, industry, the public, and governments began to realize the seriousness of this problem.

statistics The mathematical evaluation of experimental data to determine how likely it is that any difference observed is due to the variable being tested.
Montreal Protocol International treaty that laid out plans to phase out ozone-depleting chemicals like CFC.
policy A formalized plan that addresses a desired outcome or goal.
Environment Canada A Canadian federal government department responsible for environmental protection and weather monitoring.
precautionary principle A principle that encourages acting in a way that leaves a margin of safety when there is a potential for serious harm but uncertainty about the form or magnitude of that harm.
adaptive management Plan that allows room for altering strategies as new information comes in or the situation itself changes.

Canada passed its first regulation on CFCs in 1980 under the Environmental Contaminants Act. In 2001 an accelerated phase-out plan was developed. This plan is implemented by both **Environment Canada** and provincial governments.

Note that the 1987 Montreal Protocol and the international commitment to address CFCs and ozone depletion came before Susan Solomon's definitive studies were published in 1988. This is an example of applying the **precautionary principle**—acting in the face of uncertainty when there is a chance that serious consequences might occur. By 1987, though we didn't know all the details, we knew enough to take action. As more information poured in, it quickly became apparent that the Montreal Protocol targets would not be sufficient to stop ozone depletion. Amendments are still proposed and negotiated in annual meetings which strengthen the response and adjust the target dates to phase out harmful compounds. This is an ongoing process, and an example of **adaptive management**—allowing room for altering strategies as new information comes in or the situation itself changes. This same approach is needed to address other global environmental issues such as climate change. We cannot know the exact details of future climate change impacts before they occur but we do have a good idea of what we are facing. By acting now, we can lessen the severity of those impacts and adjust our response as we go along. [INFOGRAPHIC 2.6]

Fortunately, companies that produced CFCs, such as Dupont, were already researching alternative chemicals in the lab that could serve the same purposes with less damage. If companies could sell a CFC alternative just as easily, then cutting back on CFCs would not hurt them financially. Still, some of the replacements, particularly some hydrochlorofluorocarbons (HCFCs) used in refrigerants, eventually would also prove to be detrimental to ozone. In addition, illegal stockpiles and old refrigerators continue to release some CFCs, though that amount is decreasing rapidly under the Montreal Protocol. Pockets of CFCs remain because of continued legal "essential" uses, for medicines and nuclear power, for example.

Today, ozone-depleting compounds in the stratosphere are decreasing at a rate consistent with the policy changes of the Montreal Protocol. Outside of polar regions, ozone levels declined through the 1990s but seem to be holding steady more recently. The Montreal Protocol has been

Infographic **2.7** | **OZONE DEPLETION AND CFC LEVELS**

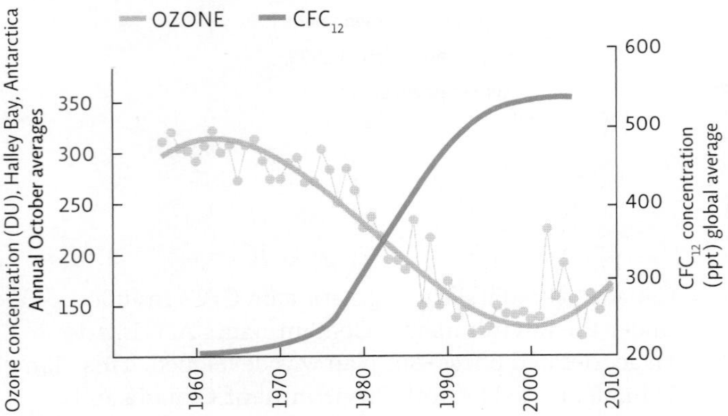
OZONE
CFC_{12}
Ozone concentration (DU), Halley Bay, Antarctica
Annual October averages
350
300
250
200
150
600
500
400
300
200
CFC_{12} concentration
(ppt) global average
1960
1970
1980
1990
2000
2010

← This graph shows a correlation between the levels of CFC_{12} and ozone over Antarctica. As CFC_{12} increased, ozone declined (a negative correlation). When CFC_{12} amounts levelled off, around the year 2000, the ozone hole over Antarctica started to recover. (Ozone concentration is expressed in Dobson units, DU, a measure of how thick the air sample would be if it were compressed at 1 atmosphere of pressure at 0°C; 1 DU = 1 mm thick.)

↓ These images show the size of the ozone hole over Antarctica in 1979 (the 1st year ozone satellite images were available), 1993 (a "mid" year), and 2006 (the worst ozone hole ever recorded). Though the 1979 image might not be "normal," since CFC-driven depletion had already begun, ozone levels are considerably higher than in subsequent years. The 1993 image shows a much larger ozone hole than that in 1979, but the record ozone hole of 2006 is wider and deeper (indicated by the purple colour—lowest ozone levels ever recorded).

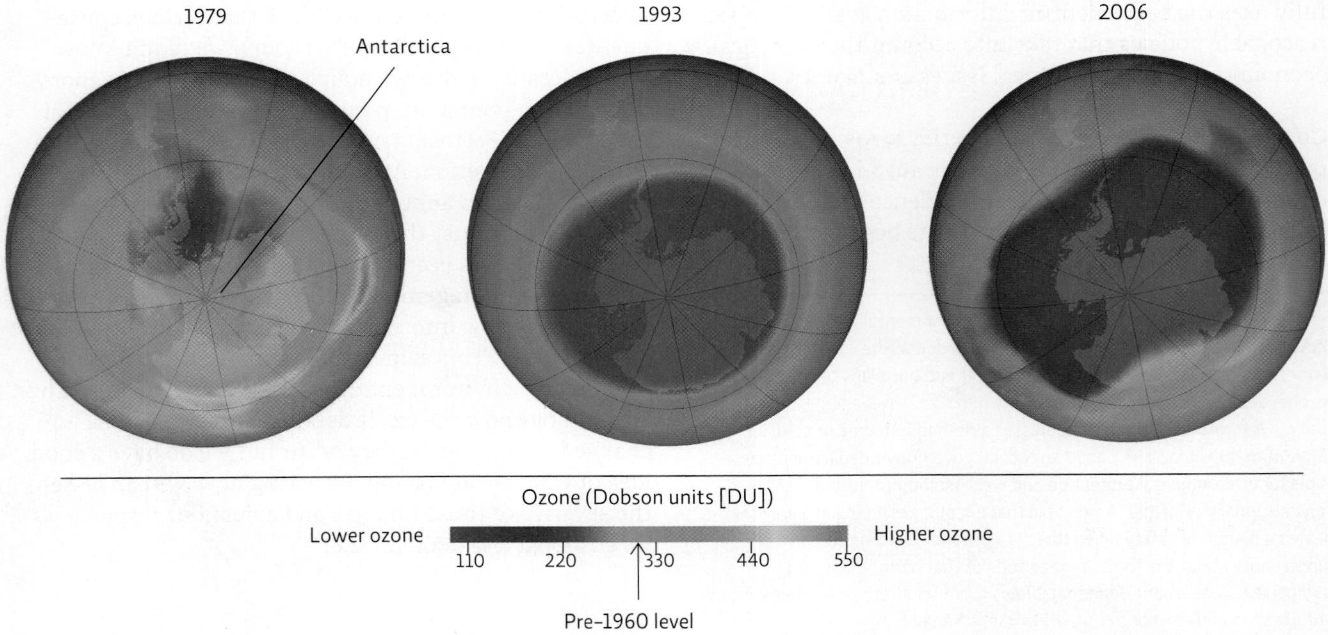
1979
1993
2006
Antarctica
Ozone (Dobson units [DU])
Lower ozone
Higher ozone
110
220
330
440
550
Pre-1960 level

hailed as the most successful international environmental agreement in history.

There is still much we don't understand about the atmospheric chemistry of ozone depletion. Just when it seemed ozone was beginning to recover, the winter of 2004–2005 in the Arctic was unusually cold and ozone loss was very high that year. This was eclipsed by a record 40% loss of ozone over the Arctic in spring of 2011. And Antarctica set two new records for ozone depletion in 2006—the widest and deepest hole ever observed, with some areas near 90% depletion. What this tells us is that even though CFCs are declining, very cold winters will still give us years of higher-than-expected ozone depletion. [INFOGRAPHIC 2.7]

Future projections estimate that midlatitude areas should be back to pre-1980 ozone levels by 2050; polar regions should be back by 2075, though the next 15 years should show periodic large declines and produce large "ozone holes" in very cold years like those seen in 2006 and 2011. Although she retired from NOAA in 2011, Solomon continues her research on CFCs, and is both hopeful and realistic. "It's clear that ozone will ultimately recover but it's also clear that it will take many decades to do so," she says. "It has been a real privilege to work on such an interesting problem and to feel that the world found it useful in making choices about the Montreal Protocol."◉

Select references in this chapter:
Abarca, J.F., & Casiccia, C.C. 2002. *Photodermatology, Photoimmunology & Photomedicine*, 18: 294–302.
Farman, J., *et al.* 1985. *Nature*, 315: 207–210.
Menzies, S.W., *et al.* 1991. *Cancer Research*, 51: 2773–2779.
Molina, M.J., & Rowland, F.S. 1974. *Nature*, 249: 810–812.
Solomon, S., *et al.* 1988. *Science*, 242: 550–555.

BRING IT HOME

PERSONAL CHOICES THAT HELP

The depletion of the ozone layer is a great example of how science documented a problem and its cause, and public action confronted the problem. All of the environmental changes we face, from rising levels of greenhouse gases to loss of biodiversity, can be addressed using science. How scientific information is or is not put into action has far-reaching consequences, making science literacy a matter of importance for every individual.

Individual Steps

→ Practise thinking like a scientist. Go outside for 10 minutes and observe the world around you. Make observations of what you see or hear. What predictions could you make from your observations; how could you test them?

→ Stay informed. Read or watch a science-related article or show once a month.

Group Action

→ Demonstrate the importance of scientific literacy with your friends and family. Develop three additional questions from the material in the chapter and discuss them over dinner.

→ Support science education. Find out about public lectures and programs in your area and attend one with your friends.

Policy Change

→ Attend a city council or county board meeting to see how policy issues are addressed in your area.

→ Make knowledgeable voting decisions on ballot initiatives.

→ Serve on local civic committees that address environmental issues in your community.

UNDERSTANDING THE ISSUE

CHECK YOUR UNDERSTANDING

1. The scientific process includes:
a. observational studies that collect data in the real world without manipulating the subject of study.
b. both testable and untestable explanations for what we have observed.
c. only provable cause-and-effect relationships.
d. just the information that can be detected with our senses: sight, hearing, taste, and smell.

2. Which of the following is NOT true about the Montreal Protocol?
a. It is a treaty that was eventually ratified by all 196 countries in the world and is considered the most successful international environmental agreement in history.
b. It is administered by the United Nations and outlines a series of deadlines for cutting back production of CFCs.
c. It is a legally binding agreement and national governments of ratifying nations have to work in lockstep with each other to reduce CFC use and production.
d. Amendments to the protocol are still proposed and negotiated in annual meetings, which strengthen the response to ozone depletion and speed up target dates to phase out the harmful compounds.

3. The statement "It is wrong to inflict pain on animals" is not scientifically testable because it is not _______ by any demonstration of empirical evidence.
a. provable
b. falsifiable
c. predictable
d. correlational

4. What is the relationship between CFCs and ozone in the stratosphere?
a. Ozone naturally breaks down in the stratosphere but substances like CFCs regulate its re-formation so that less harmful UV radiation reaches Earth's surface.
b. Ozone from the stratosphere migrates down to the troposphere where it reacts with chemicals like CFCs to produce more oxygen.
c. Ozone in the stratosphere is broken down by chemicals like CFCs, but as CFCs themselves are broken apart by UV radiation, ozone depletion slows.
d. Ozone is naturally formed and broken down in the stratosphere but substances like CFCs catalyze additional ozone breakdown, which results in increased UV radiation reaching Earth's surface.

5. Data gathered via objective observation of physical phenomena is:
a. scientific proof.
b. an inference.
c. empirical evidence.
d. a hypothesis.

6. In an experiment to study the effects of UV-B on leopard frogs, an effective control is:
a. a group of leopard frogs with skin lesions treated in the same way, but exposed to normal levels of UV-B.
b. a group of leopard frogs treated in the same way, but given a lower exposure of UV-B radiation.
c. a group of leopard frogs treated in the same way, but exposed to normal levels of UV-B.
d. a group of red-spotted toads treated in the same way, but given a lower exposure of UV-B radiation.

WORK WITH IDEAS

1. What is science? Describe the key elements of science and apply this framework to show how the story of CFCs and stratospheric ozone depletion is science.

2. What is a correlation in science? How is it different from a cause-and-effect relationship? Explain how both types of data relationships are used to understand the connection between CFCs and stratospheric ozone.

3. What are the differences between observational and experimental studies? Under what circumstances is each of these types of studies useful?

4. Define scientific hypothesis and explain how it differs from a scientific theory and the more general use of the word theory by lay people.

5. Describe the chemistry of ozone formation and breakdown in the stratosphere and explain how chemicals like CFCs interfere with this process. Explain why the chemistry of CFCs makes them so dangerous.

6. Why is the Montreal Protocol considered the most successful international environmental agreement in history? Explain what makes this treaty so successful and discuss the possible underlying reasons for this success.

ANALYZING THE SCIENCE

The graph on the following page comes from a study exploring the correlation between chlorine monoxide (ClO) and ozone. The data was collected by placing the measurement instrument on the wing of a plane and flying through the Antarctic ozone hole.

INTERPRETATION

1. What does this graph show? Describe the pattern that you see.

2. Looking at the levels of ozone relative to the levels of chlorine monoxide, which side of the graph do you think most likely represents the inside of the ozone hole? On what basis do you think this?

3. Compare this graph to the graph in Infographic 2.7 (Ozone Depletion and CFC Levels). Although both graphs show a similar pattern, they are about two different aspects of the CFC-ozone story. Compare and contrast the information presented in the two graphs.

ADVANCE YOUR THINKING

HINT: If you want to access the actual article, you can find it at:
www.undsci.berkeley.edu/article/ozone_depletion_01

4. Do the data in this graph show a correlation or a cause-and-effect relationship? Do they support the hypothesis that chlorine is the cause of ozone depletion? Explain your responses.

5. How are these data similar to and/or different from the data collected by Susan Solomon's team? And what do these data mean for the hypothesis that Solomon proposed?

6. How does this research study support the idea that science is a community enterprise?

A plot of chlorine monoxide and ozone concentrations from data collected by the aircraft.

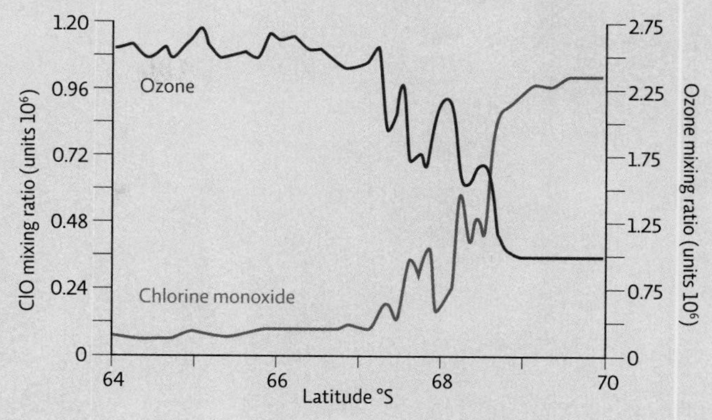
1.20
0.96
0.72
0.48
0.24
0
ClO mixing ratio (units 10⁶)
Ozone
Chlorine monoxide
2.75
2.25
1.75
1.25
0.75
0
Ozone mixing ratio (units 10⁶)
64
66
68
70
Latitude °S

EVALUATING NEW INFORMATION

You can use your understanding of the nature of science to discover the real meaning of news stories on health, science, and other national issues.

For instance, how does the Montreal Protocol affect asthma sufferers? CFCs were used in metered dose inhalers (MDIs) to deliver medication to asthma patients. Initially Canada requested, and was granted, an essential-use exemption for MDIs using CFC propellants. Having found that alternative hydrofluoroalkanes (HFA) propellants met safety guidelines, Health Canada proceeded with the transition. As of 2010, 99% of MDIs sold in Canada were CFC-free.

In Canada, The Canadian Lung Association (www.lung.ca) and The Asthma Society of Canada (www.asthma.ca) helped the public understand and accept this change. However, in the United States, where a similar process had occurred, there was resistance from the National Campaign to Save CFC Asthma Inhalers (www.savecfcinhalers.org).

Evaluate the websites and work with the information to answer the following questions:

1. Who runs each website? Do their credentials make the information presented about non-CFC MDIs reliable or unreliable? Explain.

2. What is the mission of each website? What are the underlying values? How do you know this?

3. What claim does each website make about CFCs, the ozone hole, and MDIs? How do the websites compare in providing scientific evidence in support of their position on non-CFC MDIs? Is the information provided accurate and reliable? Explain.

4. What sorts of links do the websites provide for additional information? Are these supporting links reliable and representative of scientific knowledge? Explain.

5. Whose position on non-CFC MDIs do you agree with? Explain why.

MAKING CONNECTIONS

METHYL BROMIDE AND OZONE DEPLETION

Background: In 2004, the Montreal Protocol banned widespread use of another chemical called methyl bromide (MeBr), which also destroys ozone. While some 27–42% of MeBr comes from natural sources such as oceans and soil, additional MeBr is released as a result of human activity. In Canada, MeBr is used primarily as a broad-spectrum pesticide. MeBr is not manufactured here.

Chemical companies insist that making effective substitutes for all uses of MeBr is not currently possible, and farmers maintain that relying on less effective alternatives will pose economic hardships. These groups also point out that MeBr has a much shorter atmospheric lifetime (0.7 years compared to 100 years for CFCs), a lower ozone depletion potential, and the production of MeBr has been steadily declining anyway (from 12 994 metric tons in 2004 to 1803 metric tons in 2010). In contrast, environmental advocates want a complete ban on MeBr.

Case: You have been assigned the task of developing a policy for the future use of MeBr. Your team must evaluate various options such as:

1. A complete ban on all anthropogenic sources of MeBr and switching to Environment Canada-approved alternatives (www.ec.gc.ca/toxiques-toxics/Default.asp?lang=En&n=98E80CC6-1&xml=B5655CBB-5B04-4D86-ADFC-BC480C437E12) even if this switch comes at a cost.

2. A continuance of the current ban on non-essential uses with exemptions for critical uses approved on an annual basis. (See www.ec.gc.ca/lcpe-cepa/default.asp?lang=En&n=F14675FC-1&offset=1&toc=show#_ftn2 for exemption criteria).

Research these (and possibly other) options and write a report recommending a course of action. In your report include the following:

a. An analysis of the pros and cons of each proposal including: a discussion of the consequences of each choice for stratospheric ozone depletion, human health, food availability and prices, and the economics of farming.
b. Based on the information at hand, what is the best course of action for the future of MeBr? Who should be involved in making this decision? Provide justifications for your proposal.

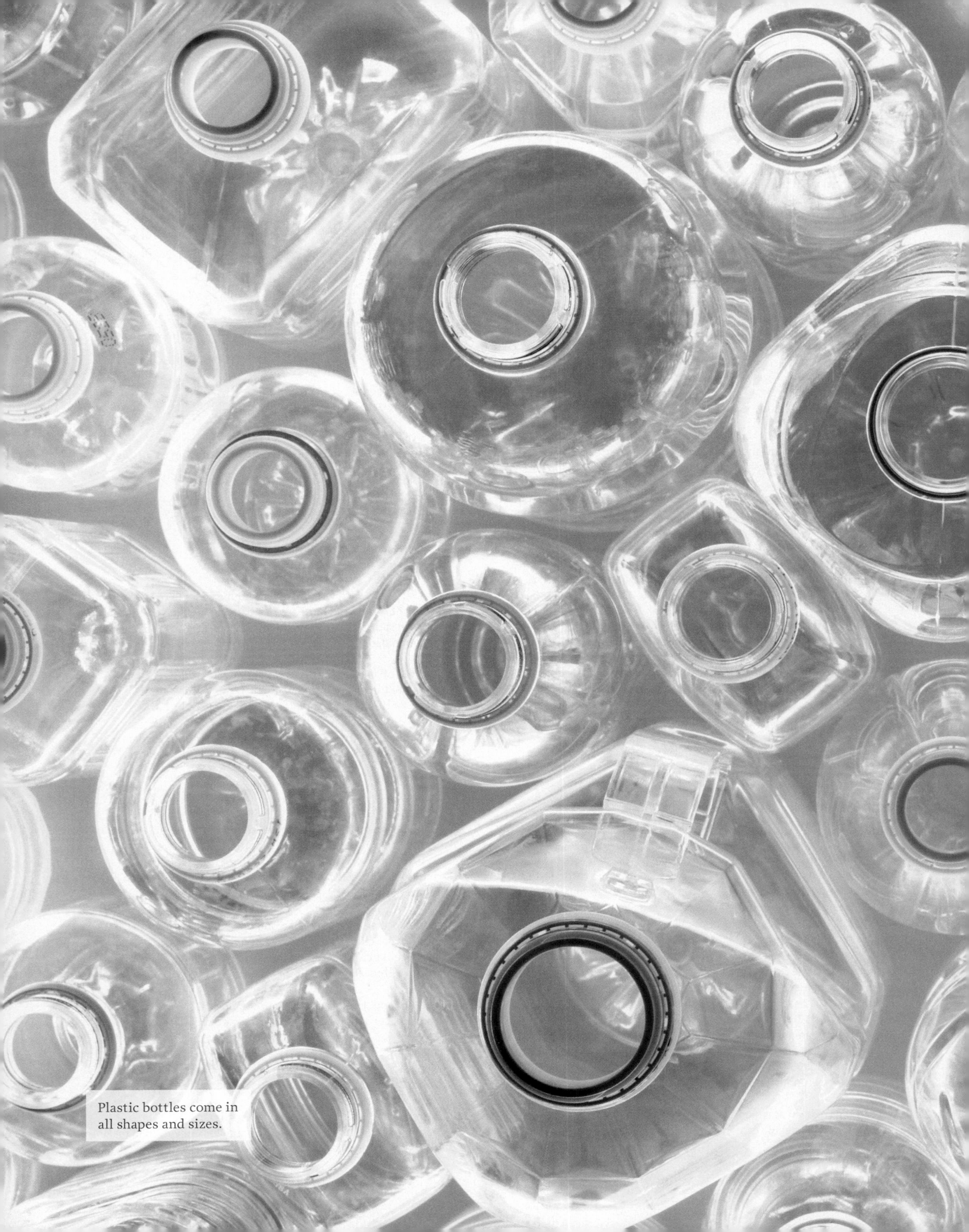

Plastic bottles come in all shapes and sizes.

TOXIC BOTTLES?

On the trail of chemicals in our everyday lives

CORE MESSAGE

Our understanding of the natural world changes almost daily as new evidence is gathered. Unfortunately, misinformation abounds in the popular press and on the Internet about the latest "science." For example, we live in an environment full of natural and synthetic chemicals, some of which are toxic, but knowing how to respond to the latest toxic scare is often difficult as we receive conflicting messages about their safety or risk. Developing information literacy skills enables us to better evaluate the usefulness and trustworthiness of various sources of information and then use the highest-quality information we can find to make reasoned decisions about how to respond.

GUIDING QUESTIONS

After reading this chapter, you should be able to answer the following questions:

→ What is information literacy and why is it important?

→ What common logical fallacies are used in presenting arguments? How can we use critical thinking skills to logically evaluate the quality of information and its source?

→ What factors influence whether or not a particular chemical is toxic and how toxic it is?

→ What kinds of scientific studies allow us to assess the potential hazard of a particular chemical and how do we determine "safe" exposures?

→ What are endocrine disruptors and why can small amounts of exposure produce big effects in an individual?

In 2008, the Canadian government did something no other countr[illegible] ever done. After reviewing reams of research and weathering contentious debate, it recommended that a staple ingredient of most food and beverage containers be banned from baby bottles. If the ban were approved, no company could legally import or sell food containers for use by children that contained the synthetic chemical *bisphenol A* (or BPA, for short). The move incited a frenzy.

Becoming the first country in the world to seek a ban on BPA from some products was "really brave," says Maggie MacDonald at Environmental Defence in Toronto. "Evidence was mounting that showed BPA could harm children, but there was no clear picture yet of just how much harm BPA could cause."

The plastic industry decried the move. In a firestorm of press releases, in newspapers across the country and on television news, industry spokespeople insisted that their own data showed BPA to be perfectly safe.

Parents everywhere were concerned and confused. On the one hand, it was hard to believe that something as commonplace as BPA could be so dangerous; if at least some studies were showing it to be safe, maybe it was. On the other hand, if there was even a chance that this chemical could harm children, shouldn't the government take every possible precaution? And in the meantime, what were average consumers to do?

To help make sense of the conflicting information, the federal government dedicated $1.7 million to study the health effects of BPA over the following three years.

Since the late 1940s, BPA has been a staple ingredient in the linings of metal cans and plastic products of every kind, including food and beverage containers and plastic baby bottles. Once in the body, however, BPA appears to act like the hormone estrogen and some, but not all, studies had linked it to a number of serious medical conditions, from impaired neurological and sexual development to cancer. Despite this uncertainty, by 2010, the European Union was also considering banning the use of BPA in baby bottles and baby food cans; that same year, a U.S. government-appointed panel of scientists agreed that there was "some concern" for unborn babies and children, but concluded that the data were still too uncertain to warrant such a drastic step.

But the lack of conclusive evidence didn't stop Canada—one month after the U.S. report was released, Canada added BPA to its list of toxic substances, officially banning it from baby bottles.

We live in an environment full of toxic substances.

Toxic substances are substances that cause damage to living organisms through immediate or long-term exposure. They fall into two broad categories: synthetic and natural. Natural toxins are not to be taken lightly: arsenic, a basic element that sometimes leaches into groundwater, can cause cancer and nervous system damage in humans.

But synthetic substances are a particular problem because there are quite a lot of them and many are persistent chemicals, meaning they don't readily degrade over time. According to Environment Canada, the agency that protects Canada's environment and natural heritage, 23 000 chemicals were in common commercial use in Canada in 1986; since then, some 800-1000 new chemicals have entered the Canadian marketplace each year.

The debate over how to regulate these substances—how to determine what quantity of any particular compound is safe for humans, and then how to ensure that our exposure levels stay well below those quantities—began in 1962 with a book called *Silent Spring*.

In this book, legendary environmental activist Rachel Carson asked her readers to imagine a world without the sounds of spring, a world in which the birds, frogs, and crickets had all been poisoned by toxic chemicals. Just 20 years had passed since the widespread introduction of herbicides and pesticides like DDT, she explained, but in that relatively short time, they had thoroughly permeated our society. These chemicals were obviously great for killing off weeds and pests; they had done an amazing job conquering mosquito-borne diseases like malaria during World War II, and combating world hunger by boosting global food production. But no one seemed terribly concerned about the effects they might have on nontarget species, or on ecosystems. After all, they were designed

D.D.T.
Powerful Insecticide
Harmless to Humans
TODD

↑ A group of men from Todd Shipyards Corporation run their first public test of an insecticidal fogging machine at Jones Beach State Park, New York, in July 1945. As part of the testing, a 6.5 kilometre area was blanketed with the DDT fog.

to kill living things. Wasn't it at least possible that what killed one organism might also harm or kill others?

Carson went on to identify three specific concerns that were being overlooked at the time: some chemicals can have large effects at small doses; certain stages of human development are especially vulnerable to these effects; and mixtures of different chemicals can have unexpected impacts. The book created an uproar, which led to much stricter regulations for chemical pesticides in general, and, in Canada and the United States, a complete ban on DDT in particular. But half a century later, we are still struggling to effectively regulate the chemicals in our world.

Regulation happens even in the face of change.

Regulation begins with **risk assessment**—a careful weighing of the risks and benefits associated with any given chemical. In an ideal world, unbiased, professional regulators would assess the safety of every new chemical before it entered our lives. They would discern all the potential consequences of excessive or continued long-term exposure and, in so doing, would protect us from any slow, unwitting poisoning. In reality, of course, a variety of factors—practicality, economic forces, sheer need—make such thorough precautions nearly impossible to implement.

The decision to ban BPA from baby bottles followed an investigation by **Health Canada**, the federal agency responsible for protecting Canadians' health. The agency had been investigating BPA as part of the **Chemicals Management Plan**, operated by Health Canada and Environment Canada.

In the case of BPA, where the data were uncertain and the effects unpredictable, the government employed a "better safe than sorry" strategy, known as the **precautionary principle**. MacDonald says that, in the face of what may be considered limited information, this approach cautions

---

toxic substances Substances that cause damage to living organisms through immediate or long-term exposure.
risk assessment Weighing the risks and benefits of a particular action in order to decide how to proceed.
Health Canada The federal agency responsible for protecting Canadians' health.
Chemicals Management Plan Joint Health and Environment Canada program to assess and research chemical substances and improve product safety.
precautionary principle A principle that encourages acting in a way that leaves a margin of safety when there is a potential for serious harm but uncertainty about the form or magnitude of that harm.

us to "stop, reduce our exposure, and do more research to find out if something is, in fact, safe before continuing to expose ourselves on such a wide basis."

This rule of thumb calls for leaving a wide safety margin when setting the *exposure limit*—the maximum quantity of a substance to which humans can safely be exposed. The width of that margin depends on the severity of the potential health effects and environmental damage.

The precautionary principle is becoming a favoured tactic in the European Union. But despite Canada's early action on BPA in baby bottles, for the vast majority of chemicals, it largely follows the lead of the United States, which takes a different approach: "innocent until proven guilty." Rather than thoroughly testing each individual compound, regulators make educated guesses about safety, based on how other, similar compounds have fared. As a result, toxic products are often discovered only after reaching the marketplace—usually when some person or group of people suffers the effects of exposure. Then these products are recalled.

As the case of BPA shows, even when concerns about safety emerge, the presence of contradictory information about the harm a substance poses can make deciding which precautions to take seem like an impossible task.

Part of the problem is that, even in our era of warp-speed communication, information is a slippery thing; this is especially true when it comes to science. We are constantly uncovering new information and gleaning new insights about the environment and our relationship to it. And as our understanding grows and changes, existing information often becomes subject to new interpretation. In fact, much of what we learn in science class today will be outdated 5 years from now—not because we are wholly ignorant in the present, but because we will know so much more in the future.

↑ A scientist in a greenhouse gathers data through first-hand accounts of her observations.

In this rapidly moving current of knowledge floats a seemingly endless array of information sources: newspapers and magazines, websites, television, scientific journals, and so on. Not all of these sources are equal. While some are carefully vetted for accuracy, others are incomplete or deliberately misleading.

Information sources vary in quality and intent.

The ability to distinguish between reliable and unreliable sources is referred to as **information literacy**. It's the key to drawing reasonable, evidence-based conclusions about any given issue or topic. It is especially useful when trying to understand the arguments for and against potentially harmful substances that are also commercially important, such as BPA. In such instances, hyperbole and misinformation abound.

Much of what we learn in science class today will be outdated 5 years from now—not because we are wholly ignorant in the present, but because we will know so much more in the future.

Primary sources are those that present new and original data or information, including novel scientific experiments published in scientific journals. Almost all of these reports, or papers, are rigorously evaluated through **peer review**—a process whereby experts in the field (a panel of the author's "peers") assess the quality of the study's design, data, and statistical analysis, as well as the soundness of the paper's conclusions. Good studies are published; bad ones are rejected.

Secondary sources present and interpret the information from primary sources, such as scientific review articles and many reports from the popular (non-scientific) press. **Tertiary sources** are those that present and interpret information from secondary sources. [INFOGRAPHIC 3.1]

Because the popular press runs on catchy sound bites and easily digestible bits of information, news outlets (both secondary and tertiary) tend to oversimplify the results

Infographic **3.1** | **INFORMATION SOURCES**

PRIMARY SOURCES

Primary sources are first-hand accounts of research and observations. They are largely restricted to peer-reviewed articles in scholarly journals that provide methods and data, but can include other kinds of reports and books based on primary research. Scholarly scientific articles contain the highest-calibre of scientific information. Review articles and editorials in scholarly journals are not primary sources, but still are considered credible.

SECONDARY SOURCES

PRIVATE AND FOR-PROFIT PUBLICATIONS

- Reviews and editorials in scholarly journals
- Not-for-profit agency publications
- Most books, including all reference books
- Magazine and newspaper articles
- Websites (including social media)

GOVERNMENT AND INTERNATIONAL AGENCIES

Reputable sources for environmental information include federal, provincial, regional, and municipal government agencies, such as Environment Canada and Ministries of the Environment and Natural Resources, and international agencies such as the United Nations Environment Programme, the World Health Organization, and the World Meteorological Organization.

UNEP

TERTIARY SOURCES

Tertiary sources can draw from, or summarize, a secondary source. Many blogs, websites, and even news shows qualify as tertiary sources: they provide additional commentary on, or foster debate over, reports from the popular press. They may be accurate; however, they may introduce errors because they do not rely on the original source for facts. They also may perpetuate any errors that appeared in the secondary source.

of individual studies, or present them as if they provided definitive answers. But there are rarely easy answers to environmental questions, and science is almost never as straightforward as we would like it to be. In fact, by its very nature, science is incremental; each study is just one small piece of a much larger puzzle, and existing hypotheses are subject to endless revision and qualification as new bits of data trickle in.

What are the dangers presented by toxic substances and how do we determine safe exposure levels?

Between 2007 and 2009, the Canadian government analyzed urine samples from a representative sample of Canadians, and found that most—91%—tested positive for BPA. These results roughly matched what other countries, such as the United States and Germany, had also found. Since there are no known natural sources of BPA, this study confirmed that the chemical was, in fact, leaching from food and beverage containers into our bodies.

But did that necessarily mean that BPA was causing problems?

Any given substance has several characteristics we must consider when evaluating its safety. One such characteristic is its **persistence**—the ability of a substance to remain in its original form, often expressed by a measure of how long it takes to break down in the environment. Chemicals with low persistence tend to break down quickly in the presence of sunlight. Chemicals with high persistence tend to linger for a long time and can affect ecosystems well after their initial release.

information literacy The ability to evaluate the quality of information.
primary sources Sources that present new and original data or information, including novel scientific experiments or observations and first-hand accounts of any given event.
peer review A process whereby researchers submit a report of their work to outside experts who evaluate the study's design and results to determine if it is of a high-enough quality to publish.
secondary sources Sources that present and interpret information from primary sources. Secondary sources include newspapers, magazines, books, and most information from the Internet.
tertiary sources Sources that present and interpret information from secondary sources.
persistence The ability of a substance to remain in its original form; often expressed as the length of time it takes a substance to break down in the environment.

↑ Water and vegetation sampling by scientists from the University of Alberta in streams near Fort McMurray, Alberta. They are testing for mercury, trace metals, and PAHs.

Another trait that must be considered is the substance's **solubility**—its ability to dissolve in liquid or gas. In some cases, *water-soluble* chemicals are safe for humans but not so good for the environment. Because we can excrete them in our urine, they don't linger in our bodies for very long (of course, at high enough doses these chemicals can still prove toxic; and at low but continual doses they can cause kidney damage). Because water-soluble chemicals are easily taken up by aquatic organisms, they can wreak slow havoc on aquatic environments, and by extension, on the ecosystems that surround them.

Fat-soluble chemicals, those that do not dissolve in water, present an extra level of complexity. Because these chemicals pass easily through cell membranes, our cells can readily absorb them. Our bodies have a hard time expelling fat-soluble chemicals once they're inside us. In some cases, the liver can covert a fat-soluble molecule into a water-soluble one, so that it can be broken down and excreted in urine. But when our livers can't work this magic, fat-soluble chemicals are stored in our fatty tissue, where they can pile up in a process known as **bioaccumulation**.

Biomagnification describes a consequence of bioaccumulation; it's what happens when animals that are higher up on the food chain eat other animals that have bioaccumulated toxic compounds. These predators consume their prey's entire lifetime toxic dose. Biomagnification means that animals higher on the food chain accumulate far more toxic substances than do those lower on the chain. The best example in our human diets is tuna or swordfish. These fish are large predators, and thus are high up on the ocean food chain. So when we eat them, we consume all the chemicals that they have picked up from preying on smaller fish. Mercury is of particular concern. [INFOGRAPHIC 3.2]

As for BPA, it has a low persistence, meaning that it breaks down rapidly in the environment. Although it is fat soluble, liver and gut cells can readily convert it to a water-soluble form, so that it's easily excreted in urine. This means it should not bioaccumulate or biomagnify.

Infographic 3.2 | **BIOACCUMULATION AND BIOMAGNIFICATION**

↓ Animals can acquire fat-soluble toxic substances through air, water, or food sources. The substances build up in the tissue of the animal over its lifetime if it has continued exposure; the lifetime accumulation is stored in fatty tissue, and sometimes in organs.

BIOACCUMULATION

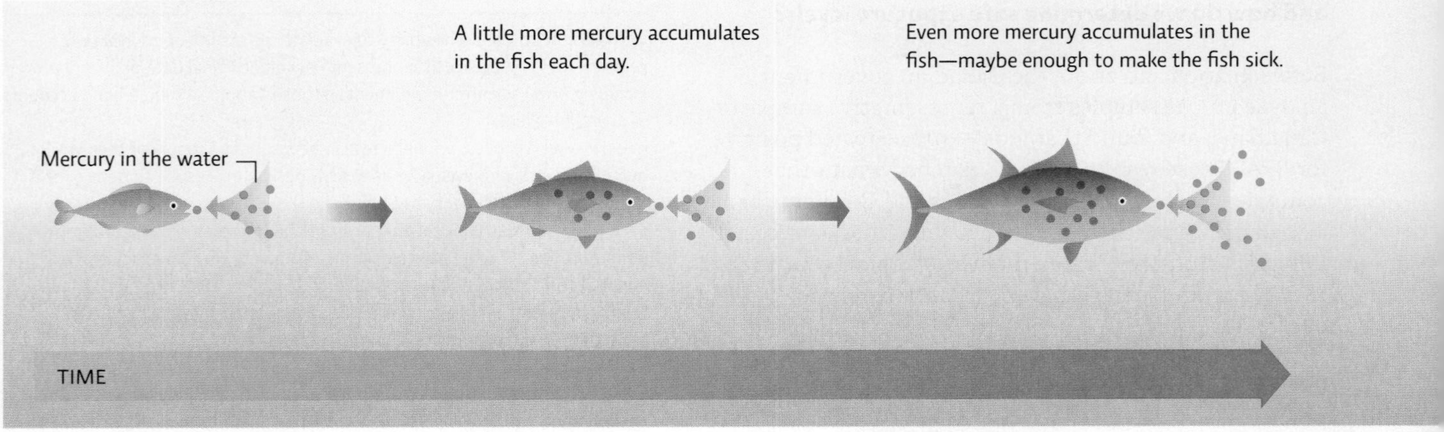
A little more mercury accumulates in the fish each day.
Even more mercury accumulates in the fish—maybe enough to make the fish sick.
Mercury in the water
TIME

But BPA is so commonplace and people are exposed to it so continuously that it remains ever present in our systems, even if it isn't accumulating; as we are breaking down and excreting some BPA, we are ingesting more. Scientists have spent the past decade trying to determine what such exposure might mean for human health.

Figuring out the cause-and-effect relationships between our bodies and the chemicals that enter them is tricky work. **Epidemiologists**—the scientists charged with this type of research—can't just expose a test group of humans to a toxic substance to see what effects it has. They must do a bit of detective work. They can start by looking for health problems in specific populations and work their way backward to find the culprit. Or they can look at groups that have been exposed to a given substance and see if any common health problems emerge. This latter approach is the one that researchers took for BPA. Epidemiologists looked at the health profiles of hundreds of individuals who had BPA in their urine; in one study of 1455 such people, they found an association between BPA concentrations and cardiovascular disease.

The task of determining exactly how BPA might go about wreaking such havoc inside human bodies falls to toxicologists. **Toxicologists** concern themselves with determining the specific properties of potentially toxic substances and how they affect cells or tissues. They do this by testing "intact" lab animals through ***in vivo*** ("in the body") studies, or by testing cells in culture, such as in test tubes or Petri dishes, in what scientists call ***in vitro*** studies ("in glass"). Toxicologists use these data to determine how toxic a substance is and what its effects on living organisms are. [INFOGRAPHIC 3.3]

Toxicity can be affected by a host of factors. Individual susceptibility varies with genetics, age, and underlying health status. When a person is exposed to a toxic substance, the type and amount of chemicals already in that person's system are important. Route of exposure (for example, inhalation, injection, or skin contact) and the dose at the time of exposure also play a role; large doses can cause immediate effects that differ from those caused by lower doses acquired over a longer time period. Some chemicals in the body might increase overall toxicity (**additive effects**); other chemicals may reduce toxicity due to interactions between the toxic substances that "cancel each other out" or at least lessen the effect

solubility The ability of a substance to dissolve in a liquid or gas.
bioaccumulation The buildup of substances in the tissues of an organism over the course of its lifetime.
biomagnification The increased levels of substances in the tissue of predatory (higher trophic level) animals that have consumed organisms that contain bioaccumulated toxic substances.
epidemiologist A scientist who studies the causes and patterns of disease in human populations.
toxicologist A scientist who studies the specific properties of any given potentially toxic substances.
in vivo study Research that studies the effects of an experimental treatment in intact organisms.
in vitro study Research that studies the effects of experimental treatment cells in culture dishes rather than in intact organisms.
additive effects When exposure to two or more chemicals has an effect equivalent to the sum of their individual effects.

↓ Top predators living in ecosystems contaminated with persistent, fat-soluble toxic substances such as mercury or DDT will have much higher levels in their tissues than organisms lower on the food chain. Fish advisories often reflect this danger.

BIOMAGNIFICATION

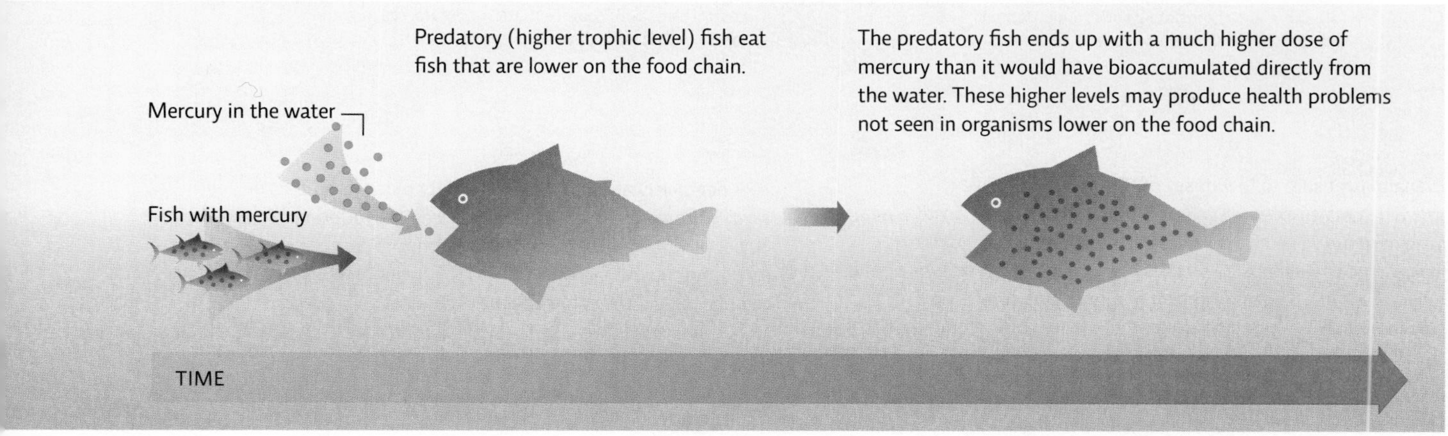
Predatory (higher trophic level) fish eat fish that are lower on the food chain.
The predatory fish ends up with a much higher dose of mercury than it would have bioaccumulated directly from the water. These higher levels may produce health problems not seen in organisms lower on the food chain.
Mercury in the water
Fish with mercury
TIME

Infographic **3.3** | **BPA STUDIES**

↓ Both *in vivo* and *in vitro* studies are used to experimentally study the effects of BPA. Note that none of the researchers discussed below concludes that they have definitive evidence that BPA is harmful, but each bit of research presents a piece of the puzzle. It is the accumulation of evidence that will be most helpful in drawing conclusions about the danger or safety of BPA in our food containers and other plastic goods.

TOXICOLOGICAL STUDIES

In vivo studies

DATA COLLECTED AND ANALYZED

Researchers Nagel and vom Saal studied the effect of low-dose BPA on the prostate size of male mice pups whose mothers were fed one of two doses of BPA when pregnant, compared to males whose mothers were not fed BPA.

RESULT

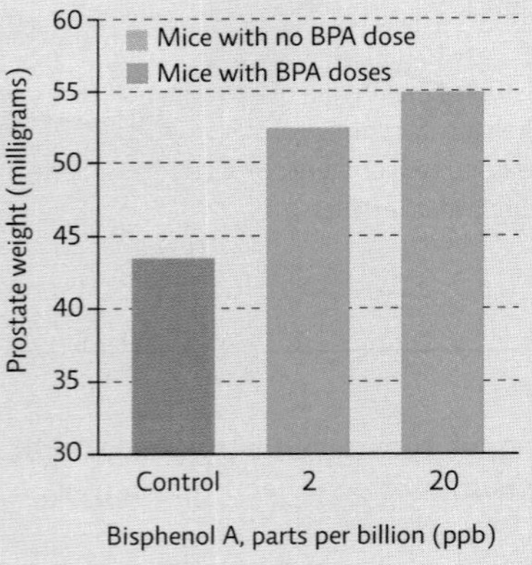
Mice with no BPA dose
Mice with BPA doses
Prostate weight (milligrams)
60
55
50
45
40
35
30
Control
2
20
Bisphenol A, parts per billion (ppb)

At both doses, male mice had significantly larger (30–35% larger) prostate glands ($p < 0.01$) but did not vary from controls in overall body size.

RESEARCHERS' CONCLUSIONS

Prenatal exposure to low doses of BPA alters prostate growth in male mice and is thus biologically active at doses seen in nature. The effect was greater than predicted for *in vivo* studies. Perhaps BPA acts synergistically with naturally occuring estrogens in the animal feed or is converted to a more active product in the body. Further studies on the processing of BPA in the body are needed.

In vitro studies

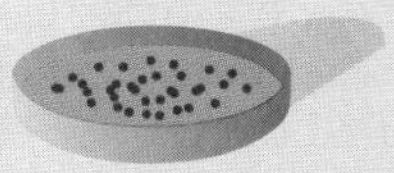

DATA COLLECTED AND ANALYZED

Researchers Ishido and Suzuki placed droplets of rat neural stem cells onto a dish treated with various doses of BPA. They then monitored the cells with a microscope. As the stem cells migrated out of the spherical droplets across the dish, the researchers measured, on average, how far the cells migrated in both the test and control groups.

RESULT

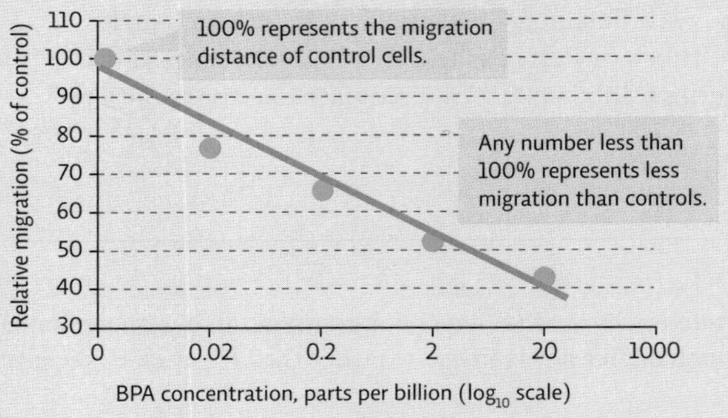
100% represents the migration distance of control cells.
Any number less than 100% represents less migration than controls.
Relative migration (% of control)
110
100
90
80
70
60
50
40
30
0
0.02
0.2
2
20
1000
BPA concentration, parts per billion (log_{10} scale)

BPA had a negative effect on cell migration compared to control cells; as BPA concentration went up, cell migration (and survival) went down.

RESEARCHERS' CONCLUSIONS

BPA negatively affects migration ability of rat brain cells *in vitro*. Since successful migration is necessary for proper brain development, low doses of BPA may impact brain development and function in these animals. Further testing is needed to determine if the same effect is seen *in vivo*.

↓ Epidemiological studies look at human populations to see if any BPA effects emerge in particular groups, such as those with a diagnosis of disease. The Lang study shown here compared the amount of BPA in the urine of subjects with and without various common health conditions to see if high BPA correlated with any of the illnesses.

EPIDEMIOLOGICAL STUDIES

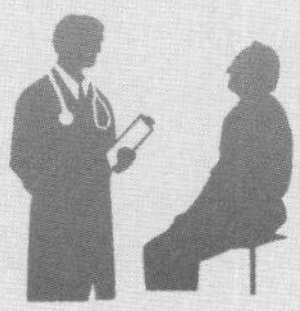

DATA COLLECTED AND ANALYZED

The Lang study evaluated 1455 subjects' urine BPA levels. Subjects were interviewed about whether they had been diagnosed with various health conditions. The data were adjusted for age and sex, and statistically analyzed to see if any groups (those with or without a certain health condition) differed in their levels of BPA.

RESULTS

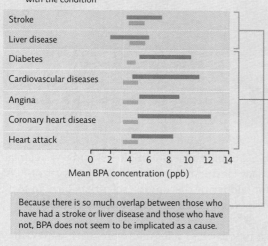

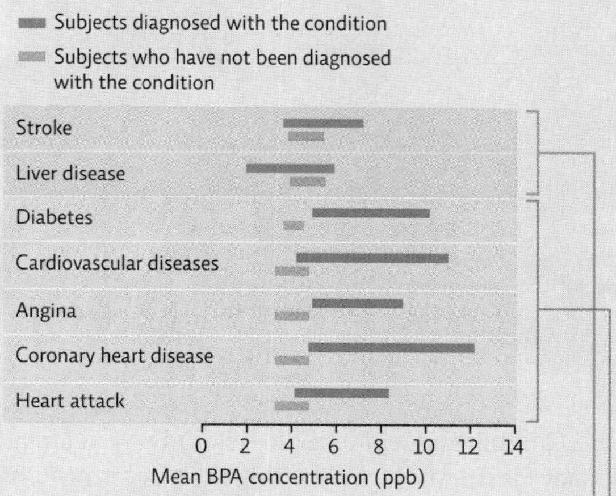
Subjects diagnosed with the condition
Subjects who have not been diagnosed with the condition
Stroke
Liver disease
Diabetes
Cardiovascular diseases
Angina
Coronary heart disease
Heart attack
0 2 4 6 8 10 12 14
Mean BPA concentration (ppb)

Because there is so much overlap between those who have had a stroke or liver disease and those who have not, BPA does not seem to be implicated as a cause.

For diabetes and all forms of cardiovascular disease reported, there is a significant difference in BPA urine levels (higher in those with the disease). Notice that the bars overlap very little, if at all.

RESEARCHERS' CONCLUSIONS

Concentrations of BPA were associated with an increased prevalence of coronary heart disease, cardiovascular disease, and diabetes. These findings add to the evidence that suggests there are adverse effects of low-dose BPA in animals. Independent replication and follow-up studies are needed to confirm these findings and to provide evidence about whether the associations are causal.

(**antagonistic effects**). Still other chemicals may work together to produce an even bigger effect than expected (**synergistic effects**). [INFOGRAPHIC 3.4]

But in general, toxicologists like to say that "the dose makes the poison." This means that almost anything can be tolerated in low enough doses; conversely, anything—even water—can be toxic if the dose is big enough. And in most cases, as the dose increases, so does the severity of the effect. This idea—that higher doses of something harmful are worse for you than lower doses—makes obvious sense. It guides both regulatory efforts and modern medicine. And it applies to many other chemicals you can think of—except for the class of which BPA is a part: endocrine disruptors.

Endocrine disruptors cause big problems at small doses.

As their name suggests, **endocrine disruptors** interfere with the endocrine system, typically by mimicking a **hormone**, or preventing a hormone from having an effect. BPA is a mimic of estrogen, a hormone that plays many roles in the body, primarily guiding reproduction and development in both males and females. BPA binds to the body's cellular estrogen **receptors**, among others, and triggers effects that natural estrogen would trigger. In animals living in the wild, chemicals like this have been shown to cause feminization of males (even sex changes), as evidenced by lower sperm counts and the production of egg proteins normally only produced by females.

In humans, we know that sperm counts are down among men and that puberty onset is earlier in both boys and girls than it has ever been before. We also know that 91% of the Canadian population has trace amounts of BPA in its urine. No one can say for sure whether one fact (changes in sexual development) is related to the other (BPA in our systems), but because hormones control the development of body organs, scientists are especially concerned about the exposure of developing fetuses, newborns, and infants to endocrine disruptors.

Endocrine disruptors are curiously different from many other chemicals, where the relationship between dose

antagonistic effects When exposure to two or more chemicals has a smaller effect than the sum of their individual effects normally would.
synergistic effects When exposure to two or more chemicals has a greater effect than the sum of their individual effects normally would.
endocrine disruptor A substance that interferes with the endocrine system, typically by mimicking a hormone or preventing a hormone from having an effect.
hormone A chemical released by organisms that directs cellular activity and produces changes in how their bodies function.
receptor A structure on or inside a cell that binds a particular molecule, such as a hormone, thus allowing the molecule to affect the cell.

Infographic **3.4** | **FACTORS THAT AFFECT TOXICITY**

↓ Some chemicals are more toxic than others due to their mode of action. Other factors also affect how toxic a particular chemical will be for an individual.

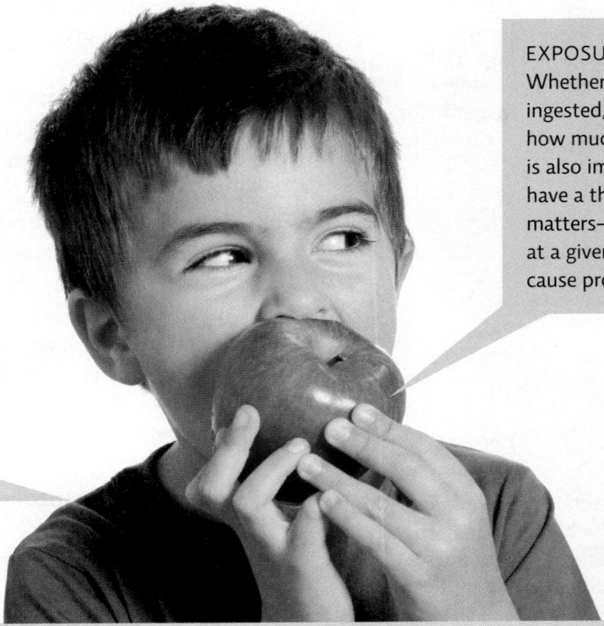

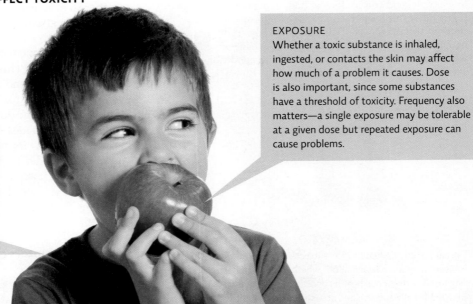

EXPOSURE
Whether a toxic substance is inhaled, ingested, or contacts the skin may affect how much of a problem it causes. Dose is also important, since some substances have a threshold of toxicity. Frequency also matters—a single exposure may be tolerable at a given dose but repeated exposure can cause problems.

INDIVIDUAL FACTORS
Factors related to the individual may affect how toxic a chemical is. Some chemicals are more of a problem for the very young or very old, or for those who are ill. In some cases, genetic differences make a person more or less vulnerable to a given chemical.

CHEMICAL INTERACTIONS
We are never exposed to just one chemical. The fact that chemicals can interact in ways that increase or decrease their toxic effects complicates our efforts to determine a "safe dose." For instance, suppose there are two chemicals (A and B); each raises one's temperature 2 degrees at a given dose. If we are exposed to both chemicals at the same time, one of three interactions is possible:

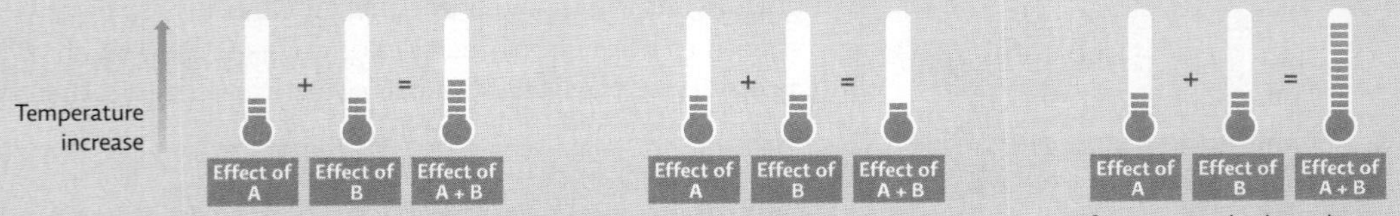
ADDITIVE EFFECT
ANTAGONISTIC EFFECT
SYNERGISTIC EFFECT
Temperature increase
Effect of A
Effect of B
Effect of A + B

If exposure to the two chemicals gives the effect we'd expect (2 + 2 = 4), then the effect is additive.

If exposure to both actually lessens the effect we would expect, then the interaction is antagonistic (2 + 2 < 4).

If exposure to both produces an effect much greater than the sum of the two individual effects, the chemicals are working synergistically (2 + 2 > 4).

and effect is linear (the more you ingest, the sicker you get—"the dose makes the poison"). They can have one set of effects at a very low dose, and then no effects (or much different effects) at higher doses.

To track the effects of a dose of a chemical, toxicologists use data from *in vitro* and *in vivo* studies to create a **dose-response curve**, from which they can calculate an **LD50** (lethal dose 50%), the dose that would kill 50% of the population. The lower the LD50, the more toxic the substance. [INFOGRAPHIC 3.5]

Because endocrine disruptors like BPA have different effects at low and high doses, LD50s and dose-response curves are much trickier to calculate, and a "safe dose" is much tougher to determine. "We don't know if BPA has a 'threshold' dose at which it starts to impact health," says Bruce Lanphear, an epidemiologist and physician at Simon Fraser University, who studies the effects of BPA. Indeed, for endocrine disruptors, scientists have to test the effects of high doses and low doses separately.

In the late 1980s, when scientists were testing BPA, they started with very high doses, given to rats, lowered the dose until they saw no adverse effects, and then stopped. As is the usual practice, a "safe dose" was established by dividing that "no effect" level by 1000, to account for the possibility that humans might be more sensitive than lab animals, or that some people, such as children and the elderly or sick, might be more vulnerable than others.

But this strategy doesn't necessarily apply to endocrine disruptors, where low doses can still cause harm. This was proven during one unfortunate chapter in our

Infographic 3.5 | **DOSE-RESPONSE CURVES**

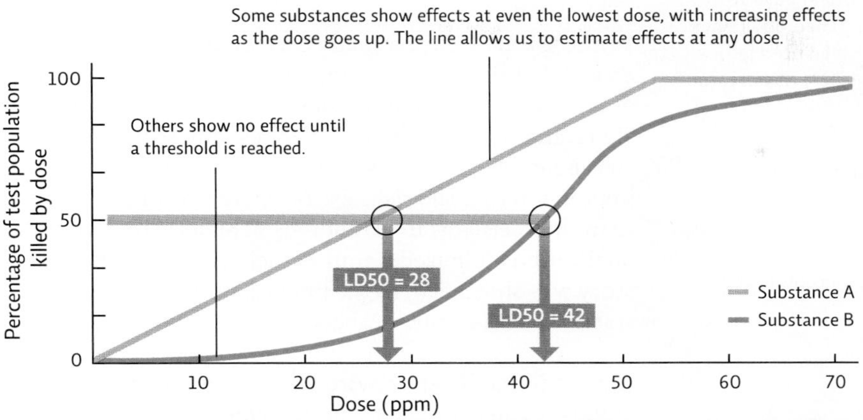
Some substances show effects at even the lowest dose, with increasing effects as the dose goes up. The line allows us to estimate effects at any dose.
Others show no effect until a threshold is reached.
Percentage of test population killed by dose
100
50
0
LD50 = 28
LD50 = 42
Substance A
Substance B
10
20
30
40
50
60
70
Dose (ppm)

← Dose-response studies evaluate the effect of a toxic substance at various doses (on animals or cells). Knowing the LD50 allows us to compare toxic substances directly, i.e., which are more toxic than others. Charting the actual change in effect as the dose increases (the "dose response") allows us to predict effects at doses other than those tested. We can also determine if there is a threshold of exposure that must be reached before harmful effects are seen.

"SAFE DOSE"
Because there is uncertainty in the determination of "safe dose," regulatory agencies factor in a safety margin of 100 or even 1000 to be on the safe side. Example: If we have evidence that says that a chemical is safe at 100 ppm, we might set the environmental standard (what is allowed) at 1 ppm (or even 0.1 ppm if children are at risk).

chemical history—between 1938 and 1971, doctors used the estrogen mimic diethylstilbesterol (DES) to prevent premature labour in expectant mothers. Thirty years later, when a higher incidence of reproductive abnormalities started showing up in the population, epidemiologists traced it back to the DES given to pregnant mothers.

Researchers actually developed BPA at the same time as DES—not as an ingredient of plastic, but as another estrogen mimic. They ended up shelving BPA when DES appeared to better replicate the effects of estrogen, says Lanphear. Only later did chemists discover that it had great sealant properties, making it useful for an infinite variety of products, from food and beverage cans to dental sealants and football helmets. Given this history, it should be no surprise that BPA has some effects on the human body, says Lanphear. "When you start to use chemicals, the first thing to ask is: does it have biological activities? And of course, BPA did, because that's what it was designed to do." [INFOGRAPHIC 3.6]

Over the years, scientists have developed a list of BPA's effects in lab animals: increased early growth in both males and females, early onset of sexual maturation in females, decreased testosterone and increased prostate size in males, altered immune function, and increased mortality of embryos. Increasingly, more experts began to worry that, at low doses—such as those quantities that leach from plastic containers into food and drinks—BPA might be particularly dangerous for pregnant mothers, developing fetuses, and young children.

Other researchers thought that developing fetuses were safe from any potential effects. They reasoned that BPA would be broken down in the mother's gut and excreted in her urine long before it had a chance to reach the womb. Even if the gut failed, they thought, BPA would never get past the placenta. A 2002 study of 37 pregnant women, however, found BPA in fetal blood and the placenta.

So where did the truth lie?

Critical thinking skills give us the tools to uncover logical fallacies in arguments or claims.

In 2010, Lanphear and a group of experts gathered in Ottawa, at a meeting convened by the World Health Organization and the Food and Agriculture Organization of the United Nations, to discuss the potential effects of BPA. By then, there had been hundreds of studies about BPA, according to Lanphear, but half of the toxicologists at the meeting said many of the studies were flawed—BPA was administered incorrectly to the animals used in the study, or the doses were wrong, or perhaps the scientists looked at the wrong health outcomes to determine its effects. As a result, the report from the meeting simply highlighted the "considerable uncertainty" in our knowledge, and recommended further study.

dose-response curve A graph of the effects of a substance at different concentrations or levels of exposures.

LD50 (lethal dose 50%) The dose of a substance that would kill 50% of the test population.

HOW HORMONES WORK

rmones like estrogen (and its mimics) pass into ere they bind to DNA and actually "turn on" genes. effects can be far reaching since the activated genes affect other genes and many cell processes, having many later effects.

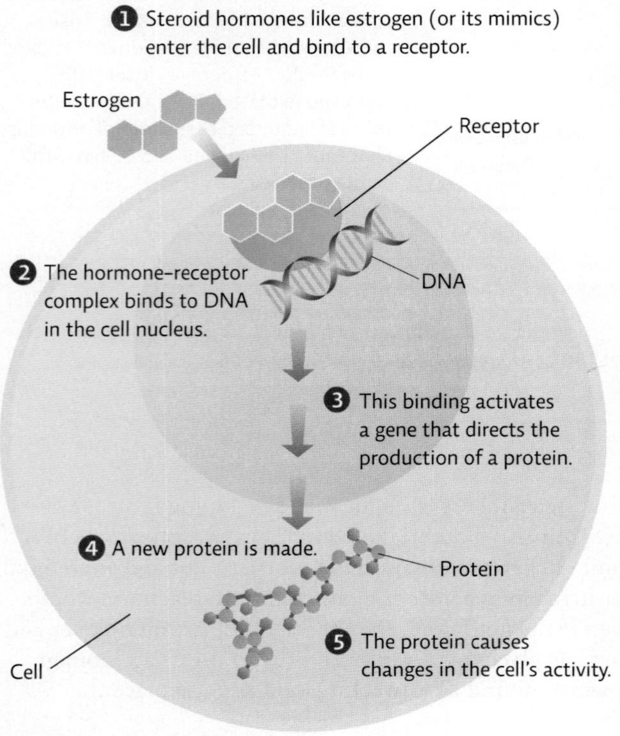
1 Steroid hormones like estrogen (or its mimics) enter the cell and bind to a receptor.
Estrogen
Receptor
2 The hormone-receptor complex binds to DNA in the cell nucleus.
DNA
3 This binding activates a gene that directs the production of a protein.
4 A new protein is made.
Protein
5 The protein causes changes in the cell's activity.
Cell

It's important for everybody—not just scientists—to think carefully about the data they are presented, advises MacDonald. "People need to develop a critical mind, and not just accept the science at face value," she says. Questions to ask yourself include: How was the research conducted? Was the study conducted by someone with a vested interest in the outcome? "Critical science literacy is more important now than ever before," says MacDonald.

Unfortunately, the media doesn't always help clarify the debate. Take, for instance, this excerpt from one of the many news stories about BPA:

Tests performed on liquid baby formula found that they all contained bisphenol A (BPA). This leaching, hormone-mimicking chemical is used by all major baby formula manufacturers in the linings of the metal cans in which baby formula is sold. BPA has been found to cause hyperactivity, reproductive abnormalities, and pediatric brain cancer in lab animals. Increasingly, scientists suspect that BPA might be linked to several medical problems in humans, including breast and testicular cancer.

While the article is factually correct, it commits several errors of omission. By failing to explain the high degree of uncertainty (similar studies reached different conclusions, and scientists had yet to reach any sort of consensus about what the risks might be) the author creates the impression that the risks are far more certain, and dire, than they actually are. Hundreds of articles in dozens of publications followed a similar tack, and before long, the story was shortened, in the public's mind, to one simple statement: BPA causes cancer.

Almost immediately, environmental and consumer groups began calling for a nationwide ban on the chemical. These groups frequently attacked the plastics industry, suggesting that because large, powerful companies profited handsomely from BPA-made products, they had an incentive to downplay or even suppress troubling data. Such attacks often employed **logical fallacies** and succeeded in stirring up the public's fears, but told it nothing about whether or not BPA was actually safe. [INFOGRAPHIC 3.7]

As a result, average consumers were left to figure out for themselves which side to trust: Were environmental activists and some scientists exaggerating, or was BPA truly dangerous? Why did some scientific studies show deleterious effects from the chemical, while others did not? More importantly, what precautions could, or should, individuals take?

Critical thinking skills are the antidote to logical fallacies; they enable individuals to logically assess and reflect on information and reach their own conclusions about it. The skill set can be broken down into a handful of components.

Be skeptical. Just like a good scientist, one should not accept claims without evidence, even from an expert. This doesn't mean refusing to believe anything; it simply means requiring evidence before accepting a claim as reasonable. For example, a *Science News* article, "Receipts a Large and Little-Known Source of BPA," quoted a researcher as saying that he believed BPA, a common component of the thermal paper of cash register receipts, "penetrated deeply into the skin, perhaps as far as the bloodstream." The actual research documented

logical fallacies Arguments which attempt to sway the reader without using evidence.

critical thinking skills Skills that enable individuals to logically assess information, reflect on that information, and reach their own conclusions.

the amount of BPA in the receipts and on the surface of fingers that held a receipt, but not the amount of BPA in the bloodstream of individuals who handled the receipts. A skeptical reader would therefore question the inference that BPA from receipts enters the body, since no evidence was presented regarding its uptake, only its presence on the skin's surface. Perhaps BPA does enter the bloodstream through receipts, but we need evidence in order to view this claim as reasonable.

Evaluate the evidence. Is the claim that is being made derived from anecdotal evidence (unscientific observations, usually relayed as secondhand stories) or from scientific studies? If it is based on actual studies, how relevant are those studies to the claim? Were they done on primates? Rodents? Cells in a Petri dish? Or did the researchers look at human populations?

Be open-minded. Try to identify your own biases or preconceived notions (most chemicals are dangerous; most people overreact to things like this, etc.) and be willing to follow the evidence wherever it takes you.

Watch out for biases. Does the author of the study or person making the claim have a position he or she is trying to promote? Is this person financially tied to one conclusion or another? Is he or she trying to use evidence to support a predetermined conclusion?

To try to make sense of some of the conflicting data, Lanphear decided to follow up on some research he heard about in Ottawa. Some toxicologists had mentioned that they'd consistently noted that young rodents exposed to BPA were more anxious than others. So he and his colleagues decided to follow nearly 250 pregnant women, testing their urine for BPA exposures, and then follow their children during the first three years of their lives. In 2012, he and his colleagues published their results in the science journal *Pediatrics*—97% of mothers and

Infographic **3.7** | **DON'T BE MISLED BY LOGICAL FALLACIES**

↓ Logical fallacies are arguments used to confuse or sway a reader or listener to accept a claim/position in the absence of evidence.

COMMON LOGICAL FALLACIES	EXAMPLES
Hasty generalization: Drawing a broad conclusion on too little evidence.	A study might show that BPA is present in the urine of babies who drink from plastic bottles; however, this is not evidence that babies are being poisoned by their bottles. All we could conclude would be that babies who drink from plastic bottles have more BPA in their urine than babies who do not.
Red herring: Presents extraneous information that does not directly support the claim but that might confuse the reader/listener.	An argument that the buildup of toxic chemicals in modern humans is significant and that many of these buildups have led to problems may be true but tells us nothing about the safety of BPA.
Ad hominem attack: Attacks the person/group presenting the opposite view rather than addressing the evidence.	Opposing the use of BPA on the grounds that the chemical industry is untrustworthy (because they profit from plastics) does not address the safety of BPA.
Appeal to authority: Does not present evidence directly but instead makes the case that "experts" agree with the position/claim.	Claiming that BPA is a health hazard (or not) because a noted toxicologist or scientific group has drawn that conclusion is not evidence in itself. If the experts agree, there should be a reason, and that evidence should be presented.
Appeal to ignorance: A statement or implication that the issue is too complex and we are not capable of understanding it.	In order to justify doing nothing about the use of BPA, one might claim there is no way to know its effect since humans are exposed to so many toxic substances. We may not know everything about the effects of BPA, but we do know some things and it is on that evidence that we should draw our conclusions.
False dichotomy: The argument sets up an either/or choice that is not valid. Issues in environmental science are rarely black and white, so easy answers (it is "this" or "that") are rarely accurate.	The claims that BPA must be *completely avoided* or is *totally safe for everyone* is a false dichotomy. The evidence so far suggests that the effects of BPA may be negligible for healthy adults but problematic for fetuses and very young children. Recommendations differ for different groups.

children tested positive for BPA, and the more BPA children were exposed to in the womb, the more they appeared anxious or depressed. The findings—supported by funding from the U.S. government—were particularly true for young girls.

In the paper, Lanphear and his team acknowledged that the study was small, and children may have been exposed to other endocrine disruptors that might have influenced the results. This makes it difficult to know the specific effects of BPA. Steven Hentges, a representative of the American Chemistry Council, which represents the chemical industry, told the media in a statement about the paper: "For parents, the most important information from this report is that the authors themselves question its relevance." This statement, however, misrepresents the authors' self critique. Most research papers end with caveats about the study's limitations, and Hentges, who is paid by the industry that profits from selling BPA, cannot be expected to provide an objective perspective.

Risk assessments help determine safe exposure levels.

When Canada declared its intention to ban BPA from baby bottles in 2008, declaring the substance toxic—a next logical step—was not so easy. That's because in 2009, the American Chemistry Council launched a formal objection.

By 2008, BPA was already a $3 billion industry unto itself; and, in some cases at least, it had no obvious substitute. That made employing the precautionary principle, or banning BPA more widely, almost impossible. "This is the central dilemma the government faces," says Lanphear. Government has to keep us safe, and make us *feel* safe, all without causing a major disruption to the economy. "If the government declared that BPA in all sources was toxic, and it must be eliminated within a year, it would put our economy into a tailspin." Companies would have to scramble to replace BPA from the myriad of products that contain it.

To make such decisions, we need to know two things: whether a given chemical has the potential to harm us, and how great that harm might be. It's crucial to consider both. Say, for example, a new study reports that BPA exposure doubles one's chances of developing a particular disease. Before hitting the panic button, we must first ask what the chances are of getting the disease in the first place, without BPA exposure. If it's a rare condition, with a 1% probability, then BPA increases our chances to a mere 2%. If it is more common, with, say, a 20% probability, then BPA doubles our chances to 40%—a significantly greater risk. How important that increase is, and thus, what we should do in light of it, depends on the seriousness of the disease.

So, what to do about BPA? Let's take a look at what we've learned so far. We know that chemicals like BPA can have effects at low doses. We know that developing fetuses and young mammals are particularly susceptible to these low-dose effects. We know that BPA is, in fact, leaching into our systems from food and beverage containers, and that it can indeed cross the human placental barrier.

But we also know that BPA is water-soluble and can be excreted in our urine. This means that, in most cases, it may not be building up, or bioaccumulating, in our tissues.

What we don't know—what we may never know—is how well the effects seen in rodents correspond to the risks faced by humans. And because of all the additive and synergistic effects BPA is likely to have with all of the countless other chemicals we encounter, it will be difficult to tell, even in the future, how much of any given health condition can be specifically attributed to BPA.

The bane of risk assessment is that we can't really wait for those facts to come in.

In the end, Canada's federal government decided to declare BPA toxic in 2010, despite the American Chemistry Council's objections. In announcing the decision, Canadian Minister of the Environment Jim Prentice said: "I am of the view that [the ACC's] notice does not bring forth any new scientific data or information."

Soon after Canada declared BPA toxic, the European Union also banned BPA from baby bottles; in 2012, the United States did, as well.

By then, the American Chemistry Council *was* asking the U.S. government to ban the substance from baby bottles, because manufacturers had stopped including BPA in bottles and sippy cups due to the public's concern. They weren't the only industry to voluntarily phase out the chemical—in 2007, outdoor gear and clothing seller Mountain Equipment Co-op, headquartered in Vancouver, was the first Canadian retailer to remove BPA-laden containers from its shelves. That same month, Lululemon Athletica, which sells yoga and running gear, said it would no longer sell plastic bottles that contain BPA. The next year, a host of Canadian companies followed suit—Walmart Canada, Canadian Tire, and Hudson's Bay Co., among others.

Since declaring BPA toxic in 2010, Health Canada has yet to ban it from anything other than baby bottles and sippy cups, concluding that the amount of BPA in canned food and beverage containers "posed no health or safety concerns to the general public." Meanwhile, France became the first country to ban BPA from all food containers; its plan will be implemented by 2015.

The purpose of the Canadian ban was to reduce exposures for children, who appear most vulnerable to BPA's effects, but children do more than drink out of bottles and sippy cups, says Lanphear. Isn't canned soup a children's product? "Some important exposures were not considered," he says. And if children are at risk, should the government consider banning BPA from products that pregnant women come in contact with? Since women don't always know when they're pregnant, the ban would have to extend to the packaging of foods that all women eat. And since women generally eat the same things as men, BPA would essentially have to be banned from all food packaging—leading to major disruptions to the economy. But restricting the ban to baby bottles simply "gives us an illusion of safety," says Lanphear.

However, a widespread ban could create even more problems. Since companies have phased out BPA, they've replaced it in many products with a very similar compound: *bisphenol S* (BPS). Scientists are still studying its properties, but early research suggests it may mimic estrogen more potently than BPA, and break down more slowly in the environment.

"Most of us think that our government is out there, testing these things before they're released into the marketplace," says Lanphear. But even in countries like Canada, where the government made that first courageous decision to ban BPA from baby bottles, products are full of chemicals that have unknown effects on the body.

"It may turn out that BPA is perfectly safe, and it may be that some other chemical—flame retardants, or whatever—isn't," says Lanphear. "But the point is we aren't doing these studies on laboratory rats. We're doing them on ourselves, by releasing these products into the marketplace without knowing their effects. Yet how can we test everything thoroughly without bringing our economy to a grinding halt? This is one of the central ethical dilemmas we're dealing with in our time."◉

Select references in this chapter:
Braun, J.M., *et al.* 2011. *Pediatrics*, 128(5): 873-882.
Ishido, M., and Suzuki, J. 2010. *Journal of Health Science*, 56: 175–181.
Lang, I.A., *et al.* 2008. *Journal of the American Medical Association*, 300: 1303–1310.
Nagel, S.C., *et al.* 1997. *Environmental Health Perspectives*, 105: 70–76.
Raloff, J., 2010. *Science News*, 175(5): 5-6.
Schönfelder, G., *et al.* 2002. *Environmental Health Perspectives*, 110: 703–707.
Statistics Canada. 2012. Bisphenol A concentrations in the Canadian population, 2007 to 2009. *Statistics Canada Catalogue* no. 82-625-X.

BRING IT HOME

PERSONAL CHOICES THAT HELP

What do some air fresheners, nail polishes, and plastic storage containers have in common? They are all potential sources of chemicals that people are exposed to every day. Many of the products designed to improve our lives actually contain chemicals that may harm us in the long run. With just a few changes, you can dramatically reduce your long-term chemical exposure.

Individual Steps

→ Check to see if the body products you use contain potentially harmful chemicals at www.ewg.org/skindeep.

→ Avoid microwaving or storing hot food in plastic containers; chemicals such as BPA can leach into your food when heated. Always use microwave-safe glass or ceramic containers when microwaving food.

→ Check with your city's or municipality's waste management services for guidelines about how to correctly dispose of household wastes including paint, medication, and cleaning products.

Group Action

→ Talk to your friends and family, especially those you live with, to see if you can switch to products with fewer harmful chemicals.

Policy

→ There are many policies governing the use of toxic substances in Canada. Visit the Government of Canada website (www.canada.gc.ca) and determine which federal departments have legislation relating to toxic substances, and how many policies there are. Contact your local representative if you feel that these polices are not effective.

UNDERSTANDING THE ISSUE

CHECK YOUR UNDERSTANDING

1. Which of the following is true of all toxic substances?
a. They cause damage through immediate or long-term exposure.
b. They don't readily break down through natural means.
c. They are synthetic chemicals; no natural substances are considered toxic.
d. They will biomagnify up the food chain.

2. The test data for a new pesticide provided mixed results about its effects on mice. The pesticide seemed harmless on its own, but when the new pesticide was combined with another pesticide some mice got sick. However, this new pesticide was very effective against an invasive insect ravaging local trees. In deciding to approve the pesticide for sale, the chemists were applying:
a. the precautionary principle.
b. logical fallacies.
c. information literacy.
d. risk assessment.

3. Pesticides increase agricultural yields but some people oppose their use because of their inherent toxicity. An argument against pesticide use that attacks the pesticide maker on the grounds that he or she is simply profit driven is:
a. an ad hominem attack.
b. an appeal to authority.
c. an appeal to ignorance.
d. a false dichotomy.

4. All of the following are secondary sources of information except:
a. an encyclopedia.
b. a newspaper article.
c. a peer-reviewed article.
d. a government website.

5. Pesticides such as DDT are known to biomagnify in food chains. This means that:
a. organisms in lower trophic levels accumulate lethal doses of toxic substances.
b. organisms at higher trophic levels have more concentrated levels of toxic substances.
c. toxic substances build up in the tissue of an organism over the course of its lifetime.
d. the environment has higher concentrations of toxic substances than organisms in the food chain.

6. The toxicity of a chemical is evaluated using animal models by creating a(n) ______; a high LD50 indicates ______.
a. *in vitro* study; a safe dose
b. dose analysis; a threshold dose
c. *in vivo* study; high toxicity
d. dose-response curve; low toxicity

WORK WITH IDEAS

1. What potential dangers do toxic substances present and what characteristics of a toxic substance make it more or less harmful?

2. What are the differences between *in vitro* and *in vivo* studies? How are these types of studies useful in toxicology?

3. Why is the debate over the safety of synthetic chemicals contentious?

4. What is the precautionary principle? In what way does the story of BPA reflect the precautionary principle?

5. What factors influence the toxicity of a given chemical? How do regulators setting "safe exposure" standards deal with the uncertainty associated with these factors?

ANALYZING THE SCIENCE

The graphs on the following page come from a 2005 study by the Toxic-Free Legacy Coalition and the Washington Toxics Coalition to identify chemical residues in the human body (http://pollutioninpeople.org/results). Samples of hair, blood, and urine of ten individuals from Washington D.C. were tested for a variety of toxic substances including mercury and DDT. Data for each individual (identified by his or her initials) are shown for p,p'–DDE (a breakdown product of DDT) and for mercury.

INTERPRETATION

1. Identify the people with the highest and lowest levels of mercury and DDT.

2. Compare the levels of mercury and DDT in each study participant to the national median. Which study participants have levels of DDT and mercury that are at least twice the national median?

3. The EPA "safe dose" for mercury is 1100 ppb (parts per billion) for women in their child-bearing years, as mercury levels above this value may impair neurological development in the fetus. Which study participant is above the safe level and by how much?

ADVANCE YOUR THINKING

4. Who conducted this study and for what purpose? What type of study do these data represent? Are the data reliable? Explain your responses.

5. The pesticide DDT has been banned in Canada and the United States since 1972. How do you explain the presence of DDT in 8 of the 10 study participants?

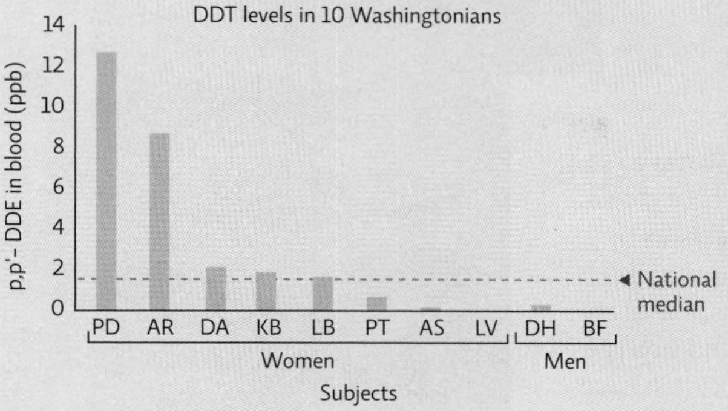
DDT levels in 10 Washingtonians
p,p'- DDE in blood (ppb)
14
12
10
8
6
4
2
0
National median
PD AR DA KB LB PT AS LV DH BF
Women
Men
Subjects

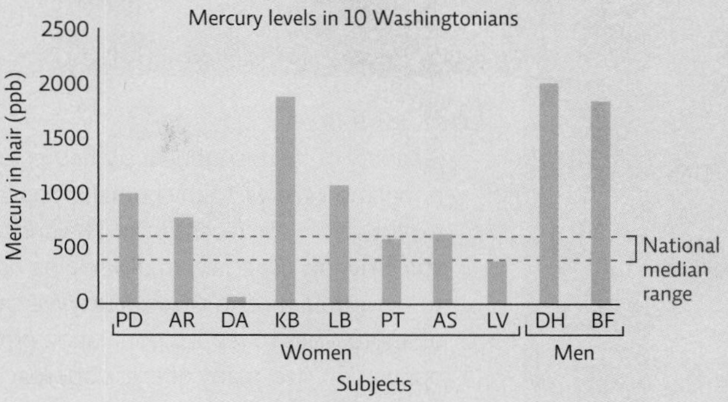
Mercury levels in 10 Washingtonians
Mercury in hair (ppb)
2500
2000
1500
1000
500
0
National median range
PD AR DA KB LB PT AS LV DH BF
Women
Men
Subjects

EVALUATING NEW INFORMATION

Some sources estimate that there are more than 12 000 chemical ingredients used in cosmetics, of which less than 20% have been reviewed for safety. While Health Canada maintains a list of substances whose use in cosmetics comes under some form of restriction (www.hc-sc.gc.ca/cps-spc/cosmet-person/indust/hot-list-critique/hotlist-liste-eng.php), most cosmetic ingredients are only subject to labelling requirements.
The Environmental Working Group's Skin Deep cosmetics database (www.ewg.org/skindeep) has ratings for more than 68 000 products from approximately 3000 brands. Each product is rated on a scale of 0–10 for the level of hazard (low, moderate, high) and data availability for the ingredients.

Evaluate the Skin Deep website and work with the information to answer the following questions:

1. Is this a reliable information source? Does it have a clear and transparent agenda?
   a. Who runs this website? Do the credentials of the organization running the site make the information presented reliable or unreliable? Explain.
   b. What is the mission of this website? What are its underlying values? How do you know this?
   c. What data sources does EWG rely on and what methodology does the organization employ in constructing its databases? Are the sources EWG uses reliable?
   d. Identify a claim made on this website. Is there sufficient evidence provided in support of this claim? Are there any logical fallacies used? Explain.
   e. Do you agree with EWG's assessment of the problems with and concerns about cosmetics and other personal care products? What about its solutions? Explain.

2. Select one of your favourite personal care products that is also on the EWG website:

   a. Check out the company website for that product. What sort of information about the product and its ingredients does the company website offer? Does the company employ any logical fallacies in talking about the product or its ingredients? Is the information provided useful to you as a consumer? Explain your responses.
   b. How do the two websites (Skin Deep and the company website) compare in providing useful and reliable information about your particular product? Explain.
   c. Search the Skin Deep website for alternative brands that have a safer rating than your personal care product. Are there other choices? Would you consider switching your brand? Explain.

MAKING CONNECTIONS

TRICLOSAN—A COMPOUND OF EMERGING CONCERN?

Background: Triclosan (2,4,4'–trichloro-2'-hydroxydiphenyl ether) is a synthetic compound used as a broad-spectrum antimicrobial agent in a variety of consumer products such as toothpaste, antibacterial hand soaps, deodorants, and detergents. It has also been incorporated into plastics such as children's toys and kitchen utensils, as well as many other industrial and household items. Triclosan was first registered as a pesticide in 1969. According to one estimate, the annual use of triclosan is more than 300 000 kg/year.

Triclosan's efficacy and value are under question. The Canadian Medical Association does not consider triclosan either necessary or efficacious in personal care products (see www.cma.ca/multimedia/CMA/Content_Images/Inside_cma/Office_Public_Health/HealthPromotion/Antimicrobial-IssueBriefing_en.pdf). Physicians are particularly concerned about the development of antibiotic resistance. They also assert that soap and alcohol-based products are just as effective in preventing infection. In addition, there are concerns about the impact of triclosan and its breakdown products on the environment.

Case: You have been assigned the task of critically evaluating the research on triclosan and developing a position paper for the future use and regulation of this chemical.

1. In your report, include the following:
   a. An assessment of the costs versus the benefits of triclosan in various products. Be sure to discuss health, economic, and environmental costs and benefits.
   b. An analysis of any logical fallacies in how stakeholders (both those for and those against the use of triclosan) present their arguments.
2. Based on the information at hand, what is the best course of action for the future of triclosan? Who should be involved in making this decision? Provide justifications for your proposal.
3. Based on the story of BPA and triclosan, what kind of balance should we strike between the economic well-being of chemical companies and the health of humans and the environment? Develop a set of guiding principles for the development and use of chemicals in our society. Discuss the principles you develop and explain the underlying rationale.

CORE MESSAGE

A variety of factors influence whether and how fast a population grows. Many human populations have grown explosively in the recent past. The human population is still growing, especially in developing nations, which leads to overpopulation in those areas. We can pursue a variety of approaches to reduce population growth and stabilize population size; many of these approaches focus on issues of social justice. Population size has increased our impact on the environment but that rising impact is also being caused by an increasing per capita use of resources and generation of waste.

GUIDING QUESTIONS

After reading this chapter, you should be able to answer the following questions:

→ How and why has human population size and growth rate changed over time? What is the size and distribution of today's population?

→ What cultural and demographic factors influence population growth in a given country? How do they differ between developed and developing nations?

→ What is the "demographic transition" and why is it important?

→ What do population growth rates look like today and how can we achieve zero population growth?

→ What impact does our current world population have on the Earth and can it support us all?

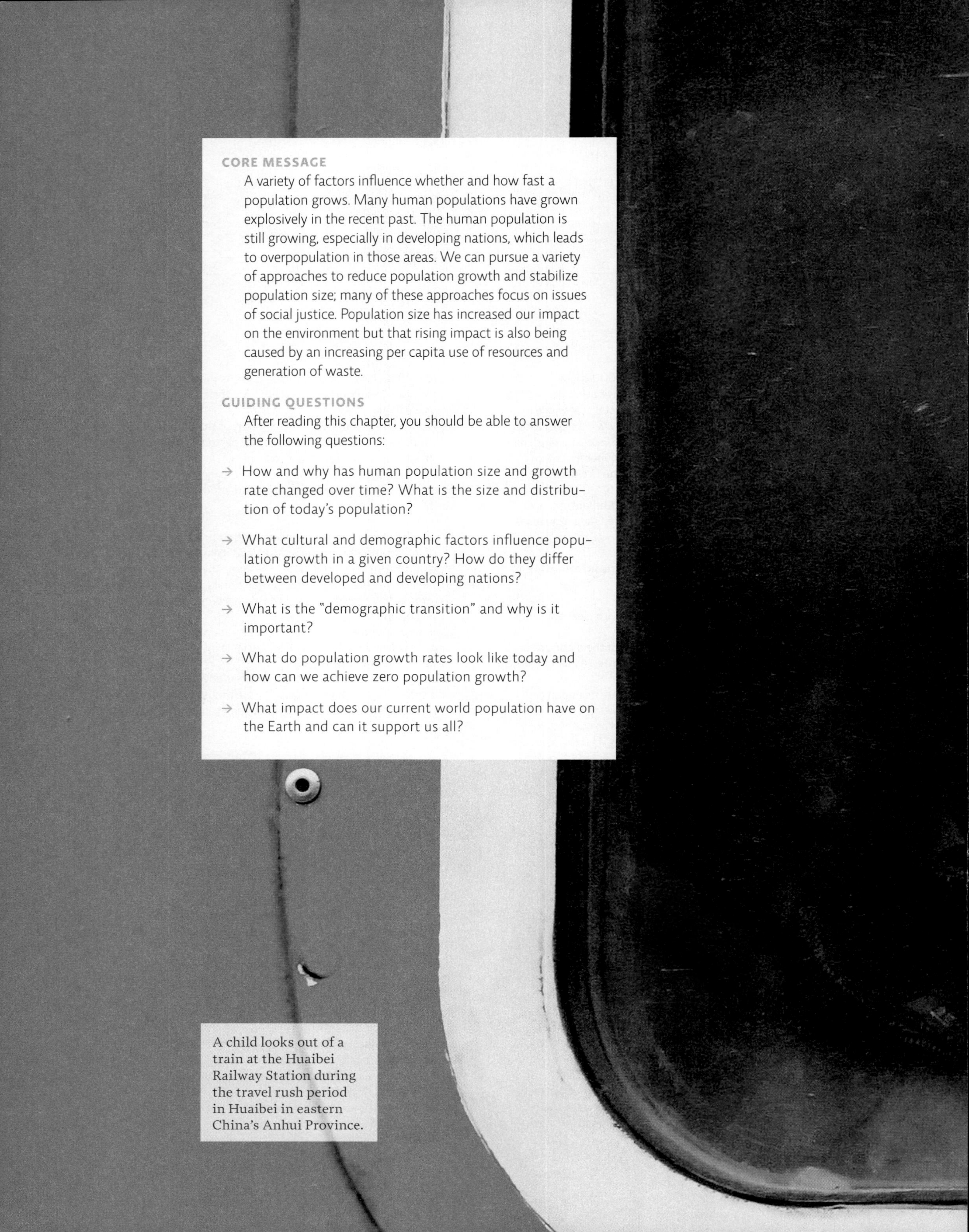

A child looks out of a train at the Huaibei Railway Station during the travel rush period in Huaibei in eastern China's Anhui Province.

ONE-CHILD CHINA GROWS UP

A country faces the outcomes of radical population control

In most countries, including Canada, it's not hard to imagine a life without siblings—plenty of families have just one child. But what about a life without cousins or aunts and uncles? Imagine not just a single family, but an entire country of single children and you'll begin to get a sense of what China is like for the generation now entering adulthood.

Their elders call them the "Little Emperors"—a title that is meant to reflect the spoiled life and haughty temperament we sometimes associate with only children; they are the result of a colossal social experiment set in motion by the Chinese government 30-some years ago in an effort to curb population growth: one child per family. For 30 years.

Human populations grew slowly at first and then at a much faster rate in recent years.

At least twice in the course of human history, global human population has hit a dramatic growth spurt. The first came around 10 000 years ago, when the Agricultural Revolution dramatically increased the amount of food we could grow, and thus the number of mouths we could feed. As more food translated into healthier people, lower death rates, and longer life spans, the **growth rate** increased and global population raced toward the 100 million mark. The next growth spurt came in the 1700s, when the Industrial Revolution ushered in a rapid succession of advances in both sanitation and health care—including vaccines, cleaner water, and better nutrition. Once again, death rates fell, life expectancy rose, and the population swelled. [INFOGRAPHIC 4.1]

In China, this cycle played out most dramatically during the latter half of the 20th century, when the country experienced a spectacular improvement in its standard of living. Between the 1950s and 1970s, **life expectancy** increased from 45 to 60. But as the **crude death rate** fell,

growth rate The percent increase of population size over time; affected by births, deaths, and the number of people moving into or out of a regional population.
life expectancy The number of years an individual is expected to live.
crude death rate The number of deaths per 1000 individuals per year.
crude birth rate The number of offspring per 1000 individuals per year.
population density The number of people per unit area.
overpopulation More people living in an area than its natural and human resources can support.
population momentum The tendency of a young population to continue to grow even after birth rates drop to "replacement rates"—2 children per couple.

Infographic **4.1** | **HUMAN POPULATION THROUGH HISTORY**

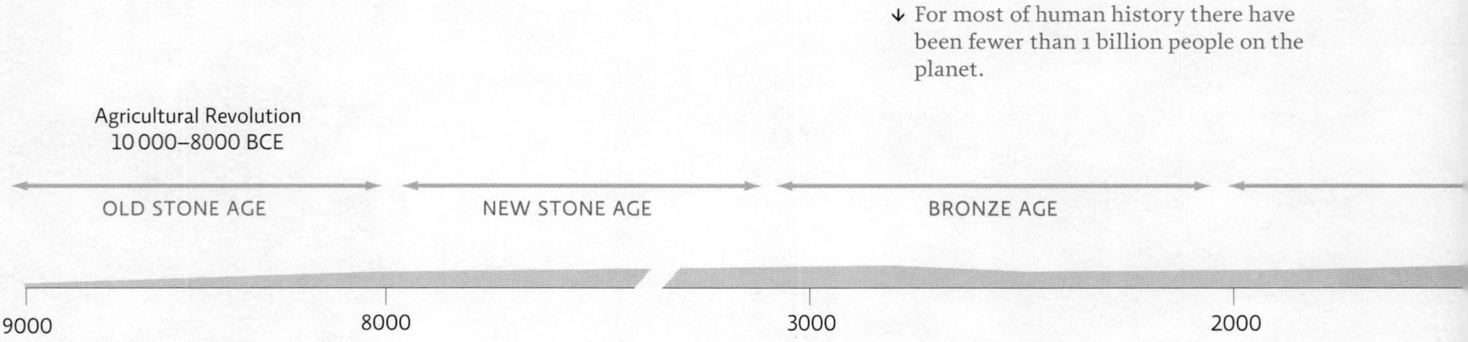

For most of human history there have been fewer than 1 billion people on the planet.
Agricultural Revolution 10 000–8000 BCE
OLD STONE AGE
NEW STONE AGE
BRONZE AGE
9000
8000
3000
2000

the **crude birth rate** held steady. By 1949, China was home to 500 million people—37 times the population of Canada at that time—making it the most populous country on Earth. By 1970, population had swelled to almost 900 million: roughly a quarter of the world's people crammed onto just 7% of the planet's arable land.

Today, China is the most populous nation, with more than 1.3 billion people, but it will soon be surpassed by India, which is projected to have 1.5 billion by 2030. Like many nations around the world, more and more of China's people live in big cities, with many urban areas reaching record **population densities**. [INFOGRAPHIC 4.2]

In China, such growth became a political issue. In the late 1950s, a famine claimed 30 million lives; and during the 1970s, a severe shortage in consumer goods—soap, eggs, sugar, and cotton—led to strict rationing. The government blamed the country's woes on **overpopulation**. Today, experts say that the real culprit was botched state planning. "China was not unlike other parts of the world at that time, in terms of higher life expectancies and growing population," says Wang Feng, Director of the Brookings-Tsinghua Center for Public Policy. "We could have managed, as other countries did, with better agricultural and economic policies."

Maybe so. But in the late 1970s, with two-thirds of the population under the age of 30, and the baby boomers of the 50s and 60s just entering their reproductive years, China's leaders had good reason to panic. A population with a lot of young women has **population momentum**; it will continue to increase for another generation, even if each couple only has two children, just enough to replace them when they die. If the government could not feed the mouths it already had, how in the world would it feed any more?

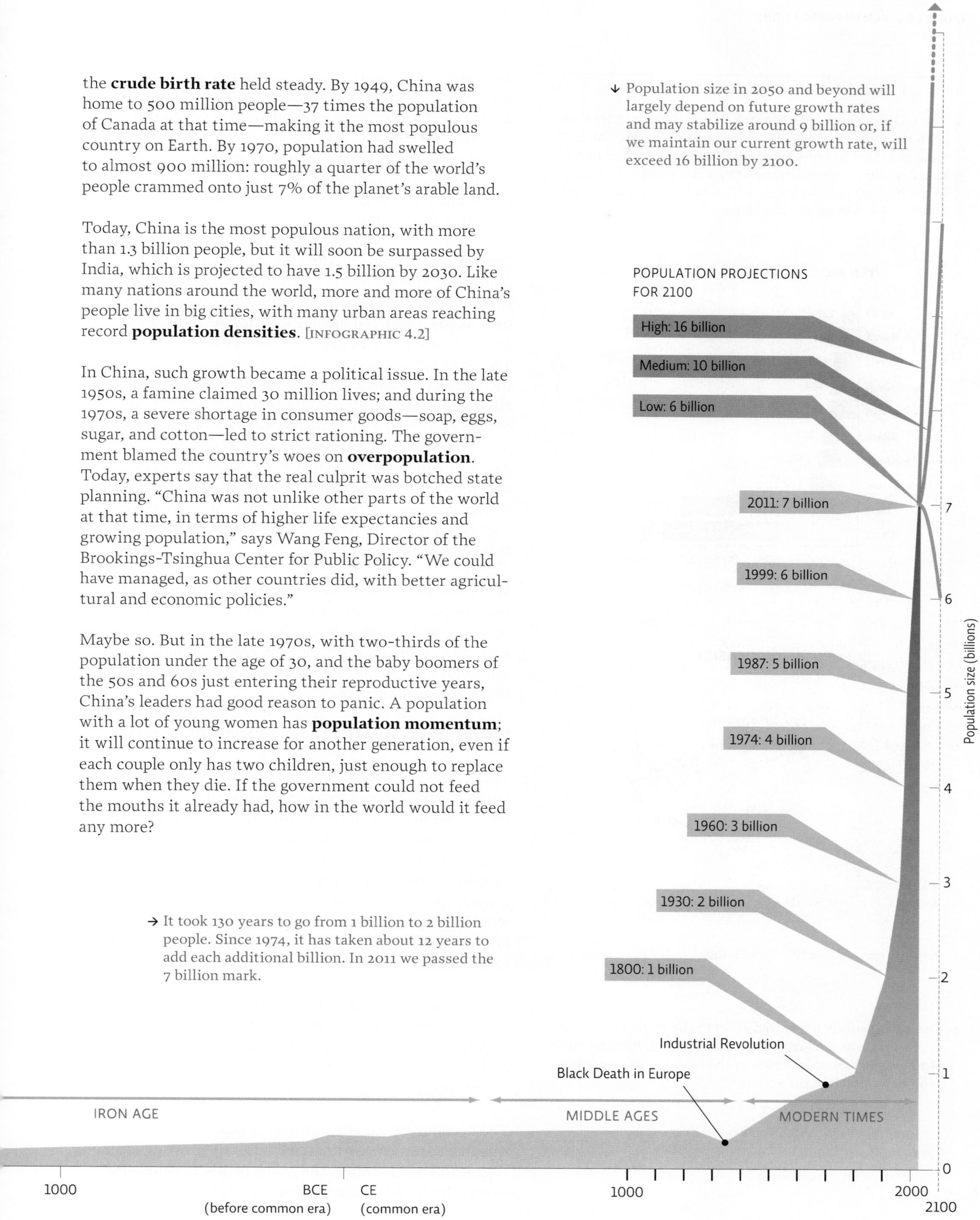
POPULATION PROJECTIONS FOR 2100
High: 16 billion
Medium: 10 billion
Low: 6 billion
2011: 7 billion
1999: 6 billion
1987: 5 billion
1974: 4 billion
1960: 3 billion
1930: 2 billion
1800: 1 billion
Industrial Revolution
Black Death in Europe
IRON AGE
MIDDLE AGES
MODERN TIMES
Population size (billions)
0
1
2
3
4
5
6
7
1000
BCE
(before common era)
CE
(common era)
1000
2000
2100

↓ Population size in 2050 and beyond will largely depend on future growth rates and may stabilize around 9 billion or, if we maintain our current growth rate, will exceed 16 billion by 2100.

→ It took 130 years to go from 1 billion to 2 billion people. Since 1974, it has taken about 12 years to add each additional billion. In 2011 we passed the 7 billion mark.

Infographic **4.2** | **POPULATION DISTRIBUTION**

↓ The ten most populous countries are not necessarily those with the largest land mass. China and the United States have roughly the same area—so China has a higher population density than the United States. Bangladesh has one of the highest population densities in the world at almost 1000 people per square kilometre.

TEN MOST POPULOUS COUNTRIES

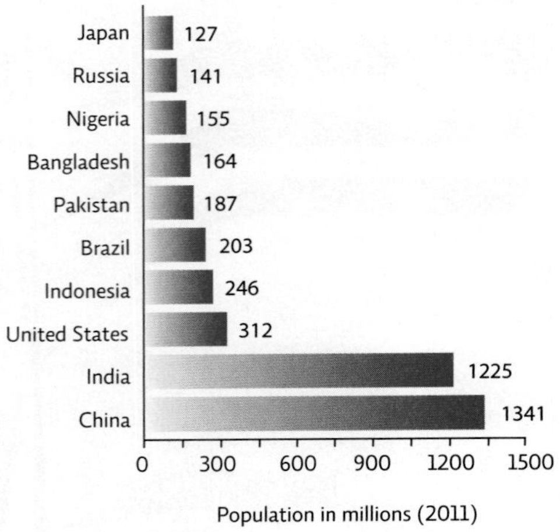

Japan 127
Russia 141
Nigeria 155
Bangladesh 164
Pakistan 187
Brazil 203
Indonesia 246
United States 312
India 1225
China 1341
0 300 600 900 1200 1500
Population in millions (2011)

POPULATION DENSITY

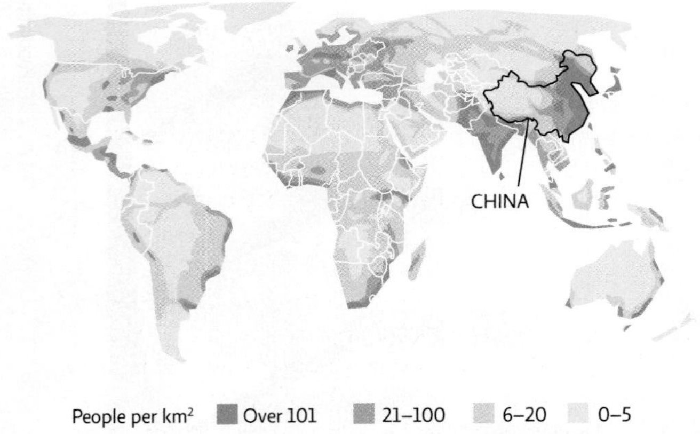

CHINA
People per km²
Over 101
21–100
6–20
0–5

↑ The most densely populated areas in the world tend to be in coastal areas or close to major waterways. Roughly 90% of all people live on just 10% of the land surface, and most people live north of the equator.

So in 1979, the country's leaders issued a mandatory decree: no family could have more than one child. The government vowed to deny state-funded health care and education to all but the first-born. Parents who didn't comply also risked losing their jobs and faced severe fines (often several times their annual income) and penalties (including a loss of government subsidies for baby formula and other foodstuffs).

Dramatic as it sounded, the idea of strictly enforced population control was not entirely new. In fact, modern societies have fretted over our swelling ranks as far back as the Industrial Revolution (that second growth spurt); that's when an English priest by the name of Thomas Malthus first noticed that food supplies were not increasing in tandem with population. Unless something changed, he warned, we would soon have more mouths to feed than food to feed them with. And when that happened, disease, famine, and war would surely follow. According to Susan Greenhalgh, Harvard University anthropologist and author of the book *Just One Child: Science and Policy in Deng's China*, such catastrophic thinking was a big part of how the Chinese government arrived at its one-child policy.

When China's leaders enacted the policy, they promised that the policy would be a short-term one, set to expire in 30 years. By then, they said, Chinese society would have adapted to a small-family culture, and with hundreds of millions of births prevented, the quality of life would have improved substantially.

Fertility rates are affected by a variety of factors.

China was once a culture built around large, extended families, and shaped by a constellation of *pronatalist pressures*—cultural and economic forces that encourage women to have more children. Parents and grandparents lived under the same roof; aunts, uncles, and cousins remained close by, usually in the same village. And couples had as many children as possible so that there would be enough hands to work the family farm, tend to household chores, and most importantly, care for parents as they aged.

Desired family size (how many children a couple wants) is one of the best predictors of actual fertility (how many children a couple has); therefore, any factors that increase or decrease one's desire to have children will significantly impact population growth rates.

The need for labour is a common pronatalist pressure in agrarian societies—so are the status and prestige associated with large families, and a lack of other options for women (in many countries women's rights and freedoms

为四化只生一个孩子好
THE 4 MODERNIZATION REQUIRES "ONE CHILD FAMILY"

↑ A poster promoting China's one-child policy.

are still strictly limited). In many developing countries, especially in Africa, a high **infant mortality rate**—the number of infants who die in their first year of life, per every thousand births—also contributes to higher fertility. Under these conditions, couples tend to have more children, which increases the odds that at least some will survive to adulthood.

For centuries—right up until the 1970s, in fact—China's **total fertility rate (TFR)**, or the average number of children a woman has in the course of her lifetime, hovered between five and six. Ironically enough, the most dramatic decline in that rate took place in the decade before China implemented its "one-child" policy. Between 1970 and 1979, a largely voluntary mandate known as "late, few, long" managed to cut the total fertility rate in half—from 5.9 to 2.9—just by encouraging couples to marry later, delay childbearing, and space conceptions as far apart as possible.

Around the world we see big differences in the kinds of factors that influence how a population changes in size and composition. **Demographic factors** such as health (especially that of children), education, economic conditions, and cultural influences are, on average, quite different for **developed** and **developing countries**. [INFOGRAPHIC 4.3]

In China, cutting the average family size from six children down to three seemed easy. But making the leap from three children down to one proved considerably more difficult. One child meant no siblings, and in subsequent generations, no cousins or aunts and uncles, either. Who would help tend the farms? If a couple's child died or was disabled, who would care for them when they aged? "Many Chinese families in the countryside found

---

infant mortality rate The number of infants who die in their first year of life per every thousand live births in that year.
total fertility rate (TFR) The number of children the average woman has in her lifetime.
demographic factors Population characteristics such as birth rate or life expectancy that influence how a population changes in size and composition.
developed country A country that has a moderate to high standard of living on average and an established market economy.
developing country A country that has a lower standard of living than a developed country, and has a weak economy; may have high poverty.

Infographic **4.3** | **WE LIVE IN TWO DEMOGRAPHIC WORLDS**

↓ Values for demographic factors differ tremendously between developing nations like many African nations and developed nations like Canada. Most of the world's population, as well as most population growth, occurs in developing nations, but most wealth is in developed nations. The higher death rate in developed nations is due to an aging population; the much higher infant mortality rates in developing nations reveal the differences in quality of life and health care between the two categories.

Demographic Factor (2011 data)	World	Developed Countries	Developing Countries	Niger: An Extreme Example
Population size	7 billion	1.24 billion	5.76 billion	16 million
% growth rate	1.2	0.17	1.44	3.5
Crude birth rate	20.3	11	22	48
Crude death rate	8	10	8	12
Total fertility rate	2.5	1.7	2.6	7.0
Infant mortality rate	44	5	48	88
Life expectancy	70 yrs	78 yrs	68 yrs	55 yrs
Wealth (per capita GNP; 2008 U.S.$)	$10 240	$32 470	$5440	$680

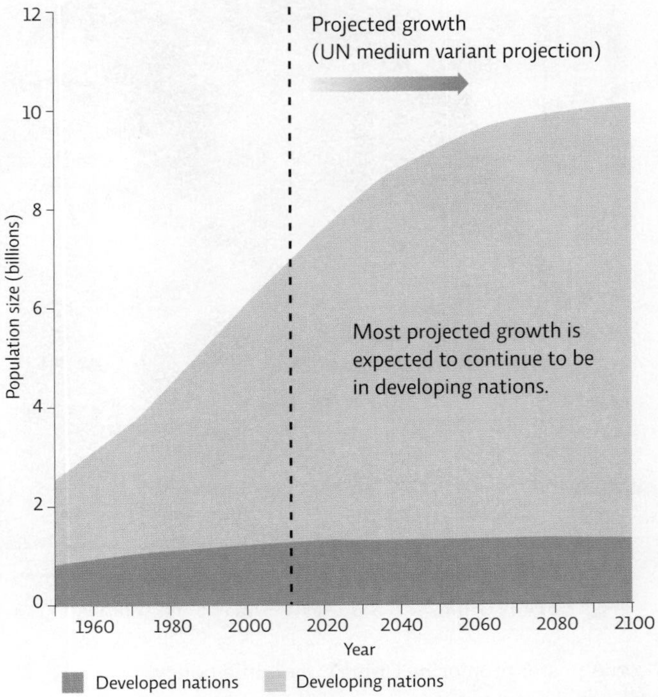
Projected growth
(UN medium variant projection)
Most projected growth is expected to continue to be in developing nations.
Population size (billions)
0
2
4
6
8
10
12
1960
1980
2000
2020
2040
2060
2080
2100
Year
Developed nations
Developing nations

it unthinkable at the time to have just one child," says Greenhalgh. So much so, that even with the penalties they faced, they resisted.

The policy's most horrendous consequences remain the source of great speculation and worry. Many experts have said that in the early years, enforcers carried out numerous sterilizations and abortions—some estimates range as high as the tens of millions—many of them by force. But because official numbers are so difficult to come by, it's nearly impossible to say whether such figures are grossly overestimating or underestimating the problem. Ultimately, the true numbers may be unknowable.

In 1998, however, a former Chinese family planning official shed some light on the inner-workings of the one-child policy, when she testified before the U.S. Congress that women as far along as 8.5 months were forced to abort by injection of saline solution, and women in their ninth month of pregnancy, or even women in labour, were having their babies killed in the birth canal or immediately after birth. The policy's defenders say that such practices have abated, but as recently as 2001, the British newspaper *The Telegraph* reported that a single county in Guangdong Province was tasked with an annual quota of 20 000 abortions and sterilizations, after some residents disregarded the one-child policy.

Chinese officials have argued that their focus on population control has actually improved women's access to health care—both contraception and prenatal classes are free to all—and dramatically reduced the incidence of childbirth-related deaths and injuries.

But critics say that the human rights violations engendered by the one-child policy far outweigh any possible health gains and that, at any rate, health care is wildly imbalanced. One study, conducted by public health researchers from Oregon State University in 1990 in the rural Sichuan Province, found that women whose pregnancies were unapproved were twice as likely to die in childbirth as those whose pregnancies were sanctioned by the government. And while nearly 90% of all married women in China use contraception—compared with

↑ An overcrowded train approaches as other passengers wait to board at a railway station in Dhaka, Bangladesh.

roughly one-third in most other developing countries—most women are not given a choice about which type of contraceptive to use; in a 2001 survey conducted by the Chinese family planning commission, 80% of women said they were compelled to accept whatever method their family planning worker chose for them. Overall, IUDs and sterilizations account for more than 90% of contraceptive methods used since the mid-1980s.

Still, there are signs of improvement. The number of sterilizations has declined considerably since the early 1990s. And in 2002, facing strong international pressure, the Chinese government finally outlawed the use of physical force to compel a woman to have an abortion or be sterilized. (Experts say that prohibition is not entirely enforced. The policy is largely exercised at the local level. And in many provinces, local governments still demand abortions.)

In any case, by 2011, China's fertility rate had plummeted to 1.54, spurring China's leaders to proclaim that the one-child policy had succeeded in preventing some 400 million births, and all the calamities those births would have brought—disease, more famine, overtaxed social services, and so on.

But while the drop in fertility was plain, not everyone would agree that the one-child policy deserved the credit. Critics argue that by the time the first generation of "Little Emperors" reached adulthood, many other forces had reshaped their society and the world.

Factors that decrease the death rate can also decrease overall population growth rates.

When Qin Yijao's mom was pregnant with him back in 1981, the one-child policy was being strictly enforced, and she already had a son. "People from the Communist Brigade said I couldn't have a second baby," she told National Public Radio (NPR) in a 2005 interview. "But I was determined. I needed a son to work our farm." So taking care to evade the birth police, she snuck off and delivered her son in a cave. When she was found out, she was penalized. "I had to pay double the price for grain, and they confiscated our television and sewing machine." In the end, she got to keep and raise her son. But at 30, he has never worked a single day on the farm. "I don't even know where my mom's farm plot is," Qin also told NPR, adding that while he appreciates the risks his mother took, he and his wife have no desire for a second child themselves. The anecdote underscores a key point: Chinese culture has changed dramatically in the years since one-child was enacted, becoming much more urban and thus lessening the need for farmhand children. Many demographers say that such social changes may be more responsible for smaller families than the policy itself.

The **demographic transition** holds that as a country's economy changes from pre-industrial to post-industrial, low birth and death rates replace high birth and death rates. The reasons are cultural as much as economic. As health care improves, infant mortality declines, and with it, the need for couples to have many children just to ensure that some make it to adulthood. And as cities grow and jobs materialize, the need for children to work the farm decreases. Meanwhile, as women find themselves with more opportunities, better health care, and greater access to education, they come to want fewer children. [INFOGRAPHIC 4.4]

Developed countries, including Canada, the United States, and those of the European Union, have already undergone the demographic transition. Demographers are split over whether or not other countries are likely to do the same. But that uncertainty has not stopped world leaders from trying to cull lessons from the demographic transition model. Around the world, countries from Bangladesh to Brazil are working to reduce pronatalist pressures, increase gender equality, and combat poverty and infant mortality, all in an effort to improve standards of living, which in turn should lower fertility rates. The ultimate goal is to achieve **zero population growth**, when population size is stable, neither increasing nor decreasing. This occurs when the number of people born equals the number of people dying; in other words, **replacement fertility rate** is reached.

In most developed countries, the replacement rate is 2.1 (rather than the expected 2.0 to *replace* the parents), since a few children die before maturity and not every female has children. In countries with high infant mortality rates, the replacement rate is higher. Canada currently has a total fertility rate below replacement at 1.7. Without immigration, total population size would be shrinking.

There are at least a handful of clues suggesting that the demographic transition is as responsible for reductions in total fertility rates as the one-child policy is. Clue number

Infographic **4.4** | **DEMOGRAPHIC TRANSITION**

↓ Some industrialized nations have gone through the demographic transition from high birth and death rates to low birth and death rates, giving them a stable population. It is unclear whether industrialization will produce the same pattern in all developing countries. Steps to improve the quality of life and decrease death rates are important worldwide, even if other measures are needed to slow birth rates.

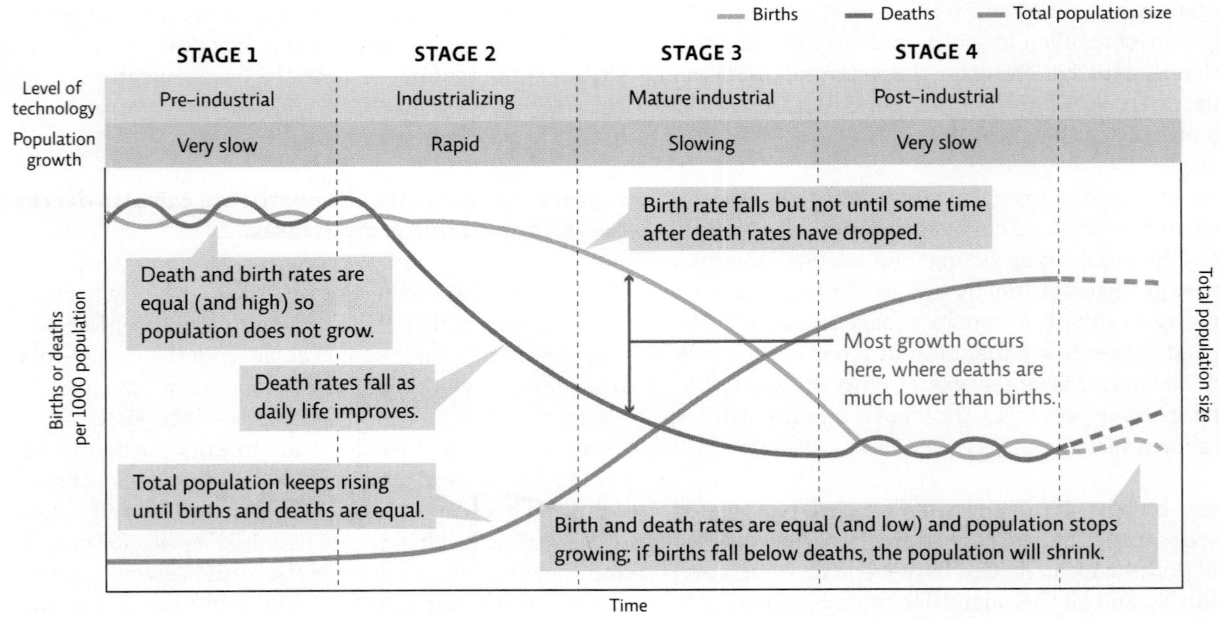

Births
Deaths
Total population size
STAGE 1
STAGE 2
STAGE 3
STAGE 4
Level of technology
Pre-industrial
Industrializing
Mature industrial
Post-industrial
Population growth
Very slow
Rapid
Slowing
Very slow
Birth rate falls but not until some time after death rates have dropped.
Death and birth rates are equal (and high) so population does not grow.
Death rates fall as daily life improves.
Most growth occurs here, where deaths are much lower than births.
Total population keeps rising until births and deaths are equal.
Birth and death rates are equal (and low) and population stops growing; if births fall below deaths, the population will shrink.
Births or deaths per 1000 population
Total population size
Time

one: the most significant fertility declines in the policy's history coincided with the largest gains in the Chinese economy, not with periods of strict enforcement. Clue number two: many other countries have had substantial declines in fertility during the past 25 years without the use of strict population-control policies, including Japan, Singapore, and Thailand, whose fertility rates mirror those of China's almost exactly—falling from 5 children per family in 1970 to 1.8 children per family in 2011.

"The parallels are hard to ignore," says Therese Hesketh, a global health professor at University College London. "It suggests that China could have expected a continued reduction in its fertility rate just from continued economic development." Feng, Director of the Brookings-Tsinghua Center for Public Policy, puts it more bluntly. "The claim that 'one-child' prevented 400 million births is bogus for sure," he says. "Many other countries experienced nearly identical fertility declines over the same time frame. So unless you say Chinese people are necessarily different from other populations, it's very hard to argue the decline is due to government policy."

Worldwide, population growth rates are declining, but overall the number is still positive, so we are still increasing. [INFOGRAPHIC 4.5] The United Nations estimates future population size leveling off somewhere between 6 and 16 billion people by 2100, depending on how quickly we reduce growth rates (see Infographic 4.1).

But if the predominant causes of declining fertility remain a source of debate, the consequences are becoming clearer and more stark by the day. Because the reasons for population growth are likely to be different in different regions, reaching zero population growth requires that the pronatalist pressures in each region be addressed. A variety of measures that don't include government-restricted family size have proven successful.

Programs that address the needs of a given population and work within the cultural and religious traditions are likely to be most successful. Many demographers believe that addressing the social justice issues associated with overpopulation (poverty, high infant mortality, lack of education and opportunities for women) will prove the

demographic transition Theoretical model that describes the expected drop in once-high population growth rates as economic conditions improve the quality of life in a population.
zero population growth The absence of population growth; occurs when birth rates equal death rates.
replacement fertility rate The rate at which children must be born to replace those dying in the population.

Infographic **4.5** | **DECLINING POPULATION GROWTH RATES**

↓ Although global growth rates are decreasing, the overall world population is still on the rise, just more slowly. Only when growth rates reach zero will our population stop growing. The growth rate of a country as big as China can strongly influence global growth rates, as evidenced here by the substantial dip in the late 1950s.

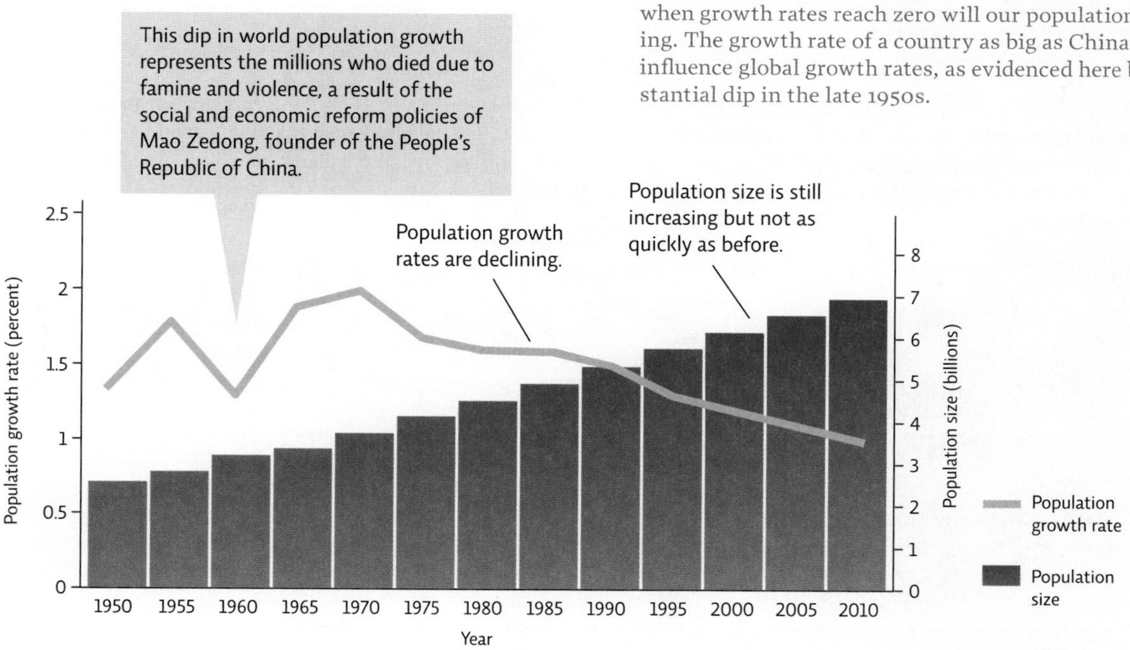
This dip in world population growth represents the millions who died due to famine and violence, a result of the social and economic reform policies of Mao Zedong, founder of the People's Republic of China.
Population growth rates are declining.
Population size is still increasing but not as quickly as before.
Population growth rate (percent)
2.5
2
1.5
1
0.5
0
1950
1955
1960
1965
1970
1975
1980
1985
1990
1995
2000
2005
2010
Year
Population size (billions)
8
7
6
5
4
3
2
1
0
Population growth rate
Population size

Infographic **4.6** | **REACHING ZERO POPULATION GROWTH**

↓ Reaching zero population growth takes two steps: identifying why birth rates are high (what are the pronatalist pressures?) and then taking steps (education, birth control, social campaigns) to address those pronatalist pressures and reduce birth rates.

STEP 1: IDENTIFY PRONATALIST PRESSURES

INFANT MORTALITY RATE IMPACTS BIRTH RATE

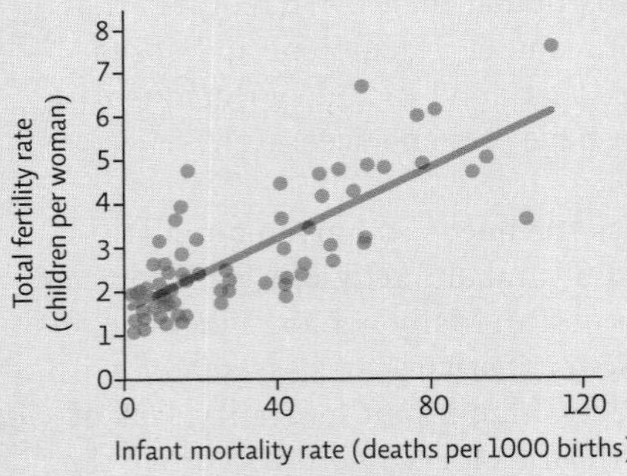

Total fertility rate (children per woman)
0
1
2
3
4
5
6
7
8
0
40
80
120
Infant mortality rate (deaths per 1000 births)

High infant mortality is closely correlated with poverty: total fertility rate (TFR) goes up as infant mortality and poverty increase.

DESIRED FERTILITY AND ACTUAL TOTAL FERTILITY

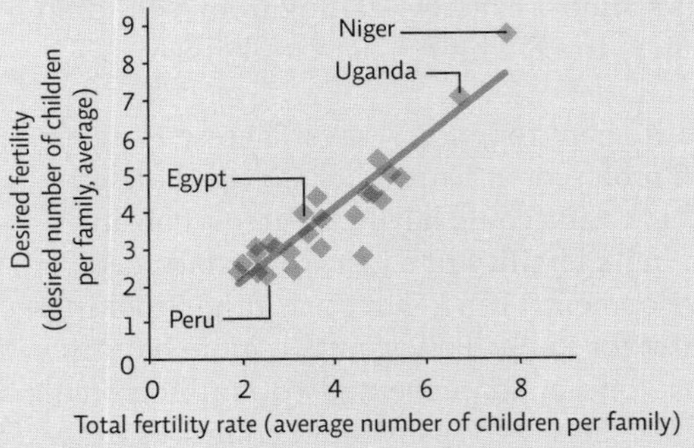

Desired fertility (desired number of children per family, average)
0
1
2
3
4
5
6
7
8
9
Niger
Uganda
Egypt
Peru
0
2
4
6
8
Total fertility rate (average number of children per family)

TFR closely aligns with the desired fertility of families—families usually have the number of children they would like to have.

STEP 2: STRATEGIES TO REDUCE TFR

AS EDUCATION OF WOMEN INCREASES, TFR DECREASES

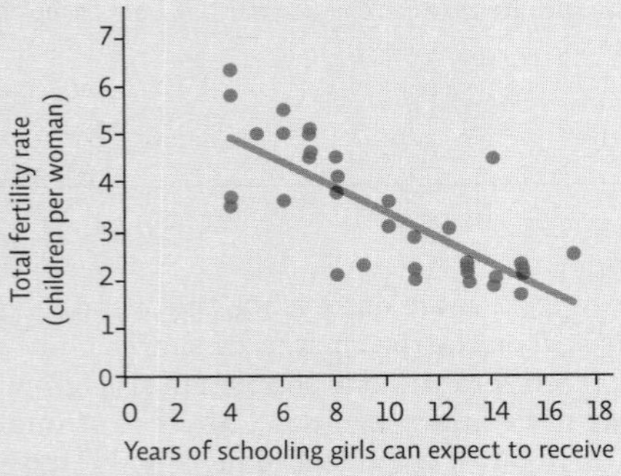

Total fertility rate (children per woman)
0
1
2
3
4
5
6
7
0
2
4
6
8
10
12
14
16
18
Years of schooling girls can expect to receive

Fertility declines as educational opportunities for girls and women increase. This means that funding for education and job opportunities for women may be more effective at lowering TFR than other approaches. It also improves the survival of children, another factor that leads to a reduced TFR.

FAMILY PLANNING: THERE IS STILL UNMET NEED FOR ACCESS TO CONTRACEPTIVES

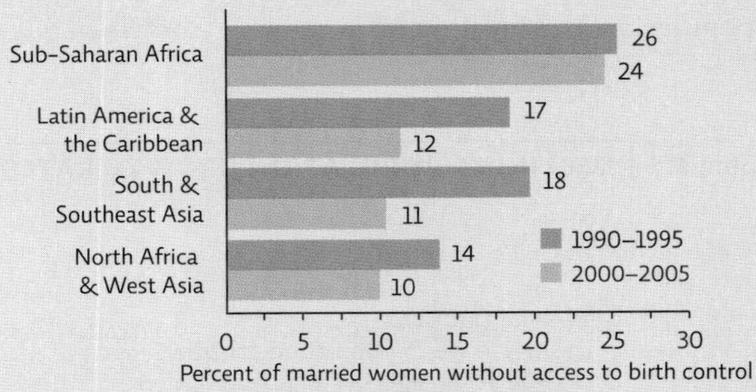

Sub-Saharan Africa
26
24
Latin America & the Caribbean
17
12
South & Southeast Asia
18
11
North Africa & West Asia
14
10
1990–1995
2000–2005
0
5
10
15
20
25
30
Percent of married women without access to birth control

Family planning programs have been effective in many areas of the world. However, many women still have "unmet need" (they would like to use contraceptives but do not have access to them), which is as high as 39% in rural areas of some African nations.

DIFFERENT SOLUTIONS FOR DIFFERENT REGIONS

Thailand's very successful family planning program started with better maternal and infant health care, family planning programs, and access to contraceptives, but also reflected the culture's love of humour—from condom distribution at public gatherings to free vasectomies on the king's birthday, to "ads" painted on the sides of water buffalo.

Kerala, India, a state in the world's most populous country, adopted a population-control plan centred around the three "e's": education, employment, and equality. As a result, Kerala's school system boasts more than a 90% literacy rate, which is almost identical between boys and girls. Educated women join the workforce before deciding to have children, and some 63% (compared to 48% in India as a whole) of women there use contraceptives. The state's birth rate has stabilized at around 2.0.

In Brazil, the dropping birth rate over the past 40 years is attributed in part to pop culture. Families in popular TV shows are small—this coupled with rising opportunities for women and crowded living spaces in large cities, has produced a generation of young women whose desired family size is only one or two children.

Infographic **4.7** | **AGE STRUCTURE AFFECTS FUTURE POPULATION GROWTH**

↓ The fastest-growing regions are those with a youthful or very young population.

Very young
Youthful
Transitional
Mature
No data

↓ Age structure diagrams show the distribution of males and females of a population in various age classes. The width of a bar shows the percent of the total population that is in each gender and age class. The more young people in a population, the more population momentum it has—it will continue to grow for some time. The more people of reproductive age, the higher the growth rate, but this measure is also influenced by income—growth rates decline as income increases.

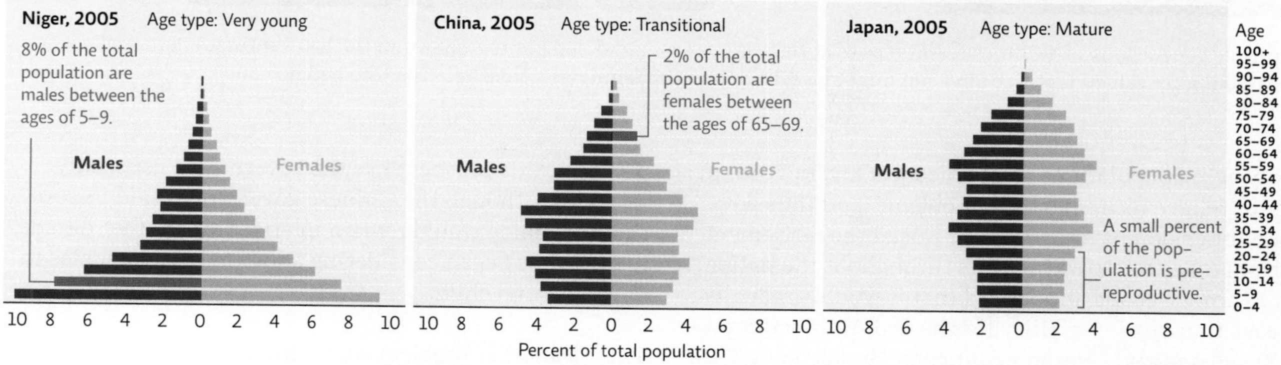
Niger, 2005 Age type: Very young
8% of the total population are males between the ages of 5–9.
Males
Females
China, 2005 Age type: Transitional
2% of the total population are females between the ages of 65–69.
Males
Females
Japan, 2005 Age type: Mature
Males
Females
A small percent of the population is pre-reproductive.
Age
100+
95–99
90–94
85–89
80–84
75–79
70–74
65–69
60–64
55–59
50–54
45–49
40–44
35–39
30–34
25–29
20–24
15–19
10–14
5–9
0–4
10 8 6 4 2 0 2 4 6 8 10
Percent of total population

In Niger's very young population, most people are under 30 and there is a very high capacity for growth (high population momentum).

China's transitional population is growing more slowly than Niger's because China's pre-reproductive and reproductive age cohorts are smaller. The slight skew toward males is noticeable.

Japan's mature population has a fairly even distribution among age classes, with two slight bulges seen around 30 and 60. This population is fairly stable or may even be decreasing slowly as deaths start to outnumber births.

most instrumental in enabling countries with high TFRs to confront what lies ahead. [INFOGRAPHIC 4.6]

The age and gender composition of a population affects its potential for growth.

Demographers create a diagram that displays a population's size as well as its **age structure** (the percentage of the population that is distributed into various age groups, called cohorts) and **sex ratio**. These **age structure diagrams** give insight into the future growth potential of a population. [INFOGRAPHIC 4.7]

When the majority of the people in a population are young, the diagram looks very much like a pyramid; there are nearly equal portions of men and women and fewer and fewer people at each higher age group. Because so many people are young, and many have not yet reached reproductive age, the population will continue to grow rapidly over time.

Currently, many industrialized nations—including Canada, the United States, and many European countries—have top-heavy age structure diagrams with many older people. In many cases, the struggle to care for these

age structure The part of a population pyramid that shows what percentage of the population is distributed into various age groups of males and females.

sex ratio The relative number of males to females in a population; calculated by dividing the number of males by the number of females.

age structure diagram A graphic that displays the size of various age groups, with males shown on one side of the graphic and females on the other.

↑ Multiple generations of a Chinese family pose at their home in an undated photograph from the last century. Historically, sons were so highly valued that a second son might be sent to join the family of a close relative who had no sons.

rapidly aging populations has turned prickly. In 2010, in order to relieve retirement age problems (and thus government pension payouts), France raised the retirement age from 60 to 62, sparking riots throughout the nation. And in 2012, Canada announced that it would gradually increase the age of eligibility for the Old Age Security (OAS) pension and Guaranteed Income Supplement (GIS) from 65 to 67.

China's age structure diagram has lost the bottom-heavy shape, reflective of its changing population age structure. Falling birth rates and rising life expectancy have tipped the balance between old and young. In 1982, just 5% of China's population was older than 65. In 2004, that percentage had climbed to 7.5; by 2050, it is expected to exceed 15%. These figures are lower than those in most industrialized countries (especially Japan, where the proportion of people over the age of 65 years is 20%). But because elderly people are still dependent on their children for support, this situation could spell disaster.

Demographers call it the 4-2-1 conundrum: as they settle into middle age, the members of each successive generation of only children will find themselves responsible for two aging parents and four grandparents. If that single child fails, family elders would be left with virtually no options—no second children, or nieces and nephews, or even friends and neighbours who could help out, since most other families would be in an equally precarious situation. In a country without an extensive pension program, this prospect has the government especially worried. Although the Chinese government said in 2008 that the policy would remain in effect for another decade, in 2011 they began considering allowing select couples to have a second child.

There's another problem with China's evolving age structure: the workforce. Economists predict that between 2010 and 2020, the annual size of the labour force aged 20–24 will shrink by 50%. This is a forecast with global implications. "The deep fertility declines have plunged China into a demographic watershed," says Feng. "It means the days of cheap, abundant Chinese labour are over." For countries like Canada, which have come to depend on a steady influx of cheap goods from China, that's not such good news. But for the only-children of China, it almost certainly means higher wages, better working conditions, and more jobs to choose from in the coming decades.

An out-of-whack age structure isn't the only unintended consequence of the one-child policy. The sex ratio of men to women has also grown alarmingly high. Almost everywhere in the world, there are slightly more males than females, though this differs according to age; because women tend to live longer, there are more women than men in older age groups. In most industrialized countries, the sex ratio tends to hover close to 1.05. However, in Canada it is on the low side at 0.99. In China, the ratio has soared from 1.06 in 1979 to 1.17 in

2011; in some rural provinces, it's as high as 1.3. What does that mean in actual numbers? The State Population and Family Planning Commission estimates that come 2020, there will be 30 million more men than women in China.

How did this happen?

Like most Asian countries, and many African ones, China has a long tradition of preference for sons. Boys are thought of as being better-suited to the rigours of manual labour that drive China's rural, and even industrial, economies. As they grow into men, they often become the main financial providers for retired and aging parents. Daughters, on the other hand, are often tasked with housework and the care of younger siblings and aging grandparents, tasks that are unpaid. When they marry, daughters traditionally become part of the groom's family, and his parents are generally taken better care of than her own.

Sex-selective abortions—terminating a female pregnancy—have been outlawed in China, but most experts say they are still common, thanks in large part to private-sector health care and the ability of a growing number of Chinese citizens to pay the high cost of a gender-determining ultrasound.

> "The deep fertility declines have plunged China into a demographic watershed."—Wang Feng

A recent study by Chinese researchers from Zhejiang Normal University shows that the sex ratio is high (favouring males) in urban areas where only one child is allowed but in rural areas where a second child is allowed, if the first is a girl, sex ratios are even higher for the second child—as high as 160 males for every 100 females. The researchers attribute this high male bias to sex-selective abortion. Although most demographers agree that outright infanticide is increasingly rare, subtler forms of **gendercide**—the systematic killing of a specific gender—are known to occur. If a female infant falls ill, for example, she might be treated less aggressively than a male infant would be.

These days, however, it is much more common for girl babies to be put up for adoption. Officially registered adoptions increased 30-fold, from about 2000 in 1992 to 55 000 in 2001. According to a 2012 *Globe and Mail* article, Chinese adoptions in Canada reached a maximum in 1998, with about 900 adoptions, likely due to a high availability of infants from Chinese adoption agencies. Even though Canadian restrictions on international adoption have increased since then, in 2007, 655 of Canada's 1710 international adoptions were still from China. Almost all children adopted from China continue to be female.

The result of this skewed sex ratio, 30 years out, seems to be lots of prospective husbands in want of wives. Recent census data shows that in a growing portion of rural Chinese provinces, one in four men are still single at 40. "The marriage market is already getting more intense and competitive," says Feng. "Men of lower social ranks—who make up a very significant chunk of the population—are being left out. And it's only going to get worse."

A growing number of social scientists say that the dearth of women will ultimately threaten the country's very stability. Experts worry that such large numbers of young men who can't find partners will prove a recipe for disaster. "The scarcity of females has resulted in kidnapping and trafficking of women for marriage and increased numbers of commercial sex workers," writes Therese Hesketh, "with a potential rise in human immunodeficiency virus infection and other sexually transmitted diseases."

Carrying capacity: is zero population growth enough?

The population size of a given area is not just determined by its birth and death rates but also by the migration of people in and out of its borders. Migration can be over large or small distances, across or within national boundaries, permanent or temporary (e.g., seasonal workers or refugees), and in large groups or small. **Immigration** refers to the movement of people into a given population; **emigration** refers to their movement out of a given population. Taken together, annual migration data and birth and death rates can tell us how quickly a population is changing in size. In China today, migration is mostly internal—from rural areas to urban ones, expanding those burgeoning populations and stressing the ability of the cities to meet the needs of the residents.

Every given environment has a **carrying capacity**—the maximum population size that the area can support. How large that population can be is determined by a range of forces, including the supply of nonrenewable resources, the rate of replenishment for renewable resources on which we

---

gendercide The systematic killing of a specific gender (male or female).
immigration The movement of people into a given population.
emigration The movement of people out of a given population.
carrying capacity The population size that an area can support for the long term; it depends on resource availability and the rate of per capita resource use by the population.

depend, and the impact each individual person has on the environment (a degraded ecosystem can sustain far fewer people than a well-maintained, healthy one).

There are roughly 7 billion people on the planet today. Whether we stabilize at 9 or 10 billion or more depends on how quickly we lower TFR to achieve replacement fertility worldwide. Even at 7 billion people, we may have already exceeded the carrying capacity of Earth. One problem we face is our dependence on nonrenewable energy sources; they will not last indefinitely. But we are also overusing our biological resources. There are differences in how well different regions of the world live within the *biocapacity* (the ability of the ecosystem's living components to produce and recycle resources, and assimilate wastes) of their own environments. North America in particular uses far more resources than its land area can provide, whereas Latin America and the Caribbean use less than half of the biocapacity of their region. This means nations like Canada and many others are importing resources from other regions.

The problem, then, is twofold: on the one hand, we are increasing, rapidly, in sheer numbers. On the other, we are consuming more resources per person than ever before. [INFOGRAPHIC 4.8]

In general, population increase is a problem of the developing world—the highest fertility rates tend to be in the least developed countries, those that have yet to complete the demographic transition. Overconsumption, meanwhile, is a problem of more developed countries like Canada, where a single person might consume more food, fuel, and other resources (and thus place more strain on the environment) than a whole group of people in a less developed country like Sudan or Bangladesh. The impact humans have on Earth is thus due to a combination of factors—population size, affluence, and how we use resources, especially our use of technology, which tends to increase our overall use of resources. This is discussed more fully in Chapter 5.

In the years since the one-child policy was enacted, China has found itself careening from one end of this spectrum to the other. The problems that come with overpopulation—slums, epidemics, overwhelmed social services, and the production of high volumes of waste, for example—have been mitigated by a reduction in the fertility rate. But as fertility has declined, relative affluence has increased, and the Chinese are now straining the environment in a different way: by overconsuming.

It's important to understand that overconsumption in one region or country can strain carrying capacity in other, disparate regions. In Bangladesh, for example, rising sea levels, spurred by global climate change, are pushing that nation's carrying capacity to the brink. Most experts agree that a major driver of current climate change is rampant fossil fuel consumption, not by Bangladesh itself, but by developing nations, many of which are on the other side of the world. China now leads the way in the release of fossil fuel-based carbon emissions linked to climate change—because of its huge population and its growing affluence.

Ultimately, the question of how many people the planet can support may not be the right one. Trying to pinpoint

Infographic **4.8** | **HUMAN IMPACT AND CARRYING CAPACITY**

↓ Many ecologists and economists think that, at our current rate of consumption, human population size has already surpassed Earth's long-term carrying capacity. For long-term sustainability, we should leave behind enough resources so that nature can replenish what we harvest. If we had one-and-a-half Earths, the resources left behind could replace what we take—but of course we do not.

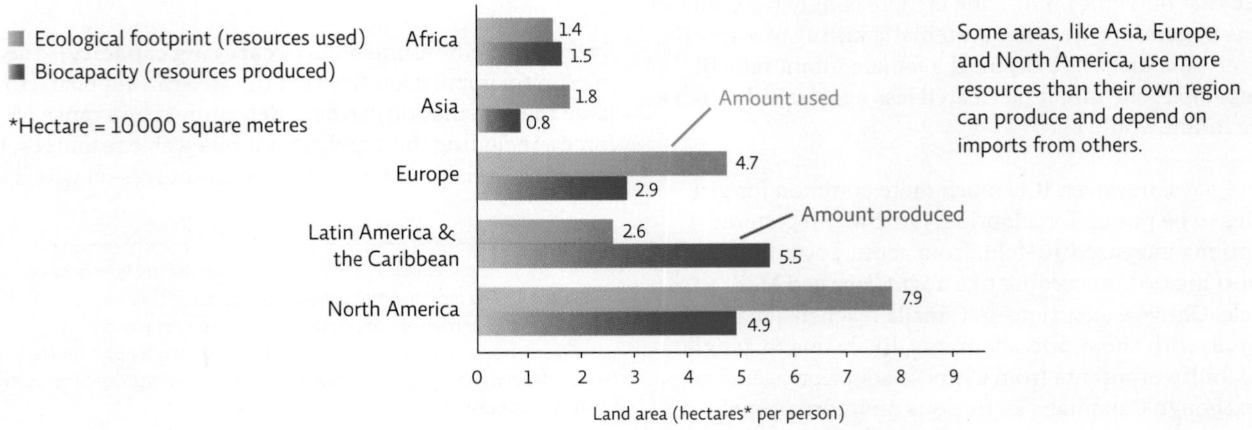
Ecological footprint (resources used)
Biocapacity (resources produced)
*Hectare = 10 000 square metres
Africa
1.4
1.5
Asia
1.8
0.8
Europe
4.7
2.9
Latin America & the Caribbean
2.6
5.5
North America
7.9
4.9
Amount used
Amount produced
0 1 2 3 4 5 6 7 8 9
Land area (hectares* per person)
Some areas, like Asia, Europe, and North America, use more resources than their own region can produce and depend on imports from others.

whether Earth's carrying capacity is 7 billion or 9 billion or even 16 billion is meaningless unless we identify what an acceptable quality of life is. And while overpopulation may be an issue with global significance and causes, it is one that ultimately plays out at the regional level: as land becomes degraded and water supplies depleted or polluted, people (and other organisms) live in an increasingly impoverished environment.

What awaits China's generation of Little Emperors?

Evidence suggests that despite the plight of the latest generation, China is indeed becoming a small-family culture. The National Family Planning and Reproductive Health Survey found that 35% of women prefer having only one child; 57% prefer having two children; and only 5.8% want more than two. True to the demographic transition's predictions, educated, urban women wanted fewer children than their rural, farm-dwelling counterparts. That's good news for those striving to bring China toward zero population growth: in the last three decades, the federal government has allowed hundreds of millions of Chinese people to move to cities in search of work. Those families will no longer need lots of children to work on the farm.

In recent years, the government has softened its stance and added several significant exceptions to the one-child rule: couples made up of two only-children, rural families with land to farm, and several groups of ethnic minorities are all allowed to have more than one child if they so choose.

Still, it's unlikely that the one-child policy will officially expire any time soon. In a 2008 interview with the state newspaper, *China Daily*, the country's minister of State Population and Family Planning, Zhang Weiqing, said that loosening the one-child policy would unleash a new baby boom. "Given such a large population base, there would be major fluctuations in population growth if we abandoned the one-child rule now. It would cause serious problems and add extra pressure on social and economic development."

For better or worse, China is now a country of small families, and of ever-fewer children. That may mean more resources per child, but it also means fewer workers, taxpayers, and innovators. Ultimately, it means that even as they bask in the spoils of one-child, the "Little Emperors" are burdened with the hopes and fears of an entire country.◉

Select references in this chapter:
Ewing, B., *et al.* 2010. *Ecological Footprint Atlas 2010*. Oakland, CA: Global Footprint Network.
Greenhalgh, S. 2008. *Just One Child: Science and Policy in Deng's China*. Berkeley: University of California Press.
Hesketh, T., *et al.* 2005. *New England Journal of Medicine*, 353: 1171–1176.
Ni, H., and A.M. Rossignol. 1994. *Epidemiology*, 5: 490–494.
Pearce, T. Feb. 17 2012. *The Globe and Mail*. Toronto.
Pritchett, L. 1994. *Population and Development Review*, 20: 1–55.
Sedgh, G., *et al.* 2007. *Women with an Unmet Need for Contraception in Developing Countries and Their Reasons for Not Using a Method*. Guttmacher Institute.
Statistics Canada. 2012. Annual Demographic Estimates: Canada, Provinces and Territories. Government of Canada. Statistics Canada Catalogue no. 91-215-XWE.
Yin, Q. 2003. *Theses Collection of 2001 National Family Planning and Reproductive Health Survey*. Beijing: China Population Publishing House. 116–26.
Zhu, W., *et al.* 2009. *British Medical Journal*, 338: b1211.

BRING IT HOME

PERSONAL CHOICES THAT HELP

The impact of humans on the planet is created by a combination of population size and our resource use. The issue of population and carrying capacity is complex. We cannot have a truly sustainable society until key components such as poverty, lack of education, and basic human rights are addressed.

Individual Steps

→ Buy Fair Trade Certified products. These products provide a livable wage to workers and are often linked to education and community development.

→ Research the products you buy to make sure that you are not supporting child slave labour or sweatshop facilities.

Group Action

→ Raise money and invest it in a socially responsible project. Kiva is a non-profit organization that provides micro-loans to help people start small businesses in less developed countries (www.kiva.org/team/canada).

→ Join an organization such as Habitat for Humanity, which builds housing for low income families.

Policy Change

→ We all know that our tendency as humans is to build up and out; however, revitalizing our current older downtown areas is important to prevent new habitat destruction as well as to make use of current infrastructure. Urge your local government to keep shopping and dining establishments in historic downtowns. If needed, work to develop a community partnership that starts clean-up programs and community gardens in abandoned lots.

UNDERSTANDING THE ISSUE

CHECK YOUR UNDERSTANDING

1. _______ is the number of children a couple must have to ensure the population neither increases nor decreases, but the most useful measure for projecting future population change is _______.
   a. Replacement fertility; total fertility rate
   b. Infant mortality rate; pronatalist pressures
   c. Crude death rate; crude birth rates
   d. Demographic transition; population density

2. Population momentum inherent in the populations of developing countries refers to the fact that:
   a. people in these countries are living longer.
   b. infant mortality rates in these nations are on the decline.
   c. populations will continue to grow even at replacement fertility rates.
   d. economic growth in these nations is unstoppable.

3. An age structure diagram can be used to:
   a. measure potential population migration to cities.
   b. see the historical growth of a population.
   c. predict the future growth of a population.
   d. measure the quality of life of a population.

4. The term "demographic transition" refers to:
   a. slower growth as the population size approaches carrying capacity.
   b. the decline in death rates and then birth rates as a country becomes industrialized.
   c. the requirement for a population to reach a specific size before it becomes stable.
   d. migration from the overpopulated countryside to urban centres.

5. Which of the following is NOT a pronatalist pressure?
   a. The child is part of the family labour pool.
   b. There is a high infant mortality rate.
   c. Contraceptives are not available.
   d. Women have many opportunities to participate in the work force.

6. Which statement illustrates human overconsumption?
   a. Fertility rate of 1.4 in Japan is below replacement-level fertility.
   b. Average per capita daily water usage in Canada is 329 litres, while the amount needed for a reasonable quality of life is 80 litres.
   c. At 176 deaths per 1000 live births, Angola has the highest infant mortality rates in the world.
   d. Per capita use of fossil fuels is decreasing in Denmark as a result of the carbon tax.

WORK WITH IDEAS

1. How has human population changed over time? What factors account for this change?

2. Compare and contrast different tactics to reduce population growth. What are the method, effectiveness, drawbacks, and cultural constraints (why a particular method might not be appropriate everywhere) of each?

3. What is a demographic transition and why is it considered important for human populations? Where have demographic transitions yet to occur and what can we predict about whether these countries will make this transition?

4. Some countries like Japan are dealing with large aging populations, while others like Niger are dealing with large numbers of youth. Why does such a demographic divide exist and what problems arise in populations that are either "very young" or "mature"?

ANALYZING THE SCIENCE

The graphs on the following page come from the United Nations Population Division and are used in its population projections.

INTERPRETATION

1. What are the population projections for each fertility variant in 2050? In 2100? What do you predict the population will be in 2150 for the constant fertility variant?

2. What is the trend in the numbers in graph A for each line?

3. How does the information in graph A relate to that in graph B? Explain.

ADVANCE YOUR THINKING

4. The three projections in graph A are based on different fertility variants. The medium fertility variant assumes global fertility rate will drop to replacement (2.1); the high fertility variant assumes global fertility rate will be 2.6; and the low fertility variant assumes global fertility rate will be 1.6. Explain why the U.N. might have three population projections based on different fertility rates.

5. Why is Africa projected to have such large growth if population projections assume that gobal fertility rates will be at replacement? Use information from the chapter to support your conclusion.

6. The following table shows 2010 population data for various world regions. Using this information as well as the data from graph B, calculate the total population in each region in 2100 as well as the distribution of the world population in each region for 2010 and 2100. According to your calculations, what are the significant changes in the distribution of the world population from 2010 to 2100?

Region	Population in 2010 (in millions)	Population in 2100 (in millions)	Population change 2010–2100	% of world population in 2010	% of world population in 2100
Africa	1022				
Asia	4164				
Europe	738				
Latin America & Caribbean	590				
North America	345				
Oceania	37				
World total	**6896**				

GRAPH A Estimated and projected world population according to different variants

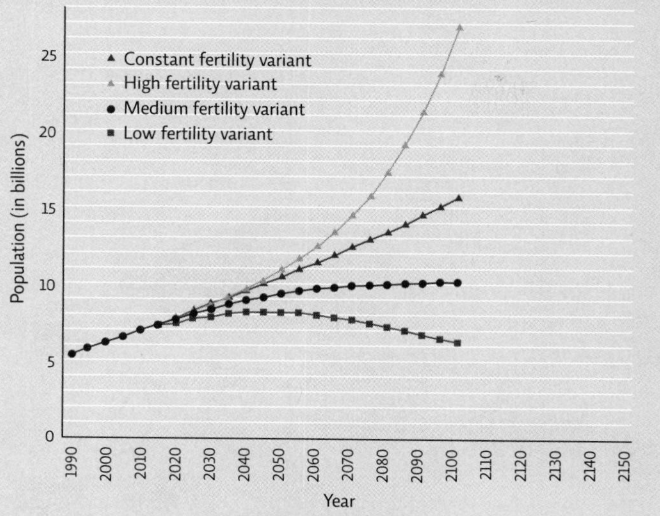
Constant fertility variant
High fertility variant
Medium fertility variant
Low fertility variant
Population (in billions)
Year
0
5
10
15
20
25
1990
2000
2010
2020
2030
2040
2050
2060
2070
2080
2090
2100
2110
2120
2130
2140
2150

GRAPH B Projected population change between 2010 and 2100 by major region

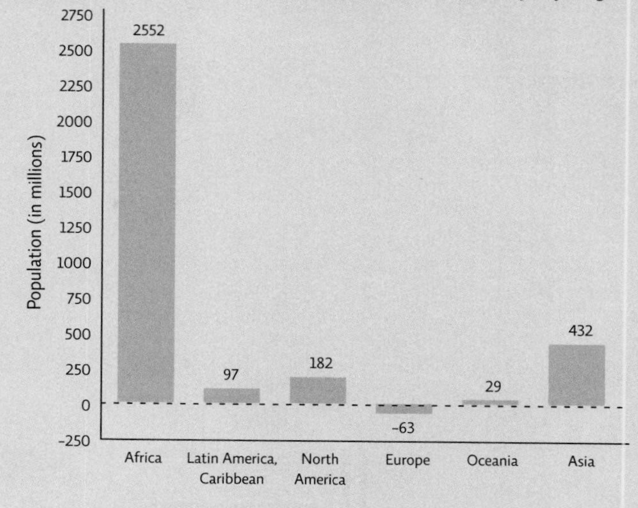
Population (in millions)
2552
97
182
-63
29
432
Africa
Latin America, Caribbean
North America
Europe
Oceania
Asia
-250
0
250
500
750
1000
1250
1500
1750
2000
2250
2500
2750

EVALUATING NEW INFORMATION

Providing education and job opportunities for women is considered important by many, not just for their effects on fertility, but also as a matter of social justice. One organization that works on this is the Foundation for International Community Assistance (FINCA), which provides microfinance services to low-income people, mostly women. Explore the FINCA website (www.finca.org).

Evaluate the websites and work with the information to answer the following questions:

1. Is this a trustworthy organization? Does it have a clear and transparent agenda?
   a. Who runs this organization? Do his or her credentials make the work of the organization and the information presented reliable/unreliable? Explain.
   b. What is the mission and vision of this organization? What are its underlying values? How do you know this?
   c. Do you agree with FINCA's assessment of the problems and concerns about poverty? Explain.
2. Explore the "About FINCA" and "Microfinance Programs" links.
   a. What are microfinance and village banking? Does the website provide supporting evidence for FINCA's claims about how these programs can help the poor? Is the evidence reliable?
   b. What is the Smart Microfinance Campaign and why is it important to microfinance clients (www.smartcampaign.org; www.finca.org/site/c.6flGIXMFJnJ0H/b.6660523/k.A0C1/Keeping_Clients_First_in_Microfinance.htm#.USaj7KWGHpg)?
   c. Do you think FINCA's business model is valid and effective? Explain.
3. Select the "Get Involved" link.
   a. Who can be a part of FINCA's solution? How can they get involved?
   b. Identify a specific solution and the strategy to accomplish it. Is the solution sufficient and the strategy reasonable? Explain.
4. How might the FINCA model influence cultural, economic, and demographic factors that influence population growth?

MAKING CONNECTIONS

CAN THE UN MILLENNIUM DEVELOPMENT GOALS ENABLE A DEMOGRAPHIC TRANSITION?

Background: The world population reached 7 billion in October 2011. This last billion was added in record time—12 years—as was the billion before it. It is also very likely that the next billion will be added in another 12 years, making this the most rapid period of human population expansion in history. This reality seems counterintuitive, given that fertility rates have been falling more rapidly than ever before. In fact, 40% of the world population lives in nations where birth rates are below replacement level. Most population growth is concentrated in the world's poorest nations.

Case: You have been given the task of analyzing the United Nations Millennium Development Goals (MDG)—eight goals for addressing the various dimensions of extreme poverty—in the context of the demographic transition. There are two basic questions:

1. Are the MDGs sufficient to enable developing countries to make the demographic transition?
2. Based on what is achieved by 2015, what should the next step be?

In your report, include the following:

a. A description of the UN MDGs, why they were selected, and what they are intended to achieve.
b. An analysis of the progress toward the MDGs. Will they meet their intended targets by 2015? Why or why not?
c. There has been much criticism of the MDGs, not only based on the assessments of progress toward their achievement, but also regarding how and why they were chosen in the first place. Who are some critics (discuss two) and what are their concerns? Discuss whether you agree with their assessment.
d. In what ways can such global goal setting be useful despite its limitations? Provide some specific examples of how such goals have been/can be used to address population matters.

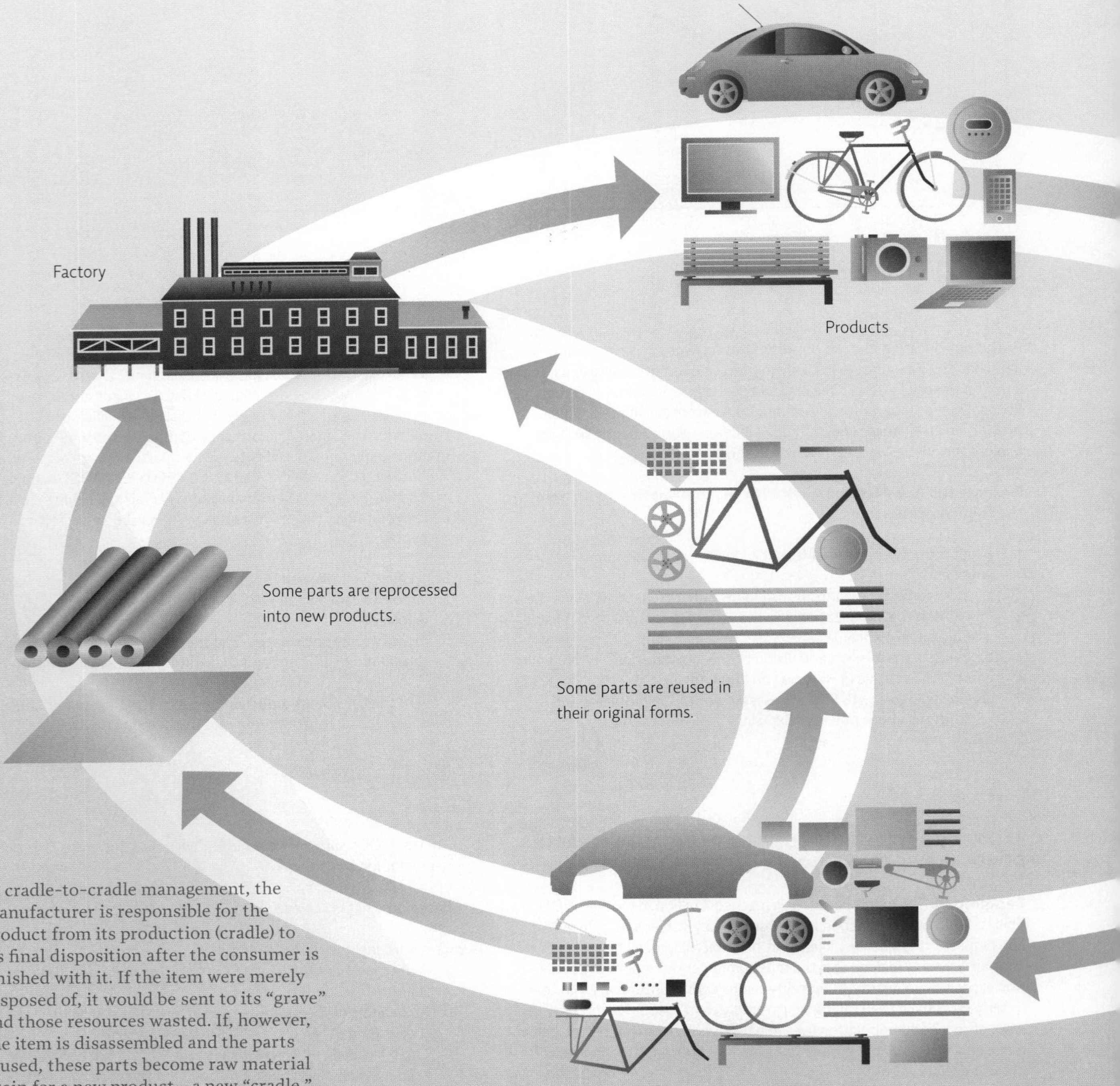
Factory
Products
Some parts are reprocessed into new products.
Some parts are reused in their original forms.
Products are disassembled.

↑ In cradle-to-cradle management, the manufacturer is responsible for the product from its production (cradle) to its final disposition after the consumer is finished with it. If the item were merely disposed of, it would be sent to its "grave" and those resources wasted. If, however, the item is disassembled and the parts reused, these parts become raw material again for a new product—a new "cradle." This provides an incentive to produce the product in a way that uses durable, reuseable parts and that minimizes toxicity, since the manufacturer is responsible for dealing with those toxic substances.

WALL TO WALL, CRADLE TO CRADLE

Successful businesses take a chance on going green

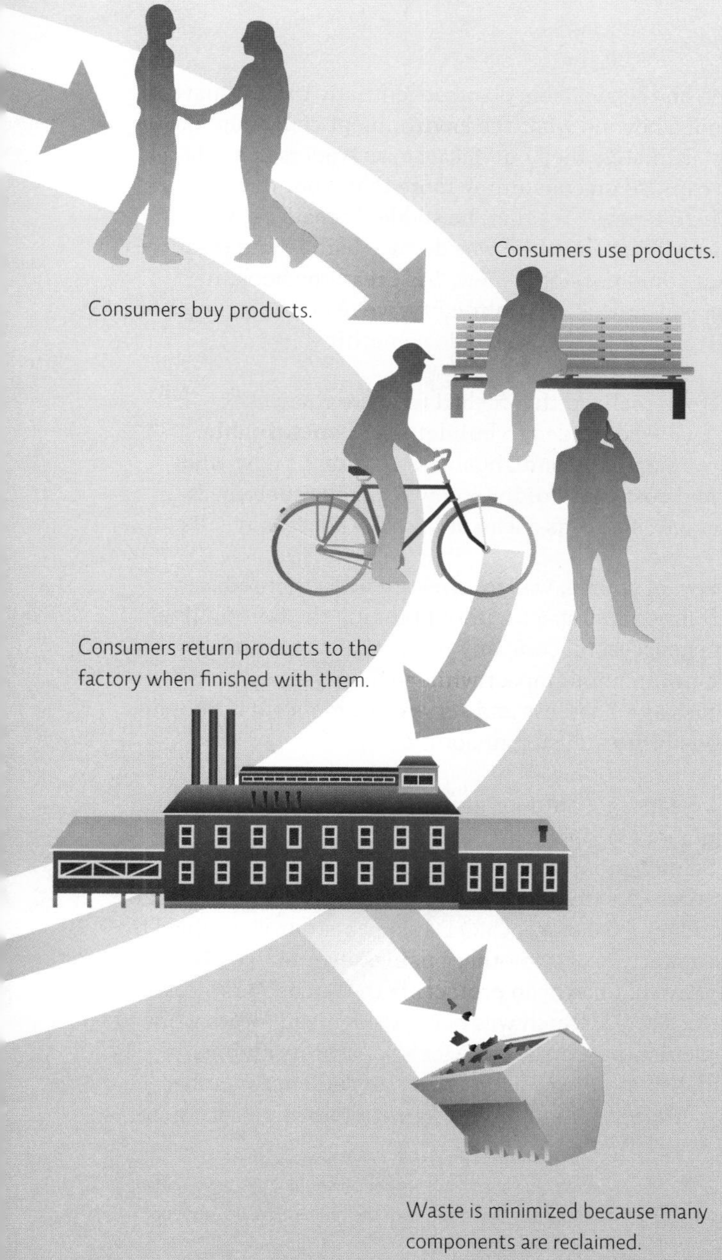
Consumers buy products.
Consumers use products.
Consumers return products to the factory when finished with them.
Waste is minimized because many components are reclaimed.

CORE MESSAGE

Human impact on Earth can be measured in terms of our ecological footprint, which is closely tied to the way we use resources. Our economic choices, both corporate and individual, tend to focus on short-term gain rather than long-term sustainability, but we can make better and more informed decisions by taking all the costs—economic, social, and environmental—of a given action into account. Using nature as a model can help us make more sustainable choices while still supporting a viable economy.

GUIDING QUESTIONS

After reading this chapter, you should be able to answer the following questions:

→ What are ecosystem services? How can it be useful to place a monetary value on these services even if we know it will not be accurate?

→ What is the concept of an ecological footprint, and how does it relate to our use of natural interest and natural capital?

→ What factors influence how much human actions impact the environment, and how can we reduce that impact?

→ What are externalities and internalities in the business world and how do these concepts relate to the concept of true costs?

→ How does ecologically based economics differ from mainstream economics, and how might ideas from ecological economics help industry and consumers make better choices?

It was the summer of 1994, and Ray Anderson was feeling pretty good. His Atlanta-based company, Interface Carpet, was the world's leading seller of carpet tiles—small square pieces of carpet that are easier to install and replace than rolled carpet—and was raking in more than a billion dollars per year. One day, though, an associate from his research division approached him with a question. Some customers apparently wanted to know what Interface was doing for the environment. One had told Interface's West Coast sales manager that, environmentally speaking, Interface "just didn't get it."

Anderson was dumbfounded. The carpet industry was not generally eco-conscious; after all, synthetic carpet is made from petroleum in a toxic process that releases significant amounts of air and water pollution, along with solid waste. Indeed, Interface used more than 450 000 metric tons of oil-derived raw materials each year, and its plant in LaGrange, Georgia, released more than 5 metric tons of carpet-trimming waste to landfills every day. "I could not think of what to say, other than 'We obey the law, we comply,'" he recalls—in other words, his company did things by the book, in terms of the environment. Wasn't that enough?

Desperate for inspiration, Anderson began leafing through *The Ecology of Commerce*, a book by environmental activist, entrepreneur, and writer Paul Hawken, which one of his sales managers had lent him. The book told the story of a small island in Alaska to which the U.S. Fish and Wildlife Service had introduced a population of reindeer during World War II. Although the reindeer thrived for a time on the available plants, eventually the population exploded beyond what the environment could support. They ultimately died out because, as Anderson explains, "you can't go on consuming more than your environment is able to renew." Yet that, he suddenly realized, was precisely what Interface was doing—using more resources than it could possibly renew. "As I read the book, it became clear that, God almighty, we're on the wrong side of history, and we've got to do something."

↑ Ray Anderson, founder of Interface. The background displays sample pieces of his carpet tiles.

Anderson realized that he had to make changes to Interface—he needed to build it into a **sustainable**, environmentally sound business. "I didn't know what it would cost, and I didn't know what our customers would pay, so it was a leap of faith," Anderson recalls.

The key, of course, was to invest in tools that reduced environmental impact without hurting the bottom line. A truly successful company is one that can reduce its environmental impact while making money, so it survives as a business and serves as a model of corporate responsibility and sustainability.

A few years ago, outdoor gear company Mountain Equipment Co-op (MEC), headquartered in Vancouver, was being honoured at the World Wilderness Congress for its decades of attention to the environmental impact of its products. On stage, CEO David Labistour was joined by representatives of two other businesses—a large cement company that has been protecting the Amazon rainforest, and Coca-Cola, which was being recognized for its work in replenishing aquifers in Mexico. After receiving his award, Labistour was accosted by an audience member who criticized him for standing with two large corporate

sustainable Capable of being continued without degrading the environment.
economics The social science that deals with the production, distribution, and consumption of goods and services.
ecosystem services Essential ecological processes that make life on Earth possible.

entities. "You shouldn't be up there with these huge businesses—it's greed that destroys the environment," he told Labistour. The CEO asked the man if he had a retirement account. Yes, came the answer. Did he expect it to grow over time? Yes, again. "Then you can't fault companies for making money," Labistour told the man.

But in the pursuit of profits, the choices businesses (and by extension, consumers) make have tremendous impacts on the environment. The amount and type of energy and water they use, the way they handle the waste they produce, the raw materials they use—these decisions affect not only business operations themselves, but also Earth as a whole, especially considering the magnitude of the resources and waste that some large businesses use and produce.

Environmentally mindful businesses like MEC and Interface, whether consciously or not, are applying principles that have always been at work in nature: since resources are finite, it's important to recycle and to maximize energy efficiency. In this way, nature is an ecological model that can inspire economic models that enhance sustainability. After all, **economics**—the social science that analyzes the production, distribution, and consumption of goods and services—is not just about money. All of the resources on which we depend come from the environment. Environmental resources like timber and water can be considered ecosystem goods. Ecological processes like water purification, pollination, climate regulation, and nutrient cycling are essential **ecosystem services**. Some of these resources are invaluable because there are no substitutes—consider the oxygen produced by green plants, which we need in order to survive. [INFOGRAPHIC 5.1]

mec

↑ Mountain Equipment Co-op is a successful business with a commitment to sustainability.

Infographic **5.1** | **VALUE OF ECOSYSTEM SERVICES**

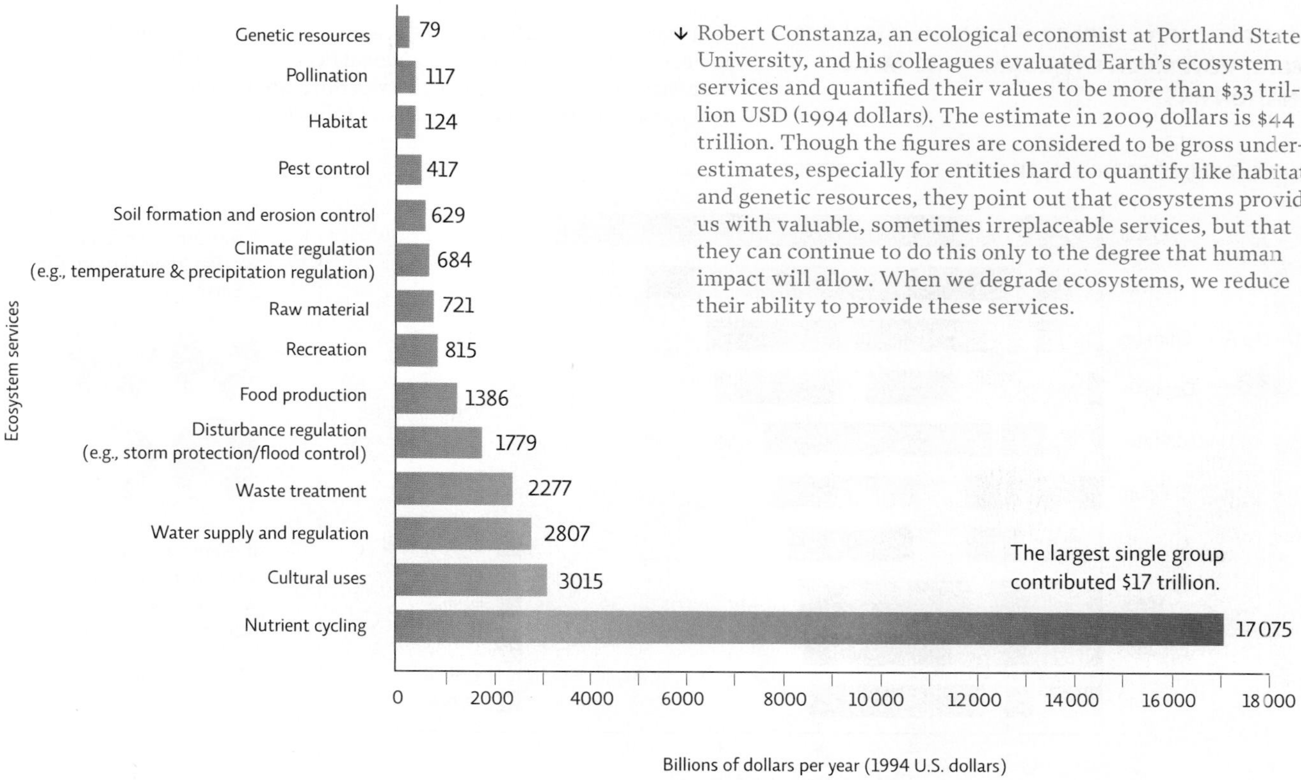
Genetic resources
79
Pollination
117
Habitat
124
Pest control
417
Soil formation and erosion control
629
Climate regulation (e.g., temperature & precipitation regulation)
684
Raw material
721
Recreation
815
Food production
1386
Disturbance regulation (e.g., storm protection/flood control)
1779
Waste treatment
2277
Water supply and regulation
2807
Cultural uses
3015
Nutrient cycling
17075
The largest single group contributed $17 trillion.
Ecosystem services
0
2000
4000
6000
8000
10000
12000
14000
16000
18000
Billions of dollars per year (1994 U.S. dollars)

↓ Robert Constanza, an ecological economist at Portland State University, and his colleagues evaluated Earth's ecosystem services and quantified their values to be more than $33 trillion USD (1994 dollars). The estimate in 2009 dollars is $44 trillion. Though the figures are considered to be gross underestimates, especially for entities hard to quantify like habitat and genetic resources, they point out that ecosystems provide us with valuable, sometimes irreplaceable services, but that they can continue to do this only to the degree that human impact will allow. When we degrade ecosystems, we reduce their ability to provide these services.

Infographic **5.2** | **ECOLOGICAL FOOTPRINT**

↓ The ecological footprint is the land area needed to provide the resources for and assimilate the waste of a person or population and may extend far beyond the actual land occupied by the person or population; it is usually expressed as a per capita value (hectares/person or square metres or kilometres/person). The current world footprint would require about 1.5 Earths to maintain, but obviously we just have one to work with.

Raw materials used in the city are imported from elsewhere.
Some waste is assimilated by areas outside the city.
Physical footprint of the city
Ecological footprint

BOTH PER CAPITA IMPACT AND POPULATION SIZE AFFECT HOW MUCH ENVIRONMENTAL IMPACT A NATION HAS

↓ Canada has the eighth largest per capita ecological footprint of all the countries in the world at 6.4 global hectares. (A global hectare represents the average productivity per hectare worldwide.) Fossil fuel use accounts for more than half of that footprint.

ECOLOGICAL FOOTPRINTS: THE TOP TEN COUNTRIES

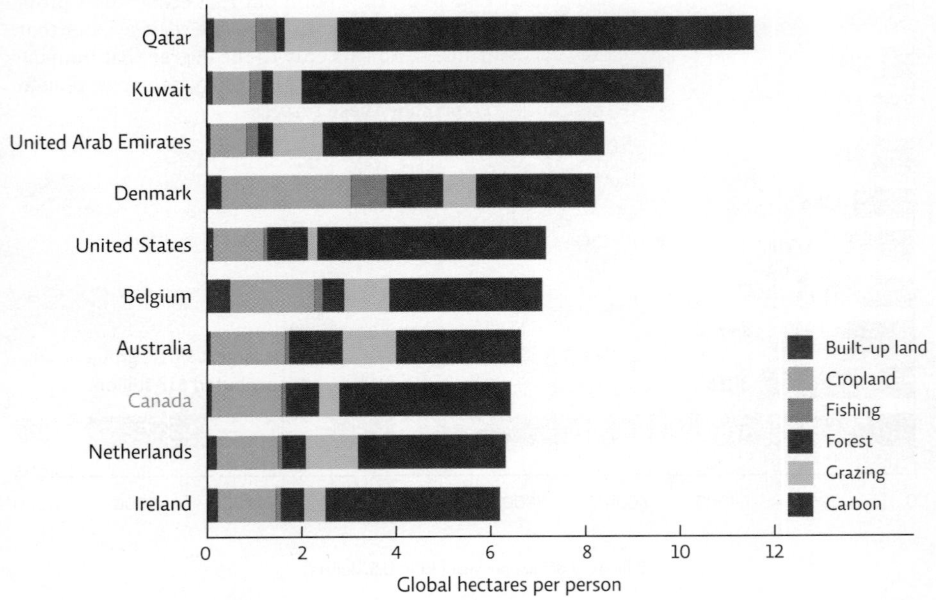
Qatar
Kuwait
United Arab Emirates
Denmark
United States
Belgium
Australia
Canada
Netherlands
Ireland
Built-up land
Cropland
Fishing
Forest
Grazing
Carbon
0
2
4
6
8
10
12
Global hectares per person

If everyone on Earth had a footprint like Canada, we would need more than 3 Earths.

If everyone on Earth had a footprint like Kenya, we would have land and resources to spare.

When ecosystems are intact, they are naturally sustainable: they rely on renewable resources and also provide services that help to renew and recycle these resources. But ecosystems will only be able to provide us with their valuable goods and services as long as we let them. As Interface and MEC have realized, when we degrade ecosystems by using more from them than can be replenished, we threaten our planet's ability to provide the services we need, which ultimately threatens our own future.

Businesses and individuals impact the environment with their economic decisions.

Every business—including Interface and MEC—has an **ecological footprint**, that is, the land needed to provide its resources and assimilate its waste (typically expressed as hectares (ha) or square metres (m^2) per person or population). Ecological footprint is a value that businesses, individuals, and populations use to quantify their impact on the environment. Just as petroleum is needed to make synthetic carpet, MEC uses petroleum as a raw material for some of its fabrics. The company also depends on energy to manufacture and sell its goods—even a simple cotton T-shirt requires farming of cotton, mechanics to gin and spin it, dyes to colour it, then fuel to ship it to factories and, finally, a retail store. [INFOGRAPHIC 5.2]

Canada has a particularly high per capita (per person) footprint, in that it requires much more land area to support each person than Canada actually possesses. In fact, if the 7 billion people who populate the planet all lived like the average Canadian, we would need the landmass of 3 1/2 Earths to sustain everyone. According to the World Wildlife Fund's *2012 Living Planet Report*, humans currently use 50% more resources than is ultimately sustainable. This means it would take 1.5 years to replenish what we take in a single year—a clearly unsustainable course.

What kinds of essential resources does Earth provide us? Considered in financial terms, our **natural capital** includes the natural resources we consume, like oxygen, trees, and fish, as well as the natural systems—forests, wetlands, and oceans—that produce some of these resources. Our **natural interest** is what is produced from this capital, over time, just like the interest you earn with a bank account. Natural interest might be represented by an increase in a fish population, for instance, or new growth in a forest—basically, the extra that is added in a given time frame. [INFOGRAPHIC 5.3]

If we only withdraw resources equivalent to (or less than) the natural interest, we will leave behind enough natural capital to replace what we took. We would be operating within our ecosystem's **biocapacity**, or its ability to produce resources and assimilate our waste. But by continually taking more than the equivalent of natural interest, we diminish resources and can potentially eliminate them. This can be an especially big problem with commonly held resources like water: once we remove it from wells or rivers, we have to wait for the next rainfall to replenish it. When many users are accessing the resource, it can quickly become depleted or degraded if they do not work together to manage it—a tragedy of the commons (see Chapter 1).

When we dip into our natural capital as humans are doing today, we are decreasing future interest potential. Essentially, we are taking resources away from the future, in what eco-architect Bill McDonough calls *intergenerational tyranny*. By liquidating our natural capital more quickly than it can be replaced and calling that "income," the question becomes this: where will future income come from?

Researchers often use the **IPAT model** to estimate the size of a population's ecological footprint, or impact (I), based on three factors: population (P), affluence (A), and technology (T). The premise is that as population size increases, so does impact. More affluent and technology-dependent populations use more resources and generate more waste than do less affluent and technology-dependent populations (technology allows us to build more things, dig deeper, and fly higher, all of which drain the environment). [INFOGRAPHIC 5.4]

One caveat with regard to this model is that the right kind of technology can have the opposite effect: it can decrease, rather than increase, environmental impact. In 2006, for instance, after deciding to become sustainable, Interface invented a new technology called TacTiles—small squares of adhesive tape that join carpet tiles together. In contrast with traditional "spread on the floor" adhesives, Interface's new tape has a 90% smaller environmental footprint, and does not contain any

ecological footprint The land area needed to provide the resources for, and assimilate the waste of, a person or population.
natural capital The wealth of resources on Earth.
natural interest Readily produced resources that we could use and still leave enough natural capital behind to replace what we took.
biocapacity The ability of land or aquatic ecosystems to produce resources and assimilate our waste.
IPAT model An equation ($I = P \times A \times T$) that measures human impact (I), based on three factors: population (P), affluence (A), and technology (T).

Infographic **5.3** | **CAPITAL AND INTEREST**

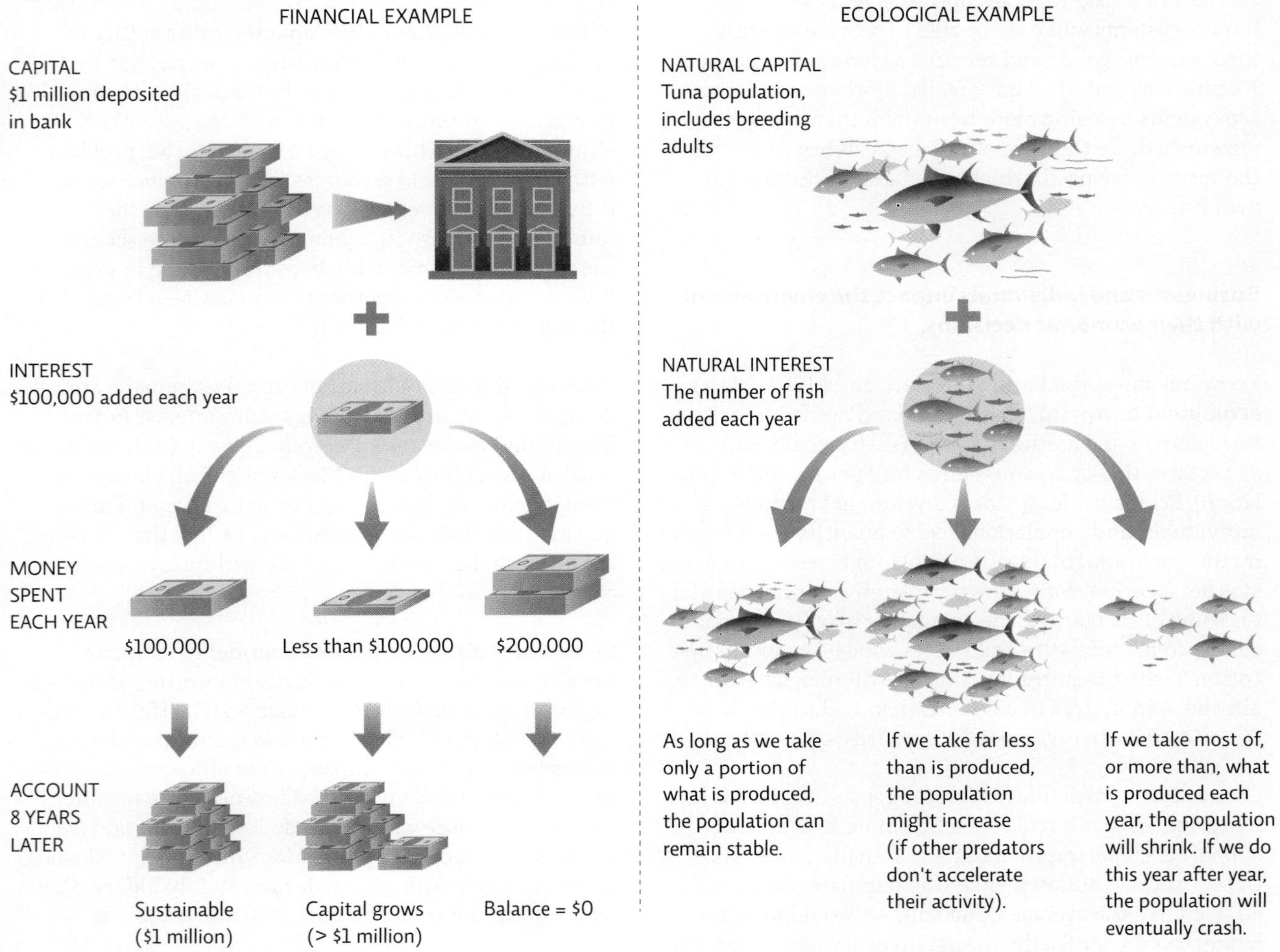
FINANCIAL EXAMPLE
CAPITAL
$1 million deposited in bank
INTEREST
$100,000 added each year
MONEY SPENT EACH YEAR
$100,000
Less than $100,000
$200,000
ACCOUNT 8 YEARS LATER
Sustainable ($1 million)
Capital grows (> $1 million)
Balance = $0
ECOLOGICAL EXAMPLE
NATURAL CAPITAL
Tuna population, includes breeding adults
NATURAL INTEREST
The number of fish added each year
As long as we take only a portion of what is produced, the population can remain stable.
If we take far less than is produced, the population might increase (if other predators don't accelerate their activity).
If we take most of, or more than, what is produced each year, the population will shrink. If we do this year after year, the population will eventually crash.

↑ Natural resources can be compared to the financial concepts of capital and interest. Natural capital is the wealth of resources on Earth and includes all the natural resources we use as well as the natural systems that produce some of those resources (e.g., forests, wetlands, oceans). Natural interest is the amount produced regularly that we could use and still leave enough natural capital behind to replace what we took.

volatile organic compounds, a group of chemicals that Environment Canada recognizes as a health risk. TacTiles also make it possible for customers to replace single carpet tiles easily, when, for instance, there has been a spill. When technologies such as TacTiles reduce the environmental footprint rather than increase it, the equation used to describe their impact changes to: I = (P × A)/T.

In 2007, to further reduce its impact, Interface launched a carpet-recycling initiative called ReEntry 2.0. More than 2 billion metric tons of carpet are pulled up and discarded globally each year, and less than 5% of that has historically been reused or recycled. With ReEntry 2.0, Interface developed a way to recycle carpets—both its own and those made by its competitors—to make new carpet, keeping more than 90 000 metric tons of material out of landfills. Interface has promised to eliminate *any* negative impact it has on the environment by 2020, in a plan it calls "Mission Zero."

internal costs Those costs—such as raw materials, manufacturing costs, labour, taxes, utilities, insurance, and rent—that are accounted for when a product or service is evaluated for pricing.

external costs Costs that are associated with a product or service, but are not taken into account when a price is assigned to that product or service.

Infographic **5.4** | **THE IPAT EQUATION**

↓ Population (P), affluence (A), and technology (T) all affect how much of an impact an individual or population has on the environment. As any or each of these factors increase, so does the population's overall impact, as indicated by the model's equation: I = P × A × T. The right kind of technology can actually lower overall impact, in which case the equation becomes I= (P × A)/T.

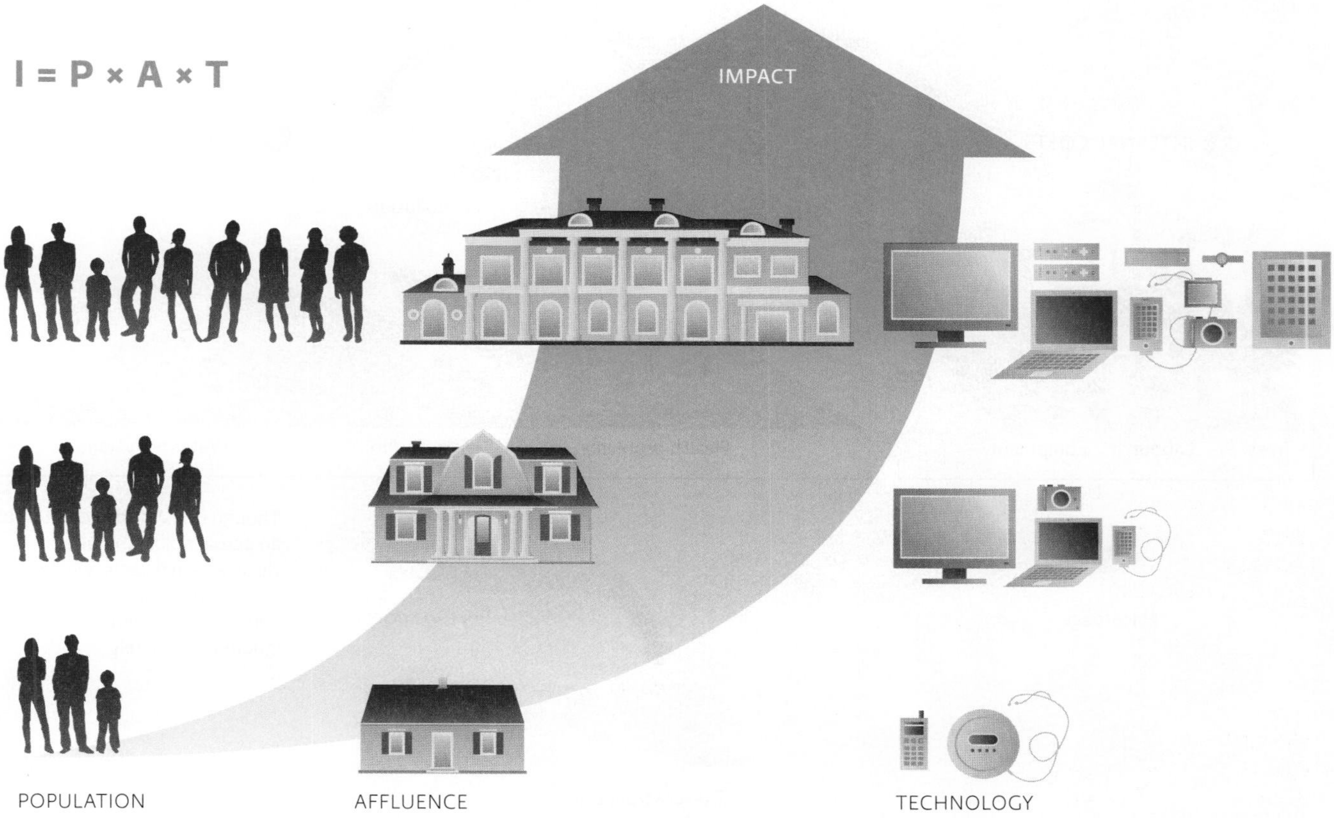
I = P × A × T
IMPACT
POPULATION
AFFLUENCE
TECHNOLOGY

Zero waste is also the ultimate goal for MEC, which in 2007 diverted 92% of its waste by recycling, donating, or composting it. Sometimes this requires creative thinking—in the Edmonton store, for instance, the staff turned old packaging into notebooks. These initiatives are not just smart for the planet, they're smart for the purse—sending 1 metric ton of waste to a landfill costs $325 in disposal charges, while recycling it runs an average of only $68.

Mainstream economics supports some actions that are not sustainable.

One of the limitations of mainstream economics is that it doesn't take into account *all* potential costs. For instance, a carpet tile might require a certain amount of material that has a particular monetary cost—but what about the environmental costs associated with drilling enough oil to make that material in the first place, or the costs associated with cleaning up the pollution it creates? The direct cost of the material is an **internal cost**—a cost that is accounted for when a product or service is priced—but it is often incomplete. There are also **external costs**, such as the health costs associated with the waste produced by making the carpet tiles, the use of chemicals to dye textiles, or the environmental damage caused by pollution. Historically, economists have regarded these as external to the business cost (the business doesn't pay for them) and they aren't reflected in the price the consumer pays for the good or service. But if the business doesn't pay for the costs or pass those costs on to the consumer, who does pay? Other people, present and future, and other species do. They pay in the form of degraded health, ecosystems, or opportunities.

An assessment of the cost of a good or service should include more than just the economic costs but also the

Infographic **5.5** | **TRUE COST ACCOUNTING**

What does it take to produce the paper you use every day?

↓ Many environmental and health costs of our goods and services are externalized (not included in the price the consumer pays). But if consumers don't pay all the costs to produce a product, such as paper, who does?

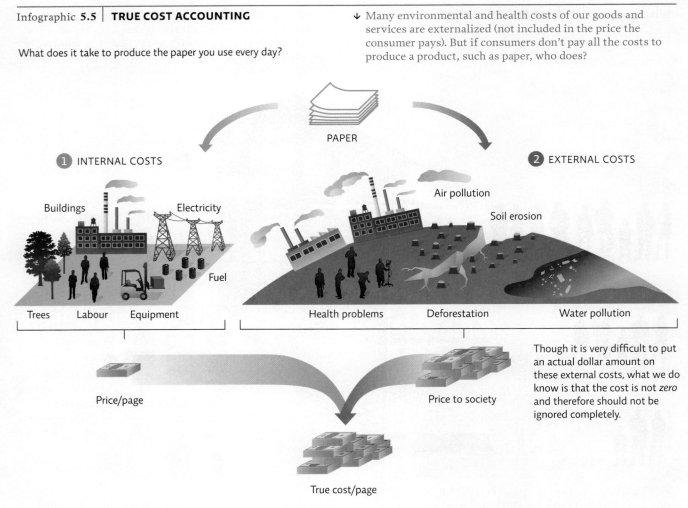

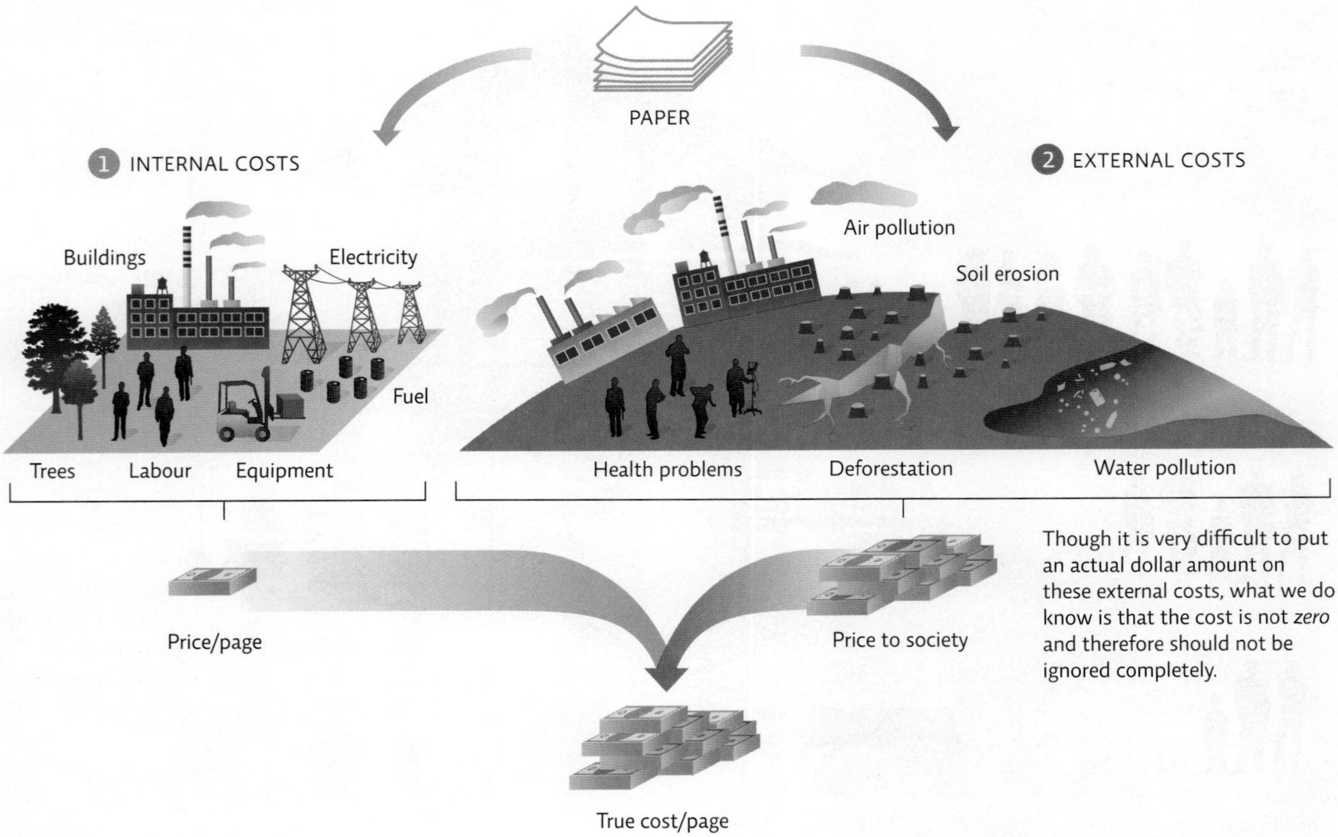
PAPER
1 INTERNAL COSTS
Buildings
Electricity
Fuel
Trees
Labour
Equipment
2 EXTERNAL COSTS
Air pollution
Soil erosion
Health problems
Deforestation
Water pollution
Though it is very difficult to put an actual dollar amount on these external costs, what we do know is that the cost is not *zero* and therefore should not be ignored completely.
Price/page
Price to society
True cost/page

social and environmental costs—the **triple-bottom line**. By ignoring the external costs, economies create a false idea of the true and complete costs of particular choices. To pay the **true cost** of a product or service, the external costs must be internalized. For example, a customer may pay $8 for every 50 square centimetre carpet tile, but the true cost for that piece of carpet would be much higher if it included the cost of greenhouse gas emissions and, for instance, the cost of treating people for asthma who might fall ill as a result of the particulate matter released during the carpet's production. The inadequate valuation of a product could eventually lead to the exploitation or overuse of resources needed to produce it, an example of market failure.

Because we are so accustomed to not paying true costs, we would most likely be appalled at how much goods and services would really cost if all externalities were internalized. Although it sounds discouraging, any time we purchase products that were made in a more environmentally or socially sound manner, we come a little closer to bearing the consequences of our choices. We also create a demand for these products in the marketplace. [INFOGRAPHIC 5.5]

Mainstream economics also assumes that natural and human resources are either infinite or that substitutes can be found if needed. While this may be true for some, it is probably not true for all resources. For instance, fossil fuels are finite and will run out, even with technological advances that allow us to access more of the fuel that is left. It remains to be seen whether or not we can replace fossil fuels with sustainable alternatives at current levels of use. Additionally, our actions can degrade air and water resources faster than nature can restore them; crop productivity also has its limits.

Another assumption of mainstream economic theory is that economic growth can go on forever. But, since there are inherent limits to what Earth can provide, unlimited resource-based growth is not, in fact, possible. We have to work within the limits of available resources in ways that allow essential ecosystem services to continue.

Most models of production follow a linear sequence: raw materials come in, humans transform those materials into some kind of product, and then they discard the waste generated in the process. Eventually the product itself becomes waste. But because some resources are finite—and waste in the form of pollution can damage natural capital like air, water, and soil—linear models of production will eventually fail. Even though only 8% of MEC's waste ends up in landfills, it adds up to 70 metric tons per year. So MEC conducts an annual waste audit for each facility to try to reduce waste even further, looking at both volume and contents. As the company says, they go "dumpster diving," periodically retrieving items from their own dumpsters and determining what and how much is going in the garbage and whether it could be diverted to recycling bins or composting initiatives.

Even better, Interface's ReEntry 2.0 program uses old carpet tiles to make new ones, and old carpet backings to make new carpet backings, in an effort to make the production process more cyclical. This is an example of a **closed-loop system**, in which the product is returned into the resource stream when consumers are finished with it, or is disposed of in such a way that nature can decompose it. By leasing carpet rather than selling it, Interface manages its product in a **cradle-to-cradle** fashion—the carpet needs to be durable and recyclable, because Interface is responsible for the impact of its use at every stage of the process. (See the diagram that opens this chapter on pages 74–75.) [INFOGRAPHIC 5.6]

Another problem with traditional economics is that it **discounts future value**: it tends to give more weight to short-term benefits and costs than it does long-term ones. In other words, something that benefits or harms us today is usually considered more important than something that might do so tomorrow. For instance, we value the tuna we can harvest today more highly than tuna we might harvest 10 years from now, so the value of taking a large harvest of tuna today seems to outweigh the benefits of taking less now to ensure there is still some later. If the money we could earn by using the resource now is more than that possible by sustainably harvesting, modern economics tells us it is more profitable to use it now and invest the resulting money in another venture. But how might the loss of tuna affect the ecosystem and other populations? And what about that "other venture"? Will there always be another fish population to harvest?

Energy usage must also be considered in a company's quest for sustainability. Thirty percent of Canada's greenhouse gas emissions come from the energy used to construct and maintain buildings. Since the 1980s, MEC has adopted an "aggressive green building vision," says Labistour, to cut down on the energy, water, and materials used in its retail buildings. For instance, the retail site in Winnipeg—dubbed by industry experts the most energy-efficient outlet in the country, according to MEC—has a structure made entirely of reclaimed materials, thus keeping these materials out of landfills. Extra insulation and double (or even triple) paned windows keep the indoor climate comfortable. Furthermore, lining the roof with vegetation to create a "green roof" provides cooling in summer, as water evaporates from the soil, and extra insulation in the winter. The roof vegetation is fed by a composting toilet, which also cuts down waste water by 50%. This store and the company's Ottawa retail location were the first two retail buildings in Canada to comply with the C2000 Green Building Standards that require facilities to use 50% less energy than conventional structures.

> A sustainable approach would be more cyclical, where "waste" becomes a raw material once again and can be used to make products.

Interface has also maximized energy efficiency in its facilities, installing skylights and solar tubes to replace artificial, electricity-dependent lighting, and using more energy-efficient heating, ventilation, and air conditioning systems. In one of its factories, Interface also installed a real-time energy tracker that displays energy use prominently for its employees to see, inspiring them to think of new ways to conserve energy. These and other changes enabled the company to cut the amount of energy it derived from fossil fuels by 55%, and reduce its total energy use by 43%.

Ecologically minded economists are considering how to incorporate environmental considerations into economic decisions. **Ecological economics** is a discipline that

---

triple-bottom line The environmental, social, and economic impacts of our choices.

true cost The sum of both external and internal costs of a good or service.

closed-loop system A production system in which the product is returned to the resource stream when consumers are finished with it, or is disposed of in such a way that nature can decompose it.

cradle-to-cradle Management of a resource that considers the impact of its use at every stage from raw material extraction to final disposal or recycling.

discounting future value Giving more weight to short-term benefits and costs than to long-term ones.

ecological economics Branch of economics that considers the long-term impact of our choices on people and the environment.

considers the long-term impact of our choices on human society and the environment. These economists support actions such as improving technology to increase production efficiency and reduce waste; valuing resources as realistically as possible; moving away from dependence on nonrenewable resources; and shifting away from a product-oriented economy. They look to natural ecosystems as models for how to efficiently use resources and live within the limits of nature. But, ecological economists feel that our ingenuity will only take us so far and that economic growth does have limits; therefore, we must significantly change the way we do things in order to become sustainable as a society.

There are tactics for achieving sustainability.

Since 1994, Interface has reduced greenhouse gas emissions from its manufacturing facilities by 35%, and now relies on recycled or bio-based ingredients for 40% of its raw materials. This drive toward sustainability has not hurt profits: over the same time period, the company has increased sales by two-thirds and doubled its earnings.

Running a **green business**—doing business in a way that is good for people and the environment—is also profitable for MEC. In 2011, the company emitted 46% less greenhouse gases from facilities than it did in 2007 while

Infographic **5.6** | **ECONOMIC MODELS**

↓ Mainstream economics assumes that resources will always be available and that waste can be disposed of in a linear (one-way) system. Ecological economics recognizes that natural ecosystems provide our resources and assimilate our wastes. If companies could fold "waste" back into production or make sure it can be decomposed by nature, we could reduce our extraction costs and be operating in a sustainable closed-loop system.

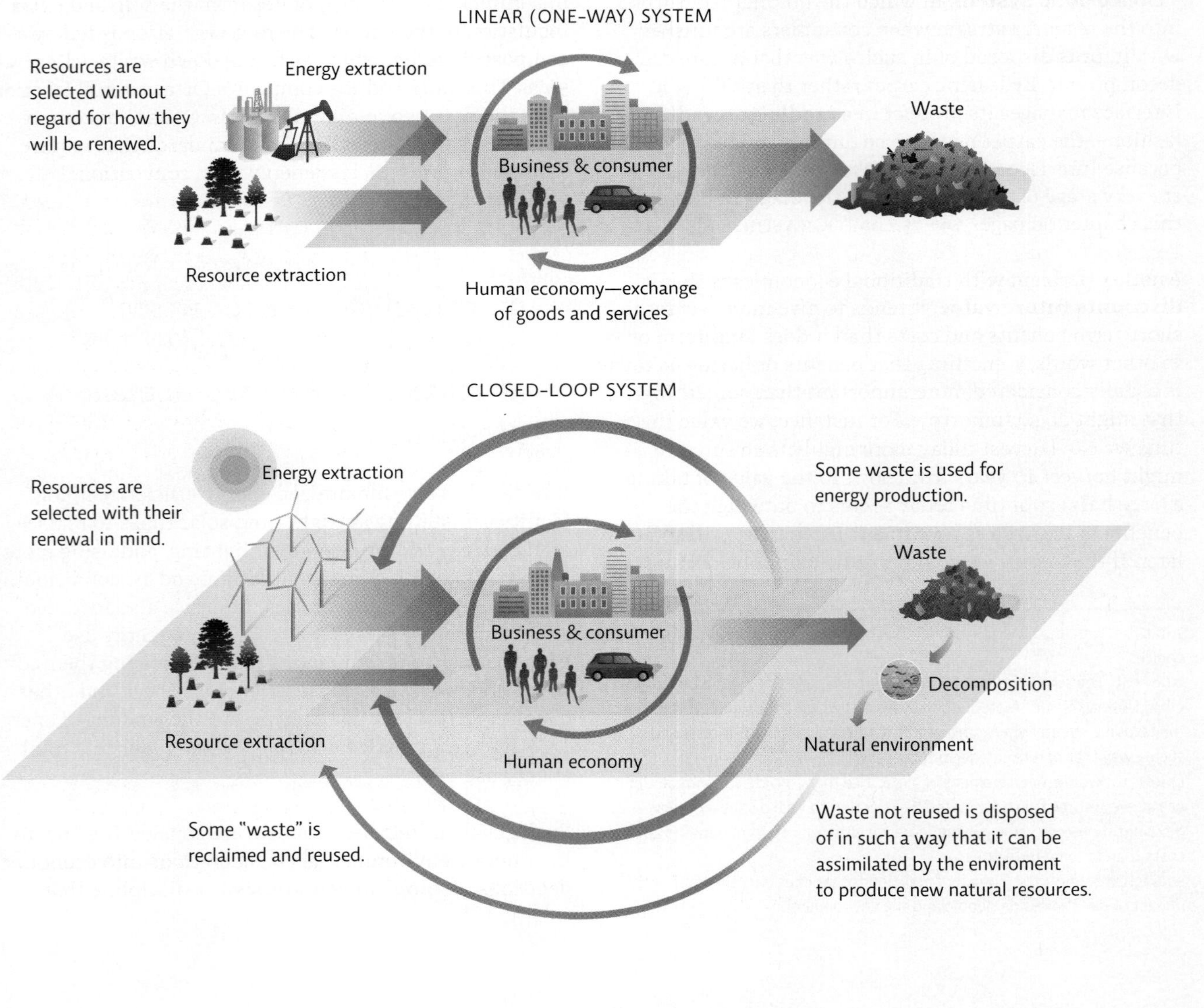
LINEAR (ONE-WAY) SYSTEM
Resources are selected without regard for how they will be renewed.
Energy extraction
Waste
Business & consumer
Resource extraction
Human economy—exchange of goods and services
CLOSED-LOOP SYSTEM
Energy extraction
Resources are selected with their renewal in mind.
Some waste is used for energy production.
Waste
Business & consumer
Decomposition
Resource extraction
Natural environment
Human economy
Some "waste" is reclaimed and reused.
Waste not reused is disposed of in such a way that it can be assimilated by the enviroment to produce new natural resources.

also reporting \$270 million in sales, a nearly \$10 million increase from the year before.

The ultimate goal is to run a fully sustainable business, with zero waste and no extra greenhouse gases. By definition, **sustainable development** must meet present needs without preventing future generations from meeting their needs. Anderson vowed in 1994 that Interface Carpet would become the world's first sustainable business. Despite these ambitions, no business has yet succeeded in this goal. "We're not there yet," says MEC's Labistour. "It's a journey."

How does a company find inspiration on how to become sustainable? One way is to look to natural ecosystems—a perfect example of sustainable resource use and waste minimization—and engage in **biomimicry** (see Chapter 1). For instance, a new fabric named *Greenshield* is modelled after a lotus leaf *(Nelumbo nucifera),* one of the most water-repellent structures in nature. Like the leaves, the fabric contains tiny crevices that trap air, so water droplets (and attached dirt) roll off easily, making it as stain- and water-resistant as conventional fabrics, but using a significantly smaller amount of harmful chemicals.

At Interface, Anderson was strongly inspired by biomimicry. For instance, the TacTiles technology that the company developed to replace glue was based on the physics that explains how a gecko lizard clings to walls and ceilings. The microscopic hairs on a gecko's foot bond to the molecular layer of water that is present on nearly every surface, allowing its feet to cling. Interface used this information to develop tiles that bond to one another rather than to the floor, making a kind of "floating" carpet that stays in place due to gravity rather than being actually glued to the floor. This makes carpet installation and removal much faster and easier. Interface also revolutionized its operations by considering itself part of a service economy; it focuses on selling a *service* rather than a *product.* Customers "lease" the carpet; Interface maintains it and replaces it as needed. This encourages Interface to produce carpet that is durable and recyclable and also easily replaceable. [INFOGRAPHIC 5.7]

Another sustainable business practice involves *take-back programs,* particularly for products with a defined lifespan, such as electronics: customers return the product to the producer when they are finished or when they need an upgrade. This provides an incentive to the producer to make a durable, high-quality product that can be reused or recycled.

Changing the way we do business is not going to be easy. Start-up or upgrade costs can be substantial, and even though they may pay for themselves in the long run, many businesses simply do not have the capital to fund the improvements. Plus, they may find themselves at a competitive disadvantage with businesses that are not trying to internalize costs. Although MEC wanted to "green" all of its retail buildings, it simply couldn't afford to make all of the changes it wanted to at those facilities that were leased, rather than purchased. "We had to prioritize which energy upgrades to do," says Labistour. "These were tough choices."

Consumers also have a role to play. We can all decrease our impact by making more sustainable choices and by consuming less. Thinking about sustainability doesn't necessarily mean "doing without," but it does mean being mindful of our choices and opting for sustainable or low-impact choices whenever possible.

Share programs are a useful option for items that people need infrequently, such as a car, bicycle, or power tool for those who live in a large city. Rather than buying, owning, and then storing the product for a large part of the time, consumers buy memberships in a sharing company and only use the product when they need it.

When we need to purchase something, it's important to consider the item's true cost. Recycled paper often costs more, but that higher cost may more closely reflect the true cost of paper. But if consumers are not willing to put their buying dollars behind their environmental ideals, businesses that make and sell paper from trees will still be more successful. Knowing a product's true cost requires transparency from the industries that produce and sell us goods and services. That is hard to come by, often because the businesses themselves don't know all the external costs associated with their products. In other words, it will take changes by both consumers and producers to put business and industry on the path to sustainability. But there are things that can be done to level the playing field.

Governments can encourage sustainability by providing incentives for businesses to account for true costs rather than just internal costs. This could be accomplished by taxing companies based on how much pollution they

---

green business Doing business in a way that is good for people and the environment.

sustainable development Economic and social development that meets present needs without preventing future generations from meeting their needs.

biomimicry Using nature as a model to inspire sustainable solutions to environmental problems.

Infographic **5.7** | **PRODUCT VERSUS SERVICE ECONOMY**

↓ In a service economy model, rather than buying a product, the customer pays for the service (i.e. the ability to photocopy pages, walk on comfortable carpet), and the vendor makes sure that the service is always available. This allows companies to decrease resource drain and lessen waste while still earning a profit.

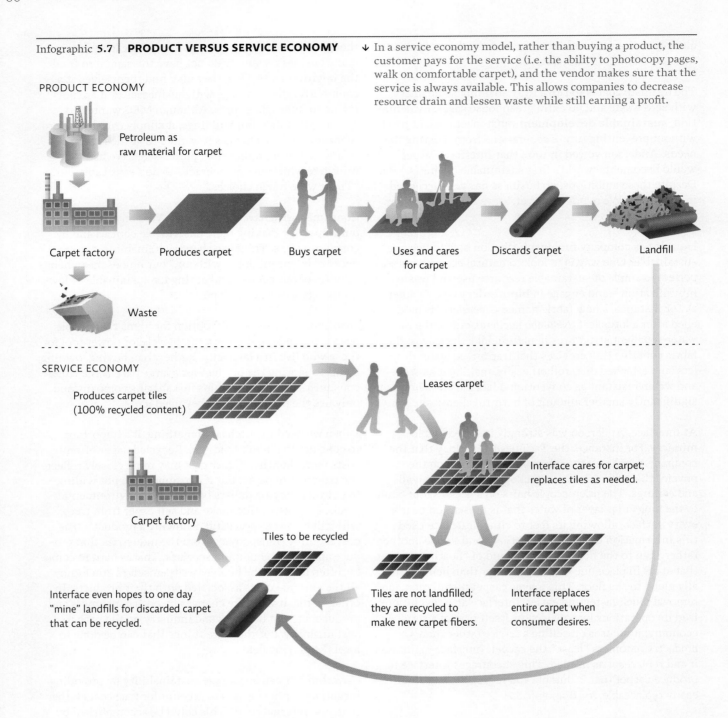

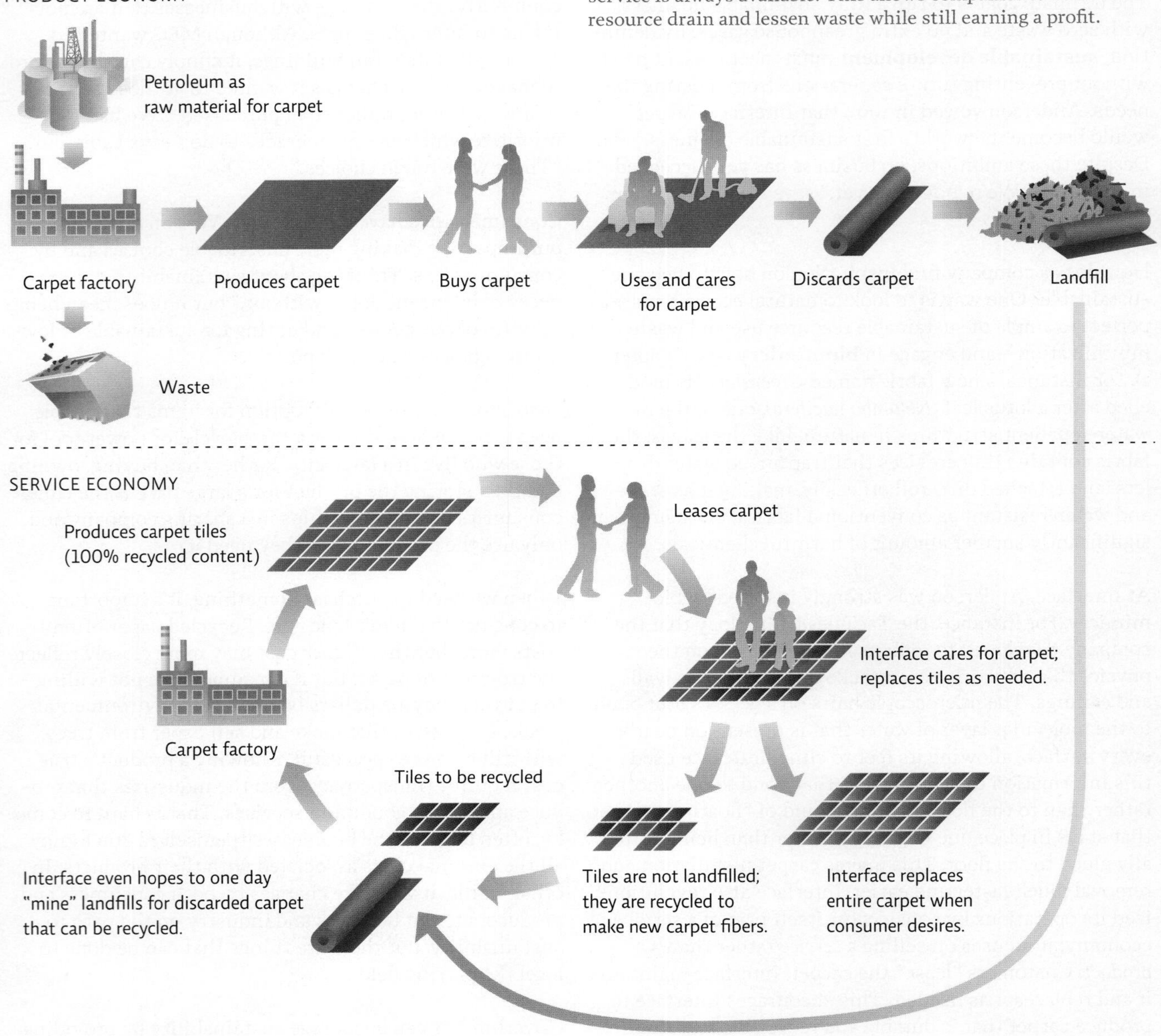
PRODUCT ECONOMY
Petroleum as raw material for carpet
Carpet factory
Produces carpet
Buys carpet
Uses and cares for carpet
Discards carpet
Landfill
Waste
SERVICE ECONOMY
Produces carpet tiles (100% recycled content)
Leases carpet
Interface cares for carpet; replaces tiles as needed.
Carpet factory
Tiles to be recycled
Interface even hopes to one day "mine" landfills for discarded carpet that can be recycled.
Tiles are not landfilled; they are recycled to make new carpet fibers.
Interface replaces entire carpet when consumer desires.

generate, subsidizing environmentally friendly processes, or giving out pollution "permits" that companies could sell if they release less pollution than they are allowed (*cap-and-trade*—see Chapter 21). For instance, if there is a pollution tax, it will be passed on to the consumer, who then decides whether or not to buy the product. The manufacturer that minimizes waste production and relies on fewer fossil fuels than its competitors might be able to meet the regulations at the lowest cost, have lower prices, and as a result, have a major market advantage. While promising, these policies can be complex to implement, because external costs are hard to quantify (How is a pollution tax fairly assessed?), and because they may put a burden on smaller manufacturers who have less ability to absorb the cost of upgrades; passing these costs on to the consumer also stresses low-income households.

To MEC, being sustainable is not just about cleaning up the environmental damage it creates—it's about finding ways to conduct business without causing environmental

Infographic **5.8** | **PURSUING SUSTAINABILITY AT MEC**

↓ Mountain Equipment Co-op's vision is to be a successful business enterprise that sells quality products in "a way that does the least harm and the most good – environmentally, socially, and economically." To meet these goals MEC works to minimize its ecological footprint in each facet of its operation, a process that continues to evolve and grow each year.

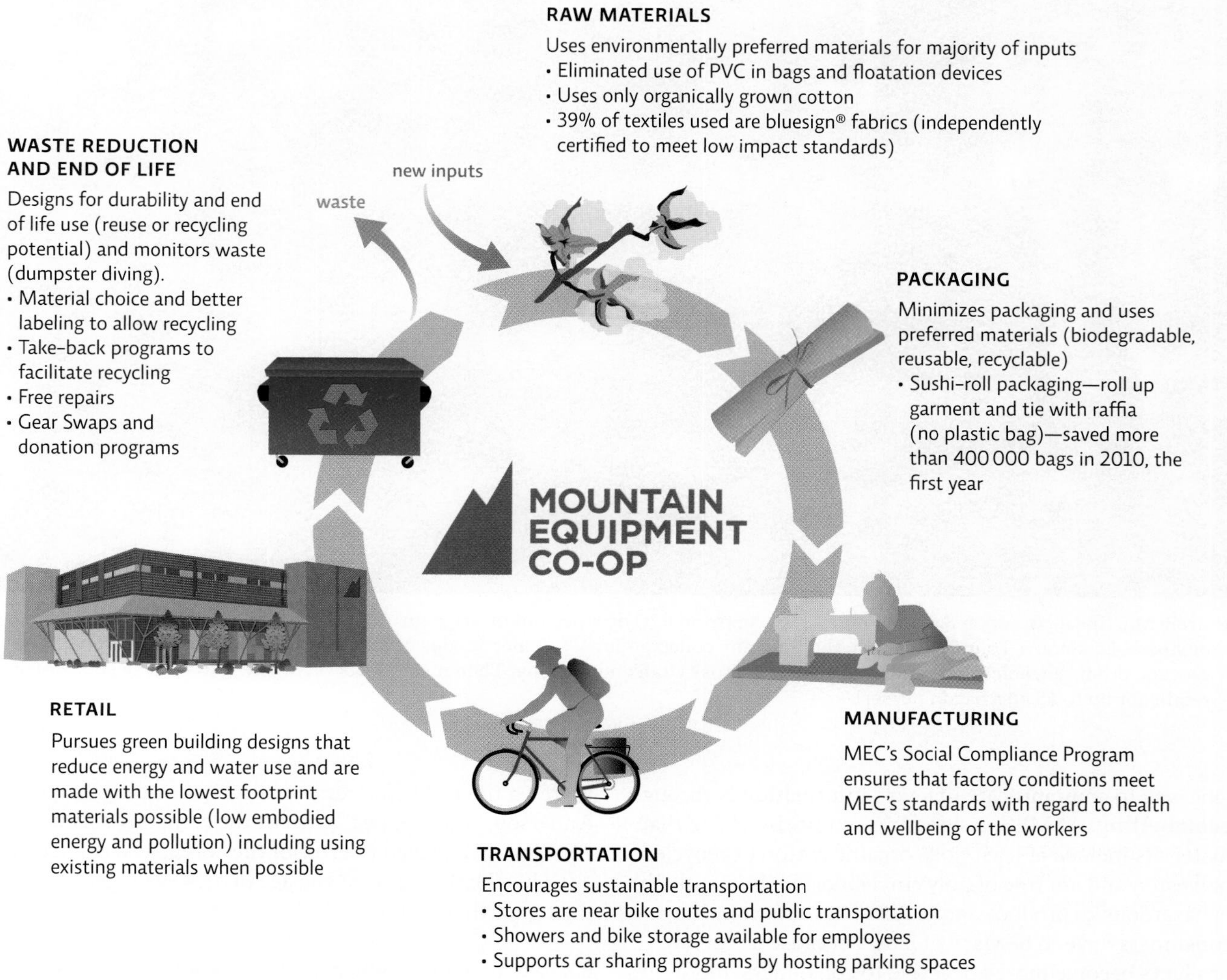
RAW MATERIALS
Uses environmentally preferred materials for majority of inputs
• Eliminated use of PVC in bags and floatation devices
• Uses only organically grown cotton
• 39% of textiles used are bluesign® fabrics (independently certified to meet low impact standards)
WASTE REDUCTION AND END OF LIFE
Designs for durability and end of life use (reuse or recycling potential) and monitors waste (dumpster diving).
• Material choice and better labeling to allow recycling
• Take-back programs to facilitate recycling
• Free repairs
• Gear Swaps and donation programs
new inputs
waste
PACKAGING
Minimizes packaging and uses preferred materials (biodegradable, reusable, recyclable)
• Sushi-roll packaging—roll up garment and tie with raffia (no plastic bag)—saved more than 400 000 bags in 2010, the first year
MOUNTAIN EQUIPMENT CO-OP
RETAIL
Pursues green building designs that reduce energy and water use and are made with the lowest footprint materials possible (low embodied energy and pollution) including using existing materials when possible
MANUFACTURING
MEC's Social Compliance Program ensures that factory conditions meet MEC's standards with regard to health and wellbeing of the workers
TRANSPORTATION
Encourages sustainable transportation
• Stores are near bike routes and public transportation
• Showers and bike storage available for employees
• Supports car sharing programs by hosting parking spaces

damage in the first place. A company that adopts cradle-to-cradle resource management is responsible for the resource or the impact of its use at every stage of the process. It can do this by considering not just the immediate impact of using a product but also the upstream and downstream impacts—a *life-cycle analysis*. This can lead to better material choices (less toxic, more sustainable) and better process choices (reusable materials, less waste and pollution). "We're trying to understand the system that creates damage—from the chemicals, manufacturing process, to packaging—to understand where our organizational impacts lie," says Labistour. MEC is now collaborating with similar businesses to establish common environmental standards they all follow, such as best practices to reduce the use of hazardous chemicals to make textiles. The company is also a supporter of bluesign®, an independent auditor of textile mills that helps facilities reduce their harm to the environment. [INFOGRAPHIC 5.8]

↑ Craig Martineau, Brandon Sargent, and Dan Blake (from left) dropped out of Brigham Young University to pursue their green business, EcoScraps. Founded in 2010, the company collects roughly 18 metric tons of food waste a day from more than 70 grocers, produce wholesalers, and Costco stores across Utah and Arizona. Then it composts the waste into potting soil, which retails for up to $8.50 a bag in nurseries.

One way to communicate lifecycle information is through **ecolabelling**. At MEC, a signature symbol indicates that materials include at least 50% organic cotton or recycled polyester, and are free of polyvinyl chloride (PVC), which is hazardous to produce and is difficult to recycle. But consumers have to be wary of labels because as "green" products become more attractive to consumers, more companies engage in *greenwashing*—claiming environmental benefits for a product when they are minor or nonexistent. *Fair trade* items are, however, more likely to be sustainably produced. In addition, for a product to be considered fair trade, workers must be paid a fair wage and work in reasonable conditions to produce the goods or services.

In June 2011, Interface began producing its first 100% non-virgin fibre carpet tiles, made from reclaimed carpet, fibre derived from salvaged commercial fishnets, and post-industrial waste. It remains the world's leading manufacturer of commercial carpet tiles. "I think we're on the right track and we'll keep on going," says Anderson, who stepped down as the company's CEO in 2001 but still played the role of the company's conscience until his death in 2011 at the age of 77. "We'll get to the top of that mountain."

It's easy to wonder why companies would go to so much trouble to reduce their environmental impact, when competitors who avoid the time and effort may be able to offer the same goods at lower prices. But to Labistour, reducing waste and pollution isn't just good for the environment—it's good for business. As an outdoor equipment company, MEC needs healthy ecosystems for co-op members to enjoy its products, he says. And resources are already becoming scarce, he adds, so

ecolabelling Providing information about how a product is made and where it comes from. Allows consumers to make more sustainable choices and support sustainable products and the businesses that produce them.

companies have to learn to use them wisely so that they can continue operating when shortages occur. "It's only a matter of time before every organization is going to have to pay far more attention to the social and environmental contexts in which they operate," says Labistour. "We focus on sustainability so the company can continue operating as it is." The fact that this focus also helps preserve natural resources for future generations is a huge bonus, he adds. "The reason I am doing these things is both for the future of my organization, and my children."◉

Select references in this chapter:
Constanza, R., *et al.* 1997. *Nature*, 387: 253–260.
World Wildlife Fund. 2012. *2012 Living Planet Report* [Online], wwf.panda.org/about_our_earth/all_publications/living_planet_report/2012_lpr/.

↑ Steve Ells, the founder, Co-CEO, and Chairman of the Chipotle Mexican Grill is committed to doing business in a way that is good for people and the environment, the definition of a Green Business. The company's environmental efforts include using organic vegetables and ethically raised meat as much as possible and constructing new restaurant buildings to be energy and water efficient.

BRING IT HOME

PERSONAL CHOICES THAT HELP

You have an impact on creating a sustainable society. Every time you buy a product or service, you are telling the manufacturer that you agree with the principles behind the product. You can use your purchasing power to show companies that people are interested in quality products that support environmental and social values.

Individual Steps

→ Reduce the amount of stuff you accumulate by buying fewer items and by choosing products that are well made and last longer.

→ Use the Good Guide app on your smartphone to scan product bar codes and see how the products rank on different scales of environmental impact, social responsibility, and health.

→ Ask your local food store or pharmacy to stock fair trade certified products if they don't already.

→ Instead of buying a new or used car, join a car-share program like Zipcar.

Group Action

→ Get together with family and friends and write a letter to your favourite companies, asking them to reduce their ecological footprint. You can ask them to become more transparent by publishing how their business practices impact the environment.

Policy Change

→ Start a blog or Facebook page to chronicle the changes you make in your buying habits and encourage others to do the same. Discuss the companies whose environmental policies you agree with.

World Salts & Sea Vegetables
Organic Spices

UNDERSTANDING THE ISSUE

CHECK YOUR UNDERSTANDING

1. **Water would be an example of _______, while the water cycle would be an example of _______.**
   a. an ecological footprint; the IPAT model
   b. an ecosystem good; an ecosystem service
   c. natural capital; natural interest
   d. an external cost; an internal cost

2. **The land needed to provide the resources for and assimilate the waste of a person or population is referred to as:**
   a. an ecological footprint.
   b. natural interest.
   c. sustainable development.
   d. true cost accounting.

3. **What does the term "cradle-to-cradle" mean when talking about product management?**
   a. Product materials must be tracked from production to disposal.
   b. The product is potentially more dangerous to children.
   c. Current legislation is too restrictive on new product development and causes the early demise of new businesses.
   d. Production is cyclical: "waste" becomes the raw material once again and can be reused.

4. **"Natural interest" refers to:**
   a. activities of human society that can be continued without degrading the environment.
   b. the rules that deal with how we allocate scarce resources.
   c. essential ecological processes that make life on Earth possible.
   d. readily produced resources that can be used and still leave enough natural capital behind to replace what we took.

5. **Which of the following statements about sustainability is FALSE?**
   a. We could make all our industrial processes sustainable if we could transform linear processes into circular ones.
   b. The Canadian rate of consumption is not sustainable; if the world population consumed as much as the average Canadian citizen, we would need more than three Earths.
   c. In its current form, mainstream economics is the optimal model for building sustainability because external costs are built in.
   d. Technology can be used to promote sustainability and decrease human environmental impact.

WORK WITH IDEAS

1. What is the IPAT model? How is the equation $I = P \times A \times T$ similar to and/or different from the equation $I = (P \times A)/T$?

2. What are the differences between internal and external costs? How do these types of costs relate to true cost accounting?

3. What are the limitations of mainstream economics that make it an unsustainable model? How does ecological economics propose to address these limitations?

4. Compare the sustainability efforts of Interface to those of MEC. What similar steps is each company taking? How do their efforts differ?

5. How did Interface incorporate biomimicry designs into their products and what are the advantages of these designs over traditional products?

ANALYZING THE SCIENCE

China and the United States are two of the world's largest consumers of resources due to sheer population size and per capita consumption rates. The figures on the following page come from the Earth Policy Institute's article "Learning from China: Why the Existing Economic Model Will Fail."

INTERPRETATION

1. What do these figures show? How many different fuel types are included in the figures?

2. How do China and the United States compare in their use of coal and oil in 2010? Use the data to explain your responses.

3. What is the projected oil consumption for China in 2035 and how does this compare to what is projected for the United States? What is the potential percentage change in oil consumption for each country? How does this trend compare to coal consumption for the two countries?

ADVANCE YOUR THINKING

Hint: Access the actual data highlight at www.earth-policy.org/data_highlights/2011/highlights18.

4. China's ecological footprint is much smaller than that of the United States. What explains the trend in coal consumption in China and the fact that it exceeds that of the United States today? How might China and the United States compare using the IPAT model?

5. The world produces 86 million barrels of oil a day. What percent of this oil production did the United States consume in 2010? China? How does this translate into per capita consumption, if the current population of China is 1.3 billion and that of the United States is 300 million? Assuming that we hold to this level of oil production, what proportion of it will China need in 2035? What about the United States? How does this translate in per capita terms, assuming that population size in China and the United States does not change significantly? Discuss the environmental consequences in terms of the IPAT model and the ecological footprint.

6. Except for oil, among the key commodities—grain, meat, coal, oil, and steel—China consumes more of each than the United States. For instance, China currently consumes twice as much meat and four times as much steel as the United States. Considering the key concepts of the chapter, what are the lessons to be learned from China's economic growth? What are some potential solutions? Discuss in terms of the IPAT model, ecological footprints, and economic models.

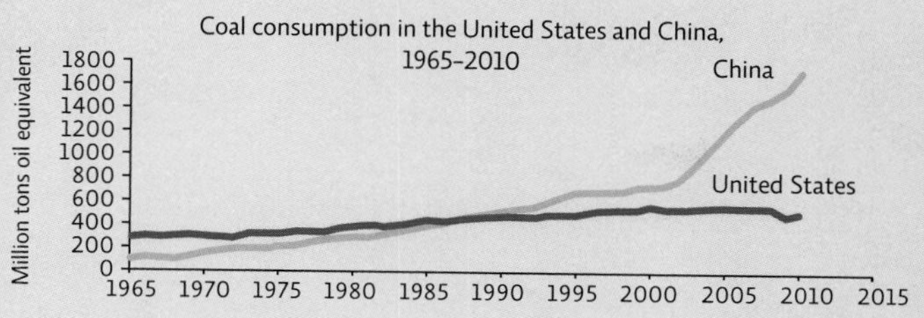
Coal consumption in the United States and China, 1965–2010
Million tons oil equivalent
1800
1600
1400
1200
1000
800
600
400
200
0
1965 1970 1975 1980 1985 1990 1995 2000 2005 2010 2015
China
United States

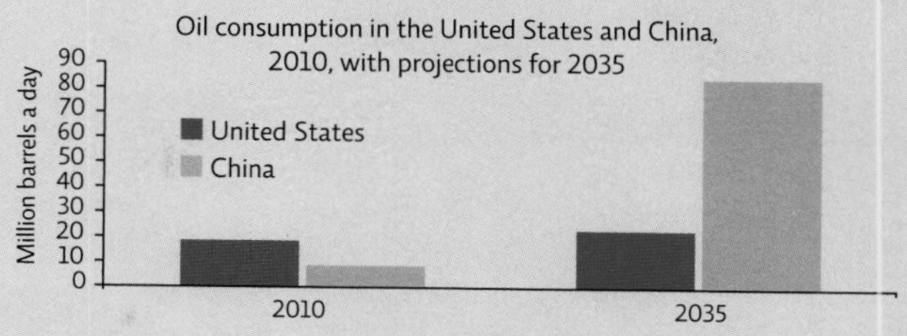
Oil consumption in the United States and China, 2010, with projections for 2035
Million barrels a day
90
80
70
60
50
40
30
20
10
0
United States
China
2010
2035

EVALUATING NEW INFORMATION

As consumers, we can rely on ecolabelling to some extent, but we need to remain vigilant to greenwashing, given that there are no clear labelling guidelines. One helpful source is the Good Guide, which has ratings for over 120 000 consumer products. Each product is given a summary score on a scale of 0–10, which is compiled from three sub-scores that address the product's health, environmental, and social impacts.

Explore the Good Guide website (www.goodguide.com).

Evaluate the website and work with the information to answer the following questions:

1. Is this a reliable information source? Does it have a clear and transparent agenda?
   a. Who runs this website? Do the credentials of the individual or organization make the information presented reliable/unreliable? Explain.
   b. What is the mission of this website? What are its underlying values? How do you know this?
   c. What data sources does Good Guide rely on and what methodology does it employ in calculating its rating? Are its sources reliable?
   d. Do you agree with Good Guide's assessment of the problems with and concerns about consumer products? Explain.
   e. Do you agree with Good Guide's solutions (e.g., its rating system)? Do you think the criteria it uses for rating products are sufficient and reasonable? Which criteria are most important to you as a consumer? Explain.

2. Select one of your favourite products that is also on the Good Guide website.
   a. How is your product rated by the Good Guide? Discuss both the overall score as well as the details of the three sub-scores.
   b. Check out the company website for your product. What sort of information about the product does the company website offer? How does it compare to the information on the Good Guide? Which information source is more useful to you as a consumer? Explain your responses.
   c. On the Good Guide website, search for alternative brands that have a better rating than your preferred product. Are there other choices? Would you consider switching your brand? Explain.

MAKING CONNECTIONS

THE EARTH CHARTER: A VISION FOR THE FUTURE

Background: The Earth Charter Initiative is an international effort whose stated mission is "to promote the transition to sustainable ways of living and a global society founded on a shared ethical framework that includes respect and care for the community of life, ecological integrity, universal human rights, respect for diversity, economic justice, democracy, and a culture of peace." The Earth Charter document (download it at www.earthcharterinaction.org) that is the basis of this initiative is promoted as a global consensus statement on what sustainability should mean and the principles by which sustainable development should be achieved at all levels of society.

Case: You have been invited to participate in the Earth Charter Initiative—that is, you are being asked to encourage people in your school, community, or workplace to endorse the Earth Charter and to apply its principles as part of their daily operations so they can join in the effort toward developing a sustainable society.

Write a letter to the people in your school, community, or workplace to describe the Earth Charter and explain why they should become part of this initiative.

1. In your letter include the following:
   a. An assessment of the importance of sustainability and the challenges to achieving it.
   b. An analysis of the Earth Charter as a framework for reaching sustainability—that is, how can the Earth Charter be used to make choices that take all the costs (economic, social, and environmental) of a given action into account and allow a school, community, or workplace to strike a balance between the current and future well-being of humans and the environment?

2. Propose some specific sustainability initiatives that your school, community, or workplace could engage in that reflect the Earth Charter principles.

3. Identify the potential challenges to implementing the Earth Charter and discuss how these challenges could be addressed.

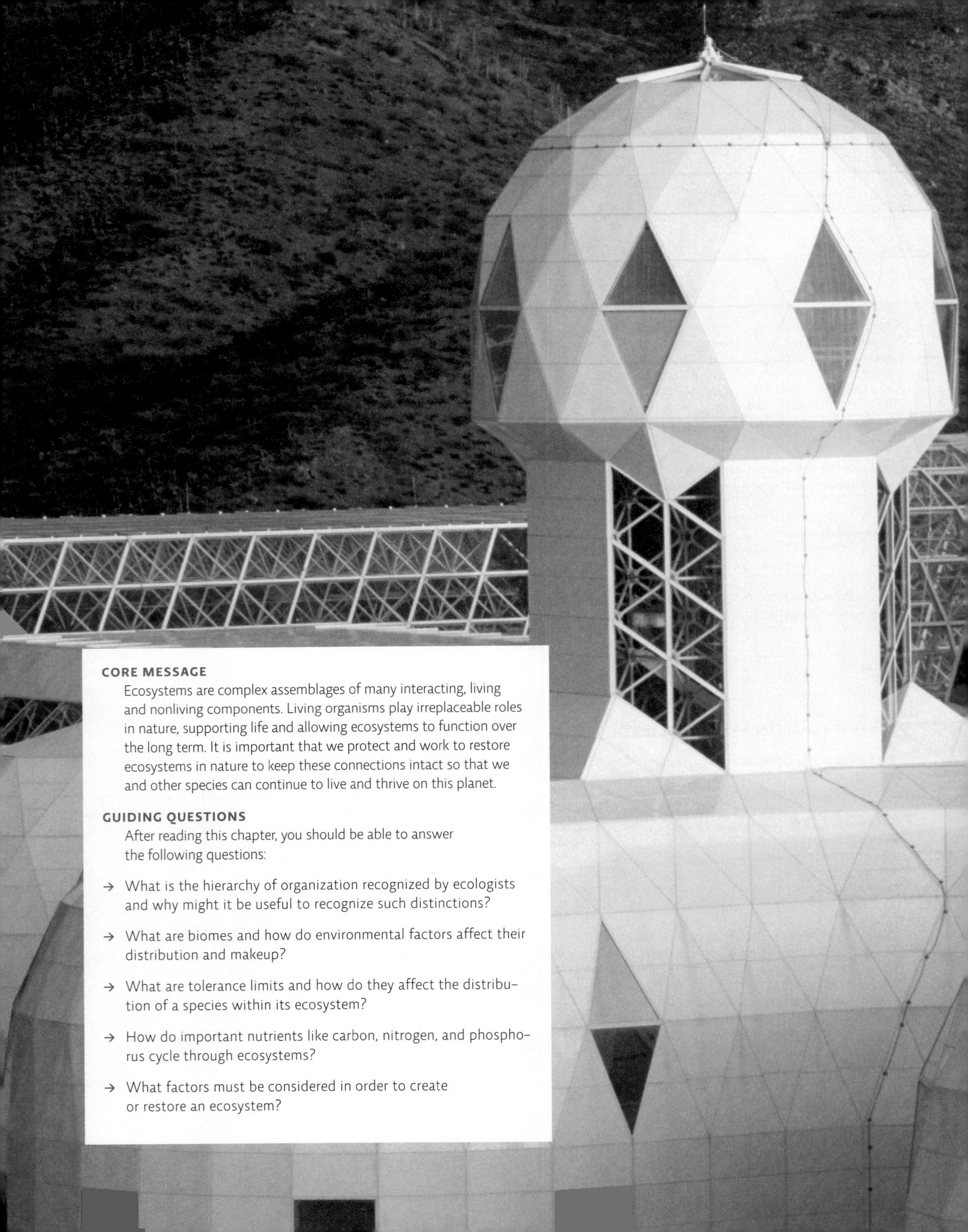

CORE MESSAGE

Ecosystems are complex assemblages of many interacting, living and nonliving components. Living organisms play irreplaceable roles in nature, supporting life and allowing ecosystems to function over the long term. It is important that we protect and work to restore ecosystems in nature to keep these connections intact so that we and other species can continue to live and thrive on this planet.

GUIDING QUESTIONS

After reading this chapter, you should be able to answer the following questions:

→ What is the hierarchy of organization recognized by ecologists and why might it be useful to recognize such distinctions?

→ What are biomes and how do environmental factors affect their distribution and makeup?

→ What are tolerance limits and how do they affect the distribution of a species within its ecosystem?

→ How do important nutrients like carbon, nitrogen, and phosphorus cycle through ecosystems?

→ What factors must be considered in order to create or restore an ecosystem?

CHAPTER 6 ECOSYSTEMS AND NUTRIENT CYCLING

ENGINEERING EARTH

A desert experiment gone awry

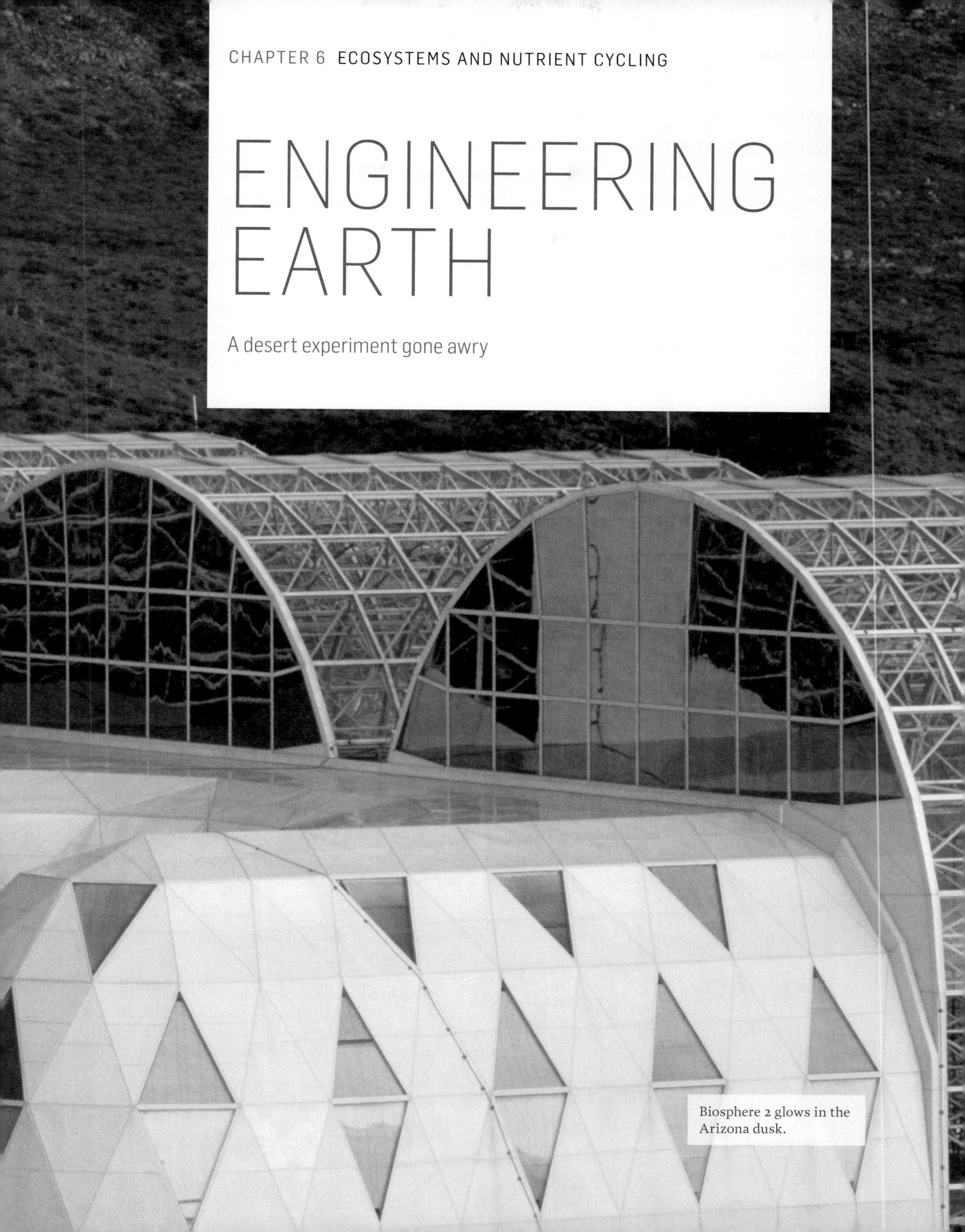

Biosphere 2 glows in the Arizona dusk.

On September 26, 1993, with their first mission complete, four men and four women emerged from Biosphere 2—a hulking dome of custom-made glass and steel—back into the Arizona desert, where throngs of spectators stood cheering. They had been sealed inside the facility, along with 3000 other plant and animal species, for exactly 2 years and 20 minutes; it was the longest anyone had ever survived in an enclosed structure.

The feat was part of a grand experiment, the goals of which were twofold. First, scientists wanted to prove that an entirely self-contained, humanmade system—the kind they might one day use to colonize the Moon or Mars—could sustain life. Second, they hoped that by studying this mini-earth, which could be controlled and manipulated in ways the real Earth could not, they might better understand our own planet's delicate balance and how best to protect it.

Despite the fanfare surrounding the biospherians' emergence, it was tough to say whether the mission had been a success or a failure. More than one-third of the flora and fauna had become extinct, including most of the vertebrates and all of the pollinating insects. Morning glory vines had overrun other plants, including food crops. Cockroaches and "crazy ants" were thriving. Too little wind had prevented trees from developing stress wood—wood that grows in response to mechanical stress and helps trunks and branches shift into an optimal position; without stress wood, the trees were brittle and prone to collapse. And too many sweet potatoes had turned the biospherians themselves bright orange (a string of plant diseases had decimated other crops).

◉WHERE IS BIOSPHERE 2?

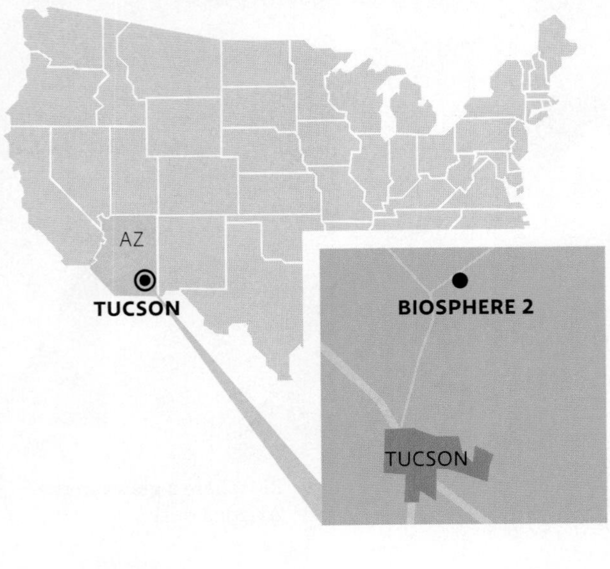
AZ
TUCSON
BIOSPHERE 2
TUCSON

On top of that, nitrous oxide (laughing gas) had grown concentrated enough to "reduce vitamin B12 synthesis to a level that could impair or damage the brain," according to one interim report. And oxygen levels had plummeted from 21% (roughly the same as Earth's atmosphere) to 14% (just barely enough to sustain human life). To fix this, project engineers had been forced to pump in 17 000 cubic metres of outside air, violating the facility's sanctity as a closed system.

Worst of all, missteps and course corrections had been mired in secrecy—each one leaked to the press only months after the fact. Rumours had begun to circulate that the eight people sealed inside—not to mention the ones they took their orders from—were more interested in creating a futuristic utopia than in conducting rigorous scientific research. As evidence for this theory mounted, the scientific community grew suspicious. Was Biosphere 2 legitimate science, a publicity stunt, or some bizarre mix of the two?

To be sure, the eight biospherians had survived, and many experts agreed that, in principle at least, the facility still held enormous potential as a scientific tool. But before that potential could be realized, the scientific community and the public at large would need to know exactly what had happened inside the desert dome.

To answer that question, we need to answer a few others first: what exactly is a biosphere, and just how did Biosphere 2's creators set about building one?

Organisms and their habitats form complex systems.

The term **biosphere** refers to the total area on Earth where living things are found—the sum total of all its ecosystems. An **ecosystem** includes all the organisms in a given area plus the nonliving components of the

biosphere The sum total of all of Earth's ecosystems.
ecosystem All of the organisms in a given area plus the physical environment in which, and with which, they interact.

← The eight biospherians emerge from Biosphere 2 after living there for two years.

↓ Visitors can tour the facility. Here they view the desert biome.

Infographic **6.1** | **ORGANIZATION OF LIFE: FROM BIOSPHERE TO INDIVIDUAL**

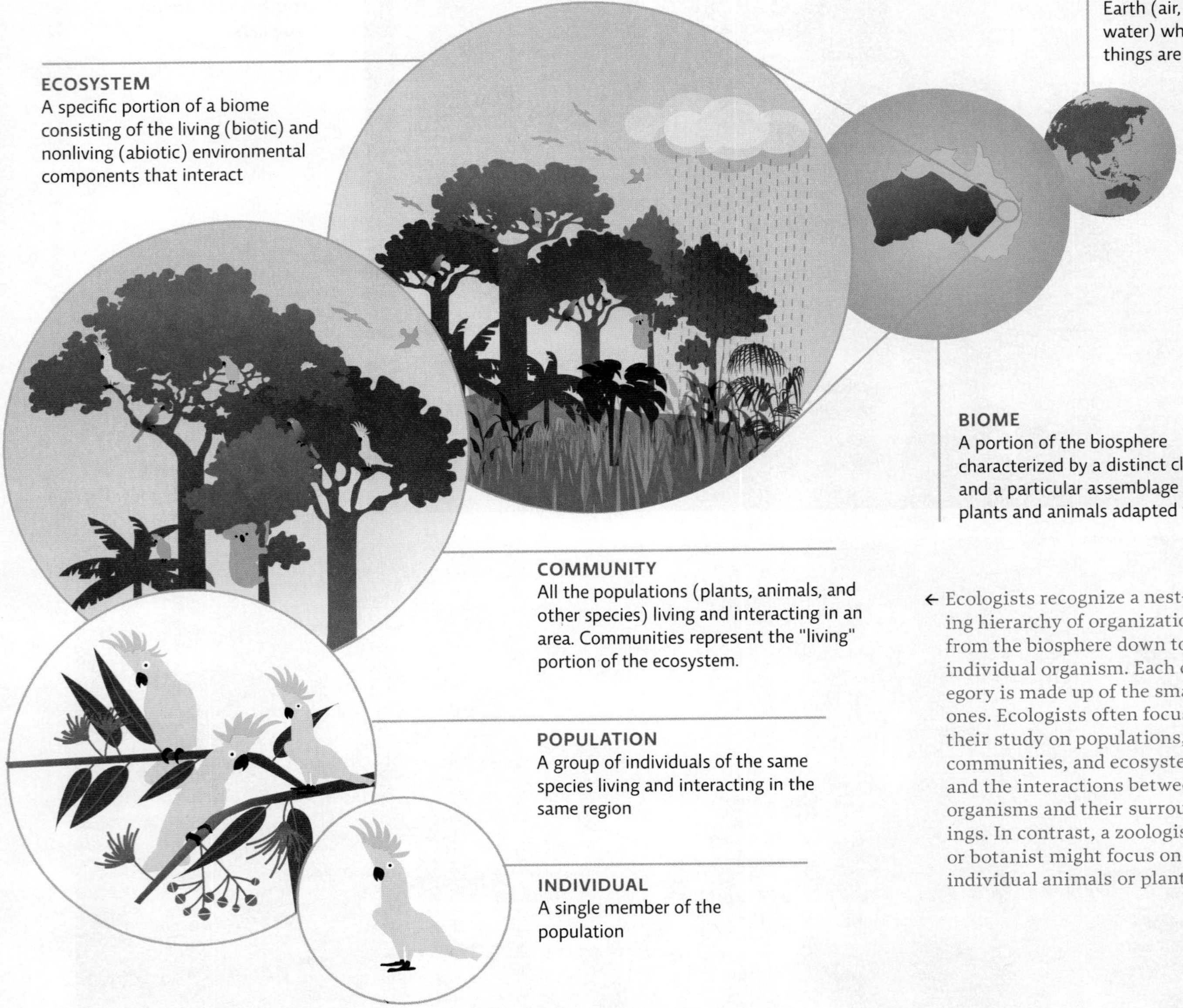
BIOSPHERE
The total area on Earth (air, land, or water) where living things are found
ECOSYSTEM
A specific portion of a biome consisting of the living (biotic) and nonliving (abiotic) environmental components that interact
BIOME
A portion of the biosphere characterized by a distinct climate and a particular assemblage of plants and animals adapted to it
COMMUNITY
All the populations (plants, animals, and other species) living and interacting in an area. Communities represent the "living" portion of the ecosystem.
POPULATION
A group of individuals of the same species living and interacting in the same region
INDIVIDUAL
A single member of the population

← Ecologists recognize a nesting hierarchy of organization from the biosphere down to the individual organism. Each category is made up of the smaller ones. Ecologists often focus their study on populations, communities, and ecosystems, and the interactions between organisms and their surroundings. In contrast, a zoologist or botanist might focus on individual animals or plants.

Infographic **6.2** | **HABITAT AND NICHE** ↓ Species depend on suitable habitats in which to live. Species fill a specific niche in their community.

SPECIES
A group of plants or animals that have a high degree of similarity and can generally only interbreed among themselves
HABITAT
The physical environment in which individuals of a particular species can be found
NICHE
The role a species plays in its community, including how it gets its energy and nutrients, its habitat requirements, and what other species and parts of the ecosystem it interacts with

physical environment in which they interact. This physical environment is often referred to as an organism's **habitat**. Ecologists study how ecosystems function by focusing on **species**' interactions with their surroundings and with other species in their communities. They also study interactions between members of the same species within a population. Individuals of each species in a community occupy a specific ecological **niche** shared by no other species in that community. In the natural world, ecosystems assume a range of shapes and sizes—a single, simple tide pool qualifies as an ecosystem; so does the entire Mojave Desert. [INFOGRAPHIC 6.1, 6.2]

All ecosystems function through two fundamental processes that are collectively referred to as ecosystem processes, namely nutrient cycling and **energy flow**. **Nutrient cycles** are biogeochemical cycles that refer specifically to the movement of life's essential chemicals or nutrients through an ecosystem. Energy, on the other hand, enters ecosystems as solar radiation and is passed along from organism to organism, some released as heat, until there is no more usable energy left. Therefore we can say that matter *cycles* but energy *flows* in a one-way trip.

Earth—or "Biosphere 1," as the creators of Biosphere 2 liked to call it—is materially closed but energetically open. [INFOGRAPHIC 6.3] In other words, the plants and other organic material that make up an ecosystem, called **biomass**, cannot enter or leave the system, but energy can: some leaves as heat or light and new energy is absorbed from outside. In fact, plant biomass is produced with energy from the Sun through photosynthesis.

Biomes are specific portions of the biosphere determined by climate and identified by the predominant vegetation and organisms adapted to live there. Biomes can be divided into three broad categories—marine, freshwater, and terrestrial. Within those three categories are several narrower groups, and within those, a variety of subgroups. An entire biome itself may be considered an ecosystem, as are the smaller groups and subgroups. For example, forests, deserts, and grasslands are the three main types of terrestrial biomes. Within the forest biome category are different types of forests, such as tropical, temperate, or boreal forest, and within each of those groups are subgroups (for example, dry tropical forest and tropical rain forest.) [INFOGRAPHIC 6.4]

Infographic **6.3** | **EARTH IS A CLOSED SYSTEM FOR MATTER BUT NOT FOR ENERGY**

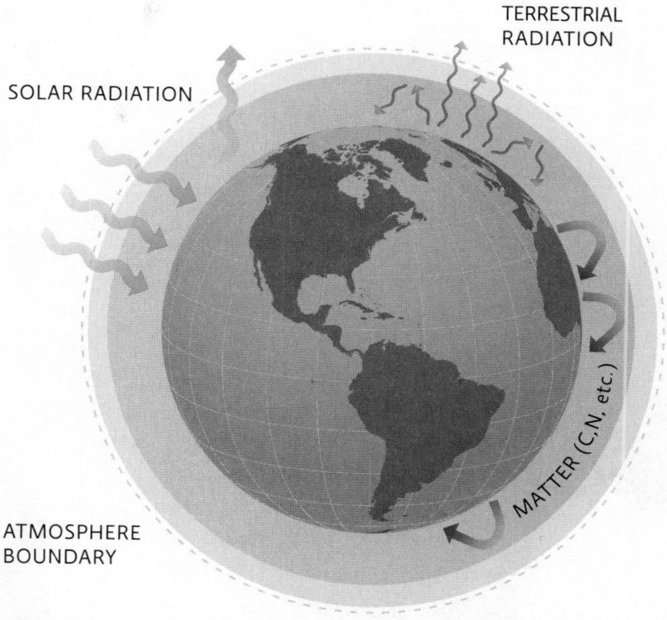
TERRESTRIAL RADIATION
SOLAR RADIATION
MATTER (C,N, etc.)
ATMOSPHERE BOUNDARY

↑ Energy can enter and leave Earth as light (solar radiation) and heat (terrestrial radiation) but matter stays in the biosphere, cycling in and out of organisms and environmental components.

When ecologists study entire ecosystems, they are limited to making observations and trying to discern cause and effect from those observations. This is no small challenge; even the simplest phenomenon is impacted by a myriad of factors. Precise, systemwide measurements are exceedingly difficult to come by. And unlike laboratory science, field research doesn't often accommodate rigorous controls. Exceptions include whole-ecosystem research, such as that conducted at Canada's Experimental Lakes Area.

Biosphere 2 would offer ecologists an unprecedented research tool: a mini-planet where a variety of environmental variables—from temperature and water availability to the relative proportions of oxygen and carbon dioxide (CO_2) at any given moment—could be tightly controlled and precisely measured. "Manipulating

habitat The physical environment in which individuals of a particular species can be found.
species A group of plants or animals that have a high degree of similarity and can generally only interbreed among themselves.
niche The role a species plays in its community, including how it gets its energy and nutrients, what habitat requirements it has, and what other species and parts of the ecosystem it interacts with.
energy flow The one-way passage of energy through an ecosystem.
nutrient cycles Movement of life's essential chemicals or nutrients through an ecosystem (they are key examples of biogeochemical cycles).
biomass The sum of all organic material—plant and animal matter—that make up an ecosystem.
biome One of many distinctive types of ecosystems determined by climate and identified by the predominant vegetation and organisms that have adapted to live there.

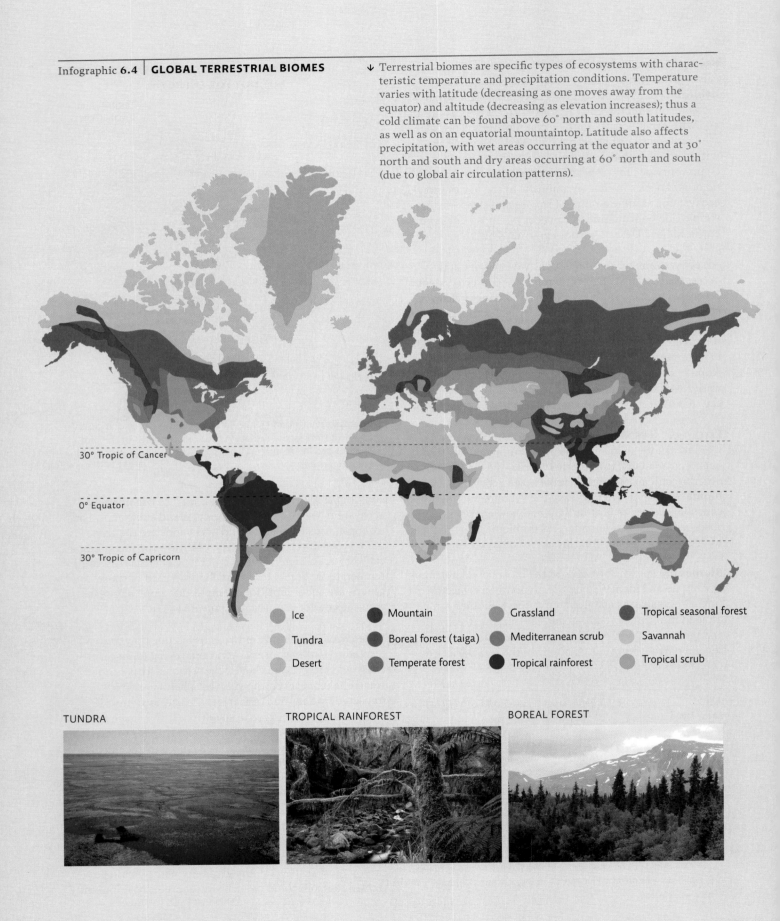

Infographic **6.4** | **GLOBAL TERRESTRIAL BIOMES**

↓ Terrestrial biomes are specific types of ecosystems with characteristic temperature and precipitation conditions. Temperature varies with latitude (decreasing as one moves away from the equator) and altitude (decreasing as elevation increases); thus a cold climate can be found above 60° north and south latitudes, as well as on an equatorial mountaintop. Latitude also affects precipitation, with wet areas occurring at the equator and at 30° north and south and dry areas occurring at 60° north and south (due to global air circulation patterns).

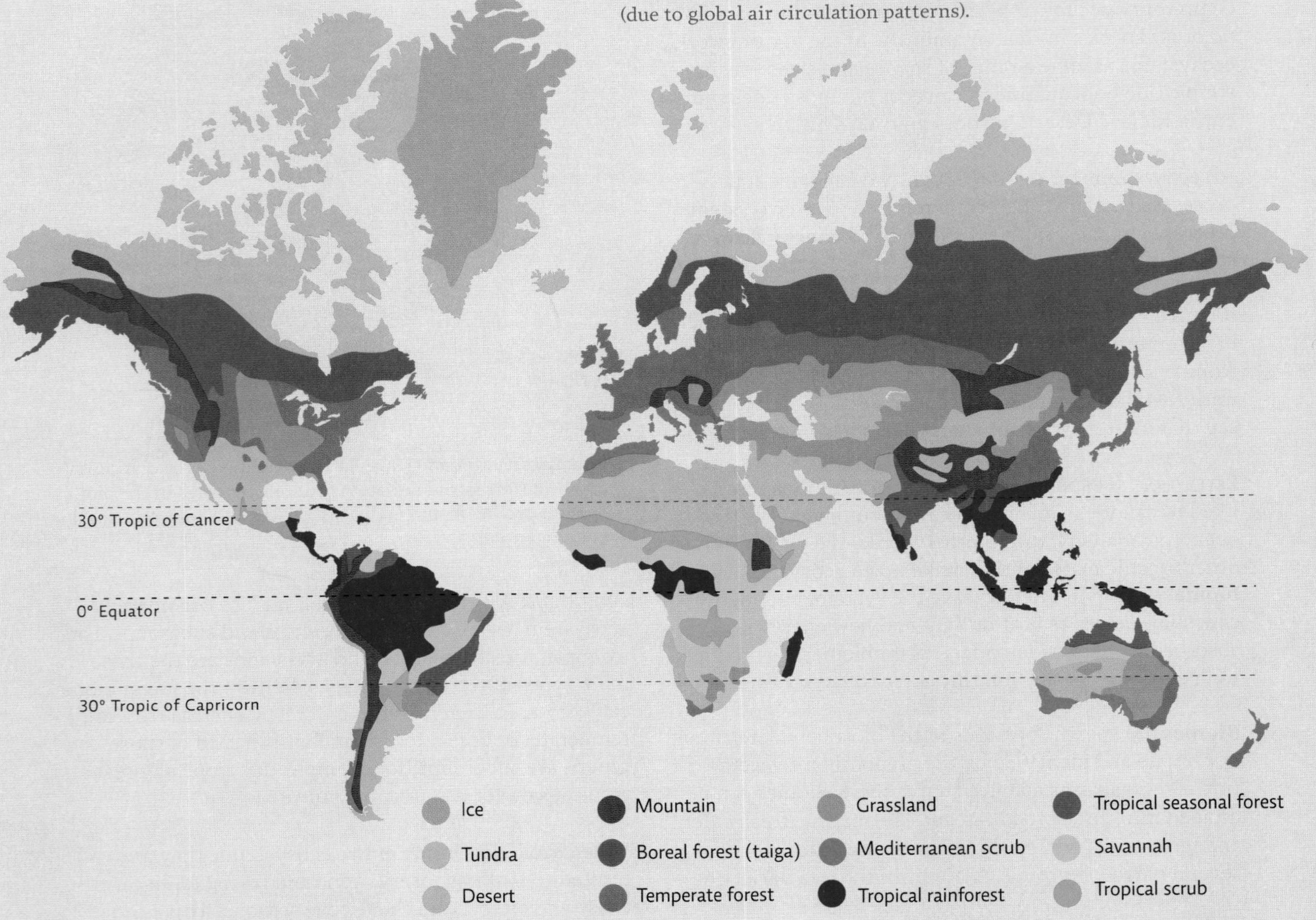
30° Tropic of Cancer
0° Equator
30° Tropic of Capricorn
Ice
Tundra
Desert
Mountain
Boreal forest (taiga)
Temperate forest
Grassland
Mediterranean scrub
Tropical rainforest
Tropical seasonal forest
Savannah
Tropical scrub

TUNDRA

TROPICAL RAINFOREST

BOREAL FOREST

↓ This biome climograph shows the approximate distribution of terrestrial biomes with regard to annual precipitation and temperature. As precipitation increases and more plant life is able to be supported, deserts, scrublands, or grasslands give way to forests. The temperature also influences what type of desert, grassland, or forest is present (for example, temperate forest versus boreal forest).

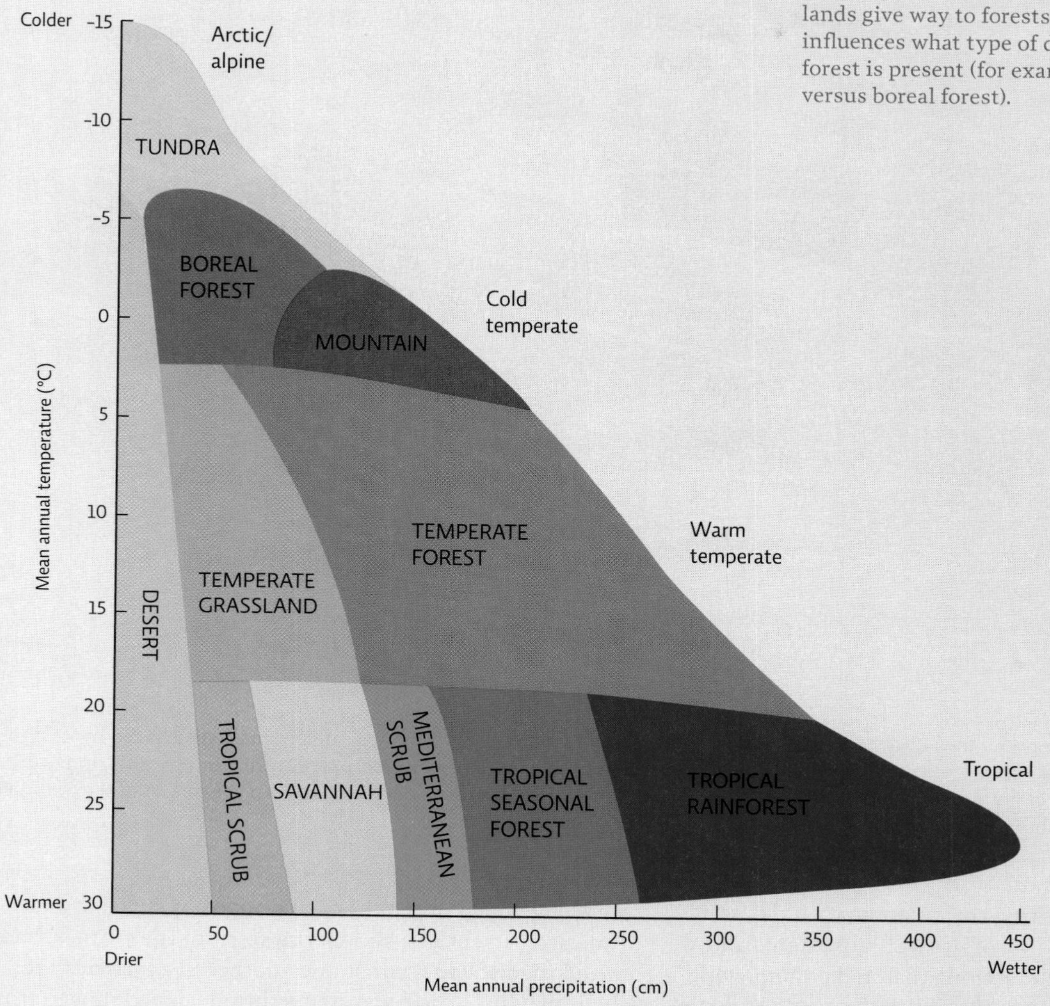

Colder
Arctic/ alpine
TUNDRA
BOREAL FOREST
MOUNTAIN
Cold temperate
TEMPERATE FOREST
Warm temperate
TEMPERATE GRASSLAND
DESERT
TROPICAL SCRUB
SAVANNAH
MEDITERRANEAN SCRUB
TROPICAL SEASONAL FOREST
TROPICAL RAINFOREST
Tropical
Warmer
Mean annual temperature (°C)
-15 -10 -5 0 5 10 15 20 25 30
0 50 100 150 200 250 300 350 400 450
Drier
Wetter
Mean annual precipitation (cm)

MEDITERRANEAN SCRUB

DESERT

SAVANNAH

Infographic **6.5** | **MAP OF BIOSPHERE 2**

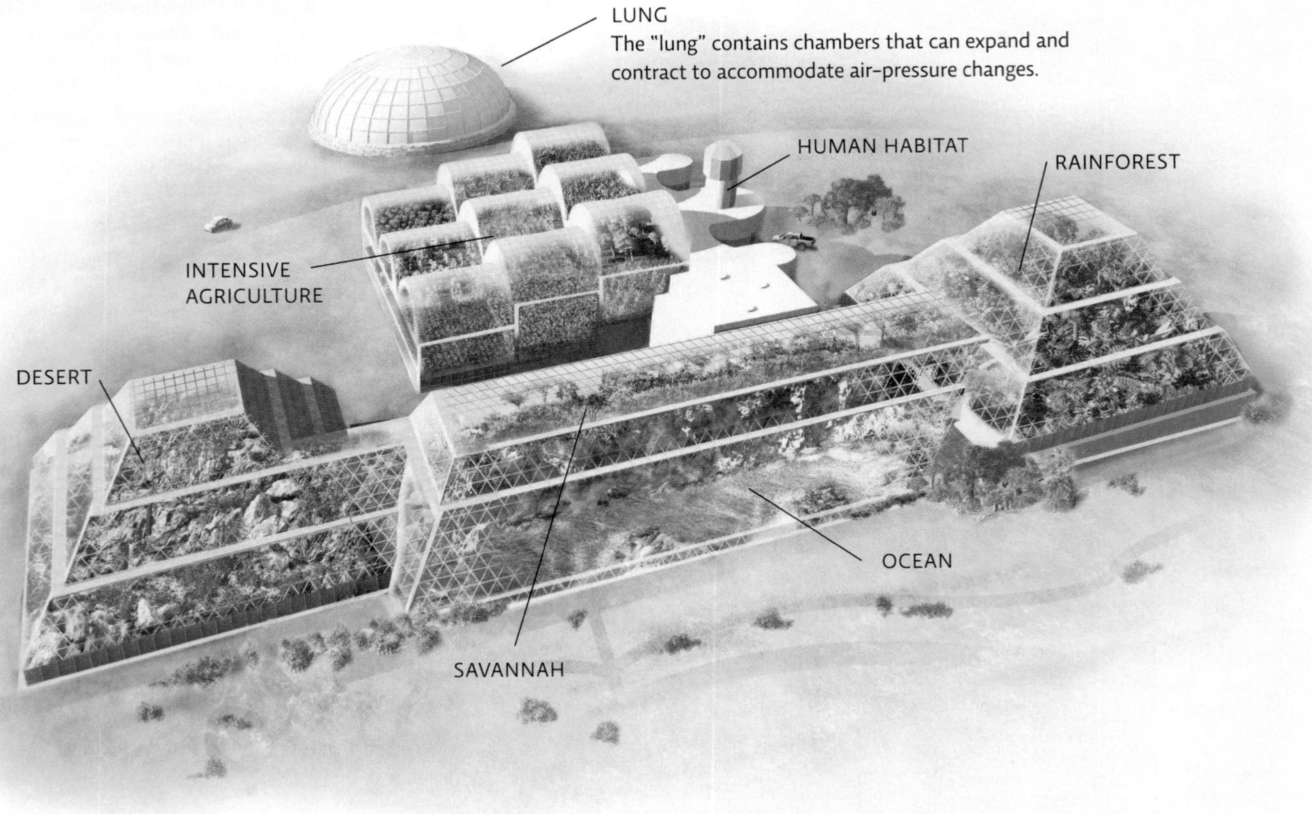
LUNG
The "lung" contains chambers that can expand and contract to accommodate air-pressure changes.
HUMAN HABITAT
RAINFOREST
INTENSIVE AGRICULTURE
DESERT
OCEAN
SAVANNAH

↑ Biosphere 2 houses several biomes under one roof, each contributing to overall function. One of the challenges faced by designers was how to include a variety of biomes in the close quarters of the 1.2 hectare Biosphere 2 structure. For example, in nature, a tropical rainforest would not be next to an arid desert. To deal with this, an ocean was placed between the desert and rainforest to serve as a temperature buffer.

these variables and tracking the outcomes could greatly advance our understanding of natural ecosystems and all the minute, complex interactions that make them work," says Kevin Griffin, a Columbia University plant ecologist who conducted research at the Biosphere 2 facility. "The plan was to use that knowledge to figure out how to repair degraded ecosystems in the real world, so that they continue to provide the services so essential to our survival."

On top of all that, proving humans could survive in a completely enclosed, manufactured system would take us one giant step closer toward colonizing space.

But would it work?

The concept of enclosed ecosystems was not a new one. Since before they put a person on the moon, astronauts and engineers had been tinkering with their own artificial ecosystems—systems they hoped could one day be used to colonize space. The earliest versions of this technology were developed in the 1960s and 1970s by Russian and American scientists who, in the spirit of their times, had pitted themselves against one another in a mad dash to the finish line. The ecosystems they designed ranged from small, crude structures in which a single person might last for a single day, to larger, more sophisticated enterprises that could sustain a few people for a few months.

Biosphere 2 is by far the most elaborate. At just under 13 000 square metres, it remains the largest enclosed ecosystem ever created and the only one to house several biomes under one roof. [INFOGRAPHIC 6.5] From a mountain under the dome's 28-metre zenith, a stream rushes down through a tropical rainforest before snaking southward into a savannah. From there, the stream wends its way through a mangrove swamp into a 3.8 million-litre ocean complete with a coral reef. On the other side of the ocean lies a desert. Biosphere 2 also includes a human habitat and an agricultural biome.

Infographic 6.6 | **RANGE OF TOLERANCE FOR LIFE**

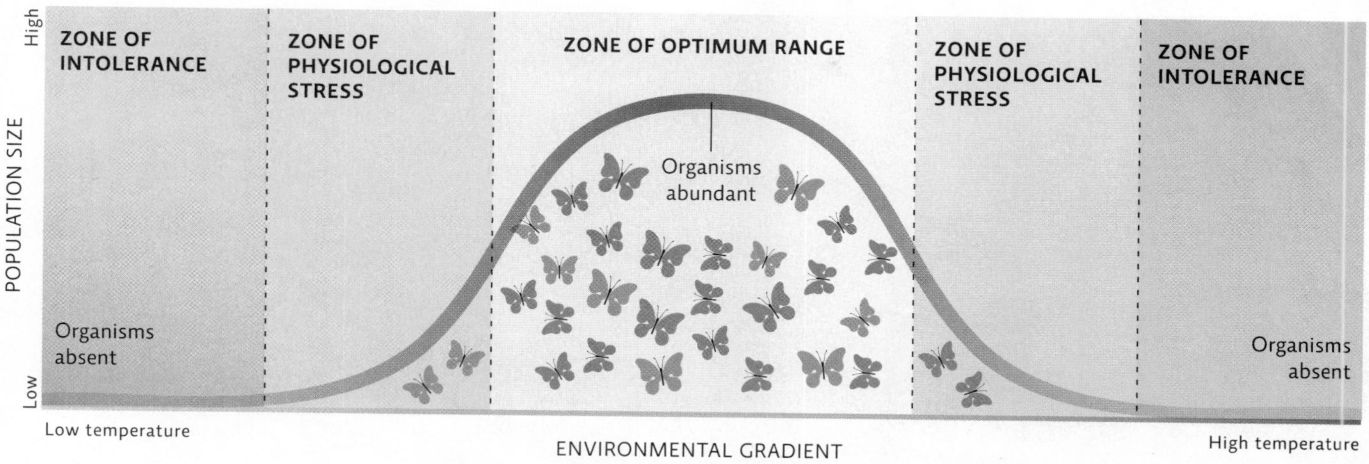
POPULATION SIZE
High
Low
ZONE OF INTOLERANCE
ZONE OF PHYSIOLOGICAL STRESS
ZONE OF OPTIMUM RANGE
ZONE OF PHYSIOLOGICAL STRESS
ZONE OF INTOLERANCE
Organisms abundant
Organisms absent
Organisms absent
Low temperature
ENVIRONMENTAL GRADIENT
High temperature

↑ Populations have a range of tolerance for any given environmental factor (such as temperature). Every species has an upper and a lower limit beyond which it cannot survive (in this example, temperatures that are too cool or too hot). Most individuals, like the butterflies in this population, can be found around the optimum temperature, though what is "optimum" for each individual may differ slightly because of genetic variability. Some individuals may find themselves in areas of the habitat that are warmer or colder than the optimum. They can tolerate these conditions but may be physiologically stressed and not grow well or successfully breed. Genetic differences that allow some individuals to tolerate or even thrive at the edges of the population's tolerance offer the population a chance to adapt to changing conditions (such as a warmer climate) if needed. The more narrow the range of tolerance and the less genetically diverse the population, the less likely it will be to survive a change in conditions.

Living things survive within a specific range of environmental conditions.

Each biome required a mind-boggling array of considerations—not only how diverse plant and animal species would interact within and across biomes, but also the nutrient requirements of each organism they planned to include. Termites, for example, would need enough dead wood at the beginning of closure to sustain them until some of the larger plants began dying off. Termites live in the soil, stirring it and allowing air to penetrate soil particles. If the termites ran out of dead wood and starved to death, organisms living in the soil would not get enough oxygen and the entire desert would be jeopardized. Hummingbirds, on the other hand, would need nectar-filled flowers. "Try figuring out how many flowers a day a hummingbird needs," says Tony Burgess, a University of Arizona ecologist who helped design the biomes in Biosphere 2 and remained involved until 2004. "From there you need to know what the blooming season is, and then what the nectar load per flower is. And then you have to translate all of that into units of hummingbird support. Now imagine doing that sort of thing about 3000 times."

Biome-level planning was equally complicated. A rainforest would need consistently warm temperatures, but most desert biomes fluctuate wildly between day and night. In the summer, the grassland biome might literally starve the rainforest by gobbling up all the CO_2 needed for photosynthesis.

In this context, dead wood, flowers, and CO_2 are all examples of **limiting factors**—resources so critical that their availability controls the distribution of species and thus of biomes. The principle of limiting factors states that the critical resource in least supply is what determines the survival, growth, and reproduction of a given species in a given biome. Living things can only survive and reproduce within a certain range (between the upper and lower limits that they can tolerate) for a given critical resource or environmental condition, referred to as their **range of tolerance**. [INFOGRAPHIC 6.6]

limiting factor The critical resource whose supply determines the population size of a given species in a given biome.

range of tolerance The range, within upper and lower limits, of a limiting factor that allows a species to survive and reproduce.

↑ The Biosphere 2 ocean, shown here, is still used for marine research on the effect of increasing CO_2 levels on coral.

Ecologists routinely monitor a variety of limiting factors as a way of assessing ecosystem health, but anticipating what each individual organism would need to survive before the fact proved daunting.

An all-star team of scientists—oceanographers, forest ecologists, and plant physiologists—spent 2 years sorting through these challenges. Drawing on their combined expertise, they set about choosing the combination of soils, plants, and animals that seemed most capable of working together to recreate the delicate balances that had made Biosphere 1 such a spectacular success. A summer-dormant desert, like the ones found in Baja California, was chosen because it would reduce the desert's CO_2 demands when the savannah's productivity was at its highest. The ocean was situated between the desert and rainforest so that it could serve as a temperature buffer between the two. And each biome was created from a carefully selected array of species: the marsh biome was composed of intact chunks of swampland harvested from the Florida Everglades, and the savannah was composed of grasses from Australia, South America, and Africa. Well water mixed with aquarium salt filled the ocean, to which were added coral reef sections culled from the Caribbean.

But it wasn't long before the rigour and pragmatism of good science began to clash with the idealism of Biosphere 2 financiers. And when that happened, critics say, science lost out.

Some scientists worried that the ocean wouldn't get enough sunlight to support plant and animal life. Others opposed the use of soil high in organic matter; soil microbes decompose the soil's organic carbon and release it into the air as CO_2. Although this organic soil might eliminate the need for chemical fertilizers, there were concerns that it would provide too much fuel for the soil microbes, and would thus send atmospheric CO_2 concentrations through the roof. Despite these concerns, scientific advisors to the project were overruled.

The first few months of Mission 1 went smoothly enough, but eventually, plants and animals started dying. Humans grew hungry and mysteriously sleepy. And before long, they turned on each other.

Like Earth, Biosphere 2 was designed as a materially closed and energetically open system: plants would conduct photosynthesis with sunlight that streamed through the glass, but no biomass would enter or leave. Temperature, wind, rain, and ocean waves would be controlled mechanically. "But the facility would have to be self-sustaining," says Burgess. "Everything would die, unless the biota met its most fundamental purpose—using energy flow for biomass production." Humans would have to survive exclusively on what they could grow or catch under the dome. The system as a whole would have to continuously recycle every last bit of nutrient that was in the soil on day one.

At first, the carefully constructed agricultural biome seemed well suited to the challenge. Carrots, broccoli, peanuts, kale, lettuce, and sweet potatoes were grown on broad quarter-hectare terraces that sat adjacent to the sprawling six-storey human habitat. A bevy of domestic animals also provided sustenance—goats for milk, chickens for eggs, and pigs for pork. Indeed, eating only what they could grow made the biospherians healthier. Everyone lost weight. Bad cholesterol and blood pressure went down; so did white blood cell counts. And slowly, the biospherians say, their relationship to food changed. "Inside, I knew exactly where my food came from, and I totally understood my place in the biosphere and how it impacted the food I ate," says Jayne Poynter, the biospherian responsible for tending the farm. "When I breathed out, my CO_2 fed the sweet potatoes that I was growing. When I first got out, I lost sense of that. I would stand for hours in the aisles of shops, reading all the names on all the food things and think 'where does this stuff come from?' People must have thought I was nuts."

Infographic **6.7** **CARBON CYCLES VIA PHOTOSYNTHESIS AND CELLULAR RESPIRATION**

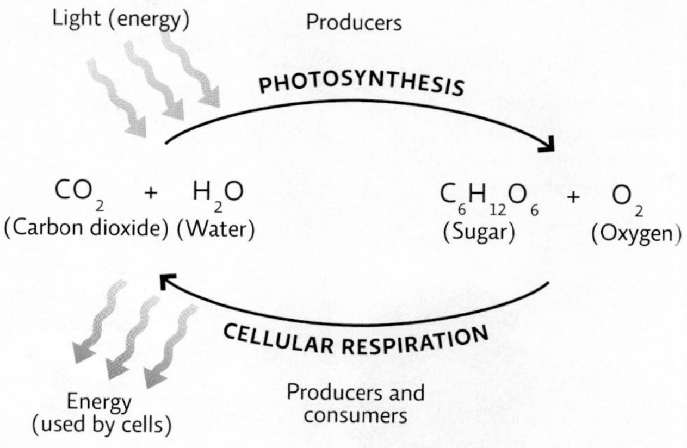
Light (energy)
Producers
PHOTOSYNTHESIS
CO2 + H2O
(Carbon dioxide) (Water)
C6H12O6 + O2
(Sugar) (Oxygen)
CELLULAR RESPIRATION
Energy (used by cells)
Producers and consumers

↑ In photosynthesis, producers use solar energy to combine CO_2 and H_2O to make sugar, releasing O_2 in the process. When producers (or any consumer who eats another organism) need energy, they break apart the sugar via the reverse reaction, cellular respiration. Oxygen is required for this reaction (which is why it is called "respiration").

But the glazed glass of the dome admitted less sunlight than had been anticipated. Less sunlight meant less biomass production. And that meant less food. Mites and disease also cut crop production. Biosphere 2's size precluded the use of pesticides and herbicides: in that small an atmosphere, the toxins would build up rapidly and could have had a deleterious impact on air quality and human health.

Now, a few months in, food was starting to run out. And that wasn't the only problem.

The humans were so tired they couldn't work. Nobody knew why, but scientists on the outside suspected it had something to do with nutrient cycles.

Nutrients such as carbon cycle through ecosystems.

On Earth, nutrients cycle through both **biotic** and **abiotic** components of an ecosystem—organisms, air, land, and water. They are stored in abiotic or biotic parts of the environment called **reservoirs**, or sinks, and linger in each for various lengths of time, known as *residence times*. Organisms acquire nutrients from the reservoir, the chemical cycles through the food chain, and eventually the chemical is returned to the reservoir. For carbon, the atmosphere—where carbon is stored as CO_2—is the most important reservoir. (Oceans and soil are also abiotic reservoirs for carbon. Oceans absorb CO_2 directly from the atmosphere and soils accumulate it during decomposition.) Plants and other photosynthesizers use carbon molecules from atmospheric CO_2 to build sugar, releasing oxygen in the process. Because they "produce" sugar, an organic molecule, from inorganic atmospheric CO_2, they are called **producers**.

This sugar molecule represents stored chemical energy that the producer can use. A **consumer**, the organism that eats the plant (or that eats the organism that eats the plant), also uses the chemical energy of sugar. This energy is released to the cell via the process of **cellular respiration**. [INFOGRAPHIC 6.7] All organisms—producers

biotic The living (organic) components of an ecosystem, such as the plants and animals and their waste (dead leaves, feces).

abiotic The nonliving components of an ecosystem, such as rainfall and mineral composition of the soil.

reservoirs (or sinks) Abiotic or biotic components of the environment that serve as storage places for cycling nutrients.

producer An organism that converts solar energy to chemical energy via photosynthesis.

consumer An organism that obtains energy and nutrients by feeding on another organism.

cellular respiration The process in which all organisms break down sugar to release its energy, using oxygen and giving off CO_2 as a waste product.

Infographic 6.8 | **THE CARBON CYCLE**

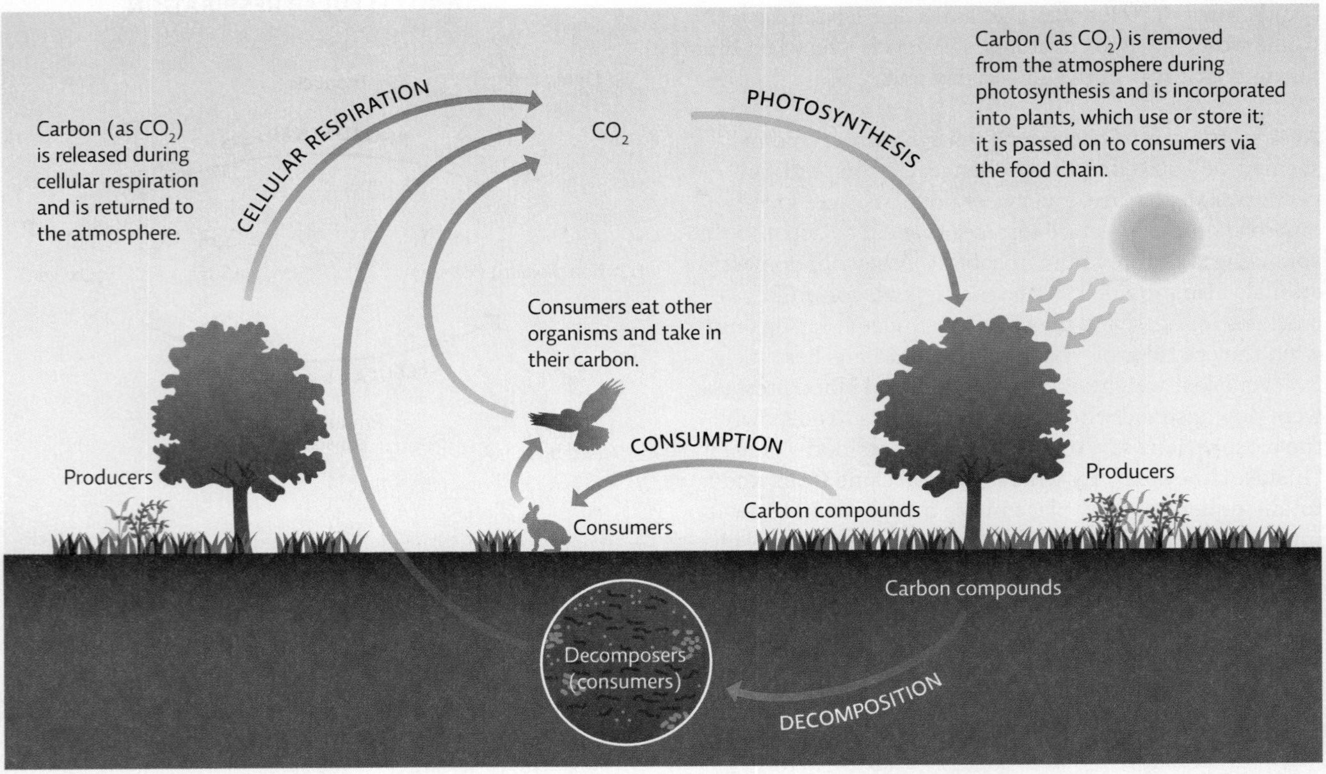
Carbon (as CO_2) is released during cellular respiration and is returned to the atmosphere.
CELLULAR RESPIRATION
CO_2
PHOTOSYNTHESIS
Carbon (as CO_2) is removed from the atmosphere during photosynthesis and is incorporated into plants, which use or store it; it is passed on to consumers via the food chain.
Consumers eat other organisms and take in their carbon.
CONSUMPTION
Producers
Consumers
Carbon compounds
Producers
Carbon compounds
Decomposers (consumers)
DECOMPOSITION

↑ Carbon cycles in and out of living things during photosynthesis and cellular respiration. As consumers (including decomposers) eat other organisms, carbon is transferred. Some carbon is stored in the bodies of organisms and in soil, but over the long term, the carbon cycle is balanced between photosynthesis and respiration.

Humans unbalance the carbon cycle via activities that increase the amount of CO_2 in the atmosphere.

Burning fossil fuels
Forest fire
Deforestation

and consumers—perform cellular respiration. (For more information on producers, consumers, and the food chain, see Chapter 8.)

From its initial incorporation into living tissue via photosynthesis, to its ultimate return to the atmosphere through respiration or through the burning of carbon-based fuels, carbon cycles in and out of various molecular forms and in and out of living things as it moves through the **carbon cycle**. [INFOGRAPHIC 6.8] Even though a single carbon atom might cycle in a few weeks or years, decades can pass before changes in a Brazilian rainforest impact a farm in Ontario. Inside Biosphere 2, the same cycle took approximately 3 days, which meant that changes in one biome could be felt in another biome much more quickly than on Earth.

Still, Biosphere 2's carbon cycle was not that different from Earth's—carbon moved from living tissue to the atmosphere and back in the same predictable manner. Or at least it should have. As the biospherians' energy waned, it became clear that something had gone terribly wrong.

It turned out that oxygen levels had fallen steadily—from 21% down to 14%. At such low concentrations, the biospherians were unable to convert the food they consumed into usable energy. "We were just dragging ourselves around the place," Poynter says. "And we had sleep apnea

Infographic **6.9** | **OXYGEN DEPLETION IN BIOSPHERE 2**

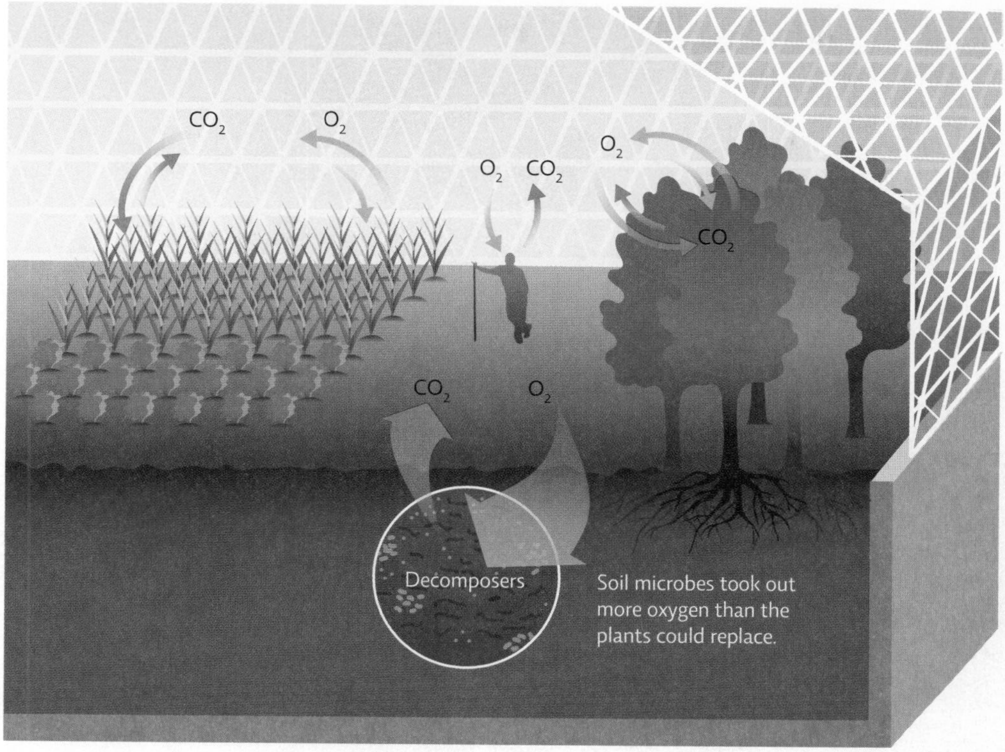
CO_2
O_2
O_2
CO_2
O_2
CO_2
CO_2
O_2
Decomposers
Soil microbes took out more oxygen than the plants could replace.

← The plants in Biosphere 2 were producing oxygen, but the excessive growth of soil microbes used oxygen faster than the plants could replace it via photosynthesis. This caused the oxygen levels in the air of Biosphere 2 to fall from 21% to 14%, causing health problems for the biospherians.

at night. So we'd wake up gasping for air because our blood chemistry had changed."

In just a few months, more than 6 metric tons of oxygen —enough to keep six people breathing for 6 months—had gone missing. As scientists from Columbia University later discovered, soil microbes were gobbling up all that O_2 and converting it into CO_2 as they decomposed the organic matter in the soil. [INFOGRAPHIC 6.9]

The biospherians responded by filling all unused planting areas with morning glory vines, a pretty and fast-growing (but as it turned out, invasive) species they hoped would maximize the amount of CO_2 converted back into O_2 by photosynthesis. But even with an abundance of plants and enough CO_2, photosynthesis was still limited by the availability of sunlight; not even morning glories could keep up with the soil microbes in their warm, well-watered, highly organic soil.

carbon cycle Movement of carbon through biotic and abiotic parts of an ecosystem. Carbon cycles via photosynthesis and cellular respiration as well as in and out of other reservoirs such as the oceans, soil, rock, and atmosphere. It is also released by human actions such as fossil fuel burning.

nitrogen cycle Continuous series of natural processes by which nitrogen passes from the air to the soil, to organisms, and then returns back to the air or soil through decomposition or denitrification.

Biosphere 2 is not alone with regard to a disrupted carbon cycle. Human activity has greatly altered carbon amounts in Earth's atmosphere. Many of our actions (such as burning fossil fuels) increase the amount of carbon normally released into the atmosphere or degrade natural ecosystems so that less carbon is removed from the atmosphere (as in the case of deforestation). Just like with Biosphere 2, this extra atmospheric carbon causes problems such as global climate change, acidification of oceans, and alterations of communities worldwide.

Adding to the confusion, concrete used to build parts of Biosphere 2 was absorbing some of the CO_2 and converting it into calcium carbonate, trapping some of the carbon and oxygen in this unexpected sink.

Besides carbon, other chemicals essential for life, such as nitrogen and phosphorus, cycle through ecosystems. Nitrogen, the most abundant element in Earth's atmosphere, is needed to make proteins and nucleic acids, but plants cannot utilize nitrogen in its atmospheric form (N_2). All plant life, and ultimately all animal life too, depends on microbes (bacteria) to convert atmospheric nitrogen into usable forms as part of the **nitrogen cycle**.

Infographic **6.10** | **THE NITROGEN CYCLE**

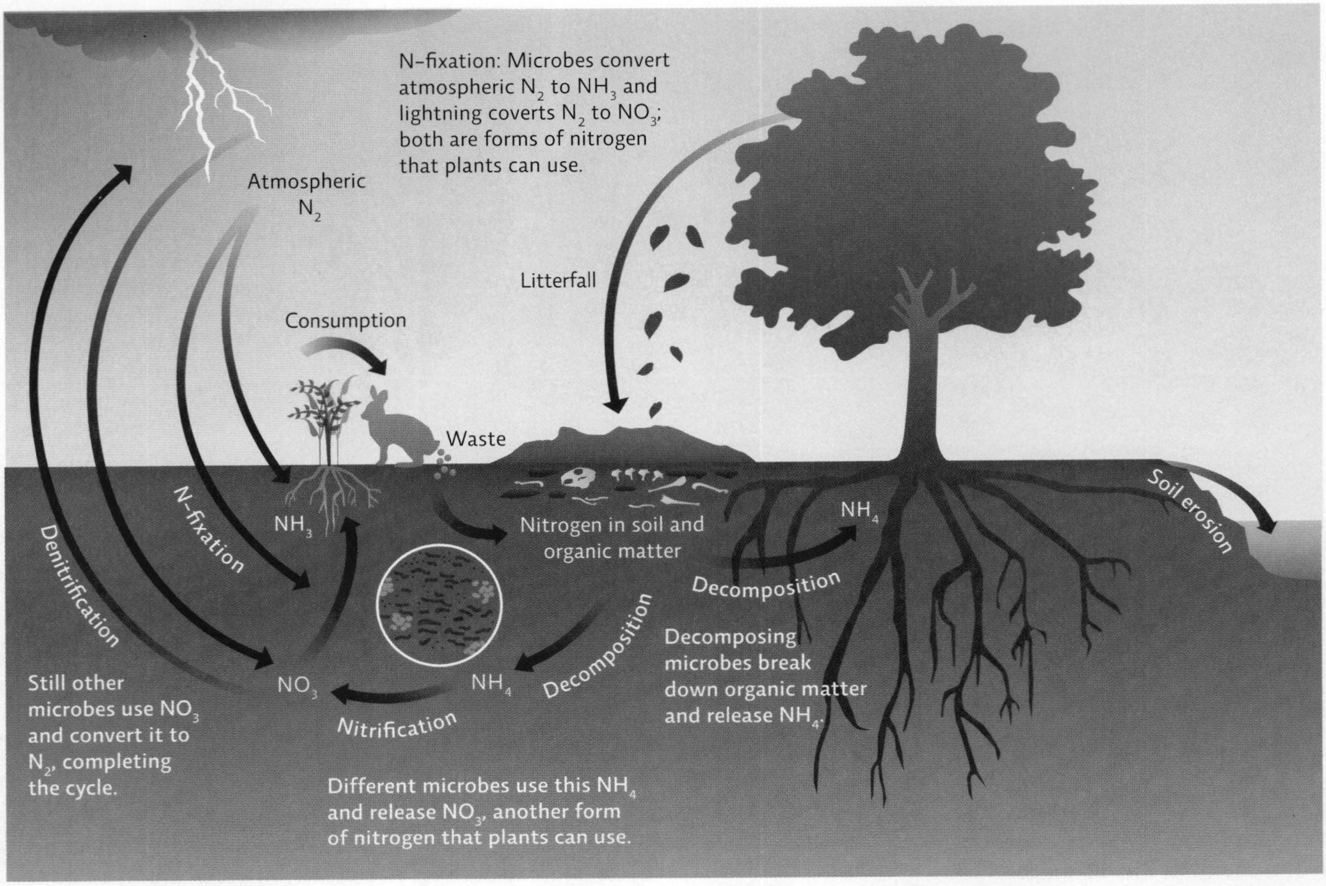
N-fixation: Microbes convert atmospheric N_2 to NH_3 and lightning coverts N_2 to NO_3; both are forms of nitrogen that plants can use.
Atmospheric N_2
Litterfall
Consumption
Waste
Soil erosion
NH_3
N-fixation
Denitrification
Nitrogen in soil and organic matter
NH_4
Decomposition
Decomposition
Decomposing microbes break down organic matter and release NH_4.
Still other microbes use NO_3 and convert it to N_2, completing the cycle.
NO_3
NH_4
Nitrification
Different microbes use this NH_4 and release NO_3, another form of nitrogen that plants can use.

↑ Nitrogen, needed by all living things to make biological molecules like protein and DNA, continuously moves in and out of organisms and the atmosphere in a cycle absolutely dependent on soil bacteria.

Nitrogen fertilizers promote plant growth but this depletes other soil nutrients; they can also leach out of soils and pollute aquatic ecosystems.

Fertilizing

Burning fossil fuels contributes to nitrogen pollution such as smog and acid rain.

Burning fossil fuels

In a process called **nitrogen fixation**, atmospheric nitrogen (N_2) is converted by bacteria into ammonia (NH_3) which plants take up through their roots; consumers take in nitrogen via their diet. A small amount of N_2 is fixed by lightning, producing nitrate (NO_3). In other steps of the nitrogen cycle (decomposition, nitrification, and denitrification) various types of bacteria feed on nitrogen compounds in organic matter or the soil, eventually returning it as N_2 to the atmosphere. [INFOGRAPHIC 6.10] The nitrous oxide (N_2O), or laughing gas, that gave the biospherians trouble is a by-product of denitrification that normally exists in trace amounts in Earth's atmosphere (it is also becoming a dangerous greenhouse gas, as it is produced in ever-higher concentrations by some human activities).

Unlike nitrogen and carbon, phosphorus—which is needed to make DNA and RNA—is found only in solid or liquid form on Earth, so the **phosphorus cycle** does not move through the atmosphere, but passes from inorganic to organic form through a series of interactions with water and organisms. [INFOGRAPHIC 6.11]

In Biosphere 2, the nitrogen and phosphorus cycles were disrupted. Thanks to an overabundance of soil microbes,

Infographic **6.11** | **THE PHOSPHORUS CYCLE**

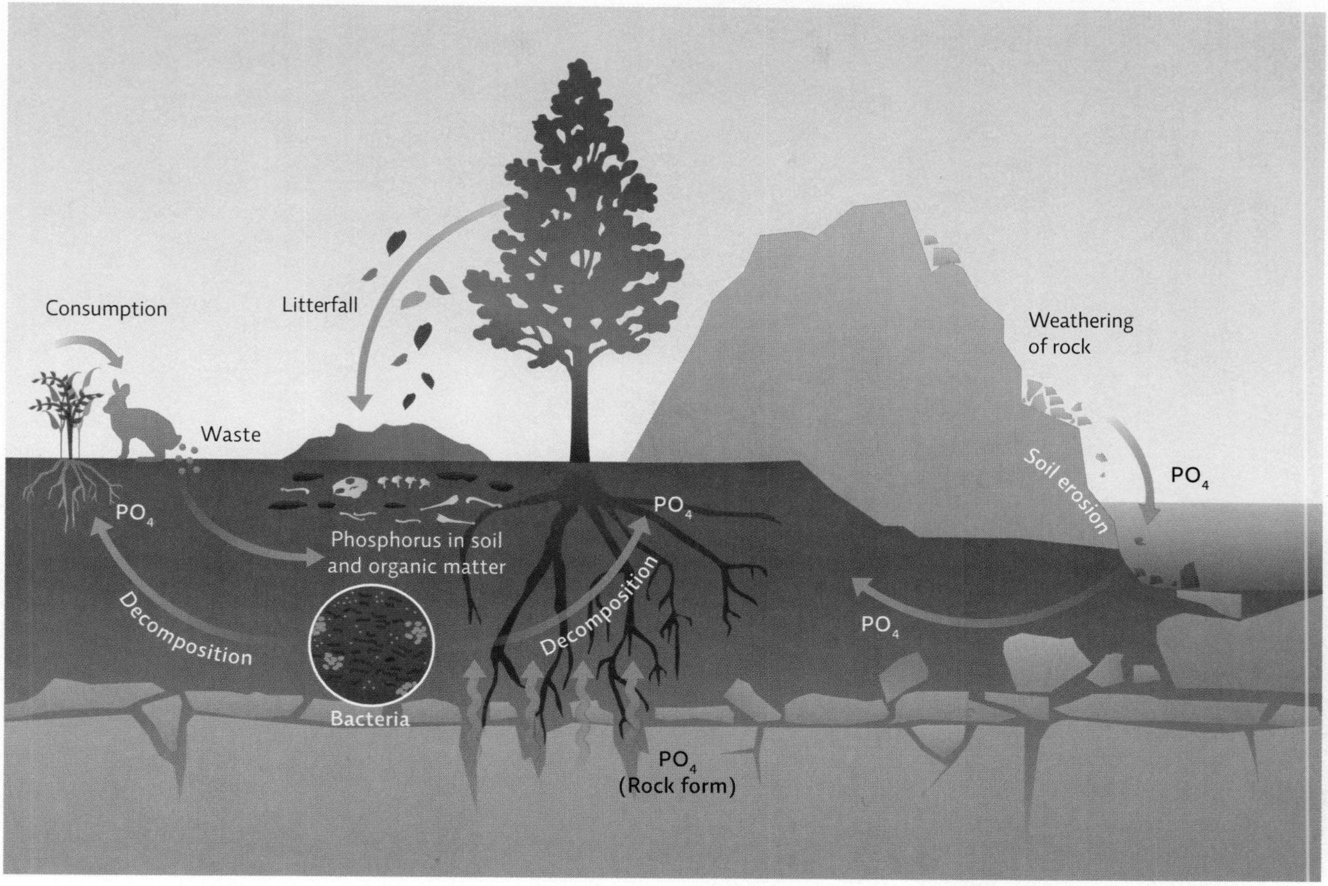
Consumption
Litterfall
Waste
Weathering of rock
Soil erosion
PO_4
PO_4
PO_4
Phosphorus in soil and organic matter
Decomposition
Decomposition
PO_4
Bacteria
PO_4 (Rock form)

↑ Phosphorus, needed by all organisms to make DNA, cycles very slowly. It has no atmospheric component but instead depends on the weathering of rock to release new supplies of phosphate (PO_4) into bodies of water or the soil, where it dissolves in water and can be taken up by organisms. Microbes also play a role when they break down organic material and release the phosphate to the soil.

Dust released through mining or in eroded areas can introduce phosphorus into the environment much more quickly than it would normally enter.

Open-pit mining

Fertilizers and animal waste (including sewage) can alter plant growth and nutrient cycling, especially in aquatic ecosystems where phosphorus is usually a limiting nutrient.

Animal waste

nitrous oxide reached levels high enough to interfere with the metabolism of vitamin B12, which is essential to the brain and nervous system.

Phosphorus got trapped in the water system, polluting aquatic habitats. The underwater and terrestrial plants of Biosphere 2 were dying off too quickly to complete this cycle. Biospherians removed excess nutrients from their water supply by passing the water over algal mats that would absorb the nutrients and could then be harvested, dried, and stored.

As food reserves dwindled, the eight biospherians split into two factions. One group felt that scientific research was the top priority, and wanted to import food so that they would have enough energy to continue with their experiments. The other group felt that maintaining a truly closed system—one where no biomass was allowed to enter or leave—was the project's most important goal;

---

nitrogen fixation Conversion of atmospheric nitrogen into a biologically usable form, carried out by bacteria found in soil or via lightning.

phosphorus cycle Series of natural processes by which the nutrient phosphorus moves from rock to soil or water, to living organisms, and back to the soil.

↑ Justin Peterson, a Biosphere 2 undergraduate intern who assists PhD candidate Henry Adams with his Pinon Pine Tree Drought Experiment. The experiment's goal is to predict the effects of climate shifts on the trees.

proving that humans could survive exclusively on what the dome provided would be essential to one day colonizing the Moon or Mars. To them, importing food would amount to a mission failure. "It was a heartbreaking split," Poynter says. "Just 6 months into the mission, and two people on the other side of the divide had been my closest friends going in." Eventually, Poynter snuck food in. That wasn't the only breech. To solve the various nutrient cycle conundrums, the project's engineers had installed a CO_2 scrubber, and pumped in 17 000 cubic metres of oxygen.

Biosphere officials hid these actions from all but a few key people. When one reporter finally broke the story, the public was outraged. Spectators of every ilk—seasoned scientists, skeptical reporters, and casual observers alike—became convinced that other data was also being fudged. "Secrets are like kryptonite to the scientific process," says Griffin. "Once you find out something has been deliberately overhyped or downplayed, or just plain lied about, all the data from that research becomes suspect. And data that can't be trusted has no scientific value."

Three years after the first mission was completed, the editors of the respected journal *Science* deemed the entire project a failure. "Isolating small pieces of large biomes and juxtaposing them in an artificial enclosure changed their functioning and interactions, rather than creating a small working earth as originally intended," they wrote. For the $200 million dome to survive as a scientific enterprise, they concluded, it would need dramatic retooling.

Ecosystems are complicated, but learning how they function will help us restore degraded ones.

Biosphere 2 taught scientists that Earth is far too complex, that ecosystem components intertwine in far too many complicated ways, for humans to recreate. Each is governed by a countless array of interacting factors and a change in one can set off a whole chain of events that degrade the system's capacity to sustain life.

Three years after the first mission was completed, the editors of the respected journal *Science* deemed the entire project a failure.

From a scientific point of view, the fact that Biosphere 2 was not able to sustain life as hoped does not mean the project was of no value. Negative results can be just as informative as positive ones—in some cases even more so because they uncover gaps in our knowledge and help us decide how to move forward. In fact, Biosphere 2's greatest liability—its skyrocketing CO_2 levels—proved to be its most valuable asset. "Now it's like a time machine," says Griffin who points out that it is allowing us a look at the consequences of elevated atmospheric CO_2 levels, the main contributor to climate change today. Recent research by Griffin has uncovered some of the complexities of carbon cycling. His group saw unexpected fluctuations in carbon release at various levels in the tree canopy, telling him there is much we still don't know about how carbon cycles—data that could only be gathered in an enclosed forest such as that found in Biosphere 2. Today, scientists from all over the world still use the facility to study the effects of an atmosphere loaded with carbon dioxide. Ultimately, though, the most valuable lesson Biosphere 2 has provided is how irreplaceable Biosphere 1 is.◉

Select references in this chapter:
Cohen, J.E., and Tilman, D. 1996. *Science*, 274: 1150–1151.
Griffin, K.L., *et al.* 2002. *New Phytologist*, 154: 609–619.

BRING IT HOME

PERSONAL CHOICES THAT HELP

Nutrient cycling is critical for maintaining Earth's ecosystems, but we interfere with nutrient cycles through our daily activities. Driving a car interferes with the carbon cycle by releasing carbon from fossil fuel reservoirs. Applying synthetic chemical fertilizers to food crops interferes with the nitrogen cycle by adding soluble nitrogen compounds to aquatic ecosystems through runoff. The challenge is to figure out ways to work with nutrient cycles rather than against them—in other words, to return nutrients to the reservoirs from which they come. How might this be done in our daily lives and in our own communities?

Individual Steps

→ Reduce your fossil fuel use to curtail carbon, nitrogen, and sulphur emissions. Take public transportation, walk, ride a bike, and drive a fuel-efficient vehicle.

→ Compost food and yard waste. Then use this material to fertilize flowerbeds, trees, and garden plots. Composting will reduce or eliminate the need for inorganic fertilizers in your yard.

→ If you cannot compost your organic waste at home, participate in your municipal organic waste program, often called a "green bin" program. Over 65% of organic waste composted at waste management facilities in Canada comes from households.

Group Action

→ Participate in or organize an event to plant trees or native grasses. By doing so, you can help recapture the carbon put into the atmosphere by driving a car.

→ Many urban areas welcome individuals who are willing to plant native trees, shrubs, and wildflowers in school grounds, parks, and other public spaces.

Policy Change

→ Support legislation to increase fuel efficiency of vehicles and subsidies for clean, renewable energy. More fuel-efficient cars and cleaner energy sources will reduce our carbon outputs from fossil fuel use.

UNDERSTANDING THE ISSUE

CHECK YOUR UNDERSTANDING

1. Biosphere 2:
 a. wasted taxpayers' money.
 b. was poor science and is ignored.
 c. was a utopian dream that did not work.
 d. was a learning experience for the scientific community.

2. Which of the following is the correct organization of life?
 a. Individual, community, biome, ecosystem, population, biosphere
 b. Individual, population, community, ecosystem, biome, biosphere
 c. Biosphere, community, biome, ecosystem, population, individual
 d. Biome, biosphere, population, ecosystem, community, individual

3. Which biome description is correct?
 a. Grasslands receive less rainfall than forests, but more than deserts.
 b. Forests have freezing temperatures regularly.
 c. Grasslands are much warmer annually than forests.
 d. Deserts are hot year round, much hotter than forests.

4. The range of tolerance for an organism might be:
 a. for a field mouse, how cold it gets.
 b. for a robin, the competing species in the area.
 c. for an earthworm, the amount of rainfall.
 d. for a new wolf pack, the number of other wolf dens in the area.

5. Carbon dioxide in the atmosphere is:
 a. captured by plants during photosynthesis.
 b. released by plants during photosynthesis.
 c. captured between water molecules.
 d. captured by plants during cellular respiration.

6. In order for nitrogen to be usable for other life it must first be:
 a. taken up by plants through photosynthesis.
 b. taken up by soil bacteria through nitrogen fixation.
 c. taken up by animals through respiration.
 d. dissolved in water and taken up through plant roots.

WORK WITH IDEAS

1. Using an example such as spring wildflowers and rainfall, explain the terms *environmental gradient* and *range of tolerance*.

2. List the steps involved in the nitrogen cycle. What would happen if a wildfire burned so hot that it sterilized the soil, killing all the microbes?

3. You are designing a habitat for living on another planet, about the same size as Biosphere 2. Do you want several small biomes or one large biome? Justify your answer, using terms such as *habitat*, *nutrient cycles*, and *biomass*.

4. You are studying a small island with the following characteristics. It has one type of grass and one type of tree. It has only four types of small invertebrates, one type of lizard, and a seed-eating mouse. It has a worm that feeds on decaying leaves and some soil bacteria for decomposers. Draw your island biome, showing how carbon will move.

5. A large hawk reaches the island described in question 4. Given your biome's size, and the amount of carbon-based biomass, is there enough carbon available for the hawk to live on the island? Explain.

6. You are studying a pocket mouse which lives in an underground burrow where it stores seeds. Make a list of biotic and abiotic things that might be limiting factors to its survival.

ANALYZING THE SCIENCE

The following table shows six tropical forests, arranged from least to most fertile soils, estimated from total soil nitrogen.

Parameter	Lower montane rainforest, Puerto Rico	Amazon caatinga, Venezuela	Oxisol forest, Venezuela	Evergreen forest, Ivory Coast	Dipterocarp forest, Malaysia	Lowland rainforest, Costa Rica
Above-ground biomass (metric tons/ha)	**228**	**268**	**264**	**513**	**475**	**382**
Root biomass (metric tons/ha)	**72.3**	**132**	**56**	**49**	**20.5**	**14.4**
Total soil nitrogen (kg/ha)	--	**785**	**1697**	**6500**	**6752**	**20 000**
Turnover time of leaves (years)	**2**	**2.2**	**1.7**	--	**1.3**	--

INTERPRETATION

1. Which two areas have the poorest soils, in terms of nitrogen availability?

2. Look at the root biomass in the two areas from Question 1. What is the relationship between root biomass and poor soils? Propose an explanation.

3. Can you determine soil nutrient levels by looking at the above-ground biomass? Explain.

ADVANCE YOUR THINKING

4. Notice how long it takes leaves to break down ("Turnover time of leaves"). Speculate as to why trees in poorer soils produce leaves that last longer.

5. If you removed the trees through logging or burning, would these areas be good for agriculture? (Hint: Cleared areas are usually used for small farms that do not have funds to purchase farm equipment, fertilizers, or other chemicals.)

EVALUATING NEW INFORMATION

Biosphere 2 covers a mere 1.2 hectares with finite resources. Although Earth is vastly larger, its resources are also finite. People have to decide, for each biome, whether to leave some untouched, or use all of an area and its resources for human purposes. One example would be plowing under an entire prairie and growing wheat, displacing all of the native plants and animals. The rainforests are another such biome. Left alone, they produce huge amounts of oxygen for the whole planet, while supporting millions of species. Many humans want to use the trees for lumber and the land to grow crops or raise cattle.

Some groups advocate using rainforests in sustainable ways: Manufacturing herbal products, or raising shade-grown coffee under the canopy. Will these tactics help preserve the biome?

Go to the Rainforest Alliance's website (www.rainforest-alliance.org) and read about Rainforest Alliance-certified products such as coffee, tea, and cocoa.

Evaluate the website and work with the information to answer the following questions:

1. Is this a reliable information source?
   a. Does the organization give supporting evidence for its claims?
   b. Does it give sources for its evidence?
   c. What is the mission of this organization?

2. Note the range of publications available on the Rainforest Alliance website and view the Research Report on cocoa certification (www.rainforest-alliance.org/publications).
   a. Do you agree that certification is an adequate tool to ensure sustainability in cocoa farming?
   b. Identify a claim the organization makes and the evidence it gives in support of this claim. Is it convincing?

3. The Global Canopy Programme (www.globalcanopy.org) is a UK-based group that has joined with the Carbon Disclosure Project to ask companies to disclose their "forest footprint" (www.cdproject.net/en-US/Programmes/Pages/forests.aspx). Should companies be required to participate?
   a. Should some intergovernmental group investigate companies and estimate impacts?
   b. Is it important to allow a "free market" for goods and services, assuming businesses will take care of the lands they use but do not own?

MAKING CONNECTIONS

LAKE ERIE: A NOT SO GREAT LAKE

Background: The Great Lakes are located in a highly populated area of North America and are susceptible to pollution from both urban and agricultural sources. In the 1960s and 1970s, Lake Erie became notorious for being heavily polluted by industrial contaminants, agricultural runoff, and raw sewage. Rivers flowing into the lake were known to catch fire owing to oil contamination, and elevated phosphorus and nitrogen led to algal blooms and massive fish kills. Following the signing of an International Water Quality Agreement in 1972 by Canada and the United States, pollution regulations were developed. By the late 1990s there was significant improvement in Lake Erie. However, a dead zone persists in Lake Erie, where hypoxic conditions prevail every year in late summer.

Case: Environment Canada is very concerned about the persistence of the Lake Erie dead zone and the threat it poses to this Great Lake ecosystem. The company where you work wants to apply for a grant from Environment Canada to investigate possible solutions to the annual dead zones.

Research the issue and write a report that includes:

1. An introduction explaining the causes of the dead zone. Be sure to take into account the following factors which can affect nutrient cycling in Lake Erie:
   a. Agricultural activities, including both crop and livestock farming in Ontario and Ohio
   b. Large cities, such as Detroit and Cleveland
   c. Industrial activities in Ontario, Ohio, and Michigan

2. A possible course of action that you think your company might want to pursue, such as greatly slowing or eliminating the pollution generated by one of the three sources listed above.

THE WOLF WATCHERS

Endangered grey wolves return to the American West

CORE MESSAGE

Population size and makeup can fluctuate or remain stable. The stability of populations is often dependent on regulation by the environment or by other species like predators. If human impact has killed or removed predators, we may have to step in to fill their role.

GUIDING QUESTIONS

After reading this chapter, you should be able to answer the following questions:

→ What is a population and why is it an ecologically significant concept? What factors do ecologists use to describe and monitor natural populations?

→ What types of population growth patterns are seen and what factors affect population growth?

→ What are the reproductive strategies of *r*- and *K*-selected species and how do these relate to population growth patterns and potential?

→ What is carrying capacity and why do some populations slowly approach it and then stabilize, whereas others overshoot it and then crash?

→ What role do predators play in regulating populations and how can the presence or absence of predators affect entire communities?

A grey wolf on the prowl at Yellowstone National Park.

At least a half a dozen times each winter, Doug Smith climbs into a helicopter, gun in tow, and hunts wolves in Yellowstone National Park. He's not looking to kill them—just to put them to sleep for a little while so that he can outfit them with radio collars to track the size of their packs, what they eat, and where they go over the course of the following year. Smith, a biologist, spends about 200 hours per year on these "hunting" expeditions, which are part of the Yellowstone Wolf Restoration Project that he leads. The project has been responsible for reintroducing many wolves to Yellowstone, after their numbers had dwindled to dangerous levels.

Sometimes, though, things go awry on Smith's radio collar missions. For instance, the tranquilizer dart doesn't fully sedate big wolves—some of which can reach 80 kilograms. Smith is forced to approach the animals while they're still awake. "I have to grab them on the scruff of the neck and manhandle them until I get the collar on," he explains. "They're typically not dangerous then—they've had enough drug to be kind of out of it—but they're still able to walk around. It's a wild experience." Sometimes the wolves—who are typically frightened of the helicopter—try to attack it while it's hovering with the doors open a couple metres above the ground. "I've had two females turn and run at the helicopter, teeth gnashing, jumping up trying to get me," Smith recalls. "I'm hovering above a wolf going back and forth, and I can't get a shot because all I'm seeing is face and teeth." Despite these adventures, Smith says a wolf has never actually bitten him—and he has tranquilized and collared more than 300 of them.

Understanding how wolf populations interact with biotic and abiotic forces in their environment, through programs like the Yellowstone Wolf Restoration Project, is key to preventing this vital species from disappearing from Yellowstone Park—and the United States—forever. That's because so many factors influence the livelihood of a species and its ability to survive and reproduce. In the early 20th century, Americans in western states not only threatened wolves by hunting them, but they also destroyed the animals' natural habitat when they cut down forests to build farms. They also starved the animals by hunting elk, deer, and bison, wolves' usual sources of food. At the time, people didn't think that this combination of changes would very nearly cause wolves to go extinct. But now that scientists like Smith have spent years watching how the animals live, they have a much better understanding of what needs to be done to keep them alive and maintain them at their natural population numbers. Today, such efforts are more crucial than ever, with more than 20 000 plant and animal species categorized as threatened on the International Union for Conservation of Nature's (IUCN's) Red List of Threatened Species.

WHERE IS YELLOWSTONE NATIONAL PARK?

MONTANA
YELLOWSTONE NATIONAL PARK
WY
IDAHO
WYOMING

Before a U.S. government-sponsored wolf control program began in the early 1900s, the exact numbers of grey wolves living in Yellowstone were unknown, but estimates range from 300 to 400. Chances are, their numbers fluctuated over time. Canadian grey wolves are faring reasonably well. With an estimated 50 000 wolves living across the country, Canada has the second largest wolf population in the world after Russia. A **population**—all the individuals of a species that live in the same geographic area and are able to interact and interbreed—frequently fluctuates in size and distribution. So, although the Canadian wolf population is relatively stable, it does fluctuate with changes in such factors as

↑ Doug Smith and Yellowstone Delta wolf 487M, the largest wolf collared in the winter of 2005.

access to food, water, and den sites, the presence of predators (including humans, who affect population stability through hunting and fishing), as well as the population's innate characteristics. Ecologists, and population biologists (like Smith), who study changes in population size and makeup (**population dynamics**), find that the population size of some species increases and decreases rather predictably (barring a catastrophic event), while others tend to fluctuate more randomly. The size of a population in a given area is determined by the interplay of factors that simultaneously increase the number of individuals in a population (births and immigration), and those that decrease numbers (deaths and emigration). An understanding of these factors helps us predict population size at any given time.

Populations fluctuate in size and have varied distributions.

In the early 20th century, the U.S. government began its Predator Control Program, for which the U.S. Congress allocated $125,000 for predator and rodent control through poisoning—and later through hunting—to eradicate or control wild animals that might harm crops or livestock. Wolves were a targeted predator because they preyed upon livestock, and wolf eradication was further supported because it boosted deer populations for human hunting. The U.S. National Park Service herded wolves out of parks so they could be hunted; moving them outside made them more accessible to hunters and made the parks more welcoming for visitors. "It was park policy to kill all predators, and wolves were their biggest objective," Smith explains. Between 1914 and 1926, at least 136 wolves were killed within Yellowstone. The last known Yellowstone wolf was killed in 1944.

Every population has a **minimum viable population**, or the smallest number of individuals that would still allow a population to be able to persist or grow, ensuring long-term survival. This is an important concept when considering how to conserve endangered or threatened species. A population that is too small may fail to recover for a variety of reasons. For example, the courtship rituals of some species require a minimum number of individuals for success. Other activities that depend on numbers—like flocking, schooling, and foraging—fail below certain population sizes. Genetic diversity (inherited variety among individuals in a population) is also important: a population with little genetic diversity is less able to evolve new traits that enable them to adapt to environmental changes and is therefore more vulnerable. A small population is also subject to inbreeding, which allows harmful genetic traits to spread and weaken the population.

population All the individuals of a species that live in the same geographic area and are able to interact and interbreed.
population dynamics The changes over time of population size and composition.
minimum viable population The smallest number of individuals that would still allow a population to be able to persist or grow, ensuring long-term survival.

In the American West, the Predator Control Program was so successful that it eventually became cause for concern and criticism as populations were close to or below minimum viable thresholds. In 1973, wolves in Montana and Wyoming became protected under the U.S. Endangered Species Act. In 1987, the U.S. Fish and Wildlife Service proposed reintroducing an "experimental population" of Canadian wolves into Yellowstone through its Northern Rocky Mountain Wolf Recovery Plan.

Not everyone was in favour of reintroducing wolves to the landscape of the American West, and opposition remains today. Ranchers in particular were (and still are) wary of the damage wolves might do to livestock. In some years, wolves were responsible for thousands of herd deaths— although not nearly as many as are caused by coyotes, according to the U.S. Fish and Wildlife Service. Current programs compensate ranchers for lost livestock, but they must first prove that the animal was killed by a wolf, which is not always easy to do. Sport hunters worry that wolves will decimate game species like deer and elk; however, these animals succumb to many factors, including bad weather and other predators—not just wolves—so it is unclear how much pressure wolves actually put on these species.

The U.S. Congress funded the U.S. Fish and Wildlife Service to conduct an **environmental impact assessment** on the local environment should wolves be reintroduced to Yellowstone. By considering all aspects of wolf ecology, scientists were able to organize the reintroduction to maximize survival of the species. For instance, scientists believed that it would be best to "softly" release wolves in the park by holding them temporarily in packs in areas that would be suitable for them to live, instead of doing a "hard" release in which the wolves could immediately disperse anywhere in the park. The soft release would curtail the wolves' movement and help them survive and acclimate to the move. In addition, it would allow the park staff, who would bring road-killed deer, elk, moose, and bison to the wolves twice a week, to become familiar with the wolves so that they could identify them once they were fully released. In January 1995, scientists captured 14 wolves from different packs across Canada, and softly released them into Yellowstone.

The number of wolves distributed across Yellowstone Park is their **population density**—the number of individuals per unit area. Population density is an important feature that varies enormously among species, or even among populations of the same species in different ecosystems. If a population's density is too low, individuals may have difficulty in finding mates, or the only potential mates may be closely related individuals, which can lead to inbreeding, loss of genetic variability, and ultimately, extinction. Density that is too high can also cause problems, such as increased competition, fighting, and spread of disease. Deer and elk populations in the United States and Canada, whose density had increased in recent years because of exploding numbers combined with shrinking habitats, now frequently suffer from an infectious disease known as chronic wasting disease, which weakens and eventually kills the animals.

In addition to size and density, another important feature is **population distribution**, or the location and spacing of individuals within their range. A number of factors affect distribution, including species characteristics, topography, and habitat makeup. Ecologists typically speak of three types of distribution. In a **clumped distribution**, individuals are found in groups or patches within the habitat. North American examples include social species like prairie dogs or beaver that are clustered around a necessary resource, like water. Wolves travel in packs and therefore have a clumped distribution. Elk, one of their prey species, congregate as well; living in herds offers some protection against the wolves.

In a **random distribution**, individuals are spread out in the environment irregularly with no discernible pattern. Random distributions are sometimes seen in homogeneous environments, in part because no particular spot is considered better than another. Species that disperse through random means and germinate where they fall—like wind-blown seeds—also often distribute randomly. **Uniform distributions**, rare in nature, include individuals that are spaced evenly, perhaps due to territorial behaviour or mechanisms for suppressing growth of nearby individuals (seen in some plant species). [INFOGRAPHIC 7.1]

Populations display various patterns of growth.

Researchers involved in species' restoration projects observe factors such as distribution, genetic diversity,

environmental impact assessment An evaluation of the positive and negative impacts of a proposed environmental action, including alternative actions that could be pursued.

population density The number of individuals per unit area.

population distribution The location and spacing of individuals within their range.

clumped distribution Individuals are found in groups or patches within the habitat.

random distribution Individuals are spread out over the environment irregularly with no discernible pattern.

uniform distribution Individuals are spaced evenly, perhaps due to territorial behaviour or mechanisms for suppressing the growth of nearby individuals.

Infographic **7.1** | **POPULATION DISTRIBUTION PATTERNS**

↓ The distribution of individuals in populations varies from species to species and is influenced by biotic and/or abiotic factors.

CLUMPED

↑ Elk stay in herds, which offers some protection against predators.

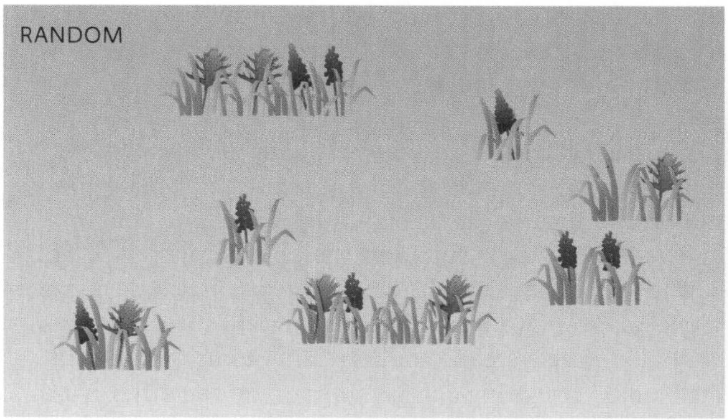
RANDOM

↑ The seeds of many flower species are distributed randomly and germinate where they fall.

UNIFORM

↑ The creosote bush of the U.S. desert Southwest produces toxins that prevent other bushes from growing close by.

initial population size, and density as well as prey availability and habitat structure, and evaluate how they influenced the size and dynamics of a vulnerable or reintroduced population. In order to predict population dynamics, scientists use some simple mathematical models that describe population growth over time. The annual **population growth rate** is determined by **birth rate** (the number of births per 1000 individuals per year) minus the population **death rate** (number of deaths per 1000 individuals per year), or by taking a simple census of population size at two time points. Between 2009 and 2010, the Yellowstone wolf population went from 320 to 343, an increase of 23 animals. This is a growth rate of 7% (23/320 = 0.07).

When there are no environmental limits to survival or reproduction, a population will reach its maximum per capita rate of increase *(r)*, called its **biotic potential**. This occurs, theoretically, when every female reproduces to her maximal potential and every offspring survives. A population increasing in this manner will quickly grow to fill its environment. This period of growth, which can't go on indefinitely, is referred to as **exponential growth**, named for the mathematical function it represents.

Infographic **7.2** | **EXPONENTIAL GROWTH OCCURS WHEN THERE ARE NO LIMITS TO GROWTH**

↓ Because deer mice have a high biotic potential, even a single pair could produce more than 31 000 descendants in their lifetime.

BIOTIC POTENTIAL OF DEER MICE

Assume each pair produces 10 pups/litter and none die: $r = 5$ (r is expressed as surviving offspring per adult; we divide the litter size by 2 since there are 2 parents per litter).

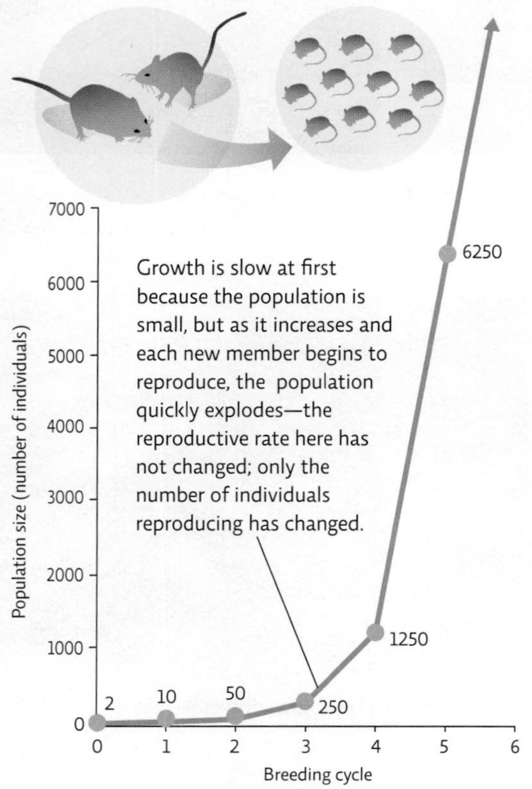
Growth is slow at first because the population is small, but as it increases and each new member begins to reproduce, the population quickly explodes—the reproductive rate here has not changed; only the number of individuals reproducing has changed.
Population size (number of individuals)
7000
6000
5000
4000
3000
2000
1000
0
2
10
50
250
1250
6250
0
1
2
3
4
5
6
Breeding cycle

Populations that have a high biotic potential have high fecundity (females typically produce lots of offspring), and reach reproductive maturity quickly. The higher the biotic potential, the faster the population of a given species will grow under ideal conditions. North American species such as deer mice and the invasive spotted knapweed have higher biotic potential than species such as grizzly bears and spruce trees.

Exponential growth is typically seen when a species first enters a new environment, or if there is an influx of new resources. The population must have a high birth rate—most individuals must have access to enough food, water, and habitat in which to reproduce—and a low death rate. The loss of predators can also lead to exponential growth among their prey species.

A population that is growing exponentially will have a *J-shaped curve* if plotted on a graph with time on the *x*-axis and population size on the *y*-axis. The J curve shows a slight lag at first and then a rapid increase. This is due to the fact that the larger the population, the faster it grows, even at the same growth rate. Think of it this way: doubling a small number yields a number that is still small. Doubling a large number, on the other hand, produces a very large number.

As an example of how profound exponential growth can be, imagine someone gave you a dollar one day and then, each subsequent day for a month, doubled the amount given to you the previous day. On day 1 you would have 1 dollar; on day 2 you would be given 2 dollars; on day 3 you'd be given 4 dollars, and on day 4, you would be given 8 dollars. By day 31, you would have over $2.1 billion. Exponential growth can create large populations quickly. [INFOGRAPHIC 7.2]

Exponential growth can't last forever, however. As a population begins to fill its environment, its growth rate decreases, because as more individuals use the available resources, the resources become scarce. Some individuals starve or are unable to find habitat in which to reproduce, and crowding may also bring about an increase in disease and aggression. Predation pressure may increase as the more numerous prey are easier to track and capture or simply because the predator population itself has

Infographic **7.3** | **LOGISTIC POPULATION GROWTH**

→ Exponential growth turns into logistic growth (S curve) as population size approaches carrying capacity (*K*) and resistance factors begin to limit survival.

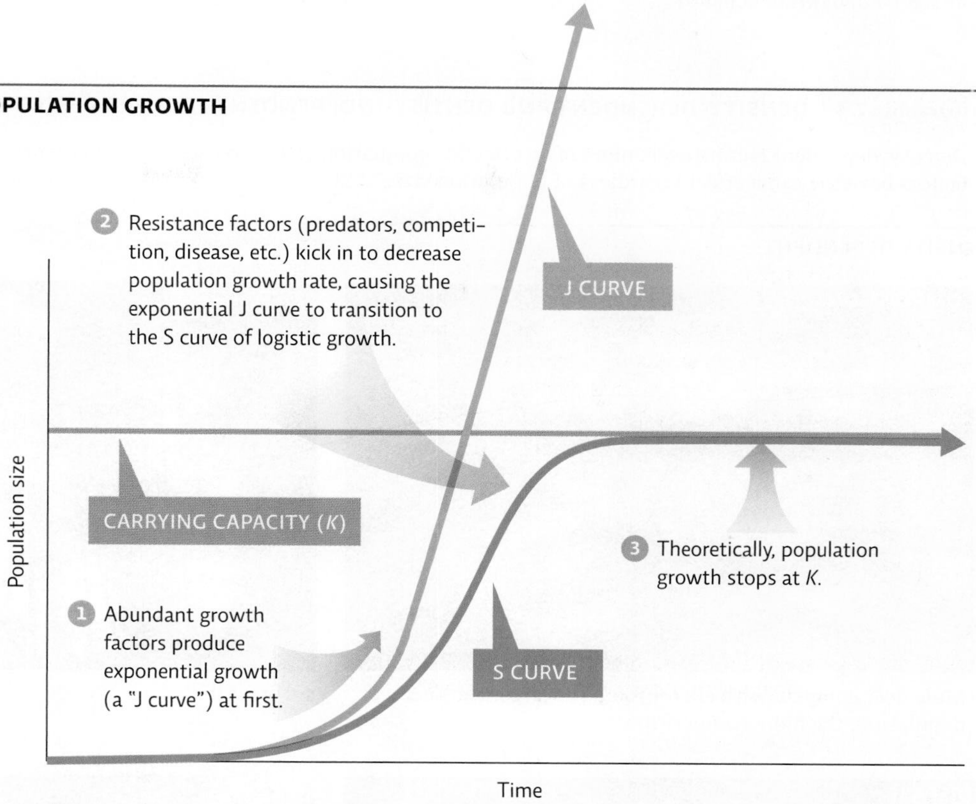
2 Resistance factors (predators, competition, disease, etc.) kick in to decrease population growth rate, causing the exponential J curve to transition to the S curve of logistic growth.
J CURVE
CARRYING CAPACITY (K)
3 Theoretically, population growth stops at K.
1 Abundant growth factors produce exponential growth (a "J curve") at first.
S CURVE
Population size
Time

increased. This kind of growth—in which as population size increases, growth rate decreases—is called **logistic growth**. A population that grows logistically will produce an *S-shaped curve* if plotted on a graph with time on the *x*-axis and population size on the *y*-axis. The S is created by the initial exponential growth (just like the first part of the J-shaped curve), followed by decelerating growth as the species approaches its maximum sustainable population size, where it levels off.

The population size that a particular environment can support indefinitely—without long-term damage to the environment—is called its **carrying capacity**, signified as *K* in population models. Carrying capacity depends on the presence of *growth factors* (resources needed for species to survive or reproduce) and it varies among species; the same environment can support many more mice than wolves, for example. Over time, a population's carrying capacity can change. If resources are diminished at a faster rate than they are replenished, the carrying capacity will drop. If, on the other hand, new resources are added or become available, perhaps due to the loss of a competitor, the carrying capacity for a species may rise. [INFOGRAPHIC 7.3]

A variety of factors affect population growth.

Populations will grow as long as growth factors are available, but as the population gets larger, resources start to decline. The *limiting factor*, that resource that is most scarce, tends to determine carrying capacity. *Resistance factors,* which include predators, competitors, and disease, also control population size and growth. The effects that predators have on populations can vary widely, in part because predators, along with disease and competition, are **density-dependent** factors—their effects all increase as population size goes up. On the other hand, some factors, such as droughts, storms, and floods, affect a population no matter how large or small it is. These **density-independent** factors don't routinely regulate population size, but they can certainly decrease it. [INFOGRAPHIC 7.4]

population growth rate The change in population size over time (births minus deaths over a specific time period).
birth rate The number of births per 1000 individuals per year.
death rate The number of deaths per 1000 individuals per year.
biotic potential (*r*) The maximum rate at which the population can grow due to births if each member of the population survives and reproduces.
exponential growth Population size becomes progressively larger each breeding cycle; produces a J curve when plotted over time.
logistic growth The kind of growth in which population size increases rapidly at first but then slows down as the population becomes larger; produces an S curve when plotted over time.
carrying capacity (*K*) The population size that a particular environment can support indefinitely without long-term damage to the environment.
density dependent Factors, such as predation or disease, whose impact on the population increases as population size goes up.
density independent Factors, such as a storm or an avalanche, whose impact on the population is not related to population size.

Infographic **7.4** | **DENSITY-DEPENDENT AND DENSITY-INDEPENDENT FACTORS AFFECT POPULATION SIZE**

↓ Density-dependent factors exert more of an effect as population size increases. On the other hand, density-independent factors have the same effect regardless of population size.

DENSITY DEPENDENT

COMPETITION

↑ Mule deer compete with elk for food; the larger the deer population, the more competition.

DISEASE

↑ Infectious diseases, such as chronic wasting disease, which weakens and eventually kills the animal, spread more easily in large populations of elk, deer, or moose.

PREDATORS

↑ Predators are more successful at capturing prey from larger populations than from smaller ones.

DENSITY INDEPENDENT

FIRE/FLOOD

↑ Forest fire

STORMS

↑ Row of trees flooded by a river

AVALANCHE AND OTHER NATURAL DISASTERS

↑ Snow avalanche

Infographic **7.5** | **REPRODUCTIVE STRATEGIES**

↓ Different species have different potentials for population growth, known as reproductive strategies. A species' biology may place it anywhere along a continuum between two extremes—the *r*- and *K*-selected species.

CHARACTERISTICS OF *r*-SELECTED SPECIES

1. Short life
2. Rapid growth of individual
3. Early maturity
4. Many, small offspring
5. Little parental care
6. Adapted to unstable environment
7. Prey
8. Niche generalists

CHARACTERISTICS OF *K*-SELECTED SPECIES

1. Long life
2. Slower growth of individual
3. Late maturity
4. Few, large offspring
5. High parental care
6. Adapted to stable environment
7. Predators
8. Niche specialists

r-SELECTED SPECIES
K-SELECTED SPECIES
DANDELIONS
DEER MICE
ELK
BEARS
SPOTTED KNAPWEED
SPRUCE TREES

A species' biology also affects its growth potential. For instance, ecologists recognize a continuum of **reproductive strategies** among species. Those that reproduce quickly, with high biotic potential, are known as ***r*-selected species**—so named because of their high rate of population increase (*r*). Species that are *r*-selected, such as deer mice and spotted knapweed, are well adapted to exploit unpredictable environments, and are able to increase quickly if resources suddenly become available.

On the other hand, ***K*-selected species**, which in Canada include bears, wolves, and slow-growing trees like spruce, are found at the other end of the continuum and have much lower reproductive rates. Because they reproduce slowly, their population growth rates are responsive to environmental conditions; they decrease or increase slowly if resources decrease or increase in availability.

reproductive strategies How quickly a population can potentially increase, reflecting the biology of the species (lifespan, fecundity, maturity rate, etc.).
r-selected species Species that have a high biotic potential and that share other characteristics such as short lifespan, early maturity, and high fecundity.
K-selected species Species that have a low biotic potential and that share characteristics such as long lifespan, late maturity, and low fecundity; generally show logistic population growth.

This responsiveness tends to keep population sizes close to carrying capacity (*K*). [INFOGRAPHIC 7.5]

K- and *r*-selected species often experience different types of population change. For instance, population sizes tend to be stable for *K*-species in undisturbed, mature areas. On the other hand, *r*-species sometimes have sudden, rapid population growth, characterized by occasional surges to high peaks, which may overshoot carrying capacity, followed by sudden crashes, especially in response to seasonal availability of food or temperature changes. Their high rate of reproduction does not allow the population the time to adjust and produce fewer offspring as resources become scarce. When this occurs, the population that has exceeded carrying capacity will drop below carrying capacity and then increase again. Some populations will eventually level off close to carrying capacity while others continue to overshoot and crash.

Predator and prey also respond to each other; predator populations increase as their prey population increases. But an increase in predators eventually reduces the prey population. Fewer prey means less food, and eventually reduces the number of predators, allowing the prey species to recover. In some cases, the fluctuations in

Infographic **7.6** | **SOME POPULATIONS FLUCTUATE IN SIZE OVER TIME**

OVERSHOOT AND CRASH

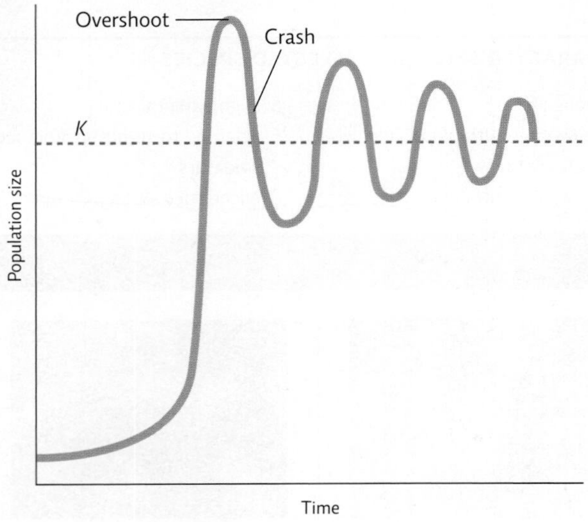
Overshoot
Crash
K
Population size
Time

↑ Not all populations show logistic growth that levels off nicely at carrying capacity. Some populations might overshoot carrying capacity, drop below it, and increase and overshoot it again until they settle down close to carrying capacity.

SNOWSHOE HARE AND CANADA LYNX POPULATION SIZE ESTIMATES AT KLUANE LAKE, YUKON

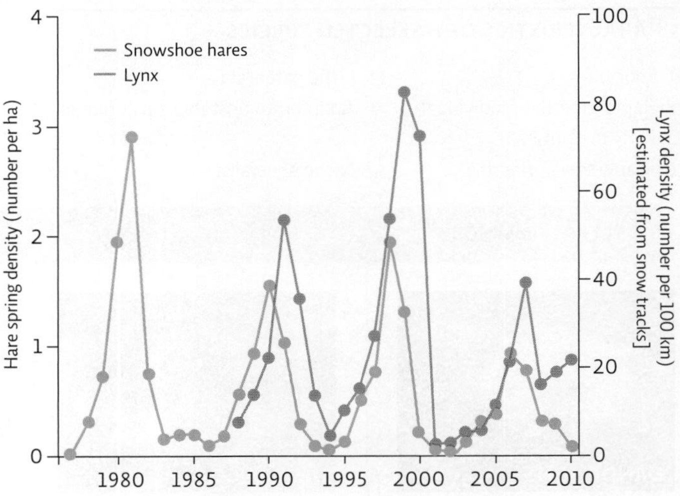
Snowshoe hares
Lynx
Hare spring density (number per ha)
Lynx density (number per 100 km) [estimated from snow tracks]
0
1
2
3
4
0
20
40
60
80
100
1980
1985
1990
1995
2000
2005
2010

↑ Some predator–prey populations, like that of the snowshoe hare and Canada lynx, regularly go through boom-and-bust cycles as part of their natural population cycle.

population size are large enough to result in **boom-and-bust cycles**, with the predator population peaks lagging behind those of its prey or food source. A classic example is that of the snowshoe hare and Canada lynx. Long-term data, based on the fur trade and going as far back as the 1800s, show that snowshoe hare populations undergo cyclic oscillations. According to recent studies by University of British Columbia researcher Charles Krebs, these mirror the size of predator populations, such as lynx and great-horned owls. The lowest point in hare population size occurred in 2006 after the rise of an additional hare predator, the marten, whose population has increased significantly since 2000. [INFOGRAPHIC 7.6]

The loss of the wolf emphasized the importance of an ecosystem's top predator.

Thanks in part to Smith's determination, the Yellowstone Gray Wolf Restoration Project has been incredibly successful. There are now more than 5000 of these wolves living in the contiguous United States, approximately 100 of which are in Yellowstone National Park. But one chief lesson hammered home by observing wolves in Yellowstone is that populations do not exist in isolation. "Work in central Yellowstone clearly demonstrated that the addition of a keystone species, such as wolves, can result in observable changes in the behaviour of its prey," says Claire Gower, a wildlife biologist at Montana Fish, Wildlife, and Parks. "These consequences may not stop at the prey individual, but may have cascading effects on other community-level processes." After wolves were reintroduced to Yellowstone, scientists noticed that coyote populations in the area shrank, in part because the wolves were hunting them. This has had a cascade of effects on the other smaller mammals upon which coyotes typically prey. For instance, scientists speculate that the elusive red fox—whose nocturnal behaviour makes these foxes difficult to track—has probably become more common in the park, because fewer coyotes are around to hunt them, and because foxes do not share any prey with wolves.

The elk population has also been affected. According to the U.S. National Park Service, the elk population in Yellowstone doubled between 1914 and 1932 as a direct result of wolf **extirpation** (local extinction). These elk then began exerting pressure on their food sources. Willow trees were overgrazed because elk like to eat the tender young shoots; this removed an important resource for other animals like birds (habitat) and beaver (food and building material). The decline of the beaver population was especially significant because beaver dams change the flow of water, creating lakes that support a wide

↑ The bare trunks of these aspen trees show the winter browse line from herbivores such as elk and deer, which eat leaves as high as they can reach during winter. As forage becomes more and more scarce, these animals will even strip off the bark, damaging the trees.

variety of fish, amphibian, and plant species that would not frequent faster-flowing streams. The loss of these dams allowed streams to return to their original flow, altering the community that lived in the area.

> "Work in central Yellowstone clearly demonstrated that the addition of a keystone species, such as wolves, can result in observable changes in the behaviour of its prey."—Claire Gower

As the wolf population increased again after reintroduction, the elk spent less time in willow stands, preferring to be out in the open where they could see their predators. As the willows recovered, so did songbird populations. Beavers returned as well, and their dam-building restored lakes and fish populations. Damming the river also slowed the flow of water enough to increase the amount that soaked into the ground, recharging groundwater supplies and providing more water for the deep roots of trees. Populations interact with their ecological community in complex and often unpredictable ways. [INFOGRAPHIC 7.7]

Ecologist and environmentalist Aldo Leopold, who lived in the early 20th century, was one of the first conservationists concerned with the effect of eliminating the wolf from the American West. He recognized that the problems resulting from interference with ecological communities could be far-reaching. "I now suspect that just as a deer herd lives in mortal fear of its wolves, so does a mountain live in mortal fear of its deer," he wrote in his book *A Sand County Almanac*. "And perhaps with better cause, for while a buck pulled down by wolves can be replaced in two or three years, a range pulled down by too many deer may fail of replacement in as many decades."

Many populations of animals other than wolves are declining worldwide due to human impact, especially from habitat loss, the introduction of non-native species, and predator removals. The number one reason that species become endangered is habitat destruction. People alter or destroy habitats or resources, breaking needed connections within ecosystems, and populations respond. Some species may benefit from the change and

---

boom-and-bust cycles Fluctuations in population size that produce a very large population followed by a crash that lowers the population size drastically, followed again by an increase to a large size and a subsequent crash.

extirpation Local extinction of a species.

Infographic **7.7** | **THE PRESENCE OR ABSENCE OF WOLVES AFFECTS THE ENTIRE ECOSYSTEM**

↓ Wolves impact the entire ecosystem by keeping elk populations in check. Whether elk are low or high in number affects the entire plant community, which then affects which animal species are present.

WITHOUT WOLVES

↑ When willow thickets are abundant, beavers thrive and build dams that create lakes and ponds. With no wolves to look out for, elk stay in willow thickets and overgraze willow needed by beavers.

↑ Without willow thickets, the beavers leave. Dams are not built or maintained and water flow changes, reducing bird and other wildlife populations.

WITH WOLVES

↑ Willows regrow because elk now feed in meadows to watch for wolves.

↑ With more willows, the beavers return and build dams. The area floods again and wildlife populations recover.

their population could increase (purple loosestrife, a non-native plant, spreads quickly in wet roadside ditches across Canada). Other species may decline in number because they are displaced by others, or because needed conditions for growth are no longer present (songbirds may lose habitat if the woodland patches they nest in are cut down). Understanding community interactions and population dynamics helps managers monitor, protect, and restore populations.

Often these dynamics are complex. For instance, while many people believe that the Wolf Restoration Project is the sole cause of Yellowstone's recent elk population decline—the reintroduced wolves, they say, are simply eating all the elk—in reality, many factors have played a role in the decline. "Wolves have come back, cougars have come back, the state was managing for fewer elk by hunting them when they left the park, and grizzly and black bears increased in number, which eat elk calves too," Smith explains. "There are many reasons why they've declined." Some studies suggest that winter weather may be one of the main factors that control elk and deer population sizes. To ensure that a population survives, it is crucial that managers assess all of the possible reasons for a population's changes.

The success of the reintroduction program has led to the wolf being "delisted" in Montana and Idaho (its status has been changed from "endangered" to "threatened"), though not without opposition from some conservation groups. This delisting has the potential to allow wolf hunts with quotas set by the U.S. Fish and Wildlife Service, similar to those allowed in Canada. But even if wolf populations become stable, biologists will have to continually manage them so that they stay that way. Many provinces and territories have wolf management plans that focus not only on maintaining healthy populations, but also on controlling their numbers to prevent conflicts with humans and on protecting prey such as caribou. The Yukon Wolf Management Plan, for instance, recognizing that wolves are a renewable and currently stable resource in that area, identifies hunting and trapping as legitimate activities (except when nursing pups are present); it reports that 2–3% are killed annually but that as much as 30% of the population could be taken without compromising the population's conservation status. The British Columbia Wolf Management Plan has no limits on hunting wolves in areas where livestock are at risk. Wolf hunting has become a major sport in Canada, but not surprisingly it has its critics, who are opposed to killing wolves and want to encourage wolf viewing as a form of ecotourism.

To ensure that the wolves reintroduced to Yellowstone are given the chance to really flourish, Smith and his colleagues diligently stay on the wolves' trail, studying their population dynamics. "We want to know their population size, what they are eating, where they are denning, how many pups they have and how many survive, and how the wolves interact with each other," he explains. Why is it so important to ensure that the wolves do well? Simply put: they were here first, he says. "We want to restore the original inhabitants to the Park."◉

Select references in this chapter:
Beyer, H.L., *et al.* 2007. *Ecological Applications*, 17: 1563–1571.
Krebs, C.J. 2010. *Proceedings of the Royal Society B*, 278(1705): 481-489.
Vucetich, J.A., and Peterson, R.O. 2004. *Proceedings of the Royal Society London Biological Sciences*, 271: 183–189.

BRING IT HOME

PERSONAL CHOICES THAT HELP

Understanding the factors that influence how populations change can help us manage species that are either becoming invasive in nature or are facing extinction. How people view species and their connection to our world has a large impact on how management plays out.

Individual Steps

→ Learn more about wolves at the International Wolf Center (www.wolf.org).
→ Use the Internet and books on wildlife to research what your area might have been like prior to human settlement. Which species have been extirpated; which ones have been introduced? How have wildlife populations changed as a result of human action?
→ See if you can recognize distribution patterns in the wild. Do some flowers or trees grow in clumps or patches? Can you find species that appear to have a random or uniform distribution?

Group Action

→ Explore organizations that support predator preservation, such as Defenders of Wildlife and Keystone Conservation, for suggestions on how you can help educate others about the importance of predators.
→ Join a local, regional, or national group that works to monitor, protect, and restore wildlife habitats, such as the Nature Conservancy of Canada's Conservation Volunteers.

Policy Change

→ Caribou *(Rangifer tarandus)* is an iconic Canadian species that is currently listed as threatened under Canada's Species at Risk Act. In Banff National Park, falling caribou numbers are linked to rising wolf populations brought on by high elk populations. Reintroduction of caribou has been proposed using captive-bred animals. Research the issue and write a letter to Parks Canada that reflects your position on this proposal.

UNDERSTANDING THE ISSUE

CHECK YOUR UNDERSTANDING

1. Consider the roach. Why does its population size never reach its biotic potential?
   a. Females produce few offspring at a time.
   b. Its tolerance limits are too broad.
   c. Overharvesting by humans.
   d. Resistance factors limit population size.

2. The concept of minimum viable population:
   a. explains how many individuals can fit into a habitat.
   b. describes the potential number of individuals if there are no predators.
   c. describes the potential number of individuals if there is no competition.
   d. describes the smallest number of individuals needed to ensure the long-term continuation of a particular population.

3. Population density is:
   a. the number of individuals there are in a given area, such as a hectare or square kilometre.
   b. the number of members there are in each pack or group of animals.
   c. complicated by the home range of each organism.
   d. the minimum number of individuals that must be in an area in order to ensure long-term continuation of that population.

4. Groups of fish called blue tang swim in large schools around coral reefs in the ocean, while barracuda, a fierce, predatory fish, prowl the waters alone, looking for prey throughout the area. Regarding distribution:
   a. blue tang are random and barracuda are solitary.
   b. blue tang are random and barracuda are uniform.
   c. blue tang are clumped and barracuda are random.
   d. blue tang are uniform and barracuda are random.

5. Exponential growth of a population:
   a. is often seen if a population reaches a new environment that is favourable.
   b. is a J-shaped curve on a population graph.
   c. occurs when every female reproduces at every opportunity and every offspring survives.
   d. All of the above are true.

6. Carrying capacity, or *K*, is:
   a. determined primarily by the number of predators in the area.
   b. the number of individuals that can live in a sustainable fashion in a particular environment.
   c. the total number of individuals that could possibly live in a particular environment for a short time during an exponential growth phase.
   d. determined only by the potential birth rate of a population.

WORK WITH IDEAS

1. Compare the reproductive strategy of a deer mouse with that of a bear.

2. Kangaroo rats eat seeds and are eaten by coyotes. Under what conditions might the kangaroo rat population increase exponentially? Logistically? Identify at least two factors that could lead to a decrease in the kangaroo rat population size.

3. Mountain gorillas live in extended families and search daily for favourite tree types that have tasty leaves. Contrast the *population distribution* of mountain gorillas with that of the trees they eat.

4. Lesser goldfinches are small, seed-eating birds. In cities, they are eaten by wild hawks and by domestic cats. Discuss several density-dependent and density-independent factors, including both critical factors and resistance factors, that could affect their carrying capacity.

5. Lemmings, the smallest mammal in the Arctic, feed on the roots, shoots, and blades of grasses. The stoat is a weasel that feeds exclusively on lemmings. Describe how these two populations undergo boom-and-bust cycle population fluctuations.

6. Ranchers in your area want to impose a bounty on wolves and mountain lions; they fear the predators are eating young calves. As a wildlife biologist, explain to the ranchers why this may or may not be a good idea.

ANALYZING THE SCIENCE

The following graph shows the numbers of Kaibab deer on the isolated Kaibab Plateau in northern Arizona. President Theodore Roosevelt declared it a National Game Preserve in 1906 and predator removal was encouraged to protect the deer. Between 1907 and 1917, more than 600 mountain lions were removed; 200 more were removed over the next 20 years. Wolves were exterminated by 1926 and more than 7000 coyotes and 500 bobcats were also removed. The deer population increased, and by 1915 there was significant damage to the grasses, shrubs, and trees that were being eaten by the deer.

Kaibab deer population, Kaibab Plateau, Arizona

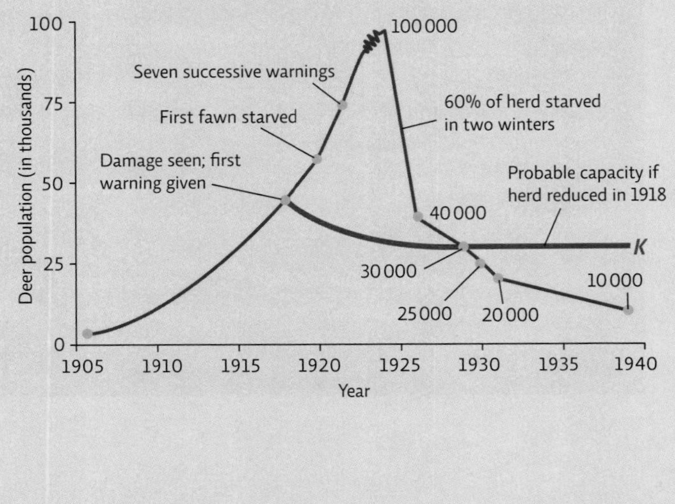
100000
Seven successive warnings
First fawn starved
60% of herd starved in two winters
Damage seen; first warning given
Probable capacity if herd reduced in 1918
40000
K
30000
25000
20000
10000
Deer population (in thousands)
0
25
50
75
100
1905
1910
1915
1920
1925
1930
1935
1940
Year

INTERPRETATION

1. With the initial level of predators, what was the probable carrying capacity (*K*) of the Kaibab Plateau for deer at the start of this story?
2. Once the population had passed its probable *K*, would it be able to sustain itself at the higher numbers? Explain your answer.
3. In what 2 to 3 year period did the deer population have the highest reproductive growth rate?

ADVANCE YOUR THINKING

4. Explain the factors—critical, resistance, density-dependent, density independent—that accounted for the changes in population numbers for the deer in 1905, 1915, and 1930.
5. Decide if Kaibab deer behaved like an *r*-selected species or a *K*-selected species, and justify your answer.
6. Compare the probable look of the habitat of the Kaibab Plateau (295 000 hectares, 2745 metres elevation) in 1905 with the way it probably looked in 1935 after 30 years of damage.

EVALUATING NEW INFORMATION

Canada has the second largest grey wolf population in the world—second only to Russia—with an estimated 50 000 animals. The species is not considered threatened, and wolf hunting is allowed outside of national and provincial parks. Many provinces and territories do have wolf management plans. The 2012 Yukon Wolf Management Plan, for example, has as its goal the conservation and management of the wolf and includes steps to reduce wolf population sizes as needed through hunting, trapping, and sterilization of individuals. However, not everyone agrees on how best to manage wolf populations in Canada.

Visit the website for the Canadian Wolf Coalition (www.canadianwolfcoalition.com) and read about the organization and its "Vision and Goals." Then take a look at the Yukon Wolf Management Plan (www.env.gov.yk.ca/publications-maps/documents/yukon_wolf_conservation_and_management_plan.pdf).

Evaluate the websites and work with the information to answer the following questions:

1. Is the Canadian Wolf Coalition a reliable information source?
   a. Does the organization give supporting evidence for its claims?
   b. Does it give sources for its evidence?
   c. What is the mission of this organization?
2. Choose one of the articles posted on the Canadian Wolf Coalition home page and read it.
   a. What is the main point of the article?
   b. Are supporting evidence and sources for this evidence given for the main point(s)?
   c. Do you detect any bias in the presentation? Explain.
3. Evaluate the Yukon Wolf Management Plan:
   a. What organization authored the plan?
   b. Do the authors give supporting evidence and the sources of this evidence for their claims?
   c. Do you detect any bias in the presentation? Explain.
   d. How do the goals of this plan compare to those of the Canadian Wolf Coalition?
4. Do you believe that wolves should be managed, and populations controlled through hunting, trapping, or other population reduction methods? Support your position.

MAKING CONNECTIONS

PRAIRIE DOGS AND THE GRASSLANDS OF SOUTHERN ARIZONA

Background: Early in the 20th century, the valleys of southern Arizona were filled with tall grasses, which attracted cattle ranchers. Summer rainfall on the mountain ranges washed down into the valleys and permeated the soil, nourishing the grasslands. In the late 1800s, the ranchers had steadily killed off large predators, such as wolves, mountain lions, coyotes, and eagles. By the early 1900s, hundreds of thousands of prairie dogs lived in vast underground "towns" (interconnected tunnels and burrows) throughout the grasslands. Prairie dogs are social, very active animals, constantly digging and moving their tunnels. They kept the soil soft and porous, and filled with numerous holes. Grazing cattle would fall in the holes, break a leg, and die, leading to the ranchers' collective decision to remove the prairie dogs.

Ralph Morrow was a game warden who received permission and assistance from the U.S. federal government to trap and poison the prairie dogs east of the Chiricahua Mountains. Within 10 years, he had removed the prairie dogs from this area. By the 1930s, the grasslands were gone as well, replaced by desert scrub vegetation. Summer floods were eroding large gullies, and the topsoil was gone, leaving hard-packed sandy clay with little organic matter.

Case: A group in Idaho is interested in eliminating prairie dogs because it is trying to farm buffalo commercially. The group members have come to your agency, which does environmental research and recommends strategies for remediation of local problems. Write a report detailing the ecological events relating to the extermination of prairie dogs east of the Chiricahua Mountains, and explaining what happened to change the area so dramatically. Make sure to consider the following in your report:

a. predator removal
b. prairie dog populations
c. prairie dog removal and its effects on soils, rainfall runoff, and erosion
d. ecology and effects on the grasslands of prairie dog removal and soil changes
e. effects of cattle (and buffalo) on soils
f. your advice regarding the group's wish to eliminate prairie dogs

BRINGING A FOREST BACK FROM THE BRINK

A forest is more than just its trees

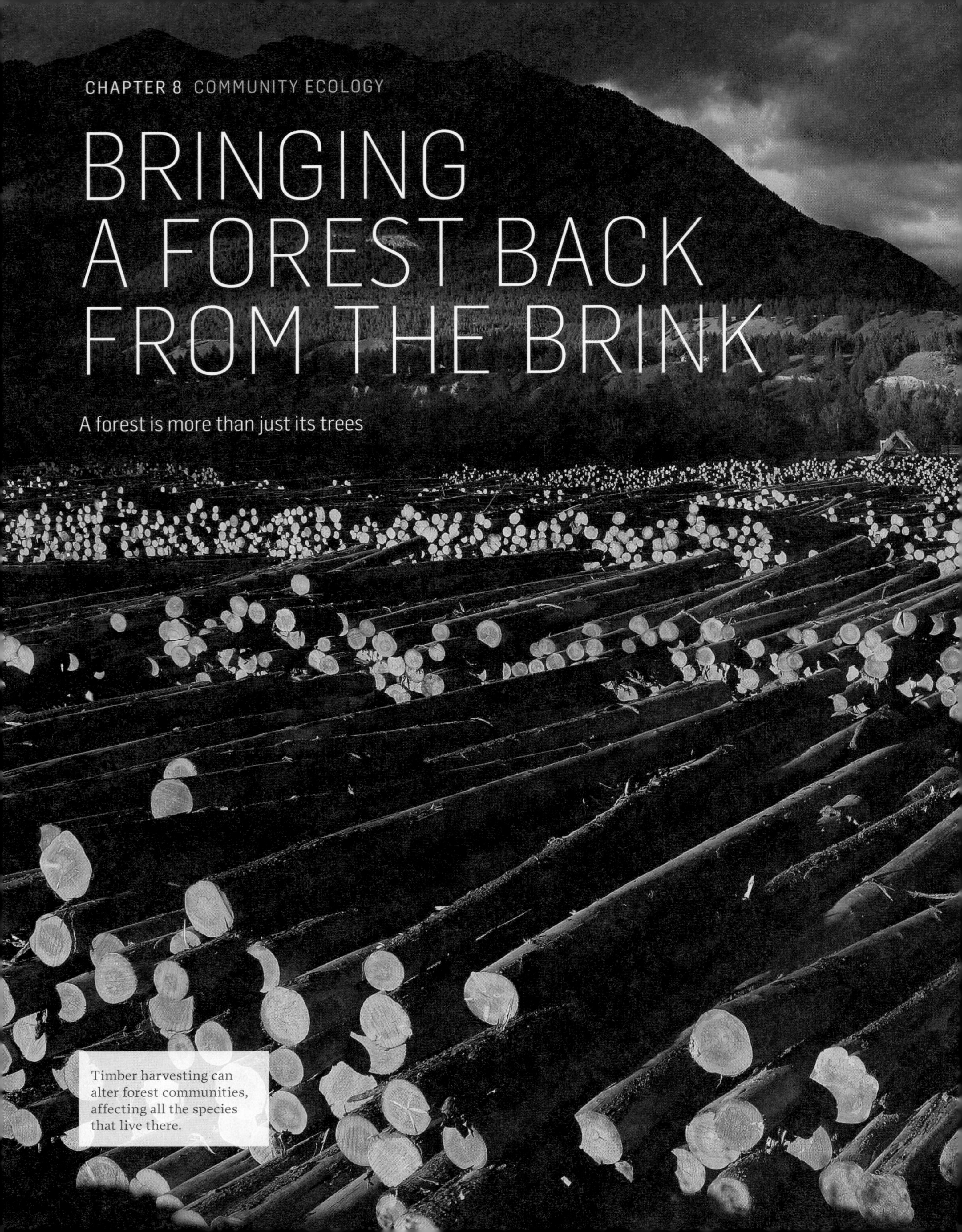

Timber harvesting can alter forest communities, affecting all the species that live there.

CORE MESSAGE

Ecological communities are complex assemblages of all the different species that can potentially interact in an area. All the pieces of the ecological community are connected; change one thing and many others are affected. This means ecosystems are often negatively affected by human impact. Understanding the interconnections within the communities may allow us to better protect and even help restore damaged ecosystems.

GUIDING QUESTIONS

After reading this chapter, you should be able to answer the following questions:

→ How do matter and energy move through ecological communities?

→ How do biotic and abiotic factors affect community composition, structure, and function?

→ How do species interactions contribute to the overall viability of the community?

→ In what ways do human actions affect ecological communities and how can we take steps to help restore damaged ecosystems?

→ How do ecosystems change over time through ecological succession? How can we use this knowledge to assist in ecosystem restoration?

Recently, conservation biologist Josh Noseworthy and a friend decided to take a winter hike through New Brunswick to Ayers Lake, one of the most pristine places in the world. To get there, they had to walk up a long forest road, which seemed to take forever—in part, because the scenery was so depressing. All they could see were the same types of trees—the same height, width, and age. It was hard evidence that this land had once been stripped by lumber companies harvesting trees, mostly for construction or paper, which were then replaced with row upon row of the same type of species, all planted at the same time. There was nothing natural about this silent forest; the only colours were the dark trunks of the monotonic trees, separated by starkly white, snowy spaces that reflected the bright sunlight that forced the hikers to occasionally squint to see. Finally, after what seemed like an interminable climb, Noseworthy came to the top of a hill, looked down at the vista below, and exhaled. This, he thought, is what a forest should look like.

Below him lay one of the last remaining parts of a crucial ecosystem known as the Acadian forest. The difference between what they had just passed through and what lay before them was remarkable. To descend into the Acadian forest, Noseworthy and his fellow hiker had to carefully make their way through the thick, diverse vegetation, winding their way around trees of all shapes and sizes. After being in the bright sunlight, it took their eyes a while to adjust to the darkness of the native forest, where tall trees create a canopy that barely lets any light through. They passed through a massive, dense stand of trees whose cavities would house the spring nests of birds or squirrels—the same species that help disperse the trees' seeds. They stepped over huge, moss-covered fallen trunks that were slowly decaying and replenishing the soil, while serving as rich habitat for invertebrate communities that recycle the trees' nutrients, and the elusive salamanders that feed on the worms and insects. It was winter, so the forest was relatively quiet, but in the spring it would come to life with the multi-layered calls of different species of birds, the chatter of squirrels, and the rhythmic carpentry of the woodpecker.

◉WHERE IS AYERS LAKE?

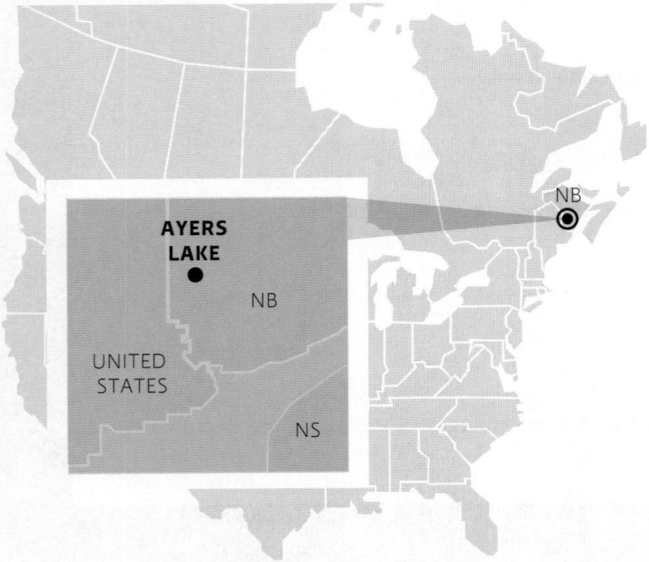
NB
AYERS
LAKE
NB
UNITED
STATES
NS

Finally, using the shards of light penetrating the treetops to deftly navigate through the diverse vegetation, the hikers reached Ayers Lake, one of the few remaining water bodies without a single cottage, farm, or humanmade structure alongside it. The sight made Noseworthy, a conservation biologist at Nature Conservancy Canada (NCC), stop in his tracks. "It was really quite an experience for me to see a lake surrounded by only forest," he says. "I've read about the lake for years, and been to many Acadian forests before, but never one that had not been altered by human activity in some way."

ecosystems All of the organisms in a given area plus the physical environment in which, and with which, they interact.
habitat The physical environment in which, and with which, individuals of a particular species can be found.
community ecology The study of all the populations (plants, animals, and other species) living and interacting in an area.

↑ A young Acadian forest in Fundy National Park, New Brunswick, shows regrowth but lacks the larger trees that would be part of the old growth forests of Canada's past.

That's because less than 5% of the Acadian forest remains in its natural state.

All the pieces of the ecological community puzzle are connected.

In the years before Europeans settled in the Maritimes, as much as 50% of the land was covered in Acadian forest—a highly diverse **ecosystem** made up of many plant, animal, and fungal species, not to mention the millions of species of bacteria that inhabit the soil. These forests, found throughout southeastern Canada and the northeastern United States, provided vital functions. These include serving as **habitat** for innumerable species, providing a natural sink for carbon dioxide that helps regulate climate, and creating a dense root network that pulls up thousands of litres of water per day in the summertime, contributing to the forest's humidity levels and its rainfall. These roots also provide other benefits: they protect surface waters by reducing the flow of sediments into nearby streams during rainstorms, allowing the water to soak into the ground to replenish groundwater supplies. (See Chapter 11 for more on the services forests provide.)

But once Western settlers arrived, they quickly began clearing the land to create farms and harvest timber. Where sections of the forest remained, lumber companies began to remove the trees they needed (hardwood for paper, softwood for lumber). If they could work around the remaining species, these less desirable trees were left behind; if not, everything was removed. Companies then replanted species that grew the fastest, such as spruce trees, so they could be reharvested as quickly as possible. Now, only pockets of the natural Acadian forest remain, mostly in small, isolated areas surrounded by steep gorges that make them inaccessible to harvesting, such as the area around Ayers Lake.

Since so little of the Acadian forest remains untouched, some of its characteristics are somewhat hard to determine. In general, a healthy Acadian forest has a mix of 30 or so different species of trees, living interdependently with associated plants and animals. The well-being of any one of those species depends on the health of the entire ecosystem—a concept that falls within the field of **community ecology**. This is the study of how a given ecosystem functions—how space is structured, why certain species thrive in certain areas, and how individual species in the same community interact with one another and their habitat. This includes understanding how various species contribute to ecosystem services like pollination, water purification, and nutrient cycling.

Infographic **8.1** | **CANADA'S FOREST REGIONS**

↓ Forests cover about half of Canada's land area and, as this map shows, boreal forests are the most expansive forest found here, stretching from coast to coast and up to the tundra biome.

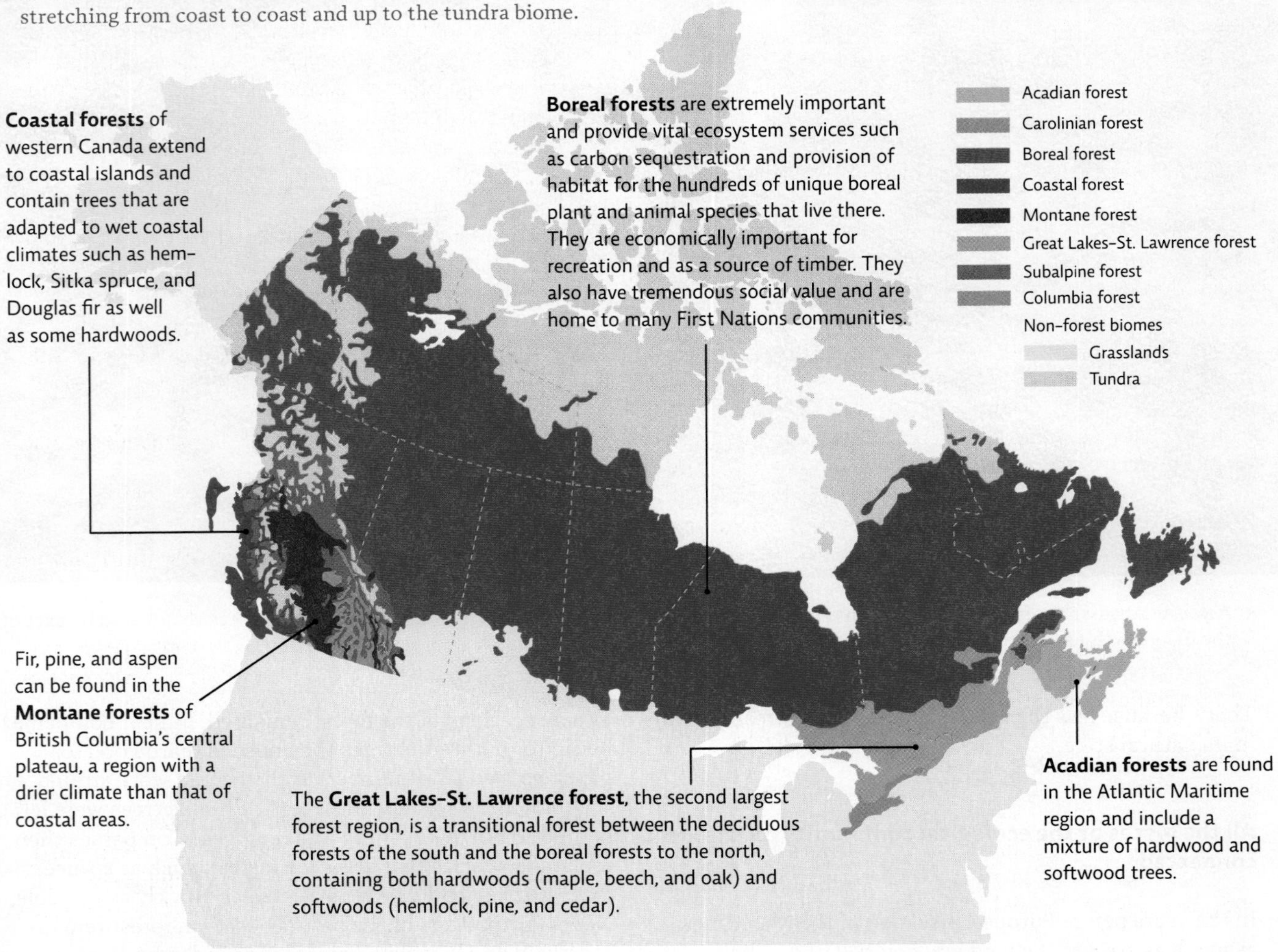
Coastal forests of western Canada extend to coastal islands and contain trees that are adapted to wet coastal climates such as hemlock, Sitka spruce, and Douglas fir as well as some hardwoods.
Boreal forests are extremely important and provide vital ecosystem services such as carbon sequestration and provision of habitat for the hundreds of unique boreal plant and animal species that live there. They are economically important for recreation and as a source of timber. They also have tremendous social value and are home to many First Nations communities.
Acadian forest
Carolinian forest
Boreal forest
Coastal forest
Montane forest
Great Lakes–St. Lawrence forest
Subalpine forest
Columbia forest
Non-forest biomes
Grasslands
Tundra
Fir, pine, and aspen can be found in the **Montane forests** of British Columbia's central plateau, a region with a drier climate than that of coastal areas.
The **Great Lakes–St. Lawrence forest**, the second largest forest region, is a transitional forest between the deciduous forests of the south and the boreal forests to the north, containing both hardwoods (maple, beech, and oak) and softwoods (hemlock, pine, and cedar).
Acadian forests are found in the Atlantic Maritime region and include a mixture of hardwood and softwood trees.

Many community ecologists are fascinated by the Acadian forest, because it contains a mix of tree species from both boreal and temperate zones, whose edges overlap to create a highly diverse mix of wildlife. The boreal forest—also known as *taiga*—is particularly important to Canada, as it covers more than half of the country. It is one of the world's richest resources, serving as the breeding site for 30% of all North American birds, among them millions of ducks and waterfowl, and containing much of the world's supply of liquid fresh water.

Although boreal forests are dominated by coniferous trees, approximately 30% of boreal land mass is covered by wetlands, mainly peatlands. Peatlands are distinct, moss-dominated ecological communities, which develop deep organic soils that store large amounts of carbon. Because mosses and other peatland plants decompose very slowly, these soils accumulate over thousands of years, making peat soils a relatively nonrenewable resource, and making damaged peatlands impossible to restore within human lifespans. Wetlands are not only highly biodiverse; they also provide extremely important water management services, including filtration and replenishment of groundwater reserves.

When the boreal forest transitions to the temperate forests found further south, you find a kaleidoscope of different organisms that rarely meet anywhere else in the world—aside from the mix of trees of many types,

---

indicator species The species that are particularly vulnerable to ecosystem perturbations, and that, when we monitor them, can give us advance warning of a problem.

food chain A simple, linear path starting with a plant (or other photosynthetic organism) that identifies what each organism in the path eats.

ages, and sizes, the Acadian forest contains a wide range of insects, worms, fungi, birds, rodents, and large predators such as hawks, bobcats, and black bears. The forest contains a wide range of habitats, from the dense tree cover Noseworthy hiked through to marshes and bogs (peatlands), which are the homes of fish, frogs, and the smaller aquatic organisms they feed on. Different levels of the forest itself, from forest floor to the upper canopy, offer different niches, occupied by different species. [INFOGRAPHIC 8.1]

The Acadian forest is a prime example of how every piece of an ecosystem is interdependent—change one thing, and many others are affected. Ecologists often say "you can't do one thing"; the loss of even one species can disrupt an entire ecosystem. For instance, cutting down trees reduces habitat for lichens, which drape from those trees; without these hanging lichens, there are fewer Northern Parula warblers (a small migratory songbird), which use those lichens for their nests. Fewer of these nests mean fewer warblers to feast on forest insects, and so more insects survive. It also means less food for predators that live off the eggs and small birds. Lichens, one of the few foods available in the winter, also help sustain deer and other year-round residents.

In the Acadian forests (and many areas worldwide) lichens are considered an **indicator species**—those that are particularly vulnerable to perturbations of the ecosystem. Most lichens are very vulnerable to the effects of air pollution, acting like sponges that soak up chemical pollutants in rainwater. If lichens start to disappear, this lets us know there is a problem and gives us time to act before other species are affected.

An ecosystem as diverse as the Acadian forest has several indicator species. Take the brown creeper, a tiny (7–8 grams), lanky songbird with a fairly narrow niche—it feeds on invertebrates found in the crevices formed by the deep bark furrows of large, live trees. "As soon as the density of those trees goes down, so does the density of the brown creeper," says Marc-André Villard, a biologist at the Université de Moncton in New Brunswick.

Another indicator species of the Acadian forest is the ovenbird. With its striped belly and loud, characteristic call, it spends most of its time on the forest floor, where it feeds on invertebrates found in dead leaves. When trees are cut down and dragged through the soil, this reduces and disturbs the amount of leaf litter—essentially depriving the ovenbird of its main food source. "This small bird is sensitive to even small changes in tree density," says Villard. "And when their numbers drop, it can take them years to recover."

Matter and energy move through a community via the food web.

Ecological communities are complex assemblages of all the different species that can potentially interact within an area. Understanding these interconnections allows us to better protect and even help restore damaged ecosystems.

Energy is the foundation of every ecosystem; it is captured by photosynthetic organisms and then passed from organism to organism via the **food chain**—a simple, linear path that shows what eats what. Any given

← A Northern Parula warbler perches on a lichen-covered branch. Many warbler species migrate to Canada in the spring to breed and are important predators of insects as well as being prey themselves for a wide variety of Acadian forest predators who feed on the birds and their eggs.

Infographic **8.2** | **THE ACADIAN FOREST FOOD WEB**

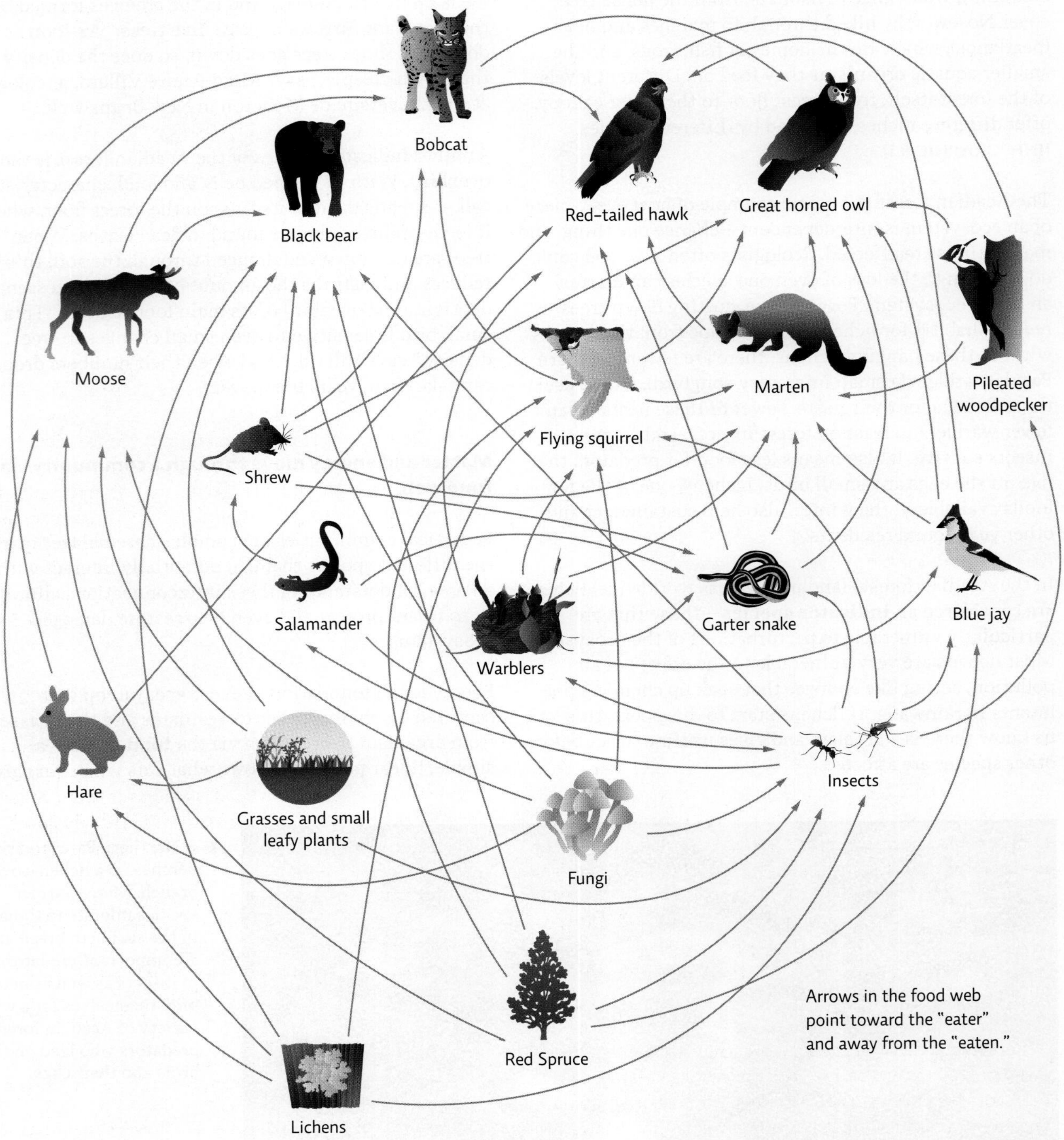
Bobcat
Black bear
Red-tailed hawk
Great horned owl
Moose
Pileated woodpecker
Marten
Flying squirrel
Shrew
Salamander
Warblers
Garter snake
Blue jay
Hare
Grasses and small leafy plants
Insects
Fungi
Red Spruce
Lichens
Arrows in the food web point toward the "eater" and away from the "eaten."

↑ The Acadian forest is very complex, containing many members. A variety of trees and understory plants are the base of the food web, which supports a wide array of vertebrate and invertebrate animals. Diets may change seasonally and this web represents only a small portion of the actual species and their connections.

ecosystem might have dozens—even hundreds—of individual food chains. Linked together they create a **food web**, which shows all the many connections in the community. Both food chains and webs help ecologists track energy and matter through a given community. They can vary greatly in length and complexity between different types of ecosystems. But most share a few common features and all are made up of the same basic building blocks—namely, consumers and producers. [INFOGRAPHIC 8.2]

Animals, fungi, and most bacteria are **consumers**—they gain energy and nutrients by eating other organisms. Some of the organisms they consume are **producers**—photosynthetic organisms such as plants that capture energy directly from the sun and convert it into food (sugar) via photosynthesis. These different feeding levels are known as **trophic levels**, into which consumers are organized based on what they eat. *Primary consumers* eat producers; *secondary consumers* eat primary consumers; *tertiary consumers* eat secondary consumers, and so on. At the top of the food chain sit the apex predators, in the last trophic level. Of course, some organisms feed at more than one trophic level: for instance, when the ovenbird eats berries or seeds found in leaf litter, it is feeding at trophic level 2 (a primary consumer); when it eats the invertebrates that live in that litter, such as beetle larvae and snails, it is feeding at trophic level 3 (a secondary consumer).

In an ecosystem as diverse and complex as the Acadian forest, there are many different organisms at each trophic level, making for a wide variety of food chains. For example, ovenbirds provide food for other birds such as blue jays, which eat their eggs; chipmunks and red squirrels also feast on ovenbird eggs and young. Both adult ovenbirds and blue jays are, in turn, the prey of martens and red-tailed hawks, which sit at the top of that particular food chain. The black bears that live in Acadian forests eat whatever they can find—anything from tree buds to ants, fruits, nuts, bird eggs, or young deer and moose, making them a versatile consumer in many Acadian forest food chains.

When any of these organisms die, they are eaten by an army of consumers known as **detritivores**—animals like worms and insects, which feed on dead plants and animals—and **decomposers**—organisms like bacteria and fungi that break decomposing organic matter all the way down into its constituent atoms and molecules.

As one moves up the food chain, energy and *biomass* (all the organisms at that level) decrease, creating a trophic pyramid. The reason is simple: nearly every organism uses up the vast majority of its energy and matter in the act of living. So when it is killed and consumed by a predator, it only passes on about 10% (some species pass on more or less than others) of all the energy and matter it consumed during its lifetime. [INFOGRAPHIC 8.3]

A pyramid's ultimate size is determined by the size of its first trophic level—the one made up of photosynthesizing producers, usually plants. More plant growth means more food for primary consumers, which in turn means more food for those organisms above them, and so on up the food chain. The end result is a larger pyramid with more and larger trophic levels. The Acadian forest supports a wide array of animal species in part because of the abundance and variety of plant life found there, particularly during the warmer seasons, which supports life at the higher trophic levels.

Productivity—the amount of energy trapped by producers and converted into organic molecules like sugar—is limited by sunlight and nutrient availability. **Gross primary productivity** is a measure of total photosynthesis. But plants only use a portion (usually less than 50%) of this energy to fuel their daily needs. Scientists use the term **net primary productivity (NPP)** to describe how much energy is left over after that. The "net" is a measure of energy available to higher trophic levels—in other words, the amount that's converted to new growth.

In the Acadian forest, along with most temperate and boreal forests, less sunlight and cooler temperatures limit photosynthesis during winter months, causing plant productivity to slow or shut down completely, resulting in less new growth, and thus less food for other organisms. This is why bears hibernate and birds fly south for the winter: the lack of productivity drives them to these

---

food web A linkage of all the food chains together that shows the many connections in the community.
consumer An organism that eats other organisms to gain energy and nutrients; includes animals, fungi, and most bacteria.
producer A photosynthetic organism that captures solar energy directly and uses it to produce its own food (sugar).
trophic levels Feeding levels in a food chain.
detritivores Consumers (including worms, insects, crabs, etc.) who eat dead organic material.
decomposers Organisms such as bacteria and fungi that break organic matter all the way down to constituent atoms or molecules in a form that plants can take back up.
gross primary productivity A measure of the total amount of energy captured via photosynthesis and transferred to organic molecules in an ecosystem.
net primary productivity (NPP) A measure of the amount of energy captured via photosynthesis and stored in photosynthetic organisms.

Infographic 8.3 | **THE TROPHIC PYRAMID**

↓ Energy enters at the base of the food chain in the first trophic level (TL) via photosynthesis and is passed on to higher levels as consumers feed on other organisms. This is shown as a pyramid (smaller on top) because only about 10% of the energy is passed on to each subsequent level, with the other 90% being "lost" to the environment (usually as heat from the energy that the organism burns in day-to-day life before it is eaten). Most food chains have only 4 or 5 levels due to this progressive loss of energy.

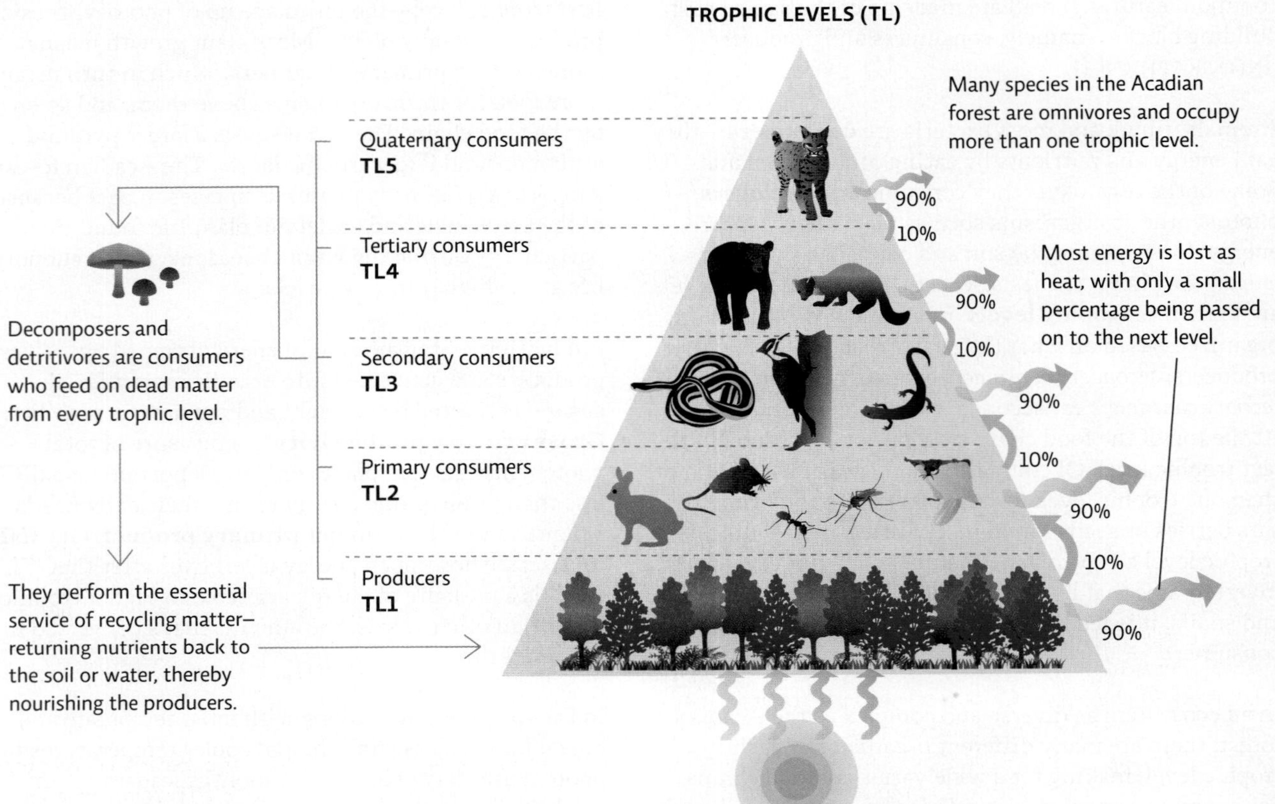

TROPHIC LEVELS (TL)
Many species in the Acadian forest are omnivores and occupy more than one trophic level.
Quaternary consumers
TL5
Tertiary consumers
TL4
Secondary consumers
TL3
Primary consumers
TL2
Producers
TL1
Decomposers and detritivores are consumers who feed on dead matter from every trophic level.
They perform the essential service of recycling matter—returning nutrients back to the soil or water, thereby nourishing the producers.
Most energy is lost as heat, with only a small percentage being passed on to the next level.
90%
10%
90%
10%
90%
10%
90%
10%
90%

extremes. NPP can be a window into the health of an ecosystem. If it rises or falls unexpectedly, ecologists can look for the cause of the change, which might be an invasion by a non-native plant or a sudden drop in a producer population. Anything that alters NPP can potentially affect organisms at every other level of the trophic pyramid.

The Acadian forest is shaped by biotic and abiotic forces.

A defining feature of the mature Acadian forest is its mix of over 30 tree species, notably red spruce. And as in any mature forest, these trees are of different ages—some young, just beginning to branch; others in their prime; and still others in a state of decay, their dead branches and decaying trunks scattered across the soil.

It turns out the variety of trees and even the deadwood and leaf litter is vitally important to the health of a forest. In a 1986 article in the magazine *Tennessee Wildlife*, Maurice Simpson, a Tennessee wildlife biologist, recounted a conversation he had with a visiting German forester as they walked through a forest in a U.S. recreational area. His visitor commented on the sounds and signs of life in the forest—birds and squirrels chirping or chattering in the treetops, insects industriously

niche The role a species plays in its community, including how it gets its energy and nutrients, what habitat requirements it has, and what other species and parts of the ecosystem it interacts with.
resilience The ability of an ecosystem to recover when it is damaged or perturbed.
species diversity The variety of species in an area; includes measures of species richness and evenness.
species richness The total number of different species in a community.
species evenness The relative abundance of each species in a community.

↑ Forest animals, like this blue jay, are important dispersers of seeds. Fungi and a wide variety of invertebrates help break down leaf litter and dead wood, returning nutrients to the soil and atmosphere for uptake by plants.

working on a decaying, fungus-covered log that housed a mouse and her family. The visitor explained that forest managers in Germany discovered just how important this "dead" material was when they carefully manicured their forests, removing dead branches, fallen logs, and standing deadwood—even leaf litter—in an effort to increase the growth of their main timber-harvesting tree, the Norway spruce. They also removed some of the competing hardwood trees to make room for more spruce. As the decaying matter and many of the native tree species were lost or removed, the animals left. And as the numbers of insects, birds, and small mammals declined, there were fewer animals to spread seeds and cycle nutrients, so the forest itself suffered. Squirrels were particularly important as dispersers of fungi that live in association with the trees' roots, helping the plants take in vital nutrients. Without these forest partners and the nutrients that the decaying material provided, the forests were dying.

These connections among species, and between species and their environment, give rise to *ecosystem complexity*. Ecosystem complexity is a measure of the number of species at each trophic level, as well as the total number of trophic levels and available niches. The Acadian forest has many food webs, some of which have thousands of species at the lower trophic levels, making it a highly complex ecosystem. Each species occupies a unique **niche**—that is, its unique role and set of interactions in the community, how it gets its energy and nutrients, and its preferred habitat. Greater ecosystem complexity means more niches and thus more ways for matter and energy to be accessed and exchanged. This increases a community's **resilience**—its ability to adjust to changes in the environment and to return to its original state rather quickly.

Species diversity, which refers to the variety of species in an area, is measured in two different ways: species richness and species evenness. **Species richness** refers to the total number of different species in a community. **Species evenness** refers to the relative abundance of each individual species. In general, populations of organisms at a higher trophic level will have fewer members than those at a lower trophic level, but populations of organisms within the same trophic level often have relatively similar numbers. If they do, the community is said to have high species evenness. If, on the other hand, one or two species dominate any given trophic level, and there are few members of other species, then the community is said to have low species evenness. In such uneven communities, the less abundant species are at greater risk of dying out.

Both richness and evenness have an impact on diversity. In general, higher species richness and evenness make for a more complex community and a more intricate food web. It enables more matter and energy to be brought into the system and also makes the community less likely to collapse in the face of calamity. The low evenness of

Infographic **8.4** | **SPECIES DIVERSITY INCLUDES RICHNESS AND EVENNESS**

The species diversity in an area is a measure of species richness (the total number of species) and species evenness (a comparison of the population size of each species). Each forest plot shown here contains 15 trees but they differ in terms of species richness and evenness.

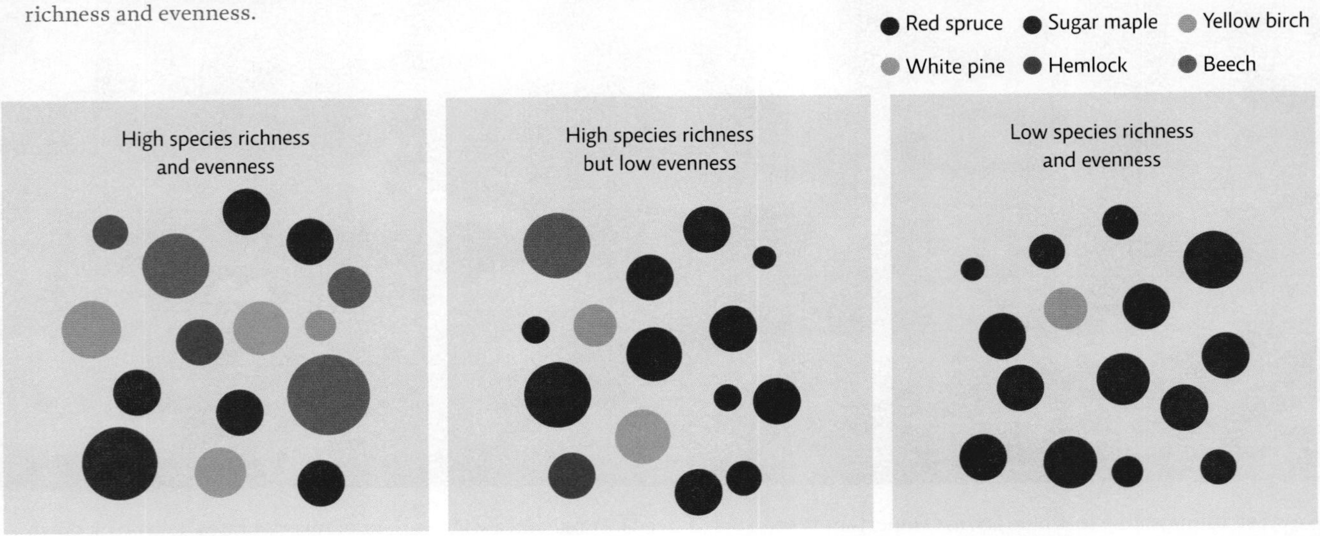
Red spruce
Sugar maple
Yellow birch
White pine
Hemlock
Beech
High species richness and evenness
High species richness but low evenness
Low species richness and evenness

the forest Noseworthy hiked through to get to Ayers Lake mirrored what has happened to much of the former Acadian forest, as well as many forests elsewhere (such as Germany)—a conversion to one species of tree, such as spruce or pine, creating an ultimate loss in species richness and a "silent forest." [INFOGRAPHIC 8.4]

A community's composition, and thus complexity, is also heavily influenced by its physical features. As physical features like temperature and moisture change, so does community composition. This often happens in **ecotones**, places where two different ecosystems meet—like the edge between forest and field, or river and shore. The different physical makeup of these edges creates different conditions, known as **edge effects**, which either attract or repel certain species. For example, it is drier, warmer, and more open at the edge between a forest and field than it is further into the forest. This difference produces conditions favourable to some species but not others. Species that thrive in edge habitats like this are called **edge species**.

There are many different examples of ecotones and their effects on various species. Throughout most of their North American range, for instance, white-tailed deer prefer areas that are close to both forest and field—they find better food in the fields, but need the forest for refuge from predators and places to bed down during the day. Creating more edges or opening up old growth forests by removing some trees improves the habitat for deer (the new growth is an excellent food source for deer). Sometimes this is done intentionally, to help the deer populations—but when forests are cleared for lumber or to make way for new subdivisions, shopping malls, and roads, we fragment habitats and create more edges, which benefit the deer. (Deer were not historically found in the Acadian forest, but since humans have begun clearing the forest, they have become more frequent inhabitants.)

Other species that can only be found deep within the centre of a habitat are called **core species**. Some birds, such as many warbler species, travel to deep, undisturbed forests to breed during spring and summer months. When storms, timber harvesting, or any other factors create new openings in the forest, this shrinks the birds' habitat and makes it easier for predators to find them. Something as small as a one-lane dirt road can isolate populations of small birds and mammals who will not cross the opening. (See Chapter 10 for more on habitat fragmentation and edge/core species.) [INFOGRAPHIC 8.5]

Relationships are crucial in an ecological community.

Although the Acadian forest is brimming with biological activity, some species have more of an impact than others. In the last remaining patches of Acadian forest, beavers are using material from a variety of tree species, old and young, to construct dams and their lodges.

The beavers' actions have many effects on neighbouring species, both positive and negative. Their dams can completely alter the flow of a river to create ponds and wetlands, which serve as new habitat for a variety of

Infographic **8.5** | **EDGE EFFECTS**

↓ Habitat structure influences where species live. Edge species, like deer, prefer habitats with forest and field edges whereas core species, like warblers, prefer the inner areas of forest and do not readily venture into edge regions.

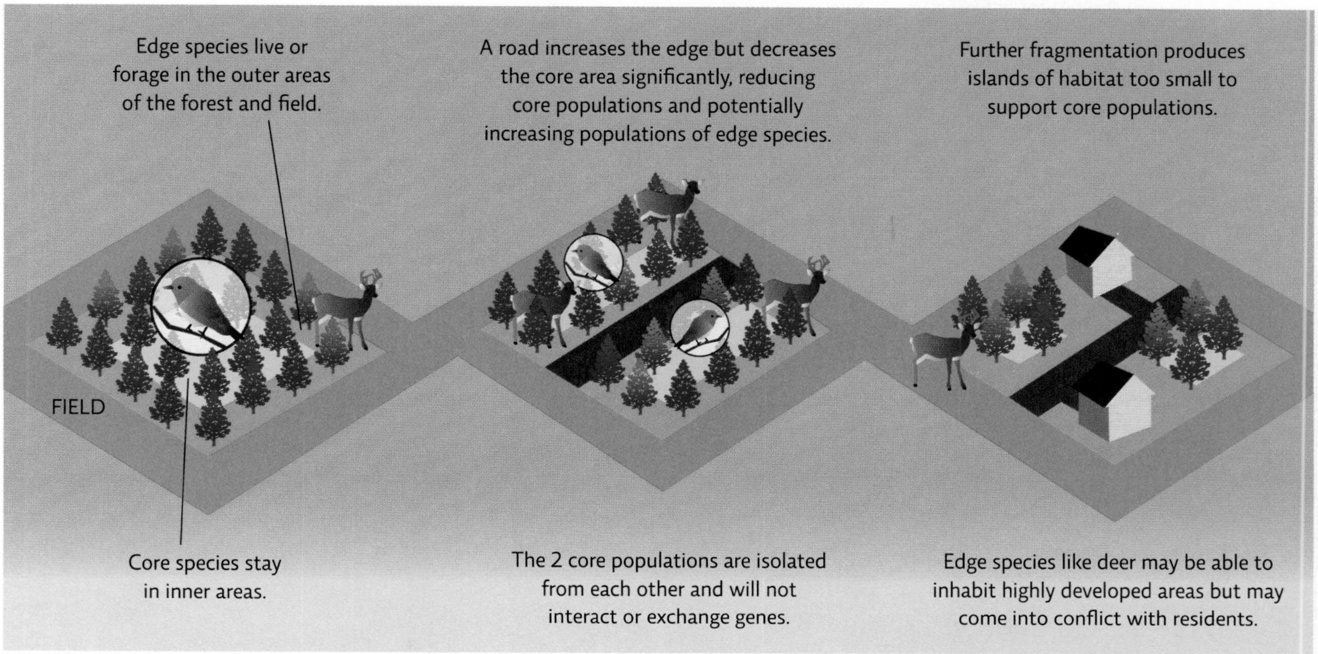
Edge species live or forage in the outer areas of the forest and field.
A road increases the edge but decreases the core area significantly, reducing core populations and potentially increasing populations of edge species.
Further fragmentation produces islands of habitat too small to support core populations.
FIELD
Core species stay in inner areas.
The 2 core populations are isolated from each other and will not interact or exchange genes.
Edge species like deer may be able to inhabit highly developed areas but may come into conflict with residents.

organisms from frogs to plants. But for fish that need to cross a dam to get to their breeding grounds, the hurdle can be insurmountable. And when heavy rains come, low-lying areas near the dam become flooded. "Beavers are one species—like humans—that have many downstream effects on their environment," says Villard.

This makes the beaver a **keystone species**—one that impacts its community more than its mere abundance would predict. It's a species that many other species depend on, and one whose loss creates a substantial ripple effect, disrupting interactions for many other species, and ultimately, altering food webs. So when the Acadian forests are cleared, and beavers can't build as many dams, many other species suffer.

An ecosystem as complex as the Acadian forest has more than one keystone species. The pileated woodpecker is another keystone species, as it excavates crucial cavities in trees used by many other species, such as songbirds, owls, flying squirrels, martens, and wood ducks, who cannot create these cavities on their own. [INFOGRAPHIC 8.6]

The multiple and varied effects of keystone species show just how important species interactions are for community viability. Communities are all about relationships. Successful communities are those where a certain balance has evolved between all the organisms living there. Species interactions serve many purposes; for example, they control populations and affect carrying capacity. Biodiversity (lots of species and lots of variety within a species) is important, because more diversity means more ways to capture, store, and exchange energy and matter. But it is not sheer numbers that matter most; it is all the connections between species—how they help or hurt one another—that determine how, and how well, an ecosystem works. Each species is unique and thus interacts in its own unique way with all the species around it.

Many species have adaptations that bind them to others or that allow them to coexist. For instance, in some years, trees will release large numbers of seeds, which

---

ecotones Regions of distinctly different physical areas that serve as boundaries between different communities.
edge effects The different physical makeup of the ecotone that creates different conditions that either attract or repel certain species (for instance, it is drier, warmer, and more open at the edge of a forest and field than it is further in the forest).
edge species Species that prefer to live close to the edges of two different habitats (ecotone areas).
core species Species that prefer core areas of a habitat—areas deep within the habitat, away from the edge.
keystone species A species that impacts its community more than its mere abundance would predict, often altering ecosystem structure..

Infographic **8.6** | **KEYSTONE SPECIES ARE KEY PLAYERS IN THEIR ECOSYSTEMS**

→ Some species are especially important to their ecosystem. If a keystone species is lost or declines in number, the ecosystem could change drastically and the other species who depend on it may suffer or be lost.

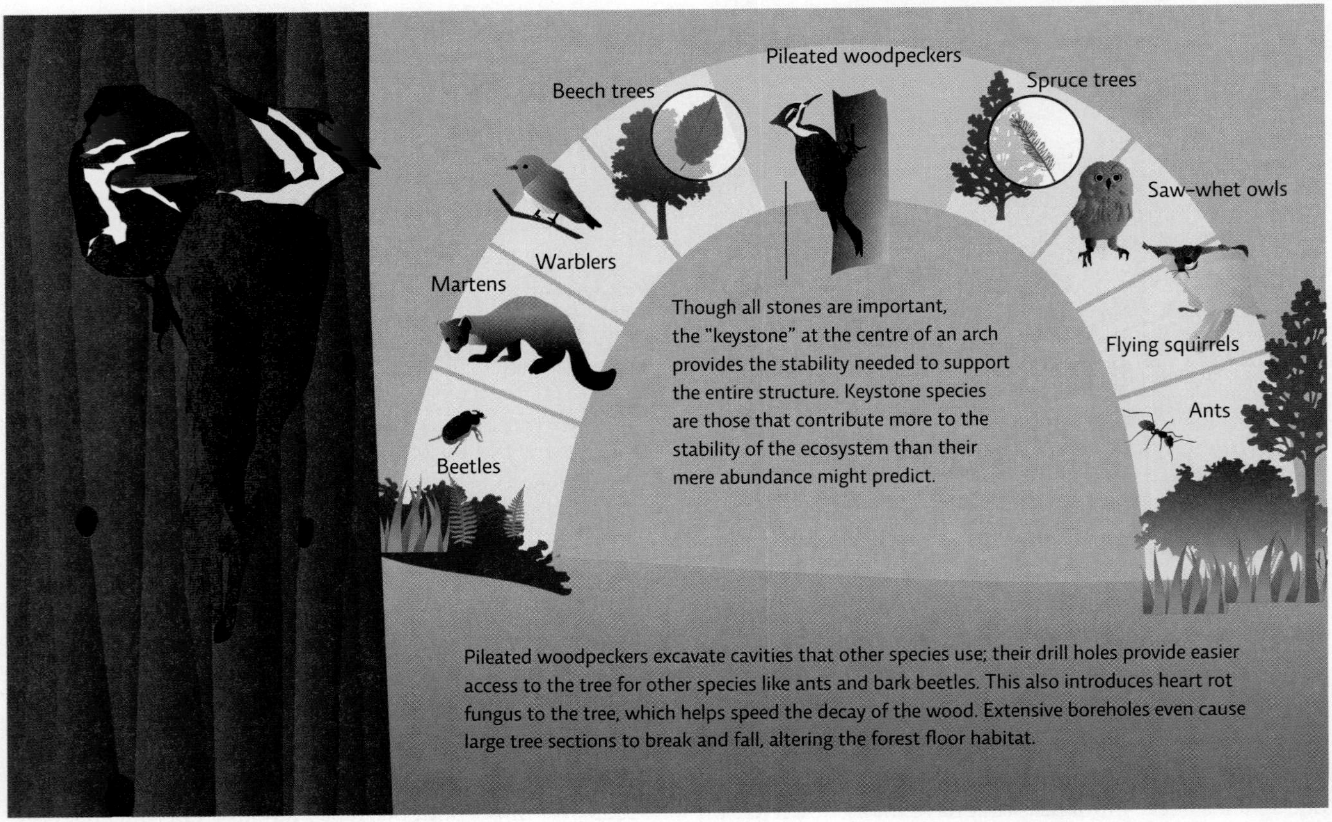
Pileated woodpeckers
Beech trees
Spruce trees
Saw-whet owls
Warblers
Martens
Flying squirrels
Ants
Beetles
Though all stones are important, the "keystone" at the centre of an arch provides the stability needed to support the entire structure. Keystone species are those that contribute more to the stability of the ecosystem than their mere abundance might predict.
Pileated woodpeckers excavate cavities that other species use; their drill holes provide easier access to the tree for other species like ants and bark beetles. This also introduces heart rot fungus to the tree, which helps speed the decay of the wood. Extensive boreholes even cause large tree sections to break and fall, altering the forest floor habitat.

spur increases in seed-eating rodents such as chipmunks, leading to increases in the populations of their predators. There are fewer seeds the following year, so rodents are forced to look for other food sources, such as bird eggs or their young, which affects those population numbers, as well.

Known as *mast yearing*, the tree's strategy is an adaptation to deal with very effective seed-eaters. If a beech tree produced the same amount of nuts (or mast) every year, the populations of the animals that eat beechnuts would stabilize at whatever level the nut crop would support and thus decrease the chance that many beechnuts would escape being eaten. But if the tree produces a low crop for a few years, that lowers the predator population; so in the years when the trees produce a bumper crop, they are essentially flooding the market with more nuts than the residents can eat. It is in these mast years that the trees experience the most reproductive success. The erratic nature of the mast yearing (varying from 2 to 7 years), tied in with the short lifespan of most nut-eaters, makes this an effective adaptation for the tree.

Competition—the vying between organisms for limited resources—is another way that species interact. Competition between individuals of the same species is generally stronger than competition that might occur between individuals of different species. This is because members of the same species share the exact same niche and thus compete for all resources in that niche, whereas members of different species may compete for only a single resource, like a certain food source or water.

If two different species try to occupy the same exact niche, the less successful species will die out or be forced to leave the area if it cannot switch to use another available resource. Some competing neighbours "switch niches" by finding a way to partition resources—that is, they divvy up the goods in a way that reduces competition and allows several species to coexist. In the Acadian forest, different species of warblers that feed on tree insects often occupy different regions of the same tree. Cape May warblers, for instance, tend to feed in the uppermost regions of trees, toward the outside of branches, while Bay-breasted warblers pick spots closer to the middle of the tree, and Myrtle warblers hang close to the bottom. This strategy—known as **resource partitioning**—increases species diversity, which benefits the entire community.

There are other strategies, too, that keep an ecosystem functioning and strong. Some of these interactions show a tremendous interdependency on the part of the participants. Known as **symbiosis**, these relationships can take one of several forms: **mutualism**, where both parties benefit; **commensalism**, where one benefits from the relationship but the other is unaffected; or **parasitism**, where one benefits from the relationship and the other is negatively affected (this is actually a form of predation).

By ensuring that all populations persist, even as individuals die, these delicate checks and balances allow more energy to be captured and exchanged, and thus increase the amount of biomass the ecosystem is able to produce. [INFOGRAPHIC 8.7]

When individual species are lost, or when a landscape is physically altered, the balance is tipped. In the case of keystone species, the forest can withstand small reductions in the numbers of beavers and pileated woodpeckers—but if too many disappear, the entire ecosystem suffers. "I sometimes compare keystone species to rivets in airplanes," says Villard. "A plane can lose many rivets and still fly—but if it loses one too many, it goes down. The same thing is true in ecosystems."

And when that happens, things can fall apart. Fast.

↓ Competition abounds in all ecosystems, not just species-rich forests. Here, a mockingbird and golden-fronted woodpecker battle over the ripe fruit of a prickly pear cactus in Texas.

Restoration ecologists can help ecosystems recover.

Given that nearly half of Canada is covered in trees, forests provide crucial resources to the environment and economy. But that fact was not foremost in the minds of the European settlers when they chopped down the Acadian forest to make way for farms, roads, and other development. Fixing this is the work of restoration ecologists. **Restoration ecology** is the science that deals with the repair of damaged or disturbed ecosystems. It requires a special blend of skills—biology, chemistry, and sometimes engineering—and a heavy dose of political will.

A series of policies established in the 1980s are designed to help restore some balance to Canadian forests and to protect and maintain these vital resources. Nationally, forests and forest resources are overseen by the *Canadian Forest Service*, a branch of Natural Resources Canada. Similar agencies exist in each province, employing foresters, economists, scientists, policy makers, and other experts to develop guidelines for maintaining forests while sustainably using the resources these ecosystems have to offer, such as timber.

Forest regulations are implemented at several levels of government, and how that happens depends on the individual province. Ecosystems may be managed by different agencies, for example Natural Resources Canada, Environment Canada, and the Ontario Ministries of Natural Resources and the Environment. In Ontario, conservation agencies often restore habitats using funding from both the government and fundraising from private donors. Although this system shares responsibility, it also can create problems if the levels of government don't communicate properly with each other or work toward the same goals.

And government can only make policies on land owned by the Crown. In most provinces, private companies own only a small proportion of land—but in Nova Scotia and

---

competition Species interaction in which individuals are vying for limited resources.
resource partitioning When different species use different parts or aspects of a resource, rather than competing directly for exactly the same resource.
symbiosis A close biological or ecological relationship between two species.
mutualism A symbiotic relationship between individuals of two species in which both parties benefit.
commensalism A symbiotic relationship between individuals of two species in which one benefits from the presence of the other but the other is unaffected.
parasitism A symbiotic relationship between individuals of two species in which one benefits and the other is negatively affected (a form of predation).
restoration ecology The science that deals with the repair of damaged or disturbed ecosystems.

Infographic 8.7 | **SPECIES INTERACTIONS**

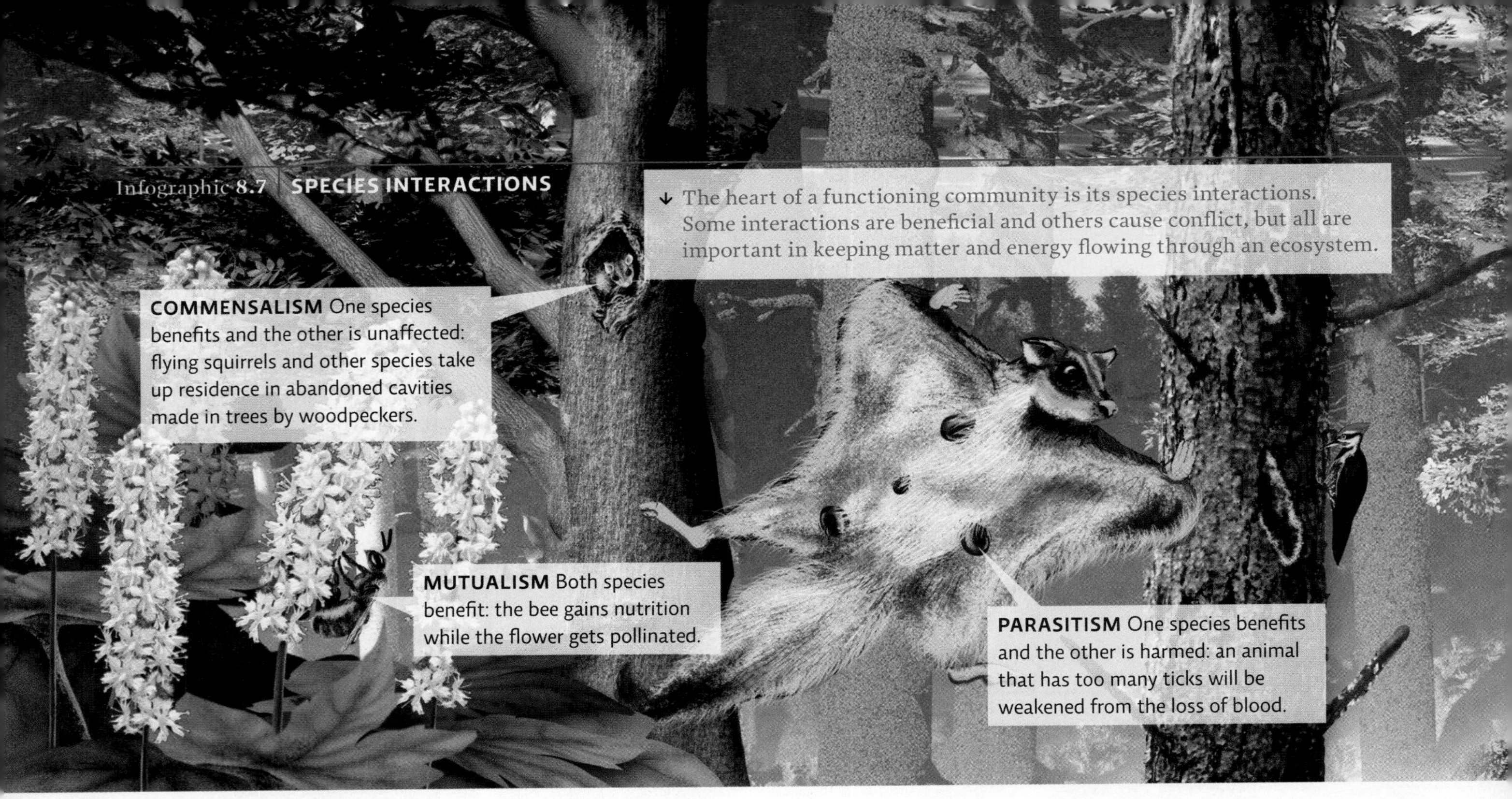
↓ The heart of a functioning community is its species interactions. Some interactions are beneficial and others cause conflict, but all are important in keeping matter and energy flowing through an ecosystem.
COMMENSALISM One species benefits and the other is unaffected: flying squirrels and other species take up residence in abandoned cavities made in trees by woodpeckers.
MUTUALISM Both species benefit: the bee gains nutrition while the flower gets pollinated.
PARASITISM One species benefits and the other is harmed: an animal that has too many ticks will be weakened from the loss of blood.

New Brunswick, where much of the Acadian forest was once located, nearly half of all land is in private hands, simply because these areas contain much inhabitable and farmable land that was sold to the first European settlers. Over time, companies have bought up that land from individual owners to manage for timber.

Forest policy in Canada is evolving, says Stephen Wyatt, a professor of forest policy and social forestry at Université de Moncton. "Before, forest policy was dictated by the government and industry—they were the two groups that decided what happened to the forests," he says. "Over the last 20 years, that's been changing, and public opinion has become much more important."

For instance, forest management policies increasingly include the input of First Nations communities. Just how much input, however, depends on the province—some provinces give First Nations communities a parcel of forest to manage as they wish; others require that they use the land to produce timber. In New Brunswick, the original site of much of the Acadian forest, First Nations groups have simply been given a quota of timber they can cut and sell to industry each year, without any land to manage.

Including First Nations communities in forest management makes sense, says Wyatt, since they often understand the ecosystem better than anyone else. "People who live in forests often have knowledge that goes back hundreds of years," he says. "They know where moose populations are, or the critical sites that geese come through each year. It just makes sense to talk to these people, to do a better job of forestry." Importantly, it's also a matter of what's right, Wyatt adds. "They have a right to be involved in forest policy in Canada, because they have been here since time immemorial."

Community composition naturally changes over time.

On a breezy day that threatened rain, a group of NCC staff and volunteers gathered along the Northumberland Strait coast for a six-hour barbeque that had one purpose: to get volunteers to plant trees. On a property that bordered both Nova Scotia and New Brunswick, people systematically scooped out more than 2000 pockets of soil, and filled them with the delicate roots from a range of species, such as white ash, yellow birch, red pine, and red spruce. They were careful to plant species of a variety of ages as well, from tiny saplings to two-year-old specimens.

Presumably, the property had once been home to a lush, vibrant Acadian forest. But it was cleared decades ago for farming, and then used as a retreat for the Girl Guides of Canada for more than 50 years.

Without any intervention from restoration ecologists, predictable transitions can sometimes be observed in which one community replaces another, a process known as **ecological succession**. **Primary succession** begins when **pioneer species** move into new areas that have not yet been colonized. In terrestrial ecosystems, these pioneer species are usually lichens—a symbiotic

RESOURCE PARTITIONING Different species of warblers forage in different areas of the tree when others are present.
PREDATION Red-tailed hawks eat a variety of prey such as rabbit.
COMPETITION Organisms that vie for the same resource are in competition with each other. The greatest competition for an individual is from a member of its own species, since both individuals are competing for all the same resources.

combination of algae and fungus. Lichens can tolerate the barren conditions. As time goes by and they live, die, and decompose, they produce soil. As soil accumulates, other small plants move in—typically sun-tolerant annual plants that live one year, produce seed, and then die—and the plant community grows. Gradually, the plant growth itself changes the physical conditions of the area—sun-loving shrubs and trees begin to cover the regions with broad, shady leaves, for example. Since these conditions are no longer suitable for the plants that created them, new species move in and those changes beget even more changes until the pioneers have been completely replaced by a succession of new species and communities.

Secondary succession describes a similar process that occurs in an area that once held life but has been damaged somehow; the level of damage the ecosystem has suffered determines what stage of plant community moves in. For example, a forest completely obliterated by fire may start close to the beginning with pioneering lichens, whereas one that has suffered only moderate losses may start midway through the process with shrubs or sun-tolerant trees moving in. The stages are roughly the same for any terrestrial area that can support a forest: first annual species, then shrubs, then sun-tolerant trees, then shade-tolerant ones. Grasslands follow a similar pattern, with different species of grasses and forbs (small leafy plants) moving in over time. [INFOGRAPHIC 8.8]

Intact ecosystems have a better chance at recovering from, and thus surviving, perturbations. Ecosystems that recover quickly from minor perturbations are said to be resilient—they can bounce back. More complex communities, such as the Acadian forest, tend to be more resilient than simpler ones with fewer species because it is less likely that the loss of one or two species will be felt by the community at large—even if some links in the food web are lost, other species are there to fill the void. For instance, if a warbler species failed to return to the Acadian forest one spring, another warbler might step in to fill its niche. However, over time, if too many species are lost, leading to the loss of keystone species, the community will start to feel the effects.

The work of the NCC and other groups greatly accelerates the process of secondary succession, providing the mix of species and different aged plants that will help restore the Acadian forest. Without their intervention on land that's been nearly totally cleared, the only trees that would initially thrive would be shade-intolerant species, such as pines, aspens, and birches. But a true Acadian forest needs shade-tolerant tree species, many of which grow to

---

ecological succession Progressive replacement of plant (and then animal) species in a community over time due to the changing conditions that the plants themselves create (more soil, shade, etc.).

primary succession Ecological succession that occurs in an area where no ecosystem existed before (for example, on bare rock with no soil).

pioneer species Plant species that move into an area during early stages of succession; these are often *r* species and may be annuals, species that live one year, leave behind seeds, and then die.

secondary succession Ecological succession that occurs in an ecosystem that has been disturbed; occurs more quickly than primary succession because soil is present.

Infographic **8.8** | **ECOLOGICAL SUCCESSION**

FOREST ECOLOGICAL SUCCESSION DEPENDS ON SOIL AND LIGHT AVAILABILITY

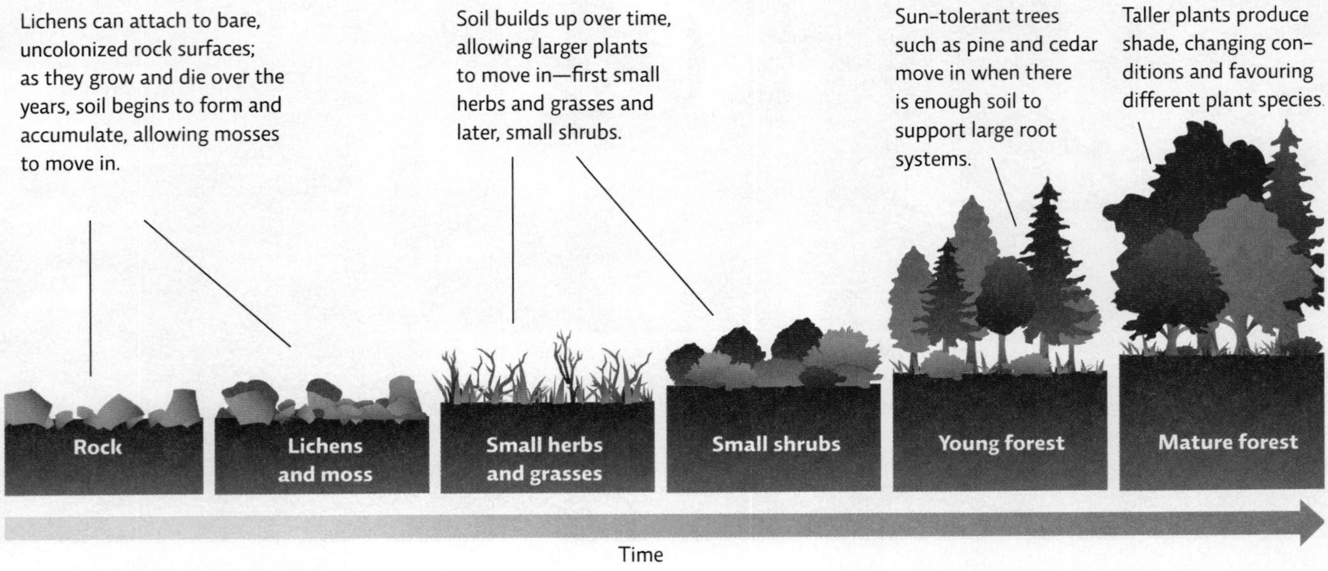
Lichens can attach to bare, uncolonized rock surfaces; as they grow and die over the years, soil begins to form and accumulate, allowing mosses to move in.
Soil builds up over time, allowing larger plants to move in—first small herbs and grasses and later, small shrubs.
Sun-tolerant trees such as pine and cedar move in when there is enough soil to support large root systems.
Taller plants produce shade, changing conditions and favouring different plant species.
Rock
Lichens and moss
Small herbs and grasses
Small shrubs
Young forest
Mature forest
Time

↑ In terrestrial ecosystems we see natural stages of succession occur whenever a new area is colonized or an established area is damaged. Sun-tolerant species give way to shade-tolerant ones as more soil is built up, supporting larger plant species.

↑ Peatlands, which include both fens and bogs, are Canada's most extensive wetland type, covering 12% of Canada's land surface (over 100 million hectares). Peatlands are important not only in terms of biodiversity and important ecosystem functions (e.g., water filtration), but also for their considerable role as an organic carbon sink (peat soil has a carbon content of 17% by weight). Peatlands are threatened by harvesting of peat soils for fuel and as a garden soil amendment (peat moss), by draining for forestry and flooding for hydroelectric development, and in some areas by habitat destruction due to such activities as agriculture and oil sands development.

great heights and provide nests for birds, small mammals, and rodents. "Some birds are very particular about the type of tree they're going to nest in," says Noseworthy. But these trees may not be able to move in on their own in the near future, so NCC staff and volunteers make sure to plant shade-tolerant species, such as eastern hemlock and sugar maple. "They won't come back if we don't plant them," says Margo Morrison, who manages the NCC's conservation science program for the Atlantic region, and is planning similar projects in other sites.

Some ecosystems remain in a constant cycle of succession; others eventually reach an end-stage equilibrium where conditions are well-suited for the plants that created them—for example, trees whose seedlings can grow in shade. These species, which can persist if their environment remains unchanged, are called **climax species**. End-stage **climax communities** can persist until disturbance restarts the process of succession—although there is debate amongst scientists over whether any community ever reaches a true end point of succession, or just continues to change and adapt, but more slowly or less obviously.

climax species Species that move into an area at later stages of ecological succession.

climax community The end stage of ecological succession in which the conditions created by the climax species are suitable for the plants that created them so they can persist as long as their environment remains unchanged.

Ideally, some parts of the Acadian forest will eventually return to the diverse, rich ecosystems they once were. But given that its primary qualities include an array of tree species of different ages and sizes, including dead and decaying trees that support a variety of organisms, such regeneration will take a long time. "This is not something we will see in our lifetime," says Noseworthy.

Planting trees to help a forest recover is relatively easy compared to what it would take for a boreal peatland to bounce back—it can take thousands of years to regenerate a metre of peatland soil depth, making it impossible to restore these ecosystems to their original, pristine state, even over hundreds of human lifetimes. "There's no way we can go back to how things were before. Definitely not," says Noseworthy. For that reason, restoration is a "goal," not an "outcome," he says—the outcome, instead, is to provide suitable enough habitats for the many organisms that once called the Acadian forest home.

"That being said, if you do nothing, a disturbed ecosystem will always show the results of human disturbances," says Morrison. "But if we take some steps now in the right direction, those disturbances will become less obvious. For many species, if we create the right conditions, they will come back."◉

Select references in this chapter:
Mosseler, A., *et.al.* 2003. *Environmental Reviews*, 11(S1):S47-S77.
Simpson, M. 1986. *Tennessee Wildlife*, 10(1): 9-12.

BRING IT HOME

PERSONAL CHOICES THAT HELP

The world is full of weird and wonderful species. Every year we discover new information about how intricate our biological communities are. By restoring habitats and increasing our understanding of the relationships between species, we can better ensure their long-term survival.

Individual Steps

→ Visit a park or nature reserve and watch for signs of species interactions. Do you hear animals or birds; can you see signs of predation or herbivory?

→ Make your backyard a safe and welcoming stopover for migrating birds in the spring and fall with bird feeders and water sources. Keep house cats indoors; they are the leading killer of songbirds in suburban areas.

Group Action

→ The Acadian forest case study is an example of a very extensive restoration project. Call your local park or nature reserve to see what restoration work is happening in your area and how you can become involved.

Policy Change

→ Follow the Nature Canada blog to learn more about wildlife and issues facing conservation (www.naturecanada.ca/).

UNDERSTANDING THE ISSUE

CHECK YOUR UNDERSTANDING

1. An indicator species:
a. helps predators keep tabs on the location of prey.
b. is an indication of the population size of a second species.
c. helps ecologists locate sensitive areas.
d. helps ecologists monitor the health of an area.

2. In ecological terms, a consumer is:
a. any plant.
b. any animal.
c. any organism that eats other organisms.
d. any animal that eats other animals.

3. Detritivores and decomposers:
a. might be earthworms and bacteria.
b. might be algae and fungi.
c. are usually insects such as flies.
d. are usually large predators.

4. Edge effects:
a. apply only to the largest and smallest members of a community.
b. occur in the areas where two or more habitats meet.
c. are beneficial for nearly all organisms.
d. are harmful for nearly all organisms.

5. In mutualism, both organisms receive an immediate benefit from the relationship. A good example of this is:
a. a dog and a flea.
b. an ant and a grasshopper.
c. a butterfly and a flowering plant.
d. a deer and a wolf.

6. An example of how secondary succession would occur in a particular area would be when:
a. one species has been outcompeted by another species, and is extirpated.
b. a flood has removed much of the vegetation.
c. hot ash from a volcano has completely burned and buried the area.
d. a disease reduces the top predator's population.

WORK WITH IDEAS

1. Which is more vulnerable to disturbances, a simple food web with only a few species, or a more complex one? Explain.

2. How do pileated woodpeckers fit the definition of a *keystone species*?

3. Draw a simple food web for a natural area near you. Include producers and at least three levels of consumers, as well as detritivores, and decomposers.

4. Explain why both species richness and species evenness are important for a healthy ecosystem.

5. Choose a different ecosystem than the one discussed in this chapter and identify examples of each of these species interactions: predation, competition, mutualism, commensalism, and parasitism.

6. Cowbirds are nest parasites—they lay their eggs in the nests of small forest birds such as bluebirds and leave them for the small birds to care for. The bluebirds then spend the next several weeks caring for the huge baby cowbird, which quickly kills the bluebird's own young. Cowbirds prefer open, disturbed areas and seldom venture far into a forest for any reason, even to lay eggs. Use the concept of edge effect to explain what happens to the populations of bluebirds when humans build roads, recreation areas, homes, and businesses in a large forest.

ANALYZING THE SCIENCE

One important value of forest ecosystems is their capacity to store carbon, thereby keeping it out of Earth's atmosphere and mitigating climate change. Intact ecological communities are better at performing this ecosystem service than degraded ones. All ecosystems can both take up carbon and emit it; it is the net difference between accumulation versus emission that tells us whether the ecosystem is a net sink or source of carbon to the atmosphere. Globally, terrestrial ecosystems are a major net sink for atmospheric carbon dioxide, storing about 1 metric gigaton of carbon per year. Wetlands store even more carbon than most forests, making wetlands extremely important carbon sinks.

CARBON ACCUMULATION IN DIFFERENT WETLAND SEDIMENTS

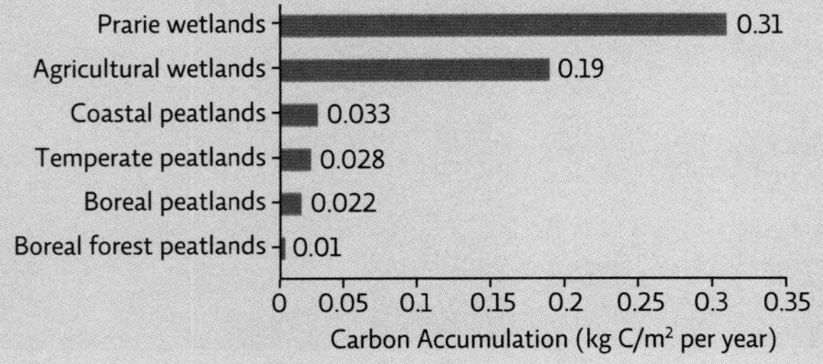
Prarie wetlands
0.31
Agricultural wetlands
0.19
Coastal peatlands
0.033
Temperate peatlands
0.028
Boreal peatlands
0.022
Boreal forest peatlands
0.01
0 0.05 0.1 0.15 0.2 0.25 0.3 0.35
Carbon Accumulation (kg C/m² per year)

INTERPRETATION

1. Which wetland type stores the most carbon? Which stores the least?

2. What is the difference in carbon accumulation (in kg C per square metre per year) between temperate peatlands and prairie wetlands? Identify two possible reasons for this difference.

ADVANCE YOUR THINKING

3. By what processes do ecosystems take up and emit carbon dioxide?

4. What do the two wetland types with the highest carbon storage have in common? (HINT: What biome is represented by both?)

5. If you had to decide what kind of wetlands to protect or restore in order to enhance carbon storage, which type of wetland would you prioritize? What challenges would you face in this effort?

EVALUATING NEW INFORMATION

Woodlands are some of the most important communities in Canada. Rich with diversity, we see them as permanent and unchanging places, filled with wildlife, which we wish to preserve. We began fighting wildfires in a systematic way around the turn of the last century; historic photographs show people building fire lines and shovelling dirt over smoldering spots in our woodlands. In the past 30 years, however, there have been increased discussions about the automatic response of immediately quelling all wildfires—perhaps some wildfires do more good than harm. Afer all, change from storms, fires, and floods is part of the natural cycle of an ecosystem.

Read the following articles about wildfires and ecology:

- An overview of wildfires (geography.about.com/od/globalproblemsandissues/a/wildlandfire.htm)
- A history of humans and wildfires (www.foresthistory.org/Education/Curriculum/Activity/activ9/essay.htm)
- The ecology of fire (www.onearth.org/blog/forest-service-appears-to-shift-controversial-and-shortsighted-firefighting-policy)
- Logging and fires (www.emagazine.com/magazine-archive/the-burning-west)
- Prescribed fires (www.bcwildfire.ca/prevention/PrescribedFire/)
- Spending on fires (www.emagazine.com/includes/print-article/magazine-archive/6631)

Evaluate the websites and work with the information to answer the following questions:

1. Are these reliable information sources?
   a. Do the authors of these articles give supporting evidence for their claims?
   b. Do they give sources for their evidence?
   c. What are the missions of these organizations?

2. Now go to the Canadian Forest Service webpage (cfs.nrcan.gc.ca/home). On the left-hand side of the page, click on *Fire*. What is the Forest Service's position on wildland fire management?

3. On the right-hand side of the Forest Service: Fire page, click on the link for the Canadian wildland fire information system (CWFIS) and then select the Interactive Map. Check the boxes for *Fire Danger* and *Active Fires* on the left-hand side. Now select a summer month in the past year and click *Update Map* on the upper left-hand side of the page.
   a. About how many active fires are shown?
   b. What locations are more prone to fires?
   c. Are there any fires in areas shown as having "low fire danger"?
   d. Repeat this for the same month in 2 additional years. Is the number and location of fires about the same? Why might there be a difference from year to year?

4. Based on the information you have found, explain your position:
   a. What should be done when a wildfire occurs?
   b. What, if anything, should be done in an area to prevent or lessen a possible wildlands fire?

MAKING CONNECTIONS

CORAL REEFS AND LOBSTER DINNER

Background: Belize is a tiny country in Central America composed of jungle, pine forests, limestone caves, Atlantic Ocean beaches, hundreds of small islands, and the major portion of the longest coral reef in this hemisphere—second only to the Great Barrier Reef in Australia. Coral reefs are found in warm, shallow waters and support huge numbers of organisms, estimated at one-fourth of all marine life. They are diverse, productive, and very fragile. The algae in coral reefs produce oxygen and the reef itself acts as a nursery for sea life, as shoreline protection from storms and erosion, and as an important feeding and breeding ground for thousands of species. Well aware of its precious resources, the Belizean government has declared over 40% of the country to be natural parks and reserves, including much of the coral reef and surrounding areas. The major businesses in Belize are tourism and commercial harvesting of fish, lobster, and conch. Can the delicate reef system withstand both uses? In July 2011, a group of 5 people were caught with nearly 300 tiny lobster tails, many weighing less than 1 ounce, and over 200 undersized Queen Conch that were also out of season. A month later, the senior marine conservationist declared that over 85% of the reef was dead, dying, or in serious difficulty.

Case: Investigate the following three important needs. How do we balance them?

1. The need for the preservation of an important world biome that gives us so many ecological services

2. The need for a productive harvest of fish, lobster, and conch each year to support thousands of fishers and to supply hundreds of thousands of consumers with seafood

3. The need for a tourism industry for the enjoyment of millions and for the kind of "clean" industry that can help a small country support itself sustainably

Write a report that focuses on one of the three needs as most important while preserving and maintaining the other two needs. In your report, be sure to use facts to justify what you consider to be the most important need and be sure to include the following:

a. The ecological services provided by a coral reef and an indication of what will happen if the reef is damaged or destroyed
b. The economics and ecology of the seafood industry in a small, rural, coastal country such as Belize
c. The damage, both to the economy and to the ecology, of poaching, and a possible solution
d. The costs and benefits of tourism in a small, rural country
e. Some possible solutions to balancing these issues

CORE MESSAGE

The variety of life on Earth is tremendous. This biodiversity provides important ecological services to ecosystems; we depend on these same services for things like food, medicine, and economic development. There are many compelling reasons to protect species and many approaches we can use.

GUIDING QUESTIONS

After reading this chapter, you should be able to answer the following questions:

→ What is biodiversity and why is it important?

→ What is the estimated total number of species on Earth and which taxonomic groups have the most species? How sure are we of these numbers? Why?

→ How do genetic diversity, species diversity, and ecological diversity contribute to ecosystem function and services?

→ What are biodiversity hotspots and why are they important?

→ What can be done to protect biodiversity?

CHAPTER 9 **BIODIVERSITY**

NATURE'S MEDICINE CABINET

Will the bark of an ordinary tree in Samoa become a cure for cancer?

A man walks along the Samoan coastline carrying bundles of vegetation in baskets made of woven leaves.

It was his mother's death in the fall of 1984, from a particularly aggressive form of breast cancer, that drove Paul Cox back to the Samoan rainforest. Cox had first visited the South Pacific island in 1973, through an undergraduate research program with Brigham Young University, where he was majoring in botany. Since then, the Utah native had earned a Ph.D. from Harvard and made a career studying plant physiology in the United States.

↑ Paul Cox and family in Samoa, 1986.

◉WHERE IS SAMOA?

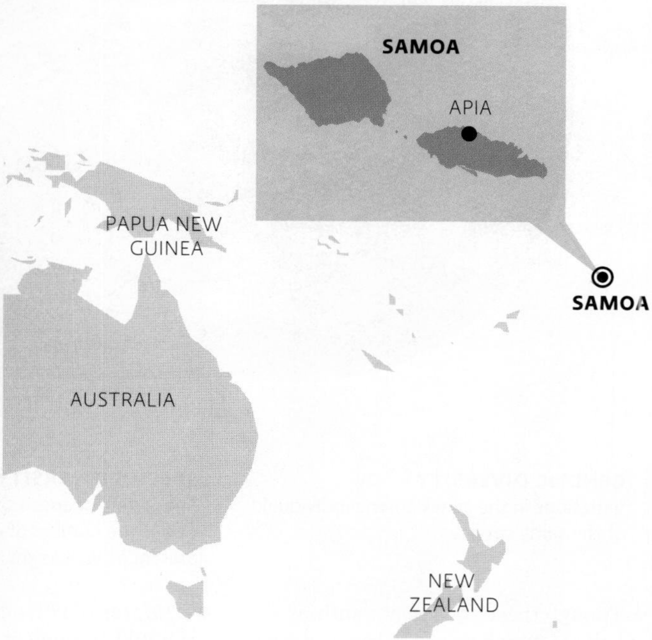
SAMOA
APIA
PAPUA NEW GUINEA
SAMOA
AUSTRALIA
NEW ZEALAND

But when the best of Western medicine failed to cure his mother, Cox remembered the Samoans—their rich folk-healing traditions, and the profound influence that jungle plants had on their health and well-being. Surely, he thought, somewhere in the lush abundance that had long sustained this impoverished island nation must be a compound powerful enough to obliterate the most insidious of tumours. "I told my friends at the National Cancer Institute, 'If there is even a 1% chance of finding something, it's worth taking a look,'" he recalls. "They said 'We think there is like a 3% chance.' So I went." Six months after his mother's funeral, with his wife and four young children in tow, Cox returned to Samoa.

biodiversity The variety of life on Earth; it includes species, genetic, and ecological diversity.
species diversity The variety of species, including how many are present (richness) and their abundance relative to each other (evenness).
genetic diversity The heritable variation among individuals of a single population or within the species as a whole.
ecological diversity The variety within an ecosystem's structure, including many communities, habitats, niches, and trophic levels.

Biodiversity benefits humans and other species.

Why did Cox suspect that Samoa might hold the key to a cancer cure? Tropical regions like Samoa—warm, lush, close to the equator—contain the greatest concentration and variety of plant and animal life forms on Earth. This variety is called **biodiversity**. Countries of this region have both high **species diversity** and high **genetic diversity**. They also usually have high **ecological diversity**, a wide variety of communities and ecosystems with

Infographic **9.1** | **BIODIVERSITY INCLUDES GENETIC, SPECIES, AND ECOSYSTEM DIVERSITY**

BIODIVERSITY

GENETIC DIVERSITY
Variations in the genes among individuals of the same species

SPECIES DIVERSITY
The variety of species present in an area; includes the number of different species that are present as well as their relative abundance

ECOLOGICAL DIVERSITY
The variety of habitats, niches, trophic levels, and community interactions

↑ Though these squarespot anthias fish are members of the same species, *Pseudanthias pleurotaenia*, they show tremendous genetic diversity.

↑ The area of Polynesia where Samoa is found has some of the highest coral reef species diversity in the world.

↑ Samoa's rainforest and coral reef ecosystems contain complex, three-dimensional structures with varied habitats and multiple niches that support high species diversity and a complex community.

different habitats, many trophic levels, and lots of niches. [INFOGRAPHIC 9.1]

It is virtually impossible to know just how many species exist on Earth. This uncertainty has given rise to a wide range of estimates—anywhere from 3 to 100 million. Further, in a 2011 paper, researchers estimated that of all the species now living on Earth, we humans have yet to discover or identify about 86% of them. [INFOGRAPHIC 9.2]

Such nearly unfathomable biodiversity brings with it many benefits. Samoa, like most other places around the world, depends on biodiversity to provide a vast array of **ecosystem services**: photosynthetic organisms (plants on land; algae and phytoplankton in the sea) bring in energy, produce oxygen, and sequester carbon. Other organisms capture and pass along important nutrients like nitrogen and phosphorus. Others still help purify the air and water and eventually become food for other creatures. And populations are kept in check by predators and competitors so that no single species grows too populous or gobbles up too many needed resources. For example, in the forests of Samoa, the tooth-billed pigeon is important for the propagation of mahogany, a species of tree found throughout Polynesia. The pigeon may be the only native animal that can open the tree's seeds so new seedlings

ecosystem services Benefits that are important to all life, including humans, provided by functional ecosystems; includes such things as nutrient cycles, air and water purification, and ecosystem goods, such as food and fuel.

Infographic **9.2** | **BIODIVERSITY ON EARTH**

↓ We have identified about 1.8 million species so far (prokaryotes and eukaryotes) but our knowledge of Earth's total biodiversity is scant. Many researchers, especially those who work with tropical species and insects, believe the estimates of 8.7 to 14 million species are far too low—there may well be 5 million insect species alone (some think the number could be much higher). Of the species we have identified, insects far outnumber any other group of organism. In fact, vertebrates (the group to which humans belong) likely make up only 1% of all creatures on Earth.

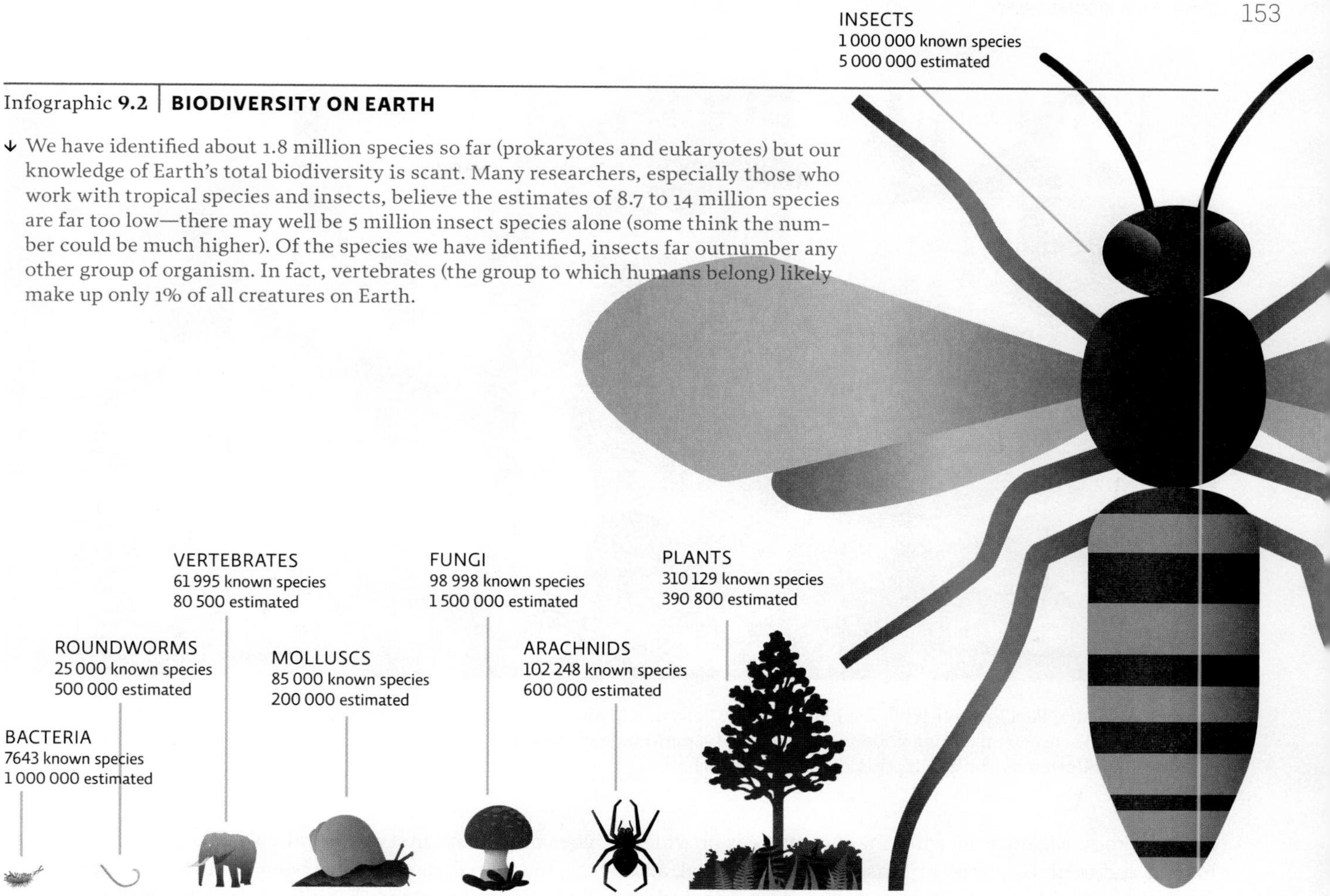
INSECTS
1 000 000 known species
5 000 000 estimated
VERTEBRATES
61 995 known species
80 500 estimated
FUNGI
98 998 known species
1 500 000 estimated
PLANTS
310 129 known species
390 800 estimated
ROUNDWORMS
25 000 known species
500 000 estimated
MOLLUSCS
85 000 known species
200 000 estimated
ARACHNIDS
102 248 known species
600 000 estimated
BACTERIA
7643 known species
1 000 000 estimated

Size roughly equals the proportion of all known species.

↓ When we compare the estimated size of different groups of species to the number within each group listed as endangered or threatened, there is a clear bias toward species of interest to us (large animals and plants). However, we are discovering that the lesser-known or -appreciated groups, such as worms and insects, perform vital ecosystem services and also warrant our attention.

PERCENT OF SPECIES LISTED AS ENDANGERED COMPARED TO THE ESTIMATED SIZE OF THE GROUP

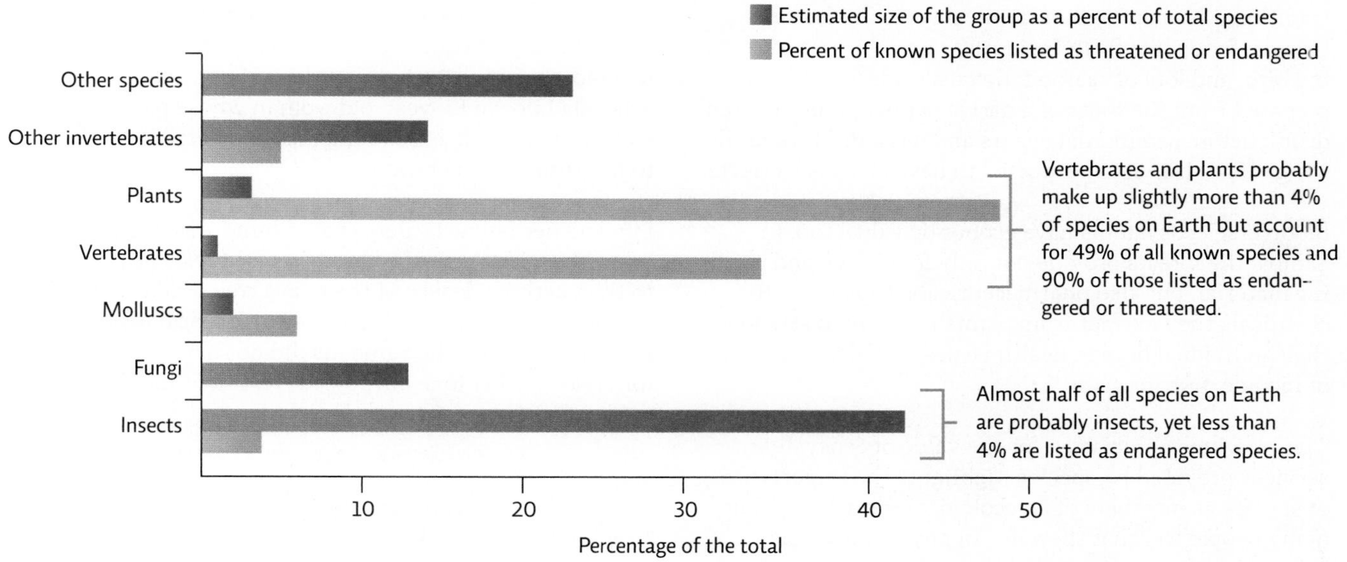
Estimated size of the group as a percent of total species
Percent of known species listed as threatened or endangered
Other species
Other invertebrates
Plants
Vertebrates
Molluscs
Fungi
Insects
Vertebrates and plants probably make up slightly more than 4% of species on Earth but account for 49% of all known species and 90% of those listed as endangered or threatened.
Almost half of all species on Earth are probably insects, yet less than 4% are listed as endangered species.
10
20
30
40
50
Percentage of the total

↑ Samoan healer, Pela Lilo, with her daughter-in-law, Fa'asaina Lamositele, use traditional medicine to treat a patient suffering from an illness seen on Samoa known as *mumu lele* (symptoms include a high fever and dermatitis). First coconut oil is rubbed on the skin, followed by the application of medicinal herbs.

can take root. Meanwhile, adult mahogany trees provide habitat and food to a variety of insect, bird, mammal, and reptile species. They are also key players in carbon cycling and soil stabilization, and serve as a windbreak in coastal areas. (See Chapter 5 for more on ecosystem services.)

Biodiversity supplies cultural benefits as well—whether it is the enjoyment of a natural area for recreation or aesthetic appreciation, or a societal tradition rooted in nature. In Samoa, native foods are an important part of traditional feasts; the Sunday meal typically features tropical foods like fresh fruit, root vegetables including taro, and lots of seafood. Kava, a traditional drink prepared from the roots of a native pepper plant, is often drunk before ceremonial events and has a mild tranquilizing effect. It is also purported to have analgesic effects.

Biodiverse ecosystems have economic value, too. In Samoa, the forests provide not only food, fuel, and building material, but also pharmaceuticals. People use the chemicals they extract from plants not only to attend to their individual human health issues, but also as a source of income. [INFOGRAPHIC 9.3]

The growing appreciation for the value of ecosystem services provided by species highlights the importance of species as members of an ecological community. But many people feel that the value of any given species goes beyond its **instrumental value** (i.e., the ecological, medicinal, or monetary benefits it can provide for humans) and contend that all species have **intrinsic value** and are therefore worth preserving.

Once they arrived in Samoa, Cox and his family settled in a thatched hut on the far western edge of the island in the village of Falealupo—just a few steps from the sea and as far as they could get from the comparatively modern neighbouring island of American Samoa. There, amid a tropical swirl of insects, humidity, and white sand, with no electricity or running water, Cox apprenticed himself to a mostly female cadre of healers. His primary teacher was Pela Lilo, an 82-year-old woman whose particular collection of plant-based remedies had been passed down to her through generations.

Lilo and her fellow healers shared hundreds of natural remedies with Cox—the bark of vavae (*Ceiba pentandra*) to treat asthma, leaves of the kuava tree (*Psidium guajaba*) for diarrhea, and root of 'Ago (*Curcuma longa*) for rashes. It turned out that the Samoans did not have a word for breast cancer (one purported treatment for "lumpy

instrumental value An object's or species' worth, based on its usefulness to humans.

intrinsic value An object's or species' worth, based on its mere existence; it has an inherent right to exist.

Infographic **9.3** | **ECOSYSTEM SERVICES**

↓ We depend on genetically diverse, species-rich communities to provide the goods and services we use every day. Impoverished ecosystems that have lost genetic, species, or ecological diversity cannot perform these tasks as well as highly diverse ecosystems can.

1. CULTURAL BENEFITS
- Aesthetic
- Spiritual
- Educational
- Recreational

2. HUMAN PROVISIONS
- Food
- Fibre products such as cotton and wool
- Fuel
- Pharmaceuticals

3. ECOSYSTEM REGULATION AND SUPPORT
- Nutrient cycling
- Pollination and seed dispersal
- Air and water purification
- Flood control
- Soil formation and erosion control
- Climate regulation
- Population control

1.
Many people enjoy spending time in nature.
2.
Many forest products can be harvested.
3.
Shoreline vegetation protects the bank from erosion.

breasts" did not prove effective against breast tumours). But they did have a treatment for "yellowing fever," a disease Westerners know as Hepatitis C; boiling the bark of the mamala tree (*Homalanthus nutans*) and drinking the extract twice a day was known to relieve symptoms. After seeing a demonstration of the potion's power, Cox gathered samples of mamala bark and sent them off (along with dozens of other promising roots, stems, and leaves) to colleagues at the National Cancer Institute (NCI), in Bethesda, Maryland, for testing.

But what he saw as a garden of medical promise, others saw as timber. And just as the mamala bark was yielding up its secrets to scientists at the NCI, loggers were negotiating with Falealupo's villagers to ***clear-cut***, or remove, all of the trees from the forests where this tree grew most abundantly. In Samoa, habitat destruction such as this had already gobbled up nearly 80% of the island's lowland forests. Cox despaired at the thought of how many undiscovered medicines would be annihilated if the loggers had their way.

Plants gain medicinal qualities as they adapt to other species.

The tropics are home to roughly 5.3 billion people, or 80% of the world's population. Because many of the countries in this region are poor, medicinal plants provide both a primary form of health care and a major source of income to local communities. Medicinal plants are one of the most prominent facets of *ethnobotany*—the study of how different cultures make use of the plants that surround them.

According to the World Bank, the international trade in medicinal plants was worth nearly $100 billion in 2011. Roughly half of all prescription drugs—including some medicine cabinet staples like aspirin, codeine, and most hypertension drugs—were originally derived from plants. And by most accounts, there are many more medical treasures just waiting to be discovered: less than 1% of the world's 265 000 known flowering species have been tested for their effectiveness against human diseases. "We are barely scratching the surface now," says Jim Miller, a scientist at the New York Botanical Garden in New York City. [INFOGRAPHIC 9.4]

Despite that untapped potential, the role of plants in Western medicine has waxed and waned throughout history. For one thing, the odds are daunting. On average, only 1 in 15 000 compounds will demonstrate the potential to treat a human condition. For most of the modern medical era, finding that 1 compound took an average of 12 years and $300 million—enough to dissuade the most tenacious of scientists.

> Roughly half of all prescription drugs—including some medicine cabinet staples like aspirin, codeine, and most hypertension drugs—were originally derived from plants.

Plants were once the biggest source of new medications in Western society. But advances in pharmacology quickly turned scientists' attention away from the forest and toward the test tube, where they could design molecules from scratch, rather than search for them like needles in a haystack. A few decades ago, scientists predicted that advances in biochemistry would eliminate the need to exploit fragile ecosystems for our own medical needs; they believed any molecule we discovered could easily be replicated in a test tube or Petri dish.

But with the exception of a few remarkable successes, like the cancer drug paclitaxel that is now synthesized in the lab rather than being extracted from the bark of Pacific yew trees where it was first found, that prediction didn't quite pan out. "An overwhelming number of plant-derived compounds have eluded all attempts to copy," says Sarah Oldfield, director of the nonprofit Botanic Garden Conservation Initiative. Because copying plant-derived molecules has proven so intractable, modern medicine continues to depend on their availability in nature. And by many accounts, evolution has bested the most apt chemists, resulting in molecules too complex to fathom, let alone replicate. "Mother Nature is the ultimate chemist," says Oldfield. "Modern chemistry has barely enabled us to copy her creations, so you can imagine the near impossibility of designing them from scratch."

It's little wonder. Plant chemicals evolve over eons, as individual plant species strive to fight off predators. Because they can't run or physically fight, the way animals can, plants employ two main anti-predator strategies: physical defence structures like thorns, thick seed coats, or bark, and chemical weapons, like the ones that ultimately find their way into our medicine cabinets. In short, it's the diversity of life itself, challenging the plant kingdom in a mind-boggling variety of ways, that spurs such a broad range of protective chemicals, and in so doing, provides us with some precious life-saving medicines. (See Chapter 10 for more on natural selection of adaptive traits.)

In 1988, the villagers of Falealupo were being pressured by the government to build a school. Knowing this, a logging company offered to pay the villagers less than $1 per hectare for the nearby 12 150 hectares forest, where the mamala tree grew most abundantly—just enough to cover school construction costs. Scientists were only beginning to study the mamala bark and Cox knew that any health or financial benefits from the tree would be decades off.

But he also knew that if the loggers had their way, hundreds of other plants with potential medicinal properties might never be found. So he negotiated with the villagers, promising them he would raise the school money himself if they agreed to protect the forest from logging. Then Cox went knocking on the doors of nonprofits and

Infographic **9.4** | **NATURALLY DERIVED MEDICINES**

Opium (flower)
Medicine: Morphine
Use: Pain

Foxglove (flower)
Medicine: Digoxin
Use: Heart disease

Madagascar periwinkle (flower)
Medicine: Vinblastine
Use: Cancer (Hodgkin's Lymphoma, breast, testicular)

Ephedra (shrub)
Medicine: Salbutamol
Use: Asthma

Curarea (woody vine)
Medicine: Tubocurarine
Use: Muscle relaxant

Cobra (snake)
Medicine: Cobroxine
Use: Pain

Pacific yew tree
Medicine: Paclitaxel, Taxol
Use: Cancer (lung, breast, ovarian)

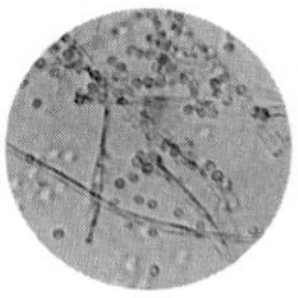

Penicillium (fungus)
Medicine: Penicillin, Metavastin
Use: Antibacterial, Cholesterol lowering

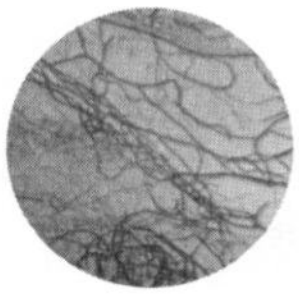

Streptomyces (bacteria)
Medicine: Ivermectin
Use: Parasites

↑ Top: Leaves from the Samoan mamala tree *Homalanthus nutans* from which the anti-HIV drug candidate prostratin was derived. Middle: Samoan healer Ake Lilo preparing a branch from the mamala tree. Bottom: The bark of the mamala tree is boiled down into an extract the Samoans use to treat Hepatitis C.

↑ More than 90% of the native species on the Hawaiian islands are found nowhere else, making these populations more vulnerable to extinction; if they are lost there, they are lost forever. The smaller the island, the smaller the population of a given species, another vunerablility for extinction. Here, botanists work through the night setting up a protective fence around the last specimen of *Delissea undulata* found in the wild. This endangered plant was growing on the side of a collapsed lava tube, but had been knocked over by wind or animals and was dangling from its roots.

conservation groups back in the United States. Before too many trees were felled, he had scrounged up $85,000—enough money to build the school and spare the forest. It was a rare victory. "It's common for indigenous people to have to choose between saving a forest or coral reef and building a school or medical clinic," says Cox. "Because they don't typically have the resources to do both, they must sacrifice one to pay for the other. And the loss when that happens is often incalculable."

This is especially true in the Pacific Islands like Samoa and Hawaii. The region has the highest percentage of threatened and endangered species (those at risk for extinction) of any region on Earth; these species are often specialists that occupy narrow niches. It also contains a high proportion of **endemic** plants and animals—those that exist nowhere else on Earth. Areas like this that contain a large number of endemic but threatened species are known as **biodiversity hotspots**. Many hotspots are in tropical regions, where biodiversity abounds. [INFOGRAPHIC 9.5]

Islands are particularly vulnerable to species loss since they are isolated from other islands or from the mainland; new individuals who might bring genetic diversity rarely arrive. In Hawaii alone, fully one-half of the indigenous flora faces immediate extinction. Such a high extinction rate threatens the fragile tapestry of life on these islands, from their soil and freshwater supplies to the health and economic future of their residents.

---

endemic Describes a species that is native to a particular area and is not naturally found elsewhere.

biodiversity hotspot An area that contains a large number of endemic but threatened species.

Infographic **9.5** | **BIODIVERSITY HOTSPOTS**

↑ Biodiversity hotspots, areas with a high number of endemic but threatened or endangered species, cover a small percentage of land and water areas but hold more than 40–50% of all plant and vertebrate endemic animal species. Most hotspots are located in tropical biomes or in isolated terrestrial ecosystems such as mountains or islands. Even small disturbances can threaten endemic species who populate specialized niches in these hotspots.

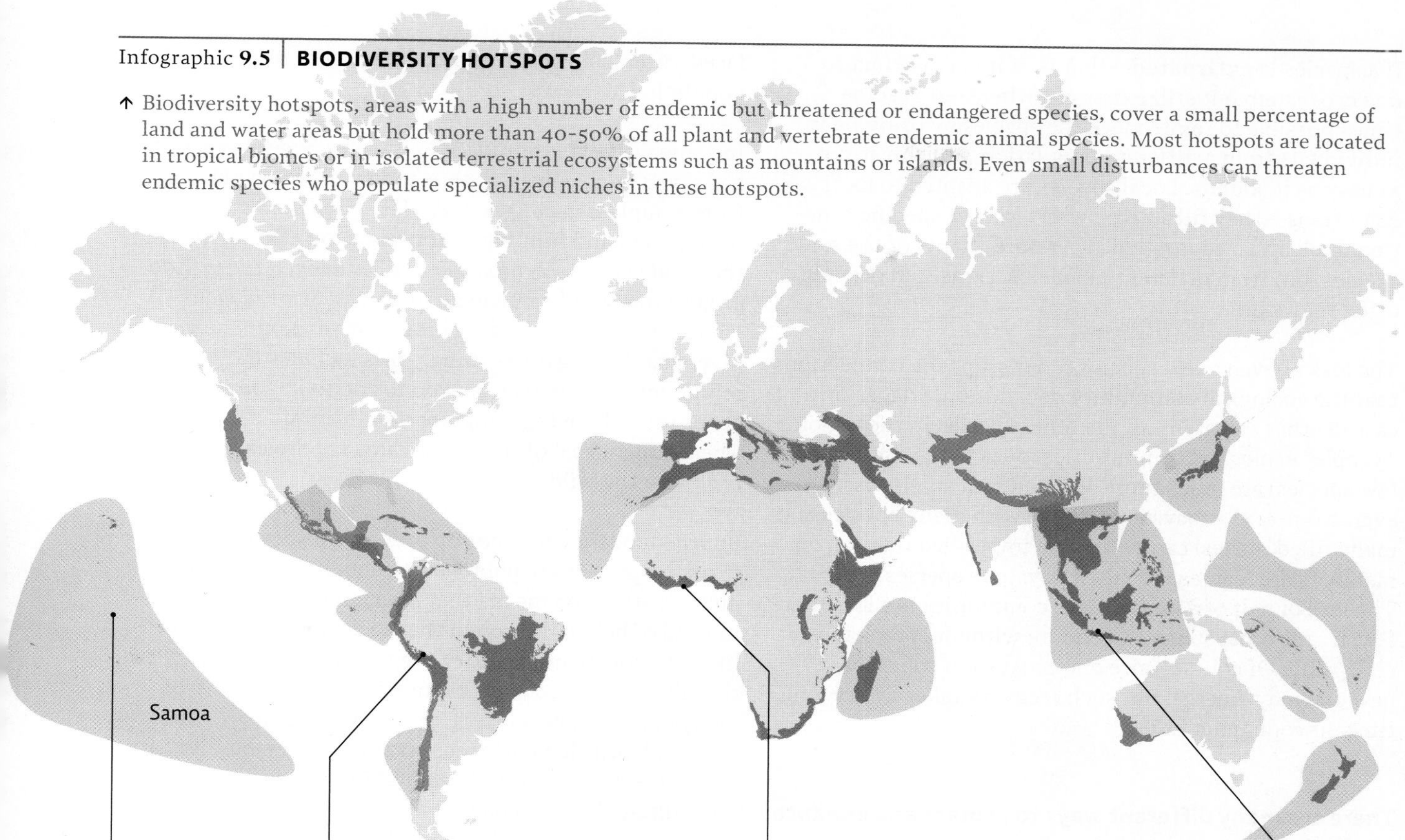
Samoa

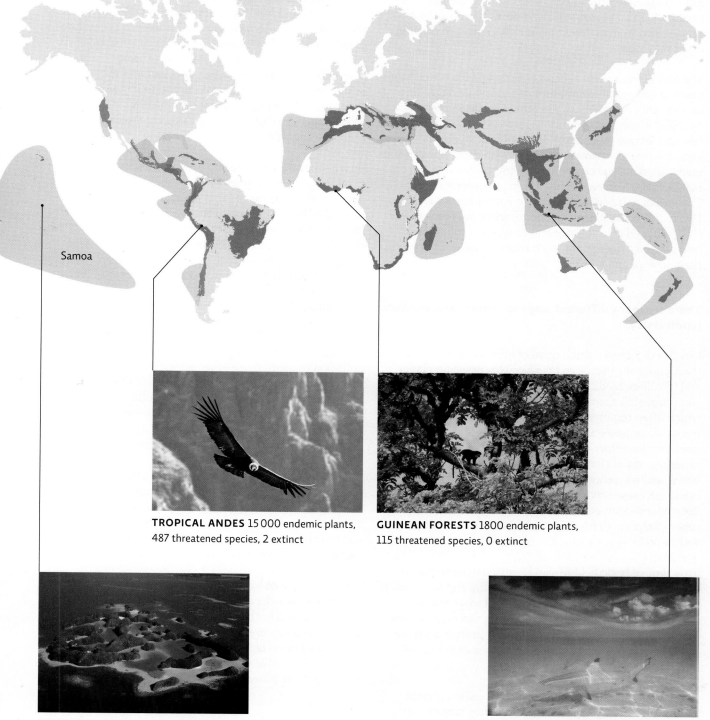

TROPICAL ANDES 15 000 endemic plants, 487 threatened species, 2 extinct

GUINEAN FORESTS 1800 endemic plants, 115 threatened species, 0 extinct

POLYNESIA-MICRONESIA 3074 endemic plants, 99 threatened species, 43 extinct

INDO-MALAYAN ARCHIPELAGO 15 000 endemic plants, 162 threatened species, 4 extinct

If a species is **extirpated**—that is, if it goes extinct in one ecosystem but still exists elsewhere—it is sometimes possible to reintroduce the species by bringing in individuals from a different population. But this is not as easy as it sounds. Local species are adapted to local conditions—they fill an *ecological niche*, or specific functional role in an ecosystem. Therefore, plants of the same species, but from different global locations, can occupy different niches.

The loss of even a few species can disrupt the connections that the ecological community depends on, triggering a cascade that threatens all the other species in the region. "Simple" ecological communities (ones with relatively few species) are much more affected by a single loss. But even "complex" ecosystems (those with many species and many filled niches) can be gravely imperilled if too many members are lost, especially if keystone species are lost. The loss of half of a native plant community is a loss that few, if any, ecosystems can endure without completely collapsing. Of course, once one ecosystem collapses, a new one may emerge. But such recovery takes a very long time. [INFOGRAPHIC 9.6]

There are many different ways to protect and enhance biodiversity.

Cox has deployed a small team of ethnobotanists throughout the Pacific to take a census of island plants and to collect seeds from those species with fewer than 100 remaining specimens. This careful accounting—which often requires scaling down cliffs and dangling from helicopters—will provide the basis of a large-scale conservation effort that scientists hope to replicate in other regions of the world. "If we know what we are losing and where we're losing it, we can plan a counter-offensive," says David Burney, from the National Tropical Botanical Garden on the Hawaiian island of Kaua'i. "The census helps us to figure out what's due to habitat loss, and so on."

This type of basic field research—evaluating an area to see what species are present, how healthy and genetically diverse that species' population is, and what threats the species face—is essential to protecting and maintaining biodiversity. After all, we can't save or protect what we don't even know exists or don't realize is in danger of extinction. But it is only the first step.

Early programs that addressed species endangerment often took the *single species approach*: they singled out well-known animals, like pandas and condors, and focused on the specific threats those individual species faced, often using captive breeding programs to increase population sizes.

In some ways, this approach has been a huge success: it has saved grey wolves, eagles, and even brown pelicans from complete decimation. And it's brought a great deal of attention and funding to conservation efforts. Indeed, the single species approach is still employed today. Zoos around the world participate in *Species Survival Plans* which use careful breeding programs to maximize genetic diversity. Among other things, this includes moving reproductive animals from zoo to zoo to introduce new genes into a breeding program or to minimize inbreeding. The ultimate goal of these programs is to release animals back into the wild.

But the tendency to focus on only those species (cute, furry, large) that are photogenic enough to capture the public's attention means that many, if not most, species in need of help fall through the cracks (people are more likely to donate money to "Save the panda!" than to "Save the white warty-back pearly mussel!"). For that reason, many experts now support an *ecosystem approach*. This means identifying entire ecosystems—often biodiversity hotspots—that are at risk, and taking steps to restore or rehabilitate damaged habitat and then to protect it from further damage or exploitation. It may involve reforestation projects, removal of non-native species, restoration of a natural river's flow patterns, or the clean-up of pollution (remediation)—the restoration goal depends on the ecosystem in question. Proponents of the ecosystem approach point out that by focusing on the ecosystem as a whole, the entire community benefits—not only the species we knew were endangered, but also those that we didn't even know were there.

In order to shift to an ecosystem focus we must know how to identify when an ecosystem is in danger. Ecologists are developing metrics (factors that are measured) to assess ecosystem quality, such as species richness and evenness, soil health, water quality, plant community composition, and the abundance of non-native species.

The protection of habitat often includes establishing preserves—which can include national parks, forests, grasslands, and in the ocean, marine protected areas—where human access is deliberately restricted. In Samoa, 5% of the land is legally protected to one degree or another; the goal of Cox and his colleagues is to increase that to 15%. In other parts of the world, the preservation

---

extirpated Describes a species that is locally extinct in one or more areas but still has some individual members in other areas.

Infographic **9.6** | **BIODIVERSITY OF SAMOA**

↓ Samoa has high biodiversity and is part of the diverse but threatened Polynesia/Micronesia hotspot. Some of the species that live in Samoan ecosystems are shown here.

TOOTH-BILLED PIGEON The endangered endemic tooth-billed pigeon (*Didunculus strigirostris*) specializes in eating mahogany seeds, using its sawtooth lower beak (like those seen in fossil birds) to open the hard seeds. Habitat loss due to deforestation threatens its survival.

FRUIT BAT Flying foxes (*Pteropus samoensis*) are fruit bats that pollinate large, tree climbing Freycinetia flowers, the roots of which are used by Samoans to make fishing baskets. The bats are endangered due to habitat loss; as the bats declined in number so did the flowers and the ability of the Samoans to make the baskets.

SAMOAN MOORHEN The endemic Samoan moorhen (*Gallinula pacifica*) may be extinct (last sighted in 1984) due to habitat loss from deforestation.

RED GINGER (*Alpinia purpurata*) is the national flower of Samoa (known as *teuila*) and a popular garden plant with its own national festival.

SAMOAN SWALLOWTAIL BUTTERFLY This endemic butterfly (*Papilio godeffroyi*) is critically endangered due to forest clearing for farming.

RACCOON BUTTERFLYFISH Samoan seas have at least 250 species of coral and almost 1000 species of fish. These raccoon butterflyfish (*Chaetodon lunula*) are one of 30 or so species of butterflyfish found there.

Infographic **9.7** | **CONSERVATION PRACTICES TO PROTECT AND ENHANCE BIODIVERSITY**

Biodiversity monitoring	Monitoring that allows us to see which species are present and how robust their populations are; also helps identify specific threats to populations' well-being.
Species Survival Plan (SSP)	A program developed by the Association of Zoos and Aquariums (AZA) that includes captive breeding plans to selectively breed individuals and maximize genetic diversity. The ultimate goal is to release individuals back into the wild to build up wild populations.
Ecosystem restoration	The repair of natural habitats back to (or close to) their original state. This may include the reintroduction of native species in an attempt to reestablish community connections. The restoration not only improves ecological diversity, it also increases or helps maintain species diversity.
Ecosystem remediation	The clean-up of pollution that is often a part of ecosystem restoration. Physical, chemical, or biological methods might be employed to remove or decontaminate an area and make it suitable for wildlife.
Preserves/parks	Protected areas that have been set aside and that limit human impact and provide sanctuaries for wildlife. These may be public areas (such as national parks, forests, or marine protected areas) and may be funded by ecotourism or debt-for-nature swaps.
Laws and treaties	Many nations have laws that protect endangered species, such as Canada's Species at Risk Act and the U.S. Endangered Species Act. International treaties include the Convention on International Trade in Endangered Species, which forbids trade of any endangered species or any product made from one, and the Convention on Biological Diversity, which promotes sustainable use of ecosystems and biodiversity.

of natural areas can be funded by **debt-for-nature swaps** in which a wealthy nation forgives part of the debt of a developing nation in return for the developing nation's pledge to protect certain ecosystems. So far, almost $1 billion in debt has been forgiven in these programs.

Laws and international treaties also offer species more formal protection, often by mandating the steps outlined above (namely monitoring, restoration, and protection). The **Species at Risk Act (SARA)**, passed in 2002, provides provisions for officially identifying a species as endangered (a *listed species*) and mandates that listed species and their habitats be protected in Canada. In many cases, this law has been very successful, but its effectiveness is perpetually hampered by inadequate funding and by conflicts with economic interests and a slow, political approval process. International treaties such as the Convention on International Trade of Endangered Species and the Convention on Biological Diversity offer protection of endangered and threatened species beyond the borders of an individual country. [INFOGRAPHIC 9.7]

debt-for-nature swap A wealthy nation forgives the debt of a developing nation in return for a pledge to protect natural areas in that developing nation.
Species at Risk Act (SARA) The primary law under which biodiversity is protected in Canada.

Biodiversity is proving invaluable in the search for cures.

For its part, the mamala tree turned out to be a good save. NCI researchers boiled down the bark and purified the remaining substance, which they named prostratin. Although prostratin had no effect against cancer, as Cox had hoped, it did prove to be a valuable weapon against HIV, the virus that causes AIDS. The first human trials of prostratin began in 2008. Researchers expect the drug to be commercially available in a few years.

So far, mamala bark provides the only known source of prostratin. While small amounts of the compound have been enough to conduct preclinical animal studies, much larger quantities will be needed for large-scale drug manufacturing. Stripping the Samoan rainforest of its mamala trees would be environmentally destructive, not to mention inefficient. So synthetic biologists are looking for better options. By locating the genes responsible for prostratin synthesis and cloning them into a simple organism, like bacteria, they hope to spare the mamala tree from overexploitation. The genetically altered bacteria would then serve as mini prostratin factories—quickly and efficiently producing mass quantities of the molecule which could then be used to make medication. This approach has already been successfully used to scale up production of the plant-derived antimalaria drug artemisinin.

Cox is quick to point out that without the Samoans' help, he would never have found prostratin. The forest he scoured housed two varieties of the tree; only one contained prostratin and within that variety, only trees of a certain size seemed to work (thus emphasizing the value of genetic diversity). The Samoans freely shared this knowledge with Cox. In exchange, he negotiated a profit-sharing agreement with the people of Falealupo. The U.S. government has promised that half of all royalty income from prostratin will go back to them. This represents one of the first formal legal recognitions of indigenous intellectual property rights. In addition to the profit sharing, any commercialized drug developed from prostratin will be supplied to developing countries for free.

While Cox and his colleagues believe ardently in each species' intrinsic value, they also agree that ethnobotany offers one of the best chances to preserve as many plant species as possible. "There is a strong link between the health of forests and the health of humans," he says. "If people understand that a rainforest might contain the best cures for diseases that plague us, they will care a whole lot more about saving it."◉

Select references in this chapter:
Cox, P.A. 1993. *Journal of Ethnopharmacology*, 38: 181–188.
Mora, C., *et al.* 2011. *PLoS Biology* 9(8): e1001127.doi:10.1371/journal.pbio.1001127.

BRING IT HOME

PERSONAL CHOICES THAT HELP

Species and habitats provide numerous benefits to people, including water and air purification, food sources, recreation, and medicine. Unfortunately, many species are facing threats at ever-increasing levels. The good news is that we as a society have a direct impact on these threats and can make changes to ensure the survival of many of our at-risk species.

Individual Steps

→ Don't buy products made from wild animal parts such as horns, fur, shells, or bones. Only buy captive-bred tropical aquarium fish, not wild-caught fish.
→ Research the policies of a not-for-profit organization that protects biodiversity, such as the Canadian Wildlife Federation (CWF), the Canadian Parks and Wilderness Society (CPAWS), and the Nature Conservancy of Canada. Is it worth donating money to their cause?
→ Make your backyard friendly to wildlife, using suggestions from www.cwf-fcf.org/en/what-we-do/habitat/gardening-for-wildlife/how-to-garden-for-wildlife.
→ Install an Audubon Guide app on your smartphone, or buy a field guide to learn the plant and animal species in your area.

Group Action

→ Work with faculty and other students to organize a bioblitz for a protected area in your region. A bioblitz, which is an intensive survey of all the biodiversity in the area, can generate a large amount of data to be used for habitat management and species protection.
→ Join a citizen science program monitoring wildlife. Many regional conservation groups have monitoring opportunities and provide training. For national programs, visit www.naturewatch.ca, a citizen science monitoring program managed by Nature Canada (www. naturecanada.ca).

Policy Change

→ Canada's Species at Risk Act was established in 2002 to protect species diversity. To learn more, visit the Committee on the Status of Endangered Wildlife in Canada at www.cosewic.gc.ca.

UNDERSTANDING THE ISSUE

CHECK YOUR UNDERSTANDING

1. Which of the following explains why there are more endangered tropical species than temperate species?
   a. More people live in the tropics.
   b. Tropical species are more likely to be generalists with broad niches.
   c. Global climate change is warming the tropics more rapidly.
   d. Both a. and c. are correct.

2. Why are many of the biodiversity hotspots around the world on islands?
   a. Islands accumulate species from many different areas.
   b. Populations of island species are isolated.
   c. Islands have more diverse habitats.
   d. There are more niches on islands.

3. The 'single species' conservation approach _______ the whole ecosystem conservation approach.
   a. has never been as successful as
   b. focuses more on less conspicuous species than
   c. has the same ultimate goals as
   d. None of the above.

4. If you live in a suburb or housing development and put out bird feeders, what kind of species are you likely to attract?
   a. Core species
   b. Edge species
   c. Extirpated species
   d. Both b. and c. are correct.

5. Whereas an individual with an anthropocentric worldview would stress the _______ value of species as a reason to conserve biodiversity, most environmental scientists would stress their _______ value.
   a. intrinsic; medicinal
   b. instrumental; ecological
   c. ecological; instrumental
   d. intrinsic; cultural

6. Which of the following does NOT directly contribute to biodiversity loss?
   a. Habitat loss in tropical hotspots
   b. Invasive species in areas with high endemism
   c. Synthetic drugs derived from plants
   d. Fragmentation of habitat

WORK WITH IDEAS

1. Which is the most convincing reason for protecting endangered species: the intrinsic value of biodiversity, the possibility of finding medicines, or preserving ecosystem function? Why?

2. What are the ethical issues involved in preserving biodiversity in tropical areas?

3. Some agricultural researchers are interested in finding the wild ancestors of our domesticated animals and plants. They are concerned that modern food production uses only a few varieties of a given species, although there are many varieties that could be used. How is the protection of genetic diversity within a species both similar and dissimilar to the conservation of threatened species?

4. Of the different organizations, laws, and treaties that protect biodiversity (the Species at Risk Act, the Convention on International Trade of Endangered Species, etc.), which one do you think is the most powerful and effective globally? Why?

5. Describe how biodiversity and national security may be connected. Why do some people make the argument that maintaining biodiversity is critical to national and international security?

6. How can you, as an individual, help maintain biodiversity worldwide? Choose at least two actions that you can take; justify your choices as they relate to biodiversity.

ANALYZING THE SCIENCE

Biodiversity hotspots are regions of the world that are of particular importance to global biodiversity both because of the total number of species they contain and because of the number of *endemic species* they have.

The map on the following page shows the location of biodiversity hotspots around the world as well as the human population density.

INTERPRETATION

1. How many of the 25 hotspots include islands?

2. Describe the relationship between population density and biodiversity in one or two sentences.

3. Can you assume that biodiversity hotspots in highly populated areas are the most at risk? What other factors would you look for?

ADVANCE YOUR THINKING

4. Based on what you've already learned about human population growth, do you think that any hotspots that are not being threatened by development will become threatened in the next 50 years?

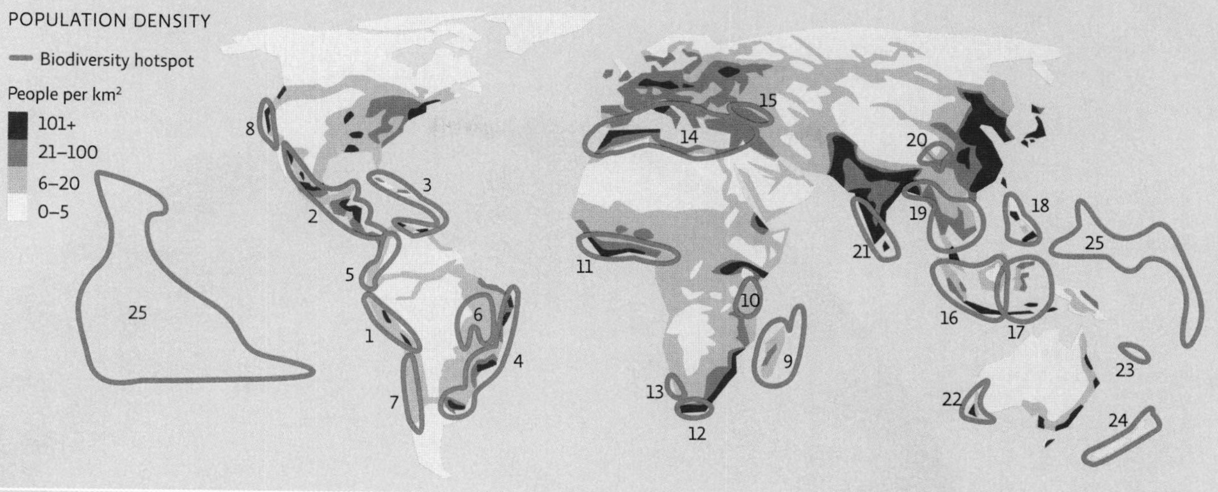
POPULATION DENSITY
Biodiversity hotspot
People per km²
101+
21–100
6–20
0–5
1
2
3
4
5
6
7
8
9
10
11
12
13
14
15
16
17
18
19
20
21
22
23
24
25
25

EVALUATING NEW INFORMATION

Visit the World Wildlife Fund website (www.worldwildlife.org). Start your exploration of the site by selecting the "What We Do" menu and reading the overview under the "Protect Wildlife" menu; note the conservation approach that WWF takes using flagship species as representatives of ecosystems.

Evaluate the website and work with the information to answer the following questions:

1. From the Protect Wildlife page, click on the "Flagship Species" link.
   a. What is the difference between a flagship species, a priority species, an indicator species, and a keystone species?

2. Does WWF provide evidence that its flagship species approach works?
   a. What scientific evidence does the organization provide?
   b. Does it mention any alternative conservation approaches?

3. Do you agree with WWF's approach to species conservation?
   a. How would you approach conservation?
   b. Do you think the WWF could get as many donations if it did not use charismatic species?

4. Select the "Conserve Places" menu and choose a conservation region to read about.
   a. What species are used to represent this region?
   b. What are the main threats to diversity in this region?
   c. What strategy is the WWF using to conserve the region?
   d. Would you do anything differently? Why or why not?

MAKING CONNECTIONS

ECONOMIC INTERESTS VERSUS PROTECTING A SPECIES

Background: COSEWIC, the Committee on the Status of Endangered Wildlife in Canada, was established in 1977 to identify species at risk of extinction using a sound, scientific approach (www.cosewic.gc.ca). Twenty-five years later, the Canadian Species at Risk Act (SARA) was passed in 2002 as part of Canada's commitment to meet its obligations under the International Convention on Biological Diversity. SARA's aim is to protect endangered and threatened species and their habitats in Canada. In 2003, COSEWIC was designated as the body that would identify and assess species status under SARA.

Although this sounds good, SARA implementation has been criticized since its inception. The problem lies with how species get final approval for inclusion on the list of species at risk. While COSEWIC, a group of scientists and wildlife experts, makes recommendations for "at-risk" designation based on scientific study, all recommended species must be approved by a panel of elected government officials under the guidance of the Minister of the Environment. Species that are approved get added to Schedule 1, which gives the species and their habitat protection, and they become subject to a federal management plan. Many species have remained under nearly perpetual "consideration" by the Minister due to conflicting influences, often economic interests. This is particularly true in the case of the Atlantic Cod, once a major source of income.

Despite the listing of Atlantic Cod as endangered by the World Wildlife Fund and the collapse of the Canadian Cod fishery, Atlantic Cod remains of indeterminate status under SARA.

Case: You are a COSEWIC member who must reassess the status of Atlantic Cod in Canada. Write a brief history of the Cod evaluation process thus far, and make a recommendation for their status now. Research the following problems in your argument:

1. How often has Atlantic Cod been recommended for protection under SARA, and what status has been approved to date?

2. How much consideration should be given to habitat protection given the existence of other fisheries in the same area?

3. Do we know enough about the different Cod populations worldwide to be able to make a sound decision?

4. What are the implications of not protecting a species on the grounds of public welfare or economic stability?

Be sure to support your recommendation with specific facts about Cod populations and habitat and the economic value of the Cod fishery. Note whether you agree or disagree with the current SARA status of Cod in Canada.

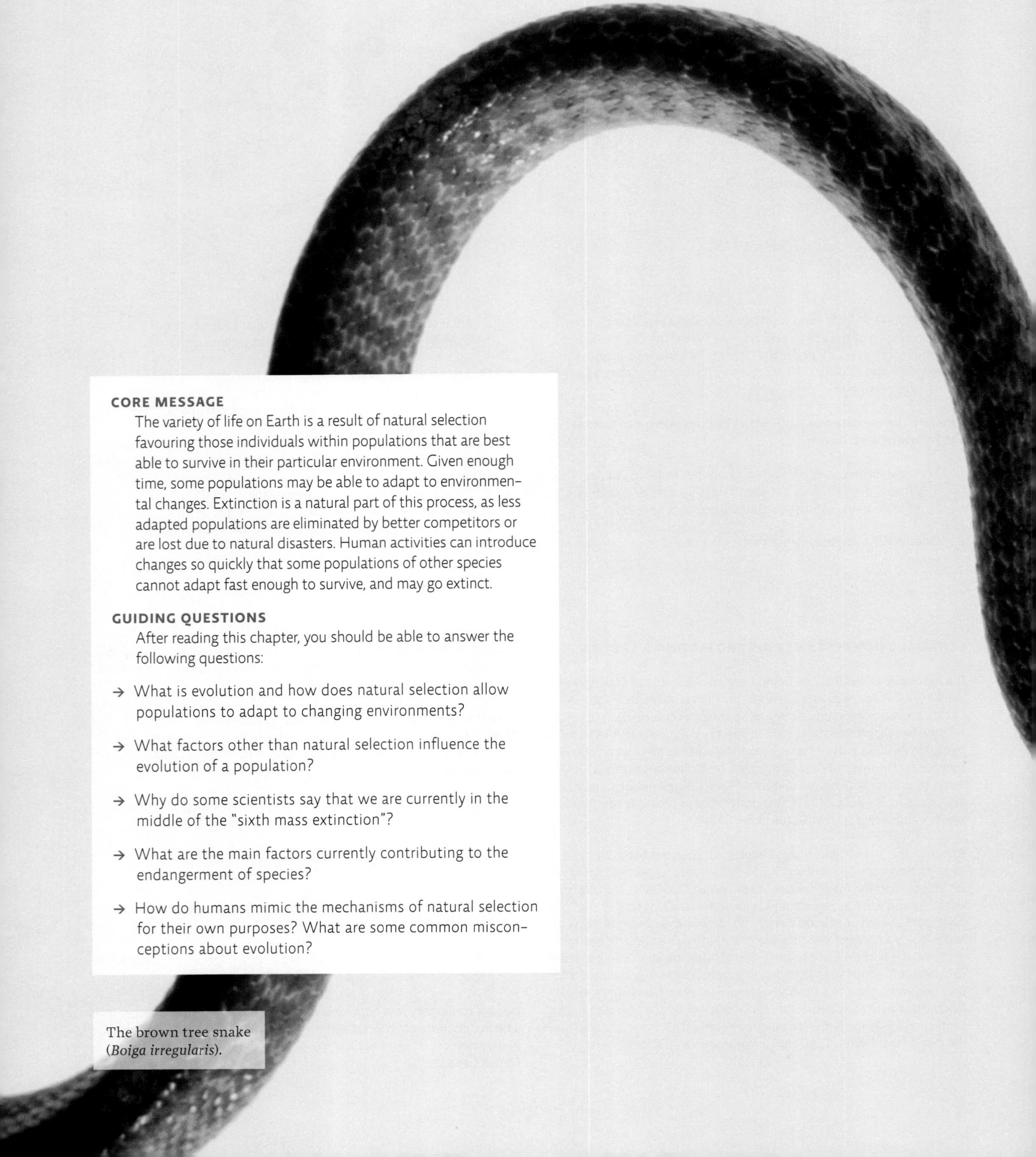

The brown tree snake (*Boiga irregularis*).

CORE MESSAGE

The variety of life on Earth is a result of natural selection favouring those individuals within populations that are best able to survive in their particular environment. Given enough time, some populations may be able to adapt to environmental changes. Extinction is a natural part of this process, as less adapted populations are eliminated by better competitors or are lost due to natural disasters. Human activities can introduce changes so quickly that some populations of other species cannot adapt fast enough to survive, and may go extinct.

GUIDING QUESTIONS

After reading this chapter, you should be able to answer the following questions:

→ What is evolution and how does natural selection allow populations to adapt to changing environments?

→ What factors other than natural selection influence the evolution of a population?

→ Why do some scientists say that we are currently in the middle of the "sixth mass extinction"?

→ What are the main factors currently contributing to the endangerment of species?

→ How do humans mimic the mechanisms of natural selection for their own purposes? What are some common misconceptions about evolution?

CHAPTER 10 EVOLUTION AND EXTINCTION

A TROPICAL MURDER MYSTERY

Finding the missing birds of Guam

↑ A brown noddy (*Anodus stolidus*) guards an egg. The brown noddy nests in large numbers on nearby Cocos but has not successfully nested on Guam since snake populations peaked in the 1970s and 1980s.

→ Dr. Julie Savidge holding a Mariana Fruit-Dove. This species only occurs on certain islands within the Mariana Islands and the last sighting on Guam was in 1985. This bird was caught as part of an early blood sampling effort to see if exotic diseases might be causing the bird decline on Guam.

Something bizarre began happening on Guam, the southernmost island in the Mariana island chain nestled halfway between Japan and New Guinea, in the late 1960s. The island (a U.S. territory), where once some 18 native avian species filled the tropical forests with song, became eerily quiet as the nation began losing its bird populations. By the early 1980s, four species of birds had gone extinct; the first documented loss was the Guam bridled white-eye, a small bird that flocked in tree canopies, feeding on nectar and small insects. Ten others were in danger of extinction. Worse, researchers and wildlife experts did not have a clue why they were dying.

In the United States, in the winter of 1980, biologist Julie Savidge was gearing up to begin Ph.D. work at the University of Illinois when she by chance attended a lecture in Redding, California, organized by the Wildlife Society. A biologist from Guam had been invited to talk about the islands' bird disappearances, and he noted—rather grimly—that no one had yet solved the devastating mystery. Savidge was immediately enthralled. "I went up to him afterwards and said, 'This sounds fascinating. Is there any way that I might be able to get out to Guam?'" Savidge recalls.

Two summers later, after much back-and-forth between Savidge and the Guam Division of Aquatic and Wildlife Resources, Savidge was hired to investigate the bird disappearances as part of her Ph.D. dissertation.

At the time, Guam's scientists were convinced that diseases like blood parasites or pesticides were to blame for the avian losses. So Savidge began looking into those possibilities. But as soon as she started talking to Guam locals, she became skeptical of the idea. "When I would interview the natives, I'd be asking all these disease questions, and they'd say, 'Why are you asking about this? The real problem is the snakes,'" she explains.

The locals were certain that brown tree snakes (*Boiga irregularis*), 1 to 2 metre-long, non-native snakes that had been accidentally introduced to the islands in military shipping cargo in the late 1940s and 1950s, were responsible for the birds' demise. Non-native species that cause ecological, economic, or human health problems and are hard to eradicate are considered **invasive species**, and they can cause significant damage in areas they invade.

Oddly, Savidge's research was starting to show that, in fact, diseases weren't playing a factor in the bird deaths at all. When she and her colleagues sampled a variety of birds for bacteria, viruses, and parasites between 1982 and 1985, they found that the native birds were basically healthy. Curious about the tree snake hypothesis (an idea other scientists had dismissed), Savidge began to investigate, on her own time, whether these reptiles might be causing the **extinctions**. Surely it was something else—could a few snakes really obliterate a whole island of birds?

◉WHERE IS GUAM?

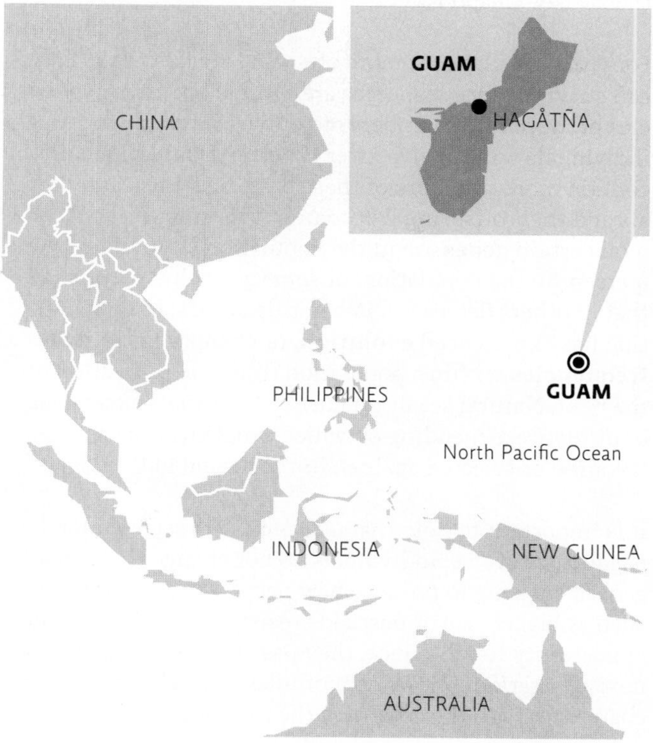

GUAM
HAGÅTÑA
CHINA
PHILIPPINES
GUAM
North Pacific Ocean
INDONESIA
NEW GUINEA
AUSTRALIA

invasive species A non-native species (a species outside of its range) whose introduction causes or is likely to cause economic or environmental harm or harm to human health.

extinct/extinction The complete loss of a species from an area; may be local (gone from a specific area) or global (gone throughout the world).

Natural selection is the main mechanism by which populations adapt and evolve.

Before they started disappearing, the birds of Guam were a diverse and resplendent bunch. The island, about a tenth the size of Price Edward Island, was home to 18 native species of birds, each specially suited to life on the island.

Populations usually contain individuals that are genetically different from one another. According to the evolutionary theory first put forth by Charles Darwin and Alfred Russel Wallace, and subsequently supported by a tremendous amount of evidence from a wide variety of scientific disciplines, **selective pressure** on a population—a nonrandom influence affecting who survives or reproduces—favours individuals with certain inherited traits over others (such as better camouflage, tolerance for drought, or enhanced sense of smell). These individuals have *differential reproductive success*: they are best suited for their environment and leave more offspring than those who are less suited for their environment. The traits that an environment favours are called **adaptations**, and the process by which organisms best adapted to the environment survive to pass on their traits is known as **natural selection**. Evolutionary biology helps us understand the diversity of life on Earth and how populations change over time. It is one of the pillars of biological science, and the vast amount of evidence in support of both the occurrence of evolution and the mechanisms by which it happens has elevated this explanation to the level of scientific theory (see Chapter 2).

For most populations, more offspring are born than can survive, since resources are limited and many species produce large numbers of young. Since only some individuals will survive, over time, the population will contain more and more of these better-adapted individuals and their offspring. Ultimately, this changes how common certain **genes** are in the population—the frequency (percent in the population) of some genes increases and that of others decreases. When this occurs, the population has experienced **evolution**, or changes in the **gene frequencies** within a population from one generation to the next. Natural selection may be *stabilizing, directional*, or *disruptive*, depending on which genetic traits are favoured or selected against. [INFOGRAPHIC 10.1]

It is important to note that *individuals* are selected for, but *populations* evolve; individuals do not change their own genetic makeup to produce new necessary adaptations, such as bigger size or pesticide resistance. If they get the opportunity to reproduce, they pass their traits on to the next generation. If they cannot tolerate environmental changes, as was the case with the first bird species to disappear from Guam (the bridled white-eye), they die or fail to reproduce and do not pass on their genes. Individuals may be able to adjust their behaviour to accommodate environmental changes, but if a trait is not genetically controlled, and therefore is not heritable, it will not influence the composition of the next generation.

Populations need genetic diversity to evolve.

The ability of a population to adapt is a reflection of its tolerance limits, which largely depend on **genetic diversity**—different individuals having different versions of genes (called *alleles*). A population that is highly diverse (has individuals with many different traits) is likely to have wider tolerance limits, which increases the population's potential to adapt to changes. This means it is more likely that some individuals will exist that can withstand (or even thrive in) the changes, and that the population as a whole will survive. If a change occurs that produces a condition outside of the range where individuals can survive and reproduce (for instance, the climate becomes too warm, or too dry), the population will die out. Similarly, if a new challenge is presented, such as the introduction of a new predator or competitor, the survival of the population will depend on whether there are any individuals in the population who can effectively deal with the new species. If the snakes on Guam were indeed responsible for killing off the birds, any birds that happened to have effective snake-avoidance behaviours would have had a greater chance of survival.

There are two main sources of variation that can increase genetic diversity in a population. The ultimate source of new variability is genetic *mutation*, a change in the DNA sequence in the sex cells that alters a gene, sometimes to the extent that it produces a new protein, and possibly a new trait. Mutations are rare, but because DNA replication and repair occurs all the time, rare events do happen. When a mutation produces traits that are beneficial, they

---

selective pressure A nonrandom influence affecting who survives or reproduces.
adaptation A trait that helps an individual survive or reproduce.
natural selection The process by which organisms best adapted to the environment (the fittest) survive to reproduce, leaving more offspring than less well-adapted individuals.
genes Stretches of DNA, the cell's hereditary material, that each direct the production of a particular protein and influence an individual's traits.
evolution Differences in the gene frequencies within a population from one generation to the next.
gene frequencies The assortment and abundance of particular variants of genes relative to each other within a population.
genetic diversity The heritable variation among individuals of a single population or within the species as a whole.

Infographic **10.1** | **NATURAL SELECTION AT WORK**

↓ When the environment presents a selective force (a new predator, changing temperatures, change in food supply), natural selection is the primary force by which populations adapt. The survivors are those who were lucky enough to have genetic traits that allowed them to survive in their changing environment (others who did not possess the trait were not as likely to survive to reproduce). Because survivors pass on those adaptations to their offspring, the gene frequencies of the population change in the next generation, which means some traits are more common and others are less common than they used to be. When this happens, the population is said to have evolved.

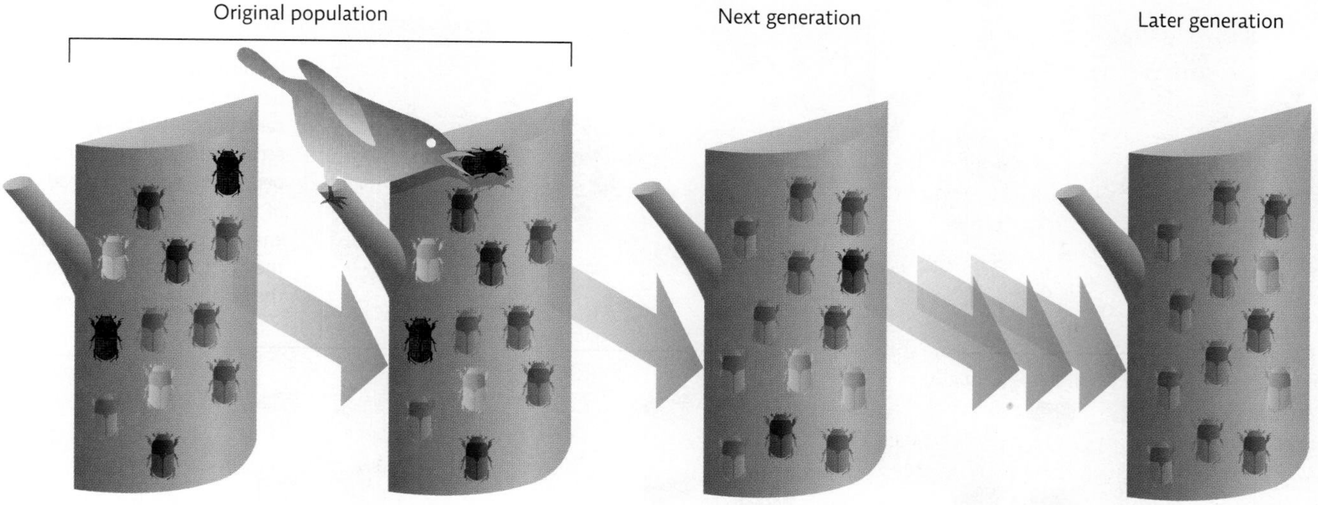
Original population
Next generation
Later generation

Beetles resting on this tree vary in colour.

Genetic variation exists in the population: individuals possess inherited differences.

Individuals with the less favourable trait (colouration that makes them stand out on a tree trunk) are more likely to be eaten.

Differential reproductive success: not everyone will survive to reproduce.

Fewer dark individuals are born (though recombination might produce some from light or tan parents).

Gene frequencies have changed: the population is evolving.

Over time, the population may be mostly or solely made up of tan individuals.

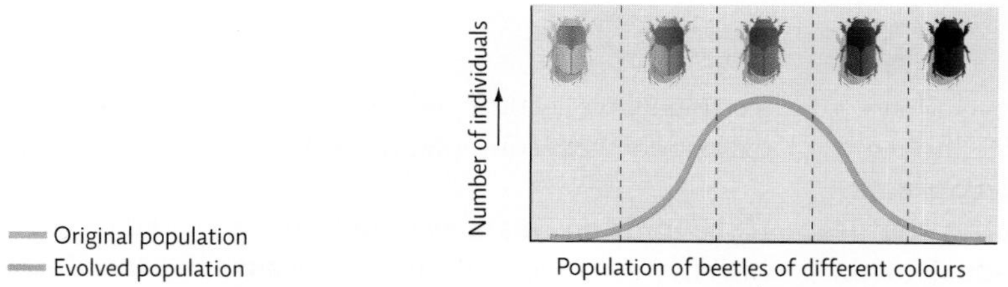
Number of individuals
Original population
Evolved population
Population of beetles of different colours

← Natural selection can have different outcomes, depending on what varieties in the population the environment favours or selects against.

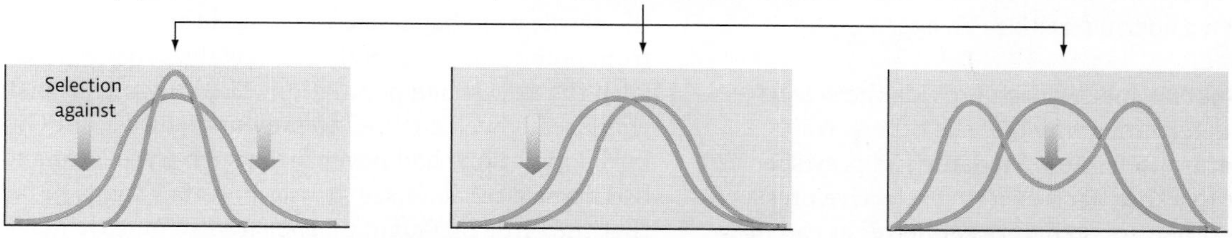
Selection against

Stabilizing selection favours the norm and selects against extremes.

All trees are tan; tan beetles are favoured.

Directional selection continually favours a particular extreme of the trait (bigger, darker, etc.).

In areas with trees darkened from pollution, darker beetles are favoured.

Disruptive selection favours the extremes but selects against the intermediate forms.

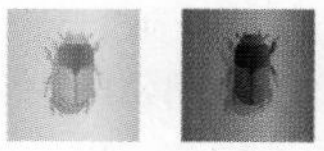

In a forest with light and dark trees but no tan trees, tan beetles (the intermediate colour) are not favoured.

Infographic 10.2 | **COEVOLUTION ALLOWS POPULATIONS TO ADAPT TO EACH OTHER**

↓ As selection favoured beetles closest to the tree colour, only birds with the keenest eyesight feed well enough to survive and reproduce.

↓ Any beetle with an even better camouflage would escape predation and pass on its genes.

↓ This then favours birds with even keener eyesight who would feed well and pass on the sharp eyesight trait to their offspring.

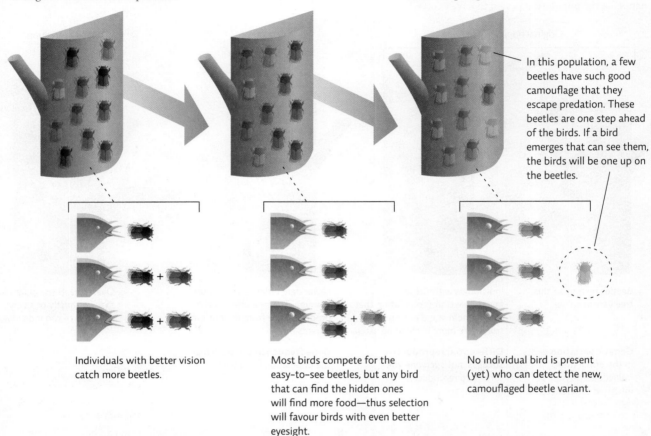

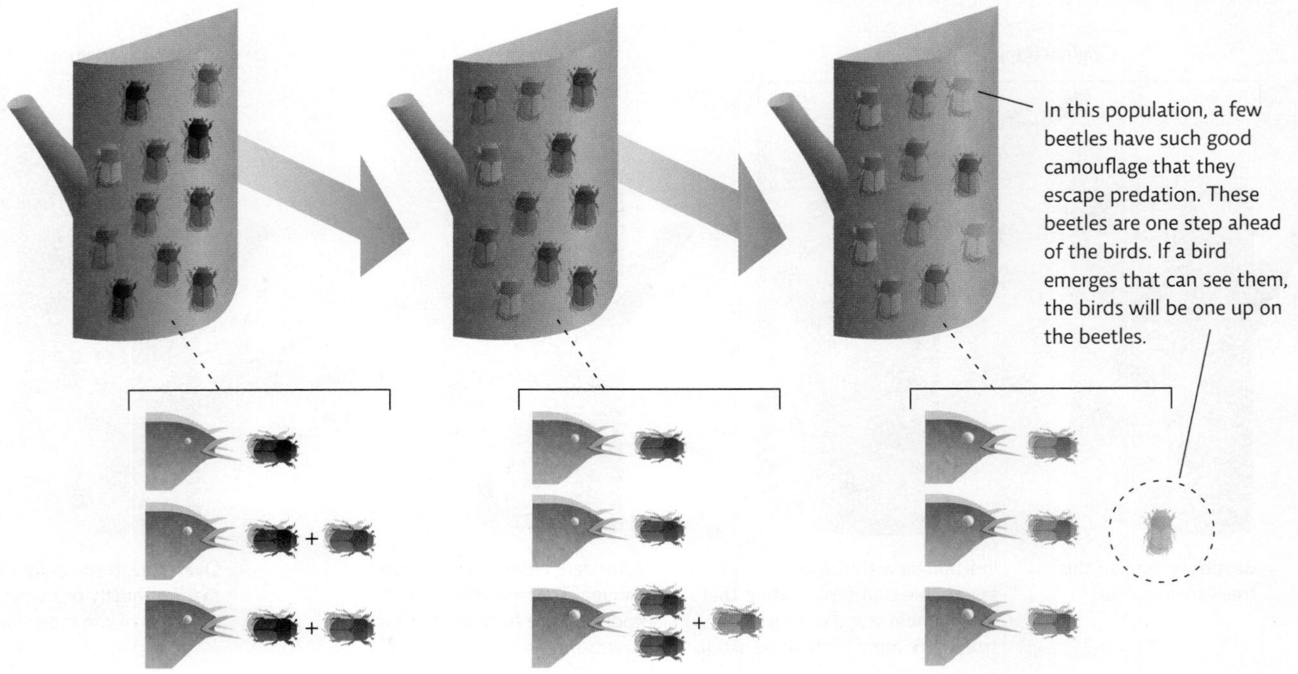
In this population, a few beetles have such good camouflage that they escape predation. These beetles are one step ahead of the birds. If a bird emerges that can see them, the birds will be one up on the beetles.
+

Individuals with better vision catch more beetles.

Most birds compete for the easy-to-see beetles, but any bird that can find the hidden ones will find more food—thus selection will favour birds with even better eyesight.

No individual bird is present (yet) who can detect the new, camouflaged beetle variant.

can quickly be passed on to the next generation, allowing the population to evolve to be better adapted to its environment. A second source of genetic variety occurs as eggs and sperm are made: *genetic recombination* shuffles alleles around and sometimes produces individuals with new traits when a sperm fertilizes an egg.

In **coevolution**, two species each provide the selective pressure that determines which of the other's traits is favoured by natural selection. Predator and prey species usually evolve together, each exerting selective pressures that shape the other. As predators get better at catching prey, the only prey to survive are those a little better at escaping, and it is these individuals that reproduce and populate the next generation. This game of one-upmanship continues generation after generation, with each species affecting the differential survival and reproductive success of the other. The result can be a predator extremely well-equipped to capture prey, and prey extremely well-equipped to escape. [INFOGRAPHIC 10.2]

If the birds on Guam were indeed eradicated by the invasive snake species, it was because the speed at which the eradication happened prevented the bird populations from potentially coevolving survival strategies to deal with the new snake population. The brown tree snake was already well adapted to prey upon birds. But Guam's bird populations had never faced such a predator and had no natural defenses. It was an unfair fight. In fact, that the birds on Guam disappeared so quickly made it difficult to tease out the cause of their extinction at all. Savidge had to work backwards to solve the mystery,

coevolution Two species each provide the selective pressure that determines which traits are favoured by natural selection in the other.

endemic Describes a species that is native to a particular area and is not naturally found elsewhere.

Infographic **10.3** | **ENDANGERED AND EXTINCT BIRDS OF GUAM**

THE SPREAD OF THE SNAKE MAPPED WITH BIRD EXTINCTIONS OVER TIME

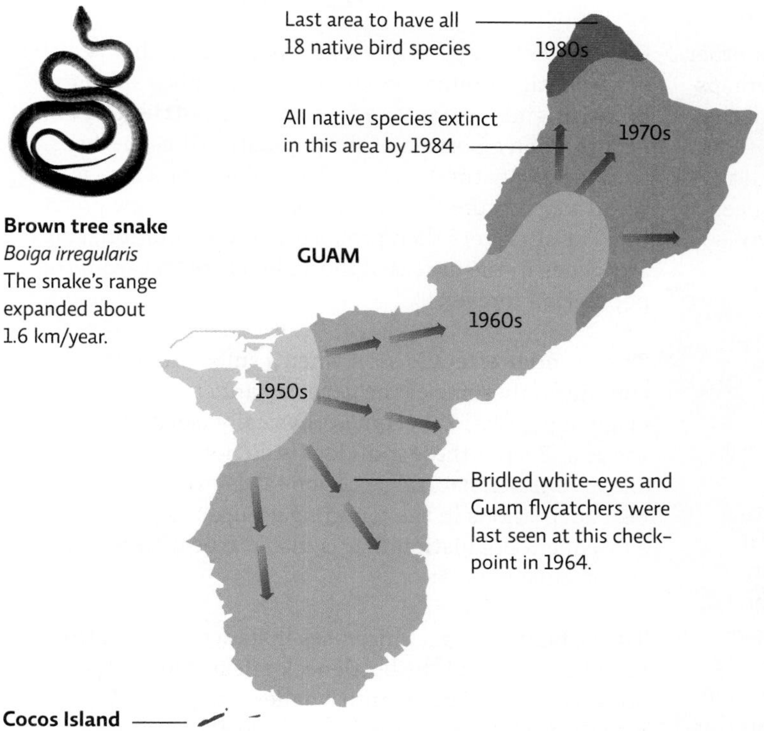
Brown tree snake
Boiga irregularis
The snake's range expanded about 1.6 km/year.
GUAM
Last area to have all 18 native bird species
1980s
All native species extinct in this area by 1984
1970s
1960s
1950s
Bridled white-eyes and Guam flycatchers were last seen at this checkpoint in 1964.
Cocos Island
Boiga is absent from Cocos Island; all bird species are present.

↑ The loss of birds followed the spread of *Boiga* as predicted. The last area to have all 10 species of native birds was the last area *Boiga* invaded.

↓ Of the 18 original native species on Guam, many are endangered or already extinct.

Extinct
Bridled white-eye
The endemic Guam bridled white-eye was last seen in 1983 and is now extinct.

Extinct
Rufous fantail
Once common all over Guam, the endemic subspecies of the Rufous fantail is now extinct.

Extinct
Guam flycatcher
The endemic Guam flycatcher was last seen on Guam in 1985 and is now extinct.

Only Captive
Guam kingfisher
By 1986, only seven Guam kingfishers existed in the wild; a captive breeding program is underway.

Only Captive
Guam rail
The endemic Guam rail is extinct in the wild; a captive breeding program is underway.

In other islands
Mariana fruit-dove
The Mariana fruit-dove is extirpated on Guam but is still found on other islands.

putting the missing pieces together, experiment by experiment. Her first step was to compare whether the distribution of the snakes on Guam matched up with the areas where birds had disappeared.

Based on what the residents told her, and what historical records like newspaper clippings reported, Savidge found that birds had begun to disappear first from southern Guam, and that their disappearances matched up perfectly with when the brown tree snakes had begun populating the area. Ritidian, an area on the very northernmost tip of Guam, on the other hand, was the last area to have lost its birds, and it was also the last to be colonized by the snakes. "I found a close correlation between the bird decline and the expansion of brown tree snakes around the island," she says, which suggested to her that brown tree snakes really might be the culprit. [INFOGRAPHIC 10.3]

Some of Guam's bird species went extinct sooner than others. For instance, the **endemic** bridled white-eye, the gregarious bird species that was extinguished first, happens to be very small, raising the possibility that the small size of these birds might have put them at a disadvantage. Larger species like flycatchers survived longer, though they, too, eventually disappeared. Other bird species experienced **extirpation**; the Guam rail, for instance, is gone from Guam but other populations still live on the nearby island of Rota.

Populations can diverge into subpopulations or new species.

If even a few individuals of the bridled white-eye or other extinct species had been able to avoid the snake (perhaps due to a heritable trait that made them more wary of the predator), they might have given rise to new populations that could cohabitate with the snake. Or, had they not been surrounded by sea, the Guam birds may have been able to relocate and find a safe haven from the brown tree snake.

> Predator and prey species usually evolve together, each exerting selective pressures that shape the other.

When populations diverge because of isolation, food availability, new predators, or habitat fragmentation such that their members can no longer freely interbreed, new species may arise (*speciation*). This increases species diversity in a community and sometimes produces specialists who can exploit open niches. This separation may be physical (for example, geographic boundaries the individuals won't cross) or may arise when something prevents some individuals from choosing others as mates, as may happen when individuals spend their time in different parts of their habitat.

However, not all evolution is driven in this manner. Random events play a role, too, typically by decreasing genetic diversity rather than increasing it.

In any population some individuals will survive and mate, and others won't. For no other reason than pure chance, this random mating may increase or decrease the frequency of a particular trait, a process known as **genetic drift**. The loss of this trait doesn't have anything to do with any inherent advantage or disadvantage the trait offers; it is merely the chance loss of some gene variants because the individuals with these genes did not happen to mate. This can produce significant changes, especially in small populations where traits that may be present in just a few could be lost in a single generation if these individuals don't happen to reproduce.

Genetic drift can produce a new population that is different from the original population when only a subset of the original variants reproduce. The **bottleneck effect** occurs when a portion of the population dies, perhaps because of a natural disaster like a flood, or a strong new selective pressure like the introduction of a new predator. The survivors then produce a new generation, and any genes in the deceased individuals are lost from the population forever.

The **founder effect** is seen when a small group that contains only some of the gene variants found in the original population becomes physically isolated from the rest. Even if these individuals do not have the most adaptive traits, natural selection takes over, favouring the best adaptations in the founding group, and producing a subsequent population that is likely to be different from the original population.

Today, human impact increases instances of both the founder effect and the bottleneck effect. Much of what we do isolates populations into smaller groups, forcing them into these situations. [INFOGRAPHIC 10.4]

The pace of evolution is generally slow but is responsive to selective pressures.

In addition to genetic diversity, the size of the population also makes a difference in how quickly natural selection can produce a change in a population: beneficial traits can spread more quickly in smaller populations simply because it is more likely that the individuals with the trait will find each other and mate. Reproductive rate and generation time also influence how quickly a population can adapt to changes. Many problem species, like insect pests, are *r* species and have fast generation times, which means they can often stay one step ahead of our efforts to control them. Many endangered species, on the other hand, are *K* species, with longer generation times; therefore they take longer to recover if population numbers fall (for more on *r* and *K* species, see Chapter 7).

The strength of the selective pressure also affects how quickly natural selection might produce a change in a population. One of the reasons the demise of birds

---

extirpation Local extinction in one or more areas, though some individuals exist in other areas.

genetic drift The change in gene frequencies of a population over time due to random mating that results in the loss of some gene variants.

bottleneck effect When population size is drastically reduced, leading to the loss of some genetic variants, and resulting in a less diverse population.

founder effect When a small group with only a subset of the larger population's genetic diversity becomes isolated and it evolves into a different population, missing some of the traits of the original.

Infographic **10.4** | **RANDOM EVENTS CAN ALTER POPULATIONS**

GENETIC DRIFT

↓ Random mating can eliminate some gene variants from the population not because these individuals were poorly adapted to their environment, or less attractive to the opposite sex, but because they were just unlucky and didn't mate.

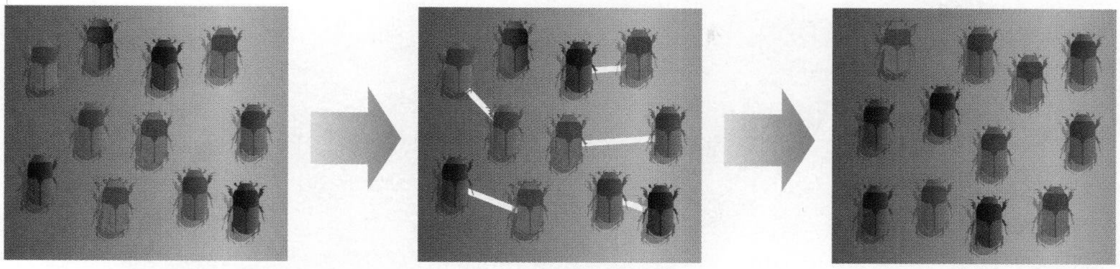

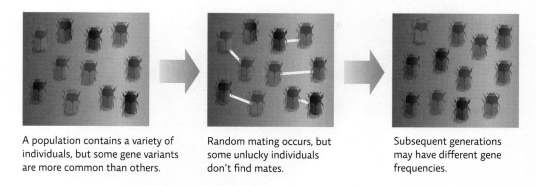

A population contains a variety of individuals, but some gene variants are more common than others.

Random mating occurs, but some unlucky individuals don't find mates.

Subsequent generations may have different gene frequencies.

BOTTLENECK

↓ If something causes a large part of the population to die, leaving the survivors with only a portion of the original genetic diversity, the population may recover in size but will not be as genetically diverse as the original population.

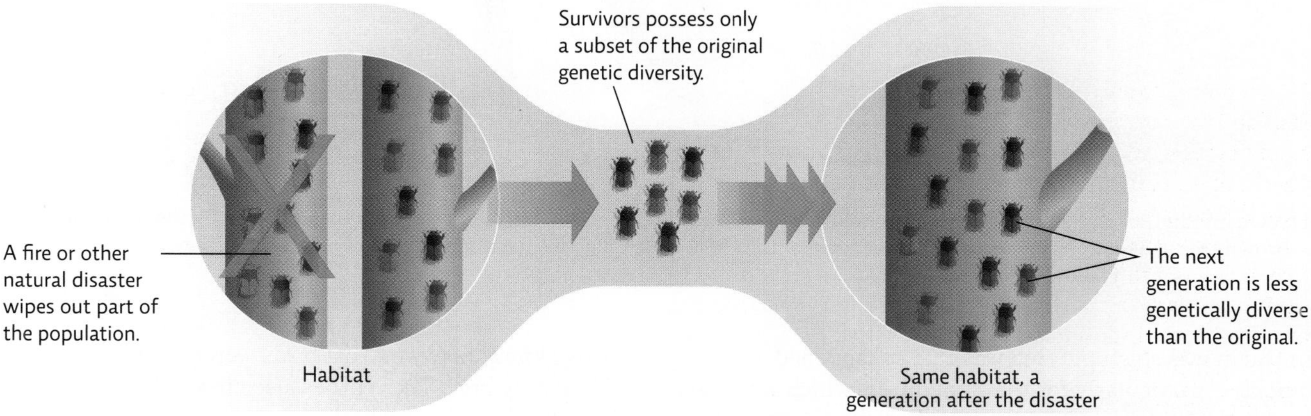
Survivors possess only a subset of the original genetic diversity.
A fire or other natural disaster wipes out part of the population.
Habitat
The next generation is less genetically diverse than the original.
Same habitat, a generation after the disaster

FOUNDER EFFECT

↓ If only a subset of a population colonizes a new area and becomes completely isolated from the others, the new population will be less genetically diverse than the original.

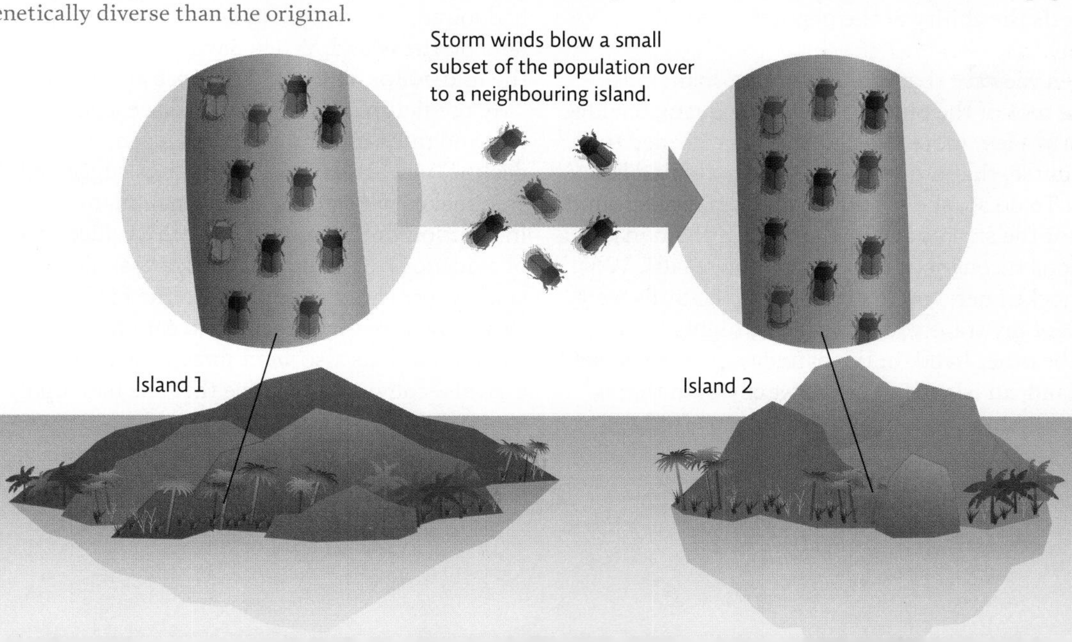
Storm winds blow a small subset of the population over to a neighbouring island.
Island 1
Island 2

↑ A Guam rail (*Rallus owstoni*), a critically endangered species. The species' decline was caused by predation by the introduced brown tree snake. It is currently considered extinct in the wild, though reintroduction programs are underway.

in Guam was so stupefying was that it happened so quickly—particularly for the small birds, which were easiest for the snakes to eat. Larger birds disappeared later, when the snakes started eating their nestlings and eggs. While speciation can take thousands or millions of years, extinction can occur much more quickly if the rate of change exceeds the ability of the population to adapt.

To strengthen the case that the brown tree snake was indeed at the root of the bird extinction in Guam, Savidge had to perform a few more experiments. She needed to be sure, of course, that brown tree snakes actually *liked* eating birds. To do so, she set bird-baited traps all around the islands for the snakes and waited. "I put them in half a dozen locations throughout the island," she recalls. What she found shocked her: "In one area where the birds were extinct, 75% of my traps got hit within 4 nights," she recalls. On the other hand, in the baited traps she had set on Cocos Island, an island off the coast of Guam that is not populated by the snakes, all the birds survived.

Savidge knew, however, that birds weren't brown tree snakes' only prey. The reptiles also ate small mammals, so she checked to see whether these animals were also being adversely affected by the snakes' presence. In the 1960s, before the tree snakes had arrived, scientists on Guam had done a survey of small mammal density and had found, on average, 40 small mammals per hectare of land on the island. When Savidge did the same thing in the mid-1980s, she found only 2.8 animals per hectare. "My prediction was that I would see a decline in these rats and mice and shrews, and indeed, it was like a 94% decline," she explains. The findings suggested that brown tree snakes were devastating small mammal populations in addition to the birds. All in all, Savidge's three pieces of evidence—the fact that the geographic location of the snakes correlated strongly with the birds' disappearance, that brown tree snakes liked to eat birds, and that other small mammals also went missing after the snakes' arrival—convinced Savidge that she had finally solved the mystery of Guam's disappearing birds. Brown tree snakes, she concluded, were definitely the culprit.

Infographic **10.5** | **EARTH'S MASS EXTINCTIONS** ↓ There have been five mass extinctions in Earth's history and many believe the high rate of extinction and endangerment seen today is leading to a sixth mass extinction.

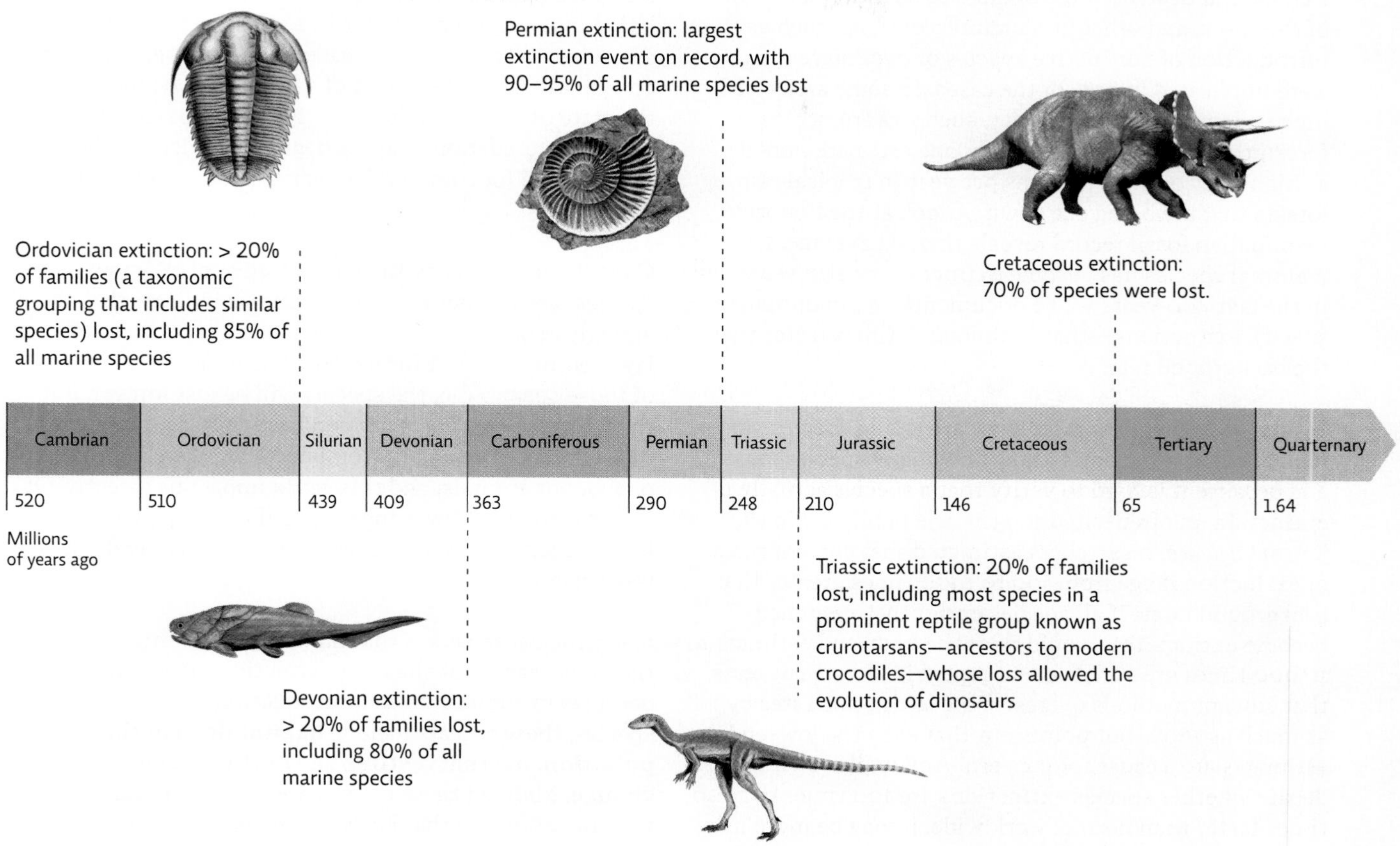
Permian extinction: largest extinction event on record, with 90–95% of all marine species lost
Ordovician extinction: > 20% of families (a taxonomic grouping that includes similar species) lost, including 85% of all marine species
Cretaceous extinction: 70% of species were lost.
Cambrian
Ordovician
Silurian
Devonian
Carboniferous
Permian
Triassic
Jurassic
Cretaceous
Tertiary
Quarternary
520
510
439
409
363
290
248
210
146
65
1.64
Millions of years ago
Triassic extinction: 20% of families lost, including most species in a prominent reptile group known as crurotarsans—ancestors to modern crocodiles—whose loss allowed the evolution of dinosaurs
Devonian extinction: > 20% of families lost, including 80% of all marine species

Extinction is normal, but the rate at which it is currently occurring appears to be increasing.

Extinction is nothing new on Earth. It is as constant and as common as evolution; by most estimates, more than 99% of all species that ever lived on the planet have gone extinct. Based on a critical analysis of the fossil record, scientists agree that there have been five *major extinction events*—when species went extinct at much greater rates than during intervening times, each event leading to the loss of 50% or more of the species present on Earth. The most infamous of these was the *K-T boundary mass extinction*, which occurred at the transition from the Cretaceous to the Tertiary period, 65 million years ago. Most scientists agree the K-T extinction event was set off by an asteroid impact in the Gulf of Mexico; 70% of all living species, including the dinosaurs, were wiped out. [INFOGRAPHIC 10.5]

But these kinds of events often lead to the emergence of new species, as other populations adapt to the open niches that are left. Cycles of extinction and evolution ultimately gave rise to the diversity of life we see on Earth today—estimates range from 3 to 100 million species. Throughout most of time, the **background rate of extinction**—the average rate of extinction that occurs between mass extinction events—has been slow. The **fossil record** tells us that, on average, 1 species out of every million species goes extinct each year. In a world with 3 million species, this would be 3 species per year; if Earth is home to 100 million species, that would be 100 species per year.

Today, most scientists in this field agree that swelling human populations have triggered a sixth major extinction event, one that we are witnessing right now. Plant and animal extinction rates are currently greater than the background rate of extinction. For example, British

background rate of extinction The average rate of extinction that occurred before the appearance of humans or occurs outside of mass extinction events.
fossil record The total collection of fossils (remains, impressions, traces of ancient organisms) found on Earth.

researchers Ian Owens of the Imperial College of London and Peter Bennett of the University of Kent analyzed the extinction risk for 1012 threatened bird species and found that habitat destruction was cited as a risk factor in 70% of the cases, and other human interventions, such as the introduction of non-native species or overharvesting were implicated in 35% of the cases. In some areas with high endemic species diversity, such as a tropical rain forest, the rate can be quite high: one estimate puts it as high as 27 000 extinctions per year in tropical rain forests that are being cut down. A look at the historic mammalian fossil record reveals that, on average, 1 mammal species has become extinct every 200 years, but in the last 400 years we've documented 89 mammalian species extinctions—that is almost 45 times faster than the background rate.

Estimates of just how rapidly we are losing species vary, in part because we don't know how many species exist and because it is hard to verify that a species actually is extinct. In an often-cited 1995 article published in the journal *Science*, researchers estimated that current rates of extinction range from 100 to 1000 times greater than background rates. If all species currently threatened become extinct, this would raise the high-end estimate to 10 000 times greater. A recent report in *Nature* suggests that current methods overestimate extinction rates by as much as 160% but points out that even the low-end estimates are a cause for concern. And while scientists debate whether species extinctions are 100 times or 1000 times faster than normal worldwide, it may be more useful to evaluate threats at a local level, like in Guam, and focus efforts on reducing species loss there. Especially on a small, isolated island, a rate that is even 10 times higher than normal is significant.

What we do know is that today's accelerated extinction is largely driven by human actions. As the human population increases, our impact is becoming much more devastating for other species. We remove the resources they need to survive, minimize their habitat ranges, introduce new predators or competitors, and strip them of their genetic diversity, all of which slowly eliminate them. In Guam, the near-total disappearance of birds between the 1960s and 1980s was a biological murder mystery of astounding proportions—one that illustrates just how vulnerable populations are to sudden changes.

Biodiversity is threatened worldwide.

The world's biodiversity faces threats on several fronts. If a species faces a very high risk of extinction in the immediate future, scientists classify that species as **endangered**. When a species is likely to become endangered in the near future, they say it is **threatened**. The International Union for Conservation of Nature (IUCN) classifies a species as endangered when it suffers a population loss of 80% or greater in a 3-year period. In Canada, a species is classified as threatened or endangered based on a qualitative assessment of several factors, including the state of its natural habitat, the threat it faces from disease or predation, and the degree to which it is being overutilized for commercial, recreational, or scientific purposes.

Globally, the IUCN estimates that 40% of all known species face extinction. Recent studies indicate that there are only about 3200 tigers, 720 mountain gorillas, and 60 Javan rhinoceros left in the wild. Once the last members of those species die, the species will be lost forever. But the fight for survival is not confined to large, furry mammals. In fact, the majority of listed endangered species are plants, not animals, and this could impact us directly. Of the 70 000 or so flowering and nonflowering plants with known medicinal value, more than 15 000 are endangered (see Chapter 9).

Scientific consensus is that human-caused threats are the main reason for the high extinction rates of recent decades. In addition to the introduction of invasive species, these threats include **habitat destruction**, **pollution**, **overharvesting**, and **anthropogenic climate change**. Many of these threats are related to overpopulation and affluence that leads to greater resource use. [INFOGRAPHIC 10.6]

endangered A species that faces a very high risk of extinction in the immediate future.

threatened A species that is likely to become endangered in the near future.

habitat destruction Altering a natural area in a way that makes it uninhabitable for the species living there.

pollution Hazardous or objectionable substances that are released into the environment; also includes noise and light.

overharvesting Human activity that removes more of a resource than can be replaced in the same time frame, such as taking too many individuals from a population.

anthropogenic climate change Alterations to climate resulting from human impact

habitat fragmentation Destruction of part of an area that separates suitable habitat patches from one another; patches that are too small may be unusable for some species.

core species Species that prefer core areas of a habitat—areas deep within the habitat away from the edge.

edge species Species that prefer to live close to the edges of two different habitats (ecotone areas).

artificial selection When humans decide which individuals breed and which do not in an attempt to produce a population with desired traits.

Infographic **10.6** | **THE MAIN THREATS TO BIODIVERSITY TODAY COME FROM HUMANS**

Climate change threatens species that cannot adapt to changing conditions or relocate to a more suitable habitat.
Habitat destruction, such as the conversion of wild habitat to lands for agriculture or the harvesting of resources such as trees, is the main cause of species endangerment. This destruction is driven by human population growth and increased affluence.
Human societies generate **pollution** that harms organisms outright or damages the other parts of their environment (such as food and water) that they depend on.
Habitat fragmentation reduces usable habitat; many species have minimum space requirements and will not move between the patches that remain.
Invasive species can drive native species to extinction by outcompeting them or by preying on them. Invasive species are hard to eradicate.
Humans can **overharvest** biotic resources and deplete species' populations.

Human activity alters the face of landscapes and the sea floor in ways that critically damage habitats needed by species for their survival. Habitat destruction, whether it be the physical destruction of an ecosystem (e.g., deforestation) or the degradation of a system such that the system is no longer inhabitable (e.g., pollution), is the number one cause of species endangerment. For some species, **habitat fragmentation** can be as devastating as total habitat destruction. For instance, a road that cuts through a forest could reduce usable habitat by half for a **core species** that will not venture close to the edge, let alone cross the open space. Isolated populations in fragmented habitats inbreed and lose genetic diversity, becoming even more vulnerable to further environmental perturbations. **Edge species** that live at the juncture of two different habitats (such as where a forest meets a field) may actually benefit from habitat fragmentation as the amount of "edge" increases, but only as long as suitable amounts of overall habitat remain. Corridors or connections between fragmented habitats are only useful for core species if the corridor is wide enough to offer protection from the edge habitat; otherwise, core species will not use the corridor. (See Chapter 8 for other examples of edge and core effects.)

Humans affect evolution in a number of ways.

The introduction of the brown tree snake to Guam was an accident: a snake hitchhiker crossed the ocean on a human tanker—unbeknownst to the crew—and landed in a veritable bird buffet. But humans also directly affect the evolution of a population through **artificial selection**. Artificial selection works the same way as natural selection but the difference is that the selective pressure is us. For many animal species, from pets to farm animals and plant species, humans choose who breeds with whom in an attempt to produce new individuals with the traits they desire. By doing this over many generations, people have accentuated certain plant and animal

Infographic **10.7** | **HUMANS USE ARTIFICIAL SELECTION TO PRODUCE PLANTS OR ANIMALS WITH DESIRED TRAITS**

→ All dogs (*Canis lupus familiaris*) are descendants of the wolf (*Canis lupus*). By only breeding those males and females with the traits desired (size, herding ability, protective instinct, etc.), humans have created more than 170 dog breeds.

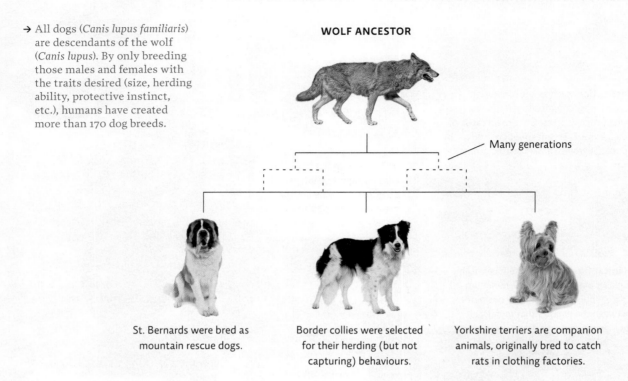

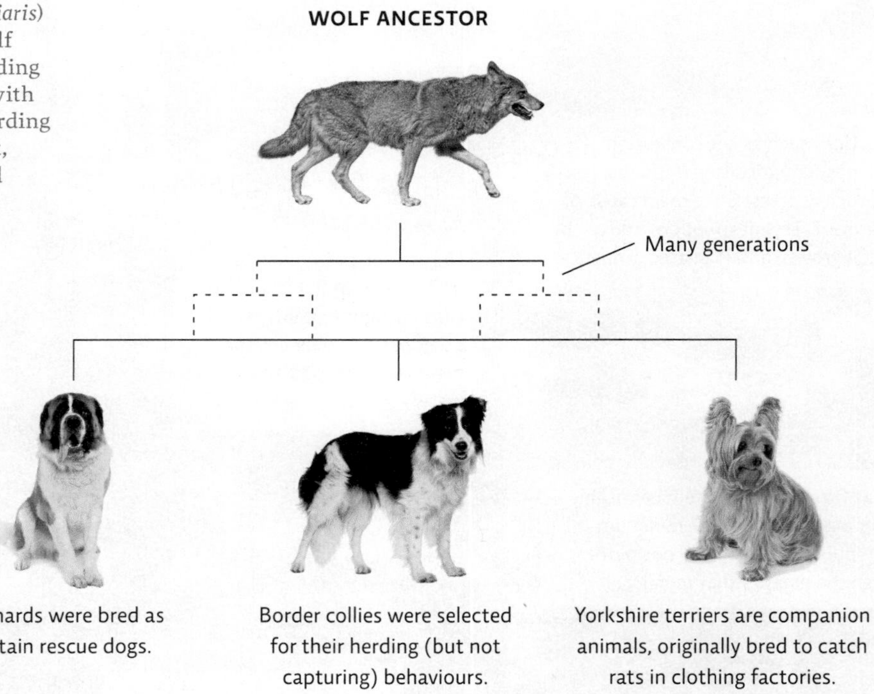
WOLF ANCESTOR
Many generations
St. Bernards were bred as mountain rescue dogs.
Border collies were selected for their herding (but not capturing) behaviours.
Yorkshire terriers are companion animals, originally bred to catch rats in clothing factories.

Infographic **10.8** | **COMMON MISCONCEPTIONS ABOUT EVOLUTION**

The study of evolution seeks to explain the creation of life	Evolution doesn't study creation or the origin of life. Evolution is the science that studies the "descent with modification" of life after it came into being.
Humans evolved from monkeys or apes	Humans did not evolve from monkeys or apes. Humans, monkeys, and apes share a common ancestor, who lived 5–8 million years ago, but each group followed different evolutionary paths to their current forms.
Evolution is goal-directed	Evolution is not working toward a particular final version or trait. It is unpredictable because we cannot know which genetic variants will be available to natural selection and we can't reliably predict selective pressures (environmental conditions and competition).
Evolution proceeds from simple to complex	Simple is not always the ancestral condition. Parasites may be simple in design but many have evolved from nonparasitic, more complex forms.
Evolution produces perfect adaptations	New traits are rarely "perfect." Species have a long history of ancestral adaptations, each giving way to the next. Evolution works with whatever is available—it doesn't scrap the anatomy of an "arm" and reform a wing for a bat. It modifies existing structures for new functions. In addition, adaptations are often compromises between differing needs. For example, the flippers of a seal are better suited for swimming but make walking on land awkward.
Evolution is just a theory	The claim that evolution is "just a theory" reveals a misunderstanding of the concept of a scientific theory. "Theories" represent our most well-accepted explanations in science—to attain the status of theory, an explanation must have substantial evidence, ideally from multiple lines of evidence.

traits, sometimes to extremes. For instance, artificial selection created domestic dogs from their wolf ancestors. [INFOGRAPHIC 10.7]

But evolution is ever at work. Pesticide- and antibiotic-resistant populations can emerge as an inadvertent human-influenced selection. When we apply a chemical that kills a pest or pathogen, some individuals survive because of their natural genetic resistance (the individuals were already resistant even though they had never encountered the chemical). These survivors are then the only individuals who reproduce, producing the next generation that is also pesticide resistant, ultimately changing the frequency of resistant genes in the population (see Chapter 18).

Evolution can be a contentious subject for some people who feel it conflicts with other views they hold. Whether or not you are convinced by the physical, scientific evidence that evolution has occurred or that natural selection is a mechanism by which it occurs, it is important that you base your criticisms of the theory on sound science and not on misconceptions of evolutionary theory. [INFOGRAPHIC 10.8]

Ultimately, by changing the environment rapidly, humans apply a number of new selective pressures on populations. Our changes have the capacity to be so great that natural selection simply cannot keep up—new needed traits are not present in the population or cannot spread quickly enough to prevent a population collapse. The accidental introduction of the predatory brown tree snake is one such example; it introduced a major change, basically overnight, that was able to eat its way through the vertebrate populations before those populations could adapt.

Now, however, things are looking up for birds in Guam. Thanks to efforts by the Guam Department of Agriculture and the United States Department of Agriculture Wildlife Services, brown tree snakes are being controlled to allow bird populations to recover. "We are using traps in northern Guam to control snakes in areas where we would like to reintroduce native species," explains Diane Vice, a wildlife biologist with the Guam Department of Agriculture. Her organization is also working to prevent the snakes from infiltrating Cocos Island, the atoll off of southern Guam that is an important haven for nesting sea birds, Micronesian starlings, and sea turtles. "The hope is to create safe habitat," Vice says, so that these beautiful native species can once again thrive.◉

Select references in this chapter:
He, F., and Hubbell, S.P. 2011. *Nature*, 473: 368–371.
Owens, I.P.F, and Bennett, P. 2000. *Proceedings of the National Academy of Sciences*, 97: 12144–12148.
Pimm, S., *et al.* 1995. *Science*, 269: 347–350.
Savidge, J.A. 1987. *Ecology*, 68: 660–668.

BRING IT HOME

PERSONAL CHOICES THAT HELP

The astonishing variety of life found on Earth is the result of natural selection favouring those individuals within populations that are best able to survive in their particular environment. Given enough time, some populations may be able to adapt to environmental changes. However, human activities may disrupt natural ecosystems so that organisms cannot adapt fast enough to survive, and may go extinct. Conservation activities can help protect vulnerable organisms and ecosystems.

Individual Steps

→ Live in an older, established area of your community. Suburban sprawl reduces habitat for wildlife, and reliance on cars causes greenhouse gas emissions that could result in species-threatening climate change.

→ Save your pocket change and at the end of every year donate the money to a land, marine, or wildlife protection agency.

→ Create a personal blog that includes photographs of wildlife, facts about current threats to plants and animals, and articles about conservation.

Group Action

→ Throw a party in support of wildlife conservation. Take a collection at the door and donate the money to an organization that supports conservation.

Policy Change

→ "Adopt an Organism." Species at Risk Public Registry maintains a list of endangered plants and animals across the country. Research this list to find wildlife that interests you. Determine what agency, conservation group, or legislator you could contact and then start your own protection campaign. See what meetings, petitions, and legislation could impact your organism and get involved.

UNDERSTANDING THE ISSUE

CHECK YOUR UNDERSTANDING

1. Which of the following is TRUE of evolutionary processes?
a. They are goal-driven.
b. New genes evolve in response to environmental change.
c. Evolution acts on existing genetic variation.
d. Random events do not influence evolution.

2. Ten thousand years ago, most members of Species A were killed by a series of volcanic eruptions. However, some members escaped; all modern members of Species A are descended from those 100 individuals. This is an example of:
a. genetic drift.
b. artificial selection.
c. mass extinction.
d. the bottleneck effect.

3. Why do most scientists think that we are in the midst of a sixth mass extinction?
a. Current extinction rates are greater than the background rate.
b. The background extinction rate is approximately 20%.
c. Most extinctions are occurring in the taiga and tundra.
d. No new species are being discovered.

4. Which of the following is an example of coevolution?
a. Moths that are preyed upon by bats can hear the ultrasonic sounds the bats use in hunting.
b. Polar bears and Arctic foxes both are white for camouflage on snow.
c. Dolphins and whales have flippers that are similar in shape and function to the fins of a fish, allowing them both to swim efficiently.
d. Humans and chimpanzees share 98% of their DNA.

5. In a population of butterflies, there used to be a wide range of sizes: small, medium, and large. But a non-native bird was introduced to the butterfly's habitat. It loves to eat butterflies, but it can only eat the medium-sized ones. Eventually, medium-sized butterflies became rare. This is an example of:
a. disruptive selection.
b. stabilizing selection.
c. artificial selection.
d. directional selection.

6. If sea levels rose 1 metre overnight, what would happen to terrestrial organisms living on the coast?
a. Their lungs would become gills, allowing the organisms to adapt to life in the water.
b. They would all die.
c. Individuals with adaptations that allowed them to live in water might survive, but the rest would die.
d. A new gill-like structure would evolve, resulting in new species adapted to living in water.

WORK WITH IDEAS

1. Identify and explain the human causes of species endangerment and extinction. Propose some actions that could help address each of these issues.

2. What kinds of predictions can you make about the effects of the current mass extinction on
a. communities?
b. ecosystems?
c. humans?

3. Based on what you've read in the chapter, what can you deduce about the relationship between extinction and islands?

4. Describe the process of natural selection, using bacterial antibiotic resistance as an example. Apply the requirements for natural selection to the development of antibiotic resistance in your answer. Include examples of human behaviour that have contributed to antibiotic resistance.

5. Describe the forces that influence evolution. Why do evolutionary rates differ? Why can't all species adjust to changes in their environment and avoid extinction?

ANALYZING THE SCIENCE

The following graph depicts the relationship between numbers of extinctions and human population size since the 19th century.

INTERPRETATION

1. Describe what is happening to:
a. the extinction rate over time.
b. human population growth over time.

2. The two curves have been graphed together. What is the implication of presenting these data in this manner?

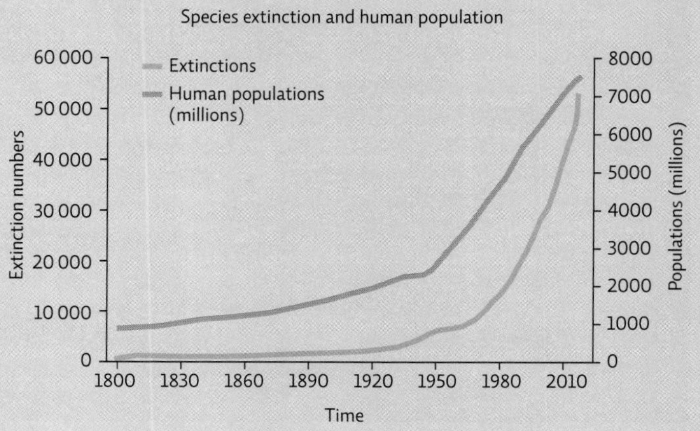
Species extinction and human population
Extinctions
Human populations (millions)
Extinction numbers
60 000
50 000
40 000
30 000
20 000
10 000
0
Populations (millions)
8000
7000
6000
5000
4000
3000
2000
1000
0
1800
1830
1860
1890
1920
1950
1980
2010
Time

ADVANCE YOUR THINKING

3. The *y* axis is labeled "Extinction numbers." What taxonomic units are being measured? What if the taxonomic unit being evaluated here had been genus or family? Would you be more concerned about the trend of extinctions or less? Explain.

4. What type of relationship is suggested by the figure (correlational or causal)? What additional data would you like to see to support the graph's main point?

EVALUATING NEW INFORMATION

Life on Earth as we know it is a result of millions of years of evolutionary processes. These processes are not immune to changes in the environment; changes in evolutionary processes will ultimately result in changes in biodiversity, which will necessarily affect life on Earth, including humans.

Go to the website www.actionbioscience.org/evolution/myers_knoll.html and read the article on the effects of the sixth mass extinction on evolution.

Evaluate the website and work with the information to answer the following questions:

1. Is this a reliable information source? Does it have a clear and transparent agenda?
   a. Who runs this website? Does this person's/group's credentials make this source reliable/unreliable? Explain.
   b. Who are the authors? What are their credentials? Do they have the scientific background and expertise that lends credibility to the article?

2. In your own words, explain why the authors think that the current mass extinction will change evolutionary processes.
   a. What changes in particular do they think will impact future evolution?
   b. What types of evidence do they provide to support their arguments? Give specific examples.

3. The sixth mass extinction is the result of human activities. Read "The First Human Caused Extinction: The Dodo" (http://catherineowen.suite101.com/the-first-human-caused-extinction-the-dodo-a85161).
   a. Why did the dodo go extinct? Be specific—there are direct and indirect consequences discussed in the article.
   b. On Mauritius, what happened as a consequence of the dodo's extinction?
   c. Do you think humans today could have a similar impact on a species? Consider both direct and indirect human impacts. Which types of species might be vulnerable? Find an example on the International Union for Conservation of Nature (IUCN) "Species of the Day" website (www.iucnredlist.org/species-of-the-day/archives).

MAKING CONNECTIONS

***Balearica regulorum:* A SPECIES WORTH SAVING?**

Background: Cranes are a group of long-legged, long-necked, wading birds that tend to prefer wetland habitats for both breeding and feeding. There are fifteen species worldwide, and they vary greatly in colouration. Many are quite tall (up to 176 cm) and several species are well-known for their beautiful mating dances. Most crane species are endangered due to habitat loss and hunting, although other threats exist. In Canada, the Whooping Crane *(Grus americana)* is listed as an endangered species under SARA (Species at Risk Act), and considerable conservation efforts have been made to save these cranes.

Case: You are a member of a philanthropic organization devoted to saving species from extinction. Your organization has a limited amount of funds, so it is imperative that the funds be given to worthwhile projects. A consortium of several African nations has approached you for funds to help save the endangered Grey Crowned Crane *(Balearica regulorum)*. Before your organization can make a decision, you need to gather information:

1. Research *Balearica regulorum* (the IUCN Red List is a good place to start). Consider the following questions:
   a. Why is this species considered endangered?
   b. What are the causes of its decline?
   c. Where does it live?
   d. Does its habitat still exist? If it were brought back from the brink of extinction, is there habitat for it to live in and food for it to eat?
   e. What plans are currently in place to save it?

2. Do some further research to answer the following questions.
   a. Why did the Whooping Crane decline? Consider both human impacts and the Whooping Crane's biology.
   b. How much money has been spent to date on Whooping Crane conservation?
   c. What is the status of Whooping Cranes that currently live in the wild?
   d. What is the outlook for the future of this species?
   e. Answer questions a. – d. for *B. regulorum* and compare your results to those for the Whooping Crane.

3. Based on your research, write a report recommending a course of action to your organization. Should you give money to save *B. regulorum*? Why or why not? Be sure to address whether it is ecologically and/or economically realistic or desirable to save *B. regulorum*.

CORE MESSAGE

Forests have great economic value but we must balance that with the value of their ecosystem services and sociocultural benefits. Sustainable management practices allow us to harvest forest products without destroying the forest or its ability to provide ecosystem services and future products.

GUIDING QUESTIONS

After reading this chapter, you should be able to answer the following questions:

- → In general what are the characteristics of a forest biome? What are the three main types of forests and what influences which forest type is found in a given area?
- → What is the three-dimensional structure of a forest and how are the plant species found there adapted to their level of the forest?
- → What ecosystem services do forests provide? Why might it be useful to put a dollar value on these services?
- → What is the current state worldwide of forest resources? What threats exist?
- → What are ways that we can act to protect and sustainably manage forest resources? What trade-offs exist with our choices?

Teams of men and women pick their way across steep, rocky slopes as they plant trees in Mahotiere, Haiti.

RETURNING TREES TO HAITI

Repairing a forest ecosystem one tree at a time

When Jean Robert was a young boy, the mountains surrounding the Haitian city of Gonaïves (pronounced go-nah-eev) were still lush and green with trees. His family's hillside farm—small though it was—provided a steady crop of mangoes and avocados in the summer, coffee and cassava in the fall, and wood for charcoal throughout the year. But today, the forests are gone and the mountains are bare. Not only have crop yields shrunk—from enough to sell in the local markets, to barely enough to feed a family—but the mountain homes and the city below have been left defenceless against the onslaught of tropical storms that pound Haiti every summer. Unencumbered by trunks or roots or shrubs, water rushes freely downward, gathering into apocalyptic mudslides that destroy homes, crops, and livelihoods. In 2004, a single storm claimed more than 2000 lives from this one city.

Now, instead of growing mangoes and cassava, Robert and his neighbours are trying to grow trees. Working as a team, moving in slow, deliberate, farm-by-farm patterns across the mountainside, the community work group, or kombit, plants a carefully planned mix of fruit and timber trees. By the end of the month, they say, each member will have his or her own saplings to tend to. If the saplings survive, and if the American ecologists they've been consulting with are right, their efforts will give rise to a new era of sustainable forestry, which strikes a balance between what people need from the forest and what the forest itself needs to survive.

◉WHERE ARE HAITI AND GONAÏVES?

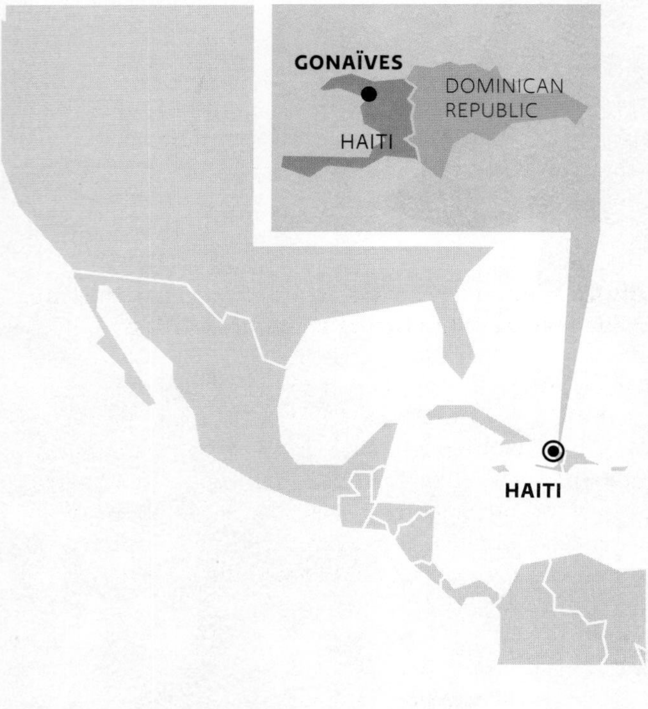
GONAÏVES
DOMINICAN REPUBLIC
HAITI
HAITI

It's a scene being replayed all over Haiti. When Europeans first arrived in the country some 500 years ago, two-thirds of the land was covered in forests so majestic that explorers dubbed this region "the Pearl of the Antilles." But as the centuries passed, the forests were cleared—sometimes with amazing breadth, to make way for coffee and sugar plantations; other times, in discrete chunks that Haitian peasants would then convert into subsistence farms. Many of the trees—along with the coffee and sugar crops that displaced them—were sold overseas. Most of the rest provided fuel for heating Haitian homes and cooking Haitian meals.

Today, less than 2% of that original forest remains; 6% of the land has no soil left at all. In these places, dead tree stumps, bleached white by the sun, protrude from the ground like bones poking through decayed flesh. And Haiti has become both the poorest and the most environmentally degraded country in the Western Hemisphere. With few trees and little soil, drinking water has grown polluted, crops have dwindled, and the people have suffered.

Farmers like Jean Robert say that forests are the key to changing all of that. They have planted thousands of trees in these mountains in recent years, and hope to plant thousands more. "Almost all of the country's

↑ South of Dajabón, Haiti's brown landscape on the left contrasts sharply with the rich forests of its neighbour, the Dominican Republic, on the right.

problems—natural disasters, food shortages, poverty—can be traced back to rampant **deforestation**," says Haitian ecologist Timote Georges, who is working with the American non-profit Trees for the Future to reforest the mountains around Gonaïves. "So if we want to fix the country, we have to put the forests back and then find a way to manage them better."

That's no small feat. Forests, after all, are one of our most contentious resources. Such is the range of economic benefits and ecosystem services they provide that any one use must be weighed against a host of others, and immediate human needs are frequently pitted against long-term conservation goals. In Haiti, where most people live on less than $2 a day, trees provide food, energy, building material, and desperately needed income. To stop them from being chopped down, Georges and his fellow Haitians will have to find alternative sources of each, not to mention a farming method that doesn't require clearing the forests.

But to really understand how trees might help alleviate poverty, we must first consider just what forests are, what they do for us, and why so many of them are being chopped down in the first place.

The location and characteristics of forest biomes are influenced by temperature and precipitation.

Forests are biomes dominated by trees. They currently cover about 30% of the planet's landmass but, thanks to their sheer concentration of biodiversity, are home to more than 50% of Earth's terrestrial life and more than 60% of its green, photosynthesizing leaves. There are many types of forest biomes around the globe, each determined by the temperature and amount of precipitation

deforestation Net loss of trees in a forested area.

Infographic **11.1** | **FORESTS OF THE WORLD**

↓ Forests are biomes whose dominant plant life is trees and other woody vegetation. There are 3 main types of forests, classified according to climate, and many subdivisions within these 3 types.

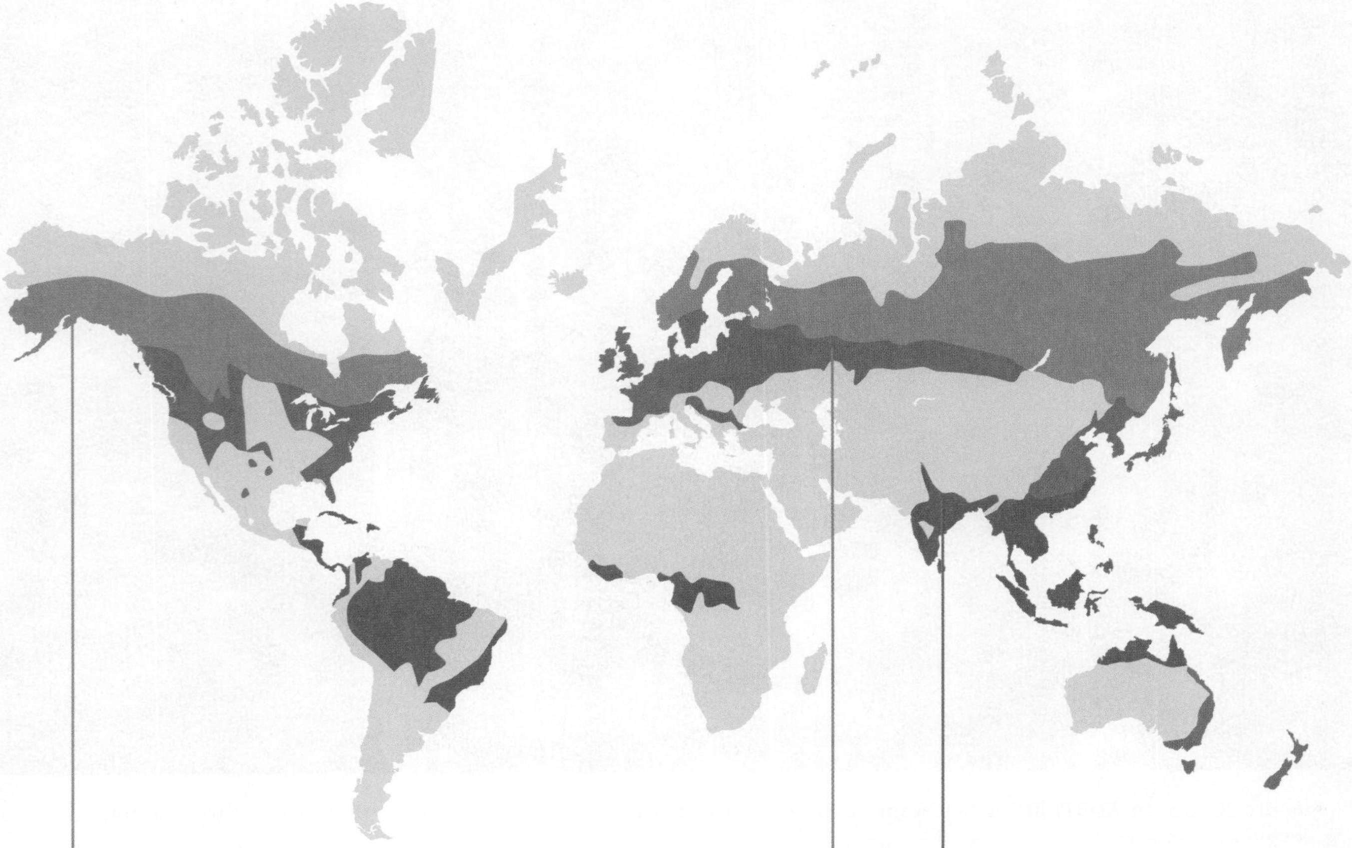

BOREAL FOREST

Boreal forests (taiga) represent the largest terrestrial biome; they stretch from Canada to Siberia and are found at higher elevations in lower latitudes. They have a short growing season with little precipitation, most of it snow. Soils here are thin and acidic and the major tree species are evergreen conifers with needlelike leaves.

TEMPERATE FOREST

Temperate forests have distinct seasons. The soil is fertile, with a thick layer of decomposing leaf litter that supports the plant life. Depending on how much precipitation an area gets, the forest may be predominantly coniferous (evergreen) or broadleafed (deciduous).

TROPICAL FOREST

Tropical forests have similar temperatures year round. Dry tropical forests have distinct wet and dry seasons, whereas tropical rainforests receive rain year round. The soils are thin, acidic, and low in nutrients. Rapid decomposition by fungi and bacteria supports the dense vegetation found in these soil-poor areas.

↑ Volunteers begin planting 25 000 donated trees in Mahotiere, Haiti, to combat soil erosion.

the area receives. **Boreal forests**, characterized by evergreen species like spruce and fir, cover vast tracts of land in the higher latitudes and altitudes and are characterized by low temperatures and precipitation levels; they represent some of the last expansive forests left on Earth. **Temperate forests**, which contain deciduous trees that lose their leaves in winter, like oak, hickory, and maple, are found in the wet areas of mid-latitudes. These forests are not as expansive as the boreal forests because they are found in latitudes with high human populations. In Canada, the world's largest timber exporter, 45% of our land area is forested, but we have lost 6% of original forest cover, mostly from temperate forest areas. **Tropical forests** contain a diverse mix of tree and undergrowth species and are found in tropical latitudes where temperatures do not vary much throughout the year. [INFOGRAPHIC 11.1]

Most forests consist of four distinct layers. The **canopy**, formed by the overlapping crowns of the tallest trees, makes up the ceiling of the forest. Some even taller trees may reach above the canopy to form an **emergent** layer. Beneath the canopy is the **understory** layer, where

boreal forests Coniferous forests found at high latitudes and altitudes characterized by low temperatures and low annual precipitation.
temperate forests Found in areas with four seasons and a moderate climate, receive 750–1500 millimetres of precipitation per year, and may include conifers and/or hardwood deciduous trees (lose their leaves in the winter).
tropical forests Found in equatorial areas with warm temperatures year-round and high rainfall; some have distinct wet and dry seasons but none has a winter season.
canopy Upper layer of a forest formed where the crowns (tops) of the majority of the tallest trees meet.
emergent The region where a tree that is taller than the canopy trees rises above the canopy layer.
understory The smaller trees, shrubs, and saplings that live in the shade of the forest canopy.

Infographic **11.2** | **CROSS SECTION OF A FOREST**

↓ Forests are stratified, having four (or more) vertical layers, each of which contains species adapted to the level of sunlight and moisture available.

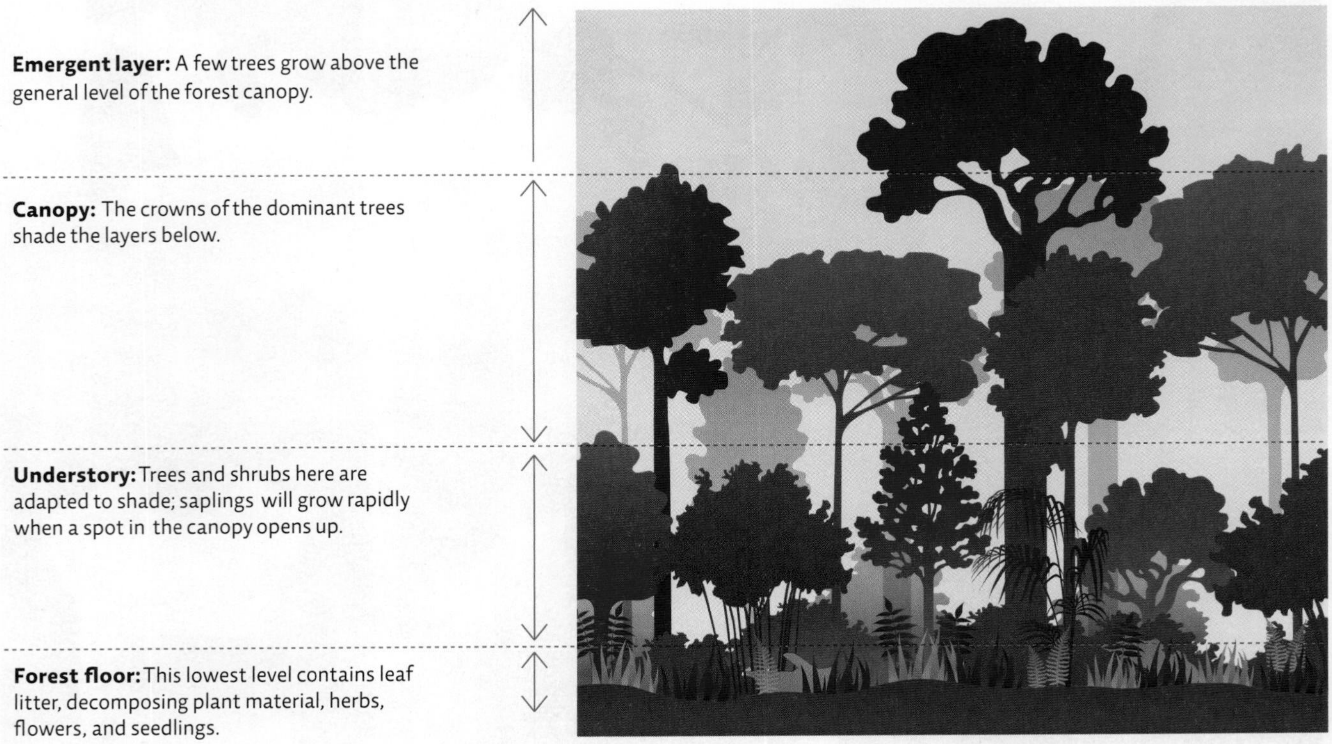
Emergent layer: A few trees grow above the general level of the forest canopy.
Canopy: The crowns of the dominant trees shade the layers below.
Understory: Trees and shrubs here are adapted to shade; saplings will grow rapidly when a spot in the canopy opens up.
Forest floor: This lowest level contains leaf litter, decomposing plant material, herbs, flowers, and seedlings.

shade-tolerant shrubs, smaller trees, or the saplings of larger trees grow. Sometimes these trees are dense enough to form a lower canopy. The lowest level is the **forest floor**, which is typically made up of seedlings, herbs, wildflowers, and ferns. The forest floor also contains soil, which is composed of leaf litter and other debris—branches, logs, and stumps—that decomposes over time.

Within each forest layer is a range of species uniquely suited to the temperature, humidity, and amount of sunlight that layer receives, and well-adapted to its particular neighbours. For instance, in a temperate deciduous forest, wildflowers on the forest floor will bloom early in the spring before the bigger trees "leaf out" and block the sun. Sunlight is a precious commodity on the forest floor and wildflowers compete for it. One wildflower species may bloom one week and another the next week. Disruption to one part of the forest (for instance, cutting down a tree that opens the canopy) can have a trickle-down effect that impacts each subsequent layer. [INFOGRAPHIC 11.2]

Most experts agree that the biggest modern contributors to Haitian deforestation are the food and fuel needs of Haitians themselves. But throughout the nation's history, there were many other culprits—from 18th-century French colonizers' coffee and sugar plantations, to the timber industry of the 19th and 20th centuries. As the population grew—from 3 million in 1940 to 9 million in 2000—rural Haitians were forced to clear ever-larger areas of mountainside to make room for subsistence crops. The trees themselves doubled as a source of fuel and cash for families who not only used the wood to cook with but also sold it as charcoal in Port-Au-Prince, the nation's densely populated, energy-starved capital city.

Charcoal—partially burned wood that ignites more easily and burns hotter than the original wood itself—is used in many developing countries that lack other reliable fuel sources. Charcoal is produced in a variety of ways but all include a slow, low-temperature "roasting" of the wood in a low-oxygen environment. This process releases both CO_2 and particulates (soot) into the atmosphere. Like Haiti, other areas dependent on charcoal as their major fuel have become severely deforested and plagued with air pollution. Many African nations including Mozambique, Malawi, Somalia, and Tanzania are

PYRAMIDcube

↑ Charcoal sellers in the Carrefour Feuilles district of Port-au-Prince, Haiti.

experiencing rampant deforestation largely to provide charcoal and fuel wood. In Mozambique, for example, charcoal is the main fuel for 80% of the population. More than 2.5 million trees are cut each year in Somalia—that is 10 trees per household, per month.

In time, Haiti's charcoal trade grew to account for 20% of the rural economy and 80% of the country's energy supply. Before long, 98% of the country's forests had been chopped down and Haitians were burning 30 million trees' worth of charcoal annually. "The trade itself became this incredibly destructive force," says Andrew Morton, a forest ecologist for the United Nations Environment Programme. "And the fact that it was based not on foreigners exploiting the land for profit, but on poor Haitians trying to earn money and feed their families and heat their homes, made it impossible to stop. There really was no other source of energy."

Of course, charcoal wasn't the only crucial service the trees provided.

Forests provide a range of goods and services and face a number of threats.

Though trees are the largest and most notable life-forms present, a forest is much more than just its trees. Together, the species inhabiting the different layers participate in a delicate symphony of chemical and physical cycles that produce an invaluable range of ecosystem services for the planet.

It starts with the soil, which the forest itself helps to form and maintain: leaves and branches die, fall to the ground, and decay, forming a thick brown layer of nutrients in which all future generations of plant life will take root.

While dead and decaying plants help form the soil, living ones hold it in place. During rain storms, soil anchored in by roots, especially by tree roots, can't flow as easily

---

forest floor The lowest level of the forest, containing herbaceous plants, fungi, leaf litter, and soil.

↑ The original forests of Haiti would have looked like this one on the Barahona Coast in neighbouring Dominican Republic.

down hillsides into nearby surface waters (lakes, streams, oceans), where it would contaminate them with sediment and chemicals it picked up from the ground's surface. By slowing this flow of rainwater across the ground's surface (called **stormwater runoff**), a well-vegetated area allows more water to soak into the ground, recharging the groundwater supplies that provide area residents with their major source of drinking water. The soil also traps chemicals that might otherwise contaminate that drinking water. (See Chapter 16 for more on runoff.)

While plants, soil, and water are playing off one another in this manner, the forest is also conducting another important cycle: pulling carbon dioxide out of the atmosphere and replacing it with oxygen. In fact, forests as a whole store more carbon in their biomass, litter, and soil than all the carbon in the atmosphere, making this biome one of the world's largest **carbon sinks** (an area that stores more carbon than it releases, such as the standing timber in a forest or organic matter in soil). And forest leaves produce so much oxygen that they are commonly referred to as "the lungs of the planet."

Last but not least, virtually every layer of forest provides food and habitat for a bevy of animals (vertebrates and invertebrates), plants, fungi, and microbes; these creatures all do their part to contribute to the functioning of the forest ecosystem as a whole. It is difficult to overstate the importance of forests to the biosphere.

In addition to all these ecosystem services, forests also provide a range of economic benefits. In fact, humans have relied on forests for millennia for a multitude of consumer goods. Wood products, including lumber, firewood, charcoal, paper pulp, and some medicines, account for some $100 billion in global trade every year. On top of that, the wildlife supported by forests provides humans with food and recreational hunting opportunities, both of which have economic value. [INFOGRAPHIC 11.3]

Infographic **11.3** | **ECOLOGICAL AND ECONOMIC VALUE OF FORESTS**

↓ All ecosystems, including forests, contribute to the ongoing functioning of the planet and the immediate well-being of humans. Some of these services we take for granted; others are recognized for their economic value. One calculation estimates the value of services provided free by ecosystems of the world at trillions of dollars per year, a value greater than the GNP of all nations of the world combined.

ECOSYSTEM SERVICES

Watershed services: Water purification and provision

Atmosphere and climate effects: A major sink for CO_2; increases rainfall in some areas; biggest contributor of oxygen to the atmosphere

Soil maintenance and protection: Soil production and recycling of nutrients; reduction of soil erosion

Disturbance regulation: Protection from storm damage, especially in coastal areas

Biodiversity and genetic resources: Food and habitat for biodiversity; a rich storehouse of genes that might prove useful to improve our crops or provide as yet undiscovered medicines

ECONOMIC VALUE

Goods: Provides many of the basic goods we depend on including:

- Food
- Fuel
- Building materials
- Other products such as rubber and cork
- Raw material for paper and other industrial products
- Medicines

Jobs: More than 10 million people make their living in and from forests

Recreation and ecotourism opportunities

SOCIOCULTURAL BENEFITS

The beauty of forests provides a place for spiritual renewal, artistic inspiration, and stress reduction. Ancient stands of trees provide a connection to the past; many indigenous people are an important part of their forest ecosystem, possessing ancestral knowledge of the forest and its inhabitants.

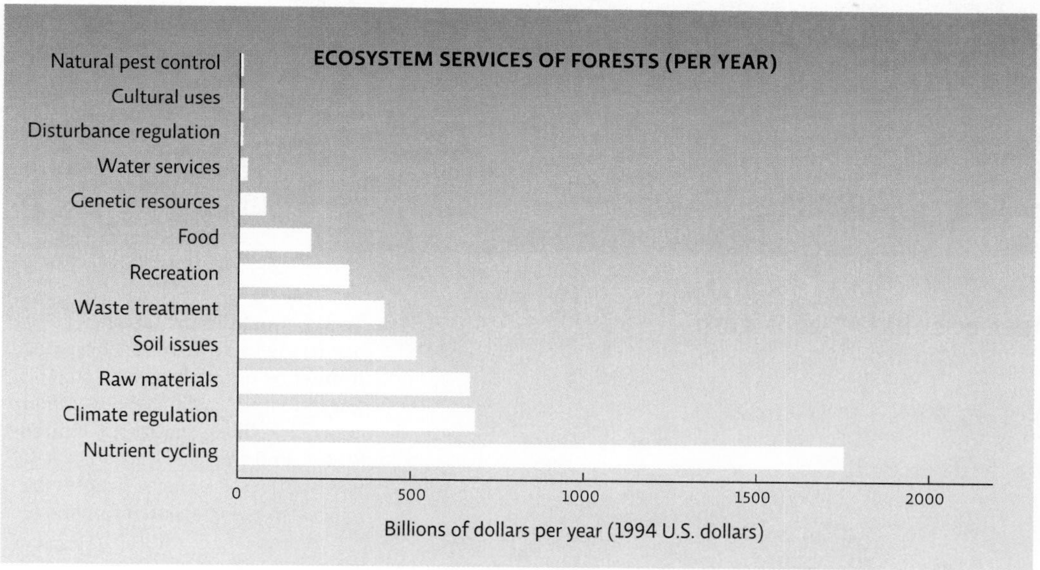

ECOSYSTEM SERVICES OF FORESTS (PER YEAR)
Natural pest control
Cultural uses
Disturbance regulation
Water services
Genetic resources
Food
Recreation
Waste treatment
Soil issues
Raw materials
Climate regulation
Nutrient cycling
0
500
1000
1500
2000
Billions of dollars per year (1994 U.S. dollars)

← Robert Constanza and his colleagues evaluated ecosystem services of forests and quantified their value (in 1994 U.S. dollars) to be $4.7 trillion.

In Haiti, the cost of deforestation spun out over decades and proved catastrophic. Soil eroded down into streams, rivers, and gullies, clogging them with sediment and disrupting aquatic ecosystems. With nothing to absorb the water, floods became more severe and groundwater sources were quickly depleted as the water flowed away rather than soaking into the ground. Slowly, crop yields shrank. As the forest habitat was fragmented, biodiversity dwindled. And as the trees vanished, the people of Haiti suffered. Unlike wealthier countries they could not afford expensive water purification systems—a service the forests once provided for free. "It took some years before we could feel the other effects of deforestation," says Georges. "The floods got worse and we lost drinking

stormwater runoff Water from precipitation that flows over the surface of the land.

carbon sinks Places such as forests, ocean sediments, and soil, where accumulated carbon does not readily reenter the carbon cycle.

Infographic 11.4 | **THREATS TO FORESTS (MAJOR CAUSES OF DEFORESTATION)**

↓ A variety of "drivers" are responsible for deforestation around the world.

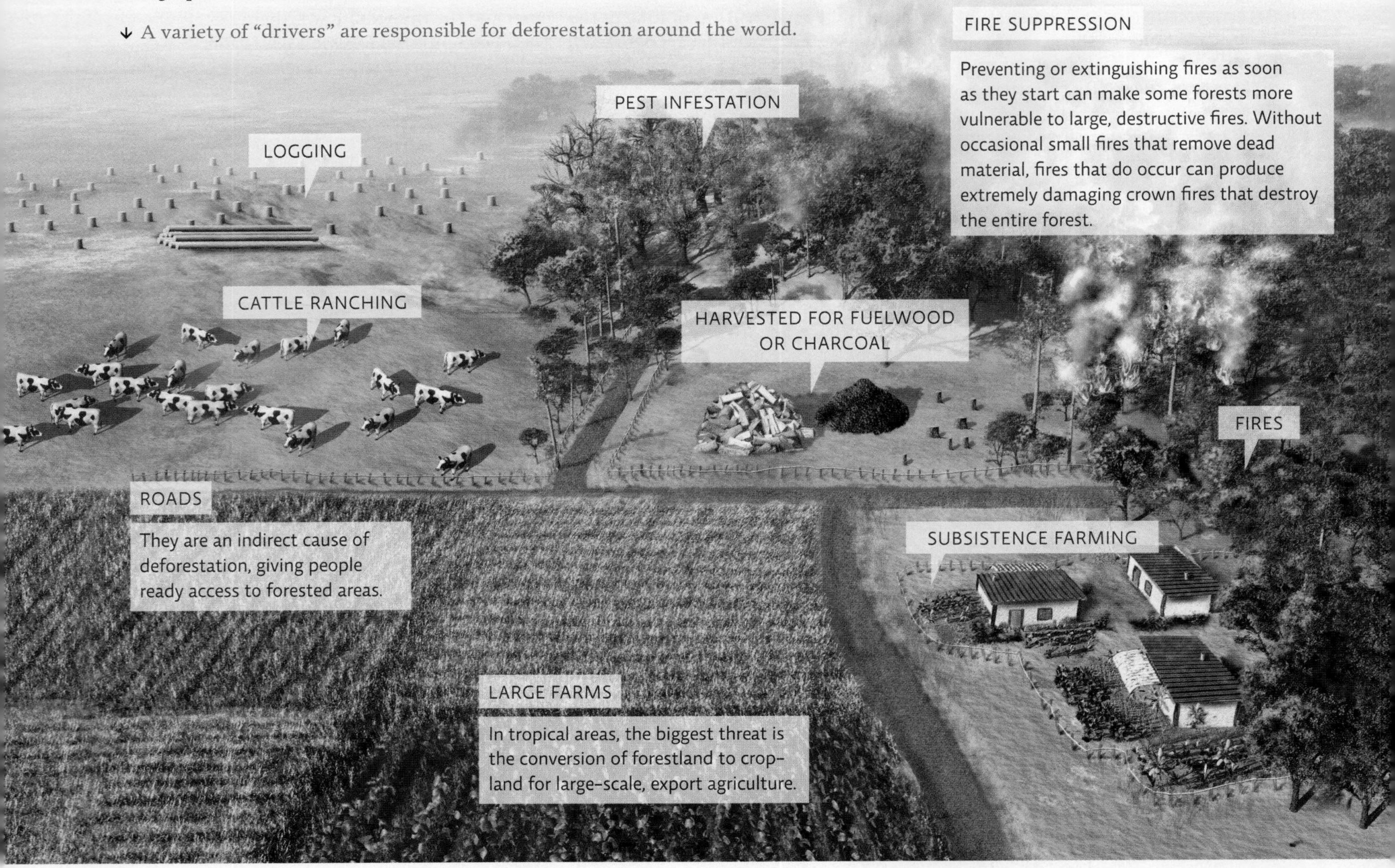
FIRE SUPPRESSION
Preventing or extinguishing fires as soon as they start can make some forests more vulnerable to large, destructive fires. Without occasional small fires that remove dead material, fires that do occur can produce extremely damaging crown fires that destroy the entire forest.
PEST INFESTATION
LOGGING
CATTLE RANCHING
HARVESTED FOR FUELWOOD OR CHARCOAL
FIRES
ROADS
They are an indirect cause of deforestation, giving people ready access to forested areas.
SUBSISTENCE FARMING
LARGE FARMS
In tropical areas, the biggest threat is the conversion of forestland to cropland for large-scale, export agriculture.

NET CHANGE IN FOREST AREA BY COUNTRY, 2005–2010

Net loss (ha/year)
More than 500 000
250 000–500 000
50 000–250 000
Small change (gain or loss)
Less than 50 000
Net gain (ha/year)
50 000–250 000
250 000–500 000
More than 500 000

← Today, most deforestation is occurring in developing countries. It must be noted, however, that the industrialization of developed countries, like Canada, the United States, and those of Europe, was supported in large part by the harvesting of their own forests.

water to runoff and pollution. And once those problems started, there was no easy way to fix them."

Eventually, everyone who could abandoned the countryside for the capital city. Even that did not stop the tide of deforestation. As the population swelled, so did the demand for fuel, which in Haiti, still comes almost exclusively from trees.

Today global deforestation has slowed considerably, from 9 million **hectares (ha)** per year in the 1990s down to 5.2 million ha in the year 2000, but the planet still has a net loss of forested land every year. There are three main culprits behind this trend: the harvesting of forests for wood and wood products, the conversion of forests into agricultural land, and urbanization. Management of fire is also a factor in forest destruction. Frequent fires remove deadwood and other flammable material but if we suppress these fires, the deadwood builds up so that when a fire does come through, it can burn so hot it catches the entire forest on fire. It turns out that fire is actually needed to maintain some forests whose trees are fire-adapted with seeds that only germinate when exposed to the heat of a fire. The nature and degree of each of these threats varies by country because forests are often used and managed differently in developing countries than they are in developed ones. [INFOGRAPHIC 11.4]

"The floods got worse and we lost drinking water to runoff and pollution. And once those problems started, there was no easy way to fix them."
—Timote Georges

In general, deforestation occurs at a greater rate in developing countries like Haiti, where dire poverty and a lack of alternatives force people to harvest their forests or remove them for other land uses. They need wood and charcoal for fuel and housing; they need more space for agriculture; and they need commodities to sell in the marketplace. In most cases, developing countries also have far greater remaining forest stands than developed countries, but many of these stands are falling fast.

In fact, many if not most developed countries, including Canada, the United States, and most European countries, became developed in part by harvesting their own forests. Today, European countries have fewer forest stands than many developing countries, but they also have much more stringent regulations in place to protect those that are left. These regulations—and the ability to enforce them—help keep deforestation in check.

Forests can be managed to protect or enhance their ecological and economic productivity.

In Canada, 7% of forests are privately owned, and 93% publicly owned. Provincial/territorial governments manage 83% of public forests, while the federal government has management jurisdiction over the other 17%. This means that forestry legislation is split over multiple levels of government, which leads to difficulties in regulating forestry practices nationally. It is also important to recognize that while almost 50% of Canada is forested, most of that area is highly inaccessible; most logging occurs near populated areas and on the west and east coasts. The most well-known logging operations in Canada are in British Columbia, where excessive clear-cutting in mountainous areas has led to large economic gains, but also to ecosystem loss and degradation. This has reduced the future economic potential for forestry, but also for the salmon industry due to excess sedimentation of spawning grounds.

Ecological knowledge can facilitate more effective forest management in developed nations. The Canadian Forest Service (CFS), a branch of Natural Resources Canada, is a science-based federal policy advisory body. Over the past century, Canada and the CFS have moved away from a timber-focused harvesting approach, towards ecosystem-based forest management. In 1992, Canada committed to **"sustainable forest management"** as part of a **National Forest Strategy.** This approach recognizes the non-timber values of forests, such as recreational uses and carbon storage, which is slowly moving forestry practices in Canada away from harvesting as much as is physically possible and economically profitable towards a **maximum sustainable yield (MSY)**, harvesting as much as *sustainably* possible (but no more) for the greatest economic benefit.

Though some timber harvesting methods are very damaging to the integrity of the forest (such as large clear-cuts on steep slopes), methods are available that reduce disruption to the ecosystem while still providing

---

hectares (ha) Metric unit of measure for area; 1 ha = 2.5 acres (ac)
sustainable forest management a forest management approach that blends ecosystem conservation with economic and social purposes.
National Forest Strategy Canada's plan to incorporate economic, social, and ecological principles into forest management.
maximum sustainable yield (MSY) Harvesting as much as is sustainably possible for the greatest economic benefit.

Infographic **11.5** | **TIMBER-HARVESTING TECHNIQUES**

→ There are many ways to harvest trees from a forest, each with its own economic and ecological trade-offs. Here are variations of 4 techniques. To evaluate the impact of a particular method, consider what the area would look like 50 years after a harvest.

ORIGINAL FOREST

CLEAR-CUTTING

All trees are cut; replanted with a fast-growing species.

IMMEDIATELY AFTER HARVEST

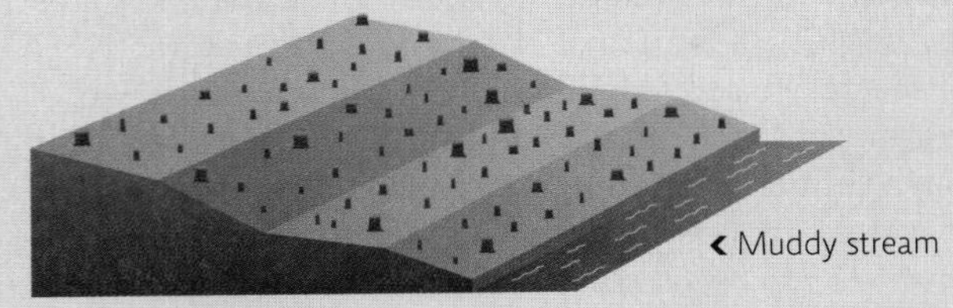
Muddy stream

- High profits at harvest, then no profits until forest regrows
- Water is polluted by heavy erosion on steep slopes.
- Biodiversity is very low after the cut.

50 YEARS AFTER HARVEST

Even-aged, single-species stand

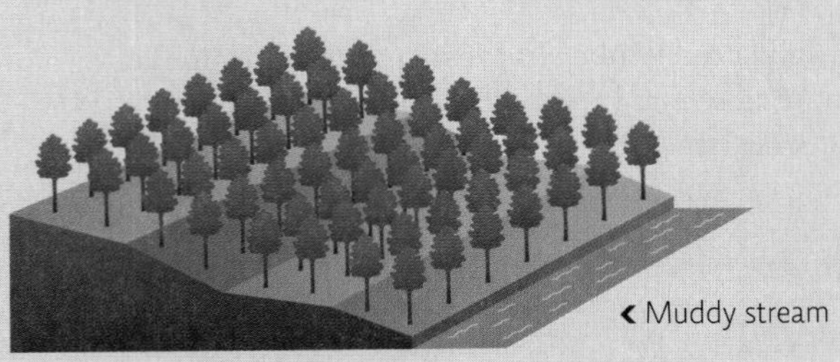
Muddy stream

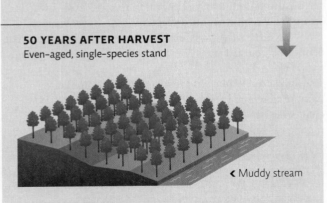

- Tree farms produce harvestable timber in a short time span.
- Water may be polluted by runoff from the open stand, but to a lesser degree than immediately after clear-cut.
- Tree farm has less biodiversity than the original forest.

STRIP HARVESTING

All trees are cut in a strip; here we show it replanted with one species but it may also be allowed to reseed from nearby trees.

IMMEDIATELY AFTER HARVEST

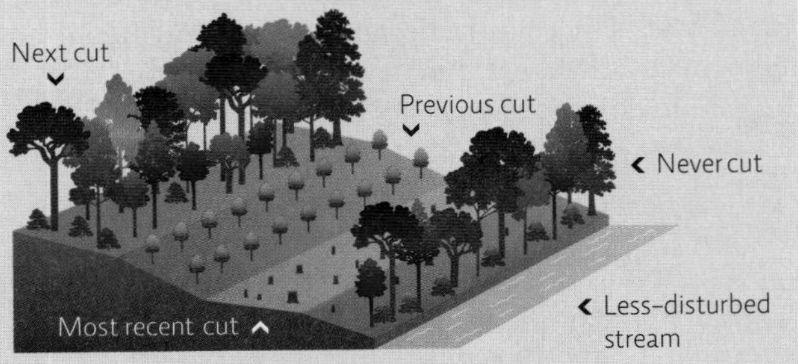
Next cut
Previous cut
Never cut
Most recent cut
Less-disturbed stream

- Profits are initially lower than for clear-cuts, but are more frequent.
- Biodiversity declines, but some organisms find refuge in uncut forests.

50 YEARS AFTER HARVEST

Even-aged stand

New cut
Less-disturbed stream

- Biodiversity is lower than in the original forest, but not as low as clear-cut lands, since a forest remains standing at all times and is still usable by some of the wildlife.

economically valuable forest products. Ideally, the stand of trees—its health, age, and species composition—and the slope of the land determine the harvesting method. Consideration is also given to the other species that reside there. When trees are harvested properly, they can provide immediate and long-term economic benefits without serious environmental damage. Even clear-cutting can be appropriate when it increases the overall health and viability of a future forest by removing unhealthy or genetically inferior trees and replacing them with better stock. Clear-cuts followed by planting of a fast-growing, high-value tree (a tree farm) can also reduce the pressure to cut other forests (though there must be enough native forests left in areas suitable for tree farms to avoid serious ecological damage due to biodiversity loss). [INFOGRAPHIC 11.5]

Forest management is not without its critics, however. Conflicting interests make it difficult to achieve a balance between multiple forest uses and ecosystem protection. For example, harvesting trees in British Columbia provides many jobs, good profits for timber companies, and useful products for our homes and businesses, but can reduce biodiversity and harm salmon runs, which are vital for ecosystem health, native cultures, and the tourism industry.

Debates are often oversimplified and shown as two competing options—owls versus jobs—however, a more accurate cost-accounting of potential forest uses requires that we factor in the economic and ecological value of ecosystem services. Forests are so important economically, ecologically, recreationally, and even

SELECTIVE HARVESTING

High-value trees are cut, leaving others to reseed the plot.

IMMEDIATELY AFTER HARVEST

Never cut
Less-disturbed stream

- Profits are intermediate between clear-cut and strip harvest levels.
- Land can be harvested again in less than 50 years.
- Biodiversity declines after harvest, but not as much as in clear-cut lands.

50 YEARS AFTER HARVEST

Multi-species, mixed-age stand

Less-disturbed stream

- Remaining large trees have been harvested for sale; since the poorer quality trees were left to reseed the area, the forest quality may decline.
- Biodiversity is higher than after the harvest, but lower than the original forest.

SHELTERWOOD HARVESTING

The best trees are left behind to reseed the plot.

IMMEDIATELY AFTER HARVEST

Never cut
Less-disturbed stream

- Profits are similar to those of selective harvest initially, but better than selective harvest on subsequent cuts, since high-quality trees remain.
- Biodiversity declines after each harvest, but may be higher than selective harvest levels.

50 YEARS AFTER HARVEST

New growth is progeny of the best trees; the large trees are now harvested.

Less-disturbed stream

- Biodiversity increases as the forest grows, and may be higher than in any of the other harvested stands.
- Harvesting the large trees left behind provides additional income.

spiritually that no matter how well managed they are, they will always be a contentious resource.

Of course, any gains made by developed countries are still largely offset by deforestation in countries like Haiti. For example, deforestation in Mexico has largely offset new plantings and reforestation in the United States and Canada. Central America has lost more than 5 million ha since 1990 and Europe has gained 12 million ha. "Industrial countries may be leading the way in conserving their own forests," says Morton. "But their demand for wood drives much of the deforestation elsewhere." Multinational corporations have simply moved deforestation operations to developing nations where regulation and enforcement are often lacking and people are desperate for income. They then export products to wealthier countries for sale.

Three months after the kombit visited his hillside farm, Jean Robert is harvesting Moringa leaves. They're tasty, packed with protein, and when harvested properly they regrow rather quickly. If everything goes as expected, he will eventually have a bevy of crops to see him through the year: mangoes in the summertime, coffee in the fall, and if he's lucky, oak and mahogany stands that will yield high prices in the timber market down the road. In the meantime, the Moringa trees provide his family with a sustainable supply of protein and fuel wood.

The trees Georges' team plants can be broken down into three main types. They start with fast-growing, multipurpose trees like the Moringa. Because it is a nitrogen-fixing plant, the Moringa helps refertilize the soil. And because it grows quickly, it can provide a sustainable source of food and fuel wood. Next, they plant fruit trees—mangoes, avocados, and citrus. These trees take longer—3 to 5 years—before they are ready to harvest, but they put down roots and thus stabilize the soil in just a few months' time. "Farmers are less likely to cut down fruit trees for charcoal, because they know it will provide the kind of food they can both eat and sell for profit," says Georges. "Fruit trees are worth something alive." The last thing the kombit will plant are the slow-growing timber trees, which may be sustainably and profitably harvested in the

↓ The nutritious leaves of the Moringa tree are high in protein, vitamins, and minerals, and contain enough iron to treat mild anemia; they can be harvested without killing the tree, eaten fresh, or dried for later use. The seeds and roots are also edible and cuttings can be planted to start new trees.

← Wangari Maathai's Green Belt Movement of Kenya has spread to more than 30 other African nations. The first Kenyan woman to earn a Ph.D., Maathai received several environmental awards, including the Nobel Peace Prize. She passed away in 2011 but the Green Belt Movement lives on.

future, but don't provide any immediate benefits to the farmers. "The goal is to mix it up," says Georges. "You can create a whole stable system that's going to provide money and food throughout the seasons and across the years."

Of course, for any of this to work, urban-dwelling Haitians will need a new, sustainable energy source.

There are several ways to protect forests, but each comes with trade-offs.

The trade-offs associated with the use of forest resources are the subject of much debate and conflict because each decision impacts both human livelihood and the health of the environment. To be sure, the economic value of wood may be dwarfed by the ecosystem services lost if the area is overharvested, and protecting the ecosystem may actually prove more profitable, even in the short term. But people still need fuel, building material, and income. "Of course trees and forests are important for the environment," says Georges, "We know that they protect us from floods, and help keep drinking water clean and plentiful. But for many Haitians, selling those same trees is the only way to feed a family. So how can you ask them not to?"

One solution, according to a growing number of experts, is to price the ecosystem services themselves. In Costa Rica, for example, higher utility bills offset the costs of maintaining rainforests that purify and replenish the water people use every day. This money is given directly to landowners who would otherwise have to chop the trees down to sell as fuel or timber or to convert the land to agriculture.

Other options include the promotion and increased availability of sustainable wood products. In developed and even in developing nations, certified sustainable wood products are becoming more available. The Forest Stewardship Council (FSC) certifies lumber and other timber products through a process that evaluates the forest itself and the timber-harvesting techniques in terms of wildlife, water, and soil quality. Worldwide, more than 6% of forests are certified by the FSC as sustainably managed. Other resources, like latex and tree nuts, can be sustainably harvested from standing forests. Alternatives for wood products also exist and can reduce the pressure

↑ Tourists on canopy walkway, Monteverde Cloud Forest Preserve, Costa Rica.

on forests. For example, lumber from old buildings can be salvaged to provide quality building materials. Paper can be made from old paper (recycled) or from fast-growing crops such as kenaf, jute, flax, and hemp. Reforestation projects are increasing in number around the world. For example, the Kenyan Green Belt Movement has planted and protected more than 30 million trees and has sparked similar movements worldwide.

For some, saving the forests may come down to recognizing their intrinsic value: their natural beauty, their inherent sacredness, and their right to exist as a living thing, regardless of what we humans might extract from them. Naturalists from John Muir to Wallace Stegner have argued for the protection and preservation of forests on these grounds alone. "We simply need that wild country available to us, even if we never do more than drive to its edge and look in," 20th-century American naturalist Wallace Stegner wrote in his famous *Wilderness Letter*. "For it can be a means of reassuring ourselves of our sanity as creatures, a part of the geography of hope."

This appreciation for the intrinsic value of a forest has been successfully translated into the economic enterprise known as **ecotourism.** Ecotourism is a viable option for many areas, especially less developed countries. These areas often possess high biodiversity precisely because they are less developed and are often found in tropical or subtropical areas with naturally high biodiversity. In the quest to develop economically, these regions may find that the highest economic value for their resources lies in keeping them intact. Ecotourism allows a way for funds to enter the country while protecting the natural areas at the same time.

But for many people, including those in Haiti, these options seem limited. Charcoal produced from wood is the only means people have to heat their homes, prepare their food, and in the cities, fuel their businesses. "We won't solve deforestation until we increase the market

ecotourism Low-impact travel to natural areas that contributes to the protection of the environment and respects the local people.

share of other energy sources," says Morton. "Right now everyone—bakers, rum makers, housewives, even factories—are dependent on charcoal." In fact, small businesses use the equivalent of 11 ha of wood each day.

"It's a vicious cycle," says Morton. "They cut down trees to make charcoal to cook and earn some money. This erodes the soil, which causes crops to shrink and floods to be more severe. Both of those render people even more impoverished, which in turn necessitates more tree-cutting." Charcoal, he says, is the fuel of the poor, and unless Haitians escape the poverty trap, they won't stop using it. In fact, declining soil fertility and falling commodity prices have led many rural Haitians to increase their charcoal production to generate more income.

In the countryside, Georges says, charcoal use will probably continue no matter how well reforestation efforts work. "It's not a matter of weaning people off charcoal," he says. "The use is ingrained into the culture. The real issue is sustainable practices." Georges notes that simple changes like harvesting only fast-growing trees for charcoal, pruning branches to encourage more growth and cutting trees 1 metre above the ground rather than at ground level for quicker regeneration could make a huge difference.

To meet intensive energy needs of the city, however, other measures will be needed. One potential alternative to charcoal that could be used in Haiti and elsewhere is Jatropha, a fast-growing plant whose oil-rich seeds have been hailed as a promising biofuel source. Even the material left after the oil is pressed out can be digested by bacteria to produce biogas, a fuel similar to natural gas. Another option is the production of composite briquettes from a variety of flammable materials such as grass, paper, or sawdust. Shredded plastic is added to make it more combustible (see Chapters 24 and 25 for more on sustainable energy).

There is a Haiti that people like Georges and Robert talk about in the quiet moments after a day's planting. It's a Haiti lush and green with trees, a country where families earn their living selling mangoes and Moringa leaves instead of charcoal and firewood. Whether this country resembles their past as much as their future will depend on an infinite number of variables—not just how well they manage their forests and reforestation efforts, but also whether they can find alternative building materials or establish a reliable energy sector based on something other than wood.

In parts of Haiti, where the top soil has long since given way to a barren landscape of rock, most experts agree that it is too late. But in much of the country there are still some trees left. And that means there is still hope. ◉

Select reference in this chapter:
Constanza, R., *et al.* 1997. *Nature*, 387: 253–260.

BRING IT HOME

PERSONAL CHOICES THAT HELP

Forests are a renewable resource that can be used sustainably for many years under proper management conditions. Using harvesting systems such as selective cutting or strip cutting can allow economic use of the forests without eliminating the ecological functions that forests provide.

Individual Steps

→ Buy paper products (toilet paper, facial tissue, and notebook paper) made of recycled content to decrease the unnecessary cutting of trees and to encourage recycling.

→ When purchasing lumber for projects, look for wood that has been certified as sustainably managed by the Forest Stewardship Council (ca.fsc.org).

→ Avoid buying noncertified furniture made from tropical wood such as rosewood, teak, and ebony; harvesting these woods contributes to tropical deforestation.

Group Action

→ Organize a workday to clear invasive species such as common buckthorn and Japanese honeysuckle to ensure that our remaining forests provide high-quality habitat.

Policy Change

→ Support legislation that protects roadless areas and old-growth forests.

→ Ask your local grocery stores, restaurants, and university and college dining facilities to offer shade-grown coffee and chocolate, which can help encourage preservation of tropical forests.

UNDERSTANDING THE ISSUE

CHECK YOUR UNDERSTANDING

1. High-latitude and high-altitude forests characterized by a short growing season, a lot of snowfall, and thin, acidic soils are:
   a. temperate forests with trees that drop their leaves in winter, such as oak and maple.
   b. boreal forests with evergreen trees that bear needlelike leaves, such as spruce and fir.
   c. savanna forests with trees that have long taproots and thick, fire-resistant barks, such as acacia and eucalyptus.
   d. tropical forests with trees that are highly valued as timber for their rot-resistant wood, such as mahogany and teak.

2. The understory of a forest is made up of:
   a. trees that push through and grow above the level of the forest canopy.
   b. the seedlings, ferns, herbs, and wildflowers that grow on the forest floor.
   c. the overlapping crowns of the tallest trees that make up the roof of the forest.
   d. shade-tolerant shrubs or the saplings of larger trees that sometimes form a lower canopy.

3. Which of the following is NOT a cause of deforestation in Haiti?
   a. Palm oil plantations for biofuel production
   b. Coffee and sugar plantations for export crops
   c. Wood charcoal production for the domestic energy market
   d. Timber harvesting for commercial sale

4. How does deforestation contribute to loss of drinking water in Haiti?
   a. Without trees, the eroded soil clogs drinking water pipes.
   b. The loss of forest habitat fragments aquatic ecosystems that provide water to the local villages.
   c. Without the trees, the water rushes away rather than soaking into the ground to recharge groundwater supplies.
   d. Clearing the forest for crops means increased irrigation, which reduces drinking water supplies.

5. The timber-harvesting system that would be most likely to cause disruption to the ecosystem services provided by a forest is:
   a. shelterwood harvesting.
   b. strip harvesting.
   c. clear-cutting.
   d. selective harvesting.

6. Which of the following activities will NOT simultaneously protect forests and provide for long-term economic well-being of local people in developing countries?
   a. Pricing ecosystem services provided by the forest and paying landowners to maintain their trees
   b. Clear-cutting the forest and planting fast-growing trees for charcoal production
   c. Promoting the harvesting and selling of wood that is certified by the Forest Stewardship Council
   d. Translating the intrinsic value of a forest into ecotourism enterprises

WORK WITH IDEAS

1. Tropical rainforests have thin, acidic soils yet they contain dense vegetation and high biodiversity. How can these tropical forests have poor soil but support such diverse arrays of plant and animal life?

2. What is an "ecosystem service"? Describe three such services provided by forests.

3. How is the current status of forests different in developing versus developed countries? What factors account for these differences?

4. How can we sustainably use forest resources?

5. What is the reforestation strategy employed by the Haitian kombit? Explain the rationale behind the approach and discuss whether it can be successful.

ANALYZING THE SCIENCE

The data in the following table comes from the most recent assessments of global forest resources conducted by the Food and Agriculture Organization of the United Nations. This table reports on the global, regional, and subregional trends in forest areas designated primarily for the protection of soil and water.

Region/subregion	Information availability		Area of forest designated for protection of soil and water (1000 ha)		
	# of countries	% of total forest area	1990	2000	2010
Eastern and Southern Africa	21	80.9	14 033	13 311	12 611
Northern Africa	7	99.1	4068	3855	3851
Western and Central Africa	22	52.5	2639	3281	3079
Total Africa	**50**	**69.2**	**20 709**	**20 447**	**19 540**
East Asia	4	90.2	24 061	38 514	65 719
South and Southeast Asia	17	100.0	55 811	57 932	56 501
Western and Central Asia	23	99.7	12 222	13 059	13 669
Total Asia	**44**	**95.8**	**92 094**	**109 505**	**135 889**
Total Europe	**45**	**99.7**	**76 932**	**90 788**	**92 995**
Caribbean	11	53.1	869	1106	1428
Central America	3	36.9	124	114	90
North America	5	100.0	0	0	0
Total North and Central America	**19**	**97.8**	**994**	**1220**	**1517**
Total Oceania	**18**	**21.6**	**1048**	**1078**	**888**
Total South America	**10**	**85.1**	**48 656**	**48 661**	**48 549**
World	**186**	**86.9**	**240 433**	**271 699**	**299 378**

INTERPRETATION

1. How much total forest area was designated for protection of soil and water globally in 2000? In 2010? (Hint: Don't forget to account for the units of measure shown in the header.)

2. Which subregion of Asia had the most land designated for protection of soil and water in 1990? Was this the same in 2010? Provide the data to support your responses.

3. Summarize the trends in forest lands set aside for the protection of soil and water:
   a. Between 1990 and 2010, how much has the total (global) area grown or decreased?
   b. Between 1990 and 2010, which major region showed the most gain and how much did it gain?
   c. Between 1990 and 2010, which major region showed the most loss and how much did it lose?

ADVANCE YOUR THINKING

Hint: To answer some of the following questions it might help to access the actual report, at www.fao.org/forestry/fra/en/ (see Chapter 6 of the report—Protective Functions of Forest Resources).

4. What might explain the patterns seen in the table?

5. Why is no forest area designated for protection of soil and water for North America? How might this affect how the data presented in the table is interpreted?

6. If the world population is 7 billlion, what is the per capita amount of forest land that is set aside for soil and water protection?

EVALUATING NEW INFORMATION

Many of the world's forests are severely degraded. Yet forests produce many consumer products that we depend on, such as food, medicine, building materials, and raw materials for industrial products like paper. So what should a conscientious consumer do?

Explore the Forest Stewardship Council (FSC) website (ca.fsc.org).

Evaluate the website and work with the information to answer the following questions:

1. Is this a reliable information source? Does the FSC have a clear and transparent agenda?
   a. Who makes up this organization? Does its membership make the FSC reliable/unreliable? Explain.
   b. What is this organization's mission? What are its underlying values? How do you know this?
   c. Does the FSC give supporting evidence for its claims about forest resources and its vision to address the problem? Does the website give sources for its evidence?
   d. Identify a claim the FSC makes and the evidence it gives in support of this claim. Is it sufficient? Explain.
   e. Do you agree with its assessment of the forest issues? Explain.

2. The FSC is currently engaged in a Regional Forest Management Standard Revision, and this is listed as a way to get involved. Go to ca.fsc.org/regional-fm-standard-revision.246.htm and download 120305 Standards Revision Process PASSED Motion.pdf.
   a. Who may participate in this revision?
   b. What is the "strategic direction" for this revision?
   c. Do you agree with this direction? Why or why not?

3. How can a consumer select FSC-certified products, according to this website? Is it easy for consumers to know if the wood products they are purchasing are FSC certified? Explain.

4. How might FSC certification help forests?

MAKING CONNECTIONS

INDONESIA'S PALM OIL DILEMMA

Background: Like those in Haiti, Indonesia's rainforests could potentially be a renewable energy source. But unlike Haiti with its charcoal production, Indonesia leads the global production of palm oil, a commodity for export as a biofuel, cooking oil, and ingredient in industrial products such as cosmetics. Palm trees can produce fruit for 30 years or more, and yield more oil per hectare than other oilseed crops. But the cost of these benefits is the loss of large stands of tropical forest to allow for plantations of palm trees. In Indonesia, the increasing demand for palm oil has accelerated seizure of indigenous peoples' land, dismantling local communities and cultures and threatening the habitat of endangered species like the orangutan.

A low-carbon alternative to fossil fuel–based gasoline, palm oil biodiesel was seen as a solution to climate change, but recent research revealed that palm oil development involving the clearing of intact tropical rainforest contributes more greenhouse gases than it helps to avoid. In fact, Indonesia already emits more greenhouse gases than any other nation besides China and the United States.

Case: You have been assigned to develop a policy for the future of the Indonesian rainforest. Your team must evaluate various options such as:

1. Protect Indonesian rainforests through the United Nations Reducing Emissions from Deforestation and Forest Degradation (REDD+) program, which offers financial incentives to developing countries to sustainably manage forests to sequester carbon. According to UN estimates, the financial value of greenhouse gas emission reductions through REDD+ could be $30 billion a year (www.un-redd.org).

2. Continue producing palm oil but seek a sustainable biofuel certification (similar to the FSC certification for wood). According to the Palm Oil Action Group, there are millions of hectares of degraded land in Indonesia that could be reclaimed for palm oil plantations.

Research these (and possibly other options) and write a report recommending a course of action. In your report, include the following:

a. An analysis of the pros and cons of each proposal, including: a discussion of the consequences of each choice on local human communities as well as on the rainforest ecosystem. What is the best option for Indonesian rainforests? Who should be involved in this selection? Provide justifications for your proposal.
b. What overall lessons can be learned from the stories of Haiti and Indonesia and applied to other forest ecosystems in jeopardy? Explain your suggestions.

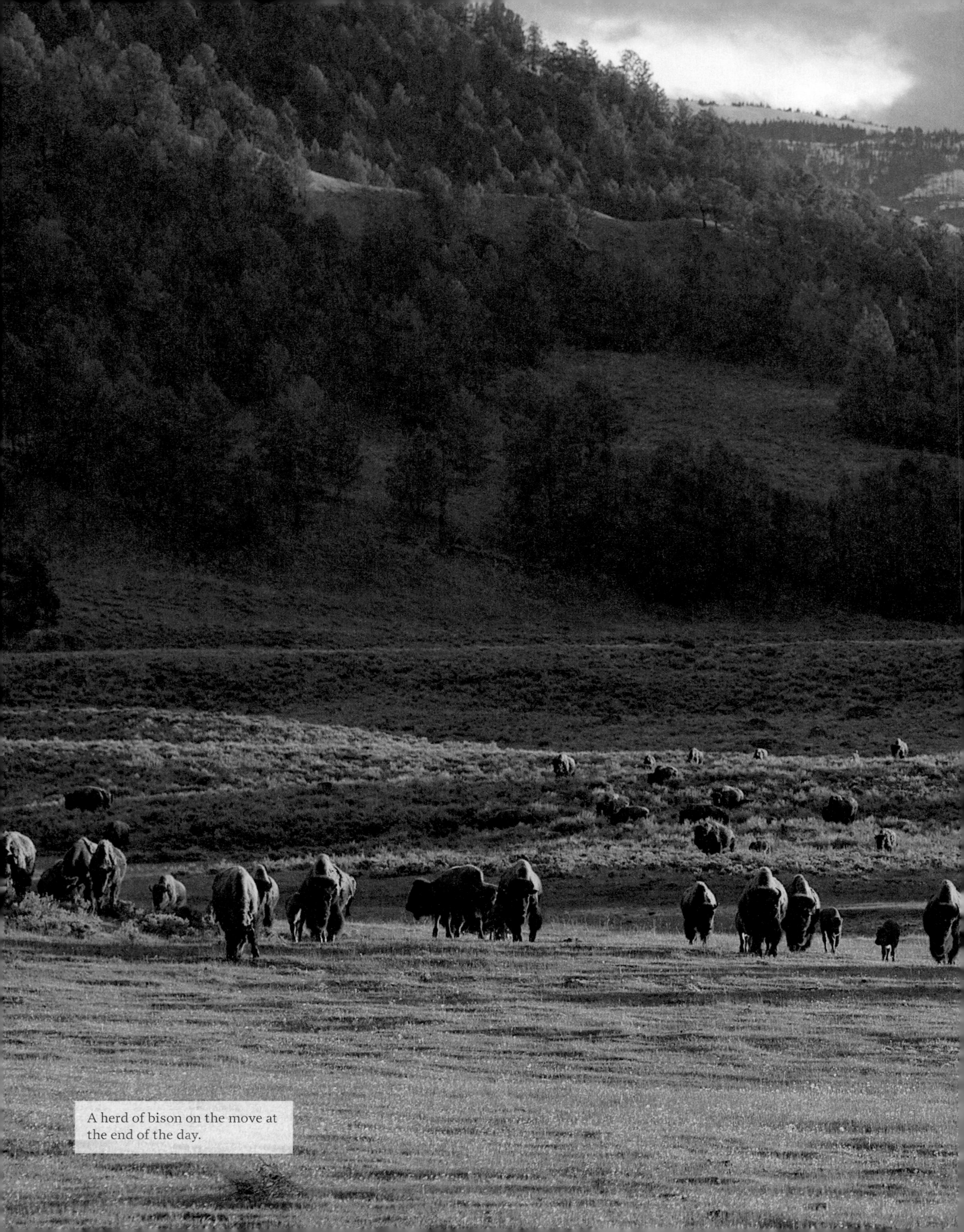

A herd of bison on the move at the end of the day.

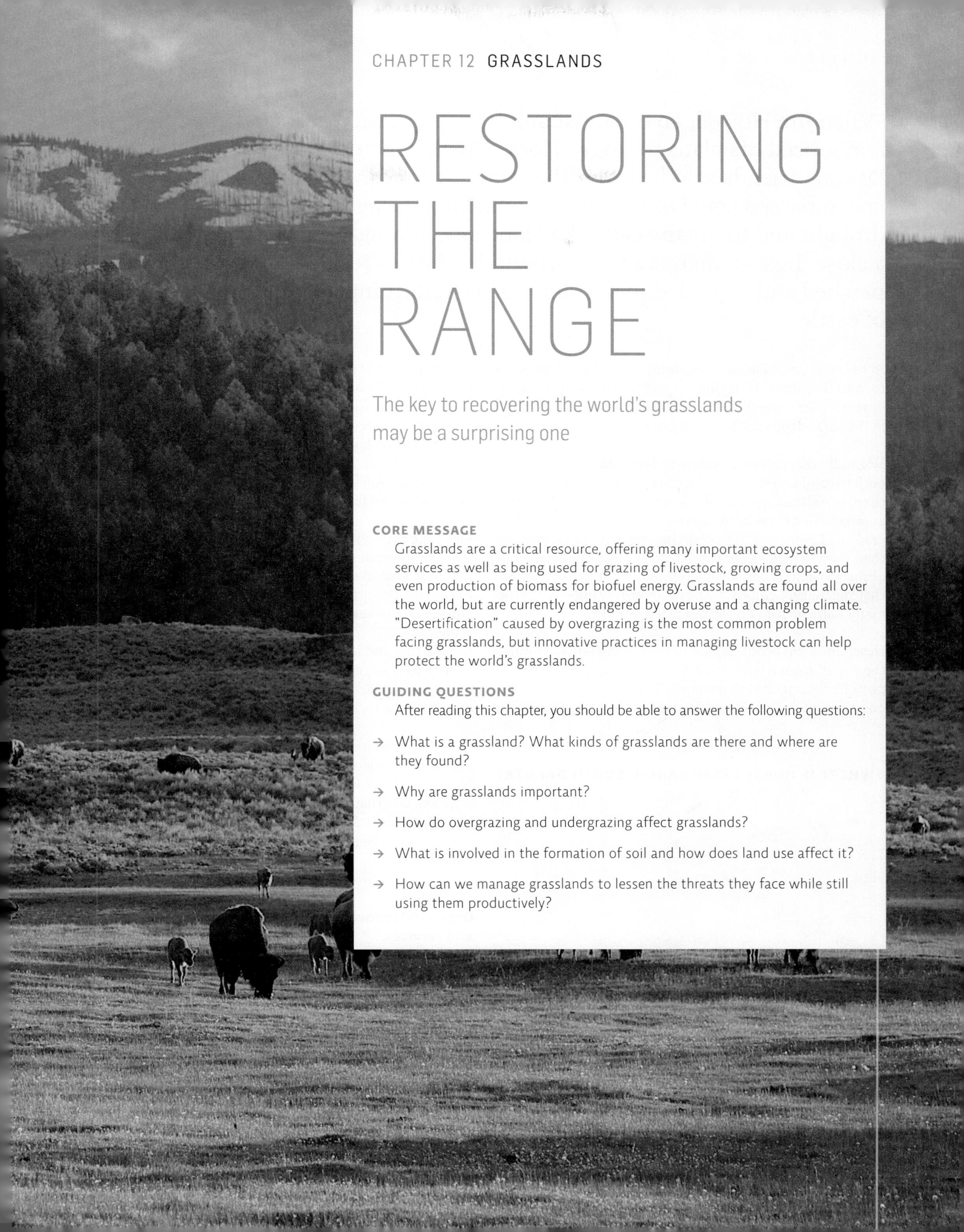

RESTORING THE RANGE

The key to recovering the world's grasslands may be a surprising one

CORE MESSAGE

Grasslands are a critical resource, offering many important ecosystem services as well as being used for grazing of livestock, growing crops, and even production of biomass for biofuel energy. Grasslands are found all over the world, but are currently endangered by overuse and a changing climate. "Desertification" caused by overgrazing is the most common problem facing grasslands, but innovative practices in managing livestock can help protect the world's grasslands.

GUIDING QUESTIONS

After reading this chapter, you should be able to answer the following questions:

→ What is a grassland? What kinds of grasslands are there and where are they found?

→ Why are grasslands important?

→ How do overgrazing and undergrazing affect grasslands?

→ What is involved in the formation of soil and how does land use affect it?

→ How can we manage grasslands to lessen the threats they face while still using them productively?

When Jim Howell, an ecologist and fifth-generation cattle rancher, first announced his plans to revive Horse Creek Ranch in Butte County, South Dakota, friends said he was either crazy or foolish. Sure, the ranch had once encompassed some of the best grazing land in the country, but persistent drought and too many cattle had long since brought those days of plenty to a close. Like so many ranches around it, Horse Creek's pastures were too parched and degraded to sustain much of anything, let alone an entire herd of cattle.

Across the Great Plains, **rangeland** (land that humans use to graze livestock) is drying out and ranchers are growing desperate. In some places, the degradation is so bad that prairie grasslands are becoming deserts.

Normally this process, known as **desertification**, is both natural and slow. As climates shift over geologic time, grasslands morph into desert and deserts back into grassland in a cycle both never-ending and imperceptibly gradual. These days, it's occurring much more rapidly than that—3 metres per year in West Texas alone. Part of the problem is climate change. But most experts agree that the biggest culprit, by far, is **overgrazing**—when too many animals feed on a given patch of land.

Desertification is not unique to the Great Plains. Around the world, from Afghanistan to Zimbabwe, 70% of the planet's rangeland is threatened. This represents roughly one-third of the world's entire land surface. And while the phenomenon has not garnered as much media attention as, say, global climate change, the consequences are no less dire. In fact, from the Fertile Crescent of ancient Babylon to modern-day developing countries like Darfur, desertification has been the stuff of wars. The cascade is both simple and devastatingly comprehensive: plants die, soil erodes, prairies fall to dust, famine sets in, economies falter, societies fail.

◉WHERE IS HORSE CREEK RANCH, SOUTH DAKOTA?

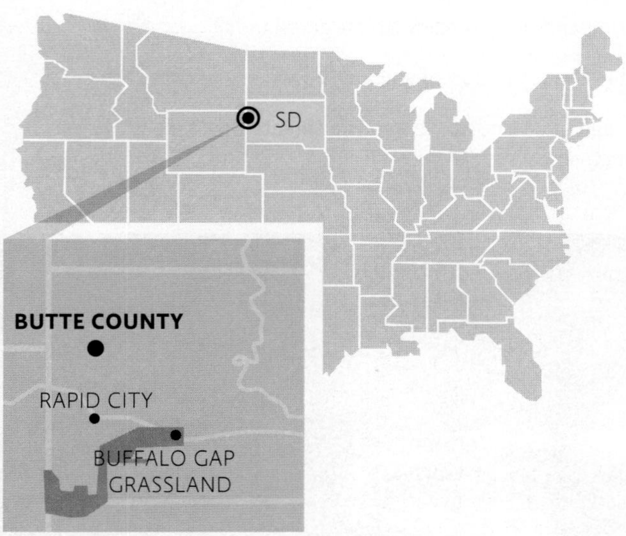
SD
BUTTE COUNTY
RAPID CITY
BUFFALO GAP
GRASSLAND

Since the early 1980s, the U.S. government has paid hundreds of thousands of ranchers to quit ranching so that vast swaths of degraded grassland can have a chance to recover. But critics say that holding the land in such **conservation reserve programs** decimates local prairie economies. "There are real environmental benefits to placing the land in trust and saying it can't be grazed or farmed over anymore," says Howell. "But it leaves farmers dependent on government cheques and does nothing to stimulate the local economy."

Starting with Horse Creek—a forgotten ranch, in a fading prairie town—Howell and his business partners, who took control of the ranch in 2008, are working on a different solution, one that aims to repair economy and environment in tandem. The bottom line, they say, is this: if you want to save the prairies, you've got to graze more cattle, not less.

Grasslands provide a wide range of important goods and services.

Grasslands are biomes that receive enough rainfall to support grass and herbaceous plants, but not enough to support forests. They may also be found in regions where rainfall is plentiful but larger plants are kept in check by periodic fires or herds of grazing herbivores. Broadly, there are several different types: *tropical grasslands,* also known as savannahs, occur in places that have both rainy and dry seasons, but are warm year-round. *Cold grasslands* like steppes are, as the name suggests, cold most of the

↑ Cattle and goats have pulverized the drought-prone Omo, Ethiopia, region into dust.

year, and are characterized by a very short growing season and ultrathin layers of soil. The Great Plains—which lies between the Rocky Mountains and the Mississippi River and stretches from Alberta, Saskatchewan, and Manitoba into the south of Texas and Mexico—is *temperate grassland*. These grasslands, known as prairies, contain many species of plants, have thick soils, and have a truly seasonal climate with cold winters and hot summers. [INFOGRAPHIC 12.1]

Though we might most often think of grasslands in terms of their human uses (pasture, farmland), they provide extremely important ecosystem services such as nutrient cycling, soil formation and protection, carbon sequestration, protection of surface waters, and provision of habitat for both year-round and seasonal wildlife. But it is also true that for thousands of years, grasslands—which cover about 40% of Earth's surface and contain some of the richest soil in the world—have provided humans with a multitude of goods and services. In fact, most major cereal crops, including wheat, rye, barley, and millet, were originally derived, thousands of years ago, from grassland seedbeds. Today, scientists trying to develop disease-resistant crops continue to rely on the genetic material found in grasslands. Meanwhile, other scientists are perfecting ways to harvest grassland plants as an energy source (see Chapter 25 for more on biofuels).

But the most widespread, human-centred use of grasslands is as food for large grazing herd animals.

In addition to supporting countless wild ungulate (hoofed animals) grazer herds, a little more than half of the world's grasslands—26% of the planet's land surface—are used to graze more than 3 billion domestic livestock (mostly cattle, goats, and sheep). This type of grazing provides the primary source of food and income for some 2 million of the world's poorest people. And while other livestock (like chicken and pigs) consume vast stores of grain that might otherwise go to humans, grazing animals

rangeland Grassland used for grazing of livestock.
desertification The process that transforms once-fertile land into desert.
overgrazing Too many herbivores feeding in an area, eating the plants faster than they can regrow.
conservation reserve program Farmers and ranchers are paid to keep damaged land out of production to promote recovery.
grasslands A biome that is predominately grasses, due to low rainfall, grazing animals, and/or fire.

Infographic **12.1** | **GRASSLANDS OF THE WORLD**

↓ Grasslands are found on every continent except Antarctica. The examples of grasslands shown here vary based on climate.

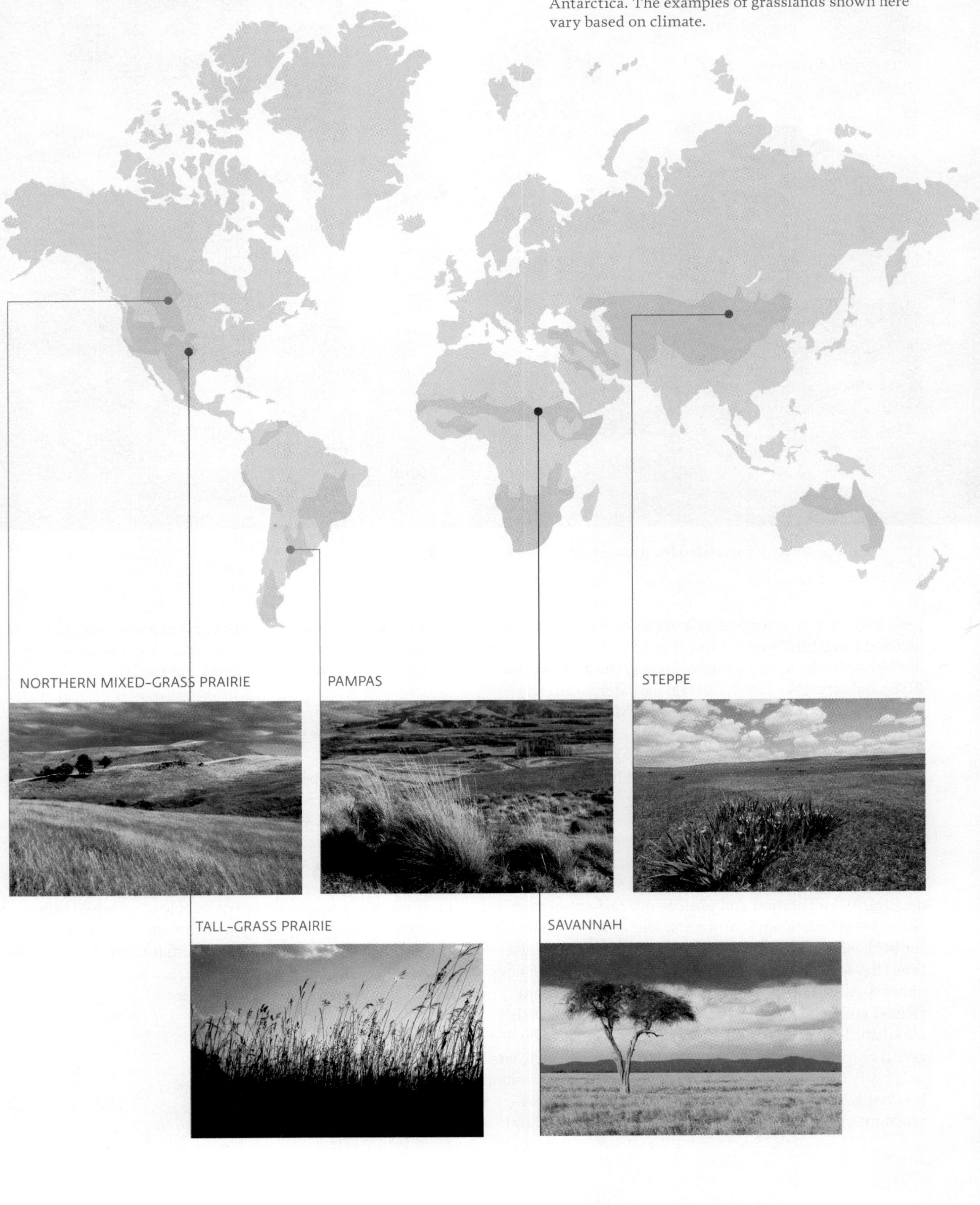
NORTHERN MIXED-GRASS PRAIRIE
PAMPAS
STEPPE
TALL-GRASS PRAIRIE
SAVANNAH

are effectively converting food we cannot eat (grass) into food we can eat (meat and dairy). [INFOGRAPHIC 12.2]

Infographic **12.2** | **GRASSLAND GOODS AND SERVICES**

WILDLIFE HABITAT
CONVERSION TO AGRICULTURE
GRAZING
BIOMASS FOR BIOFUEL

Grasslands face a variety of human and natural threats.

A trifecta of forces now threatens all of these goods and services. First, global climate change: as temperatures rise, scientists expect that shifting rainfall patterns will help push many current grasslands—including vast swaths of Canada's prairies—into desert. The second threat involves human land-use decisions: rising global populations will force us to convert more land into cities and suburbs for living, and into croplands for food. Their wide, flat expanses and nutrient-rich soil make grasslands ideal for both. The third, and many say largest, threat that grasslands face is the one posed by overgrazing.

Ironically, grazing is normally very good for grasslands. Because wild **herbivores** evolved to subsist on grasslands alone, both grazers and grasses are well adapted to grazing. Grasses can grow from the base upward, so by clipping off the top part of the blade, herbivores expose new growth shoots to the sunlight, thus stimulating the plants' growth. By breaking up the soil with their hooves, they allow water to penetrate the ground and enable seeds to germinate and take root. And as they defecate and urinate, large grazing mammals return nutrients such as nitrogen and phosphorus to the soil in a form that plants can easily absorb.

But when grass is overgrazed, or chewed down to its roots, the growth area on the blade is destroyed, the blade can no longer regenerate, and the plant eventually dies. When too many plants die at once, the soil has nothing to hold it in place. And when too many large grazing animals stomp their hooves directly onto the soil, the soil becomes compacted, which makes it harder for water to penetrate, seeds to germinate, or seedlings to grow. [INFOGRAPHIC 12.3]

Together, plant loss and soil compaction increase the rate of **soil erosion**—a process in which soil is swept away by wind and rain down into streams, rivers, and gullies, faster than it can possibly be replenished. That's no small matter. Soil formation is a slow process that requires the weathering of rock and decomposition of organic material. Under the best conditions, it takes 1 year to generate just a millimetre of the precious brown gold in which our food

---

herbivore An animal that feeds on plants.
soil erosion The removal of soil by wind and water that exceeds the soil's natural replacement.

Infographic **12.3** | **DESERTIFICATION**

↓ Every inhabited continent has grasslands vulnerable to desertification, especially in arid areas close to existing deserts.

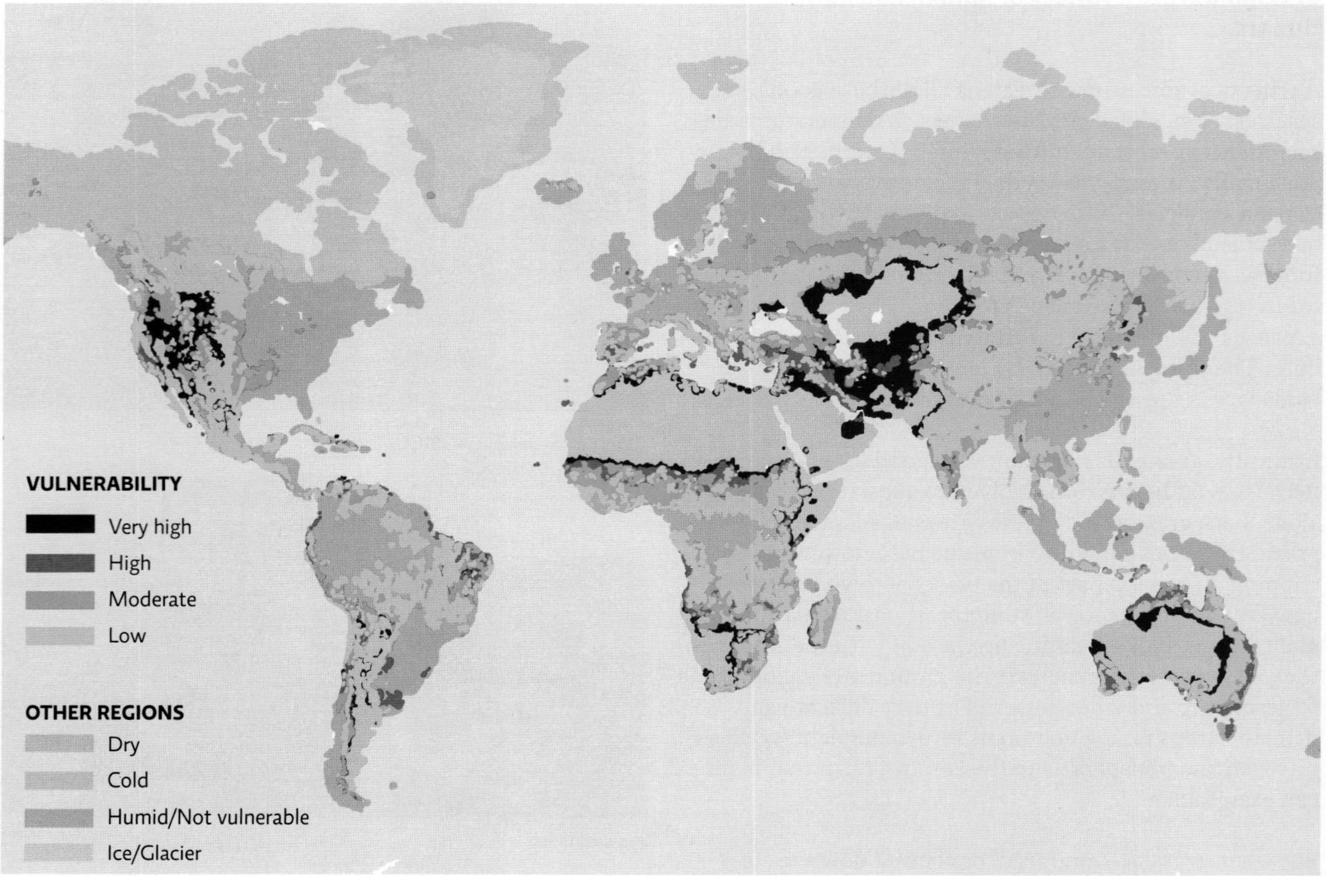
VULNERABILITY
Very high
High
Moderate
Low
OTHER REGIONS
Dry
Cold
Humid/Not vulnerable
Ice/Glacier

GROWTH AREA

HEALTHY GRASSLAND
Deep roots hold soil in place and keep the plant alive during drought or fire.

OVERGRAZED GRASSLAND
Regeneration is harder in damaged, drier soil.

↑ Grasses are adapted to grazing; cropping the grass stimulates the growth area at the base of the blade. Overgrazing may remove this growth area and kill grasses, increasing the potential for desertification.

Infographic **12.4** | **SOIL FORMATION**

↓ Soil is produced by the decay of organic material and the weathering of rock. Distinct layers are seen in healthy soils, with the topsoil (A horizon) being the most fertile for plant growth. Desertification will reduce or remove the O and A horizons and produce drier B and C horizons.

↑ Native prairie grasses have deep roots (up to 5 metres long) which allow them to access deep water supplies and to weather droughts. The native grass roots also hold the soil in place much better than do shallow-rooted annual crops like wheat.

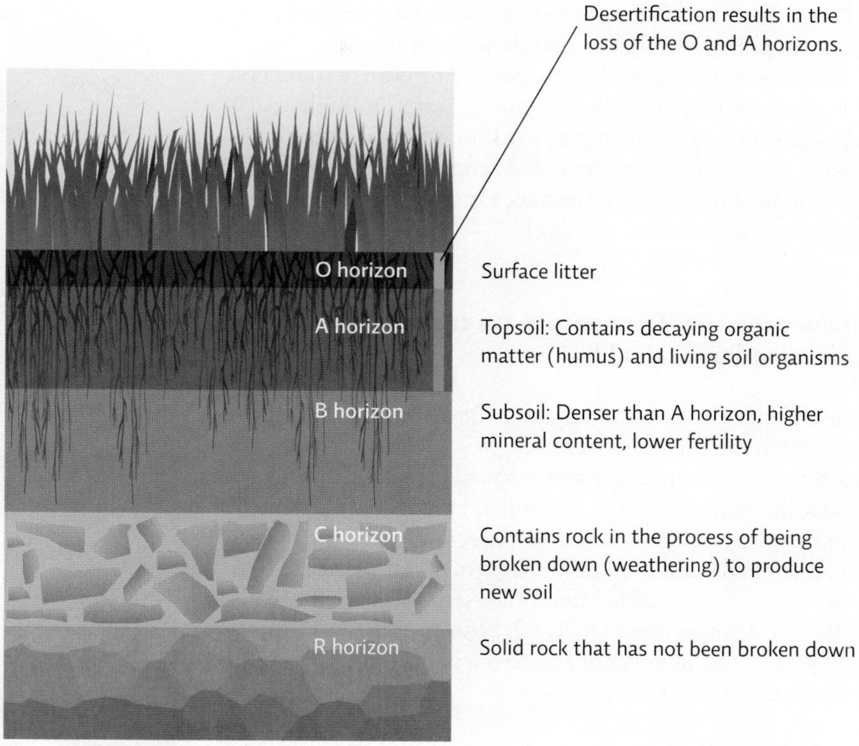
Desertification results in the loss of the O and A horizons.
O horizon
Surface litter
A horizon
Topsoil: Contains decaying organic matter (humus) and living soil organisms
B horizon
Subsoil: Denser than A horizon, higher mineral content, lower fertility
C horizon
Contains rock in the process of being broken down (weathering) to produce new soil
R horizon
Solid rock that has not been broken down

grows. A closer look at a soil's profile sometimes reveals layers (called horizons) that reflect the formation process. As plants die out and soil erodes, the denuded landscape begins to reflect incoming sunlight rather than absorb it; this touches off a cascade that ultimately alters wind and temperature patterns. Before long, grasslands give way to deserts. [INFOGRAPHIC 12.4]

At least once in North America's history, human activities so amplified the speed and scope of desertification that it triggered the largest human migration inside a decade our continent has ever seen. Beginning in 1934, clouds of dust so massive that ranchers named them "black blizzards" swirled relentlessly across the Great Plains, tearing the paint off of houses and cars and forcing some 2.75 million homesteaders from the land. History named this time and place the Great Dust Bowl, and it became a classic example of the tragedy of the commons. As each individual homesteader expanded farm and ranch for his or her own gain, the land as a whole fell victim to overgrazing. When the next drought cycled through, there was no grass left to hold the crumbling soil in place.

It took nearly three decades for the prairies to recover, but according to some critics, only half as long for us to forget the Dust Bowl's most important lessons. By the 1970s, with the market prices for agricultural goods rising steadily, many farmers were overplowing and overgrazing with the same pre-Dust Bowl fervour. Now, soil erosion is approaching Dust Bowl rates in some parts of the United States. Canada, which is not as susceptible as much of the drought-ridden United States, has held back the desert tide through regulations and farmer education on best practices.

So far, grassland degradation has cost humans roughly 12% of global grain production, not to mention $23 billion per year in global GDP (gross domestic product). All told, the food supply of more than 1 billion people is threatened. The Food and Agricultural Organization (FAO) predicts that some 50 million people will be faced with displacement in the coming decades. The majority of these will be subsistence farmers who live in the world's poorest regions and depend solely on cattle ranching for their livelihoods; but North American ranchers, like the ones who owned Horse Creek, will also suffer.

Scientists around the world have spent decades trying to prevent or even reverse desertification, to little avail. These days, most experts tend to agree that beyond a certain point, recovering grasslands that have swirled into deserts is impossible. "Most of our efforts to reverse desertification have failed dismally," says Dr. Richard Teague, a research ecologist working with ranchers in West Texas to restore degraded rangeland. "But a number of ranchers here are having success with protocols developed halfway around the world."

Infographic **12.5** | **SOUTHERN AFRICA MAP**

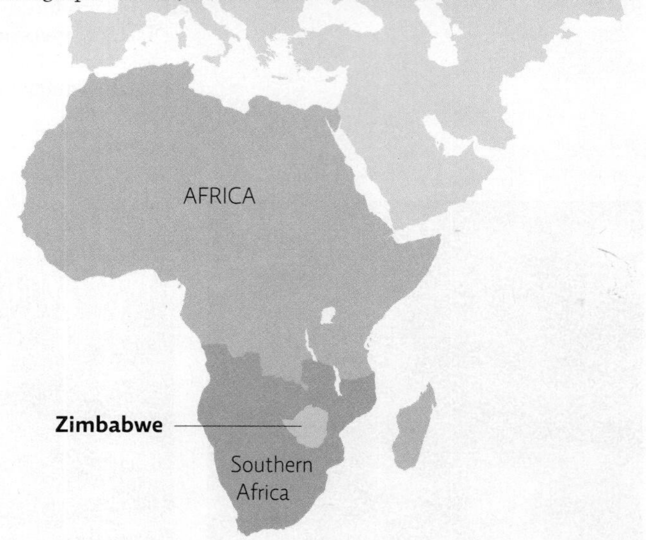
AFRICA
Zimbabwe
Southern
Africa

Taking our cues from nature, we can learn to use rangelands sustainably.

As the sun rises over Zimbabwe (formerly Rhodesia) in southern Africa, herds of antelope and zebra traverse a patchwork of temperate and tropical grasslands, feeding steadily on reedy stalks and short, fat shrubs; elephants and wildebeest splash around a precious watering hole, well fed and content. The animals may not realize it, but they have stumbled upon the Africa Centre for Holistic Management (ACHM), 2630 hectares of thriving rangeland in the heart of an otherwise parched and ailing savannah. [INFOGRAPHIC 12.5]

Perhaps nowhere else on Earth is such an oasis more urgently needed. Because the region is too arid to support much else, ranching provides the only livelihood for most of the people living there; about 75% of all land is used to graze domestic herds of cattle, goat, and sheep, and even that has not been enough. With population, and thus the number of mouths in need of food, rising steadily, farmers have crowded more and more livestock onto lands that grow sparser and drier by the day. Already stressed by climate change, that land is crumbling quickly into desert. And as viable pastures become increasingly difficult to find, neighbouring tribes have descended further into violent conflict—sometimes killing each other over a few stalks of grass.

↑ Biologist Allan Savory, squatting beside a patch of dry grasses in the desert where he teaches holistic land management.

The ACHM was established in 1992 by Allan Savory, a Rhodesian-born scientist-turned-rancher. Before then, the land had been so thoroughly desertified that neither wild nor domestic herds bothered to graze there. But in the nearly two decades since, the picture has changed dramatically. Both plants and wild herds have rebounded with surprising speed; even during the dry season, water is plentiful enough to sustain fish and water lilies. And if that's not enough, livestock has increased by 400%. Ranchers come from around the world to marvel at the turnaround and to seek counsel from Savory, who is widely credited for the dramatic recovery.

Savory came by his expertise in a circuitous way. He spent the 1950s working as a research biologist and game ranger for the British Colonial Service. At the time, the British government was culling thousands of wild herds in an effort to create more land for farming. "Zimbabwe

Infographic **12.6** | **HOW IT WORKS: GRAZING AND IMPACT ON GRASSES**

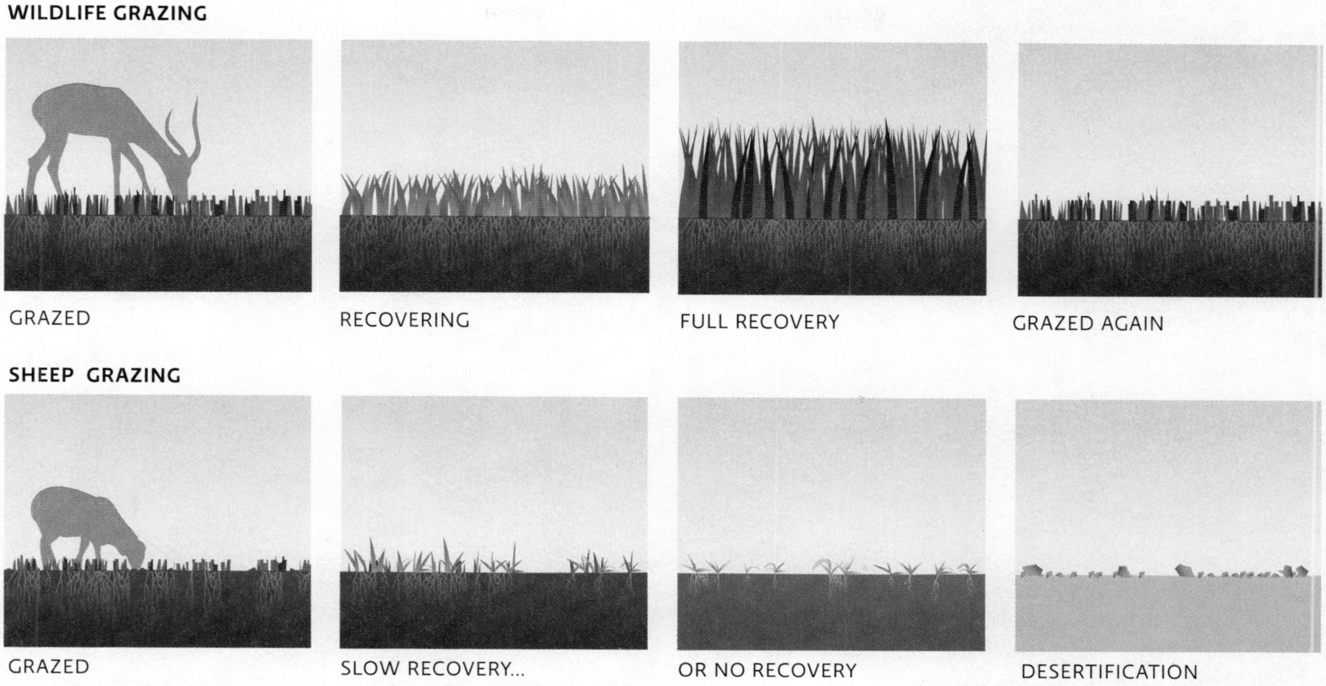
WILDLIFE GRAZING
GRAZED
RECOVERING
FULL RECOVERY
GRAZED AGAIN
SHEEP GRAZING
GRAZED
SLOW RECOVERY...
OR NO RECOVERY
DESERTIFICATION

↑ Savory noticed that wild herbivores grazed intensively and then moved on, allowing the grasses to recover. Livestock allowed to stay on a pasture too long will damage the plants, which slows or prevents recovery.

was infamous for the veterinary policy of shooting out all game over large areas," Savory says. "The unofficial slogan was, 'you cannot ranch in a zoo.'" That policy bred tension between ecologists like Savory who wanted to preserve the splendour of Africa's natural environment, and ranchers and government officials who wanted to farm right over it. Like many of his colleagues, Savory developed a healthy disdain for the cattle that were slowly replacing his beloved zebra, elephants, and antelope, and a firm belief that the land he loved was being destroyed by cattle ranchers who crowded too many animals onto too little land.

As he patrolled the vast terrain, Savory noticed that lands grazed by wild herds were healthier than those managed by cattle ranchers; plants were more abundant and diverse, rivers were cleaner and better stocked. "The wild herds were doing basically the same thing as the domestic herds," he says. "They were eating the grass. But they were having the exact opposite effect." He noticed that as they tried to protect themselves from predators and avoid feeding on their own feces, wild herds grazed in tightly bunched packs, and moved quickly from one patch of land to the next. They would stay just long enough to fertilize the ground with their waste and agitate the soil without compacting it, and they would not return until the dung had been absorbed and the land was clean again. This meant that grasses were heavily grazed over a short period, but then left for a long time to recover. The result: animals fed and grasses regenerated in an endless and mutually beneficial cycle. [INFOGRAPHIC 12.6]

Once upon a time, **pastoralists**—individuals who herd and care for livestock as a way of life—mimicked the processes laid out by Mother Nature: grazing their stock in tight herds, moving them constantly across vast swaths of rangeland. But in the 19th and 20th centuries, ranchers began partitioning their grasslands into distinctly fenced-in pastures and dividing their livestock so as to control them more easily. This made ranching easier, to be sure. But as time wore on, farmers found that their lands were being overrun by shrubs and weeds. The cattle were selectively eating only the sweetest tasting grasses and leaving all the less palatable varieties to flourish. Season by season, the plant species composition shifted,

pastoralists Individuals who herd and care for livestock as a way of life.

↑ A pastoralist, a Rabari tribal shepherd herding the sheep home in Gujarat, India.

and as it did, the land became less productive. Less sweet grass meant less cattle feed, which in turn meant thinner cattle, less food and, ultimately, thinner profits. And as scientists like Savory began to notice this, a simple idea took root: the lands were being destroyed because too many animals were feeding off of them. They decided that trimming the populations of both domestic and wild herds would be the key to salvation.

But, as scientists—including Savory himself—soon realized, undergrazing presented its own set of problems. "Where we culled too many animals, there was nothing to eat the grasses," Savory says. "So they would die standing upright, and the detritus would prevent the sun from reaching the growth buds." This also meant that instead of being returned to the soil through animals' digestive tracts, nutrients would instead be processed by soil microbes—a much slower and thus less efficient process. And because there were fewer of them, animals could be more selective about which plants to eat; left untouched, unpalatable weeds and plants began to take over the pastures. [INFOGRAPHIC 12.7]

North American ranchers were experiencing similar problems, and by the 1970s the farming industry had caught on. Land-grant universities in Texas and Arizona designed machines like the Dixon Imprinter that simulated the physical effects of large grazing herds—breaking soil crusts and laying down plant litter over vast swaths of pasture. Of course, those machines could not cycle nutrients the way animal digestive systems could, and so the colossal machine solved some problems (detritus no longer blocked the sun from growth shoots), but not others (soil quality still dropped because nutrients weren't cycled as effectively).

Around the same time back in Africa, after two decades of surveying the land and observing both ranching and wild grazing in action, Savory was certain that ranching practices were contributing heavily to land degradation. He was equally certain that humans could reverse the process by managing the land differently—but he didn't know exactly how. To figure it out, he would have to design and implement several different methods of ranching so that he could compare them to one another.

Infographic **12.7** | **INCREASING PROPORTION OF UNPALATABLE SPECIES**

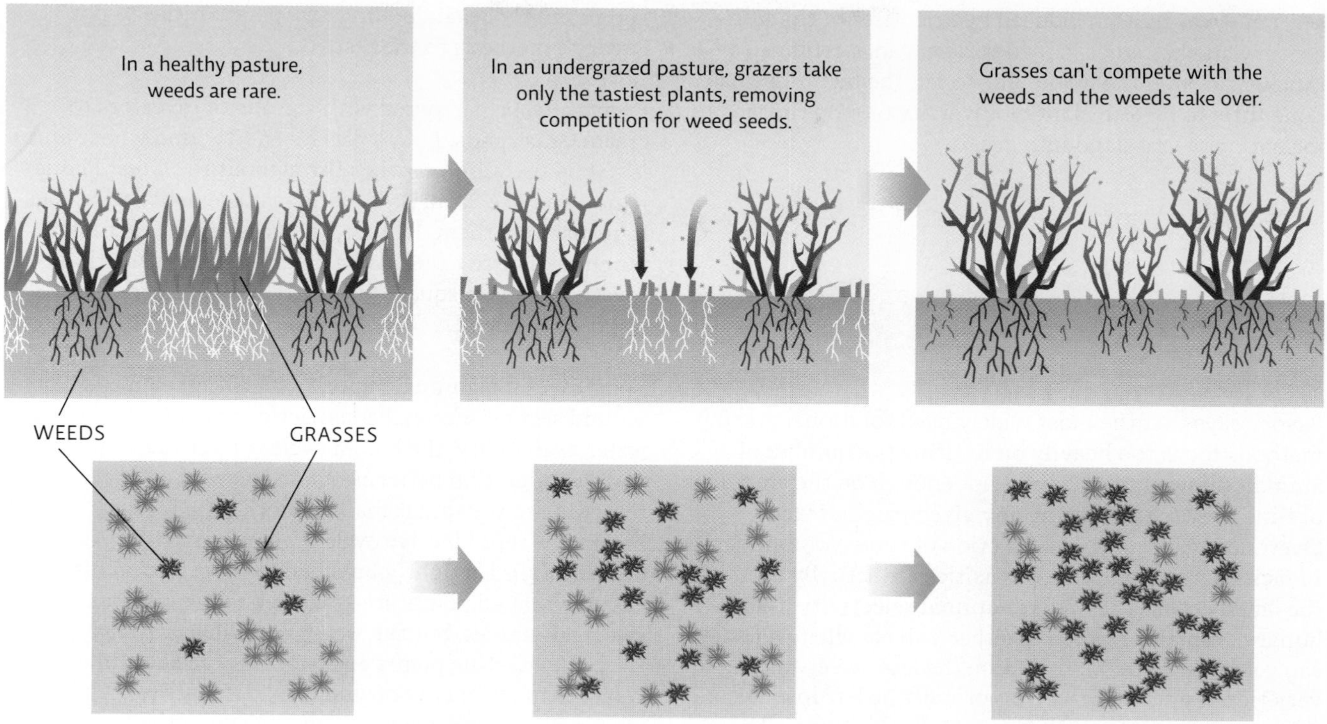
In a healthy pasture, weeds are rare.
In an undergrazed pasture, grazers take only the tastiest plants, removing competition for weed seeds.
Grasses can't compete with the weeds and the weeds take over.
WEEDS
GRASSES

↑ When lands are undergrazed, animals will only eat the tastiest plants, ignoring the less palatable weeds. This gives the weeds a competitive edge, allowing them to quickly take over the area.

And to do that, he would need to persuade the ranchers of Zimbabwe to let him experiment on their lands.

At first his imploring fell on deaf ears. Everyone agreed the lands were ailing, but government officials blamed the lack of rain, combined with overgrazing in some areas. Ranchers insisted they were already taking great care not to graze too many animals on too small a patch of land. What more could they do? "I went to the government scientists, and I tried talking to the ranchers themselves," he says. "Nobody wanted to hear that they weren't doing things in exactly the best way."

Then one night, a couple of ranchers paid him a visit. They'd been doing everything the government advised, they said, and both their ranches, which sat side by side on the open plains of Bulawayo, Zimbabwe, had enjoyed ample rainfall that season. But their grasses were not rebounding and their businesses were suffering. Could he help them? "I told them I didn't have any answers yet, just a sense that we were causing the land degradation ourselves, and that we could fix it by changing the way we ranched," he says. "We agreed to be the blind leading the blind."

Savory's plan was based on biomimicry—grazing livestock the same way wild herbivores grazed. Borrowing some lessons from his time in the British Army, he devised a technique that called for controlling herds' movements with military precision. He used electric fencing to divide the land into small paddocks, and then pulled all the livestock together into just one such paddock. Then he allowed the animals a day or two to eat everything they possibly could before releasing them into the next paddock. By that time, the first paddock had been churned into lumps of soil, dung, and freshly exposed growth buds. And so the herd would move, from paddock to paddock, devouring and fertilizing each patch of land as it went. By the time they had moved through the last paddock, they would be ready for market. Each paddock would then have an entire season—roughly 180 days, depending on how the rains fell that year—to recover.

The concept was not entirely new. **Rotational grazing**—where animals are allowed to graze on a small section of pasture for a few days before being moved to another section—was first introduced by scientists of the 18th century and has since gained widespread acceptance by ranchers around the world. But so far, the technique had done little to restore plant biodiversity or stop the transformation of grassland into desert.

> Savory's plan was based on biomimicry—grazing livestock the same way wild herbivores grazed.

Savory says that the most widely used rotational grazing methods focus too heavily on limiting the number of animals allowed to graze and not enough on the amount of time they spend grazing any given patch of land. Overgrazing, he says, is a function of time, not numbers. In fact, he says, high stock densities are actually better for the land because they reduce animal selectivity (that is, hungry animals crowded together will eat whatever they can get their teeth on, including the less sweet-tasting varieties of grass that would normally be left to overtake the pasture) and thus help preserve biodiversity. "The results we are seeing at our centre today are because of a 400% increase in the number of animals we graze, not despite it," he says. "I have to make that point to almost every scientist and rancher that passes through."

In 1960s Zimbabwe, degraded land gave way to starving people and civil unrest. Amid this turmoil, Savory and his wife were exiled, along with hundreds of others whom the government decided were activists or political dissidents. They took their work to the United States, where they found a surprisingly similar state of land-use affairs. Government agencies and land-grant colleges around the country scoffed at the idea of increasing the number of livestock as a means to combating desertification. But as individual ranchers caught wind of Savory's work, they began seeking him out.

Counteracting overgrazing requires careful planning.

One of these ranchers was Jim Howell. Howell had grown up on his family's ranch in Colorado, and was studying to be a veterinarian when he came across Savory's work in the mid-1990s. "I saw quickly that the way my family had managed the land, going back to when they first bought it in 1937, was hurting it," he says. "We were losing biodiversity, losing soil, and every year the land was yielding up less profits than it had the year before." Convinced that being a good rancher meant being a good ecologist, Howell switched majors, became an ecologist, and promptly began looking for opportunities to put Savory's methods to the test.

Eventually, he partnered with a group of investors to create Grasslands, LLC, a private equity fund whose aim is to buy up failing ranches throughout the Great Plains region and use Savory's holistic management techniques to resuscitate them. In 2010, the group made its first purchase: Horse Creek and a neighbouring ranch that together hold 60 square kilometres of degraded South Dakota rangeland.

With careful planning, Howell's ranching team can work around several seasonal constrictions at once, including water availability, the bloom cycles of poisonous plants, and the migration patterns of various wild animals. Not returning to the same small paddock for a full year also breaks the reproductive cycle of many parasites. And keeping animals tightly bunched for a set period of time ensures that all plants are grazed equally—not just the sweetest grasses, but the weeds, too. This actually gives the better-tasting plants an advantage: because they don't waste as much energy on chemical defences (which is what makes some plants so unpalatable in the first place), they regrow faster and thus colonize more area than the foul-tasting plants.

On top of that, planned grazing helps keep plant biomass levels within an ideal range, where plants capture a maximum amount of sunlight and thus grow exceedingly quickly. Below this range, plants have less leaf area, capture less solar energy, and grow much more slowly as a result; above it, excessive leaf growth blocks newer shoots from the sun, and old leaves die as fast as new ones are made. Grasses kept between these two extremes regenerate much more quickly and are thus ideal for grazing. [INFOGRAPHIC 12.8]

Howell says the end result of such careful planning is that land productivity is maximized and ranching becomes profitable once again. So far, the numbers support his claim. According to one assessment of Alberta's grazing practices, by Warren Elofson of the University of Calgary, rotational grazing decreases land needed per cow by two thirds. "You're raising twice as many animals in the same amount of space," explains Howell. "It's like getting a second ranch for free."

In fact, the financial benefits have proven so substantial, Howell and his partners are betting that the economy will rebound in tandem with the land. "Ultimately, we

Infographic **12.8** | **PLANNED GRAZING**

↓ Livestock are allowed to graze intensively on a plot and then are moved to the next plot. In this example, each plot is grazed for 1 month. By the time the livestock return to a given plot, it will have recovered.

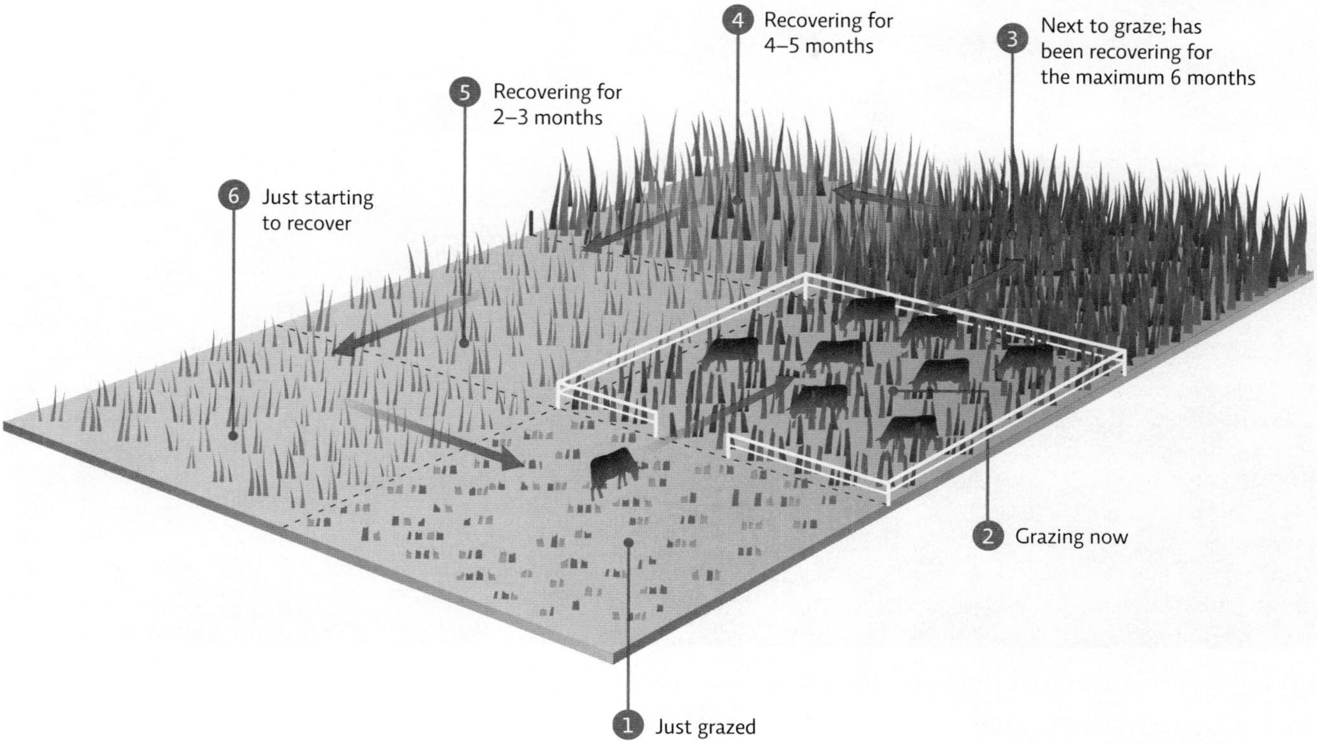
4 Recovering for 4–5 months
3 Next to graze; has been recovering for the maximum 6 months
5 Recovering for 2–3 months
6 Just starting to recover
2 Grazing now
1 Just grazed

want to sell these places back to the surrounding communities," says Howell. "At the very least, we hope to inject some energy into the region—provide jobs, attract young people."

Planned grazing is tricky work.

But planned grazing is tricky work, especially for ranchers who are set in their ways and are already anxious about the bottom line. "It's very hard to make a ranch profitable in the first place. So significant change can be very frightening," says Howell. In more traditional grazing methods, animals are left in the same pasture for long periods, sometimes for an entire growing season, which can run for as long as 180 days. With *planned grazing*—as Savory's method is called—animals are moved much more frequently. "It's really easy to screw things up when you switch from regular grazing to planned grazing," Howell, adds. "It's easy enough if you have 3 pastures and a 180-day growing season—each pasture gets about 60 days of grazing and 120 days of recovery," Howell says. "But say you have 45 pastures. That'd give you an average grazing period of only 4 days. If you're off by even a day, your animals can suffer considerably." Leave them on a given pasture too long, and they will run out of food. Do that too often, and the animals' ability to gain weight, lactate, come into heat, and reproduce will all be compromised. And that can take a huge economic toll.

Not everyone agrees the risk is worth it. Most ranchers acknowledge that **sustainable grazing**—grazing that maintains the health of the ecosystem and allows the grasses to recover before the animals return—is essential to protecting grasslands the world over. But some argue that existing methods of continuous grazing work just as well, if not better, when done properly. "There is plenty of evidence showing comparable outcomes for biodiversity, soil erosion, etc.," says David Briske, a rangeland ecologist at Texas A&M University. "Effective

rotational grazing Moving animals from one pasture to the next in a predetermined sequence to prevent overgrazing.

sustainable grazing Practices that allow animals to graze in a way that keeps pastures healthy and allows grasses to recover.

↑ Desertification is overtaking farmland at the U.S./Mexico border.

management of grazed ecosystems is sufficiently dynamic and complex that it should not be envisioned to have any one correct solution."

Grasslands are not just here for human use and anything we can do to restore or protect them will benefit us and other species as well. Savory is showing that improved grazing techniques can help protect and even restore some degraded grasslands. Another land-management technique that also reduces soil erosion and protects grasslands from degradation is the planting of *shelterbelts*—a stand of trees that blocks the wind and thus decreases wind erosion. Shelterbelt programs helped Canada and the United States recover from the Dust Bowl and are being used today in areas facing desertification, such as Inner Mongolia and China (where the shelterbelt is referred to as the "Green Wall of China"). Other approaches, such as conservation easements, reserves, and protected parks also help protect our grasslands by limiting land uses and making it easier to keep grasslands in some level of protected status. [INFOGRAPHIC 12.9]

Nevertheless, a growing number of Canadian and U.S. ranchers are following Howell's and Savory's lead. Success in Zimbabwe has been followed by successes in California, Texas, Alberta, and Ontario. "Change is tough," says Joe Morris, a rancher in central California who has been using Savory's methods with great success. "Ranching is a culture, steeped in tradition. But our lands are hurting and our communities are dying, and we know we've got to do something to fix that."

This past spring, Howell and his partners ran 2300 yearling cattle out of Horse Creek Ranch—a nearly 100% increase from years past and a number they hope to double next season. "It's going to be a long road," says Howell. "But we've already made so much progress, in just one season." Meanwhile, back on the range, tiny buds of grass have begun to poke up everywhere.◉

Select references in this chapter:
Bailey, D.W., *et al.* 1996. *Journal of Range Management,* 49:386–400.
Briske, D. D., *et al.* 2011. *Rangeland Ecology and Management,* 64:325–334.
Briske, D. D. *et al.* 2008. *Rangeland Ecology and Management,* 61:3–17.
Derner, J.D., & Hart, R.H. 2007. *Rangeland Ecology and Management,* 60:270–276.
Diaz-Solis, H., *et al.* 2003. *Agricultural Systems,* 76:655–680.
Elofson, W. 2012. *Agricultural History,* 86(4): 144-168.
Heitschmidt, R.K., *et al.* 2005. *Rangeland Ecology and Management,* 58:11–19.
Savory, A., & Parsons, S. 1980. *Rangelands,* 2:234–237.

Infographic **12.9** | **PROTECTING OUR GRASSLANDS**

MECHANISM	DESCRIPTION	BENEFITS
CONSERVATION EASEMENTS	Legal agreement that restricts how a landowner can develop and use land; allowable uses are typically agriculture, forestry, recreation, or natural preserves.	Land will be protected in perpetuity; reduced taxes to landowner
CONSERVATION RESERVE PROGRAMS	Government program that assists farmers and ranchers in land management; farmers are paid to allow sensitive areas to be taken out of production and planted with grasses or trees.	Reduces soil erosion and water pollution, and increases wildlife habitat
PARK STATUS	Designation as a national or provincial park offers protection from overuse.	Areas can be managed in ways that protect the habitat and allow recreational use, a benefit to both nature and people.
SUSTAINABLE GRAZING	Grazing only the number of animals that the area can support; rotating animals from pasture to pasture is often used to maximize the number of animals that can be grazed.	Allows grassland to recover after grazing and to benefit from the rejuvenating effects of intensive short-term grazing
SHELTERBELTS	Trees are planted at the edge of a farm or grassland to reduce wind erosion.	Prevents or slows the encroachment of nearby deserts; also protects the crop or grasses from wind damage

BRING IT HOME

PERSONAL CHOICES THAT HELP

Because of overgrazing and rampant soil erosion, the grasslands have lost a larger proportion of their habitat than the tropical rainforests. But most people have never worried about preserving the grasslands, let alone are even aware of how little remains (only about 1–2%). Grassland habitats provide many ecosystem services—they capture CO_2 and are home to an immense range of biodiversity. They can be used for both economic growth as well as recreation, but only if they are managed properly.

Individual Steps

→ Visit a remnant or restored prairie. Such prairie preserves are scattered throughout the heart of North America, ranging from Alberta, Saskatchewan, and Manitoba in the north, down through the Great Plains to southern Texas and Mexico, and from the Rocky Mountains eastward to western Indiana. There are more than 20 national grasslands and several dozen more provincial, state, and local prairie preserves.

→ Explore the difference between turf grass and native grasses. Find out what grasses are native to your area and plant some in your yard.

→ Purchase grass-fed beef or free-range bison meat as a way to support sustainable use of grassland ecosystems.

Group Action

→ If there are prairie restoration efforts underway in your community, participate in these efforts, which include seeding, brush cutting, removal of invasive species, and seed collecting.

Policy Change

→ Research national legislation designed to protect and restore native habitats like grasslands and write a letter to your provincial representative asking him or her to vote for the legislation.

→ Organize a fundraiser and donate the proceeds to research that examines the pros and cons of using grasslands for either biomass production or wind farms.

UNDERSTANDING THE ISSUE

CHECK YOUR UNDERSTANDING

1. Which of the following statements describes a grassland biome?
  a. Grasslands are biomes with abiotic conditions like fires that support the growth of grass and herbaceous plants, but not trees.
  b. Grassland biomes are found exclusively in semi-arid areas with very low levels of rainfall, most of which falls in the winter season.
  c. Grasslands have thick soils that develop due to the long, warm growing season characteristic of their tropical environment.
  d. Grasslands are only found in temperate regions that experience hot summers and cold winters, conditions that make it impossible to support tree growth.

2. Temperate grasslands:
  a. occur in places that have both rainy and dry seasons, but are warm year-round.
  b. have a seasonal climate with cold winters and hot summers.
  c. are cold most of the year and are characterized by very short growing seasons.
  d. are high-altitude grasslands that get all their precipitation in the form of snow.

3. Which of the following is NOT a cause of desertification of rangeland?
  a. Climate change, especially shifting rainfall patterns that will result in less precipitation
  b. Conversion of rangeland to urban and suburban development
  c. Soil erosion due to the loss of vegetative cover and the compaction of the soil column
  d. Overgrazing by domesticated livestock

4. Pastoralists:
  a. are herders who graze their livestock in tight herds, moving them constantly across vast swaths of rangeland.
  b. are landowners who plant fast-growing, nutrient-dense grasses to maximize livestock production in small pastures.
  c. partition their land into separate fenced-in pastures and divide their livestock between these fenced areas so as to control them more easily.
  d. are livestock owners who are paid to remove sensitive grassland areas out of production.

5. A program that pays farmers to take sensitive areas out of production and plant them with grasses or trees instead is a:
  a. conservation easement agreement.
  b. conservation reserve program.
  c. shelterbelt planting program.
  d. rotational grazing system.

6. Shelterbelts are:
  a. trees planted at the edge of a farm or grassland that protect crops or grasses from wind and prevent encroachment of nearby deserts.
  b. planned grazing systems where animals are allowed to graze on a small section of pasture for a few days before being moved to another section so as to prevent overgrazing.
  c. legal agreements, including tax breaks, that restrict how a landowner can develop and use land so that the land will be protected in perpetuity.
  d. areas managed in ways that protect the habitat but allow recreational use so as to benefit both nature and people.

WORK WITH IDEAS

1. What is a grassland? How and why do grasslands differ worldwide?
2. Describe the major threats to grasslands and explain the factors that underlie these threats.
3. How does grazing by native herbivores impact grassland ecosystems?
4. What are the differences between pastoralist grazers and ranchers? Which is better for the grassland and how do we know this?
5. What is overgrazing and how can it damage a grassland? Why is undergrazing just as problematic as overgrazing?
6. Planned grazing can make ranching activities mutually beneficial to both the livestock and the grassland ecosystem, yet it is a risky venture for ranchers in Canada and the United States. Why is this? Why are some farmers hesitant to change? What trade-offs do they face?

ANALYZING THE SCIENCE

The data in the table on the following page comes from a research document prepared by the Temperate Grasslands Conservation Initiative of the IUCN World Commission on Protected Areas. The document was prepared to build a case for the conservation and protection of temperate grassland ecosystems by evaluating their total economic value to human well-being.

INTERPRETATION

1. What does this figure show? Which of the *grassland* ecosystems is the most converted? Which one is the most protected?
2. The Conservation Risk Index (CRI) is a ratio of habitat converted to habitat protected for each biome. It is used to assess the level of threat to a particular biome.
  a. Calculate the CRI for each of the 13 biomes. Here is one example: CRI of deserts and xeric shrublands is 6.8 / 9.9 = 0.7.
  b. A CRI greater than 8 suggests a critically endangered biome (in other words 8 times as much land is being converted as is being protected). A CRI of 2 or less is considered a biome with a low risk of endangerment. According to the CRIs you calculated in part a, what is the status of each of the grasslands?

ADVANCE YOUR THINKING

3. What patterns do you see in the CRIs you calculated above? Describe the patterns and include data to support your explanation.
4. Which ecoregions are under the greatest threat?

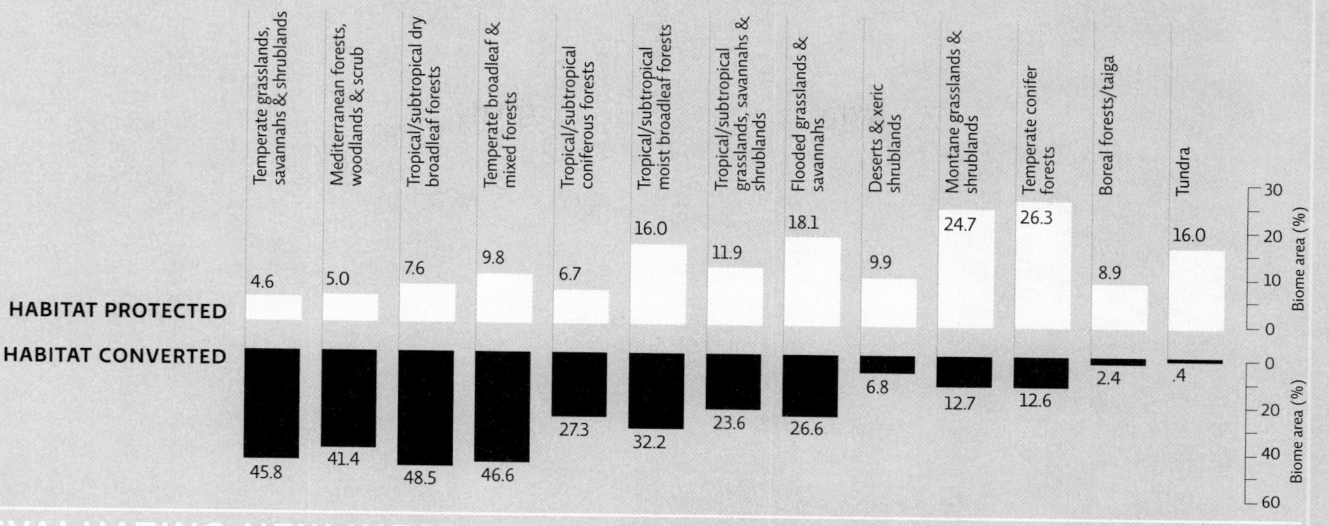

Temperate grasslands, savannahs & shrublands
Mediterranean forests, woodlands & scrub
Tropical/subtropical dry broadleaf forests
Temperate broadleaf & mixed forests
Tropical/subtropical coniferous forests
Tropical/subtropical moist broadleaf forests
Tropical/subtropical grasslands, savannahs & shrublands
Flooded grasslands & savannahs
Deserts & xeric shrublands
Montane grasslands & shrublands
Temperate conifer forests
Boreal forests/taiga
Tundra
HABITAT PROTECTED
HABITAT CONVERTED
4.6
5.0
7.6
9.8
6.7
16.0
11.9
18.1
9.9
24.7
26.3
8.9
16.0
45.8
41.4
48.5
46.6
27.3
32.2
23.6
26.6
6.8
12.7
12.6
2.4
.4
Biome area (%)
30
20
10
0
Biome area (%)
0
20
40
60

EVALUATING NEW INFORMATION

About 90% of grassland degradation is a result of intensive agriculture, including grazing practices and the conversion of croplands to grow livestock feed.

How can a conscientious consumer help protect the grasslands ecosystem? What choices can we make that are sensitive to the welfare of the ecosystem as well as the animals that we eat?

Explore the Eat Wild website (www.eatwild.com).

Evaluate the website and work with the information to answer the folllowing questions:

1. Is this a reliable information source? Does it have a clear and transparent agenda?
   a. Who runs this website? Does this person's credentials make him or her reliable/unreliable? Explain.
   b. Does the website provide supporting evidence for its claims about grass-fed versus feedlot animal products? Does it give sources for its evidence?
2. Select the link "A direct link to local farms."
   a. What information does that link provide? How can you as a consumer use this information?
   b. Select the "Criteria" link. What are the criteria for listing a farm on the website? Do you think these criteria are sufficient and reasonable? Which criteria are most important to you as a consumer? Explain your responses.
3. Examine two other websites that certify animal products: Animal Welfare Approved (www.animalwelfareapproved.org) and American Grassfed (www.americangrassfed.org).
   a. How similar/different are the criteria each website uses for certifying animal products? Which website uses criteria that are most important to you as a consumer? Explain your response.
4. How might certifications for animal products help grasslands?

MAKING CONNECTIONS

THE BUFFALO COMMONS AND THE FUTURE OF THE GREAT PLAINS

Background: Allan Savory's planned grazing methods are one approach to reverse desertification in the Great Plains. But according to some, not only is farming and ranching ecologically unsustainable, it is economically impractical as well. Several researchers have proposed that cattle and grazing be replaced by wildlife refuges, where bison (buffalo) could once again be allowed to roam freely. In 1987, Rutgers scientists Frank and Deborah Popper suggested converting about 360 square kilometres of the Great Plains, which they argued would most likely become depopulated anyway, into wildlife preserves called the Buffalo Commons.

Case: You have been assigned the task of determining the future of the Great Plains. Select between the following two options:

1. The federal government should use financial incentives to attract people to the Great Plains and should also offer subsidies to encourage sustainable grazing practices.
2. As the Great Plains become depopulated, the unoccupied land should be purchased and transformed into ecological reserves supporting activities such as ecotourism, wind farms, and free-range bison ranching.

Research these two alternatives and write a report recommending one of the two options above. In your report include the following:

a. An analysis of the pros and cons of each proposal including: a discussion of the consequences of each choice both for local human communities and for the grassland ecosystem, and a reflection on the values underlying each proposal.
b. Based on the information at hand, what is the best option for the future of the Great Plains and who should be involved in this decision? Provide justifications for your proposal.

Scientist from the Aquarius underwater ocean laboratory taking coral samples.

SCIENCE UNDER THE SEA

Aquanauts explore an ecosystem on the brink

CORE MESSAGE

Marine ecosystems contain a huge diversity of life, though we know far less about ocean ecosystems than those on land. Many ocean ecosystems are suffering as a result of pollution, overfishing, misuse of the ecosystem's resources, and global climate change. Of particular concern is the change in ocean chemistry caused by the release of CO_2 from fossil fuel use. This has the potential to alter ocean ecosystems drastically, and some effects are already being seen. Choices we make today will influence the future of ocean ecosystems and the species that inhabit them.

GUIDING QUESTIONS

After reading this chapter, you should be able to answer the following questions:

→ What is contributing to ocean acidification and why is this a problem?

→ What environmental conditions determine the location and makeup of marine ecosystems?

→ Where are coral reefs found and what is the community makeup of these complex ecosystems?

→ What threats do coral reefs and other ocean communities face?

→ How can we reduce the threats to coral reefs and other ocean ecosystems?

↑ Two goliath groupers swim past a porthole on the Aquarius Reef Base.

→ A researcher swims past the Aquarius Reef Base, the only undersea research station in the world. Scientists can spend up to 10 days on the station before they must return to the surface (to avoid decompression illness).

2

It was four o'clock in the morning and Marc Slattery could not sleep. He and his six crewmates had just settled into the Aquarius Reef Base—an artificial, undersea research station located on Conch Reef, 16 kilometres off shore from, and 15 metres below, Key Largo, Florida. Tomorrow they would begin an 8-day stretch of underwater experiments and data collection—all aimed at understanding how the physiology of various coral reef species was changing in response to changes in ocean conditions. Slattery, a scientist at the University of Mississippi and the team's principal investigator, was anxious to get started.

But it was not this eagerness keeping him awake; it was the two goliath groupers—each about 2.5 metres in length—gliding persistently past the viewport near his bunk. Sticking close to one another, they would swish up to the small circular window, stare at him briefly, and then swish away to complete another lap around the structure. Slattery was captivated. As a young kid, he had spent countless hours watching a collection of tropical fish glide around his home aquarium. Now, for the first time, he knew how the fish must have felt. "They seem to swim back and forth between the bunkroom and galley viewports looking for those fascinating creatures that walk on two legs, breathe air and eat constantly," he wrote in the expedition log. "They accept our odd behavior and even seem to enjoy our presence. Maybe they know we are here to help."

Whether they knew it or not, the groupers, and all of their underwater neighbours, were being threatened by an avalanche of forces—global climate change chief among them. Around the world, temperatures were rising, glaciers melting, and the ocean's chemistry changing in peculiar and disturbing ways. Scientists like Slattery had been on a quest to understand these changes from their land-based labs. Now Slattery wanted to dig for clues down below.

WHERE IS KEY LARGO, FLORIDA?

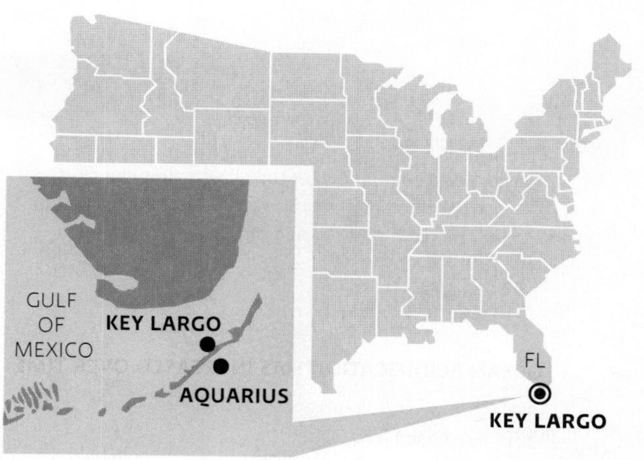
GULF OF MEXICO
KEY LARGO
AQUARIUS
FL
KEY LARGO

Acidification threatens life in the world's oceans.

From the beginning of the Industrial Revolution to the present day, we humans have burned enough fossil fuels and clear-cut enough forests to release more than 500 billion metric tons of CO_2 into Earth's atmosphere, making it higher than at any point in the past 800 000 years. Even worse than these unprecedented levels is how fast they have risen—too fast, experts say, for many organisms to adapt. Much has been made of what heat-trapping molecules do to terrestrial ecosystems. But, as scientists are now learning, their effect on aquatic environments is just as profound.

Ocean and atmosphere come into direct contact over 75% of Earth's surface, and they are constantly exchanging gases over that interface; anything emitted into one eventually ends up in the other—including CO_2. Winds, waves, and currents quickly mix CO_2 into the top one hundred metres or so of water, and as years pass, currents pull it ever deeper into the ocean. Between the early 1990s and mid-2000s, scientists around the world collected and analyzed nearly 80 000 water samples from a range of ocean environments. By their estimates, some 30% of all the CO_2 released by humans in the last two centuries has been absorbed by the world's oceans. "For terrestrial ecosystems, it's a good thing," says Slattery. "Because it

Infographic 13.1 | pH AND OCEAN ACIDIFICATION

↓ The pH scale is a measure of how acidic or basic a solution is. It is a measurement that compares the proportion of hydrogen (H^+) or hydroxide (OH^-) ions in a solution (ions are charged atoms or molecules). Water (H_2O) can dissociate into hydroxide ions (OH^-) and hydrogen ions (H^+). In pure water, there is always one H^+ for every OH^-. These solutions are neutral, with a pH of 7.0. Acids have a pH lower than 7.0 and have extra H^+ ions in the solution. Bases (alkaline solutions) have a pH higher than 7.0 and have extra OH^- ions in the solution.

pH SCALE

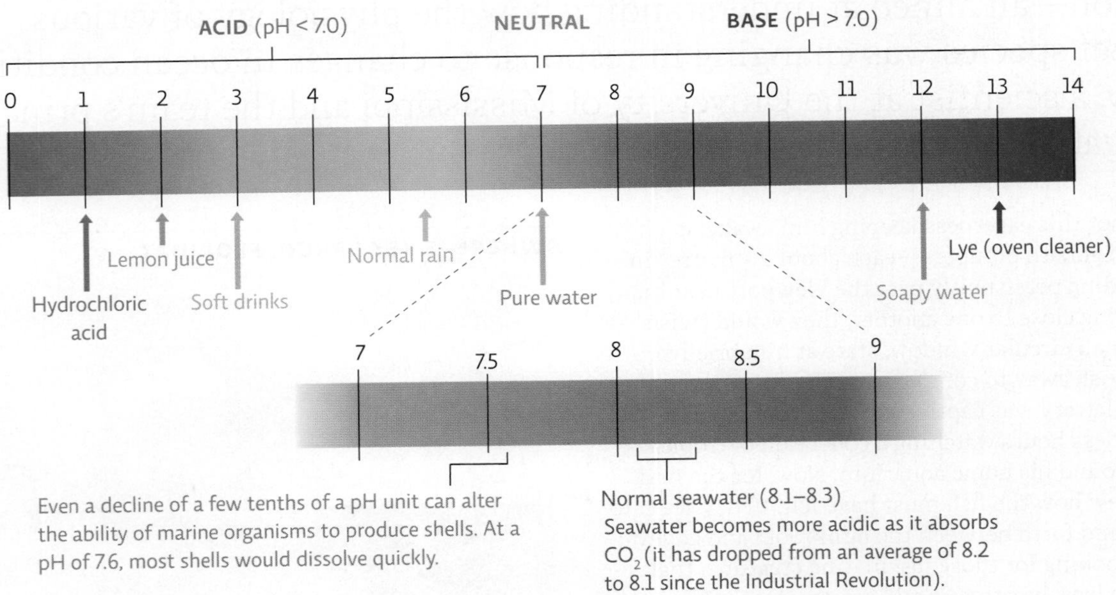
ACID (pH < 7.0)
NEUTRAL
BASE (pH > 7.0)
0 1 2 3 4 5 6 7 8 9 10 11 12 13 14
Hydrochloric acid
Lemon juice
Soft drinks
Normal rain
Pure water
Soapy water
Lye (oven cleaner)
7 7.5 8 8.5 9
Even a decline of a few tenths of a pH unit can alter the ability of marine organisms to produce shells. At a pH of 7.6, most shells would dissolve quickly.
Normal seawater (8.1–8.3)
Seawater becomes more acidic as it absorbs CO2 (it has dropped from an average of 8.2 to 8.1 since the Industrial Revolution).

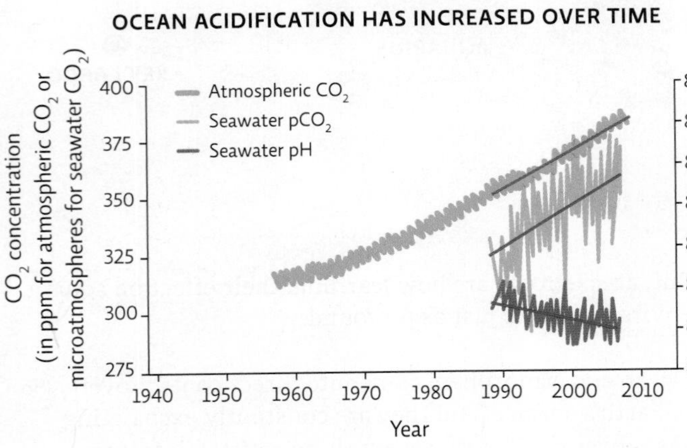
OCEAN ACIDIFICATION HAS INCREASED OVER TIME
Atmospheric CO2
Seawater pCO2
Seawater pH
CO2 concentration (in ppm for atmospheric CO2 or microatmospheres for seawater CO2)
400 375 350 325 300 275
8.38 8.33 8.28 8.23 8.18 8.13 8.08 8.03
pH
1940 1950 1960 1970 1980 1990 2000 2010
Year

↑ The increase in ocean acidity correlates well with the increases in CO_2 dissolved in seawater and atmospheric CO_2 concentrations.

→ Ocean pH will continue to drop significantly if we do not curtail fossil fuel use and our release of extra CO_2.

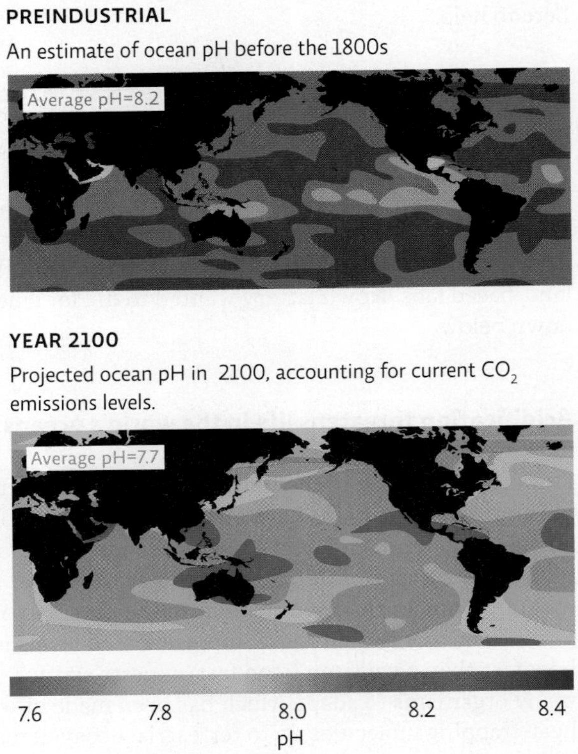
PREINDUSTRIAL
An estimate of ocean pH before the 1800s
Average pH=8.2
YEAR 2100
Projected ocean pH in 2100, accounting for current CO2 emissions levels.
Average pH=7.7
7.6 7.8 8.0 8.2 8.4
pH

↑ A healthy reef area, like this one off the coast of Castello Aragonese near Naples, Italy, is full of life, made up of a variety of coral species and inhabited by a wide array of fish and invertebrates like the red sponges that dot the coral here. A well-camouflaged fish (a tompot blenny) hugs the coral just above the black sea urchin (bottom left).

↑ Just a few hundred metres away from the scene in the photo to the left, CO_2 can be seen bubbling out of volcanic vents in the sea floor, giving us a glimpse of what future ocean communities might look like if oceans continue to acidify. This CO_2 so acidifies the water that only sparse algal mats are found here.

means that much less CO_2 is lingering in the atmosphere. But for oceans, it could be very bad."

Here's how: normal seawater has an average pH of around 8.2, meaning it is slightly basic (alkaline). So far, CO_2 emissions have reduced the ocean's pH by about 0.1. That might not sound like much, but the pH scale is logarithmic, so even small numbers represent large effects. A pH drop of 0.1 corresponds to a 30% increase in ocean water acidity. If present trends continue, by 2100 the ocean's surface waters will be about 150% more acidic than they were in 1800. In 2003, scientists adopted the phrase "ocean **acidification**" to describe this coming catastrophe. [INFOGRAPHIC 13.1]

Scientists expect such a colossal reordering of ocean chemistry to have a huge impact on marine ecosystems. "Ocean acidification is global warming's equally evil twin," says Jane Lubchenco, a marine ecologist and the head of the National Oceanic and Atmospheric Administration.

One big concern is that as pH shifts, the availability of key nutrients like nitrogen and iron will change, plummeting in some areas, soaring in others, and threatening the stability of virtually all marine ecosystems. In just one example, Michael Berman and his colleagues at the University of California have found that the rate of nitrification (a process that produces nitrate, a form of nitrogen that marine organisms need to grow) decreases in tandem with pH; as that happens, smaller species of plankton (which are more tolerant of nitrate declines), gain an advantage over larger ones. Were it pervasive enough, such a change in species composition could alter the food chain and decrease primary production throughout the oceans (see Chapter 8 for more on food chains and webs). Indeed, some researchers think this shift may already be occurring. Overall, plankton biomass may have decreased as much as 40% in the 20th century, they say, especially since 1950.

The consequences of this particular chain of events are, for now, anybody's guess. On one hand, there could be a **positive feedback** effect that amplifies ocean acidification and climate change: less plankton means that less CO_2 is taken in by the organisms, and more is left behind

acidification The lowering of the pH of a solution.
positive feedback Changes caused by an initial event accentuate that original event (i.e., changes brought on by warming lead to even more warming).

to further acidify the water or reenter the atmosphere. On the other hand, there could be a **negative feedback** effect on climate change: less nitrification means less N_2O (nitrous oxide—a potent greenhouse gas) is produced, and thus less is released into the atmosphere.

So far, the most well-documented effect of acidification seems to be on marine calcifiers—ocean organisms that make shells, plates, and exoskeletons from calcium minerals. There are thousands of different types of calcifiers—from snails to corals to plankton—dispersed widely throughout the ocean. Early research suggests that acidification may well threaten all of them. When dissolved in water, CO_2 forms carbonic acid, which not only eats away at existing calcium-based materials, but interferes with the chemical reactions by which new ones are made. [INFOGRAPHIC 13.2]

Scientists have already documented significant effects on pteropods—tiny swimming snails that are important food for whales, birds, and fish (including juvenile salmon, pollack, and other commercially important species) in the Arctic and Antarctic. Experiments show that pteropod shells grow more slowly—and even start to dissolve—in acidified seawater. At predicted ocean pHs of the near future, they will become too compromised to support all the life-forms that feed on them.

If present trends continue, by 2100 the ocean's surface waters will be about 150% more acidic than they were in 1800.

To be sure, all calcifiers play an important part in the grand choreography of ocean life. But the most substantial of them are **coral reefs**, like the one Slattery and his team were studying near Aquarius.

Aquarius is the only facility of its kind: a 73-metric ton, double-lock pressure vessel, just 13 metres long and under 3 metres in diameter. That's big enough to house six people for 8 days, sturdy enough to weather the violent storms that periodically shake the region, and unique enough that for the next week, Slattery and his five fellow aquanauts would be the only people on the planet living at the bottom of the sea.

negative feedback Reduction or reversal of an effect by its own influence on the process giving rise to it (i.e., changes brought on by warming lead to cooling).

coral reef Large underwater structures formed by colonies of tiny animals (coral) that produce a calcium carbonate exoskeleton that over time build up; found in shallow, warm, tropical seas.

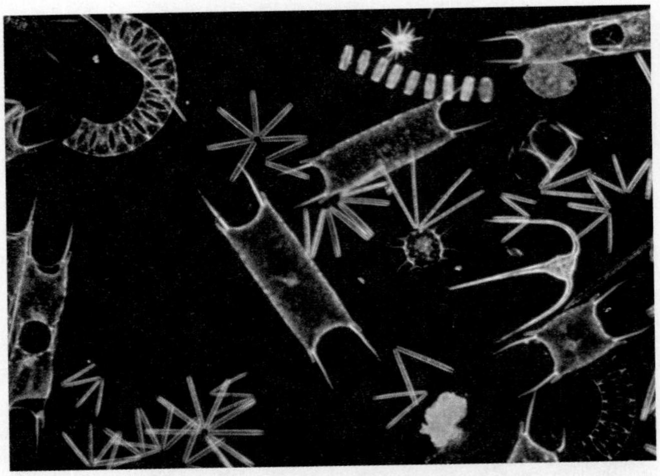

↑ Declines in plankton productivity lead to decreased carbon capture. This means less carbon, stored in the shells of dead plankton, sinks to the ocean bottom. This type of carbon capture is an important mechanism for locking away some of the extra CO_2 that enters the oceans.

To a marine biologist, the advantages of this particular perspective are innumerable. Without having to return to the surface every hour, or dive in shifts to complete a mere day's worth of work, individual team members would be able to take measurements in real time and could observe myriad reef species in all their splendour for hours on end. "Aquarius will enable us to observe and take measurements at much closer intervals than we otherwise might," Slattery says. "It will give us a much fuller picture of what's going on down there."

What he really wanted to get a picture of were the hidden crevices tucked deep within the reef. "These are areas within a coral reef landscape that are naturally acidified," Slattery explains. "They're packed with sponges and other ocean life, which means lots of respiration, and at the same time they have poor water circulation." Because CO_2 is released during cellular respiration, Slattery reasoned that CO_2 concentrations would be high and pH would be low (see Chapter 6 for more on cellular respiration). Slattery and his team hoped the creatures living in such crevices might provide some clues about how the larger reef would respond to an acidified ocean.

Once they found these acidified microhabitats, the team planned to measure individual cellular respiration rates for the creatures living in these crevices—a crucial detail that had yet to be ascertained by anyone. "Knowledge of the *in situ* rates of respiration processes and their impact on local pH has been virtually nonexistent," says Chris Martens, a marine biologist from the University of North Carolina, Chapel Hill, who has also studied acidification from the Aquarius Reef Base. "But it's hugely important." Without these data, researchers can't tell how much

Infographic **13.2** | **ACIDIFICATION ALTERS OCEAN CHEMISTRY AND CALCIFICATION**

↓ Acidic ocean water decreases the ability of aquatic organisms to form shells or exoskeletons, while also dissolving shells and coral that have already been formed.

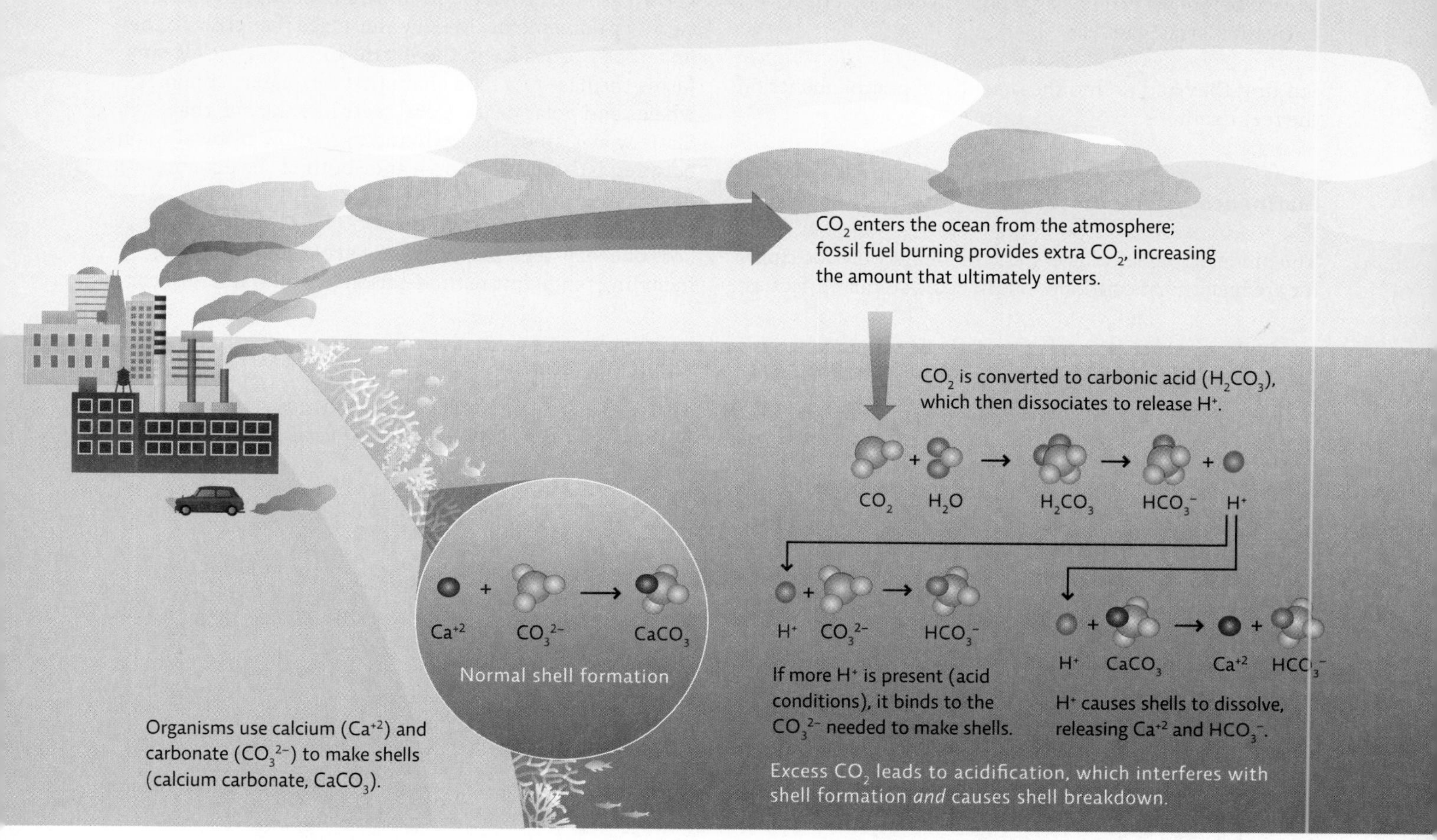

CO_2 enters the ocean from the atmosphere; fossil fuel burning provides extra CO_2, increasing the amount that ultimately enters.
CO_2 is converted to carbonic acid (H_2CO_3), which then dissociates to release H^+.
CO_2 + H_2O → H_2CO_3 → HCO_3^- + H^+
Ca^{+2} + CO_3^{2-} → $CaCO_3$
Normal shell formation
Organisms use calcium (Ca^{+2}) and carbonate (CO_3^{2-}) to make shells (calcium carbonate, $CaCO_3$).
H^+ + CO_3^{2-} → HCO_3^-
If more H^+ is present (acid conditions), it binds to the CO_3^{2-} needed to make shells.
H^+ + $CaCO_3$ → Ca^{+2} + HCO_3^-
H^+ causes shells to dissolve, releasing Ca^{+2} and HCO_3^-.
Excess CO_2 leads to acidification, which interferes with shell formation *and* causes shell breakdown.

↓ An experiment by Orr *et al.* showed the increased rate of dissolution (dissolving) of pteropod shells in water with less calcium carbonate. In this test, shells were exposed to water containing the lower levels of calcium carbonate we expect to see in 2100 if we continue to use fossil fuels at the same rate that we now do. (At lower pHs there will be less calcium carbonate in the ocean).

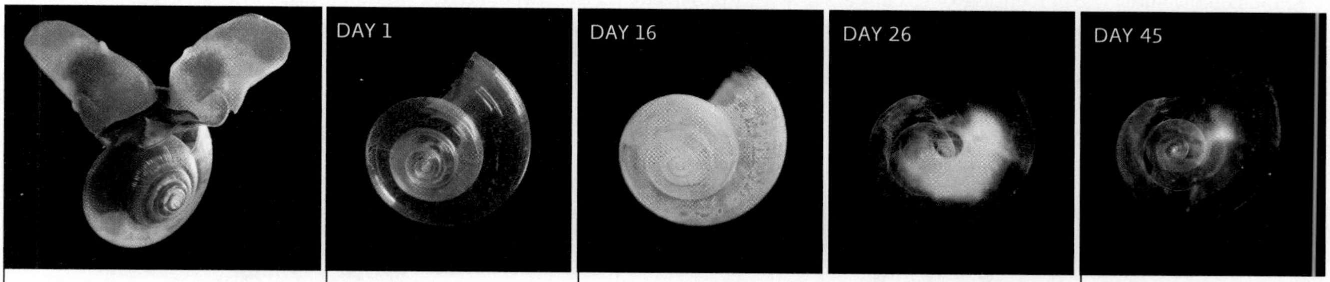

DAY 1
DAY 16
DAY 26
DAY 45

Live healthy pteropod

Normal shell

Shells become pitted (making them appear more opaque) as they begin to decalcify.

After 45 days, major dissolution has occurred.

acidification is coming from CO_2 absorption from the atmosphere and how much is coming from CO_2 respired by the infinite array of ocean life in all the various micro-environments. "We need to know the rough contributions of each before we can possibly hope to develop effective mitigation strategies."

But first they had to find these acidified microhabitats on the reef itself.

Marine ecosystems are diverse.

The hidden crevices Slattery and his team were looking for are just one of countless ocean ecosystems. In fact, the world's oceans are a wonderland of diversity. They cover about 70% of Earth's surface, and house a greater variety of flora and fauna than all land masses combined. In shallow, temperate regions, forests of tall brown seaweed known as kelp provide both food and habitat to a wide variety of organisms. Meanwhile, algae that cling to the underbelly of sea ice in the Arctic and Antarctic Oceans forms the base of a food chain that ultimately supports whales and polar bears. Coral reefs like the one that Slattery was studying are found in shallow tropical waters between latitudes 30° north and south of the equator. [INFOGRAPHIC 13.3]

Lots of ecosystems means lots of ecosystem services, including temperature moderation (ocean water absorbs

Infographic **13.3** | **CORAL REEF DISTRIBUTION, THREATS, AND DESTRUCTION**

↓ Most reef-building corals are found in warm, tropical and subtropical waters between 30° N and 30° S latitudes, shown here as red dots. Coral reefs around the world are threatened by a variety of forces, and destruction of reefs is on the rise.

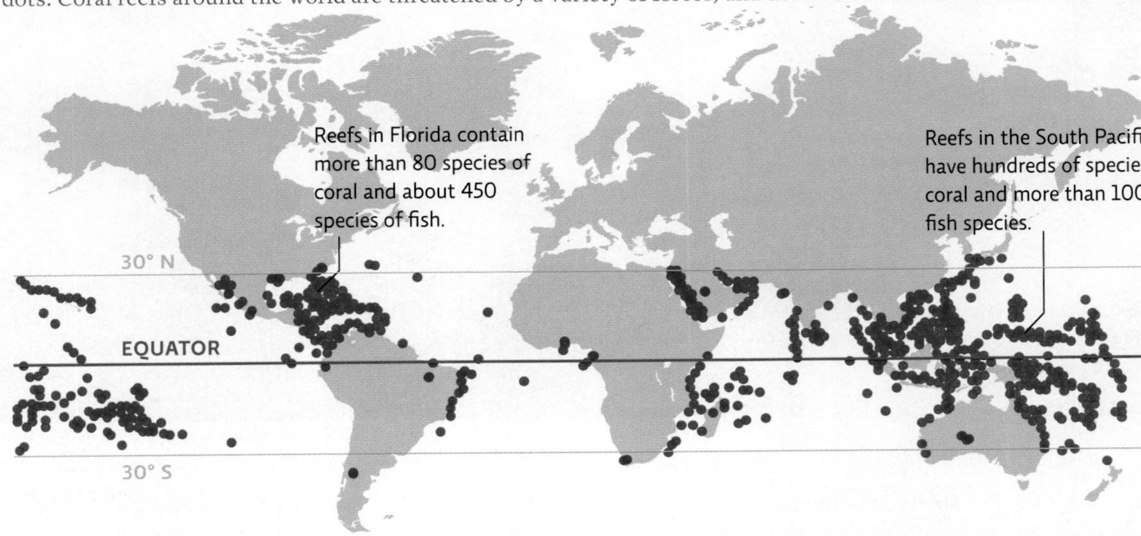
Reefs in Florida contain more than 80 species of coral and about 450 species of fish.
Reefs in the South Pacific can have hundreds of species of coral and more than 1000 fish species.
30° N
EQUATOR
30° S

↓ NOAA estimates that 75% of coral reefs worldwide are threatened by human activities or environmental changes. The two graphs below show the percentage of coral reefs already destroyed and the threats that face those that remain. Overfishing alters the community makeup, disrupting important relationships that keep the reef alive and healthy; heavy nets can also directly damage coral reefs themselves. Coastal development can add sediment to water, making it cloudy; in some areas, coral reefs are actually mined for limestone building materials. Pollution, whether from land or sea (ships), can directly harm coral as well.

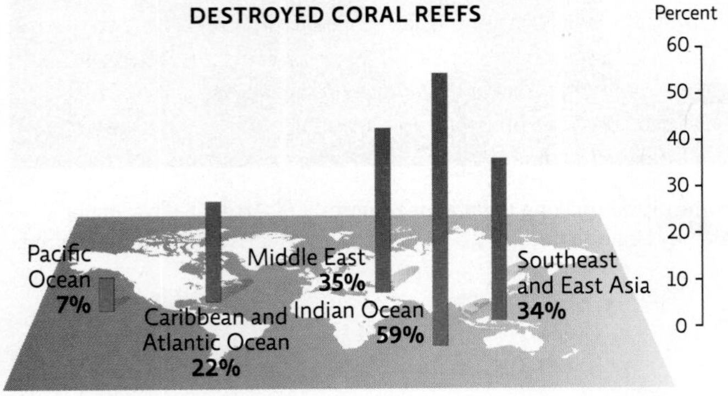
DESTROYED CORAL REEFS
Percent
60
50
40
30
20
10
0
Pacific Ocean 7%
Caribbean and Atlantic Ocean 22%
Middle East 35%
Indian Ocean 59%
Southeast and East Asia 34%

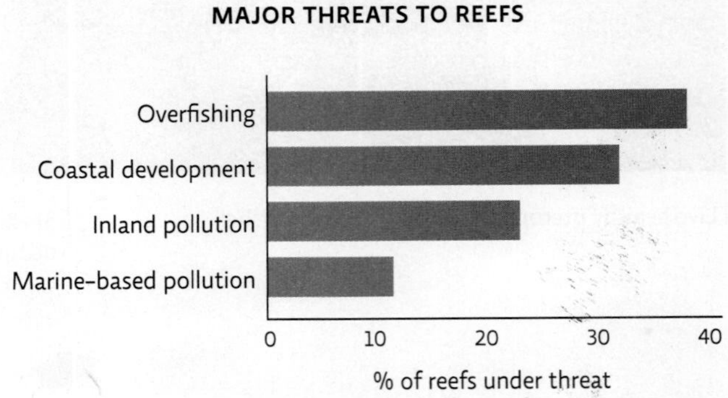
MAJOR THREATS TO REEFS
Overfishing
Coastal development
Inland pollution
Marine-based pollution
0
10
20
30
40
% of reefs under threat

↑ Researchers at the Aquarius Reef Base determine cellular respiration rates of a brown barrel sponge by measuring CO_2 production.

a lot of heat and releases it slowly), nutrient cycling, and support for large populations of commercially valuable fish. In particular, coral reefs are especially valuable for their services: protection of coastal areas from storms, purification of the water (many reef occupants are filter feeders), provision of recreational opportunities, and support of important commercial fisheries. They even serve as a source of current and potential medicines such as antibiotics and anticancer drugs. And oceans have absorbed a good bit of the CO_2 we have released due to fossil fuel burning, reducing the atmospheric warming that CO_2 would have caused if all of it had remained in the air.

Marine ecosystems, much like terrestrial ones, vary by location. In the ocean, depth is a key determinant of environmental conditions. About 80% of all sunlight is absorbed in the first 10 metres of the water column, even more than that in murky waters. Because sunlight supplies both heat and the essential energy for photosynthesis, upper layers of the ocean are much warmer and more productive than lower layers. In fact, most life-forms are found in shallow waters where light can penetrate; both heat and productivity dwindle as we move down the water column to deeper depths.

Scientists have divided the oceans into zones, based on depth. Each region differs in the amount of available sunlight, which in turn affects rates of photosynthesis and thus strongly impacts the diversity and abundance of ocean life at that level. Ocean regions are also identified by their proximity to land; the closer to land and to rivers that empty into the ocean, the more nutrient rich the area, and thus productive, the ocean community. In fact, **estuaries**, regions where rivers empty into the ocean, are known as the nurseries of the sea because so many species come to these areas to spawn (see Chapter 16 for more on nutrient enrichment of estuaries). [INFOGRAPHIC 13.4]

estuary Region where rivers empty into the ocean.

Infographic 13.4 | **OCEAN LIFE ZONES**

↓ Different communities of species are found at different depths of the ocean and at different distances from shore. Most life is found on the continental shelf, the region off each shoreline that is relatively shallow compared to the deeper regions that begin at the edge of the shelf. Communities in deeper waters have life-forms adapted to little or no sunlight, whereas the most productive communities are found in relatively clear, shallow waters with ample sunlight.

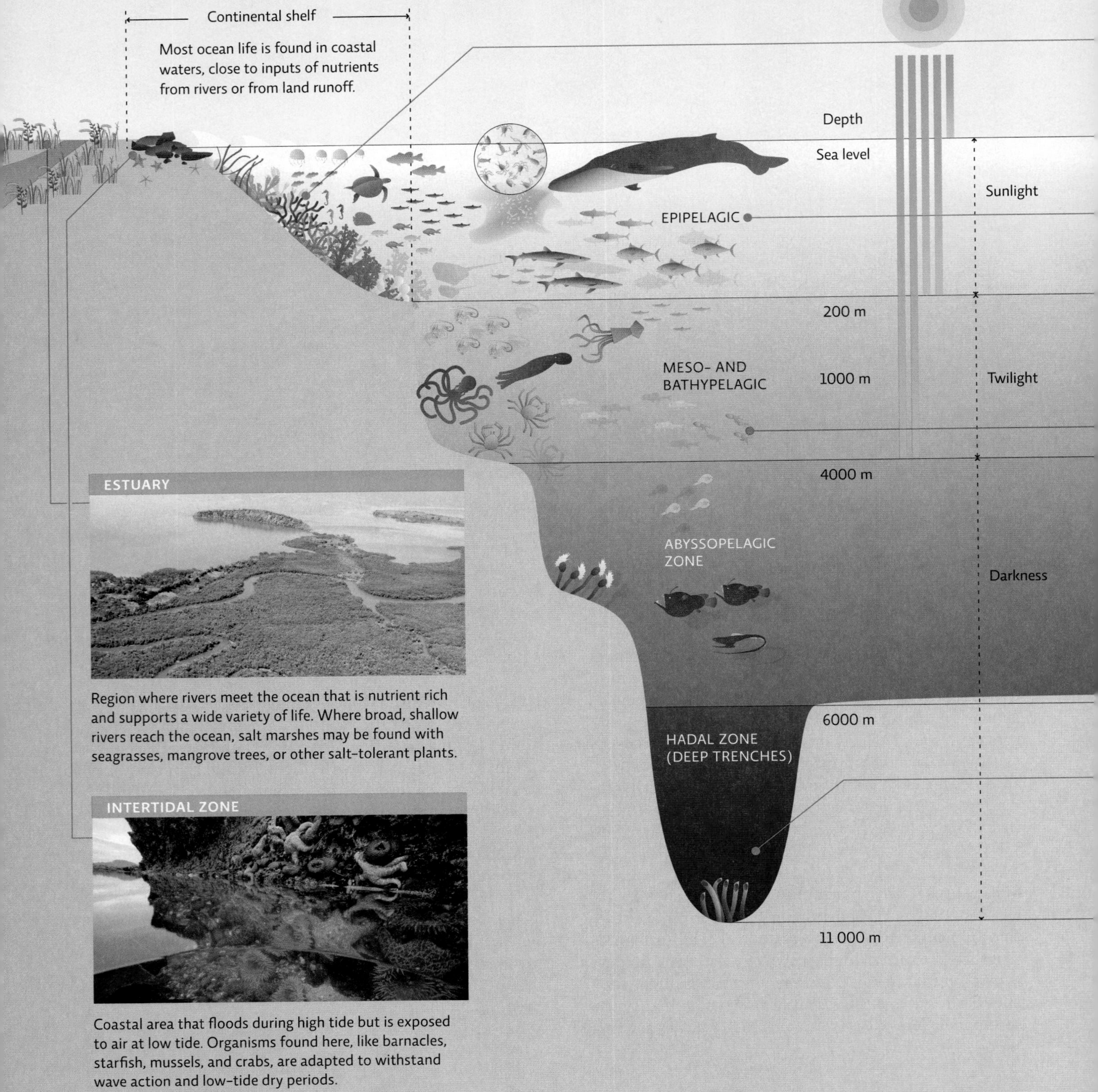

Continental shelf
Most ocean life is found in coastal waters, close to inputs of nutrients from rivers or from land runoff.
Depth
Sea level
Sunlight
EPIPELAGIC
200 m
MESO- AND BATHYPELAGIC
1000 m
Twilight
4000 m
ABYSSOPELAGIC ZONE
Darkness
6000 m
HADAL ZONE (DEEP TRENCHES)
11 000 m
ESTUARY
Region where rivers meet the ocean that is nutrient rich and supports a wide variety of life. Where broad, shallow rivers reach the ocean, salt marshes may be found with seagrasses, mangrove trees, or other salt-tolerant plants.
INTERTIDAL ZONE
Coastal area that floods during high tide but is exposed to air at low tide. Organisms found here, like barnacles, starfish, mussels, and crabs, are adapted to withstand wave action and low-tide dry periods.

CORAL REEF

Found in clear, shallow water in tropical regions; in addition to the keystone coral species, the reef supports a wide variety of fish, crustaceans, sponges, and other marine organisms.

OPEN OCEAN

Not as densely populated by ocean life, but home to many species on the move, like sharks and whales

MESO-, BATHY-, AND ABYSSOPELAGIC ZONES

Deeper regions that are home to squid and strange fish like this anglerfish; plankton make up the base of this food chain.

HADAL ZONE COMMUNITIES

Organisms such as tube worms, sea stars, and this sea cucumber live near thermal vents (areas where superheated water containing hydrogen sulphide and other chemicals is released from Earth's crust); these communities do not depend on photosynthesis but get their energy from chemicals released by the vents, captured by specialized bacteria, and passed up the food chain.

Can some organisms adapt to ocean acidification?

From high above, the Aquarius Reef Base looked something like an alien anthill. A small army of divers had descended on the structure to take care of some daily maintenance tasks, and all six aquanauts were shuttling samples and equipment back and forth from reef to station. Eventually, this would grow exhausting. "The pinnacle, where we sample, seems to get further away each day, particularly when the current gets ripping," wrote one aquanaut on the station's blog. "But then we get to ride the current back in, 'Finding Nemo' style!"

The team's main focus was the *Xestospongia muta*, a giant barrel sponge that thrives in both acidified and nonacidified microhabitats. By monitoring pH and CO_2 levels around each sponge, and taking tissue samples for later analysis in the lab, they hope to determine if those sponges that live in acidified waters have a different protein makeup that helps them adapt to, and even thrive in, these regions.

Some sponges lived in the crevices where pH was low; others lived out on the open reef, where pH was closer to normal. Slattery wanted to see whether the crevice dwellers had any special adaptations that their open-reef-dwelling cousins lacked. "It's just like humans and the common cold," he explains. "Some people go through the whole season without getting sick at all, and then others catch every little bug that's been flying around. We want to see if there's the same type of natural variation in sponges responding to acidification."

They also transplanted paired sponges—from acidified and nonacidified habitats—to sites facing additional stress, namely temperature increases. The idea was to see what impact prior exposures to low pH had on the organisms' health, and also to see what effects these environmental changes had on physiology.

Lastly, the team took a detailed census of the flora and fauna in various microhabitats, especially the acidified crevices, and collected samples of all the species they could. Like the sponge tissue, these samples would be assessed for protein expression to see if similar organisms behaved differently when they were exposed to lower pH.

Coral reefs are complex communities with lots of interspecific interactions.

Species of coral reef communities interact with and depend on each other in many ways (see Chapter 8). Clown fish and sea anemones offer each other both food

Infographic **13.5** | **CORAL BIOLOGY**

↓ Coral live in a mutualistic relationship with zooxanthellae ("zooks")—photosynthetic algae that live inside the epidermal cells of the coral. The two species share nutrients. Zooks also raise the pH of the coral cells slightly, which helps the coral lay down its coral skeleton. Though they come in a tremendous variety of shapes and colours due to species differences and even growing conditions, all coral have the same basic body plan: the polyp—a tube-like structure attached at the base to a substrate. At the top of the tube are stinging tentacles that trap food floating by, which is taken in by the mouth and sent to the digestive sac.

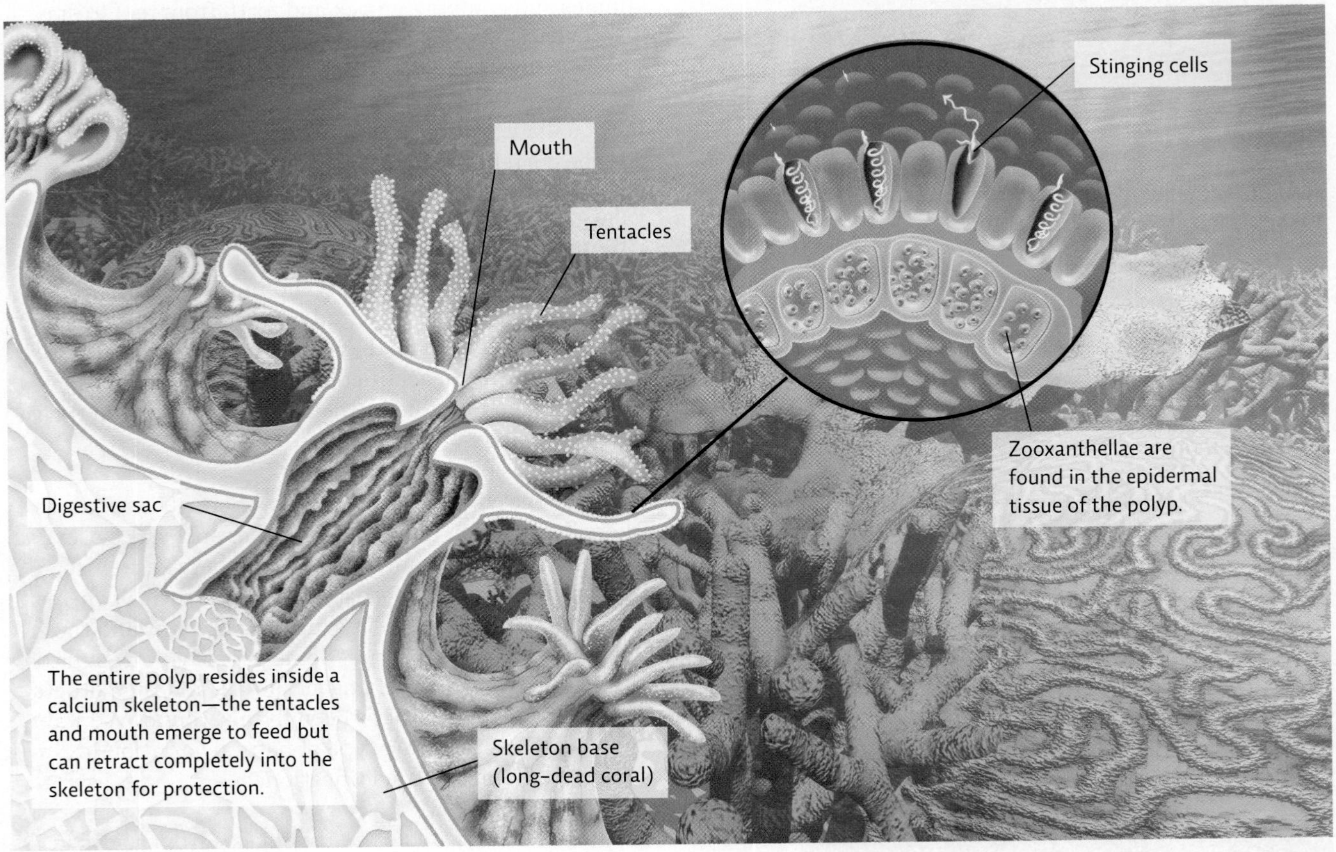
Stinging cells
Mouth
Tentacles
Zooxanthellae are found in the epidermal tissue of the polyp.
Digestive sac
The entire polyp resides inside a calcium skeleton—the tentacles and mouth emerge to feed but can retract completely into the skeleton for protection.
Skeleton base (long-dead coral)

and protection (mutualism); remora fish attach their suckers to manta rays and eat any bits of food that flow out of the ray's mouth (commensalism); and sea urchins act like lawn mowers, feeding on the reef itself in a way that enables new corals to attach and grow, thereby keeping the overall structure strong and healthy.

Like kelp, coral reefs attract and provide food and habitat for many other species. In fact, scientists estimate that 25% of all ocean species spend at least some portion of their life in a coral reef. This outsized role in marine ecology makes reefs an appealing target for oceanographers to study, and is but one of several reasons that the Aquarius station was positioned near a reef. "Coral reefs are the ultimate keystone species," says Martens. "Understanding them is the key to understanding everything else that's going wrong in the ocean right now."

Corals are tiny marine organisms that live in densely packed colonies of many individual polyps (basically a tube with one opening surrounded by tentacles). Coral larvae (immature stage) are free floating but must attach to a surface to survive. Established coral reefs actually attract the larvae via chemical cues, increasing the odds that the larvae will attach to and build on top of existing coral skeletons. Once attached to a surface, the larvae enter the polyp stage and secrete a calcium carbonate skeleton. A reef thus grows from corals building on top of other corals over the course of generations. Studies have also shown that lower pH leads to declines in fertilization, in larval development, and in a process called settlement—the dropping of coral larvae out of the water column, and their attaching to something solid—the necessary prerequisites to new coral colony formation.

Infographic **13.6** | **CORAL BLEACHING**

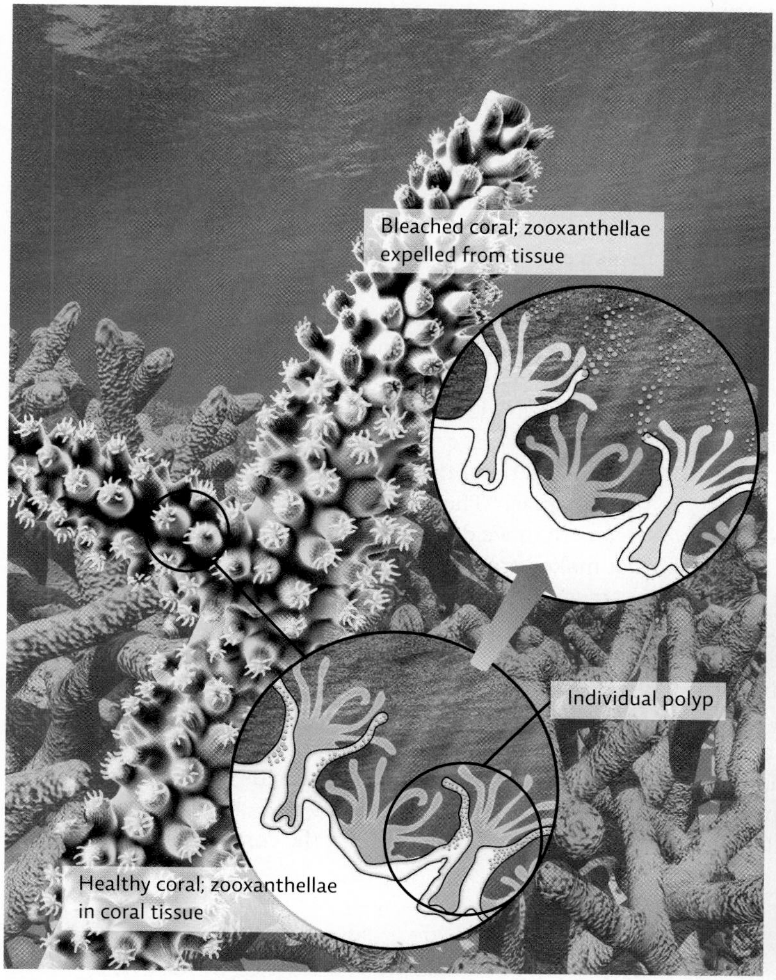
Bleached coral; zooxanthellae expelled from tissue
Individual polyp
Healthy coral; zooxanthellae in coral tissue

← Coral may expel their zooxanthellae if stressed, such as when water temperature gets too high (though the mechanism for how this expulsion occurs is poorly understood). Without their pigmented algae, the white coral skeleton can be seen through the translucent polyp—leading to the term "coral bleaching." Bleaching may be an adaptation that allows the coral to take up a different species of zooks, one that can tolerate the warmer temperatures, for instance. Coral may be able to survive for a few months without the zooks but if the polyps don't eventually take other algae back up, they will die. Multiple warming events or a warming event that persists for several weeks may stress the coral beyond its ability to survive. Pollution and even excess sun exposure are also linked to bleaching events.

↓ Bleached staghorn coral on a Fiji reef

Corals feed on decaying organic material and plankton—pretty much anything that floats by. But they live in nutrient-poor regions: tropical waters tend to remain stratified, with warm water trapped above colder, deeper water. This stratification prevents the mixing of deep water and surface water that would normally bring up nutrients from below.

Thus the corals themselves could not survive without some help from their algal partners. They live in a mutualistic relationship with zooxanthellae (known affectionately to most marine biologists as "zooks")—a photosynthetic, dinoflagellate algae that lives inside, and shares nutrients with, the coral. Corals provide their resident zooks with nutrients (from the polyps' waste) and CO_2 for photosynthesis. Zooks, in turn, provide the coral polyps with food (sugars made during photosynthesis). The zooks also give the corals their beautiful colours.

It's a well-honed evolutionary collaboration, one that enables each species to thrive in a nutrient-poor environment. [INFOGRAPHIC 13.5]

Corals can also control the number of zooks they host in their cells by adjusting the amount of sunlight or nutrients they supply. Under extreme stress, corals have been known to expel the algae. This event is known as coral bleaching because when it happens, the coral turns bone white. Though the coral can survive for a short period of time without zooks, they will die if not recolonized. Since the many species of zooks are symbiotically tied to their coral hosts, any events that lead to coral death, such as bleaching or decalcification, could also lead to the loss of the zooxanthellae species that reside in those coral. [INFOGRAPHIC 13.6]

Scientists have found that in the past, corals have gone through natural death and birth cycles in response to environmental changes such as shifts in temperature or pH. However, many scientists feel that today's accelerating rates of acidification and bleaching are likely unprecedented, as is their global scope. But CO_2-related effects are by no means the only threats.

The world's oceans face many other threats.

Rising sea levels may decrease sunlight penetration—and thus reduce photosynthesis—in the deepening coral seas. And increased flooding is almost certain to bring more pollutants from land to sea.

On top of all that, oceans everywhere are now threatened by overfishing, pollution, and invasive species.

In the ocean at large, some 90% of predators in the top trophic levels have been eliminated by overexploitation (a.k.a., fishing pressure). Such heavy losses disrupt the interdependent relationships needed to sustain each community. For example, without grazer fish to keep it in check, algae overgrows on coral. Fishing pressure inflicts other wounds as well: bottom trawling can decimate sea beds, crushing or burying organisms that live close to the beds and uncovering those that need to remain buried (see Chapter 14). Cyanide sprays, used to stun fish for aquarium collection, kill most fish and coral that encounter it. And dynamite, also used in fishing, physically destroys reefs and other ecosystems around it.

Meanwhile, sediments and high levels of nutrient runoff from agricultural areas are boosting algae production and creating algal blooms, which in turn smother corals and block sunlight, reducing photosynthesis. Trash from both land and sea, and petroleum from ships and boats are also polluting the marine environment at unprecedented rates.

Invasive species pose yet another threat; more than 80% of ocean harbours around the world now host at least one invasive. Whether they arrive in the ballast water of ships, escape from aquaculture pens, or are moved in the aquarium trade, the outcome is the same: they wreak havoc on aquatic ecosystems. "We don't even know how much damage they're doing," says Slattery, who has studied the impact of invasive lionfish on reefs near the Bahamas. "The fact is, we know much less about the oceans than we do about land. But if you think about it, that makes protection even more vital, because we need to err on the side of caution." [INFOGRAPHIC 13.7]

Scientists have been working to create marine protected areas (MPAs)—places where fishing and other human activities are restricted or completely prohibited. Evidence shows that in the right conditions, MPAs can significantly improve the marine ecosystems they encompass. Three years after a 2003 bleaching event that devastated the South Pacific coral reefs of Kanton

→ Lionfish (escapees from aquaria) are aggressive (and poisonous) predators and either kill native fish outright or outcompete them when foraging. When they are introduced to a coral reef, the survival of native fish can drop by as much as 80%. In some regions of Florida, divers who remove them are eligible for a $10,000 prize. In areas where groupers are protected, lionfish are not as big of a problem because the native groupers eat them—another reason to protect communities and try to keep them intact.

Infographic **13.7** | **THREATS TO OCEANS**

↓ Oceans are threatened by a variety of human activities.

FISHING PRESSURE

↑ A severely damaged coral reef in Indonesia from "blast fishing" with dynamite. This brings up lots of fish but kills many in the process, including many nontarget species. (See Chapter 14 for information on overfishing.)

↑ Fish collector spraying cyanide on a reef in the Philippines. This stuns some fish, which are collected, but kills most of the fish and kills the coral as well.

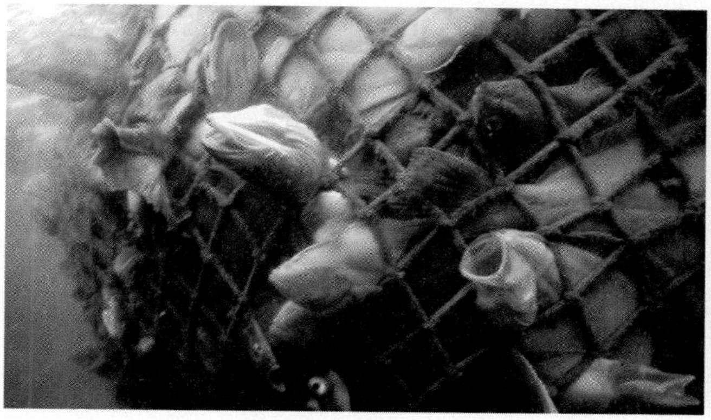

↑ Many species of commercially valuable fish, such as bluefin tuna and herring, are facing population crashes due to overfishing; others, like the Newfoundland cod population, have already crashed. (See Chapter 14 for information on commercial overfishing.)

POLLUTION

↑ Plumes of sediment pollution along coasts can introduce nutrients, toxins, pathogens, and solids that can smother and damage coral reef or other marine ecosystems. (See Chapter 16 for more on water pollution.)

↑ Debris like this plastic bag can choke ocean animals or quickly become entangled on a coral and smother it. (See Chapter 17 for more on trash in the oceans.)

↑ Oil spills from drilling operations and from tanker transport damage coastal and offshore marine ecosystems. (See Chapter 20 for more on oil spills.)

Island, the area was declared an MPA where fishing was prohibited. By 2010, the reefs showed significant recovery. Credit is given to the community of fish: because they were left as intact, undisturbed populations, the fish grazed on algae that would normally move in and prevent coral from recolonizing the area. By keeping the community connections intact, the reef appears to be well on its way to recovery, even after a serious bleaching event, showing the resiliency of intact ecosystems.

Of course, the best way to save the oceans may be to change the way we live on land. We humans now emit more than 30 billion metric tons of CO_2 each year. Steps taken now to decrease the amount of CO_2 released into the atmosphere will reduce the CO_2 available to be absorbed by the oceans. Unfortunately, even if we stopped completely, right now, it would take tens of thousands of years for ocean chemistry to return to its preindustrial state. "We can't reverse the tide at this point," says Slattery. "But if we act quickly, we can at least slow it down." [INFOGRAPHIC 13.8]

On the last day of the mission, Slattery and his team ascended back up to the sunlit surface. Slattery did not remember the sun being quite so bright. He was anxious to get back to his lab at the University of Mississippi, where he and his team would sort through the reams of data and samples, like detectives sorting through clues in an effort to solve a great mystery.

They already knew that their initial hypothesis would hold true: there were indeed major differences in pH across the reef landscape. But whether those differing conditions had any impact on the reef species' physiology remained to be seen. Back in the lab, they would measure the protein levels in each of the samples they had collected and try to determine whether species that had grown up in acidified microhabitats expressed different proteins, or responded differently to stress, than those that had grown up on the open reef.

They'd spent 9 days at the bottom of the ocean. Now the real work could begin.◉

Select references in this chapter:

Beman, J.M., *et al.* 2011. *Proceedings of the National Academy of Sciences*, 108: 208–213.

Orr, J., *et al.* 2005. *Nature*, 437: 681–686.

Infographic **13.8** | **REDUCING THE THREATS TO OCEAN ECOSYSTEMS**

STRATEGY	EFFECT
Designate vulnerable areas as marine protected areas.	Protecting ocean communities helps an area withstand or even recover from environmental damages, as evidenced by the astounding recovery of the Kanton Island reef.
Reduce use of fossil fuel.	Conservation efforts and transition to non-fossil fuel-based energy sources will reduce the amount of anthropogenic CO_2 released and slow ocean acidification and warming (see Chapter 22).
Limit development in vulnerable areas.	Reducing beachfront development and ceasing the practice of coral reef harvesting for building materials keeps the areas intact and reduces exposure to land-based pollution.
Prohibit bottom trawling for fish in vulnerable areas and reduce overfishing in general.	Nets dragged across the sea floor can destroy coral reefs and other seabed formations. Using other, safer methods in reef areas will protect the reef from these methods. Keeping populations viable by not overfishing improves an ocean community's ability to withstand perturbations.
Reduce pollution.	Trash, oil spills, and other types of pollution harm sea life. Keeping waters clear, clean, and free of nutrient (fertilizer or sewage) or toxic pollution benefits both ocean organisms and humans. In particular, fungi and bacteria, which thrive when extra nutrients are in the water, can infect and kill coral, so keeping the water free from this type of pollution will reduce infections and keep coral healthy.

↑ This healthy new plate coral in the lagoon on Kanton Island is a sign of hope. In their recovery after being hit by a severe bleaching episode, the coral has grown to a diameter of more than 1.2 metres.

BRING IT HOME

PERSONAL CHOICES THAT HELP

Marine ecosystems are vastly diverse, with many different aquatic habitats and an impressive array of species. Unfortunately, marine ecosystems are facing numerous threats due to human action. By bringing awareness to these threats and taking action to reduce negative human impacts, we can dramatically improve the health of our oceans.

Individual Steps

→ Don't leave trash on the beach; avoid disposable plastic bags and bottles, as many of them end up in the ocean where they are a danger to marine organisms.

→ If you go diving or snorkelling on a coral reef, do not touch or remove any coral; don't drop the boat's anchor on coral—find a sandy spot to place it.

→ Decrease your fossil fuel use. The less CO_2 produced, the healthier the ocean will be.

Group Action

→ Lead a book discussion on *Oceana: Our Endangered Oceans and What We Can Do to Save Them* by Ted Danson, which discusses many of the threats to the oceans and what we can do to help.

→ Follow groups like the Coral Reef Alliance , Oceana, or the Living Oceans Society on Facebook and Twitter to learn more about protecting marine ecosystems.

→ Volunteer for the Great Annual Fish Count at fishcount.org to help survey reef fish species.

Policy Change

→ Ask pet store owners and aquarium supply centres to sell only captive-bred fish to help protect coral reefs from being poisoned by cyanide.

→ Contact your members of Parliament and ask them to support legislation to reduce carbon emissions.

UNDERSTANDING THE ISSUE

CHECK YOUR UNDERSTANDING

1. Which of the following is TRUE about ocean acidification?
a. CO_2 emissions have reduced the ocean's pH by about 0.1, which corresponds to a 300% increase in ocean water acidity.
b. Ocean pH has now dropped below a pH of 7.0.
c. Ocean acidification is harmful, as CO_2 forms carbonic acid when dissolved in water, which eats away at existing calcium-based materials.
d. In a more acidic ocean, the increase in key nutrients such as nitrogen and iron stabilizes almost all marine ecosystems, counteracting the effects of global warming.

2. Organisms that live on the ocean floor are said to live in the:
a. benthic zone.
b. abyssopelagic zone.
c. hadal zone.
d. pelagic zone.

3. An estuary is:
a. a shallow coastal region affected by the tides, where organisms are adapted to strong currents.
b. the only ecosystem on Earth that does not depend on photosynthesis.
c. a pitch-black, cold, high-pressure zone where few organisms live.
d. a productive marine ecosystem in a region where rivers empty into the ocean, bringing nutrients from land runoff.

4. Which of the following threats to ocean ecosystems currently has the biggest impact on coral reefs?
a. Overfishing
b. Coastal development
c. Sediment runoff from land
d. Oil spills

5. What is/are the consequence(s) of coral bleaching for coral reefs and other ocean ecosystems?
a. Coral reefs attract coral larvae via visual cues, so when reefs lose their colour, larvae have difficulty attaching to the top of existing coral skeletons.
b. Coral bleaching occurs when corals under stress expel their zooxanthellae: corals can only survive for a short period of time without their photosynthetic algae partners.
c. Although coral reefs are home to many species, if bleaching occurs, most of these species can move to other marine ecosystems such as estuaries and kelp forests, thus tempering the impact on ocean ecosystems.
d. All of the above are consequences of coral bleaching for coral reefs and other ocean ecosystems.

6. The highest percentage of destroyed coral reefs are found in the ______ while the ______ has the lowest percentage of destroyed coral reefs.
a. Arctic Ocean; Atlantic Ocean
b. Indian Ocean; Pacific Ocean
c. Caribbean Sea; Mediterranean Sea
d. Middle East; Asian South Pacific

WORK WITH IDEAS

1. What are coral reefs and where are they found? Explain the symbiotic relationship the polyp shares with its resident zooxanthellae.

2. What is ocean acidification? Describe the causes and consequences of this phenomenon.

3. Besides ocean acidification, what threats do coral reefs and other ocean ecosystems face? Describe the causes and consequences of these threats.

4. How can we reduce the threats to coral reefs and other ocean ecosystems? Discuss three strategies.

ANALYZING THE SCIENCE

The data in the table on the following page comes from a recent assessment of coral reefs (*Reefs at Risk Revisited*) prepared by the World Resources Institute.

INTERPRETATION

1. How many different coral reef regions are the global reefs divided into?

2. How are reef areas distributed globally? What patterns do you see in the distribution of reefs and the size of the population that lives within 30 km of reefs? Provide the data to support your conclusions.

3. What are the patterns for local threats to coral reefs relative to population within 30 km of reefs? Do the patterns seem reasonable when considering that local threats consist of overfishing and destructive fishing, marine-based pollution, coastal development, and land-based pollution? Provide the data to support your conclusions.

ADVANCE YOUR THINKING

4. What is thermal stress in coral reefs? What are the patterns in past thermal stress to coral reefs relative to population within 30 km of reefs? Are they the same as with local threats? Why or why not? Provide the data to support your conclusions.

5. How might thermal stress affect reefs in the future? As discussed in the chapter, what is another threat that reefs are facing or will be facing in the future? How are these two stresses related?

6. What challenges do these different threats present for protecting coral reef systems? What is the role of MPAs in protecting coral reefs against these various threats? Based on the information about reefs in the table, which regional reef is most secure and which is least so? Explain your responses.

Integrated threat to coral reefs by region

Region	Reef area (sq km)	Reef area as percent of global	Coastal population (within 30 km of reef), 2000	% of reefs threatened		Reef area in MPAs (%)
				Local threats	Severe thermal stress (1998–2007)	
Atlantic	25 849	10	42 541	75	56	30
Australia	42 315	17	3509	14	33	75
Indian Ocean	31 543	13	65 152	66	50	19
Middle East	14 399	6	19 041	65	36	12
Pacific	65 972	26	7487	48	41	13
Southeast Asia	69 637	28	138 156	94	27	17
Global	**249 713**	**100**	**275 886**	**61**	**38**	**27**

EVALUATING NEW INFORMATION

One way for a marine enthusiast to participate in ocean conservation is as a citizen scientist. Citizen science engages individuals or teams of volunteers, many of whom have no special scientific training, in research-related tasks. This allows scientists to accomplish research objectives more easily and also promotes public engagement with science research.

Explore the SciStarter website (scistarter.com).

Evaluate the website and work with the information to answer the following questions:

1. Is this a reliable information source? Does the organization have a clear and transparent agenda?
   a. Who runs this website? Do the organization's credentials make the information presented reliable or unreliable? Explain.
   b. What is the mission of this website? What are its underlying values? How do you know this?
2. From the "Pick a Topic" drop-down menu, select "Ocean & Water."
   a. How many different projects related to oceans and water are there? What sorts of information about each project does the website offer? Provide some examples.
   b. Are there any projects directly related to coral reefs? What are they and how can a citizen scientist be involved?
3. Would you participate in a citizen science project? If so, what type? Do you think that the citizen science model is valid and effective? Explain your responses.
4. How might the citizen science model be useful as a way to influence the public understanding of environmental issues and shape environmental policy?

MAKING CONNECTIONS

DESIGNING MARINE PROTECTED AREAS

Background: The health of the oceans affects our well-being, regardless of whether or not we live on the coast. Historically, we have treated the oceans as an endless resource provider as well as a bottomless waste receptacle. Numerous scientific studies show that human activities are degrading ocean ecosystems around the world, and these changes are impairing the ocean's ability to provide the ecosystem services we rely on.

Case: One way to protect coastal ocean ecosystems is by creating effective marine protected areas (MPAs) (see Chapter 14 and the Fisheries and Oceans Canada web page on MPAs, http://www.dfo-mpo.gc.ca/oceans/marineareas-zonesmarines/mpa-zpm/index-eng.htm). You have been selected to be part of a team to evaluate current MPAs and develop guidelines for the future design and management of MPAs. Based on your research and analysis, write a position paper that includes the following:

1. An evaluation of the importance of coastal ocean ecosystems and the challenges of managing them.
2. An analysis of MPAs as a means of protecting coastal ocean ecosystems, including:
   a. an assessment of the ecological and human cost/benefits of establishing MPAs to determine what management priorities and techniques should be.
   b. an evaluation of current MPAs in Canada to determine criteria for effective design, and to assess whether Canada's MPAs meet global standards in terms of size and activity restrictions.
3. An assessment of the potential challenges to designing and implementing MPAs and a discussion about how to address these challenges.

CORE MESSAGE

Although the oceans are vast, many fisheries are in serious jeopardy due to overfishing. Aquaculture may allow us to raise fish for harvest, taking some pressure off of wild stocks.

GUIDING QUESTIONS

After reading this chapter, you should be able to answer the following questions:

→ How are fish and fisheries like that of the Atlantic cod important for humans?

→ How do technology and the tragedy of the commons interact to jeopardize global fisheries?

→ Why have fisheries declined so precipitously in the last half-century and what is the current status of the world's fisheries?

→ What are some of the ways we are trying to protect our fisheries?

→ What is aquaculture and how might it ease the strain on at-risk fisheries? What trade-offs does aquaculture involve?

Raising sea bass inland in a recirculating aquaculture system is an alternative to marine aquaculture, which can cause problems in coastal ecosystems.

CHAPTER 14 FISHERIES AND AQUACULTURE

FISH IN A WAREHOUSE?

How one fish scientist could change the way we eat

In Frenchman Bay, Maine, eight fishers huddle around their aquaculturist instructor as he explains how to feed the "fingerling" cod, each about 40–60 millimetres long, in the wired cage beneath the water. The group is part of a new program meant to turn fishers into fish farmers. Instead of harvesting wild fish from the depths of the ocean, with factory ships and huge fishing nets, the fishers will learn how to raise them, from hatchlings to full-sized adults—how to feed them, monitor their health, and ultimately, how to prepare them for sale. Today, the fishers are observing a coastal *net pen,* where fish are raised in a system of stationary, floating nets, usually positioned in coastal waters. Tomorrow they will tour an indoor fish farm where the same types of fish are raised in colossal indoor tanks.

This, say experts, is the future of fishing.

For more than 400 years, the Atlantic cod—a *demersal* fish that feeds along the ocean floor—supported not just a fishing industry, but an entire culture along the northeastern coast of North America. Huge **fisheries**—operations in which fish are caught, harvested, processed, and sold and/or shipped—sprang up in the port towns of Newfoundland, New England, and even European countries, and were sustained for generations almost exclusively by this one fish. Cod became a crucial part of the early Newfoundland economy. Up to 400 000 metric tons of cod were caught each year in the 1800s and, justifiably, cod earned the name "Newfoundland currency." But after a dramatic peak in the late 1960s, catches declined rapidly. The Canadian government imposed quotas in 1974, but by 1992 the annual catch had dropped to just 2% of its historic high and the federal government closed the Grand Banks area to fishing completely. When annual catches fall below 10% of their historic high, scientists call this a **collapsed fishery**. [INFOGRAPHIC 14.1]

More than 20 years later, with the moratorium still in place, the Grand Banks cod fishery has yet to recover. Cod population estimates still hover at around 30% of what would be needed for the fishery to survive commercial fishing pressure. Meanwhile, some 40 000 cod fishers are still out of work and reliant on government support.

How did this happen? One reason is that nobody owns the ocean's resources but everyone uses them; the collapse of the Grand Banks cod fishery is a classic example of the tragedy of the commons (see Chapter 1). Trawlers from Canada, the United States, and several western European nations all fished the Grand Banks and contributed to the cod fishery collapse.

↑ As seen in this photo, cod are big fish and the cod fishery was the economic backbone of colonial Canada and New England; in 1497, John Cabot's crew reported that cod was so abundant that one only had to lower a basket into the water to "fish" for cod.

fisheries The industry devoted to commercial fishing or the places where fish are caught, harvested, processed, and sold.
collapsed fishery A fishery in which annual catches fall below 10% of their historic high; stocks can no longer support a fishery.

Infographic **14.1** | **MEET THE COD**

ATLANTIC COD, *Gadus mordua*
Conservation status: Vulnerable

↓ Cod are cold-water fish that inhabit coastal waters in the North Atlantic (they are also found in the Pacific). They are usually found in waters 10 to 200 metres deep, and live close to the sea floor. Adults eat a wide variety of fish (including other cod) as well as mussels, squid, and crab. Their diet and early life-cycle stages depend on a seabed with a complex structure (lots of hiding places and niches for their prey); therefore, trawling methods that drag a heavy net across the sea floor seriously degrade their habitat and spawning beds. The cod fishery has declined throughout its Atlantic range (blue area), especially in the Grand Banks of the Newfoundland–Labrador Shelf region, where the fishery is seriously depleted.

GRAND BANKS

The Grand Banks are a part of North America's eastern continental shelf. This underwater plateau area supports abundant fish and other marine life because of its relatively shallow depth (24–100 metres) and its location, which is right where the warm Gulf Stream meets the cold Labrador Current. These conditions promote mixing, which brings nutrients to this area, fostering a rich food web.

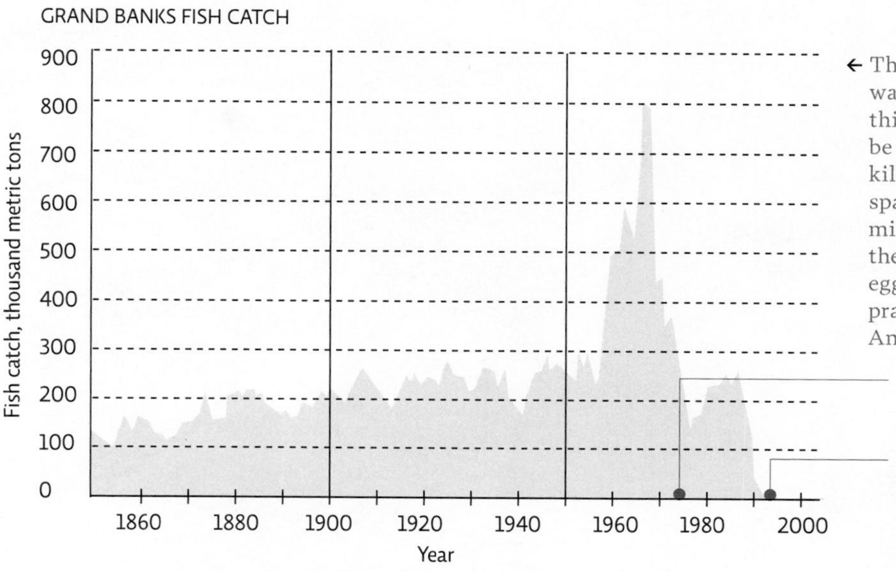
GRAND BANKS FISH CATCH
Fish catch, thousand metric tons
900
800
700
600
500
400
300
200
100
0
1860
1880
1900
1920
1940
1960
1980
2000
Year

← The collapse of the Grand Banks fishery was surprising, considering the biology of this fish. Cod are quite large—an adult can be more than 2 metres long and weigh 90 kilograms or more. When one considers that spawning schools of more than a hundred million fish have been observed and that the average female can spawn millions of eggs, it is hard to imagine that our fishing practices could devastate this population. And yet, they did.

1974 Quotas set for all cod stocks of the Northwest Atlantic

1992 Canadian government closes the cod fishery

Another reason for the fishery's collapse involves a trifecta of technological advances—steam engines, flash freezing, and trawler ships that could drag huge nets behind them—that enabled fishers to travel farther into the ocean, catch more fish, and transport those fish greater distances than ever before. At the same time, health-conscious eating patterns made fish more popular than any other source of protein. With the coalescence of new technology, rising demand, and multiple fishing nations, the once-prolific Grand Banks cod fishery was decimated. Fish populations plummeted, and the increased traffic of fuel-guzzling ships polluted the water. Bottom trawlers are particularly damaging. They destroy the sea beds, which are the cod's spawning grounds, and catch or destroy billions of coral, sponges, starfish, and other invertebrates.

Other marine species are lost as **bycatch**, meaning they are trapped in the trawler nets that are meant to capture cod. Being caught as bycatch seriously threatens many species, including small whales and dolphins (over 300 000 lost per year), sea turtles (over 250 000 lost per year), and even seabirds (over 300 000 lost per year). Seals, sea lions, and sea otter populations have declined by as much as 85% in some areas due to this unintentional capture. Non-target fish are also taken as bycatch, including sharks, rays, and juvenile fish of many species, including cod. Bycatch is also a problem with other industrial fishing techniques such as long-line fishing (where a main line holds many individual lines with baited hooks) and drift netting (where a free-floating net entangles fish; this method was banned in 1992). Fishing methods that minimize bycatch are available and their use is increasing. [INFOGRAPHIC 14.2]

Such is the story of our last wild food. It begins in the oceans and ends on our dining room tables. But as fish stocks around the world—not only of Atlantic cod, but also of Chilean sea bass, Alaskan pollock, bluefin tuna, and Atlantic herring—dwindle to unprecedented lows, scientists and fishers alike are trying to rewrite that story. Their success could mean sparing the world's fisheries

bycatch Non-target species that become trapped in fishing nets and are usually discarded. Some methods, like trawling, have very high bycatch levels, and discards often exceed the actual target species catch.

Infographic **14.2** | **BOTTOM TRAWLING**

↓ Bottom trawlers, like those that are used for cod, drag huge nets across the ocean floor, damaging the seabed where cod and other species live and breed.

Other organisms that the fishers cannot use are caught in the large nets and are discarded (unwanted or illegal species) or may drown (turtles and sea mammals). This bycatch can be huge (30–70%) and is the main threat to many whales and marine turtles. Special net modifications can decrease bycatch.
The trawl net has heavy weights to keep it on the seabed and has an opening that can be more than 60 metres wide.
BYCATCH
SEDIMENT PLUME
DAMAGED SEABED

from an unthinkable fate; but it could also mean that our grandchildren never eat a wild caught fish.

Humans rely on protein from fish.

Simply put: humans need fish. We consume more seafood every year than we do beef, pork, and chicken combined. In fact, over 15% of the world's population relies on fish as their main source of protein. In poorer nations, this preference is driven by the cheaper cost of fish compared with meat and poultry. In wealthier nations like Canada, a preference for seafood emerged as scientists uncovered an array of health benefits associated with the omega-3 fatty acids found in fish oils—especially the fish oils of cold saltwater species. In recent decades, a bevy of trendy diets have been built around fish consumption. Canada's Food Guide recommends that Canadians eat at least two servings of fish per week.

It's not just fish protein we have come to rely on. More than 200 million people around the world earn their living in the fishing industry, which generates some $130 billion annually in global revenue. Developing countries are especially dependent on fish, not only as a source of protein, but also as a source of revenue. More than half of all fish sold in the global market comes from developing countries, and fish make up the single biggest developing country export.

> With the coalescence of new technology, rising demand, and multiple fishing nations, the once-prolific Grand Banks Atlantic cod fishery was decimated.

But for all the health and economic benefits this massive industry provides, it has also exacted huge—some would say catastrophic—environmental costs.

Declining fish populations can impact an entire ecosystem, especially one with a simple food web comprised of few species, like that of the cold waters of the North Atlantic.

↑ A commercial catch of sand eel fish, which have been overfished, caught on the North Sea, Denmark.

Unbalanced food webs could irreversibly change the ocean ecosystem of the Grand Banks cod. Loss of apex predators often disrupts food chains by allowing lower-trophic-level species to increase in number. For example, declines in cod and the fish they feed on have led to an increase in populations of small jellyfish-like organisms called hydroids. Hydroids make it harder for the cod population to recover, by feeding on the same food as the very young cod and on the juvenile cod themselves. [INFOGRAPHIC 14.3]

Overfishing of upper-trophic-level fish has led humans to continually seek out new species to harvest, at lower and lower trophic levels; this is called "fishing down the food chain." Interestingly, fish lower on the food chain are seen as less desirable for human consumption. Some vendors get around this by renaming the fish once it gets to market. For example, there is not much of a market for the invasive Asian carp that is causing problems in North American waters, but marketers here have plans to bring it to market as "Silverfin."

Today, the oceans—not to mention the fish themselves—are in grave danger. The number of large fish like tuna, cod, and halibut in the ocean is only 10% of what it was in 1950. Over half of the world's fisheries are *fully exploited* and at their **maximum sustainable yield** (the amount that can be harvested without decreasing the yield in future years). Another 32% are **overexploited fisheries**, meaning they are being harvested at an unsustainable level, or **depleted fisheries**, meaning that the population size is very low compared to historic levels and there are not enough fish left to support a fishery. Unless we change our ways, scientists predict that all commercially valuable wild fish populations may collapse by the middle of the century. [INFOGRAPHIC 14.4]

Laws exist to protect and manage fisheries.

For most of human history, it seemed impossible that we could ever use up the resources in something as vast as the ocean, but in recent decades, that thinking has begun to change. As human populations swell and fish populations plummet, scientists and lawmakers around the world are struggling to reverse the damage that's already been done, and to protect our fisheries from further decline. They started by enacting legislation.

For centuries, most nations defined their territorial waters, or exclusive fishing zones, as extending 12

maximum sustainable yield The amount that can be harvested without decreasing the yield in future years.
overexploited fisheries More fish are taken than is sustainable in the long run, leading to population declines.
depleted fisheries The fish population is well below historic levels and the population's reproductive capacity is low, meaning that recovery will be slow, if at all.
exclusive economic zones (EEZs) Zones that extend 200 nautical miles (370 kilometres) from the coastline of any given nation, where that nation has exclusive rights over marine resources, including fish.

Infographic **14.3** | **FISHING DOWN THE FOOD CHAIN**

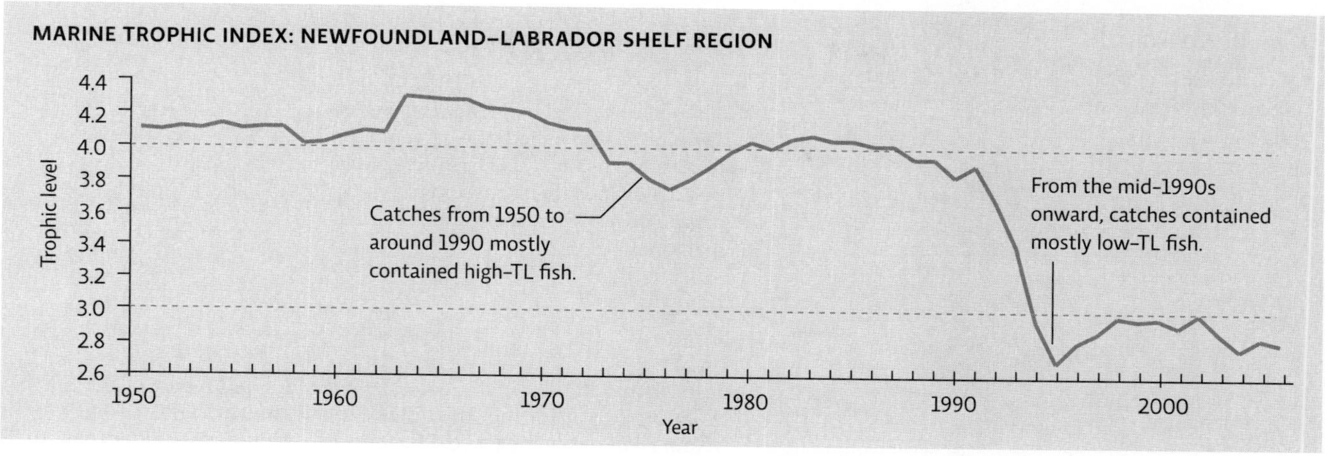
MARINE TROPHIC INDEX: NEWFOUNDLAND–LABRADOR SHELF REGION
Trophic level
4.4
4.2
4.0
3.8
3.6
3.4
3.2
3.0
2.8
2.6
1950
1960
1970
1980
1990
2000
Year
Catches from 1950 to around 1990 mostly contained high-TL fish.
From the mid-1990s onward, catches contained mostly low-TL fish.

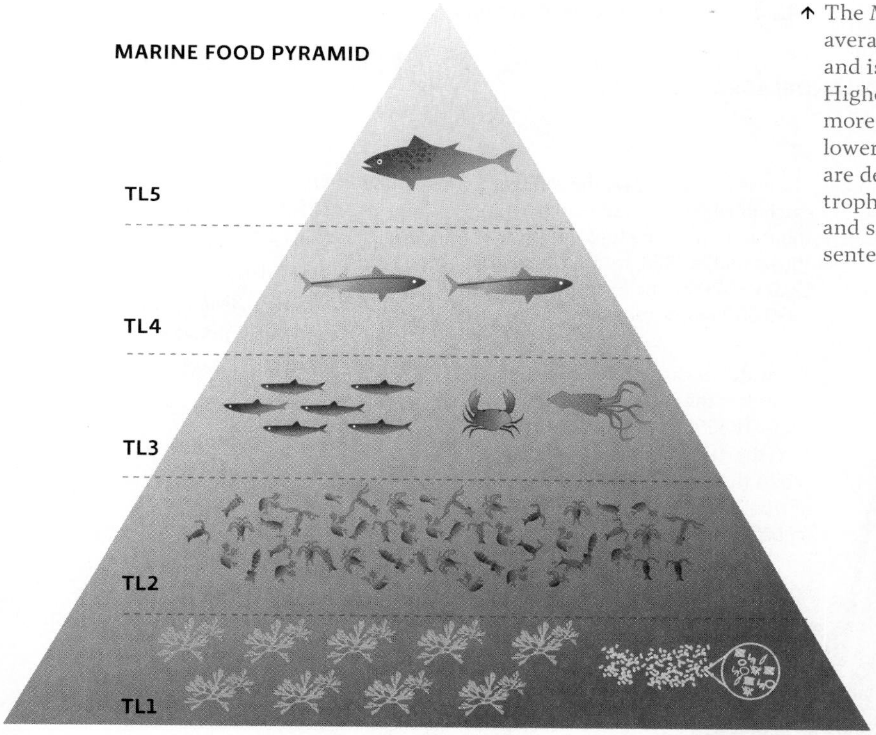
MARINE FOOD PYRAMID
TL5
TL4
TL3
TL2
TL1

↑ The *Marine Trophic Index* (MTI) is a measure of the average trophic level (TL) of fish taken in a given year and is an indicator of the status of a marine ecosystem. Higher-TL species like cod and tuna are typically the more sought-after fish. A catch that contains mostly lower-TL fish suggests that higher-TL fish populations are depleted. To determine the MTI, one identifies the trophic level of all fish taken (how many at TL2, TL3, and so on) and then calculates the average TL represented by all fish taken that year.

← One of the obstacles to the recovery of cod and other large predatory fish is the loss of their own prey. When cod are depleted, fishers pursue the herring, crabs, and shrimp at lower trophic levels, reducing the food supply for the cod, and ultimately jeopardizing the cod recovery.

nautical miles (22 kilometres) off their coastlines. In the 1960s, it became clear that such zones were inadequate. Wide-ranging industrial ships—floating factories, replete with freezing and processing technology—had long freed fishers from the shackles of time and distance. They could now pursue any fishery they liked, no matter how far from their home country. With no rule of law to constrain them, they could stay as long as they pleased and take as much as they wanted. No fishery stood a chance against a global army of such ships. One response to this problem was the creation of larger **exclusive economic zones (EEZs)** that extend 200 nautical miles (370 kilometres) from the coastline of any nation, where that nation has exclusive rights over marine resources, including fish.

For a long time, most countries, including Canada, simply chased other nations out of their fishing territories, while still allowing their own fishers to take as much as they pleased. But with fisheries around the world at or near collapse, fisheries managers started cracking down. In the

Infographic **14.4** | **STATUS OF MARINE FISHERIES**

STATUS OF GLOBAL MARINE FISHERIES (2010)

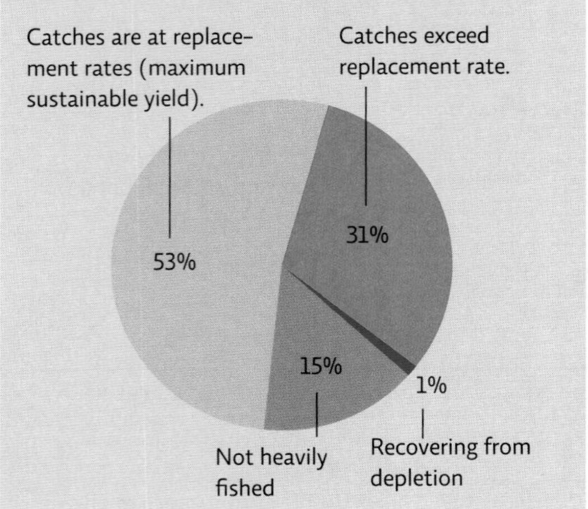
Catches are at replacement rates (maximum sustainable yield).
Catches exceed replacement rate.
53%
31%
15%
1%
Not heavily fished
Recovering from depletion

← Though 68% of global marine fisheries are sustainably fished, the percentage that is overfished (31%) has increased threefold since the 1970s and is a major concern for fishers and fisheries managers.

↓ Fish catches in the Newfoundland–Labrador Shelf region of the North Atlantic show a distinct shift in the species and amounts taken over the years. Fisheries managers called for a 50% reduction in allowable catch of cod in 1988, but political officials only reduced the amount by 10%. The cod fishery was closed in 1992.

FISH CATCH BY COMMERCIAL GROUP: NEWFOUNDLAND–LABRADOR SHELF

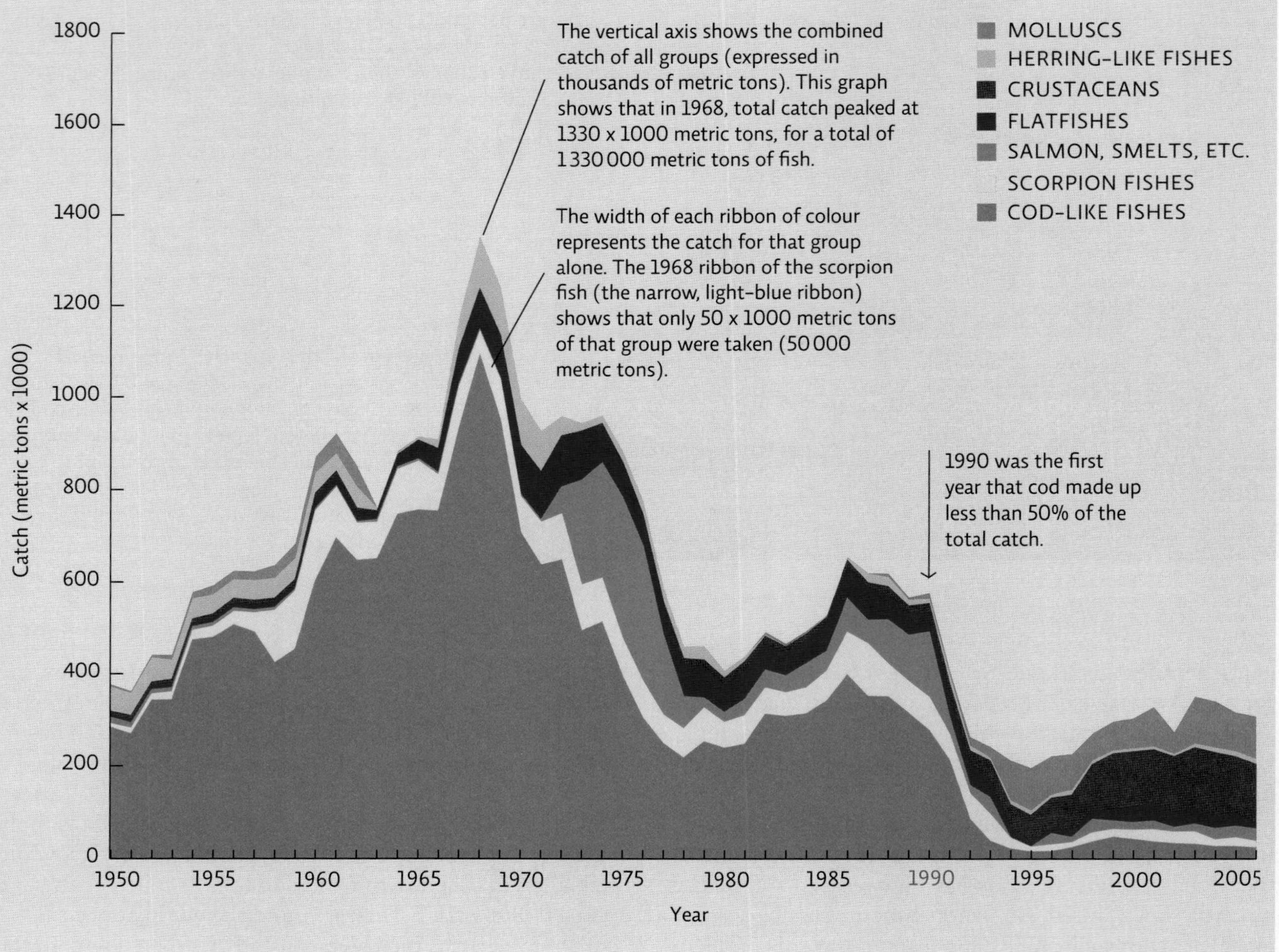
1800
1600
1400
1200
1000
800
600
400
200
0
Catch (metric tons x 1000)
1950
1955
1960
1965
1970
1975
1980
1985
1990
1995
2000
2005
Year
The vertical axis shows the combined catch of all groups (expressed in thousands of metric tons). This graph shows that in 1968, total catch peaked at 1330 x 1000 metric tons, for a total of 1 330 000 metric tons of fish.
The width of each ribbon of colour represents the catch for that group alone. The 1968 ribbon of the scorpion fish (the narrow, light-blue ribbon) shows that only 50 x 1000 metric tons of that group were taken (50 000 metric tons).
1990 was the first year that cod made up less than 50% of the total catch.
MOLLUSCS
HERRING-LIKE FISHES
CRUSTACEANS
FLATFISHES
SALMON, SMELTS, ETC.
SCORPION FISHES
COD-LIKE FISHES

Infographic **14.5** | **CANADA'S EXCLUSIVE ECONOMIC ZONE AND MARINE PROTECTED AREAS**

→ Marine protected areas (MPAs) provide varying degrees of protection for different species, depending on the area and the need. Exclusive economic zones (EEZs) extend 200 nautical miles (370 kilometres) from the coastline of any given nation, giving that nation exclusive rights over marine resources, including fish. Both MPAs and EEZs can help protect species at risk by restricting harvests and use in the area.

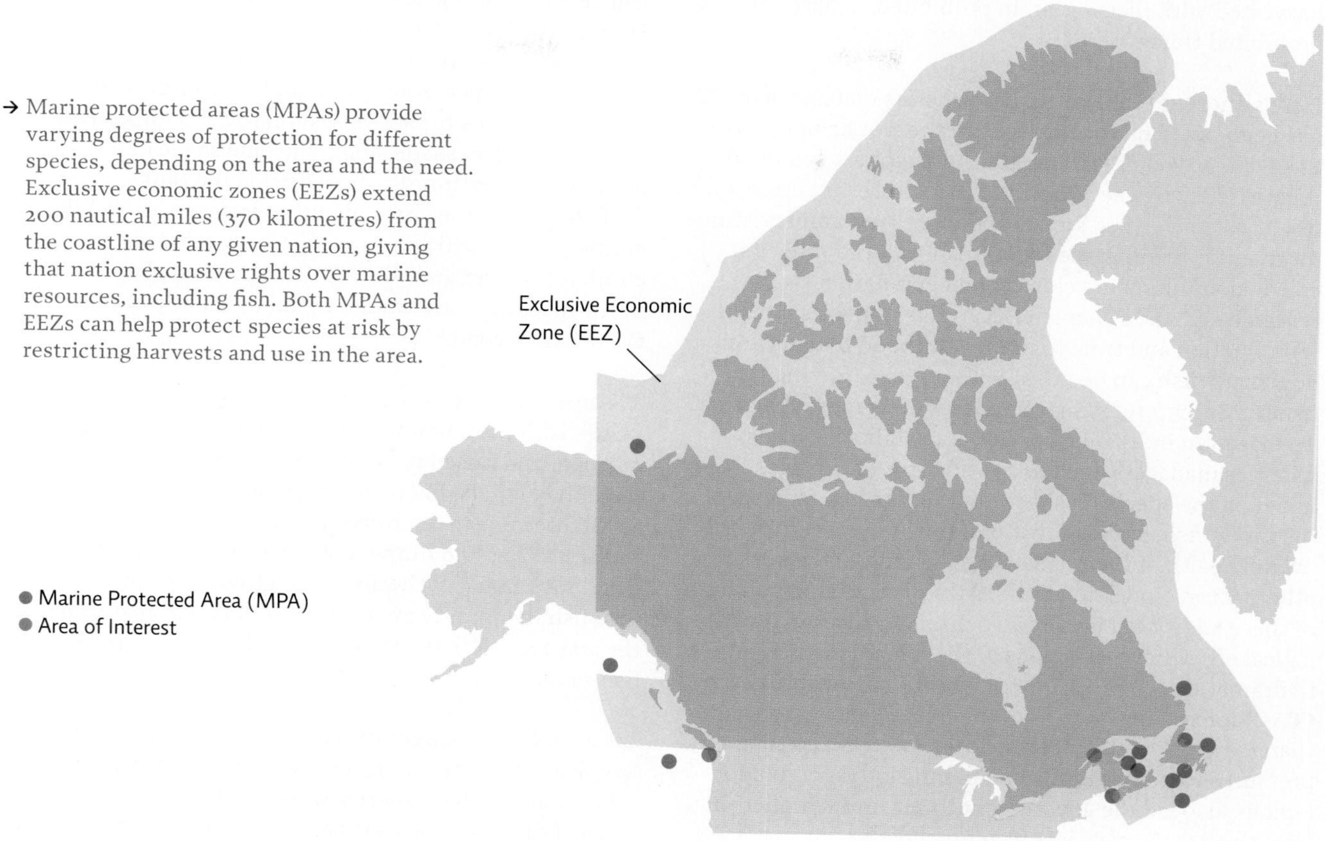
Exclusive Economic Zone (EEZ)
Marine Protected Area (MPA)
Area of Interest

Canadian cod fisheries, for example, officials set strict limits on the amount of ground (bottom-dwelling) fish of any species that any vessel may take. Once the vessel reached its predetermined limit, it was required to stop fishing for the season.

Fisheries managers worldwide have also created more than 1500 **marine protected areas (MPAs)**—discrete regions of ocean that are legally protected from various forms of human exploitation. As the country with the longest coastline in the world (including 3 oceans and the 5 Great Lakes), Canada plans to develop a national network of over 800 MPAs, in a comprehensive approach to protecting marine resources. The Oceans Act is federal legislation that allows **Fisheries and Oceans Canada** to designate existing "areas of interest" as official MPAs. Many of these MPAs protect only specific elements within the area—certain species of fish, a discrete portion of the habitat (like a spawning ground), or a cultural artifact (like a sunken ship). Some of them—albeit only a small number—are fully protected **marine reserves**, "no-take" zones, where no fishing can occur and human activities are prohibited. Sea life does recover in and near no-take MPAs, so there is reason to be encouraged. [INFOGRAPHIC 14.5]

The primary law governing marine fisheries management in Canadian waters is the **Fisheries Act**, first passed in 1867 (the same year the nation was formed). It was tweaked over the years and modernized in 1985. It regulates fishing, as well as the harvesting of marine plant life, and protects marine habitats by prohibiting destruction and pollution. The main goal of the act is to maintain a sustainable fishery for commercial, recreational, and Aboriginal uses.

But managing oceans, which make up 70% of the planet, is no small feat, and despite these efforts, fisheries have continued to dwindle. Part of the problem is the enormous amount of illegal, unreported, and unregulated fishing—between \$4 and \$9 billion worth every year.

marine protected areas (MPAs) Discrete regions of ocean that are legally protected from various forms of human exploitation.

Fisheries and Oceans Canada The federal agency that protects oceans and manages fisheries.

marine reserves Restricted areas where all fishing is prohibited and absolutely no human disturbance is allowed.

Fisheries Act The federal law that regulates fishing and the harvesting of marine plant life, and protects marine habitats.

In some fisheries, 30% or more of the annual catch is harvested with illegal gear, in prohibited areas, or at prohibited times.

Here's another problem: so far, fisheries management has been based on incomplete science. For example, *The State of the World's Fisheries and Agriculture,* a biannual report issued by the Food and Agriculture Organization (FAO), estimates fish stocks and makes recommendations for maintaining fisheries sustainably. But it bases its conclusions almost exclusively on population studies—research that uses characteristics like a species' rate of reproduction and time to reach adulthood to determine how many fish can be taken without jeopardizing the fishery. Successful management requires more than that. To maintain individual species at sustainable levels, fisheries managers need to consider the entire ecosystem in which the species resides—not just the abundance of the target fish, but also that of the fish they feed on, the overall health of their spawning grounds, and a host of other factors. Indeed, it was partly a lack of understanding of these additional factors—such as increases in technological efficiency—that caused the Canadian government to dramatically overestimate the **total allowable catch (TAC)** for cod in the 1980s, which let fishers catch more than the stock could stand to lose. Political and economic pressures to maintain the cod fishing industry in poor regions also led the government to set limits higher than even their own scientists recommended.

The Marine Stewardship Council, an international, nonprofit organization that certifies fish food products as sustainable, defines a **sustainable fishery** as one that ensures that fish stocks are maintained at healthy levels, that the ecosystem is fully functional, that fishing activity does not threaten biological diversity, and that the fishery is managed effectively. The Council currently recognizes 104 certified fisheries, and another 143 are undergoing assessment for certification.

Of course, no management strategy can reconcile limited supplies with rising demand, which is why, even as net pen operations like the ones in Frenchman Bay, Maine, proliferate around the world, more and more scientists are saying that the real solution to our fish problem won't be found in the ocean at all.

Scientists study the possibility of growing marine fish indoors.

Perched on the edge of Baltimore's Inner Harbor, the University of Maryland's Center of Marine Biotechnology (COMB) looks like a cross between a classic ship and a modern research facility. A sleek glass façade covers one side of the building, and a giant sail covers the other. Deep in the lower decks, in the lab of COMB director and marine scientist Yonathan Zohar, land and sea get blurred further still. Pipes, computers, and narrow cylindrical filters buzz and whir above several rows of tanks, each roughly the size, shape, and light-blue shade of a small above ground swimming pool. Inside the tanks, giant fin fish—mostly sea bass, many weighing over 36 kilograms—create whirlpools as they retrace the same circumference over and over; they seem blissfully unaware of the machinations around them, or the bay outside, or the ocean beyond.

The operation, known as a *recirculating aquaculture system* (RAS), is Zohar's brainchild. The fish were spawned, hatched, and raised to adulthood in the lab. If all goes to plan, they will be harvested and sold to area restaurants—as Mediterranean sea bass—and will eventually end up in the lunches and dinners of local patrons. "This is a real leap," says Zohar. "The ability to grow these large predatory fish indoors, away from the ocean, will really change the way we see these animals and the way we think about fish as food."

Zohar began his career in the 1970s on the banks of the Red Sea, when the environmental movement was in its infancy and global fisheries were just beginning to show signs of distress. Forward-thinking scientists had begun searching for ways to domesticate fish—just as humans had domesticated cattle, poultry, and pork—so that wild stocks might be spared as human populations swelled. **Aquaculture** had already been used for centuries to

↑ Dr. Yonathan Zohar works at COMB in Baltimore, Maryland.

↑ Sea bream fish farm, Shikoku, Japan

grow freshwater fish like tilapia. Canada, too, is trying its hand at aquaculture, which is projected to grow over time. Already, it's become the fourth largest producer of farmed salmon. Sadly, cod aquaculture has been less successful, as they are difficult to raise and easily substituted by other forms of white-fleshed fish, creating less of a market. But most fish farms are family or village operations—just big enough to provide a few dozen people with a steady food supply. Zohar wanted a fish farm that could feed a modern city, or even a country. He also wanted one that could raise large marine species. "With marine aquaculture, we are talking about fish that one, we are running out of, and two, are the most beneficial to human health," he says.

But there was a huge barrier to realizing this vision: most commercially important marine fish—like sea bream, sea bass, and tuna—would not reproduce in captivity. Some scientists had suggested recreating spawning grounds—specific areas where fish come every year to deposit their eggs and sperm—in captive breeding sites. But that seemed wildly impractical to Zohar. "Most of these fish travel hundreds to thousands of miles to reach their spawning grounds," he says. "They move through a wide range of temperatures, water depths, and salinities, and nobody had any clue which of these variables was the key to getting them to reproduce, let alone how to recreate all of that in a finite space." So instead he tracked several species across the ocean and measured their hormonal changes as they navigated the open waters. As it turned out, one particular hormone was produced in the wild but not in captivity. By developing a hormone supplement for the fish, Zohar and his colleagues tricked the captive fish into breeding as they would in the wild.

But just as that problem was resolved, other problems began creeping up.

Aquaculture presents environmental challenges.

Once scientists could get marine fish to reliably breed in captivity, entrepreneurs began adapting the aquaculture techniques used in freshwater operations to suit larger ocean fish. Before long, elaborate net pen operations began dotting coastlines around the world. This simple technology has enabled fish farms to out-produce traditional fishers. In 2009, aquaculture crossed the threshold of providing more than half of all seafood consumed worldwide, with Asia leading the way, and China alone producing some 63% of all farmed marine species. But, it has also led to a range of environmental problems including depletion of the populations of smaller fish harvested as food for aquaculture species, excess nutrient release into coastal waters, and ecosystem damage from the aquaculture

---

total allowable catch (TAC) The maximum amount (weight or numbers of fish or shellfish) of a particular species that can be harvested per year or fishing season in a given area; meant to prevent overfishing.

sustainable fishery A fishery that ensures that fish stocks are maintained at healthy levels, the ecosystem is fully functional, and fishing activity does not threaten biological diversity.

aquaculture Fish farming; the rearing of aquatic species in tanks, ponds, or ocean net pens.

Infographic **14.6** | **NET PEN AND POND AQUACULTURE: PROBLEMS AND POSSIBLE SOLUTIONS**

PROBLEMS	POSSIBLE SOLUTIONS
More diseases and parasites than wild fish Use of antibiotics and pesticides Large amounts of waste are released into the environment.	Decrease the density of net pens to minimize the impact. This would decrease concentrated waste (though there would still be some) and decrease/eliminate the need to use antibiotics or pesticides; pursue RAS aquaculture.
Fish farms displace mangrove swamps and other wetlands.	Move outdoor fish farms farther from the coast; robotic fish cages that can be navigated through the sea with boat-operated remote controls are being developed.
Farmed fish are fed fish meal made from wild caught fish so still exert pressure on wild stocks	Find alternative foods that do not rely on fully or overexploited wild caught fish.
High-trophic-level fish are typically raised; this requires large amounts of food (2 kilograms of fish for 0.5 kilogram of salmon; 9 kilograms for 0.5 kilogram of tuna).	Concentrate on fish lower on the trophic chain to increase the efficiency of food conversion.
Escape of non-native or genetically modified species	Only use native species if farming in natural waters.

↑ The Aquapod is a submersible net pen for open ocean aquaculture made from recycled materials. Aquapods are designed to be located several kilometres from the marine coast to lower the environmental impact while optimizing growing conditions for farmed fish.

ponds and net pens. Hundreds of thousands of hectares of tropical mangrove forests have been cleared to make way for shrimp aquaculture ponds. Mangroves are keystone species that provide habitat and protection for a wide variety of species, including humans, and protect coastlines from storm surges during hurricanes. Another problem is transfer of diseases and parasites from farmed to wild populations. In British Columbia, one study showed that farmed salmon spread sea lice to wild populations, causing up to 80% juvenile mortality. Most experts agree that these problems are not insurmountable. But how best to resolve them is a matter of some debate. [INFOGRAPHIC 14.6]

Indoor fish farming may provide a solution.

Zohar's laboratory starts with water from the same supply that feeds household taps. By adding a precise mix of salts and trace elements, the lab creates its own seawater. The water is monitored and regulated by computer as the seawater cycles through the swimming pool-sized fish tanks and the network of pipes and filters that make up these miniature, indoor oceans. Everything about the artificial salt water—from its temperature, salinity, and pH, to its CO_2 and oxygen concentrations—can be adjusted, on a tank-by-tank basis, to suit the species of fish and to mimic the changing conditions the fish would experience as they moved through both time and distance in the wild. "The entire system is biosecure," boasts Zohar. "We don't take a drop of water out of the harbour and we don't drain a drop of water into the harbour." In fact, the same thousands of litres of water are recycled over and over. Of course, in areas with water shortages, commandeering this much water to set up tanks may not be an option, so this may limit the applicability of this technology.

For Zohar, recirculating aquaculture represents a lifetime of research—each detail a carefully crafted response to the problems associated with existing aquaculture technology. Fish growing in Zohar's tanks require less food per kilogram than the same fish grown in a net pen. The reason for this is simple: because there are waves and because salinity can't be optimized, typical net pen fish expend more energy on movement and regulating their internal salt concentrations (osmoregulation) than they otherwise might. On top of that, food in the pens is not completely consumed; a portion of it sinks to the bottom of the pen, where it can't be recovered or reused. In Zohar's RAS setting, by contrast, aquaculturists can control food intake much better, and because salinity is always optimal, fish don't have to invest as much energy in osmoregulation. As a result, they convert their food into flesh at a much higher ratio.

To further reduce the dependence on smaller marine fish as a food source, the researchers are experimenting with a variety of alternative feeds. The main contender so far seems to be algae. In a room adjacent to the fish tanks, long plastic tubes—the size, shape, and colour of colossal lime green freezies—hang from thin metal racks. Each one is filled with algae that will eventually be converted into food pellets (along with several other ingredients) and given to the fish. "All of the optimal ingredients in marine fish—the omega-3s, etc.—all come from the base of the food chain," says Zohar. "So why not feed them algae?"

To manage accumulated liquid and solid waste, Zohar's team employs carefully calibrated microbial communities that function much like they do in the ocean. One community converts ammonia into non-toxic forms; other bacteria convert 96% of the solid waste into fuel-grade methane. "So on top of all the other benefits, the system produces energy, about 10 litres of biogas every day," Zohar explained, on a recent tour of the facility. That's not enough to power the entire system, he says, but it certainly offsets some of the energy costs. [INFOGRAPHIC 14.7]

As good as it all sounds, RAS facilities will have to meet a number of challenges before they can possibly hope

Infographic **14.7** | **BIOMIMICRY IN THE POOL**

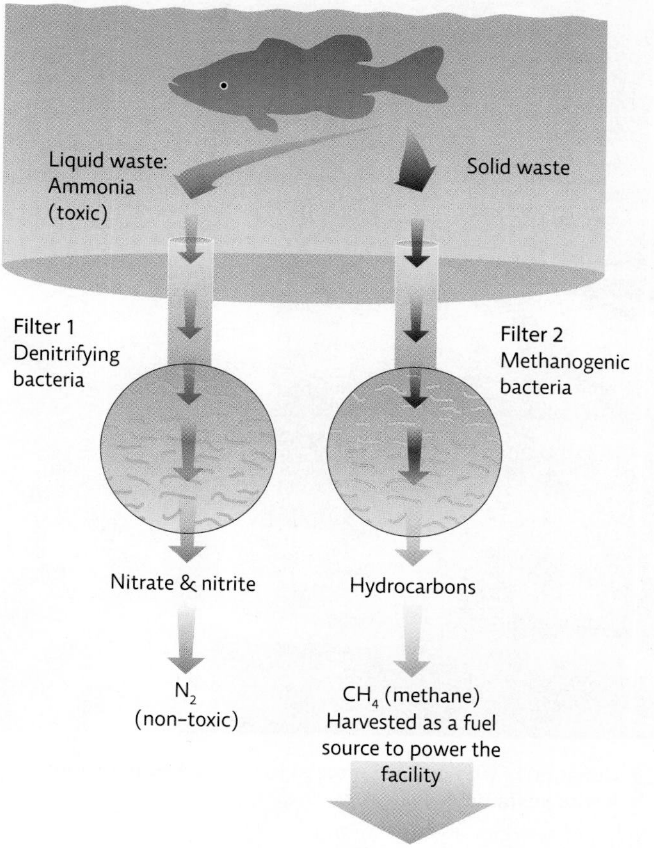
Liquid waste:
Ammonia
(toxic)
Solid waste
Filter 1
Denitrifying
bacteria
Filter 2
Methanogenic
bacteria
Nitrate & nitrite
Hydrocarbons
N_2
(non-toxic)
CH_4 (methane)
Harvested as a fuel
source to power the
facility

↑ Microbes in the RAS manage the waste in the water. Like their wild counterparts, these microbes detoxify ammonia (the same waste product that is so toxic to net pen fish) or break down solid waste. This system has the added benefit of providing an energy source for the facility.

to replace net pens. Net pens, for one, are much cheaper to set up and operate. The technology is simpler—and proven to work—and for fishers used to dealing with water and boats and navigation, growing fish in coastal or offshore net pens is less of a leap than growing them in warehouses.

"Systems like the one Yonathan has developed are definitely more environmentally friendly," says Lorenzo Juarez, deputy manager of the U.S. National Oceanic and Atmospheric Administration (NOAA) Aquaculture Program. "But in our world, everything comes down to economics. If it's going to take off, they have to prove it's profitable."

In a warehouse on the edge of Baltimore, tucked away from the harbour, Zohar and some business partners

What's DIFFERENT about our... Seafood
Pristine
Independent Fishermen
Sustainable
Our Docks
WHAT'S Special
Swordfish Steaks
TILAPIA
SEAFOOD

↑ Consumers who make choices at the market or in restaurants that support sustainable fishing practices may ultimately be the key to sustainable fisheries.

are trying to do just that. They've rented a nearly 17 000 square-metre space (that's 7 times larger than the main lab) where they plan to scale up operations—from tanks the size of small, above ground pools, to tanks the size of large, in-ground pools, and from 2 metric tons of fish per year to 400 metric tons of fish per year. "That's almost a million fish a year," says Zohar's business partner, David Wolf.

While the start-up costs are much higher than they would be for a net pen facility, RAS comes with its own set of advantages, too. For example, as Zohar and Wolf are quick to point out, the type of fish grown can be chosen based on economic opportunity rather than location. "So you can grow Mediterranean fish in Maryland or Minnesota," says Wolf. "We could literally raise these fish next to a distribution house in Kansas, and because of that proximity, our fish will be fresher, and probably cheaper, too."

Right now, roughly 85% of fish consumed in the United States is imported, often from great distances. Not only does that make the fish less fresh and more expensive, it also means the environmental footprint for Mediterranean fish consumed in U.S. restaurants—which includes fossil fuel for transportation and freezing, along with the resultant air pollution—is significant. By reducing the distance between the fishery and the consumer, RAS technology could make that footprint much smaller.

In an ideal world, aquaculture would provide the vast majority of fish we humans eat, while wild populations—along with the ecosystems they inhabit—would be left to recover, with only sustainable harvesting of healthy stocks. If RAS facilities prove commercially viable, coastal aquaculture ponds and net pen farms in developing countries could be scaled back, the mangrove swamps and coastal ecosystems they displaced could be restored, and, just maybe, fish could be produced, both profitably and sustainably, for local markets rather than shipped thousands of kilometres across the globe.

However, the start-up costs of a RAS facility make it unlikely to be feasible in poorer regions. Nor will large-scale RAS facilities be a viable option in water-starved regions, where such large quantities of fresh water can be difficult to come by. Meanwhile, the loss of net pen farms, while good for the environment, translates into a loss of income for areas that rely heavily on fish exports.

As with many environmental problems, we must consider the triple bottom line (see Chapter 1) and take steps to address the social and economic issues as well as the environmental ones.

And finally, even if RAS operations can be made cost-effective, and the fish feed problem can be solved, and the fishers retrained, there will still be one more hurdle to clear before fish become as domesticated as chicken. Consumers must accept farmed fish, must come to think of them as no different than, or even better than, those harvested from the ocean. This may be an easier sell in developing countries, where some 2 billion people rely on fish protein for their very survival. But in North America, where fish protein is still something of a luxury, and where the first RAS facilities are just coming to life, Zohar knows that it will all come down to taste. That's why, when the first harvest of Mediterranean sea bass was ready, he organized a cook-off at a local restaurant. Chefs from five nearby restaurants prepared two plates of five different dishes—one using wild caught fish and one using the fish he had grown in his laboratory fish farm. The result? "You couldn't tell them apart," says chef Damon Hersh, who participated in the contest. "They tasted exactly the same." ◉

Select references in this chapter:
FAO Fisheries and Aquaculture Department. 2010. *The State of World Fisheries and Aquaculture*. Rome: Food and Agriculture Organization of the United Nations.
Fisheries and Oceans Canada. 2013. *Marine Protected Areas*. www.dfo-mpo.gc.ca/oceans/marineareas-zonesmarines/mpa-zpm/index-eng.htm.
Krkosek, M., *et al.* 2007. *Science*, 318 (5857): 1772-1775.
Mason, F. 2002. The Newfoundland Cod Stock Collapse: A Review and Analysis of Social Factors. *Electronic Green Journal*, 1(17). http://escholarship.org/uc/item/19p7z78s#page-1.
Sea Around Us Project. 2011. *EEZ Waters of Canada*. http://www.seaaroundus.org/eez/124.aspx.

BRING IT HOME

➲ PERSONAL CHOICES THAT HELP

While many see oceans as a vast and infinite resource, you now understand some of the threats facing our oceans and fisheries. By properly managing our fisheries, using sustainable harvest methods, and working to prevent and reduce the pollutants that enter our bodies of water, we can have oceans and rivers that support both aquatic life and human life for years to come.

Individual Steps

→ Educate yourself on which fish we should be eating: you can now check to see which fish species and locations have stable population levels and are being caught using sustainable methods. For an example, see www.seachoice.org, where you can search for recommended seafood choices, download a printable guide, or download the SeaChoice app for your smart phone.

→ When purchasing prepackaged or fresh fish in a grocery store or restaurant, look for the Marine Stewardship Council seal, which labels products sourced from wild-caught fisheries that have been independently certified to the MSC environmental standard.

MARINE STEWARDSHIP COUNCIL

Group Action

→ Organize a beach or river clean-up to prevent pollution from entering waterways.

Policy Change

→ Write or talk to restaurant owners to ask them to serve only fish that are considered sustainable.

→ Write your policy makers to encourage them to establish marine protected areas and support legislation that protects aquatic health.

UNDERSTANDING THE ISSUE

CHECK YOUR UNDERSTANDING

1. Why might aquaculture be considered the "future of fishing?"

a. Systems like recirculating aquaculture systems (RAS) are so affordable that more fishers will turn to them.
b. Today's aquaculture techniques provide fish to consumers without damaging the environment.
c. It can produce large numbers of fish without depleting wild stocks.
d. It will raise fish to be released into the ocean to support commercial fisheries.

2. Why might bottom-trawling fishing techniques make it harder for the cod population to recover even after the ships have left the area?

a. No fish are left to breed.
b. The nets damage the seabed where cod spawn.
c. The cod leave the area once a trawler has come through.
d. The ships attract seabirds that eat the cod.

3. In what way is the RAS waste-treatment plan an example of biomimicry?

a. It purifies the water with biodegradable filters.
b. It depends on bacteria doing what they normally do in nature.
c. It allows the waste to naturally sink to the bottom of the pool where it remains out of the way.
d. It focuses on frequent water changes rather than on purification of water.

4. The collapse of the cod fishery is a tragedy of the commons because:

a. some fishers are not allowed to fish in the Grand Banks.
b. cod was the most common fish in the area.
c. it was the privatization of the fishery that caused its collapse.
d. fishers would take as much as possible because if they didn't, someone else would.

5. A disadvantage of net pen operations is:

a. fish that can be successfully raised this way are not popular food items.
b. these operations are expensive.
c. they release pollution into the surrounding water.
d. they cannot be done on a large scale.

6. "Fishing down the food chain" refers to:

a. taking only higher-trophic-level species and throwing back lower-trophic-level species.
b. making sure equal numbers of individuals are taken from all the trophic levels.
c. current management techniques designed to restore a depleted fishery.
d. moving on to take lower-trophic-level species once the higher levels are depleted.

WORK WITH IDEAS

1. Why do we define a fishery in terms of human use? Does that diminish it as an ecological concept?

2. Ecosystems like those found in the extreme environment of the Arctic have simple food webs, with only a few organisms at each trophic level, whereas ecosystems with more moderate climates have more robust food webs with many species. Why is a simple food web more vulnerable to collapse than a more complex one?

3. What are the advantages of RAS like the one Yonathan Zohar is developing and where (in what nations or regions) will they likely be adopted? If adopted in these areas, will they help lower the impact of eating fish? Explain.

4. Consider the triple bottom line and propose (based on current technology) a low-impact way for people in developing nations to meet their need for fish.

ANALYZING THE SCIENCE

The graph to the right comes from *The Ecological Fishprint of Nations* report by Redefining Progress. It tracks the global fishprint (a measure of the ocean area needed to produce the fish catch in hectares per metric ton of fish, weighted to reflect the ecological productivity of particular biomes) against global biocapacity (a measure of the ocean's ability to supply a steady quantity of fish based on the producer productivity in ocean ecosystems and also expressed in terms of area).

Global Fishprint and Biocapacity 1950–2003

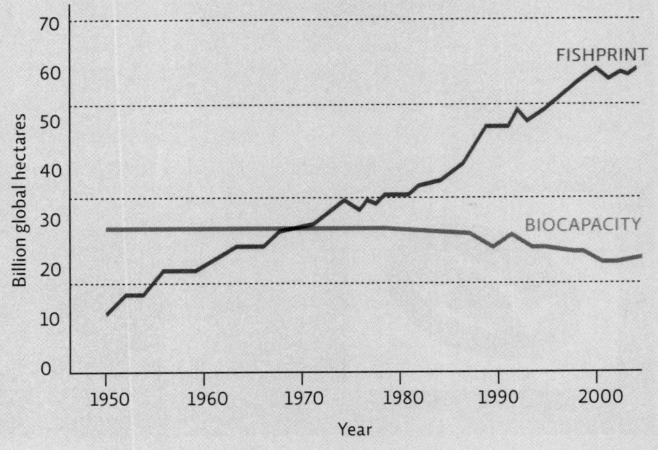
70
60
50
40
30
20
10
0
Billion global hectares
FISHPRINT
BIOCAPACITY
1950
1960
1970
1980
1990
2000
Year

INTERPRETATION

1. In one sentence, explain what has happened over time in terms of the human fishprint and the ability of global oceans to support it.

2. In what year did our fishprint exceed the biocapacity of the ocean?

3 Ecological overshoot is the difference between the fishprint and biocapacity. What was the ecological overshoot in 1980, in global hectares (gh)? In 1990? In 2003?

ADVANCE YOUR THINKING

Hint: To answer some of the questions below it might help to access the actual report.

4. Why do you suppose global biocapacity has decreased in recent years?

5. If RAS facilities like the one described in this chapter are widely adopted, what would likely happen to biocapacity and fishprint in the future? Explain why.

6. What would it take to return to a global fishprint and biocapacity as they were in the 1950s or 1960s? Describe three specific strategies using the fishprint as a tool to help accomplish this reversal.

EVALUATING NEW INFORMATION

Should one eat fish? Many of the world fisheries are seriously degraded. But fish are a major protein source for many people around the world and a healthy alternative to other forms of meat. So what is a conscientious consumer to do?

Explore the SeaChoice website (www.seachoice.org).

Evaluate the website and work with the information to answer the following questions:

1. Is this a reliable information source? Does it have a clear and transparent agenda?
   a. Who runs this website? Do the organization's credentials make the information presented reliable or unreliable? Explain.
   b. What is the mission of this organization? What are its underlying values? How do you know this?
   c. What data sources does SeaChoice rely on and what methodology does it employ in calculating its rankings? Are the sources it uses reliable?
   d. Do you agree with its ranking system? Do you think that such a ranking system is sufficient and useful to help consumers make sustainable choices in selecting fish? Explain your responses.

2. From the main page, choose the "State of Our Oceans" link.
   a. Do you agree with the organization's assessment of the ocean issues? Explain.
   b. Identify a claim that is made and the evidence given in support of this claim. Is it sufficient? Explain.

3. The website offers a wallet guide that identifies safe fish to eat based on health and ecological considerations. Open the "Recommendations" tab, select "Browse for Seafood" and type in "cod" in the search box; select Atlantic cod from the choices that appear. Why are some Atlantic cod identified as fish to avoid and others not?

4. Go back to the "Recommendations" tab and select the link for "Resources." Open the wallet guide to Canada's sustainable seafood. Look at the fish listed and find one that you eat frequently. (If you don't eat fish, choose one of the tuna species, a popular fish in Canada.) Is this a good fish to consume? If not, identify an alternate fish you could consume instead.

5. How might a program like SeaChoice help the oceans?

MAKING CONNECTIONS

THE FUTURE OF SALMON FISHERIES

Background: According to SeaChoice, Alaskan wild salmon are an ocean-friendly choice, but other salmon fisheries raise concerns. The management of salmon fisheries is challenging, as salmon require both freshwater and ocean habitats.

One solution is to switch to salmon aquaculture. But while salmon raised in tank systems are environmentally friendly operations, most salmon are farmed in open net pens and should be avoided for their environmental impact. Wastes that pollute water, transmission of disease to wild salmon stocks, and interference with life cycles of wild fish from farmed fish (which are selectively bred or non-native species of salmon) that escape are the major concerns.

Case: To ensure that the fate that befell the cod fisheries is not repeated for salmon, efforts are being made to develop a sustainable salmon aquaculture. You have been assigned the task of developing a policy for the future of salmon farming based on an evaluation of the following options:

1. The AquAdvantage salmon: a genetically modified salmon that reaches market-size in 18 months—half the time of a wild salmon.

2. Integrated multi-trophic aquaculture (IMTA) systems: these blend the farming of fish with the culture of seaweed and shellfish that process the waste the fish generate.

3. Recirculating aquaculture systems (RAS): above ground, temperature-controlled environments that raise single species of fish without any interaction with the wild.

4. The Arctic char: a salmon-like fish with a smaller footprint.

Research these four options and write a report recommending a course of action to invest in. In your report, include the following:
   a. An analysis of the pros and cons of each proposal including:
      • A discussion of the consequences of each choice from an economic, ecological, food security, and human health perspective.
      • A reflection on the values underlying each proposal.
   b. The best option for the future of salmon aquaculture. Who should be involved in this decision? Provide justifications for your proposal.
   c. A set of guiding principles that can be applied to sustainably managing any fish species as a food source, based on what you now know about fishery management and aquaculture, and their challenges. Discuss the principles you develop and explain why you consider them key for the future of sustainable seafood.

A glass of treated water from the Groundwater Replenishment System in Fountain Valley, California.

CHAPTER 15 FRESHWATER RESOURCES

TOILET TO TAP

A California county is tapping controversial sources for drinking water

CORE MESSAGE

Freshwater is a precious but limited resource, and is essential to life. Some regions consume water faster than it is replenished. And unfortunately, water is not evenly distributed across the globe; many people worldwide lack access to enough clean water. Methods are available to recover and purify otherwise dirty water, but we also need to use water more wisely.

GUIDING QUESTIONS

After reading this chapter, you should be able to answer the following questions:

→ What are the sources of freshwater on Earth and how does water cycle through the environment?

→ What are the causes and consequences of water scarcity?

→ What is an aquifer, how does it receive water, and what problems emerge when too much water is removed?

→ What are some of the ways that our wastewater is treated to make it usable again or safe to release into the environment?

→ How can conservation help us address water scarcity issues?

One of the most exciting moments in Shivaji Deshmukh's career as a water engineer came one bright, sunny day in January 2008. He had gathered with staff from the Orange County Water District (OCWD) in Anaheim, California, to watch for the first time as former sewage water, cleaned using state-of-the-art techniques, was pumped into underground drinking water sources. It was the beginning of a groundbreaking project designed to help save the region from ongoing, and frightening, water shortages.

"It's basically this drought-proof supply of water," says Deshmukh. "Nobody else has done it. Nobody thought a community could support it, because they would be too grossed out by it."

The water that Deshmukh and other engineers watched seep into the region's underground water stores that day in 2008 was treated **wastewater**—including sewage and used water from homes and industrial sites. Understandably, when many residents first heard about the project, they were concerned.

But that same month, Deshmukh and other OCWD staff attended a dedication ceremony at the water treatment plant in Fountain Valley, California, along with hundreds of other people, including various community groups, to honour the massive project. Having that support from the community was key to the project's success, says Deshmukh—but getting it hadn't been easy.

◉WHERE IS ANAHEIM, CALIFORNIA?

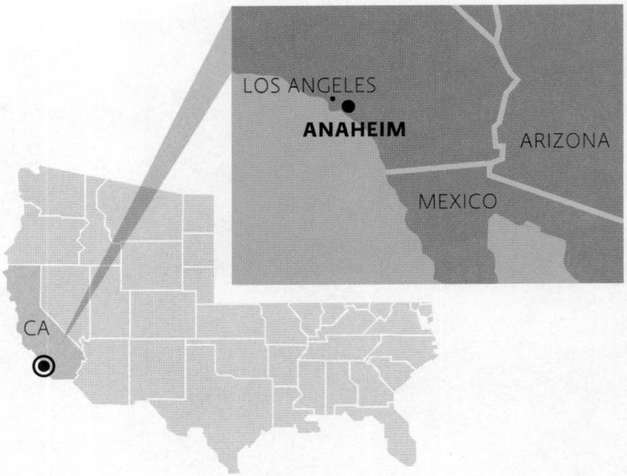
LOS ANGELES
ANAHEIM
ARIZONA
MEXICO
CA

Water is one of the most ubiquitous, yet scarce, resources on Earth.

Even though Earth is covered in more than 1400 million cubic kilometres of water—about 75% of its surface—only about 1/100 of 1% of that water is usable by humans.

Water provides many important ecosystem services that animals and plants require to live. Up to 75% of the human body, for instance, consists of water. But humans need liquid **freshwater** (which has few dissolved ions such as salt); ocean water is too salty for human consumption and is toxic in large doses.

Complicating things, nearly 80% of the freshwater on the planet is trapped in ice caps at the poles and glaciers around the world, which contain more than 35 million cubic kilometres—enough to fill 140 billion Olympic-sized swimming pools. [INFOGRAPHIC 15.1]

Wherever there is water, it is constantly moving through the environment via the **water cycle** (hydrologic cycle). Heat from the sun causes water to evaporate from **surface waters** (rivers, lakes, oceans) and land surfaces. At the same time, plant roots pull up water from the soil and then release some into the atmosphere in a process called **transpiration**. Plants with deep roots, like trees, may bring up many thousands of litres of water a year, releasing much of this to the atmosphere. Altogether, the combination of **evaporation** and transpiration—*evapotranspiration*—sends more than 64 000 trillion litres of water vapour into the atmosphere every year. Once

wastewater Used and contaminated water that is released after use by households, industry, or agriculture.
freshwater Water that has few dissolved ions such as salt.
water cycle The movement of water from gaseous to liquid states through various water compartments such as surface waters, soil, and living organisms.
surface water Any body of water found above ground such as oceans, rivers, and lakes.
transpiration The loss of water vapour from plants.
evaporation The conversion of water from a liquid state to a gaseous state.

↑ Crystal-clear purified water from the Groundwater Replenishment System is piped to the Orange County Water District's percolation ponds in Anaheim, California.

Infographic **15.1** | **DISTRIBUTION OF WATER ON EARTH**

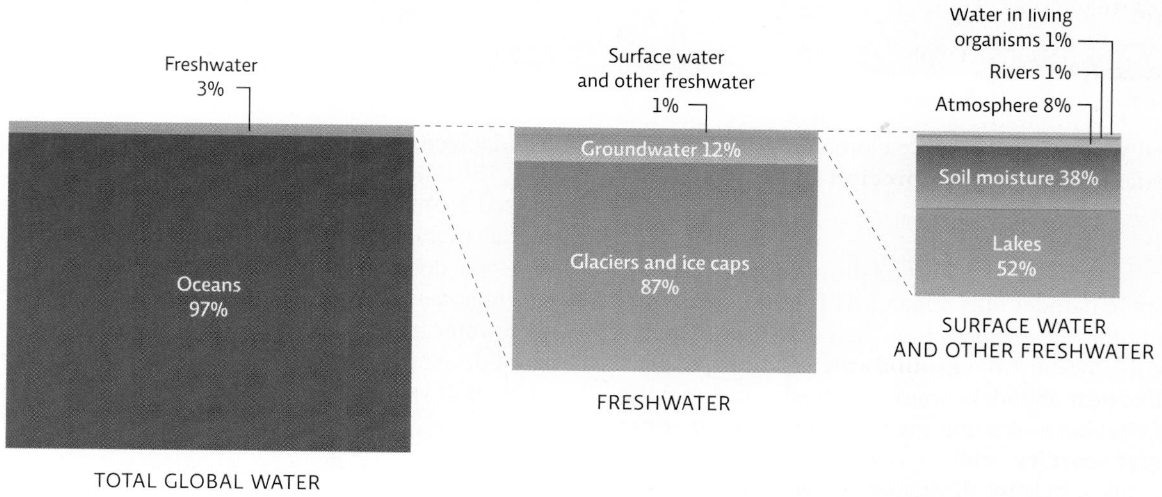

Freshwater
3%
Oceans
97%
TOTAL GLOBAL WATER
Surface water
and other freshwater
1%
Groundwater 12%
Glaciers and ice caps
87%
FRESHWATER
Water in living
organisms 1%
Rivers 1%
Atmosphere 8%
Soil moisture 38%
Lakes
52%
SURFACE WATER
AND OTHER FRESHWATER

↑ Most of the water on Earth is found in the oceans and most of the freshwater is tied up in ice and snow. Only about 0.001% of all of Earth's water is available for us to use, but with more than 1.3 trillion cubic metres of water on the planet, that is still a lot of water.

Infographic **15.2** | **THE WATER CYCLE**

↓ Water cycles between liquid and gaseous forms as it moves through space and time. Ocean water (which we cannot use) is converted to freshwater when it evaporates and falls back to Earth as precipitation, refilling freshwater surface and underground water supplies. Liquid freshwater is a renewable resource as long as we don't use it faster than it is naturally replenished.

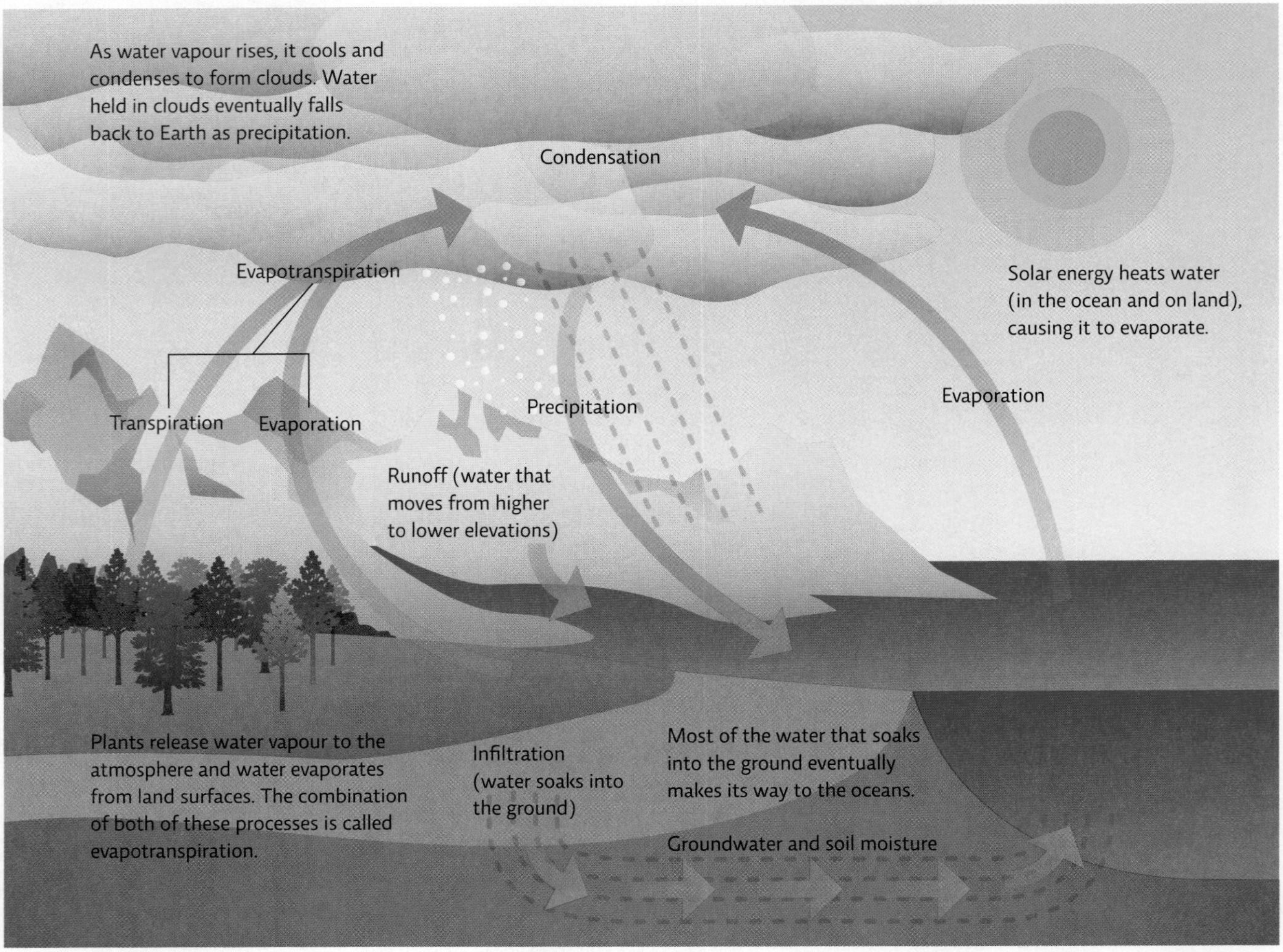
As water vapour rises, it cools and condenses to form clouds. Water held in clouds eventually falls back to Earth as precipitation.
Condensation
Evapotranspiration
Solar energy heats water (in the ocean and on land), causing it to evaporate.
Transpiration
Evaporation
Precipitation
Evaporation
Runoff (water that moves from higher to lower elevations)
Plants release water vapour to the atmosphere and water evaporates from land surfaces. The combination of both of these processes is called evapotranspiration.
Infiltration (water soaks into the ground)
Most of the water that soaks into the ground eventually makes its way to the oceans.
Groundwater and soil moisture

aloft, that water condenses into clouds (**condensation**) and may fall back to Earth as **precipitation** (rain, snow, sleet, etc.). [INFOGRAPHIC 15.2]

Almost all precipitation ends up falling on the oceans, and a tiny remainder falls on land. This latter portion is the part humans can harvest for their own use. Typically, we draw freshwater from **groundwater**. But people don't always live near abundant sources of freshwater, making access a vital issue. Around the world, many areas suffer from **water scarcity**—not having access to enough clean water supplies. In some dry regions, there is simply not enough to meet needs; many arid nations like those of the Middle East, parts of Africa, and much of Australia face water shortages as a way of life. According to the Water Stress Index, compiled by the risk assessment analyst firm Maplecroft, the Middle Eastern nations of Bahrain, Qatar, Kuwait, and Saudi Arabia are the most water-stressed countries in the world, with the lowest per capita water availability. In other areas, there is enough water but people do not have the money to purchase it or dig wells to access it. By far, per capita domestic (household) use of water in developed nations is much higher than that of developing nations—Canadians alone use up to 350 litres of water per day. A wealthy nation may even be

condensation Conversion of water from a gaseous state (water vapour) to a liquid state.
precipitation Rain, snow, sleet, or any form of water falling from the atmosphere.
groundwater Water found underground in aquifers.
water scarcity Not having access to enough clean water.

Infographic **15.3** | **AVAILABILITY AND ACCESS TO WATER WORLDWIDE**

↓ Around the world, more than 2 billion people lack access to clean water. An additional 2.6 billion have no access to sanitation.

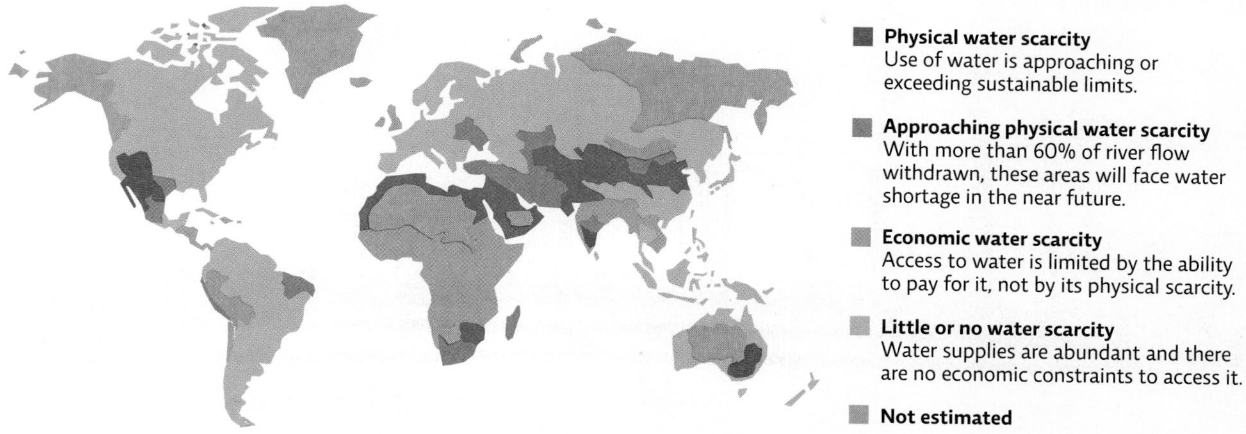
Physical water scarcity
Use of water is approaching or exceeding sustainable limits.
Approaching physical water scarcity
With more than 60% of river flow withdrawn, these areas will face water shortage in the near future.
Economic water scarcity
Access to water is limited by the ability to pay for it, not by its physical scarcity.
Little or no water scarcity
Water supplies are abundant and there are no economic constraints to access it.
Not estimated

↓ Sanitation: Access to adequate sanitation facilities is higher in urban areas than in rural areas in many regions of the world.

PERCENT OF POPULATION WHO USE IMPROVED SANITATION

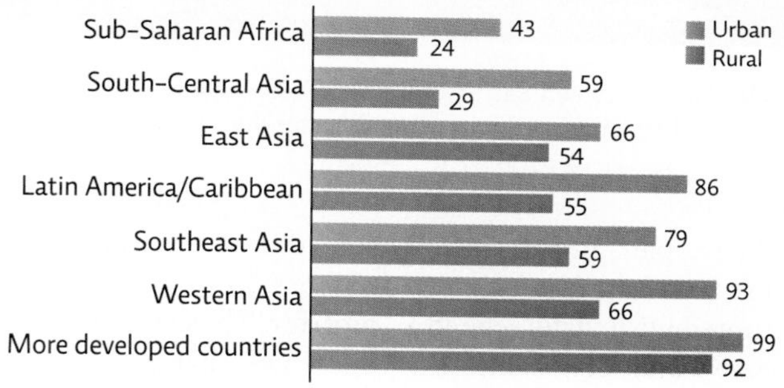
Urban
Rural
Sub-Saharan Africa 43 24
South-Central Asia 59 29
East Asia 66 54
Latin America/Caribbean 86 55
Southeast Asia 79 59
Western Asia 93 66
More developed countries 99 92

↓ Water footprint: Per capita water use is growing faster than our population. Developed countries have a larger per capita "water footprint" than developing nations. In some areas, individuals must make do with only a few litres a day.

DOMESTIC WATER USE

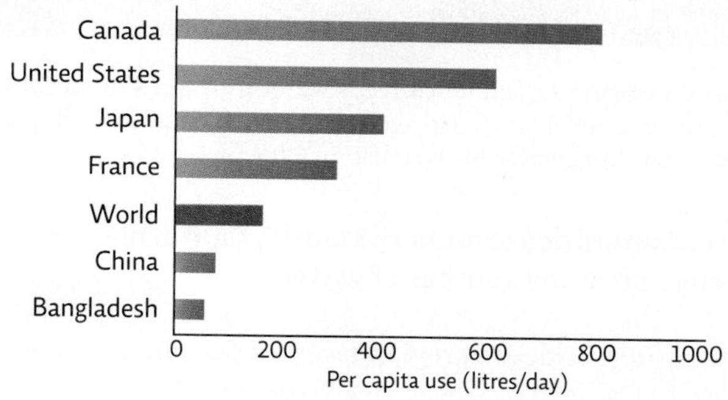
Canada
United States
Japan
France
World
China
Bangladesh
0
200
400
600
800
1000
Per capita use (litres/day)

able to invest in costly technology to access water (like facilities to remove salt from seawater), but poorer nations do not have this luxury. People in remote regions may have to resort to digging deep wells or installing catchment systems to capture rainwater.

The World Health Organization (WHO) estimates that 1 in 3 people—more than 2 billion—lack sufficient access to clean water; even more lack access to sufficient sanitation facilities (safe disposal of human waste). In developing nations where water and funding for basic sanitation are scarce, people use nearby surface waters to meet their basic cooking, drinking, and washing needs. These waters can be contaminated with raw sewage, which increases the chance for disease transmission. According to the WHO, more than 1.1 million litres of raw sewage enters the Ganges River of India every minute. In Africa, almost 3000 people die each day from water-borne diseases like cholera and typhoid fever as a result of poor sanitation and contaminated water.

The World Health Organization (WHO) estimates that 1 in 3 people—more than 2 billion—lack sufficient access to clean water.

As populations increase, so will scarcity and sanitation issues; according to the United Nations, 2 out of 3 people will face water shortages by 2025. [INFOGRAPHIC 15.3]

↑ The California Aqueduct conveys water past Kettleman City, located halfway between San Francisco and Los Angeles, to Southern California. Local agencies are working to develop a treatment facility that would tap into the aqueduct to provide clean drinking water for Kettleman City.

Like communities around the world, California depends on many sources of water.

Around the world, each region faces its own water challenges. In California, freshwater flows into the northern part of the state when the Sierra Mountain snowpack melts in the spring. But as Earth's climate changes, the state could lose much of its snowpack. Indeed, in 2009, California's precipitation was 80% of its average, and the snowpack only 60% of its average size.

Even in a good snowpack year, the state faces major water issues, explains Deshmukh, now working at the West Basin Municipal Water District in Carson, California. Two-thirds of California's water is located in the northern part of the state, but two-thirds of the state's residents live in the south, he explains. At the moment, the state ships some of the northern water to the south via pipes and canals, which costs money and uses a great deal of electricity. And if there is an earthquake or other natural disaster, that transport system could be cut off.

Many people (not just in California) draw their water from an underground region of soil or porous rock saturated with water, called an **aquifer**. These stores receive water from rainfall and snowmelt that soaks into the ground through **infiltration**. Plant roots take up some of the water along the way, but much of it continues to move downward, filling every available space in the aquifer. As the water trickles down, it becomes naturally filtered by rocks and soil, which trap bacteria and other contaminants as the water passes by. The top of this water-saturated region, referred to as the **water table**, rises and falls due to seasonal weather changes.

The depth of Orange County's groundwater varies, says Deshmukh. At the coast, the aquifer is 6 to 90 metres deep, but further inland, at its deepest, the groundwater

aquifer An underground, permeable region of soil or rock that is saturated with water.

infiltration The process of water soaking into the ground.

water table The uppermost water level of the saturated zone of an aquifer.

Infographic **15.4** | **WATER USE**

↓ Globally, agriculture remains our biggest user of water and also the sector with the largest amount of waste. There is much room for improvement in how we use water in all sectors. For the average individual in a developed nation, reducing waste is tied to reducing consumption, not just of water but of industrial and agricultural products as well. It may be surprising just how much water goes into some of the products you use.

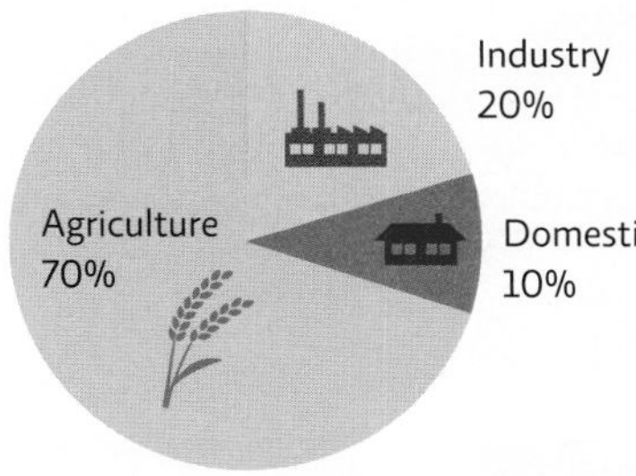
Industry
20%
Agriculture
70%
Domestic
10%

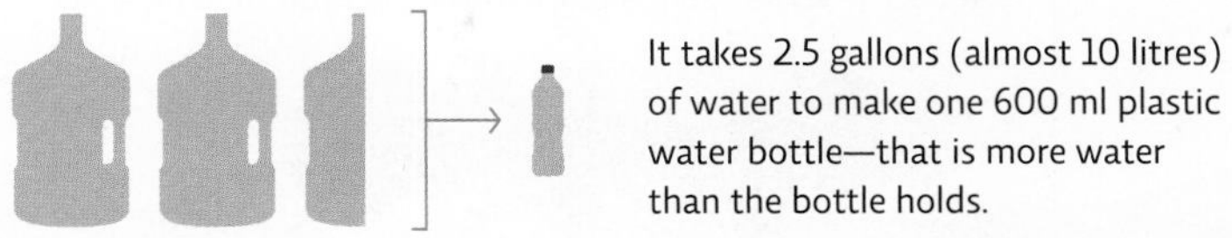
It takes 2.5 gallons (almost 10 litres) of water to make one 600 ml plastic water bottle—that is more water than the bottle holds.

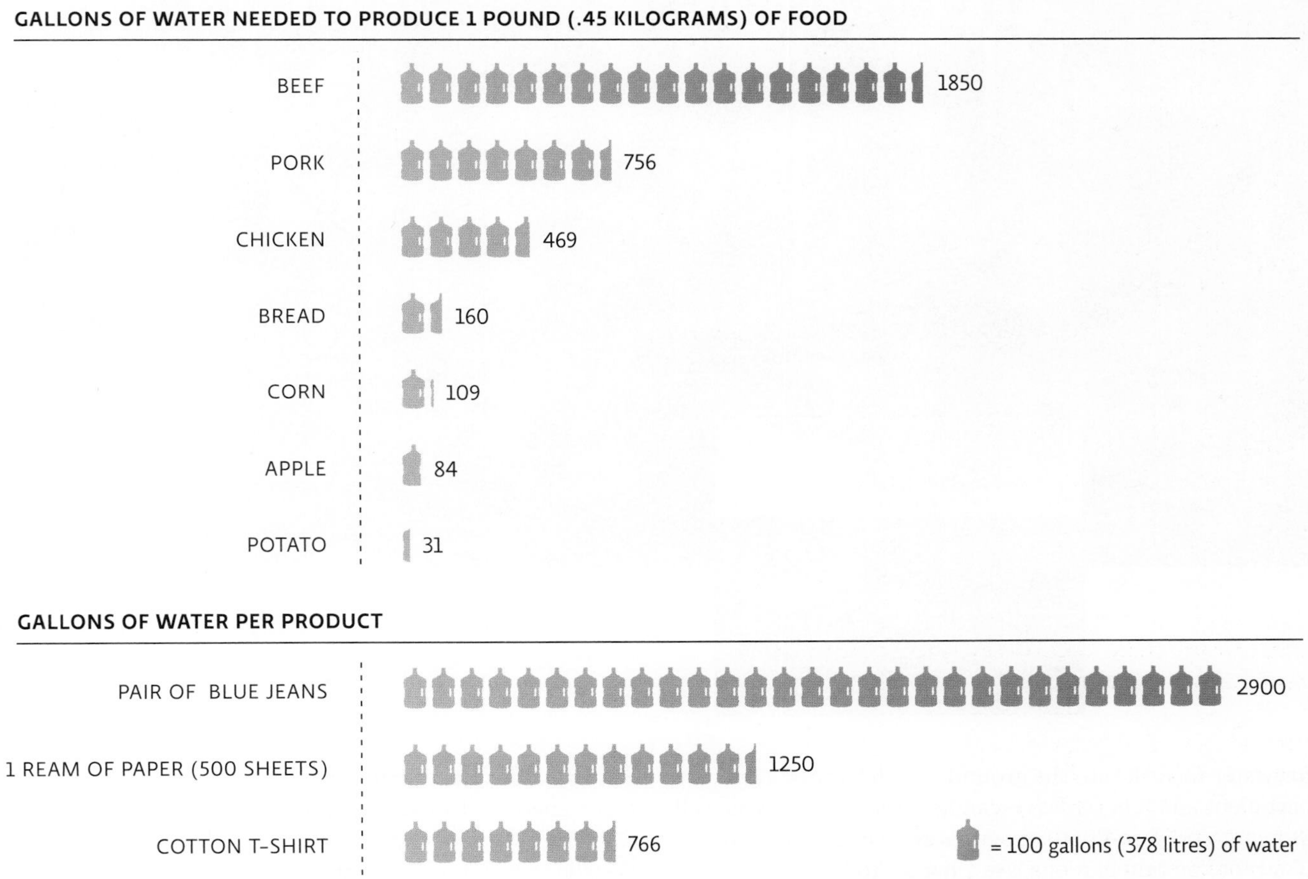
GALLONS OF WATER NEEDED TO PRODUCE 1 POUND (.45 KILOGRAMS) OF FOOD
BEEF 1850
PORK 756
CHICKEN 469
BREAD 160
CORN 109
APPLE 84
POTATO 31
GALLONS OF WATER PER PRODUCT
PAIR OF BLUE JEANS 2900
1 REAM OF PAPER (500 SHEETS) 1250
COTTON T-SHIRT 766
= 100 gallons (378 litres) of water

extends about 900 metres deep. Water quantity is usually measured in terms of cubic metres. Five hundred cubic metres of water is enough for one Canadian family for 1 year. Deshmukh estimates that over 6 million cubic metres of water is accessible from the deepest part of the aquifer. But that deep subterranean water is harder to get because it costs more to pump it out of the earth than the groundwater closer to the surface.

Aquifers remain dependable sources of water only as long as removal rates don't exceed replenishment rates. The biggest drain on freshwater supplies isn't the water used to shower, flush the toilet, and wash dishes. While the biggest user of water worldwide is agriculture (at 70%), in Canada, the major user is thermal power generation at 60%; agriculture uses only 9%, mainly due to high precipitation. [INFOGRAPHIC 15.4]

In addition, anything that reduces infiltration will reduce the rate at which the aquifer refills and thus decreases the amount of water we can sustainably remove. Infiltration is hampered in urban and suburban settings because of all of the hard surfaces, such as roads and buildings; even the typical suburban lawn is so compacted from the home construction process that very little water infiltrates the ground. Urban and suburban designs that provide ways

Infographic 15.5 | **GROUNDWATER IS FOUND IN AQUIFERS**

↓ Groundwater in aquifers is naturally replenished as water soaks into the ground. Humans can access this groundwater through wells but we can pull out water faster than it is naturally replaced. This can lead to saltwater intrusion in coastal areas or dry wells in inland areas.

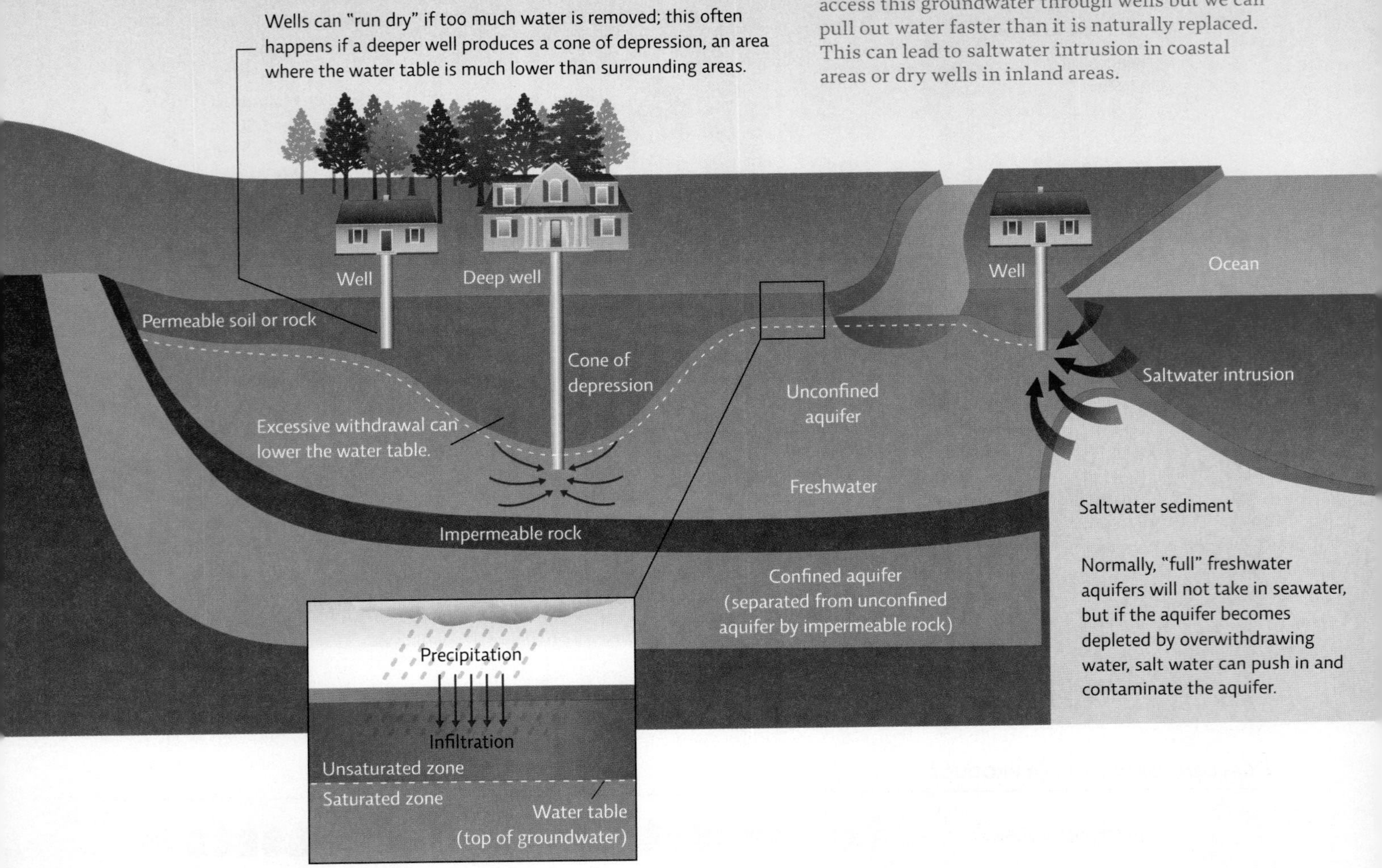

Wells can "run dry" if too much water is removed; this often happens if a deeper well produces a cone of depression, an area where the water table is much lower than surrounding areas.
Well
Deep well
Well
Ocean
Permeable soil or rock
Cone of depression
Unconfined aquifer
Saltwater intrusion
Excessive withdrawal can lower the water table.
Freshwater
Saltwater sediment
Impermeable rock
Confined aquifer (separated from unconfined aquifer by impermeable rock)
Normally, "full" freshwater aquifers will not take in seawater, but if the aquifer becomes depleted by overwithdrawing water, salt water can push in and contaminate the aquifer.
Precipitation
Infiltration
Unsaturated zone
Saturated zone
Water table (top of groundwater)

for water to soak into the ground—such as permeable pavement and *rain gardens*—can help refill aquifers as well as help to prevent flooding events after heavy rainfalls. (For more on rain gardens, see Chapter 16.)

Of course, Californians are lucky enough to have plenty of water all along the coast. But removing salt and other minerals from salt water—a process known as **desalination**—is expensive and uses a large amount of energy. Still, there are thousands of desalination plants worldwide operating today that meet some of the water needs of their regions. The largest such facilities in the world are in the Middle East, some of which are processing more than 750 million litres of water per day—about 10 times the volume of the largest U.S. plants (which are in Tampa Bay, Florida, and El Paso, Texas).

But in California, the proximity to salt water has actually threatened the freshwater supply.

Decades ago, Orange County officials discovered to their dismay that salt water was seeping into some of the region's aquifers, putting those precious freshwater stores in jeopardy. Groundwater levels are typically higher than sea level, so salt water doesn't infiltrate aquifers. But as freshwater was pumped out of the county's aquifer inland, the groundwater level dropped, and salty ocean water had started to enter the coastal edge of the aquifer to the west, where it bordered the Pacific. In Orange County, some aquifers are confined by geologic faults, which prevent ocean water from entering at some points—but not everywhere. It was in these unconfined coastal aquifers that **saltwater intrusion** was becoming a problem. [INFOGRAPHIC 15.5]

ULTRAVIOLET LIGHT

↑ The Groundwater Replenishment System in Fountain Valley, California, solves two problems at once: water scarcity and dealing with wastewater. This $480 million water treatment system converts the sewage water of Orange County into drinking water, producing more than 200 million litres of drinking water every day.

To stem the influx of salt water, in 1975 the Orange County Water District started pumping highly treated (purified) sewage wastewater into saltwater-infiltrated wells. At almost 20 million litres a day, the underground injection created a curtain of freshwater along the California coast that prevented salty ocean water from seeping into the county's aquifer. This water management program was the first to take treated wastewater and pump it into the ground, says Deshmukh.

Most residents had no idea this was going on—even those who thought about water every day. Jack Skinner is a medical doctor and lifelong surfer based in Orange County, whose house in Newport Beach is supplied with water from the county's wells. His second home is in Laguna Beach, surrounded by water; in the morning, he and his wife would watch from their oceanview apartment as the sun would light up a pole stationed about 1.6 km offshore, marking the spot where a nearby sewer treatment plant discharged its **effluent**. "Every morning when we got up, we would see that marker." In the early 1980s, he began experiencing eye infections while distance swimming along the coast. He decided to learn more about the impact of releasing wastewater into the ocean, and how the wastewater was being treated.

Sewage can carry pathogens like viruses and disease-causing bacteria, and surfers get exposed to these when they surf in sewage outflow that is untreated. Public health officials monitor drinking and recreational waters for the presence of **coliform bacteria**. Since many coliforms are intestinal bacteria, their presence could indicate fecal contamination of the water.

As a doctor specializing in internal medicine and cardiology, Skinner was also concerned about potential health problems that could come with exposure over a long period of time to some chemical contaminants that are carried in drinking water. In California, the water sometimes carries traces of toxic chemicals. In the 1990s, officials detected two toxic chemicals: N-Nitrosodimethylamine (NDMA), created by chemical reactions during wastewater treatment with chlorine; and 1,4-dioxane, which came from a computer circuit board manufacturer releasing industrial wastewater into the sewer system.

desalination The removal of salt and minerals from seawater to make it suitable for consumption.
saltwater intrusion The inflow of ocean (salt) water into a freshwater aquifer that happens when an aquifer has lost some of its freshwater stores.
effluent Wastewater discharged into the environment.
coliform bacteria Bacteria often found in the intestinal tract of animals; monitored for fecal contamination of water.

Over the years, Skinner had become a self-proclaimed "troublemaker," making sure that wastewater was highly treated before being discharged into the Pacific Ocean, and speaking publicly about water pollution.

One day, the Orange County Sanitation District (OCSD) called Skinner and asked him to come in for a meeting. They wanted to consult with him about a massive new project that they were undertaking that would need the support of the community to succeed. What they said made Skinner very concerned.

There are a variety of approaches to water purification.

In the mid-1990s, Orange County water sanitation engineers and hydrologists found that they were faced with a problem of water scarcity and, at times, the opposite problem—too much water.

As an increasing number of people moved to the region, they used more water, creating excess wastewater. The OCSD already operated an 8 km "outfall," or underground pipeline that takes treated sewage water past California's beaches and out to the Pacific Ocean. The OCSD also had another 1.6 km-long outfall to the Santa Ana River, but it was not enough to handle heavy rainfall which could wash sewage through and out of a facility before it could be adequately treated. About 375 million litres a day of partially treated sewage water was already flooding the river. Nearly five times that amount would reach the river during a major storm. The county was considering building another pipeline to carry excess water to the Pacific Ocean.

But such an endeavour would be incredibly expensive. So the agency considered a project that would solve both the problem of too much wastewater and too little freshwater. They decided to expand the existing system that used treated wastewater to protect groundwater from infiltration by the ocean. But they knew they would face community opposition.

Since the 1970s, the county had been pumping only small amounts of treated wastewater into the ground—about 19 million litres per day, and only at the coasts. This new project would expand that amount to 265 million litres per day, and take place not just at the seawater barrier. The actual amount of treated wastewater that would make its way into people's homes would range, but would ultimately average 15% in north and central Orange County, says Deshmukh. "All of a sudden, it became a significant component of the water supply."

He and the other staff at the OCSD knew they would need to consult the community about the project, known officially as the Groundwater Replenishment System (GWRS)—without the support of community members, it would never get off the ground.

Once the project was conceived, they began intensive community outreach, such as presentations in front of community groups about water scarcity and the need for new sources, and placing representatives at big events. "Any chance we got we would talk about it," says Deshmukh. Part of the multiyear initiative included outreach to community leaders, such as Skinner.

When Skinner initially learned about the project, and the fact that they had been doing a smaller version at the coast for decades, he was uneasy. "Since the late 1970s, my family had unknowingly been drinking that water," Skinner says. "I thought of my daughter, who lives in the area. I thought about my grandson, Robbie. He would be drinking the water, and is probably right now. I wanted to be sure it was safe."

So Skinner set about learning how the wastewater would be treated. The state health department appointed him to serve on a state committee that would review the treatment process for the recycled wastewater, and Skinner began speaking with experts about the techniques.

He learned that the first step of the cleaning process consisted of microfiltration, in which microscopic, strawlike fibres, 1/300 the diameter of a human hair, filter out many suspended solids, bacteria, and other viruses because only water can pass through the centre of the fibres.

Then, to render wastewater **potable**—meaning, safe for humans to drink—engineers would perform a crucial step known as reverse osmosis. During this process, they use pressure to force water through a plastic membrane. The pores of this membrane are so tight, explains Deshmukh, that salt and other contaminants (such as pharmaceutical drugs and toxic chemicals) do not pass through, but water does. After reverse osmosis, the water is exposed to ultraviolet (UV) light, which kills any remaining viruses and bacteria. Adding hydrogen peroxide helps convert traces of dioxane and some of the other toxic chemicals that might be present into harmless molecules.

Where some places only clean their wastewater with two or three treatment processes before dumping it, the Orange County water and sanitation districts made the

---

potable Water clean enough for consumption.
wetland An ecosystem that is permanently or seasonally flooded.

decision to spend the money and time to do complete reverse osmosis and complete UV-light disinfection. At completion, the final product is cleaner than state and federal regulations require.

In addition, after the region discovered dioxane and NDMA in the water supply, officials tracked down the sources of those chemicals and rerouted the discharge into wastewater plants that weren't going to recycle the water into the freshwater supply, says Deshmukh.

Engineers also treat wastewater not destined to be potable, says Deshmukh, such as for watering lawns or irrigation, using filtration and disinfection, without reverse osmosis.

The Orange County Water District was Deshmukh's first job after engineering school at the University of California, Los Angeles. His thesis was on reverse osmosis, so he had total confidence in the treatment process. During presentations with community groups, he would explain that he and his parents would be affected by the project, as well. "I live here, my family lives here, this is the water we drink too." During tours of the treatment facility, they would offer visitors a taste of the water on cups labelled "It tastes just like water!" Tasting water that was sewage only 2 hours ago, and seeing its pristine quality, "was very convincing for people," says Deshmukh. In fact, sometimes they had to stop people from taking some home in water bottles, he adds.

Another California community, Arcata, tackled its water purification problems in a different way. Rather than constructing a typical wastewater treatment facility to handle sewage that had been contaminating nearby Humboldt Bay, they repurposed a retired landfill near the coast by converting it to a **wetland**—an ecosystem that is permanently or seasonally flooded. A slow river meanders through the wetland where organisms there purify it; to them, it's not "sewage," it's food. The Arcata facility depends on nature to perform the job of water purification. The water discharged into the ocean is very clean and the health of the bay ecosystem has improved. But while the Arcata facility addresses the need to decontaminate sewage, it does not take the extra steps to produce potable water like Orange County's GWRS. [INFOGRAPHIC 15.6]

↓ This wetland marsh in Arcata, California, is actually part of a wastewater treatment system that uses nature to help purify sewage. The wetland is now an Audubon birding sanctuary.

Infographic **15.6** | **WASTEWATER TREATMENT**

↓ Sewage must be treated before it can be safely released to the environment. Most communities use chemical- and energy-intensive high-tech methods, but systems that mimic nature can also effectively purify water.

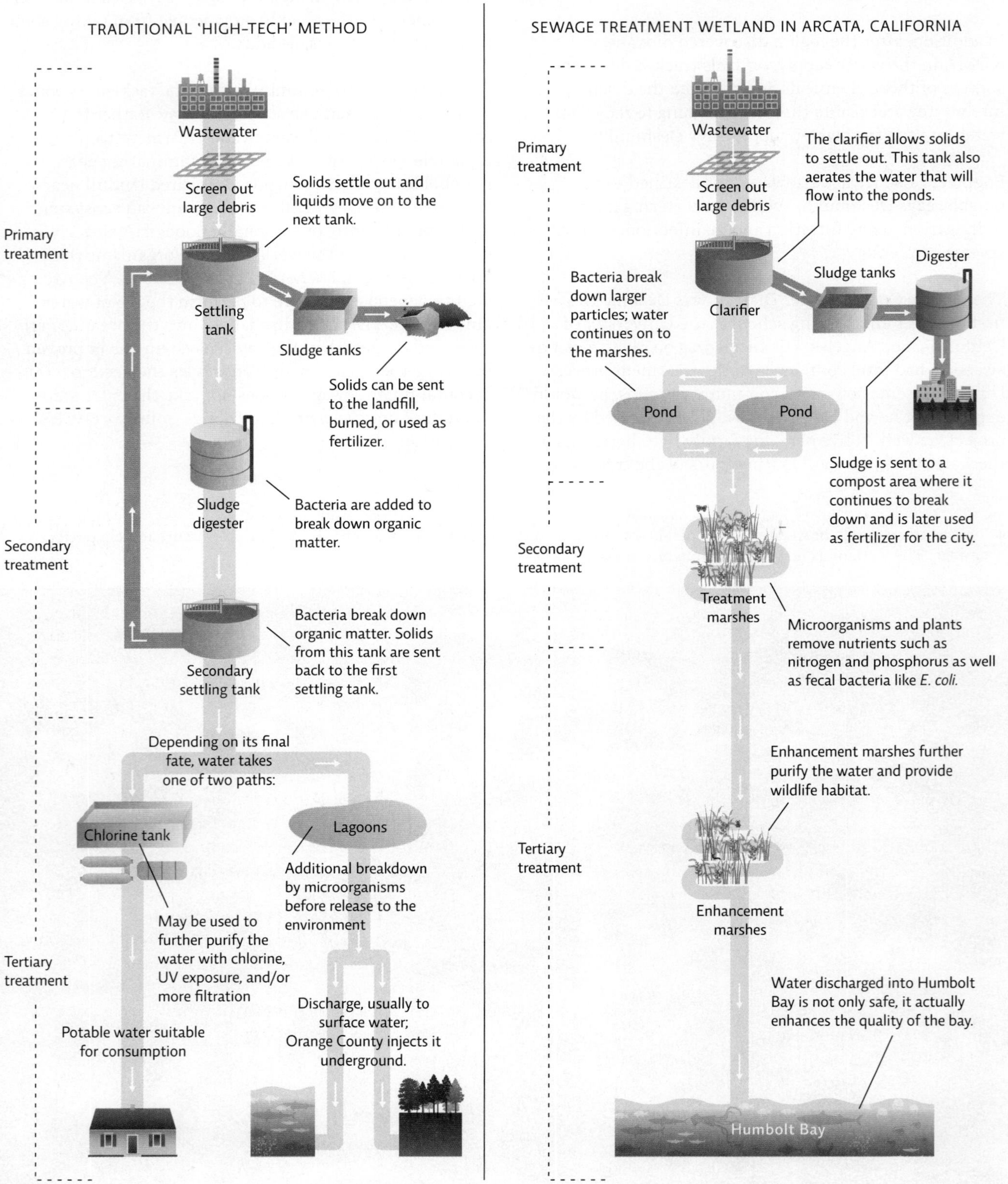
TRADITIONAL 'HIGH-TECH' METHOD
Primary treatment
Wastewater
Screen out large debris
Solids settle out and liquids move on to the next tank.
Settling tank
Sludge tanks
Solids can be sent to the landfill, burned, or used as fertilizer.
Secondary treatment
Sludge digester
Bacteria are added to break down organic matter.
Secondary settling tank
Bacteria break down organic matter. Solids from this tank are sent back to the first settling tank.
Tertiary treatment
Depending on its final fate, water takes one of two paths:
Chlorine tank
May be used to further purify the water with chlorine, UV exposure, and/or more filtration
Potable water suitable for consumption
Lagoons
Additional breakdown by microorganisms before release to the environment
Discharge, usually to surface water; Orange County injects it underground.
SEWAGE TREATMENT WETLAND IN ARCATA, CALIFORNIA
Primary treatment
Wastewater
Screen out large debris
The clarifier allows solids to settle out. This tank also aerates the water that will flow into the ponds.
Clarifier
Sludge tanks
Digester
Bacteria break down larger particles; water continues on to the marshes.
Pond
Pond
Sludge is sent to a compost area where it continues to break down and is later used as fertilizer for the city.
Secondary treatment
Treatment marshes
Microorganisms and plants remove nutrients such as nitrogen and phosphorus as well as fecal bacteria like *E. coli*.
Tertiary treatment
Enhancement marshes further purify the water and provide wildlife habitat.
Enhancement marshes
Water discharged into Humbolt Bay is not only safe, it actually enhances the quality of the bay.
Humbolt Bay

↑ Revelstoke Dam is one of four dams on the Columbia River in British Columbia. A 5th penstock turbine was recently added adjacent to the original four shown here (the long tubes on the front face of the dam), giving the dam a generating capacity of around 2500 megawatts. The 6th penstock may be installed and operational by 2019, making this the most powerful dam in British Columbia.

The GWRS went online in January 2008 when Orange County's water district began tapping its artificially replenished groundwater to deliver clean freshwater to people for drinking, irrigation, and other uses. As of fall 2011, it had recycled more than 270 billion litres of water. Every day, approximately 265 million litres of recycled water are pumped into wells or percolation basins in Anaheim, where sand and gravel naturally purify the water as it trickles down into the region's aquifers—enough to meet the needs of nearly 600 000 residents.

Solving water shortages is not easy.

A major benefit of the Groundwater Replenishment System, Deshmukh and his colleagues explained to local residents, is that injection into groundwater can address one of the biggest sources of water loss.

The majority of Earth's liquid freshwater is found in lakes. To ensure an ongoing freshwater resource, communities often invest in **dams**. These barriers stop the flow of rivers and create **reservoirs**, large bodies of water that hold freshwater for a variety of uses (freshwater source, flood control, electricity production). In Canada and the United States, these often become recreation sites and fishing resources. While reservoirs are a valuable resource, they lose an enormous amount of water every day through evaporation.

Depending on temperatures, atmospheric conditions, and the surface area of a reservoir or lake, wide-open bodies of water can evaporate tens of thousands of litres of water a day in desert settings where it's hot and dry. The California Department of Water Resources calculated that fresh-water reservoirs in the South Coast region lost more than 200 million cubic metres to evaporation in 2000 alone. Worldwide, reservoirs lose more water to evaporation than that used for industry and domestic purposes combined.

The construction of dams can also spark political conflicts. In the Middle East, Turkey's plans to build 22 dams that pull water from the Tigris and Euphrates rivers for agriculture and electric power will impact its neighbours Syria and Jordan downstream. With too little water available for too many people, this hotspot may be the site of future conflict. Similarly, a proposal to build two new dams on the Churchill River in Labrador to provide electricity to Newfoundland, and potentially as an export to the United States, is being met with opposition by residents and native Inuit who counter it is not worth the destruction to the natural area that the dams will cause.

dam Structure that blocks the flow of water in a river or stream.
reservoir Artificial lake formed when a river is impounded by a dam.

Conservation is an important "source" of water.

The GWRS project was expensive: the total price tag to build the system came to about $487 million from federal, state, and local funding.

An easier and cheaper way to maintain water supplies is simply not to waste so much. For example, water-saving irrigation methods limit loss to evaporation and runoff, thus significantly reducing the water that is used. This has the added advantages of protecting surface waters and of preventing soil salinization (the buildup of salt as water evaporates), a common problem in dry climates. Choosing to plant crops more suited to the environment and water availability will also decrease agricultural water use. Many industrial processes are now designed to reuse water rather than discharge it into the environment. Small individual changes in the household can also save a lot of water. [INFOGRAPHIC 15.7]

Infographic **15.7** | **WATER USAGE AND CONSERVATION**

↓ Much of our water usage can be reduced by using new water-efficient technologies and by making behavioural changes that don't waste water. This can be as easy as simply turning off a faucet when not using the water, taking shorter showers, and only using appliances like dishwashers when they are full. Buying less stuff and using less energy also saves water because much water is used to produce both.

CANADIAN HOUSEHOLD WATER USE

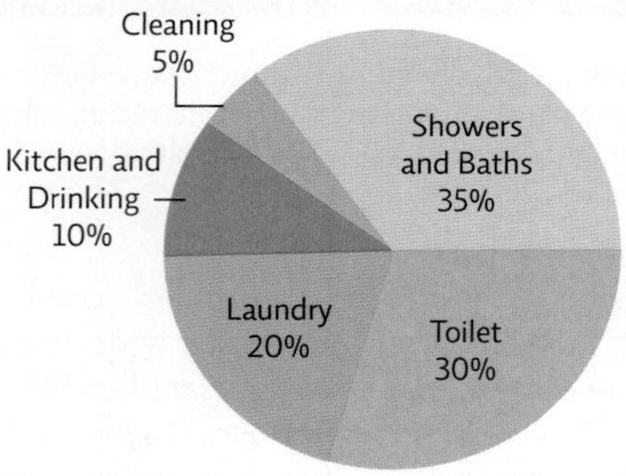
Cleaning 5%
Kitchen and Drinking 10%
Showers and Baths 35%
Laundry 20%
Toilet 30%

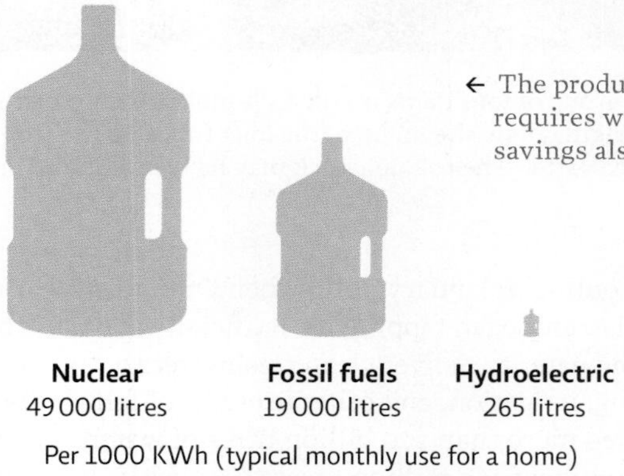
Nuclear 49 000 litres
Fossil fuels 19 000 litres
Hydroelectric 265 litres
Per 1000 KWh (typical monthly use for a home)

← The production of electricity requires water; any energy savings also save water.

OLD TECHNOLOGY	NEW TECHNOLOGY	BEHAVIOURAL CHANGE
TOILET 22.7 LITRES/FLUSH	**LOW-FLOW TOILET** 4.9 LITRES/FLUSH	Don't flush tissues—use the trash. Flush liquid waste less frequently.
SHOWER 14.3 LITRES/MINUTE **BATH** 132.5 LITRES	**LOW-FLOW SHOWER HEAD** 8.7 LITRES/MINUTE	Turn off the shower head except to rinse (some shower heads come with a convenient valve that allows you to switch off the water without turning it off at the source).
FAUCET 18.9 LITRES/MINUTE	**LOW-FLOW FAUCET** 5.7 LITRES/MINUTE	Don't leave the faucet running while brushing your teeth, shaving, or washing your face.
WASHING MACHINE 151.4 LITRES/LOAD	**WASHING MACHINE (ENERGY STAR)** 83.2 LITRES/LOAD	Don't wash a clothing item unless it needs it (those jeans can probably be worn several times before washing) and only run the washer when it is full.
DISHWASHER 34 LITRES/LOAD	**DISHWASHER (ENERGY STAR)** 15 LITRES/LOAD	Only run the dishwasher when it is full and limit the amount of rinsing you do before loading dishes into the dishwasher; if you have a new dishwasher, it isn't needed.

In the meantime, supplementing potable water supplies with recycled water is an innovative way to help ameliorate ongoing water issues, says Channah Rock, a water-quality specialist and assistant professor at the University of Arizona. It's rare to find initiatives like Orange County's, she says, but several communities—in Arizona, California, Nevada, and Florida, for instance—are reusing recycled water for nonpotable use, such as for irrigating landscapes and crops, filling fountains and fire hydrants, and flushing toilets. Communities are trying to "match the quality of water with the right use of water," she says. Since people are prohibited from drinking the water that is used for irrigation, for example, says Rock, it's not necessary to subject that water to the same advanced treatment processes as those used for potable water.

Public perception of recycled water projects actually varies, says Rock, depending on how familiar or confident people are with the treatment process and how plagued their community is by water-scarcity issues. In Arizona, which has regions affected by drought, a recent statewide survey found that Arizona residents generally supported most potential uses of recycled water and felt that it was very important that their community use recycled water to help meet its water needs. To many people, she says, what matters is that the water ultimately meets regulatory standards designed to protect public health.

In fact, Rock says, the "toilet to tap" phrase is incredibly misleading, because it leaves out the testing, treatment, and scrutiny that take place in between.

The Groundwater Replenishment Program has received many accolades, including the prestigious 2008 Stockholm Industry Water Award. In 2010, just 10 years after Deshmukh finished graduate school, his work was profiled in *National Geographic*. It's an achievement that fills Deshmukh with pride. "This is water that's normally just wasted in the ocean. For the first time, it was being added to the water basin, cleaner and at a higher quantity than we'd done before."

Once Skinner learned how the water would be treated, spoke to scientists about the effectiveness of the treatment process, and learned the project would have continuous oversight, he was reassured. "I feel they have successfully addressed my concerns," he says. Skinner and Deshmukh even worked together to create videos describing the project. Today, Skinner, his wife, daughter, and grandson consume the recycled water. "I think it's safe for my family to drink."◉

Select reference in this chapter:

Hoekstra, A., and Chapagain, A. 2007. *Water Resources Management*, 21: 35–48.

BRING IT HOME

PERSONAL CHOICES THAT HELP

Regardless of whether our water comes from an aquifer or a local reservoir, we can make those water sources last longer by taking steps to use our water as efficiently as possible.

Individual Steps

→ If you have a smartphone, download a water usage tracking app. Once you have a baseline, try to reduce it by 10%.

→ Time your shower and try to reduce it by 1 to 2 minutes.

→ Have a container by the sink or shower to catch water while it warms up—just make sure not to get soap in it. Use this water for watering plants both inside and out.

Group Action

→ Install a rain barrel at home. Rain barrels allow people to use the rain that falls on the roof of a building to water plants as opposed to letting it run off into the storm drain. If you live in a dorm or an apartment, see if you can get permission to have a rain barrel installed.

Policy Change

→ Do you know where your water comes from? Talk to a city representative to find out where your water comes from, and what steps are being taken to make sure it lasts as long as possible.

→ Encourage local policy-makers to ban the watering of lawns, or restrict the use of water for landscaping, to certain days of the week.

UNDERSTANDING THE ISSUE

CHECK YOUR UNDERSTANDING

1. Approximately 75% of Earth's surface is covered with water. The amount of that water that is drinkable by humans is:
a. nearly all of it.
b. about half of it.
c. perhaps 1/10 of it.
d. much less than 1%.

2. Most humans get their water from:
a. underground wells.
b. rivers with dams.
c. lakes.
d. desalination plants.

3. Infiltration:
a. is made easy in urban and suburban areas by all of the lawns.
b. is made harder in urban and suburban areas by roads, buildings, and lawns.
c. is the process of removing particulate matter from sewage.
d. is what happens when seawater enters a freshwater system.

4. Many people say that desalination (removing the salt from ocean water) is the obvious way around our water shortages. This is:
a. becoming relatively easy and inexpensive and will be common soon.
b. inexpensive now but it is difficult to route the water from the coasts to inland areas.
c. still very expensive and uses a great deal of energy.
d. not being done at this time.

5. If too much water is removed by a well in coastal areas:
a. the water table will rise.
b. a cone of depression will form.
c. the aquifer might collapse.
d. salt water can seep into the aquifer.

6. One creative way that some communities deal with wastewater is:
a. creating wetlands that support birds and other wildlife.
b. using microfiltration, then bottling and selling it as mineral water.
c. pumping it through a separate water system for people to use for laundry and for yard irrigation.
d. evaporating it in gigantic reservoirs, creating additional clouds and rain.

WORK WITH IDEAS

1. Why do the problems of water scarcity and unsanitary water conditions often occur together?
2. Discuss at least two of the reasons why having enough potable water is a problem for so many people on Earth.
3. Draw a flow chart of the water cycle. (Don't copy from the book; do your own small drawing.) Follow a single water molecule from a cloud through some portion of the cycle, including a living organism, and back to a cloud.
4. Describe, in your own words, the percentage of total water on Earth that is freshwater, and identify the compartments where it is located.
5. What are some different things that communities can do to deal with their wastewater?

ANALYZING THE SCIENCE

The World Health Organization's (WHO) *Global Burden of Disease* analysis provides a comprehensive and comparable assessment of mortality and loss of health due to diseases, injuries, and risk factors for all regions of the world. The overall burden of disease is assessed using the disability-adjusted life year (DALY), a time-based measure that combines years of life lost due to premature mortality and time lived in states of less than full health.

INTERPRETATION

1. According to the map, which continent appears to have more problems with the lack of sanitation?
2. What is the DALY status attributable to water and hygiene in most of the world's developed countries?
3. In your own words, describe the DALY status in Eurasia, one of the most populous areas on Earth.

DALYs ATTRIBUTABLE TO WATER, SANITATION, AND HYGIENE (DIARRHEA), 2004

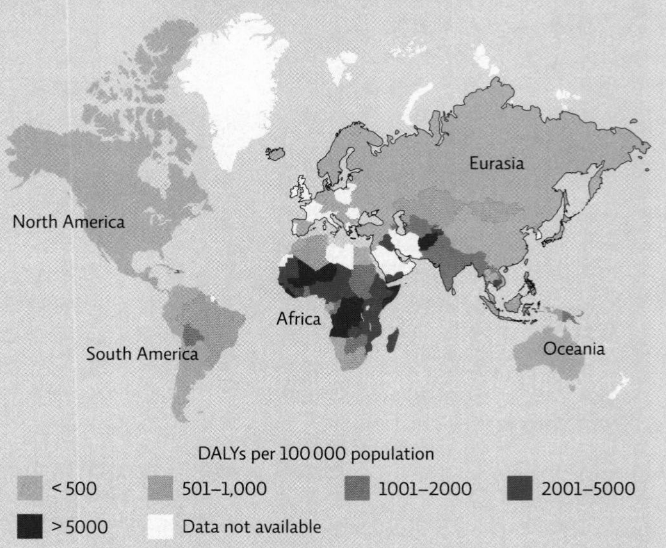
North America
Eurasia
Africa
South America
Oceania
DALYs per 100 000 population
< 500
501–1,000
1001–2000
2001–5000
> 5000
Data not available

ADVANCE YOUR THINKING

4. Explain why people in many places in North America complain of water shortages, while this map indicates that they have access to clean water and sanitation.

5. If much of Africa is grasslands and jungle, why do the inhabitants have so little access to clean water and sanitation?

6. Some countries that border oceans are able to afford desalination plants to improve their access to clean drinking water. If all countries could afford it, would this solve the worldwide problem? Explain.

EVALUATING NEW INFORMATION

In the summer of 1989, Dr. Noah Boaz and his archaeological Earthwatch crews were excavating a site of ancient human habitation along the Semliki River, which runs by the border between Zaire (now the Democratic Republic of Congo) and Uganda. They could not, however, just drink the water from the river, or even swim in it. They had to filter the water a few litres at a time and then add chemicals to it in order to remove the waterborne parasites and diseases. Bathing meant wearing shoes and keeping their eyes, nose, and mouth out of the water. The nearby villagers did drink the water, and they had endemic health problems.

More than 2 billion people do not have access to clean drinking water. The results affect all aspects of life in developing countries: according to Water.org, a child dies every 20 seconds from a water-related illness and women in some water-stressed areas spend several hours every day collecting water for their families' basic needs.

There have been many suggestions about ways to improve access to clean water. One of the problems is that in many areas, the lack of access is coupled with a lack of the electricity, developed roads, machines, and equipment necessary to be able to support digging municipal wells and providing pumping stations, reservoirs, and pipelines.

Go to the Global Water website (www.globalwater.org) and explore their current projects. Then go to the water.org website (water.org) and look at some of their featured projects. Finally, go to NEED magazine (www.knowh2o.org/assets/pdf/NEED03_FUTURE.pdf) and look at some of the solutions proposed in the featured article.

Evaluate the websites and work with the information to answer the following questions:

1. Are these authors/sponsoring groups reliable information sources?
   a. Do they give supporting evidence for their claims?
   b. Do they give sources for their evidence, as well as clear explanations?
   c. What is the mission of the organization or of the scientist/author who wrote the article?
   d. Does the author or organization appear to have a workable solution or solutions?

Now go to the bottlelessvancouver website (bottlelessvancouver.wordpress.com/2011/02/25/clean-water-solution-to-third-world-countries) and watch the video about LifeStraw. Then go to the HowStuffWorks website (science.howstuffworks.com/environmental/green-tech/sustainable/playpump.htm) to read about the PlayPump, and to the gizmag website (www.gizmag.com/mobile-bicycle-powered-water-pump/15281) to read about the bicycle-powered water pump.

2. Evaluate the proposed solutions for:
   a. price.
   b. ease of use.
   c. whether they would be portable or stationary.
   d. whether they include pumps for underground water, or items to clean surface water.

3. Do any of the proposed solutions stand out as the best possible one for remote or undeveloped areas? Explain your answer.

MAKING CONNECTIONS

RETHINKING TOILETS

Background: In the 1990s, the Arizona State Museum created a major new exhibit featuring the life of the Tarahumara people, who live in the rugged Sierra Madre Occidental mountains in Mexico, in and around the gigantic Barranca del Cobre (Copper Canyon). The museum staff invited a number of the people from this remote area, especially those who had donated their rugs, pottery, carvings, and other artifacts, to come to Tucson to attend the opening. On arriving in Tucson, they declined to stay in the high-rise hotel where museum guests usually stayed. Instead they opted to stay with someone they knew, sleeping in the backyard so they could see familiar stars. When the uses of a bathroom were explained, the group was horrified. Since they lived in the high mountains with only a few streams, water was their most precious resource. The idea of putting human waste into a porcelain bowl of clean, pure water was unthinkable.

Case: People in many areas of North America are reaching this same conclusion, that clean water should not be used for human waste, or at least that its use should be minimized. Write a position paper for a local magazine, newspaper, or online newsblog encouraging your community to vote for or against the local proposition to reduce the use of clean water for toilets. Include the following in your paper:

a. How much water is used in a standard (pre-1994) toilet? Multiply that amount times 100 000 people (or use your local population number) times an average of three flushes per person per day times 365 days to get an estimate of the amount of water that, literally, goes down the toilet every year in your community.
b. At least three ways to reduce the amount of water used for toilets, including a redesigned toilet (with or without water).
c. Examples of ways other communities are addressing this problem.

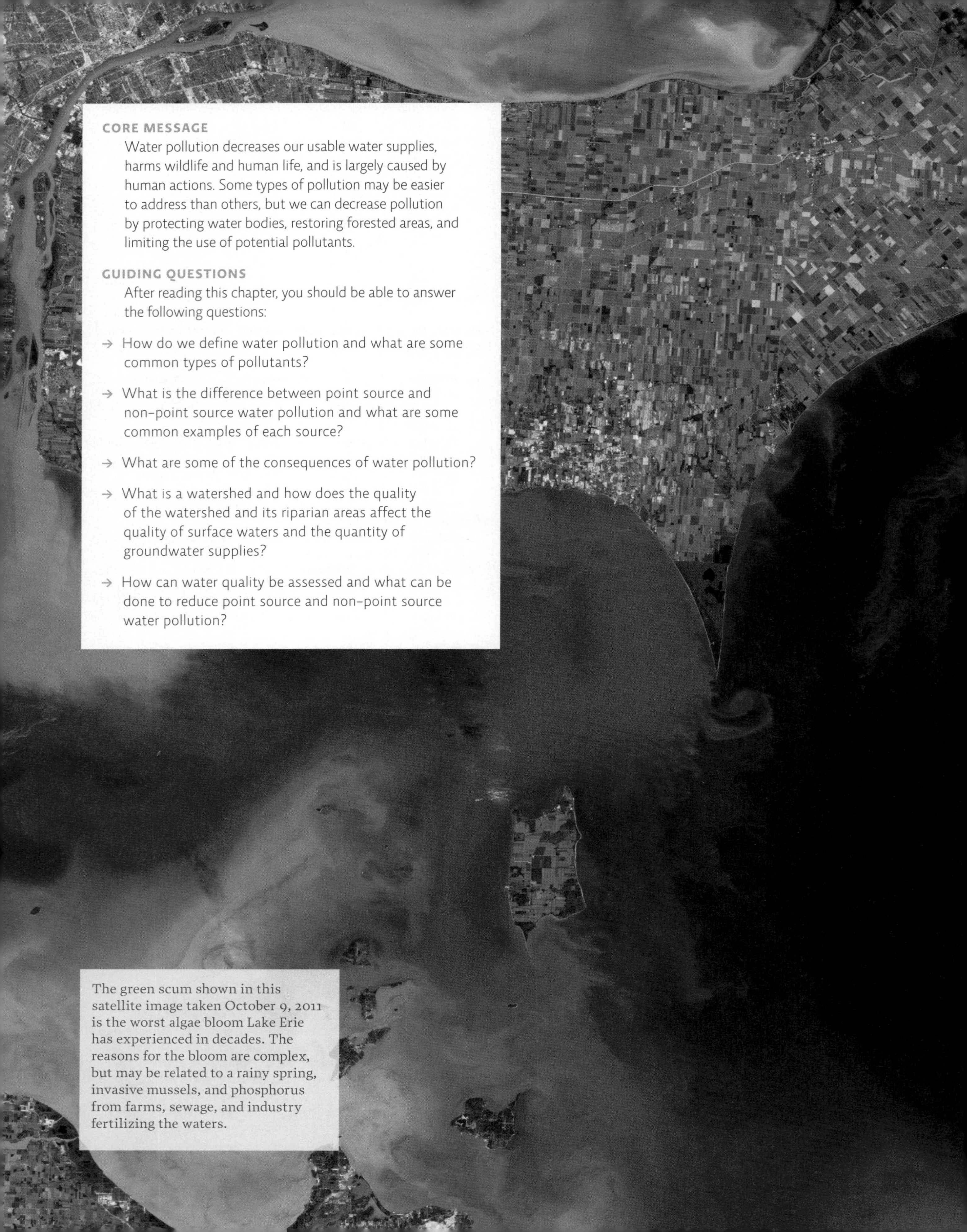

CORE MESSAGE

Water pollution decreases our usable water supplies, harms wildlife and human life, and is largely caused by human actions. Some types of pollution may be easier to address than others, but we can decrease pollution by protecting water bodies, restoring forested areas, and limiting the use of potential pollutants.

GUIDING QUESTIONS

After reading this chapter, you should be able to answer the following questions:

→ How do we define water pollution and what are some common types of pollutants?

→ What is the difference between point source and non-point source water pollution and what are some common examples of each source?

→ What are some of the consequences of water pollution?

→ What is a watershed and how does the quality of the watershed and its riparian areas affect the quality of surface waters and the quantity of groundwater supplies?

→ How can water quality be assessed and what can be done to reduce point source and non-point source water pollution?

The green scum shown in this satellite image taken October 9, 2011 is the worst algae bloom Lake Erie has experienced in decades. The reasons for the bloom are complex, but may be related to a rainy spring, invasive mussels, and phosphorus from farms, sewage, and industry fertilizing the waters.

CHAPTER 16 **WATER POLLUTION**

RESCUING THE GREAT LAKES

Brought back from the brink in the 1970s, pollution threatens Lake Erie once again

It was supposed to be an uneventful trip around Lake Ontario, collecting water samples to determine how the lake was doing. It was a trip Sue Watson and her colleagues, research scientists at Environment Canada, make several times a year, spending most of their summers on a research vessel that visits both Lake Ontario and adjacent Lake Erie, the shallowest, southernmost, and smallest of the Great Lakes. Mostly, the data they collected were routine—samples showed the water was relatively healthy. But over Labour Day in 2012, Watson got some news that made this trip anything but routine.

There had been a massive fish kill along the northern shore of Lake Erie, she learned—a rotten, putrid smell was emanating from the beaches and the lake, and decaying fish corpses were piling up on the shorelines. She asked a colleague to visit the site, and he reported back that whatever was happening was killing all types of fish, indiscriminate of age or species. And one more thing: "I'm seeing the beginning of a massive algal bloom here," he told her.

When Watson arrived at Lake Erie two weeks later, she saw a disturbing sight.

"This was the first time I'd ever seen an algal bloom that stretched from Point Stanley, in the middle of the north shore, all the way into Long Point Bay," she recalls. "It was probably 10 kilometres wide."

This was not supposed to happen in Lake Erie. Decades earlier, such a sight would have been commonplace, as unprecedented levels of algae and toxic chemicals made the lake a poster child for environmental destruction. In response, the Canadian and U.S. governments investigated the situation and developed new laws to restrict pollution. This was largely successful, and the waters of Lake Erie became clear once again. People returned to its shorelines in droves to sun, swim, and sport fish. But now the algae were back, and pollutants were once again infiltrating the lake. What was going on?

◉WHERE IS LAKE ERIE?

ON
QC
ON
QC
WI
LAKE ERIE
MI
PA
IL
IN
OH

When pollutants infiltrate water bodies, things go awry.

Most of the problems facing Lake Erie can be traced back to **water pollution**—the addition of anything that might degrade water quality. The list of potential pollutants is discouragingly long. Industrial chemicals and raw sewage get dumped directly into water bodies. Garbage, oil, pesticides, fertilizers, and sediments wash into water with runoff. Contaminants, such as mercury, sulphur, and other air pollutants from fossil fuel combustion, enter lakes with rainwater. While some types of pollutants are more difficult to address, some can be reduced more easily or even ultimately eliminated.

Water pollution is a major concern for Canada, because it is home to 7% of the entire world's supply of renewable freshwater. The Great Lakes alone supply drinking water to 8.5 million Canadians.

It's a resource that cannot be taken for granted. In the late 1960s, Harvey Shear spent his summers as an undergraduate student working for professors at the University of Toronto campus in Scarborough, and during breaks he would wander to one of his favourite spots, a creek in the nearby woods. On many days, his reverie would be ruined

water pollution The addition of anything that might degrade the quality of the water.

↑ A pipe releasing raw sewage and stormwater into Lake Ontario near Scarborough, Ontario.

← Unprotected farm fields lose topsoil as well as farm fertilizers and other potential pollutants when heavy rains occur.

Infographic **16.1** | **MAJOR CAUSES OF WATER POLLUTION**

↓ Water pollution comes from a variety of point and non-point sources. In the 1960s and 1970s, the leading contributor of nutrients that caused algal blooms was point source pollution from the release of sewage treatment plant wastewater, a pollution source that was effectively addressed. The new round of algal blooms that began in the 1990s is likely due to nutrients from non-point sources, although point sources still play a role.

POINT SOURCES

Some industrial and agricultural sources discharge pollutants directly into a body of water.

NON-POINT SOURCES

A variety of sources contribute pollutants that can run off the surface of the land during rainfall and enter the water; air pollutants can fall directly with rain or as dry deposition.

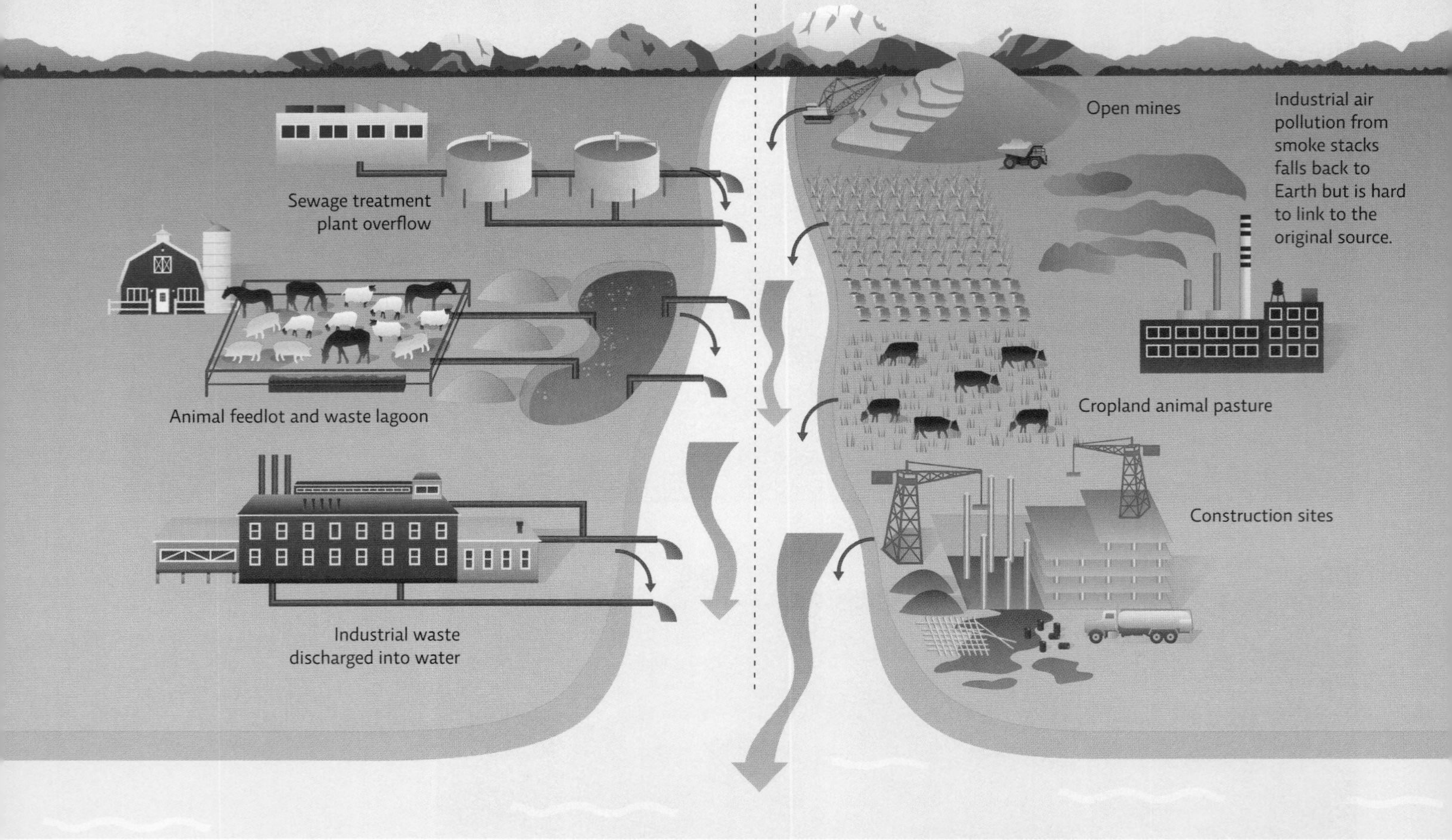
Sewage treatment plant overflow
Animal feedlot and waste lagoon
Industrial waste discharged into water
Open mines
Industrial air pollution from smoke stacks falls back to Earth but is hard to link to the original source.
Cropland animal pasture
Construction sites

by the sight of toilet paper and waste bobbing along the running water, which led right into Lake Ontario. "I thought, 'this is ridiculous,'" says Shear, now a professor at the U of T. "This is one small creek—how much more of this was going on?"

By that time, sanitation plants and companies had been dumping raw sewage and chemicals into the Great Lakes for years. Tons of unused fertilizers washed off from farm fields into the lake with rainwater. In the 1960s and 1970s, thick layers of floating petrochemicals would start fires on the water's surface. Massive algal blooms were seen from Lake Superior to the outflow of the St. Lawrence River. Lake Erie was declared "dead."

Once experts began looking for the sources of pollution, they found specific entry points, such as oil from an oil spill, or pipes emptying raw sewage or industry discharge into the Great Lakes. This is known as **point source pollution**, meaning that it can be traced back to discrete sources, which can carry toxic compounds, acids, or pathogens into the water. In contrast, **stormwater runoff** that makes its way into surface waters is known as **non-point source pollution**, because there is no

↑ Algal blooms along the shore of Simon Lake in Naughton, Ontario, are still occurring despite steps to reduce the amount of phosphorus entering the lake from various sources. Spring runoff still delivers unacceptable levels of phosphorus to the lake but lake turnover may also be releasing stored phosphorus from sediments, triggering the algal blooms.

single discharge point. For example, this type of pollution carries fertilizer from farms, golf courses, and lawns to the lake and its tributaries (rivers and streams that empty into the lake). [INFOGRAPHIC 16.1]

In the 1960s and 1970s, Lake Erie was being inundated with high loads of phosphorus, most of it coming from sewage (raw and treated) discharge, an example of point source pollution. Stormwater runoff was also an increasing contributor of nutrients and other chemicals, as more land surfaces that used to absorb water got paved over. These impermeable landscapes are more likely to be sources of problematic substances, such as fertilizers, pesticides, gasoline, antifreeze, pet waste, and motor oil. Five million metric tons of road salt are scattered across icy roads in Canada every year, making salt a major contaminant in waters close to treated roadways. Runoff water can also pick up soil and cause sediment pollution, the most common type of water pollution. Sediments can cloud the water, decrease photosynthesis, clog fish gills, and cover rocky stream beds with a thick coating of mud.

Other types of non-point source water pollution come from industry or vehicle air pollution, which enters surface water via precipitation or dry particle fallout. Lands adjacent to large rivers and lakes have always been preferred sites for industry and cities; more than 360 industrial chemicals have been found in the waters of the Great Lakes, including PCBs, dioxins, and pesticides such as DDT. Fossil fuel combustion also releases sulphur and nitrogen pollution, which can lead to acid rain and lake acidification (although the Great Lakes, because of their size and the fact that they contain minerals such as calcium carbonate that can neutralize acids, can buffer the effects of acid rain more easily than smaller lakes or bodies of water with less of these minerals).

Canada and other high-latitude areas face an additional issue—an accumulation of toxic compounds such as pesticides and heavy metals that are transported in the upper atmosphere to the polar region, where they enter ecosystems (and food chains) as particles that fall to the ground. (See Chapter 21 for more on long-range air pollution.)

---

point source pollution Pollution that can be traced back to discrete sources such as wastewater treatment plants or industrial sites.

stormwater runoff Water from precipitation that flows over the surface of the land.

non-point source pollution Runoff that enters the water from overland flow and can come from any area in the watershed.

← Firefighters stand on a bridge over the Cuyahoga River, a Lake Erie tributary in Cleveland, Ohio, to spray water on a tugboat as a fire—started in an oil slick on the river—moves toward the docks at the Great Lakes Towing Company site. This 1952 blaze, one of 13 fires on the river since the late 1800s, was the most costly, destroying three tugboats, three buildings, and the ship-repair yards.

Nutrient enrichment can lead to oxygen depletion.

The huge influx of nutrients that run off from farms, golf courses, and lawns into Lake Erie and the rivers and streams that feed into it (tributaries) is the first step in a process known as **eutrophication**, or more specifically **cultural eutrophication** (nutrient enrichment caused by human activities). Because nutrients—primarily phosphorus but also nitrogen—fuel plant growth, excess amounts trigger explosions of algae. In the 1970s, Lake Erie wasn't blue—it was a pea-green colour, full of algae. While not necessarily toxic, pollution that adds nutrients to a body of water can throw its natural ecosystem out of balance, spurring the growth of some organisms and reordering the entire ecological community. "The Great Lakes have aged over many thousands of years," explains Watson. "They developed a balanced system that functioned well, and enabled them to survive dramatic shifts in climate and other changes. Within 200 years, humans have completely altered the entire area around the Great Lakes so rapidly, they never had a chance to recover."

eutrophication Nutrient enrichment of water bodies, which typically leads to algal overgrowth and oxygen depletion, and which can occur naturally or via human activities.

cultural eutrophication Eutrophication specifically caused by human activities (most eutrophication is caused by humans).

dissolved oxygen (DO) The amount of oxygen in the water.

hypoxia A situation in which the level of oxygen in the water is inadequate to support life.

Algae aren't just a cosmetic problem—blooms may also contain one or more species of cyanobacteria (formerly called *blue-green algae*, because scientists initially thought they were algae), a type of photosynthetic bacteria. Some cyanobacteria produce potent toxins that can harm the liver and even cause death after chronic exposure. In Brazil, 46 people receiving dialysis died after the facility unknowingly used water that contained cyanobacterial toxins. In Canada, most people won't accidentally drink contaminated water because of the foul smell produced by cyanobacteria, but pets and livestock are less discriminating. Children are also more vulnerable due to their smaller size; even small amounts can make them sick.

But not every bloom produces toxins, and even when toxins are present, they are not always the source of the biggest problems. Though algae release some oxygen into the water as a by-product of photosynthesis, thick layers of surface algae also block sunlight from reaching underwater plants, causing them to die. As a result, levels of **dissolved oxygen (DO)** in the water plummet—a condition known as **hypoxia**. Water can only hold a limited amount of oxygen, much less than that found in air, so even a small decrease can have immediate effects on other aquatic life.

An influx of nutrients or an algal bloom also cause an explosion of bacterial decomposers that consume these nutrients and dead algae; as part of this process, bacteria consume even more oxygen, which reduces DO levels

Infographic **16.2** | **EUTROPHICATION CREATES DEAD ZONES**

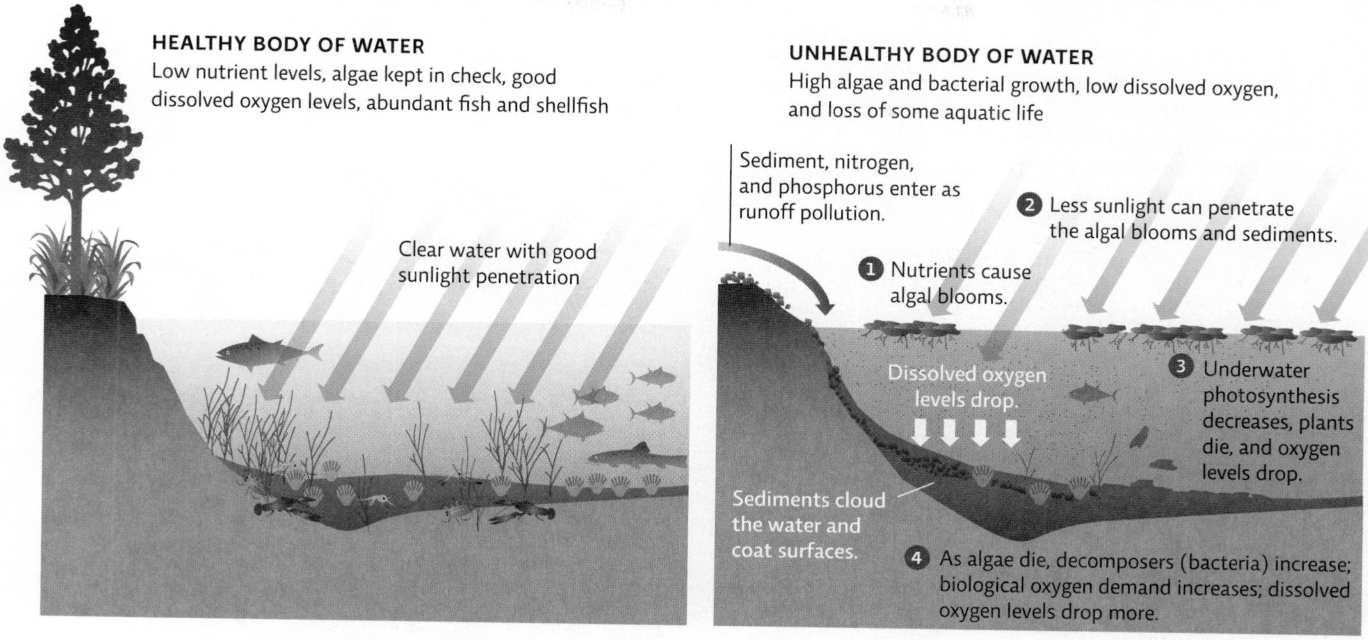
HEALTHY BODY OF WATER
Low nutrient levels, algae kept in check, good dissolved oxygen levels, abundant fish and shellfish
Clear water with good sunlight penetration
UNHEALTHY BODY OF WATER
High algae and bacterial growth, low dissolved oxygen, and loss of some aquatic life
Sediment, nitrogen, and phosphorus enter as runoff pollution.
1 Nutrients cause algal blooms.
2 Less sunlight can penetrate the algal blooms and sediments.
3 Underwater photosynthesis decreases, plants die, and oxygen levels drop.
Dissolved oxygen levels drop.
Sediments cloud the water and coat surfaces.
4 As algae die, decomposers (bacteria) increase; biological oxygen demand increases; dissolved oxygen levels drop more.

↑ Healthy water is relatively clear with abundant aquatic life. A variety of fish, plant, and invertebrate life, as well as low nutrient inputs, kept the water clean before human action introduced sediments and chemical pollution.

↑ An influx of nutrients or sediments can ultimately result in hypoxia in a body of water due to both a decrease in underwater photosynthesis and an increase in bacteria (consumers that use much of the water's oxygen). Any organism that cannot leave the area will suffer in the hypoxic waters.

even further. When people referred to Lake Erie as "dead" decades ago, they weren't so far off—hypoxic areas are often referred to as *dead zones*. "But in fact, 'dead zones' are not 'dead'," says Watson—"they are full of bacteria." [INFOGRAPHIC 16.2]

The source of pollution can be hard to pinpoint.

To restore the damaged Great Lakes, in 1972, the Canadian and U.S. governments established the *Great Lakes Water Quality Agreement (GLWQA)*. The policy affirmed both countries' commitments to cleaning up the lakes. In Canada, the GLWQA is implemented by a number of agencies working together, such as Environment Canada, Fisheries and Oceans Canada, and Health Canada. The Canadian government has the power to regulate water pollution via a series of acts of parliament—most notably, the *Canadian Environmental Protection Act*, or CEPA, which is concerned with protecting the environment and human health. But there is also the *Fisheries Act*, which includes provisions that protect fish habitats, and the *Canada Water Act*, to manage the quality of the country's water. Together, these laws enable officials to set limits for the release of nutrients and chemicals into the

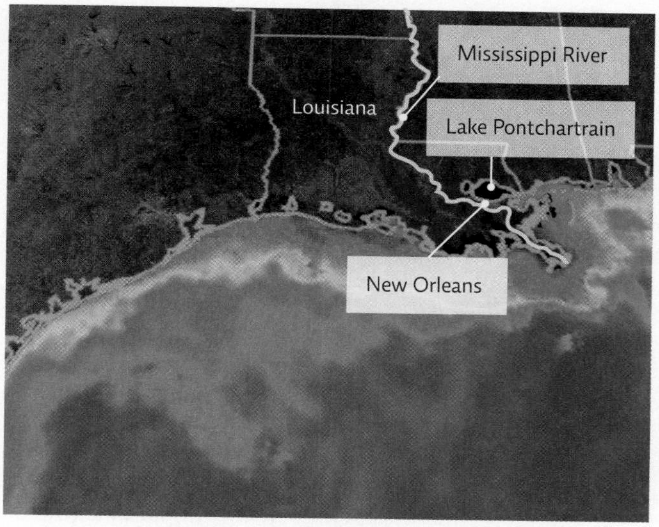
Mississippi River
Louisiana
Lake Pontchartrain
New Orleans

↑ This summertime satellite image of the Gulf of Mexico depicts water clarity; areas with high sediment loads or phytoplankton (algal) blooms are less clear (more turbid). Sediment can block sunlight and deplete oxygen through diminished photosynthesis. Phytoplankton blooms also result in oxygen depletion and dead zones. Red and orange represent the most turbid water; yellow, green, and aqua are progressively less turbid but still contain enough phytoplankton to trigger oxygen depletion.

Infographic 16.3 | **WATERSHEDS**

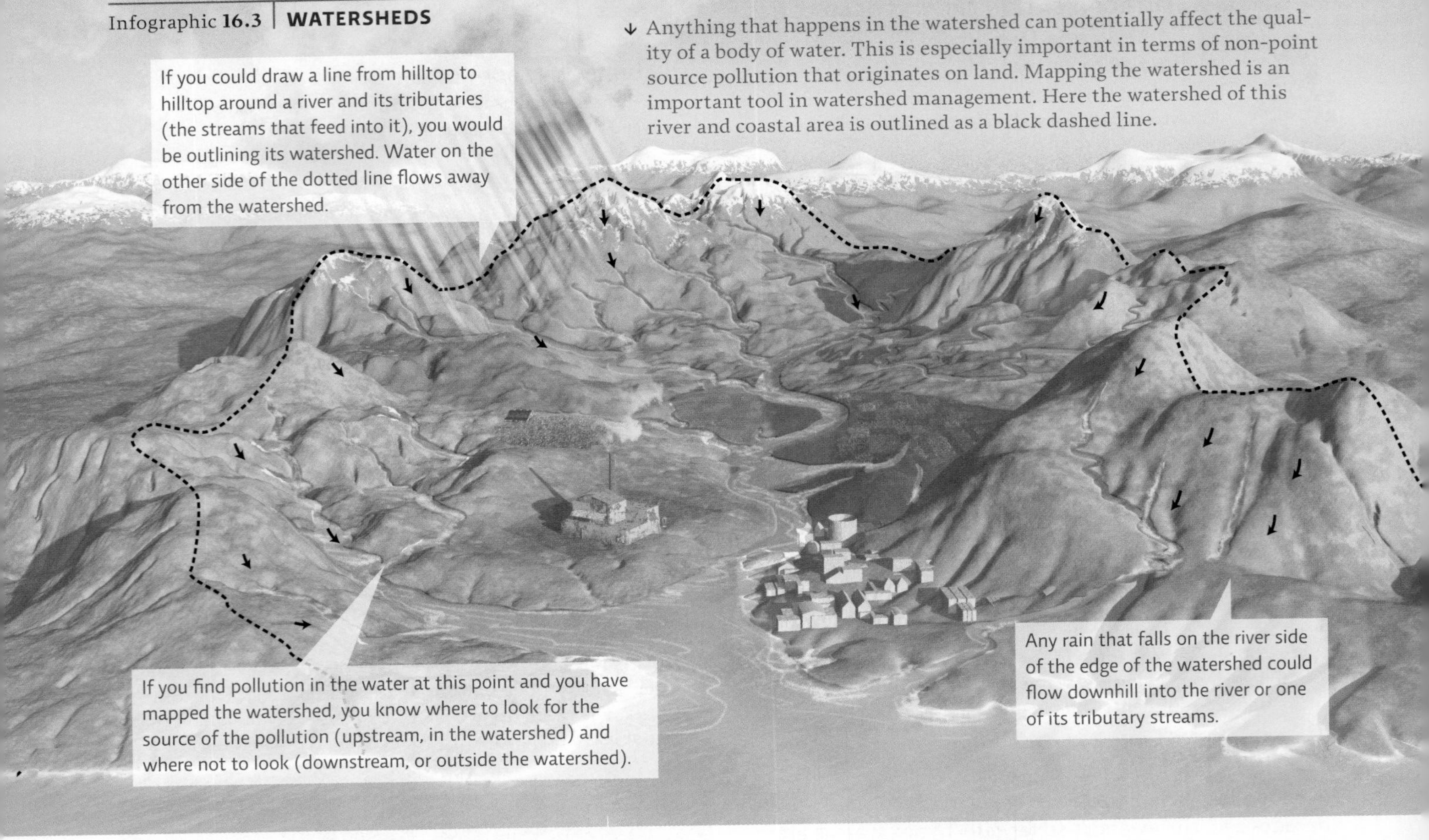
Anything that happens in the watershed can potentially affect the quality of a body of water. This is especially important in terms of non-point source pollution that originates on land. Mapping the watershed is an important tool in watershed management. Here the watershed of this river and coastal area is outlined as a black dashed line.
If you could draw a line from hilltop to hilltop around a river and its tributaries (the streams that feed into it), you would be outlining its watershed. Water on the other side of the dotted line flows away from the watershed.
If you find pollution in the water at this point and you have mapped the watershed, you know where to look for the source of the pollution (upstream, in the watershed) and where not to look (downstream, or outside the watershed).
Any rain that falls on the river side of the edge of the watershed could flow downhill into the river or one of its tributary streams.

water, known as **pollution standards**. Recent changes to these acts, however, have weakened their ability to protect water quality and aquatic habitats.

One major feature of the Canadian regulatory system is *shared responsibility*—each level of government has the ability to create laws that protect the environment. As a result, often the federal government will create a guideline—for instance, establishing which chemicals should be regulated—and then leave it up to provinces, territories, regions, and municipalities to set maximum allowable levels for those chemicals and to enforce those limits. While this ensures that different levels of government must collaborate and cooperate, problems can occur when one agency simply passes responsibility for a problem onto another.

Even if all levels of government work together well, pollution can be a tricky problem to address. Point sources are relatively easy to remedy: identify the source and make a change. Pollution from non-point sources is much more intractable because there is no single discharge point that can be pinpointed. "These non-point sources of pollution are so difficult to control," says Watson, "because we don't know exactly what all farmers are doing, how much fertilizer they use, and how they use it."

Attempts to clean up the lake in the 1970s were relatively successful—household detergents were modified to contain fewer phosphates, industries were required to reduce pollution, and billions were spent to upgrade wastewater treatment plants that once released raw sewage into the lakes. Farmers made changes to reduce runoff of soil, chemicals, and nutrients. Lake Erie was restored as a premier vacation spot for boating, fishing, and lounging on the beach; fishers and anglers tugged millions of fish from the restored lake every year.

pollution standards Allowable levels of a pollutant that can be released over a certain time period.

watershed The land area surrounding a body of water over which water such as rain could flow and potentially enter that body of water.

riparian area The land area adjacent to a body of water that is affected by the water's presence (for example, water-tolerant plants grow there) and that affects the water itself (for example, provides shade).

Then, after a few decades of relative ecosystem harmony, some funny colours began to reappear in the crystalline waters.

During her recent surveys of Lake Erie and Ontario, Watson would scoop up water samples that contained what looked like grass cuttings, tiny green blobs, or thin wisps of blue-green she had to squint to see; suspended material was sometimes a brilliant blue, a dark willowy brown, or even pink. All of this, she says, is algae.

Shear got the news in a phone call from a former colleague. The algae are back, the colleague said—would you come speak at a community meeting near Lake Ontario about the problem? There, Shear listened to local representatives talk about how they would pull sheets of some species of algae from the water, stuff them in garbage bags, and bring them to landfills. This won't work, Shear told them. "It's like cutting your grass—it's going to keep growing back. You're not addressing the problem, you're addressing the result."

Identifying the types and sources of the pollution is the first step to cleaning up our water.

Some additional sources of pollution were creeping into Lake Erie's **watershed**—the area of land from which rain and other sources of water drain into the lake or its tributaries. Pinpointing those sources was not an easy task. Lake Erie is the shallowest of the Great Lakes, but still one of the largest freshwater bodies in the world in terms of surface area. Its watershed covers more than 78 000 square kilometres and is home to a lot of activity: dense population (approximately 12 million residents), extensive agriculture, and many industries. To make things even more complicated, the lake is bordered by one province (Ontario) and four states (New York, Pennsylvania, Ohio, and Michigan). [INFOGRAPHIC 16.3]

The resurgence of algal blooms in Lake Erie was not just reminiscent of earlier times—in some recent years they have been worse. The 2012 bloom was looking like a repeat of the previous year's record bloom. Research led by Anna M. Michalak of the Carnegie Institution for Science's Department of Global Ecology concluded that the 2011 bloom was the result of a "perfect storm" of conditions—a moderate autumn allowed time to apply fertilizer and prepare the ground for the next season, a very rainy spring and summer delivered the fertilizer to the lake, and a summer with less wind than normal prevented the nutrients from being dispersed—all of which conspired to fuel the record algal bloom.

Watershed topography (flatness or steepness), natural land cover type, and how humans use the land (as parks or parking lots) influences whether pollution makes its way into surface waters. Ideally, the land near the water's edge (the **riparian area**) should have abundant plant life. Well-vegetated riparian areas stabilize banks, which reduces sediment pollution. In addition, undergrowth and deep rooted plants slow the flow of runoff which helps keep pollution out of surface waters and also increases groundwater supplies. The farther water has to trickle down the soil to reach groundwater the more infiltration acts as a filter, purifying the water. [INFOGRAPHIC 16.4]

The soil filtration system does not always work perfectly. In 2000, thousands of people fell ill in Walkerton, Ontario, when a deadly strain of *Escherischia coli* contaminated one of the wells providing water to some residents. Unfortunately, *E. coli* and other pathogens such as *Salmonella* can enter water sources from untreated sewage or animal waste. In Walkerton, heavy spring rains carried runoff with contaminated manure into a well, entering the groundwater. The incident—which claimed the lives of 7 people—was aggravated by administrative missteps, such as failure to sufficiently chlorinate the water and to warn residents of the contamination. (For more on freshwater resources, see Chapter 15.)

Clearly, healthy, well-vegetated watersheds serve many important functions within an ecosystem—they purify groundwater, prevent floods and major erosion, and protect nearby bodies of water. Any problem in the Lake Erie watershed could affect the lake, so officials from every shoreline have to work together to determine the cause of the latest problems. "There's no point in half the watershed doing something [to prevent pollution] if the other half continues" polluting the water, says Watson.

And even when wastewater is treated before being released into water bodies, it can contain the remnants of many medications we humans excrete in our waste. This treated wastewater from residential areas, such as the busy communities lining Lake Erie, contains enough drug residues to affect aquatic life. Some of these substances are endocrine disruptors, altering fish and amphibian reproduction. Downstream from wastewater treatment plants in Alberta, a research team led by Ken Jeffries has identified fish populations that are predominately female, and that have some male fish exhibiting female traits, likely a result of exposure to synthetic estrogens and other contaminants.

In this second round of pollution in Lake Erie, researchers soon realized they were also dealing with an entirely new complication that was indirectly spurring the growth

Infographic 16.4 | HEALTHY RIPARIAN AREAS PROVIDE MANY BENEFITS
Plants are nutrient sinks—they store nutrients in their tissues, reducing the amount free to enter the water.
Ground vegetation slows runoff and allows water to soak into the ground. This keeps the water and any pollutants it may carry out of the stream and increases infiltration, which replenishes groundwater.
Shade provided by overhanging trees cools the water; important because cooler water can hold more oxygen than warmer water.
Plant roots stabilize banks and prevent erosion and the deepening of the channel, which can speed water flow and lead to further erosion.
Trees provide food; leaf litter that falls in the water is a main nutrient source for aquatic organisms, including nitrifying bacteria which break down nitrates, reducing the amount sent downstream.

↑ The area next to a body of water that impacts that water (provides shade and nutrients) and is itself impacted (water-tolerant species live here) is the riparian area. A well-vegetated riparian area reduces the runoff that reaches a body of water by slowing the water's movement across the land so that it soaks into the ground rather than flowing into the stream.

↑ Tiny zebra mussels, like these found in Lake Ontario, have invaded the Great Lakes and have spread as far west as British Columbia and are beginning to move down the Mississippi River.

of algae: an invasive species. In 1988, a transatlantic freighter carrying ballast water from another region—a technique meant to balance a boat in the absence of cargo—entered into the Great Lakes system. The water contained some tiny Russian hitchhikers: zebra mussels. Once the vessel picked up new cargo, it dumped the water in Lake St. Clair, which feeds into Lake Erie. Zebra mussels are hardy creatures that breed prolifically; each female can produce 1 million eggs in a year. Young mussels are barely wider than a human hair, and can drift for kilometres before attaching onto a stationary object, such as rocks, docks, and piers; they also attach to boats or trailers, which carry them even farther. Within 10 years, they had spread to all five of the Great Lakes. "If you look at a map of North America, they're everywhere—even British Columbia," says Shear. "They've colonized just about every available habitat there is."

In some areas of Lake Erie, you could find 1 million mussels per square metre. Shells make beaches unusable; they even blocked a Michigan town's water supply when they clogged a major pipeline. Zebra mussels aren't the only invasive species problem, either. More than 140 non-native species have become established in the Great Lakes since the 1800s, a form of *biological pollution*. Other new, noxious neighbours include the round goby, a small fish that outcompetes native fish and eats the eggs of lake trout, and quagga mussels, which are similar to zebra mussels.

How could the introduction of zebra mussels kill fish and trigger algal blooms? Invasive zebra mussels can actually help remove excess nutrients or toxins since they are filter feeders, but they disrupt the entire community and outcompete some native species for food or habitat. Each mussel can filter up to 4 litres of water per day, including suspended sediments and algae, but they then excrete excess phosphorus back into the water. Water clarity increases—from 15 centimetres to 9 metres in some parts of Lake Erie—but that allows sunlight to penetrate further, fuelling an explosive growth of algae, potentially causing hypoxia.

So what can be done? Unfortunately, once zebra mussels and other invasive species have taken root, they can be very difficult (if not impossible) to eradicate. Some researchers are investigating forms of biological control, in which they release a predator of zebra mussels into the water, but Lake Erie is so large, and zebra mussels reproduce so rapidly, that no predators have kept pace. Ultimately, the most effective solution is to try to prevent more invasive species from entering and spreading—by, for instance, removing vegetation and attached mussels from boats, and thoroughly cleaning scuba gear before using it to enter new waters.

You can't prevent it if you can't see it coming.

Every time Watson boards her research vessel, she takes numerous water samples from different parts of Lake Erie and Ontario, from both surface and deeper waters. As part of this **water monitoring**, she analyzes the samples to compare them to what would be expected in healthy systems.

On board the vessel, she performs a *chemical analysis* to monitor what chemicals are present, and at what levels, and to check for quantities of phosphorus and nitrogen, the nutrients that encourage growth of algae. She also conducts a *physical assessment* of the sample, in which she measures the pH (most freshwater organisms require near-neutral pH levels), temperature, and cloudiness of the water (turbidity). Usually, the lower the temperature the better, since water can hold more oxygen when it's cold. Cloudy water limits photosynthesis, can clog the gills of aquatic organisms, and decreases visibility.

When Watson sees wispy filaments and other telltale signs of algae, she tests them to determine which nutrients trigger their growth, and examines their genetic makeup, which reveals if they are toxic species. Unfortunately, since Watson began these trips in Lake Ontario in 2000, and Lake Erie in 2005, she's found that the blooms are more likely to contain cyanobacteria than in the past, resulting in more cyanobacteria toxins in the water. Algal blooms are also more severe and occur later in the season, in areas that used to be relatively algae free.

The health of an aquatic ecosystem can also be assessed by examining what lives there. This is known as **biological assessment**. One approach relies on simply netting, identifying, and counting **benthic macroinvertebrates** such as insects and crayfish that live on the bottom of streams (the benthic zone); if the stream is unhealthy, there won't be many organisms present that are sensitive to pollutants. The abundance and diversity of pollution-tolerant and pollution-sensitive species in the sample can be used to assess stream quality. [INFOGRAPHIC 16.5]

Long-term chemical and biological monitoring of lakes and streams is important if we want to be able to detect and predict changes in water quality and ecosystem health. The Experimental Lakes Area (ELA) is an internationally recognized whole-lake research facility in northwestern Ontario, where long-term monitoring of eutrophication has been done since 1969. This kind of research and monitoring is essential to understanding how lakes respond to key pollutants.

Reducing pollution means reducing it at the source and preventing it from reaching the water.

Although officials took steps to reduce nutrient runoff from agricultural lands in previous decades, farmers everywhere can prevent chemicals and fertilizers from seeping into water bodies by using them judiciously, limiting toxic types, and planting cover crops during off-seasons, as exposed soil allows more runoff into streams during rain events. It's often better to use fertilizers in warmer months, so it can be absorbed, rather than lying on top of frozen soil only to be washed away with the rain.

---

water monitoring Collection of water samples from different parts of a body of water (surface and deeper water), particularly for comparison with what would be expected in healthy systems.

biological assessment Sampling an area to see what lives there as a tool to determine how healthy the area is.

benthic macroinvertebrates Easy-to-see (not microscopic) arthropods such as insects and crayfish that live on the stream bottom.

Infographic **16.5** | **BIOLOGICAL ASSESSMENT**

↓ While one can learn about the water quality in a given location by performing physical (i.e. pH, temperature, clarity) or chemical (i.e. salinity, levels of nitrogen or phosphorus) tests, these assessments typically require some expertise and equipment. An easier "first look" at a body of water is often a biological assessment—simply collecting a net sample of aquatic organisms to see what is living in the water. If there is good diversity and abundance of aquatic life, especially of those organisms that are sensitive to pollution, then it is reasonable to conclude the water is clean. A poor sampling suggests there is a problem, prompting one to look more closely at water quality to determine the reason for the problem.

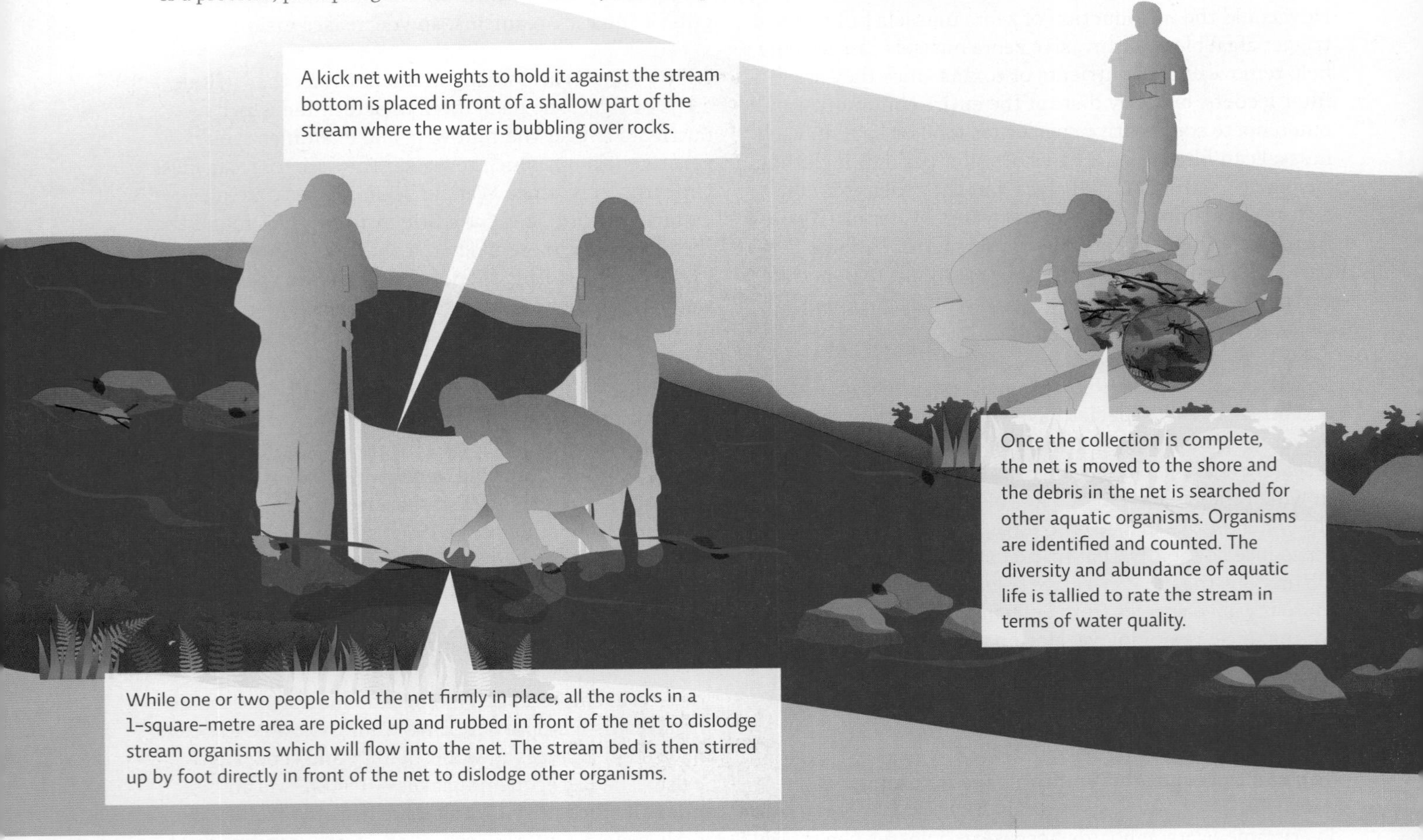
A kick net with weights to hold it against the stream bottom is placed in front of a shallow part of the stream where the water is bubbling over rocks.
Once the collection is complete, the net is moved to the shore and the debris in the net is searched for other aquatic organisms. Organisms are identified and counted. The diversity and abundance of aquatic life is tallied to rate the stream in terms of water quality.
While one or two people hold the net firmly in place, all the rocks in a 1-square-metre area are picked up and rubbed in front of the net to dislodge stream organisms which will flow into the net. The stream bed is then stirred up by foot directly in front of the net to dislodge other organisms.

BENTHIC MACROINVERTEBRATE GROUPINGS

Since all these organisms differ somewhat in their tolerance of different pollutants, the ideal collection would be one with a wide variety of pollution-sensitive and intermediate organisms. This tells you that the water quality is good.

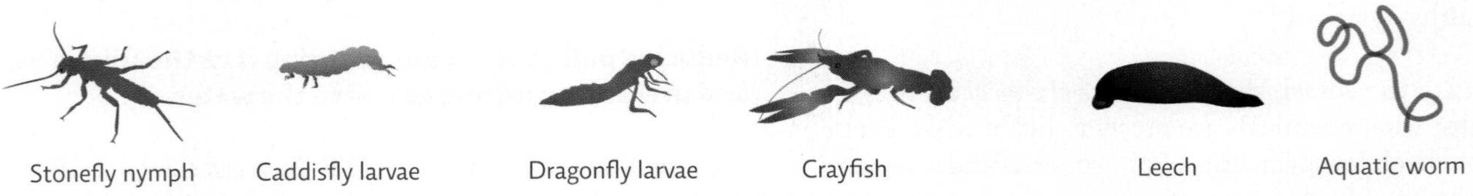
Stonefly nymph
Caddisfly larvae
Dragonfly larvae
Crayfish
Leech
Aquatic worm

POLLUTION-SENSITIVE GROUP: Benthic macroinvertebrates like stonefly nymphs and caddisfly larvae are easily killed by pollution or low oxygen levels, so they are good indicators of water quality.

INTERMEDIATE GROUP: Organisms like dragonfly larvae and crayfish are somewhat sensitive to pollution but can tolerate low levels of some pollutants or slightly diminished oxygen levels.

POLLUTION-TOLERANT GROUP: This group includes organisms like leeches and aquatic worms; they can tolerate higher levels of pollution and low oxygen levels. That is not to say they will be absent from clear, well-oxygenated water, but if you find only pollution-tolerant organisms in your net, this is a sign there is a problem.

Infographic **16.6** | **RESTORATION OF LAKE ERIE**

GOAL	STRATEGY
Reducing pollution at the source 	• Agriculture: Reduce use of hazardous chemicals and fertilizers in areas close to bodies of water; plant cover crops between seasons to hold soil in place; prevent manure runoff from farms and fields. • Implement better regulation of industrial air and water pollution. • Improve sewage handling and treatment at sewage treatment plants; upgrade private sewage systems if needed (septic systems and those used on watercraft).
Fish and wildlife protection 	• Restock fish populations to reestablish natural ecological communities. • Restore or construct wetlands to capture stormwater and provide habitat. • Monitor fish populations for toxic substances.

GOAL	STRATEGY
Reduce and manage stormwater 	• Reduce stormwater runoff by protecting stream corridors and revegetating watershed areas. • Create basins to capture, store, and treat stormwater runoff. • Increase permeable surfaces in urban areas to allow infiltration and decrease runoff; minimize land disturbance when building.
Water conservation and protection 	• Restrict use of lead shot in waterfowl hunting and lead sinkers for fishing. • Encourage household and industrial water conservation and promote use of safer chemicals.

In fields around Lake Erie, many farmers have cordoned off areas for livestock that prevent them from wandering into streams, where they damage stream banks and bottoms, and pollute the water with feces and urine. Another key step they have taken is to restore riparian areas by vegetating buffer zones to slow runoff and give the rainwater time to soak into the ground. Trees at the water's edge also provide cooling shade, which keeps water cooler and better able to hold oxygen.

In 1987, under the Great Lakes Water Quality Agreement, the U.S. and Canadian governments created *Lakewide Management Plans* for the Great Lakes, essentially plans of action to tackle the ongoing issues affecting each lake. Over the years, both governments have dedicated millions of dollars to preserving and cleaning up the lakes. Canada's federal budget in 2010, for instance, allocated $16 million to cleaning up the most damaged areas.

Many wetlands along Lake Erie have been converted into agricultural fields, even though these wetlands act as natural filters for pollutants or nutrients. Now, as part of the Great Lakes Restoration Initiative, more than 120 hectares of unused farmland will be converted back into wetlands.

One of the most important steps officials took in the 1970s to clean up Lake Erie was to upgrade sewage treatment facilities. But now that the same problems have returned, it's time to revisit those improvements to make sure they still meet the demands of the twenty-first century population, says Shear. More people have moved to the shores of western Lake Ontario, he says, which means more flow into sewage treatment plants. So pollution standards set in the 1970s may no longer suffice, he says. For instance, if facilities are allowed to release treated water containing less than 1 milligram of phosphorus per litre, the additional volume of water from the increased population means the total amount of phosphorus released will increase. "When more people move to an area, that increases the volume going to sewage treatment plants, so we have to lower the limits for pollutants even further." [INFOGRAPHIC 16.6]

Infographic **16.7** | **INCREASING INFILTRATION OF STORMWATER**

↓ Stormwater that doesn't soak into the ground can enter storm drains that flow directly to rivers and streams, or cause floods, especially in heavily built-up urban settings. Anything that increases infiltration can help avoid these stormwater problems.

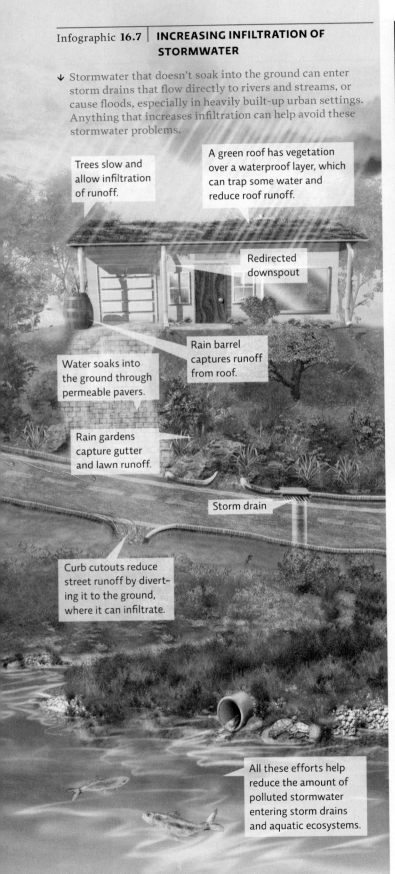

Trees slow and allow infiltration of runoff.
A green roof has vegetation over a waterproof layer, which can trap some water and reduce roof runoff.
Redirected downspout
Rain barrel captures runoff from roof.
Water soaks into the ground through permeable pavers.
Rain gardens capture gutter and lawn runoff.
Storm drain
Curb cutouts reduce street runoff by diverting it to the ground, where it can infiltrate.
All these efforts help reduce the amount of polluted stormwater entering storm drains and aquatic ecosystems.

City of Chicago
Green Alley
Richard M. Daley
Mayor

↑ "Green alleys" like this one in Chicago are those that have been resurfaced with permeable pavement to allow rainwater or snowmelt to pass through the pavement and soak into the ground rather than run off into storm drains which would eventually reach Lake Michigan.

Even though officials may be more aware of the problems water pollution can cause, these problems are far from being fully addressed. The city of Victoria dumped more than 120 million litres of raw sewage into the ocean every day in 2010; more than 181 million litres of untreated storm and wastewater are dumped every day into Halifax Harbour. Not surprisingly, both regions have reported massive dead zones as a result. But Canada's "sickest" lake is Lake Winnipeg, which is now coated in a putrid mat of algae that is twice the size of Prince Edward Island. It's even visible from space.

Steps can be taken in towns and cities to reduce sources of pollution and to decrease stormwater runoff. In suburban areas, lawns can be major non-point sources of nitrogen pollution if homeowners apply too much fertilizer. However, the grass also sequesters nitrogen because of its long growing season, preventing it from flowing to the nearby bodies of water. Homeowners are encouraged to limit fertilizer use on lawns and to plant native plants and grasses that do not need fertilizer. Planting a **rain garden** of water-tolerant plants in low-lying areas that tend to flood in rain events will also capture water and reduce runoff.

In urban areas, replacing some hard surfaces with porous surfaces such as green space or permeable pavers reduces runoff by allowing water to seep through. Green roofs can also help capture some rainwater and slow or prevent its release to the environment (see Chapter 26). [INFOGRAPHIC 16.7]

The story of Lake Erie reveals that only constant vigilance and effort will keep the world's waters clean. Complacency and cutting corners are not an option. Two years ago, Shear brought scientists visiting from Mexico to a sewage treatment plant in late winter to show them how phosphorus is removed before discharging the water into Lake Ontario. But the tanks and chemicals needed to treat the water were missing. Shear asked a technician where they were. "The technician said, 'oh, we don't use those in the winter to save money, because algae won't grow in cold water.'" Shear's jaw dropped. "Even if algae don't grow in colder months, all those nutrients are pouring into the water—it's like dumping fertilizer." And when the water warms up in the spring, a huge burst of algae will follow.

We know the sources of water pollution, and we know how to address them, says Watson. It's time to re-enact Lake Erie's success story. "The Great Lakes' contaminants are coming back to haunt us," she says. "We've got a lot of work to do."◉

Select references in this chapter:

Cruikshank, D.R. 1984. Whole lake chemical additions in the Experimental Lakes Area, 1969–1983. *Fisheries and Marine Service Data Report,* 449: iv + 23.

Jeffries, K., *et al.* 2008. *Environmental Toxicology and Chemistry,* 27(10): 2042–2052.

Michalak, A.M., *et al.* 2013. *Proceedings of the National Academy of Sciences* [Online], www.pnas.org/cgi/doi/10.1073/pnas.1216006110.

Watson, S.B., *et al.* 2008. *Canadian Journal of Fisheries and Aquatic Sciences,* 65(8): 1779–1796.

rain garden Runoff area that is planted with water-tolerant plants to slow runoff and promote infiltration.

BRING IT HOME

PERSONAL CHOICES THAT HELP

We are facing not only shrinking supplies of easily accessible water, but also the potential degradation of this resource due to pollution. By changing products and modifying common practices, we can improve our water quality for years to come.

Individual Steps

→ Read your community's water quality report to see which pollutants are prevalent in your area.

→ Decrease your use of chemicals (fertilizers, pesticides, harsh cleaners, etc.) that will end up in the water supply. For alternatives to traditional yard chemicals, see safelawns.org.

→ Always dispose of pet waste properly. In high quantities it acts as an oxygen-demanding waste and can also spread disease.

Group Action

→ Marking storm drains with "Don't Dump" symbols can remind people not to dump waste liquids down sewer drains. If the drains in your area are not marked, talk to city officials to see if you and other volunteers can mark them.

Policy Change

→ Canada Water Week happens every March. Take steps every day to reduce your water pollution, and help raise awareness in March by writing a letter to your local newspaper outlining simple steps people can take to improve water quality.

UNDERSTANDING THE ISSUE

CHECK YOUR UNDERSTANDING

1. Water pollution is:
a. found only in surface waters near cities.
b. primarily excess nutrients from lawns, farms, and animal feedlots.
c. usually from excess carbon being added to the system.
d. contaminants or excess nutrients in surface waters and in groundwater.

2. The following items are part of the process of eutrophication.
1. Algae quickly reproduce, using up oxygen and blocking sunlight to underwater plants.
2. Bacteria consume excess wastes and nutrients, using up oxygen.
3. Underwater plants die.
4. Excess nutrients enter a body of water.
What is the correct order for the process?
a. 1, 2, 3, 4
b. 4, 1, 3, 2
c. 3, 4, 2, 1
d. 2, 4, 1, 3

3. The riparian area of a lake or stream:
a. should be cleared of any natural debris like downed trees or fallen branches.
b. is the area where water infiltration is most likely to occur.
c. can protect the stream from runoff if well vegetated.
d. is a good place for cropland or livestock grazing.

4. Fertilizer from your lawn and motor oil from the leaky oil pan on your car are examples of:
a. non-point source pollution.
b. point source pollution.
c. eutrophication.
d. pathogenesis.

5. Biological pollution is:
a. invasive, non-native species that degrade an ecosystem.
b. a spill of a naturally occurring chemical.
c. any chemical pollutant that can kill wildlife.
d. the decline of native populations in an area.

6. A watershed includes:
a. only the land that would be underwater during a normal rainfall year.
b. the surface water and the underground aquifer.
c. all the uphill land surrounding a river and its streams that can feed water into that river.
d. all the land downhill from a river that could potentially be flooded.

WORK WITH IDEAS

1. A factory manufactures convenience foods such as frozen lasagna. Its wastes are filtered for large particulates and then fed into the local river. Why might this be a problem for the fish that live in the river?

2. The fish in your local lake or bay are rapidly declining in number. List the possible water problems, as discussed in the chapter, which might be causing the decline.

3. Compare and contrast three typical point source pollutants and three non-point source pollutants from the area where you live.

4. What are potential non-point sources of pathogens and drugs entering our waterways and what problems can these cause for people and other organisms?

5. As we saw in Chapter 6, in most natural ecosystems, nitrogen is a limiting factor for the growth of plants, and subsequently for the total biomass of plants and animals. How have humans changed this biogeochemical cycle?

6. You are investigating a die-off of fish in the local mountain lake, Lake Pleasant. It is fed by Hilltop Stream, and the Happy Valley River comes out of the lake and heads down toward Happy Valley. Do any of the following have something to do with the die-off? Justify your answers.
   - The new manufacturing plant just over the ridgetop in the next valley
   - The cattle feedlot on the edge of the Happy Valley River
   - The recent construction of a new subdivision on the eastern slopes above Lake Pleasant
   - The lovely older neighbourhood with wide lawns, a golf course, and a large grassy park on the western slope above the lake

ANALYZING THE SCIENCE

The application of road salt in cold climates can result in runoff pollution that can damage aquatic ecosystems. Researchers at Saint Mary's University in Halifax surveyed local roadside ponds (some were occupied by amphibians and some were not) to determine chloride concentration and species richness of amphibian species. They also investigated the vulnerability of 5 amphibian species to sodium chloride (NaCl) by determining the LD-50, the dose or concentration of chloride that kills 50% of the population.

INTERPRETATION

1. Look at Graph A on the following page. In a single sentence, explain the relationship between chloride concentration and species richness.

2. Look at Graph B. Which species of amphibian is least affected by the presence of chloride in the water? Explain your reasoning.

3. Look at Graph C. Which species is the most sensitive to chloride? Which is the least sensitive?

ADVANCE YOUR THINKING

4. Look at Graph B and Graph C. Do the LD-50s for each species shown in Graph C correlate with the data (presence or absence of each species according to chloride concentration) shown in Graph B? Explain.

5. Suggest two actions that could be taken to reduce the chance that road salt would enter ponds as runoff.

GRAPH A: Effects of chloride on amphibian species richness

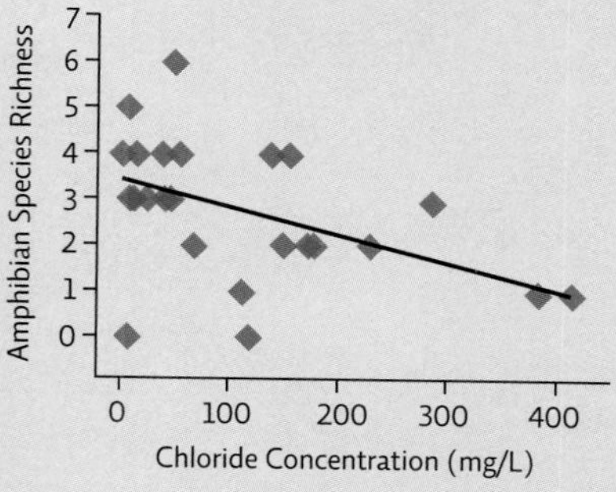
Amphibian Species Richness
0
1
2
3
4
5
6
7
0
100
200
300
400
Chloride Concentration (mg/L)

GRAPH B: Chloride concentration in occupied and unoccupied ponds for 5 amphibian species*

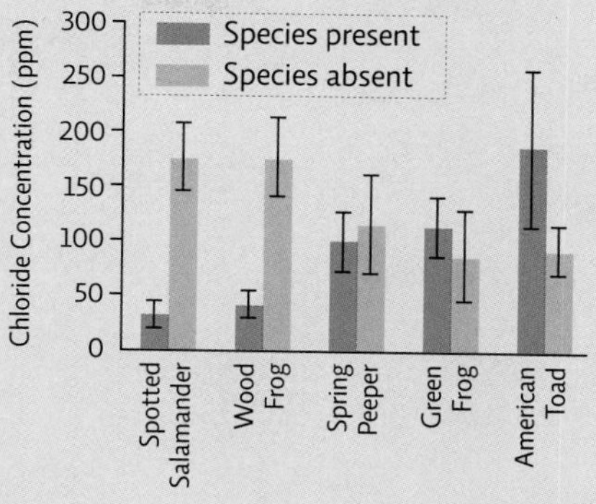
Species present
Species absent
Chloride Concentration (ppm)
0
50
100
150
200
250
300
Spotted Salamander
Wood Frog
Spring Peeper
Green Frog
American Toad

GRAPH C: Comparison of median LD-50 values in amphibians*

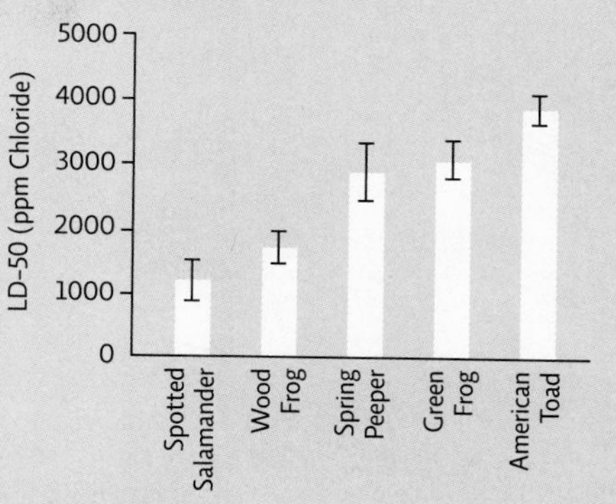
LD-50 (ppm Chloride)
0
1000
2000
3000
4000
5000
Spotted Salamander
Wood Frog
Spring Peeper
Green Frog
American Toad

*Error bars represent ±1 SE

EVALUATING NEW INFORMATION

Settling ponds, also known as holding ponds, are locations where contaminated water is allowed to stand so that particulates suspended in the water settle to the bottom of the pond. The water then evaporates or is drawn off, leaving behind the unwanted sediment, which can be from a mine, a quarry, a manure pit at an animal facility, industrial wastewater, stormwater, or other sources.

Go to www.technology.infomine.com/sedimentponds to see an overview of some of the concerns about building and maintaining a settling pond for surface mines. Then visit www.agriculture.gov.sk.ca. Choose an article to read. Summarize that article and evaluate it using all of the items under Question 1.

Evaluate the authors of each website or article and work with the information to answer the following questions:

1. Is this author/sponsoring group a reliable information source?
   a. Does the author or group give supporting evidence for his/its claims?
   b. Does he/it give sources for the evidence?
   c. What is the mission of this organization, or of the scientist/author who wrote the article?
2. Do an Internet search and read two or three news articles about the toxic sludge spill near Kolontár, Hungary.
   a. What is the most recent article you can find and what does it say about the incident?
   b. How similar is the information in the articles?
   c. Do the authors give any sources cited for the "facts" presented?
3. Now do an Internet search using Google Scholar or another scholarly database. Select a peer-reviewed article that you can access in its entirety (not just the abstract—look for a pdf link). Read the article and summarize its purpose and main points.
   a. How does this article compare to the news articles you read? Compare and contrast the types of information regarding scope of coverage, intended audience, clarity of information, and reliability of information (whether reliable sources are cited).
   b. In general, when would you go to a news source for information on a topic like this and when would you go to a scholarly source?
4. What could be done to prevent another incident like what happened at Kolontár from happening again?

MAKING CONNECTIONS

COPPER MINING IN BRITISH COLUMBIA

Background: Copper is a critical element in our modern lives—it's used in our power lines, roofing materials, water pipes, and circuit boards. A proposed open pit copper and gold mine in British Columbia (the Ajax project) would be located partly within the city limits of Kamloops. Open pit mines have the potential for significant air and water pollution. Mines must pump water constantly, both to remove excess water from the mine itself and to prevent some of the toxic compounds used in the mining process from reaching the local aquifer and tainting the water supply. Even after it has removed the valuable ore, a mine must continue to pump in order to keep from polluting the groundwater. In places where pumping has stopped, a pit lake may form—such as the one at the Anaconda Copper Mine in Montana which became a federal Superfund site that the U.S. government must clean up. Despite the potential for significant water pollution, the mining company, KGHM Ajax, says that the mining will be done in an "environmentally and socially responsible manner." Kamloops has a thriving tourism industry with year-round indoor and outdoor activities. It also has a productive farming community, known for its diverse farm products sold at local farmer's markets. What impact would the mine likely have on the city and the region's local farms, ranches, and tourism industry?

Case: You are an investigative reporter, trying to lay out the facts for your readers. Research and write an article explaining:

a. how water is used in the mining process.
b. the types of water pollution that can occur with a copper mine.
c. whether the mine and its water use will impact the people who live there.
d. whether the mine will impact the ecology and the groundwater of the region.
e. whether the mine will impact the area's tourism.

Trash litters coastal waters near the Vietnamese island of Phu Quoc. Millions of metric tons of solid waste find their way to the world's oceans every year.

CHAPTER 17 SOLID WASTE

A PLASTIC SURF

Are the oceans teeming with trash?

CORE MESSAGE

We are often unaware of the waste we produce, but it profoundly affects the environment. Waste that cannot be reused or recycled simply accumulates; some of it is toxic or otherwise disruptive to living things. We can address its impact by minimizing the waste we generate in the first place, recovering and recycling waste materials, and disposing of all waste safely.

GUIDING QUESTIONS

After reading this chapter, you should be able to answer the following questions:

→ What kinds of trash do humans generate, particularly in developed nations like Canada?

→ What options do we have for dealing with solid waste and what are the trade-offs for each option?

→ What is hazardous waste and what can the average person do to reduce his or her production of it?

→ What are some of the environmental consequences of our waste production and disposal methods?

→ How can the amount of waste produced be reduced?

From the deck of the tall ship *SSV Corwith Cramer,* the surface waters of the North Atlantic looked like smooth, dark glass. The 41-metre oceanographic Sea Education Association (SEA) research vessel had set out from Bermuda just 24 hours before on a month-long expedition aimed at tracking human garbage across the deep sea. It was a cool mid-June evening, the sun was setting, and the crew had just launched its first "neuston tow" of the trip. Giora Proskurowski, the expedition's chief scientist, stood watching as the tangle of mesh skidded along the water's surface, or *neuston layer,* keeping pace with the ship as it went. It was hard to imagine that any garbage would be found in such a flat, serene seascape.

After 30 minutes, crew members pulled the net—dripping with seagrass and stained red with jellyfish—from the water. Sure enough, when Proskurowski got closer, he could make out scores of tiny bits of plastic glistening in the mesh. The crew members counted 110 pieces, each one smaller than a pencil eraser and no heavier than a paper clip. Based on the size of the net and the area they had dragged it over, 110 pieces came out to nearly 100 000 bits of plastic per square kilometre.

Despite the water's pristine look, the yield was hardly surprising. In the past 20 years, undergraduate students on expeditions like this one had handpicked, counted, and measured more than 64 000 pieces of plastic from some 6000 net tows. That might not sound like much, given the vastness of the Atlantic and the smallness of the plastic. But the tiny bits were gathering in very specific areas, known as *gyres.* Gyres are regions of the world's oceans where strong currents circle around areas with very weak, or even no, currents. Lightweight material, like plastic, that is delivered to a gyre by ocean currents becomes trapped, and cannot escape the stronger circling currents. When scientists first evaluated all the data from the Atlantic Gyre, they discovered surprisingly dense patches of plastic—more than 100 000 pieces per square kilometre—across a surprisingly large "high-concentration zone." The press and general public would come to know this region as the "Great Atlantic Garbage Patch," cousin to a "Great Pacific Garbage Patch" discovered around the same time, and, they suspected, to several other patches scattered about the ocean.

◉WHERE IS BERMUDA?

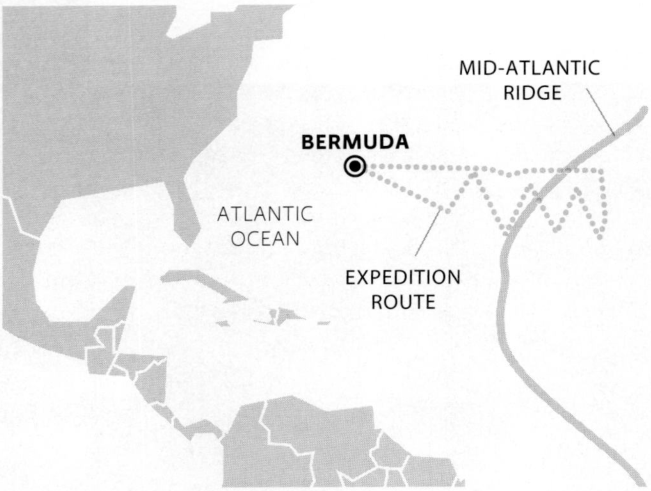
MID-ATLANTIC RIDGE
BERMUDA
ATLANTIC OCEAN
EXPEDITION ROUTE

Scientists knew where the plastic was coming from (seagoing vessels, open landfills, and litter-polluted gutters around the world). And they understood why it was being trapped in the gyres (wind patterns and ocean currents). But other questions remained. No one could say how big the Atlantic Patch actually was, or how all that plastic was affecting ocean ecosystems. Were the toxic substances found in plastic accumulating up the food chain, making their way into fish, bird, or mammal diets? Were all those tiny bits of solid surface providing transport for invasive species? And why wasn't there more of it? Despite a fivefold increase in global plastic production over 22 years, the concentration of plastic in North Atlantic surface waters had remained fairly steady.

To answer these questions, Proskurowski, his team, and their captain would sail the *Corwith Cramer* all the way out to the Mid-Atlantic Ridge, an underwater mountain range that lies about 1600 kilometres farther east than any previous plastics expedition had ever gone. "That's far enough from Bermuda that getting back will be a challenge," he wrote on the ship's blog as land faded from view. "The same forces that drew plastic into this particular part of the ocean—namely low and variable winds and currents—will make operating a tall ship sailing

↑ Team members ready a neuston net for deployment.

↓ Giora Proskurowski, chief scientist of the *Corwith Cramer*

vessel tricky, to say the least." But if all went according to plan, Proskurowski thought, he and his colleagues would be the first to reach the so-called garbage patch's easternmost edge.

Waste is a uniquely human invention, generated by uniquely human activities.

In natural ecosystems, there is no such thing as waste. Matter expelled by one organism is taken up by another organism and used again. This natural recycling is consistent with the **law of conservation of matter**, which states that matter is never created or destroyed; it only changes form. Forms of matter that are dangerous to living things (think arsenic and mercury) tend to stay buried, deep underground, and are released only during extreme events like volcanic eruptions.

law of conservation of matter Matter can neither be created nor destroyed; it only changes form.

Infographic **17.1** | **MUNICIPAL SOLID WASTE**

↓ At 777 kilograms per capita (in 2008), Canada ranks 18th out of 29 Organisation for Economic Co-operation and Development (OECD) nations—developed and developing democratic states with free market economies—in terms of amounts of municipal waste produced annually. Although municipal solid waste makes up only 3% of total waste produced here, it is the type of waste we can most directly impact with our individual day-to-day choices. Residential waste production continues to rise in Canada, but so do municipal recycling and composting programs, which now divert 25% of residential waste from landfills.

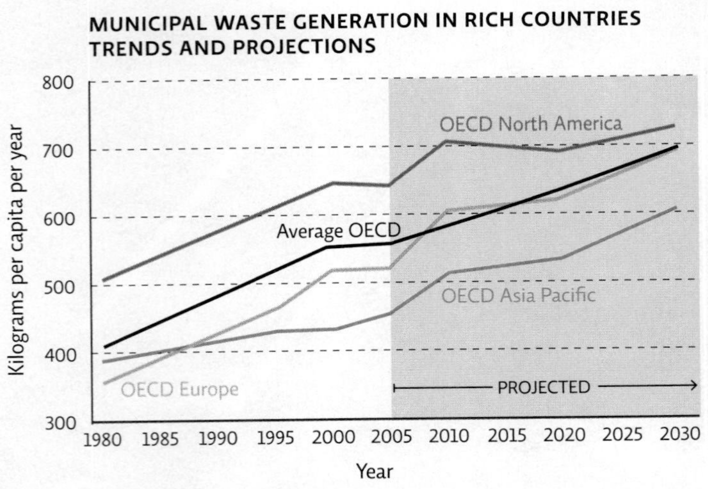
MUNICIPAL WASTE GENERATION IN RICH COUNTRIES TRENDS AND PROJECTIONS
Kilograms per capita per year
800
700
600
500
400
300
OECD North America
Average OECD
OECD Asia Pacific
OECD Europe
PROJECTED
1980 1985 1990 1995 2000 2005 2010 2015 2020 2025 2030
Year

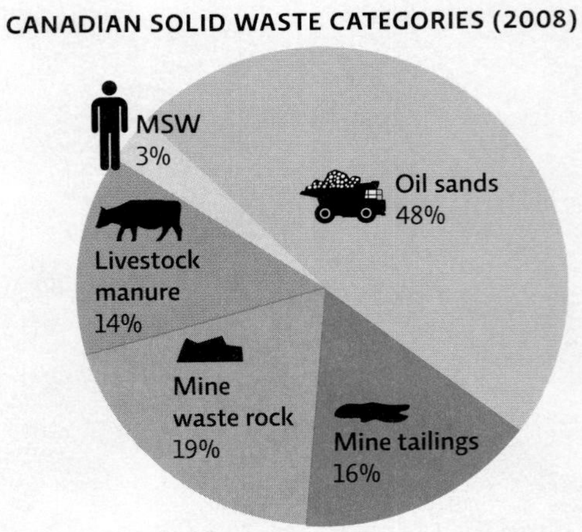
CANADIAN SOLID WASTE CATEGORIES (2008)
MSW 3%
Oil sands 48%
Livestock manure 14%
Mine waste rock 19%
Mine tailings 16%

Human ecosystems are another story. By taking matter out of the reach of organisms that can use it, we continually disrupt this natural cycle. We do this by converting usable matter into synthetic chemicals that can't easily be broken down, and by burying readily degradable things in places and under conditions where natural processes can't run their course.

For example, paper and cardboard break down easily in a compost bin, thanks to the microbes that feed on them. But we typically keep this type of waste in landfills, where the lack of water, oxygen, and microbes forces it to decompose much more slowly than it otherwise might.

Thus matter—the physical substance of which the universe is made—becomes **waste**—a uniquely human term used to describe all the things we throw away. Waste that can be broken down by microbes is considered **biodegradable**. Waste that can be broken down by chemical and physical reactions is considered degradable, even if that degradation takes a long time. Some waste—mostly synthetic molecules like the pesticide DDT and CFCs once found in aerosols—is considered **nondegradable**.

waste Any material that humans deem to be unwanted.
biodegradable Capable of being broken down by living organisms.
nondegradable Incapable of being broken down under normal conditions.
municipal solid waste (MSW) Everyday garbage or trash (solid waste) produced by individuals or small businesses.

These molecules are chemically stable and don't degrade in normal atmospheric conditions. And because they haven't been around for very long, no organism has yet developed (through mutation or genetic recombination) the enzymes to use them as food.

Almost any human activity you can think of generates some form of waste. The harvesting of coal, oil, and ores produces *mining waste,* which can pollute air, water, and soil, and makes up 83% of solid waste in Canada. Processes that produce food, consumer goods, and industrial products generate *agricultural* and *industrial waste*; in Canada, livestock manure waste alone accounts for 14% of solid waste. In built-up municipal areas, the increasingly complicated act of living, working, and producing goods generates its own steady stream of trash. This mix of residential, business, and industrial activity produces what is called **municipal solid waste (MSW)**, or at the community level, an MSW stream. [INFOGRAPHIC 17.1]

In Canada, MSW streams make up just 3% of total waste, but this is still a lot of trash: 34 million metric tons annually. In 2008 alone, each Canadian produced about 777 kilograms of solid waste (2.13 kilograms per day), up from about 769 kilograms per capita in 2002 (2.1 kilograms per day). With more than 35 million people living in Canada, this adds up to over 27 billion kilograms (27 million metric tons) of household trash per year. That's twice the per capita amount produced by the European Union and as

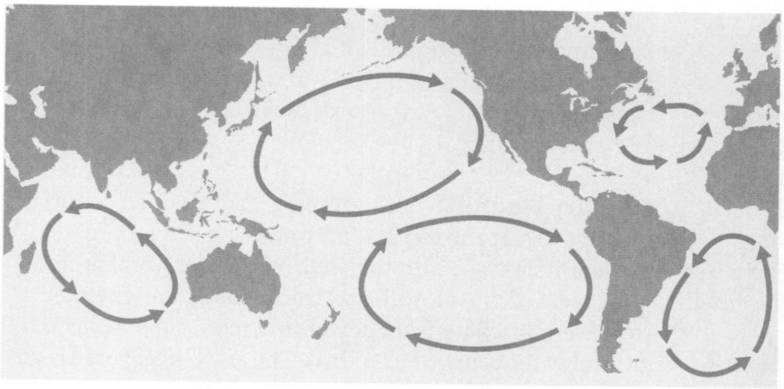

↑ Five major gyres are found in the world's oceans and there are floating bits of plastic in all of them. These "garbage patches" are not floating islands and aren't even necessarily visible when gazing at the water. Much of the debris is very small and lies just below the surface. The highest density captured in one of the 6100 sampling tows was approximately 200 000 pieces per square kilometre. It is hard to estimate the size of any garbage patch since the material is so spread out and may reach down to depths of 20 metres.

much as 10 times the amount produced by most developing countries.

The vast majority of this garbage comes from a familiar array of goods: paper, wood, glass, rubber, leather, textiles, and of course plastic—cheap enough to have become a staple of both advanced and developing societies, light enough to float, and durable enough to persist for hundreds of thousands of years across thousands of kilometres of ocean.

How big is the Atlantic Garbage Patch and is it growing?

The first garbage patch was discovered in 1997 by Captain Charles Moore, a veteran seafarer who was crossing the Pacific on his way back from an international yacht race.

As the press caught wind of what sounded like a giant plastic island in the middle of the ocean, a media frenzy ensued. Some reports said the patch was twice the size of Texas; others said that, in the densest portions, plastic outweighed plankton (plankton consists of all the organisms floating in the ocean's upper reaches—from microbes to jellyfish). Still others claimed that the patch was growing exponentially.

But none of those claims was accurate. In fact, most of them were completely unfounded.

Part of the trouble was semantic. "An oceanographer understands the term 'patch' to mean 'an uneven distribution,'" says Kara Lavender Law, a scientist at the Woods Hole Oceanographic Institute who has led several expeditions on the *Corwith Cramer*. "We say the upper ocean is 'patchy' because organisms are often observed in clumps, separated by regions with sparse populations. But when reporters and laypeople hear 'patch,' they incorrectly think of an 'island' or a 'continent' of trash."

> In 2008 alone, each Canadian produced 2.13 kilograms of solid waste per day.

A bigger problem was the data itself. MSW and global plastic production data indicated that more plastic was being produced, used, and discarded throughout the world. According to one scientific study, the amount of plastic being ingested by Subarctic seabirds had nearly doubled between 1975 and 1985. But it was impossible to tell whether the patches themselves were growing or not. Data from the Pacific Patch suggested that the concentration of plastic had risen by an order of magnitude between the 1980s and 1990s. But that comparison was of limited value because sampling methods differed widely from one expedition to the next. In the Atlantic, where sampling methods had remained consistent across the years, no such increase could be detected.

Some scientists had suggested that the nets used on most plastics expeditions were too porous to trap the smallest of particles and that, with finer nets, much more plastic would be found at the surface. Others thought that plastic was hiding below the surface, throughout the mixed layer—a layer of uniform density at the top of the water column that is saturated with sunlight, low in nutrients, and easily mixed by wind. Part of the *Corwith Cramer*'s mission was to try and figure out which hypothesis might be correct.

By July 3rd, Proskurowski's team had gleaned at least part of the answer. In the ship's daily blog, he wrote:

The results from the past several days of tucker trawls (which collect water samples from discrete depths within the mixed layer) *have been very interesting. While we've seen fairly low numbers of plastic at the surface, we have observed an almost equal amount of plastic from one meter's water depth, slightly less at ten meters and no plastic below the mixed layer...These tows show that plastic is undoubtedly being mixed down into the water column, and what we measure at the surface is, in many cases, not the sum total of what is out here. Like many environmental problems, the closer you look at the system, the worse it appears.*

It would be a few more weeks before they found out just how much worse.

How we handle waste determines where it ends up.

"It is with startling accuracy that so many tiny particles of plastic end up in such obscure but well-defined stretches of ocean," Proskurowski wrote one morning, as the *Corwith Cramer* floated under the early summer sun near the Mid-Atlantic Ridge. "Especially given how long and convoluted the journey they took to get here was."

In fact, all of our solid waste makes a similar journey, one that begins when we toss something we no longer need into a trash bag that is then carried from building to curb to garbage truck, before making its way to one of several kinds of waste facilities.

Open dumps are places where trash—both **hazardous** (waste that presents a health hazard) and nonhazardous—is simply piled up. Because they are one of the cheapest ways to get rid of human trash, they are common in undeveloped countries, where entire communities often spring up around the dumps, and people survive on what they scavenge from the waste piles. Open dumps attract pests such as flies and rats, which can be a human health hazard. Open dumps also contribute to water pollution: rain either washes pollutants away from the dump to surrounding areas or pulls it along as it soaks into the ground. If this contaminated water, called **leachate**, continues to travel downward, it can contaminate the soil and groundwater.

Sanitary landfills, more common in developed countries, seal in trash at the top and bottom in an attempt to prevent its release into the environment. Several protective layers of gravel, soil, and thick plastic are intended to prevent leachate from depositing toxic substances into groundwater below the landfill. The trash is covered regularly with a layer of soil that reduces unpleasant odours, thus attracting fewer pests.

But there is a downside to these, too. The compacting of trash under a layer of soil excludes oxygen and water so well that the aerobic bacteria (those that require oxygen to live) and other detritovores that normally decompose at least some of the waste can't survive. Newspaper that would degrade in a matter of weeks is preserved in landfills for decades. Anaerobic microbes (those that live in oxygen-poor environments) pick up some of the slack. But they are much slower and produce lots of methane, a combustible greenhouse gas that is 25 times more potent than CO_2. As a result, landfills are a significant anthropogenic contributor of methane. [INFOGRAPHIC 17.2]

Just under 5% of Canada's waste winds up in **incinerators**. Burning waste in this way reduces its volume dramatically—by about 80-90%, in fact. But it also pollutes the air and water, and produces ash, some of which is toxic, and must therefore be disposed of in a separate, specially designed landfill. Incinerators are also extraordinarily expensive to build, and tipping fees (fees charged to drop off trash) are usually much higher at an incinerator than at a landfill. [INFOGRAPHIC 17.3]

↑ At Phnom Pen, Cambodia's municipal garbage dump, people work around the clock collecting plastic, metals, wood, cloth, and paper, which they sell to recyclers.

open dumps Places where trash, both hazardous and nonhazardous, is simply piled up.
hazardous waste Waste that is toxic, flammable, corrosive, explosive, or radioactive.
leachate Water that carries dissolved substances (often contaminated) that can percolate through soil.
sanitary landfills Disposal sites that seal in trash at the top and bottom to prevent its release into the atmosphere; the sites are lined on the bottom, and trash is dumped in and covered with soil daily.
incinerators Facilities that burn trash at high temperatures.
e-waste Unwanted computers and other electronic devices such as discarded televisions and cellphones.

As local landfills reach their capacity, cities and towns begin seeking alternative locations for their waste. Some Ontario municipalities and businesses, for example, send their trash to landfills or incinerators in Michigan. Most of Canada's hazardous waste is exported to facilities in the United States for safe processing and disposal.

Electronic waste, such as old computers, cellphones, and televisions (collectively known as **e-waste)**, is often shipped to countries in Asia and Africa, despite Canada's participation in the Basel Convention, which bans e-waste exports to developing nations. There, impoverished villagers do their best to extract the precious metals within.

Infographic **17.2** | **MUNICIPAL SOLID WASTE DISPOSAL**

↓ There are a variety of options at our disposal for dealing with solid waste and a hierarchy of preference can be identified. Environment Canada promotes reducing waste at the source as the preferred option for dealing with solid waste. In 2008, MSW generation decreased almost 8% compared to 2002 levels and 25% of that was diverted to recycling or composting. But more than 70% was landfilled. There is still much room for improvement.

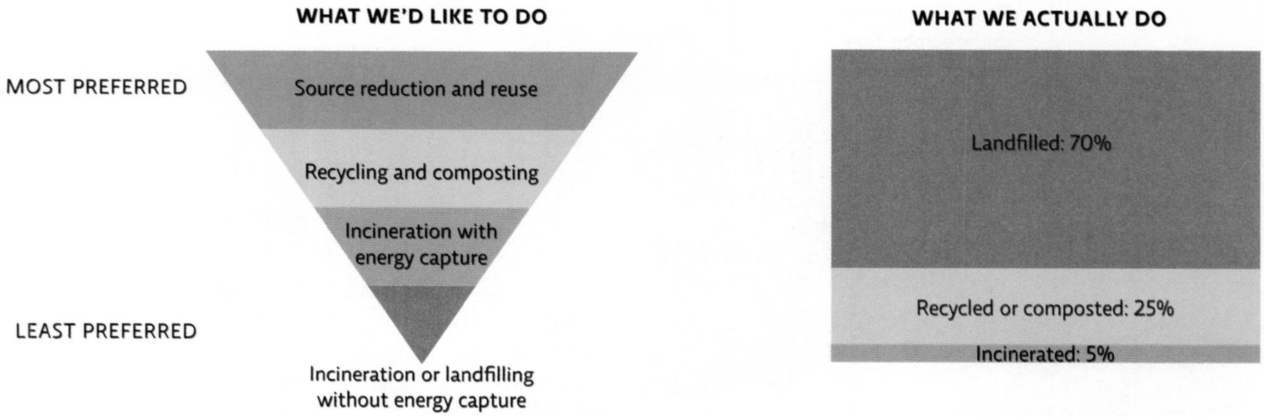
WHAT WE'D LIKE TO DO
MOST PREFERRED
Source reduction and reuse
Recycling and composting
Incineration with energy capture
LEAST PREFERRED
Incineration or landfilling without energy capture
WHAT WE ACTUALLY DO
Landfilled: 70%
Recycled or composted: 25%
Incinerated: 5%

↓ In a sanitary landfill, an area is dug out and lined to prevent groundwater contamination from leachate; trash is dumped and covered with soil frequently (this soil may take up to 20% of the landfill area.) Newer landfills have a leachate-collection system built in; older landfills can be retrofitted to collect leachate. Leachate from holding ponds is treated before being released into the environment.

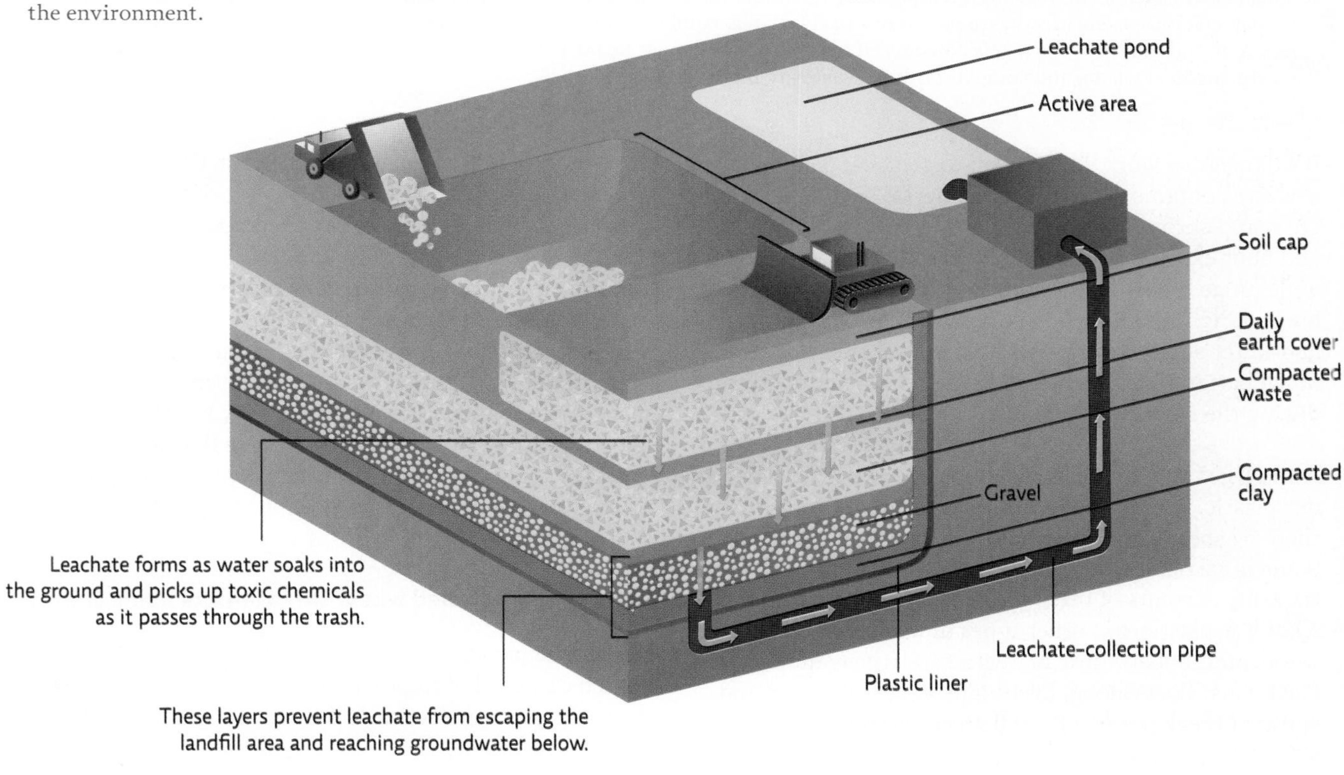
Leachate pond
Active area
Soil cap
Daily earth cover
Compacted waste
Compacted clay
Gravel
Leachate forms as water soaks into the ground and picks up toxic chemicals as it passes through the trash.
Leachate-collection pipe
Plastic liner
These layers prevent leachate from escaping the landfill area and reaching groundwater below.

Infographic **17.3** | **HOW IT WORKS: A MODERN "ENERGY-TO-WASTE" INCINERATOR**

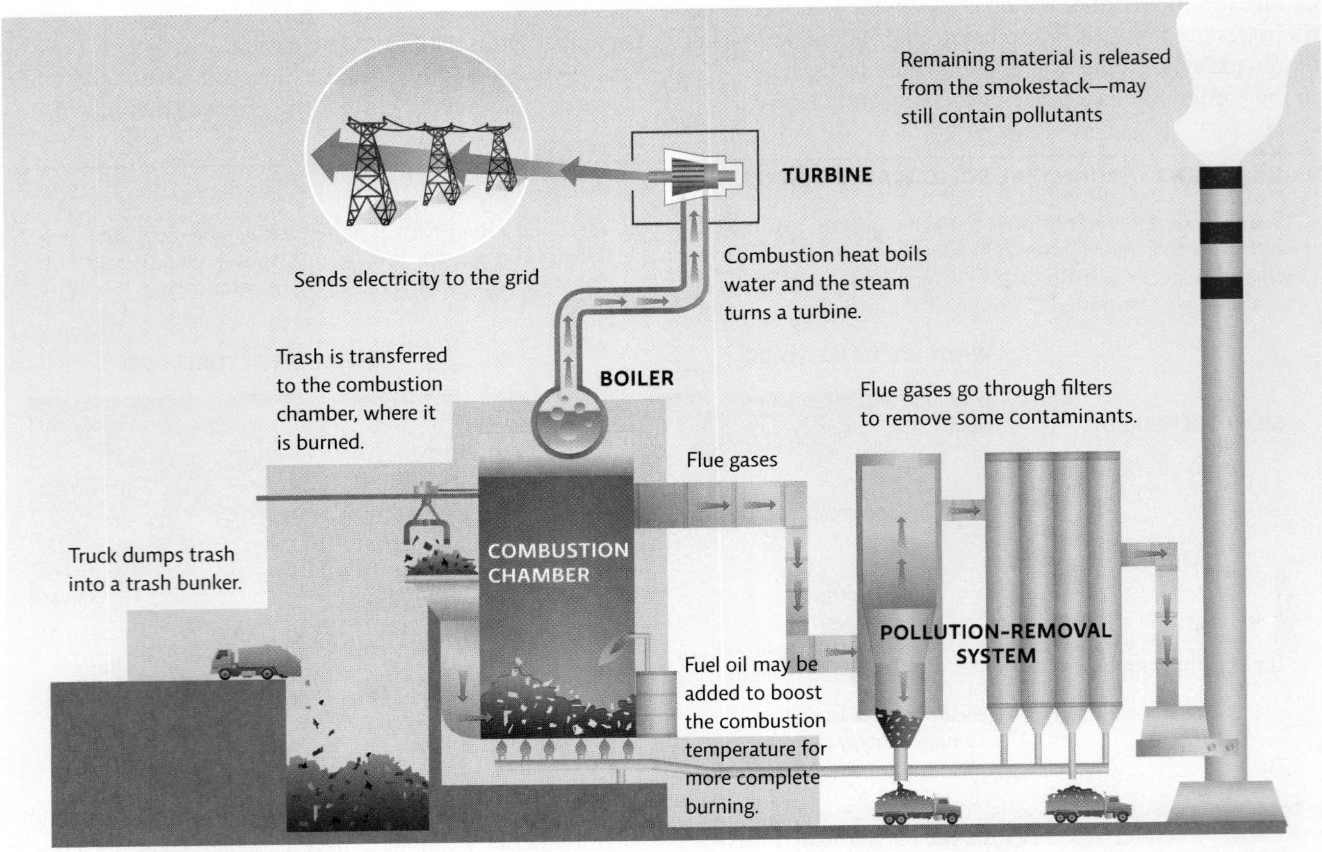
Remaining material is released from the smokestack—may still contain pollutants
TURBINE
Sends electricity to the grid
Combustion heat boils water and the steam turns a turbine.
Trash is transferred to the combustion chamber, where it is burned.
BOILER
Flue gases go through filters to remove some contaminants.
Flue gases
Truck dumps trash into a trash bunker.
COMBUSTION CHAMBER
POLLUTION-REMOVAL SYSTEM
Fuel oil may be added to boost the combustion temperature for more complete burning.

↑ Trash can be burned at very high temperatures in incinerators (some of which are designed to also generate electricity). In modern facilities, cleaning systems remove particulates, sulphur, and nitrogen pollutants, as well as toxic pollutants like mercury and dioxins. The ash is considered toxic waste and must be buried in hazardous waste landfills. Municipal solid waste, medical waste, and some hazardous waste are incinerated in Canada.

It's dangerous work. In addition to gold, copper, and zinc, e-waste contains a suite of toxic metals such as lead, mercury, and chromium. When they are released by unsafe and poor extraction methods, these substances cause a wide range of medical conditions—from birth defects to brain, lung, and kidney damage to cancer. The soil, air, and water surrounding e-waste disposal sites don't fare much better, as toxic chemicals are released into all three during the extraction process.

Amid this trash transfer, too much of our waste, especially our plastic, is escaping to the open sea. Some of it is blown there by aberrant winds from the tops of open landfills. Some is carried through faulty sewage systems, or in the trickling currents of litter-polluted gutters. To be sure, not all of it is plastic, but other types of waste—textiles, glass, wood, and rubber—sink or degrade relatively quickly. Plastic just floats along. Eventually time, saltwater, and sunlight break it down from its recognizable, everyday forms—combs, chocolate bar wrappers, CD cases—into fragments so tiny that even thousands of them together can't be seen by a naked eye trained upon a calm sea.

And as far as it has travelled, the plastic in the garbage patch still has a long way to go. It will take decades, maybe even centuries, for those fingernail-sized fragments to degrade into smaller molecules. And even then, they will not cease to pollute. In one study of the North Pacific Gyre, virtually all water samples, even those that were free of plastic debris, contained traces of polystyrene, a common chemical used in a wide range of plastic consumer goods.

Improperly handled waste threatens all living things.

The consequences of mismanaging our trash are manifold. When disposed of improperly, chemical waste can wreak havoc on plant and animal life (even some household trash is considered hazardous and should not be disposed of in a landfill or municipal incinerator); incinerators create small particle air pollution, and landfills produce methane. All of this threatens the health of ecosystems and the organisms who live there. [INFOGRAPHIC 17.4]

Aquatic life is especially vulnerable. Sea mammals get tangled in everything from discarded fishing nets to plastic six-pack rings, often with fatal consequences. On top of that, many fish and nearly half of all seabirds eat plastic, often by mistake (plastic bags floating in the open ocean look a lot like jellyfish). Some of these animals choke on the plastic or are poisoned to death by its toxicity. [INFOGRAPHIC 17.5]

Other sea animals live long enough to be consumed by predators, including humans. That's no small matter. BPA, an organic compound used in plastic, has been shown to interfere with reproductive systems, and styrene monomers, the subunits of polystyrene, are a suspected carcinogen. Plastic also absorbs fat-soluble pollutants such as PCBs and pesticides like DDT. These toxic substances are known to accumulate in the tissue of marine organisms, biomagnify up the food chain, and find their way into the foods we eat. (For more on bioaccumulation see Chapter 3).

Researchers suspect that floating bits of plastic can also serve as an attachment point for fish eggs, barnacles, and many types of larval and juvenile organisms. Thus each tiny bit of plastic could potentially transport harmful, invasive, or exotic species to new locales. "I think one of the most underrated impacts of these so-called garbage patches is the introduction of hard surfaces to an ecosystem that naturally has very few of them," says Miriam Goldstein, a PhD candidate at Scripps Institute of Oceanography who studies the Pacific patch. "Organisms that live on hard surfaces are very different than those that float freely in the ocean. And adding all that plastic is providing habitat that would not naturally exist out there."

↑ In some areas of China, the process of recovering electronic waste is not done safely and exposes workers and the community at large to toxic substances. Though officially banned, export of electronic waste from Canada continues due to legal loopholes, and Asian and African countries continue to accept the waste.

However, knowing that these things can happen is not the same as proving that they are happening in the patches. In general, as Goldstein points out, gyres tend to be areas of very low productivity, which means there are very few large fish there. It is not yet clear to scientists whether large numbers or important species of fish (or seabirds)

Infographic **17.4** | **HOUSEHOLD HAZARDOUS WASTES**

↓ Hazardous wastes are those that are toxic, flammable, explosive, or corrosive (like acids). Many chemicals that enter your home are actually hazardous and should not be discarded in the regular household trash. Contact your local municipality for information on disposing of hazardous materials. Most have websites that will tell you what products need to be taken to special drop-off collection facilities.

To protect your health and that of your family and the environment, avoid or reduce your use of hazardous chemicals such as:

Drain openers
Oven cleaners
Engine oil and fuel additives
Grease and rust removers
Glue
Pesticides
Mould and mildew removers
Paint thinners, strippers, and removers

Other materials that are also considered hazardous and may need to be disposed of as hazardous waste include:

Batteries
Fluorescent light bulbs
Mercury thermometers

Infographic **17.5** | **PLASTIC TRASH AFFECTS WILDLIFE**
COMPARISON OF FOOD INGESTED BY LAYSAN ALBATROSS CHICKS IN TWO REGIONS OF THE PACIFIC

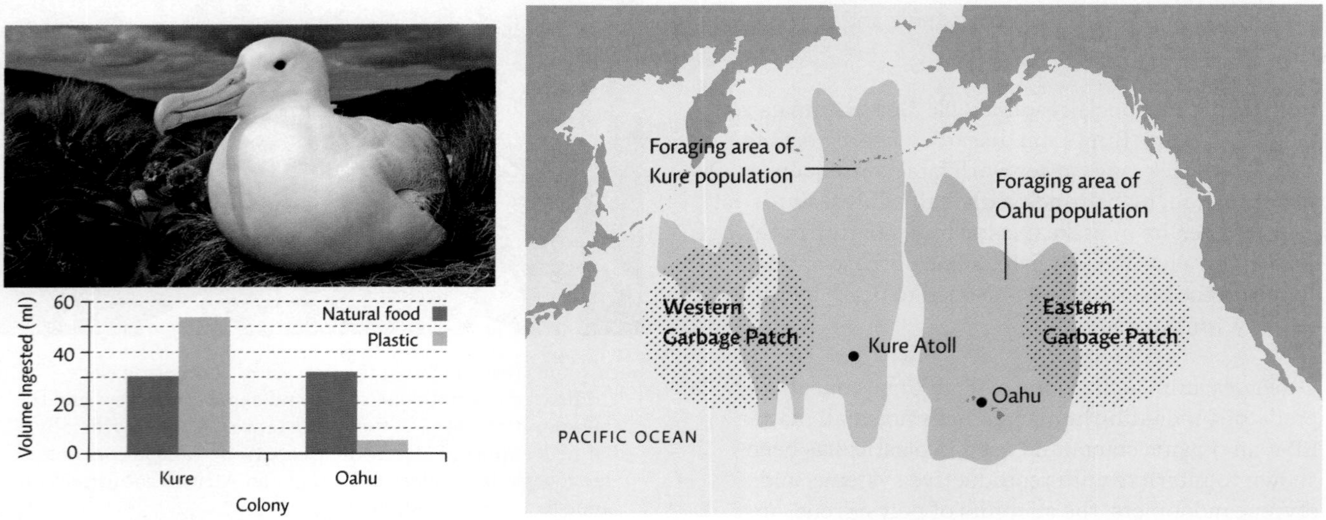
Volume Ingested (ml)
60
40
20
0
Natural food
Plastic
Kure
Oahu
Colony
Foraging area of Kure population
Foraging area of Oahu population
Western Garbage Patch
Eastern Garbage Patch
Kure Atoll
Oahu
PACIFIC OCEAN

↑ Research by Lindsay Young and her colleagues compared the food ingested by Laysan albatross chicks in two populations, more than 1900 kilometres apart in the Pacific. Their data show that, while both populations consumed roughly the same amount of actual food, chicks in the western Pacific near Kure ingested 10 times more plastic than chicks near Oahu. In addition, the Kure chicks had 4 times as many plastic pieces and these pieces were, on average, twice as large as those of the Oahu chicks. A comparison of the foraging areas for these two populations gives a clue as to why the Kure birds eat so much more plastic than the Oahu birds.

are ingesting plastic from the gyre, or if toxic chemicals from the plastic are accumulating in their tissues.

On June 21st, three weeks into their journey, Proskurowski wrote on the ship's blog:

At 0930 this morning, we sampled what I predict will be the largest amount of plastic [the program] *has ever recorded in 25 years of sampling...When the tow started, I could distinctly see a few pieces in the upper layer of water and it looked like it was going to be a "good one"—one where we got enough plastic to keep the lab busy. A couple minutes into the tow, suddenly we started to see more and more macro debris—a toilet seat, white plastic bags, oil jugs, a few bread bag fasteners, Styrofoam cups, several shoes, a few foot insoles, a loofah sponge, sunscreen, and liquor bottles—sometimes appearing to form loosely organized windrows (indicative of a special type of turbulence in the upper ocean called Langmuir circulation). While everyone was commenting on all the debris, a red 5-gallon bucket drifted by the port bow with a school of about 20 fish underneath it.* [The bucket got caught in the net.] *Fearing the bucket would tear our net, we ended the tow early and pulled the bucket with two of its associated schools of grey triggerfish and thousands of tiny fragments of plastic on deck.* [Team scientist] *Skye Moret quickly dissected one fish and found 46 pieces of plastic in its guts. This school of fish, a coastal species that typically lives on reefs, was thousands of miles from land with plastic-filled stomachs.*

He was right. At 23 000 pieces of plastic—more than 26 million pieces per square kilometre—the June 21st tow was the largest in the research program's 25-year history. It would take two lab members (rotating every hour or so) more than 14 hours of continuous work to process all of it.

When it comes to managing waste, the best solutions mimic nature.

While nondegradable trash like plastics present a challenge for disposal, much of our waste is biodegradable and we can apply the concept of biomimicry—emulating nature—to teach us how to better deal with this part of our waste stream. **Composting**—allowing waste to biologically decompose in the presence of oxygen and water—can turn some forms of trash into a soil-like mulch that can be used for gardening and landscaping. Composting can be done on a small scale (in homes, schools, and small offices) or on a large one (in municipal "digesters"). But this only works for organic waste, such as paper, kitchen scraps, and yard debris. [INFOGRAPHIC 17.6]

composting Providing good conditions for the decomposition of biodegradable waste, producing a soil-like mulch.

Infographic **17.6** | **COMPOSTING**

↓ Composting can reduce the amount of a household's trash tremendously. A simple compost pile can be started in the backyard, or a compost bin can be built or purchased. Curbside pickup of organics in green bins is available in some Canadian municipalities. The goal is to reduce or eliminate organic waste going to landfills.

HOME COMPOSTING

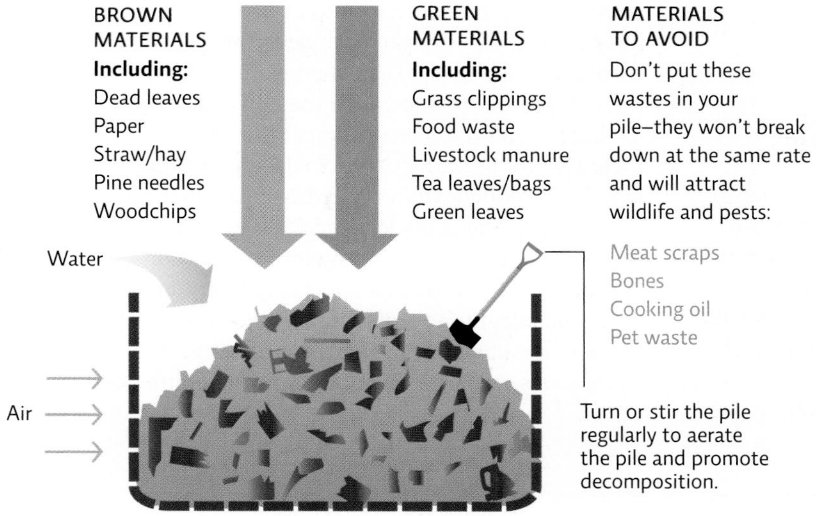
BROWN MATERIALS
Including:
Dead leaves
Paper
Straw/hay
Pine needles
Woodchips
GREEN MATERIALS
Including:
Grass clippings
Food waste
Livestock manure
Tea leaves/bags
Green leaves
MATERIALS TO AVOID
Don't put these wastes in your pile—they won't break down at the same rate and will attract wildlife and pests:
Meat scraps
Bones
Cooking oil
Pet waste
Water
Air
Turn or stir the pile regularly to aerate the pile and promote decomposition.

↑ A variety of compost bins can be used. Small tubs and countertop green bins can be used indoors.

↑ Household plant-based food scraps can be added, but meat products should not be included.

↓ Municipal composting is becoming more widely available across Canada. Curbside green bins collect this "organics stream" and decompose it in large anaerobic digesters, like the one shown below. These digesters quickly produce mulch that residents can often pick up for use in their yards.

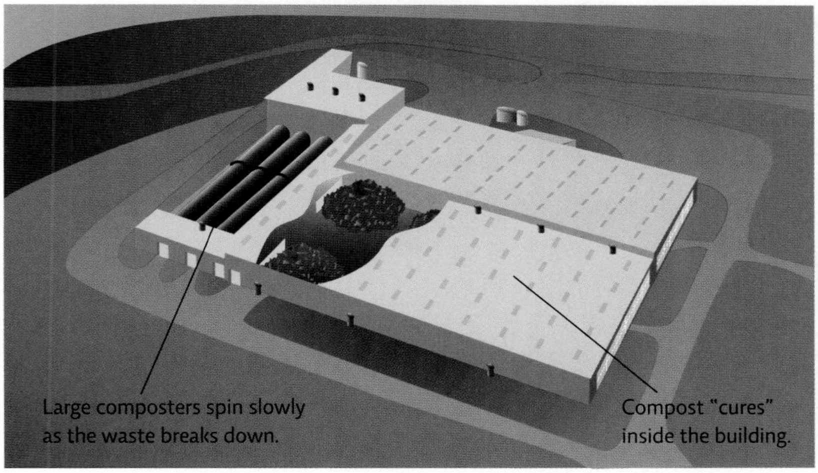
Large composters spin slowly as the waste breaks down.
Compost "cures" inside the building.

↑ The end product is a rich, soil-like mulch that can be used in gardens.

Another promising solution to some of our waste woes involves converting garbage or its by-products into usable energy. For example, the heat produced during incineration might be converted into steam energy or electricity. And the methane produced during anaerobic decomposition can also be harnessed as an energy source. Landfill methane can be captured and used to produce electricity on site and scientists are working on ways to use methane-producing microorganisms to process trash in vats or digesters.

In fact, the concept of spinning waste into energy is so appealing that it has even extended to waste found in the open ocean. Not long after the Atlantic and Pacific patches were found, a San Francisco-based environmental group tried devising ways to turn the plastic bits into diesel fuel. That plan proved impossible. "There is no way to remove all those millions of tiny pieces without also filtering out all the millions of tiny creatures that inhabit the same ecosystem," says Kara Lavendar Law. "So once it's out there, all we can really do is hope that nature eventually breaks it all the way down." That could take hundreds to thousands of years. In the meantime, she says, we need to stop adding to the patches that already exist.

Life-cycle analysis and better design can help reduce waste.

The real trick to keeping garbage out of the ocean is to go back to the beginning, before the things we throw away are even made. Environmentally friendly products continue to grow in popularity, and manufacturers have given a name to the tactics used to create these products: life-cycle analysis. By assessing the environmental impact of every stage of a product's life—from production, to use, to disposal—an increasing number of companies are trying to reduce the amount of waste generated by the things they design, make, and sell. Cradle-to-cradle analysis takes this even further as it tries to increase reuse potential and turn *waste* back into *resource* (see Chapter 5).

Part of this shift has been spurred by legislation. Canada, some European countries, and at least 19 U.S. states have implemented "take back laws," which require manufacturers to take back some of their products after consumers are finished with them. Canada's Action Plan for Extended Producer Responsibility (EPR) creates an incentive for manufacturers to design products from which components can easily be salvaged and reused. An *environmental handling fee* paid for new electronics builds the cost of eventual recycling into the purchase price. Indeed, many companies are trying to cut down on waste by *de-manufacturing*—disassembling equipment, machinery, and appliances into the component parts so that those parts can be reused. BMW, for example, designs its

Infographic 17.7 | **INDUSTRIAL ECOLOGY**

↓ The industrial park in Kalundborg, Denmark, is a prime example of industrial ecology. Here, by design, there are 24 different connections between various industries and local farms, such that the waste of one becomes the resource for another.

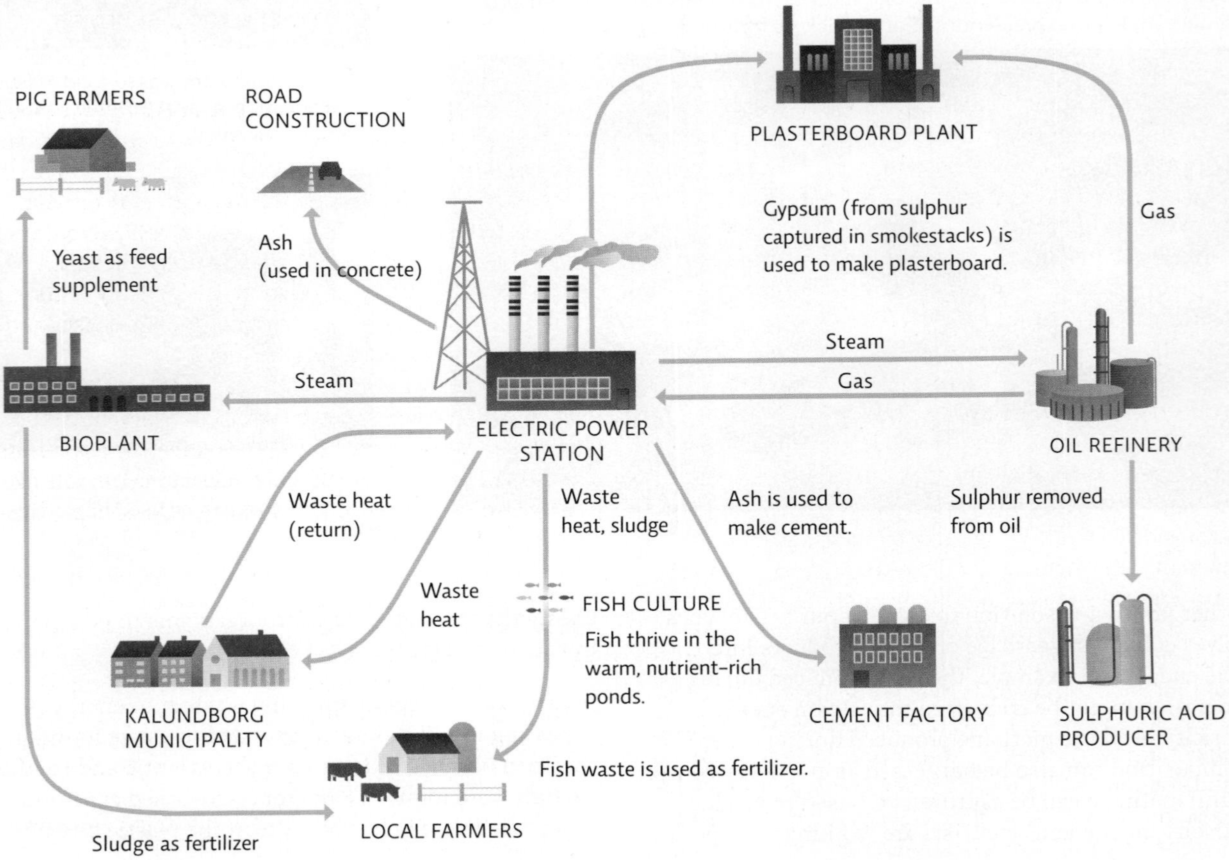
PIG FARMERS
ROAD CONSTRUCTION
PLASTERBOARD PLANT
Gypsum (from sulphur captured in smokestacks) is used to make plasterboard.
Gas
Yeast as feed supplement
Ash (used in concrete)
Steam
Gas
BIOPLANT
Steam
ELECTRIC POWER STATION
OIL REFINERY
Waste heat (return)
Waste heat, sludge
Ash is used to make cement.
Sulphur removed from oil
Waste heat
FISH CULTURE
Fish thrive in the warm, nutrient-rich ponds.
KALUNDBORG MUNICIPALITY
CEMENT FACTORY
SULPHURIC ACID PRODUCER
Fish waste is used as fertilizer.
LOCAL FARMERS
Sludge as fertilizer

↑ The Foster family of Stowe, Ohio, with their polymer-based possessions.

vehicles so that at least 97% of the components can be reused or recycled.

On a grander scale, in **eco-industrial parks**, industries are positioned so that they can use each other's waste in the same industrial area; one company's trash becomes another's raw material. This type of strategic thinking is known as *industrial ecology*; in addition to "waste-to-feed" exchanges, participants in eco-industrial parks also decrease their ecological footprint by coordinating activities and sharing some spaces, like warehouses and shipping facilities. This form of waste reduction can occur within a single company, between two companies, or between a company and a community. For example, one UK company sells the soapy water produced by its shampoo manufacturing plants to area car washes, rather than throwing it away. The same company also uses its ethanol waste—a by-product of its manufacturing processes—to heat some of its facilities. [INFOGRAPHIC 17.7]

Consumers have a role to play, too.

Advertising—a virtual staple of advanced societies—bombards us from every corner of modern life: not just on the televisions in our living rooms, but on digital monitors in waiting rooms and food courts, billboards in bus shelters, and the pop-up ads that invade our laptops. The message is surprisingly uniform: to live a happy, more fulfilling life, we simply must have more "stuff." A good deal of that stuff—though certainly not all of it—is made of plastic. "Plastics are incredibly useful and have made possible many of the greatest breakthroughs in technology and standard of living," Proskurowski says. "But they have also made our lives lazier. We now buy a bottle of water rather than refill a canteen. We buy individually wrapped bags of mini carrots, instead of buying carrots that are straight from the ground and have to be washed. There are countless other examples, and as we learn with every net tow, there are significant costs to the planet for those choices."

So how do we start making different choices? As any good environmentalist will tell you, it comes down to the "4 Rs": refuse, reduce, reuse, recycle.

The first thing we can do is simply **refuse** to use things that we don't really need, especially if they are harmful to the environment. This may be as minor as declining to take a plastic bag for a few items purchased at the drugstore or as major as biking or walking to work rather than taking the car. The logic is simple: when we save a resource by refusing to use it, that resource lasts longer, which in turn means that less pollution will be generated disposing of it and producing replacements, and more will be available for future uses. "Refusal doesn't mean never using the resource," says David Bruno, founder of the 100 Thing Challenge, a popular movement to pare down our worldy possession to 100 items or fewer. "It just means using it at a more sustainable rate."

If we can't completely refuse a given commodity, we can still try to **reduce** our consumption of it, or minimize our overall ecological footprint by making careful purchases. People who must drive to work can minimize their fossil fuel consumption by choosing a more fuel-efficient vehicle. Those who cannot drink tap water might purchase a specialized faucet or pitcher filter instead of relying on bottled water. And all consumers can greatly reduce the amount of waste they generate by paying special attention to packaging. Canada's EPR Program includes a Sustainable Packaging Strategy; however, this plan is industry-driven, which makes consumer choices an important component in compelling producers and retailers to reduce packaging waste.

And if we can't avoid using a product, our next best choice is to **reuse**—the third "R"—something consumers can do with just a little effort, by choosing durable products over disposable ones. "Products produced for limited

---

eco-industrial parks Industrial parks in which industries are physically positioned near each other for "waste-to-feed" exchanges (the waste of one becomes the raw material for another).

refuse The first of the waste-reduction "4 Rs": choose NOT to use or buy a product if you can do without it.

reduce The second of the waste-reduction "4 Rs": make choices that allow you to use less of a resource by, for instance, purchasing durable goods that will last or can be repaired.

reuse The third of the waste-reduction "4 Rs": use a product more than once for its original purpose or another purpose.

Infographic **17.8** | **THE FOUR Rs**

REFUSE: DON'T USE IT

- Avoid disposables (they are made to be trash!).
- Opt out of junk mail at www.the-cma.org/consumers/do-not-contact.
- Choose goods with no packaging.
- Bring your own cloth bags when shopping or don't accept a plastic bag if you don't need one.

REDUCE: USE LESS

- Choose goods with minimal packaging.
- Buy in bulk (but only if you will use all that you buy).
- Buy durable, well-made, and repairable goods.
- Use both sides of paper at work or school.
- Post notices on a central bulletin board or send them via email to reduce copies.
- Buy local, fresh food; it comes with less packaging.
- Compost your kitchen and yard waste.

REUSE: USE IT AGAIN

- Use products—such as shopping bags, food containers, and plasticware—multiple times.
- Rent, borrow, or lend items.
- Reuse products in different ways: use yogourt containers to hold screws, scrap paper for a note pad, and so on.
- Repair broken equipment, tools, furniture, and toys (this becomes possible when you choose items that can be repaired).
- Buy and sell old clothes and household goods or donate them to charities.
- Choose reusable containers for leftovers rather than plastic bags or plastic wrap.
- Bring your own coffee mug to work.

RECYCLE: RETURN IT FOR REPROCESSING

- Check with your local waste management service to see which recyclables it takes (just because a product or its packaging says it is recyclable doesn't mean you can recycle it in your area).
- Buy products made from recycled material (close the loop).
- Encourage family, friends, and co-workers to recycle as well.
- Help start a recycling program at your workplace or in your community.

use are really just made to be trash," Bruno says. "They pull resources out of the environment and produce pollution at every step of production, shipping, and disposal." He advises considering use and reuse each time we head to the store; whether you are purchasing clothing, razors, cups, or plates, ask yourself: How long will this last and for what other purpose might it be used?

Reusing also applies to industry. TerraCycle is a U.S. company that produces worm compost fertilizer and packages it in used pop and water bottles. The company also collects hard-to-recycle packaging like chocolate bar wrappers and juice pouches and turns them into new products like backpacks.

Once we've refused, reduced, and/or reused a given commodity as much as possible, we are left with the final R, **recycling**, the reprocessing of waste into new products. Recycling has several advantages. By reclaiming raw materials from an item that we can no longer use, we limit the amount of raw materials that must be harvested, mined, or cut down to make new items. In most cases, this helps conserve limited resources, not only trees and precious metals, but also energy. To execute this step properly, we must first have purchased items that can be recycled. We must also *close the loop* by purchasing items that are made of recycled materials to encourage manufacturers to make products from recycled materials. [INFOGRAPHIC 17.8]

Of course, even recycling comes with trade-offs. For example, while less energy and water is used making paper from paper than is used making paper from trees, there are also energy and environmental costs in collecting used paper, storing it, and shipping it to recycling plants. Once these costs are factored in, sustainably harvesting a forest of fast-growing trees may prove to be more environmentally friendly than recycling used paper. [INFOGRAPHIC 17.9]

By journey's end, the crew of the *Corwith Cramer* had spent a total of 34 days at sea, travelled more than 7000 kilometres, conducted 128 net tows, and counted 48 571 pieces of plastic along the way. Crew members found the northern and southern boundaries of the Atlantic Patch, at latitudes near Virginia and Cuba, respectively. But despite making it as far as the Mid-Atlantic Ridge, they never found the eastern edge.

As the coast of Bermuda came back into view, on July 12, 2010, Proskurowski wrote his last blog post:

recycle The fourth of the waste-reduction "4 Rs": return items for reprocessing to make new products.

It is easy to brush off the topic of plastic pollution in the ocean. It occurs thousands of miles from land, in regions of the planet that are rarely visited by humans, and relatively sparsely populated by marine life. It is also easy to pass off responsibility, to say 'I recycle,' or 'It must come from some other (developing) country,' or 'It is all fishing or marine industry waste.' While there may be kernels of truth in all those arguments, the reality of the situation is that open ocean plastic pollution occurs over incredibly large regions of the Earth, has widely distributed point sources, and—because the oceans connect the whole globe—has far-reaching consequences.◉

Select references in this chapter:
Law, K. L., *et al.* 2010. *Science*, 329 (5996): 1185 – 1188.
Statistics Canada, 2012. *Human Activity and the Environment: Waste management in Canada* (16-201-X).
Young, L. C., *et al.* 2009. *PLoS ONE*, 4: e7623.

Infographic **17.9** | **OPTIONS FOR PRODUCT DISPOSAL**

↓ We have better options than simply throwing away many products. An item like a plastic bottle can be recovered and reused, as the innovative company TerraCycle does, or the bottle may be recycled into another product.

RECYCLING REQUIRES 3 STEPS: Consumers and industry must turn in or collect materials for recycling, the material must be used to make new products, and the products must be bought by consumers.

Collection
Purchase
Production

RECYCLING PLASTIC: The number code on a plastic item indicates the type of resin it is made with. In many areas, #1 and #2 are the only plastics that can be recycled; in others, all types are taken by recyclers. Check with your community recyclers to see what is taken in your area.

1
2
3
4
5
6
7
PETE
HDPE
V
LDPE
PP
PS
OTHER

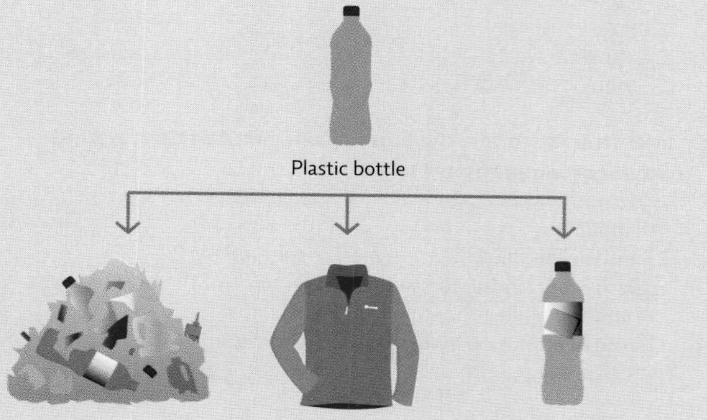
Plastic bottle

Discard
The bottle may end up in the environment such as the Atlantic Garbage Patch.

Recycle
Consumers turn in used plastic bottles to be recycled into new products like this Mountain Equipment Co-op fleece jacket.

Reuse
Companies such as TerraCycle collect used bottles and package products such as fertilizer in them.

BRING IT HOME

PERSONAL CHOICES THAT HELP

How much solid waste you produce is under your control. Reducing your waste not only reduces community costs for waste disposal, it also places less pressure on the resources used to produce consumer goods. It is also easy to do and can save you money.

Individual Steps

→ Track your trash. Record what you throw out for a week by category and weight. How could you reduce your total trash weight by a quarter? By half?

→ Use the information in Infographic 17.8 to identify five changes you can make to reduce your solid waste.

→ Consider a semester or summer session with SEA (Sea Education Association) to participate in expeditions such as the one described in this chapter. Visit www.sea.edu/.

Group Action

→ Start recycling unusual items in your community. TerraCycle is a company that takes items that usually end up in the garbage, like candy wrappers, corks, and chip bags, and recycles them into new products, like purses or backpacks.

→ Talk to friends and family about having "no gift" or "low gift" celebrations. Instead of buying lots of presents, treat friends to a dinner or a fun activity. For large families, use a grab bag or draw names and buy for only specific people.

Policy Change

→ Talk to community leaders to discuss the possibility of starting a community-wide composting program.

→ Research recycling rates of participation in your community. Advocate for recycling education and curbside recycling programs.

UNDERSTANDING THE ISSUE

CHECK YOUR UNDERSTANDING

1. All of the following are consistent with the law of conservation of matter EXCEPT:
- a. there is no such thing as waste.
- b. matter is neither created nor destroyed, but it can change form.
- c. any matter that is dangerous to living things is quickly biodegraded into a harmless form.
- d. one organism's waste matter is taken up by another organism and reused.

2. Garbage that you generate in your home, at work, or at school would be called:
- a. municipal solid waste.
- b. hazardous waste.
- c. industrial waste.
- d. nonbiodegradable waste.

3. In sanitary landfills, the primary safeguard against groundwater contamination is to:
- a. do nothing, as groundwater contamination is not a problem with landfills.
- b. line the landfill with impermeable clay and plastic.
- c. design the landfill to use the natural slope of the land so water runs downhill without being contaminated.
- d. locate landfills on industrial sites where groundwater is not used for drinking.

4. Which of the following statements about incinerators is FALSE?
- a. Incinerators reduce the volume of material that needs to be landfilled.
- b. Incinerators are expensive to build and operate.
- c. The burning of waste reduces the toxic and hazardous materials in the waste stream, as the heat neutralizes the toxic substances.
- d. Burning solid waste in incinerators can be used to produce energy.

5. The best solutions to managing waste would include:
- a. source reduction and application of natural decomposition processes.
- b. supporting the siting of a landfill in your community to reduce waste transportation costs.
- c. disposing of hazardous waste in oceans, as the water dilutes the waste.
- d. shipping electronic waste to developing countries to provide jobs there in extracting precious metals from the electronic devices.

6. An industrial ecologist's goals would NOT include:
- a. increasing resource-use efficiency by altering manufacturing processes.
- b. designing products that are durable and can be de-manufactured.
- c. identifying points in the product life cycle at which waste products can be used in other processes.
- d. identifying points in the product life cycle at which waste products can be landfilled.

WORK WITH IDEAS

1. Why is waste considered to be a human invention? Furthermore, why do we distinguish between different types of waste?

2. Compare and contrast landfilling and incineration. What are the trade-offs for each option?

3. What is biomimicry? Describe two ways this concept can be applied to reducing and managing waste.

4. What is industrial ecology? How can it be applied in managing waste? What is the consumer role in this process?

ANALYZING THE SCIENCE

The data in the following table come from a report by Greenpeace based on research published on plastic debris in the world's oceans between 1990 and 2005.

Species group	Total number of species worldwide	Number and percentage of species with entanglement records	Number and percentage of species with ingestion records
Sea turtles	7	6 (86%)	6 (86%)
Seabirds	312	51 (16%)	111 (36%)
Penguins	16	6 (38%)	1 (6%)
Grebes	19	2 (10%)	0
Albatrosses	99	10 (10%)	62 (63%)
Pelicans and cormorants	51	11 (22%)	8 (16%)
Gulls and terns	122	22 (18%)	40 (33%)
Marine mammals	115	32 (28%)	26 (23%)
Baleen whales	10	6 (60%)	2 (20%)
Toothed whales	65	5 (8%)	21 (32%)
Fur seals and sea lions	14	11 (79%)	1 (7%)
True seals	19	8 (42%)	1 (5%)
Manatees and dugongs	4	1 (25%)	1 (25%)
Sea otters	1	1 (100%)	0

INTERPRETATION

1. How many different species are included in these data and how many are affected?

2. For seabirds:
   - a. What percentage of seabird species had entanglement records? Which two groups had the highest rate of entanglements?
   - b. What percentage of seabird species had ingestion records? Which two groups of seabirds had the highest rate of ingestions?

3. For marine mammals:
   - a. What percentage of marine mammal species had entanglement records? Which two groups had the highest rate of entanglements?
   - b. What percentage of marine mammal species had ingestion records? Which two groups had the highest rate of ingestions?

ADVANCE YOUR THINKING

Hint: Access the actual report at http://bit.ly/tdQ7Kr.

4. Scientists suspect that entanglement is a significant cause of population decline for many species, but they consider the reported entanglement rates to be conservative.
   a. Based on the data in the table, which group (sea turtles, seabirds, or marine mammals) is likely to be most affected by entanglement? Why?
   b. Why might reported entanglement rates underestimate the real problem?

5. Ingestion refers to animals eating plastic. While many species of marine mammals, seabirds, and sea turtles ingest plastic, some groups ingest more than others. What might explain these differences in ingestion rates among species?

6. Not much is known about the specific consequences of aquatic species ingesting plastics. In the process of science, observation leads to more questions. List three questions about ingestion of plastics that should be studied. How would you go about answering one of these questions?

EVALUATING NEW INFORMATION

Packaging makes up a large part of the solid waste stream. Canada has taken steps to reduce packaging with its National Packaging Protocol. But are some types of packaging better than others? Paper readily decomposes, but if it is made from trees it can contribute to deforestation and increased atmospheric greenhouse gases. Plastics are lighter weight and may be more durable (and reusable) but may be made from petroleum. Glass is more environmentally benign to make and use, but it is heavier, which translates into higher fuel costs for transportation. The Canadian Plastics Industry Association (CPIA) makes the case for plastics as a sustainable packaging option.

Go to www.plastics.ca.

Evaluate the website and work with the information to answer the following questions:

1. Is this a reliable information source? Does the association have a clear and transparent agenda?
   a. Who runs the website? Do the association's credentials make the information reliable or unreliable? Explain.
   b. Do you detect a bias in the information presented? Explain.

2. Click on the "Intelligent Plastics" link and read the information there. Check out at least two of the links on this page.
   a. What are the main claims made by CPIA regarding intelligent plastics? What evidence does the association offer to back up these claims?
   b. Does this website present any information regarding the potential drawbacks of plastics? Should it? Explain.

3. Go back to the home page and click on the "Sustainability" link. Once there, select "Plastic Packaging." What reasons are presented in support of plastic as a good packaging material? What evidence does CPIA offer to back up these claims?

4. This website offers educational links and handouts. Open the "Education Tools" link on the home page and look over some of the material. Do you recommend these resources for use in a school classroom? Why or why not?

MAKING CONNECTIONS

IS IT TIME TO SKIP THE UBIQUITOUS PLASTIC BAG?

Background: Canadians use approximately 12 billion plastic bags a year. After their short use, these bags often end up as litter and many eventually make it to the ocean. Concerns over the impacts of plastic bags on marine wildlife have led to a movement to restrict or ban plastic bags. Ireland and China have instituted fees on single-use plastic bags, and in January 2011, Italy became the first country to implement a nationwide ban on them. In the United States, plastic bag use in the city of San Francisco has dropped by 5 million per month since it instituted a ban on plastic bags in March 2007. Toronto imposed a plastic bag fee in 2009 but later eliminated it and briefly considered a ban, first passing and then reversing the ban in 2012.

Case: Your community is considering some sort of restriction on plastic bags. You have been assigned the task of evaluating alternative strategies to reduce plastic bag pollution, including:

- implementing an outright ban on single-use plastic bags.
- instituting a fee for consumers to use plastic bags. The fee would be used for waste education and litter clean-up efforts.
- establishing a reusable bag credit for consumers. This credit would be paid by the businesses, as it would save them having to purchase plastic bags.

Research these (and possibly other) options and write a report recommending a course of action for your community.

1. In your report, include the following:
   a. an analysis of the pros and cons of each proposal.
   b. a discussion of the consequences of each choice from economic, ecological, and convenience/practical perspectives.
   c. a reflection on the values underlying each proposal.

2. Based on the information at hand, what is the best option for your community? Who should be involved in this decision? Provide justifications for your proposal.

3. From what you now know about the consequences of plastic pollution and the challenges of managing plastic waste, develop a set of guiding principles that could be applied to addressing the issue of plastic packaging. Discuss the principles you develop and explain why you consider them to be key to the future of reducing and managing plastic waste.

CORE MESSAGE

Modern industrial agriculture technologies and the use of genetically modified crops may help us feed the 10 billion or more people of the near future, but there are trade-offs to their use. To support the world's population, we can implement sustainable agriculture techniques as well as look to nature for farming practices that increase productivity with less environmental impact.

GUIDING QUESTIONS

After reading this chapter, you should be able to answer the following questions:

→ What modern industrial agriculture methods are used to produce food, and what are the advantages and disadvantages of these approaches?

→ Why does raising livestock often come with a higher environmental impact than plant crops?

→ How can we sustainably produce crops or raise animals for food?

→ What are the advantages and disadvantages of organic farming?

→ What is the focus of the Green Revolution 2.0, and what are the advantages and disadvantages of this technology?

In Japan, the quiet rice paddies of Takao Furuno.

FINE-FEATHERED FARMING

Creative solutions to feeding the world

If there's one thing Greg and Raquel Massa hate, it's weeds—all varieties, but especially the azolla—an insidious fernlike plant that grows on the surface of water. Each spring, azolla plants invade the couple's rice farm, snaking their way through the dense, muddy paddies that stretch for kilometres along the Sacramento River near Chico, California, strangling young rice plants and forcing Greg into an endless and tedious battle.

The rivalry—Massa versus azolla—has spanned three generations. Greg's grandfather, Manuel Massa, planted the family's (and some of California's) first rice crops on the same land that Greg and Raquel now manage, in 1916. Back then, rice farming was a hard and uncertain life; Manuel was largely powerless against the azolla, which in some seasons claimed his entire crop.

By 1962, when Greg's father, Manuel Jr., took over, human ingenuity and modern science had completely changed the nature of the fight. Heavy doses of chemical herbicides enabled him to obliterate the weed. And specially bred higher-yield rice varieties, along with modern farming equipment and a heavy dose of chemical fertilizers, made the family farm both efficient and profitable.

Of course, that modern approach had its own problems. For one thing, it relied on cheap energy and lots and lots of water, both of which are in much shorter supply these days. For another, it impaired ecosystem services, often gravely, because it involved clearing huge swaths of land to increase production, and focusing all resources on a single crop. On top of that, chemical fertilizers and pesticides were expensive and not especially good for the land.

Greg and Raquel wanted to find a better way. So when they took over in 1997, they converted a portion of their farm to **organic agriculture**. Instead of synthetic fertilizers and pesticides, organic agriculture employs more natural or "organic" techniques in the growing of crops. This type of farming uses fewer chemicals and in some cases may produce more nutritious food. For example, research by Washington State University soil scientist John Reganold showed that organically grown strawberries had a longer shelf life and higher level of antioxidants than conventionally grown berries. But organic farming forced Greg and Raquel to battle weeds much as the first Manuel had: with great difficulty.

The trick was to lower water levels enough to kill the weeds, but not so much that the rice crop also died. Each day, Greg would wade into the paddies to see how the rice plants were faring against the azolla. Some weeks, he worried his entire crop would die. After a few seasons, the Massas started to despair: How could they make their farm environmentally friendly without losing their livelihood to an army of mangy weeds?

The Massas' story is the story of North American farming, and by extension, of global agriculture; it's the story of how we feed ourselves. And on a planet where population is exploding, it's also a story of constant change.

The biggest of these changes, the ones that helped Greg's father thrive, came in a midcentury wave of scientific development and technological innovation collectively known as the **Green Revolution**. The Green Revolution—which took place between the 1940s and the 1960s—was a coordinated, global effort to eliminate hunger by improving crop performance and bringing modern agricultural technology to developing countries. Working across the globe, scientists, farmers, and world leaders introduced developing countries to technologies—such as chemical pesticides, sophisticated irrigation systems, synthetic nitrogen fertilizer, and modern farming equipment—that

◉WHERE IS CHICO, CALIFORNIA?

CHICO
—SACRAMENTO RIVER
CA

organic agriculture Farming that does not use synthetic fertilizer, pesticides, or other chemical additives like hormones (for animal rearing).

Green Revolution Plant-breeding program in the mid-1900s that dramatically increased crop yields and led the way for mechanized, large-scale agriculture.

eutrophication Nutrient enrichment of an aquatic ecosystem that stimulates excess plant growth and disrupts normal energy uptake and matter cycles.

Infographic **18.1** | **THE USE OF FERTILIZER COMES WITH TRADE-OFFS**

BENEFITS

INCREASE PRODUCTIVITY

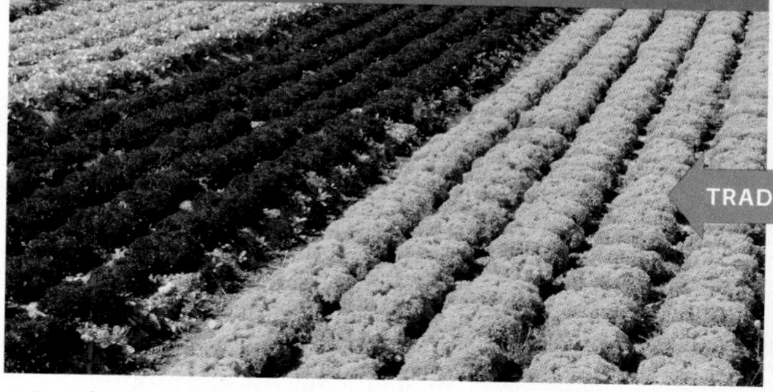

↑ Fertilizer can greatly increase productivity of the soil and is required for many of the high-yield varieties now under industrial cultivation.

GROW CROPS IN MARGINAL SOILS

↑ Fertilizers help crops grow in areas that may not otherwise be able to support agriculture. This may be the only way to farm in many areas of the world and would help increase local food supplies.

PROBLEMS

EUTROPHICATION

TRADE-OFFS

↑ Runoff pollution that contains fertilizer can cause algal blooms that block sunlight and prevent underwater photosynthesis, resulting in oxygen deprivation and the death of many aquatic organisms.

DEVELOP A DEPENDENCE ON FERTILIZER

TRADE-OFFS

↑ Addition of a fertilizer can cause the soil to become more depleted of nutrients. The addition of a limiting nutrient boosts growth but this extra growth pulls other nutrients out of the soil, requiring even more fertilizer in the future.

most industrialized nations had already been using for decades. They also introduced some novel technology, namely new high-yield varieties of staple crops like maize, wheat, and rice. High-yield varieties (HYVs) are those plants that have been selectively bred to produce more than the natural varieties of the same species, usually because they grow faster or larger or are more resistant to crop diseases.

The combination of these forces (HYVs plus existing technology) resulted in a 1000% increase in global food production and a 20% reduction in famine between 1960 and 1990. Thanks to this pivotal chapter in our history, most experts agree that we are now producing enough food to feed every one of the planet's 7 billion or so mouths.

As impressive as these gains are, they have come at a huge environmental cost. It turns out that while heavy doses of fertilizer can indeed boost crop production, the excess fertilizer (whatever is not used by the plants) is easily washed from fields by rain and modern irrigation practices. When this excess enters waterways, it causes algae, aquatic plants, and cyanobacteria to grow into massive blooms. These blooms, which are at the water's surface, block sunlight, shutting down underwater photosynthesis and ultimately causing hypoxia, oxygen-poor conditions that threaten aquatic life. This process, known as **eutrophication**, has been a significant problem in North America, where nutrient runoff from farms in Ontario, Ohio, Pennsylvania, New York, and Michigan has led to eutrophication and an intermittent "dead zone" in Lake Erie (See Chapter 16). [INFOGRAPHIC 18.1]

Pesticide application has also proved problematic. To be sure, pesticides kill pests and thus dramatically reduce the amount of crops lost each year to infestation. But because they are toxic, pesticides also pose a threat to human and ecosystem health. And as scientists quickly discovered, pest populations can develop resistance to almost any pesticide that we invent. When this happens, farmers must either use more of the original pesticide (which is even more toxic and spurs additional pesticide resistance), or find a different pesticide altogether. As resistance to that next pesticide develops, the cycle repeats itself. It's like an arms race between humans and pests, with the deck perpetually stacked in favour of the pests (see Chapter 10). [INFOGRAPHIC 18.2]

Environmental damage is not the Green Revolution's only shortcoming. Even though we are producing enough food to feed the planet, many people still go hungry each day. In 2010, the United Nations estimated that some 925 million people were underfed (not getting enough calories) or malnourished (getting enough calories but not enough essential nutrients). Such poor nutrition is a root cause of more than half of the world's diseases and more than half of all child deaths (roughly five million children per year).

> Poor nutrition is a root cause of more than half of the world's diseases and more than half of child deaths (roughly five million children per year).

Why? Because even though we are producing enough food, our production is unevenly distributed around the globe. Some countries, like Canada the United States, produce much more food than they need. In other

Infographic **18.2** | **EMERGENCE OF PESTICIDE-RESISTANT PESTS**

↓ Exposure to a pesticide will not make an individual pest resistant; it will likely kill it. However, if a few pests survive because they happen to be naturally resistant, they will breed and their offspring (most of which are also pesticide resistant) will make up the next generation. Over time, the original pesticide will no longer be effective and will have to be applied at a higher dose or a different pesticide will have to be used. Application of a pesticide might even increase the size of the pest population by killing the predators that eat the pests.

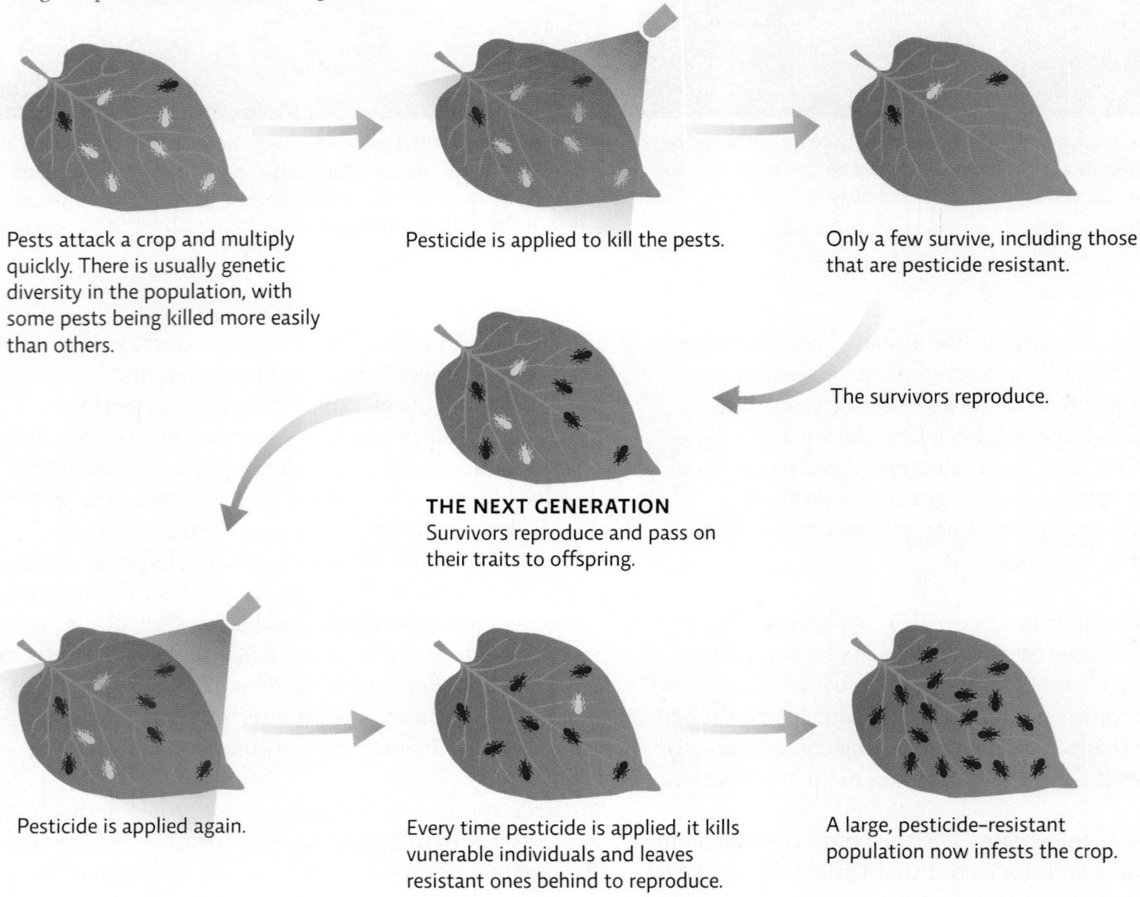
Pests attack a crop and multiply quickly. There is usually genetic diversity in the population, with some pests being killed more easily than others.
Pesticide is applied to kill the pests.
Only a few survive, including those that are pesticide resistant.
The survivors reproduce.
THE NEXT GENERATION
Survivors reproduce and pass on their traits to offspring.
Pesticide is applied again.
Every time pesticide is applied, it kills vunerable individuals and leaves resistant ones behind to reproduce.
A large, pesticide-resistant population now infests the crop.

↑ Greg Massa in his rice fields.

countries, like Haiti, degraded land, natural disasters, and a lack of other resources make food self-sufficiency impossible. Armed conflict is another factor, in Africa especially. It often destroys existing crops and prevents new ones from being planted.

Meanwhile, food has become a global commodity, instead of a locally produced good, meaning that it can be shipped huge distances for the right price. Most food now travels some 1500 miles (2400 kilometres) from producer to consumer, using up an unsustainable amount of fossil fuel. The more **food miles** a product travels before reaching the consumer, the greater the ecological footprint of that food.

Into this constellation of problems, an additional 3 billion people will soon be born. Global population is expected to reach 10 billion by 2050; experts say that to feed that many mouths we will need to produce twice as much food as we are now producing. That will mean either farming more land or devising new production-boosting technologies. But we are already farming most of the cleared, arable land out there; to farm more, we would have to clear more forests, and thus destroy more natural habitats, species, and ecosystem services. And while genetic engineering offers hope that we can produce more food on the same amount of land, most experts agree that no such technology will enable as dramatic an increase as is needed.

As land degrades, population soars, and global climate warms, the story of how we feed ourselves is changing once again. We are already growing crops on most of the available land and using all that modern technology has to offer. This time, technology may not be enough. Instead of turning to the lab, we may have to turn back toward the natural world. In searching for a new weapon against the azolla, Raquel found a Japanese rice farmer who was doing just that.

The natural world holds answers to some environmental problems.

Takao Furuno was, by most standards, a very successful industrial rice farmer, with annual yields among the highest in southern Japan. But it was a tough grind. Each year he was forced to put all his earnings back into the next year's crop—pesticides, herbicides, irrigation, and fertilizer —so that despite his success, he and his family were left with very little for themselves at the end of each season.

In searching for a better way, he turned, as he often did, to his forebears to see what he could learn from their knowledge. He was surprised to discover that they used to keep ducks in their rice paddies. Most rice farmers consider

food miles The distance a food travels from its site of production to the consumer.

Infographic **18.3** | **INTEGRATED FARMING: THE DUCK/RICE FARM**

↓ Takao Furuno's farm is a self-regulating, multiple-species system that naturally meets the needs of the farm ecosystem. The species present all play a role in the system and help each other and overall production.

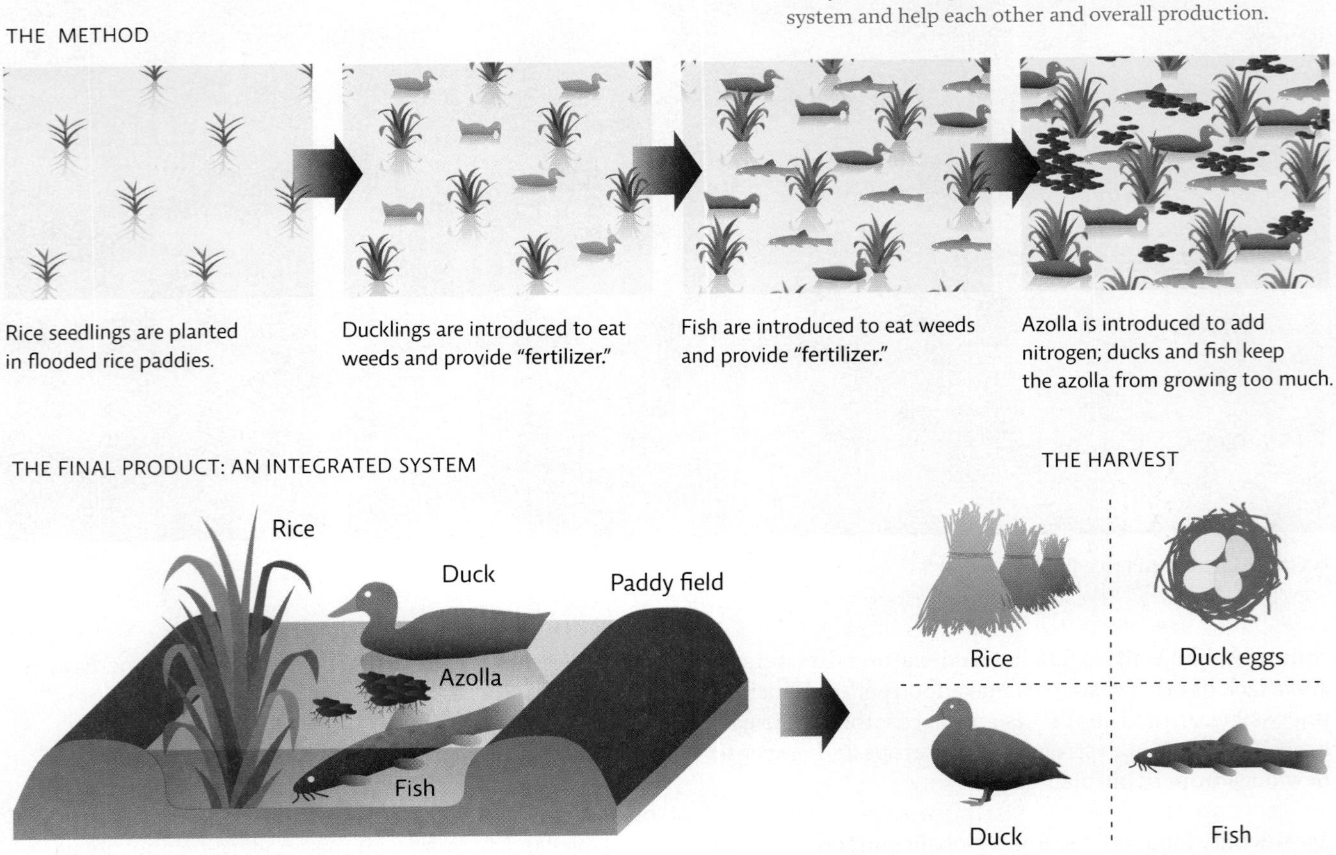
THE METHOD
Rice seedlings are planted in flooded rice paddies.
Ducklings are introduced to eat weeds and provide "fertilizer."
Fish are introduced to eat weeds and provide "fertilizer."
Azolla is introduced to add nitrogen; ducks and fish keep the azolla from growing too much.
THE FINAL PRODUCT: AN INTEGRATED SYSTEM
Rice
Duck
Paddy field
Azolla
Fish
THE HARVEST
Rice
Duck eggs
Duck
Fish

ducks a pest, albeit a slightly cuter one than the azolla. Adult ducks eat rice seeds before they have a chance to grow, and as they forage, trample young seedlings into the mud. This creates open patches of water, which in turn, invite only more ducks. "If you're not careful, you end up with a big problem pretty quickly," says Raquel.

But ducklings, Furuno soon realized, were too small to do such damage; for one thing, their bills were not big or strong enough to extract seeds from mud. Instead, they ate bugs and weeds. Azolla was one of their favourites.

Furuno's forebears also grew loaches—a type of fish—in their paddies. The loaches would eat azolla and could be harvested and sold as food.

Together, the ducklings and loaches would keep the weed from strangling the rice crop; but they would not completely eliminate the azolla the way a heavy dose of pesticides would. Furuno quickly discovered that, when kept at this benign concentration, the azolla (which contain symbiotic bacteria that produce a usable form of nitrogen) actually fertilize the rice. In fact, between the nitrogen from the azolla, and the duck and fish droppings, he soon found that he no longer needed to spend money on synthetic fertilizer.

There were other financial gains, too. Duck eggs, duck meat, and fish all fetched a good price in the market. And because he was no longer using pesticides, he could also grow fruit on the edges of his rice field. (He opted for fig trees, a crop he could harvest yearly without having to replant).

When raising ducks, fish, and rice crops together, Furuno discovered the root crowns (where the root meets the stem) of rice plants increased to about twice the size that they had been in his old industrial system. A larger root crown meant more rice. "We're not exactly sure why the crowns grew," Furuno told an audience of American farmers at a recent convention in Iowa. "But the ducks seem to actually change the way the rice grows. It's got something to do with the synergy of the whole system." [INFOGRAPHIC 18.3]

In the decade and a half since Furuno began duck/rice farming, his rice yields have increased by 20-50%, making his among the most productive farms in the world, nearly twice those of conventional farmers. Independent researchers have verified his results, including the Bangladesh Rice Research Institute which recommends the technique to Bangladeshi farmers. And by now, some 10 000 Japanese farmers have followed his lead; the Furuno method is also catching on in China, the Philippines...and California.

Some industrial agricultural practices have significant drawbacks.

By presenting an alternative way to grow both rice crops and duck meat, Furuno's methods confront two of the Green Revolution's biggest legacies: monoculture crop production, and concentrated animal feeding operations (CAFOs) for livestock.

In **monoculture** farming, a single variety (genetically identical individuals) of a single crop (rice or corn or soybeans, for example) is planted over huge swaths of land. The uniform crops of a monoculture are easier to plant, maintain, and mechanically harvest. This makes mass production easier, which in turn makes for a more bountiful harvest.

In a **concentrated animal feeding operation (CAFO)**, livestock are raised in confined spaces, with a focus on raising as many animals in a given area as possible. They are fed a nutrient-rich diet of grain and soybean (in the case of cattle, this diet is supplemented with some hay) and are generally not permitted to graze or roam free. CAFOs are highly productive; they minimize the amount of land that is used; and because animals are so concentrated and confined, it is easy to harvest the animals' manure and sell it as fertilizer. Since about the mid-1900s, high-density feedlot operations like this have been the most common method of raising cattle in Canada and the United States. Today, poultry (chickens and ducks) and pigs are also predominantly raised in CAFOs.

But for all their benefits in efficiency and production volume, monocultures and CAFOs present a number of disadvantages.

monoculture Farming method in which one variety of one crop is planted, typically in rows over huge swaths of land, with large inputs of fertilizer, pesticides, and water.

concentrated animal feeding operation (CAFO) Meat or dairy animals being reared in confined spaces, maximizing the number of animals that can be grown in a small area.

↑ Cattle in a feedlot pen

↑ Harvesting wheat with four John Deere combines

For starters, both monocrop agriculture and CAFOs contribute heavily to global warming. Clearing so much land to grow food reduces the amount of carbon that can be sequestered by natural vegetation through photosynthesis. Also, fertilizer and livestock emit greenhouse gases. In fact, according to the UN Food and Agriculture Organization, livestock is responsible for some 18% of anthropogenic greenhouse gases.

In monocrop agriculture, the crop that is chosen is not necessarily locally adapted. Instead of focusing on which crop is best suited to the existing ecosystem, farmers focus on those crops with the highest market demand, and thus the highest dollar value. This shift, combined with the exponentially higher volume of plants, has made most farms heavily dependent on external inputs—water, pesticides, and fertilizer—that are added to the farm from outside its own ecosystem. "In the 1920s, half of Iowa's farms produced 10 commodities each," says

Infographic **18.4** | **GROWING LIVESTOCK: FEED AND WATER NEEDS**

↓ Feed conversion rates show how much usable product is produced from the food an animal is fed. We never see 100% conversion because the animal has already used most of the energy it has consumed in its life in day-to-day activities; however, some species are better at converting feed to usable product than others are. It may also be surprising how much water is needed to produce meat and dairy product; much of it goes to grow the feed.

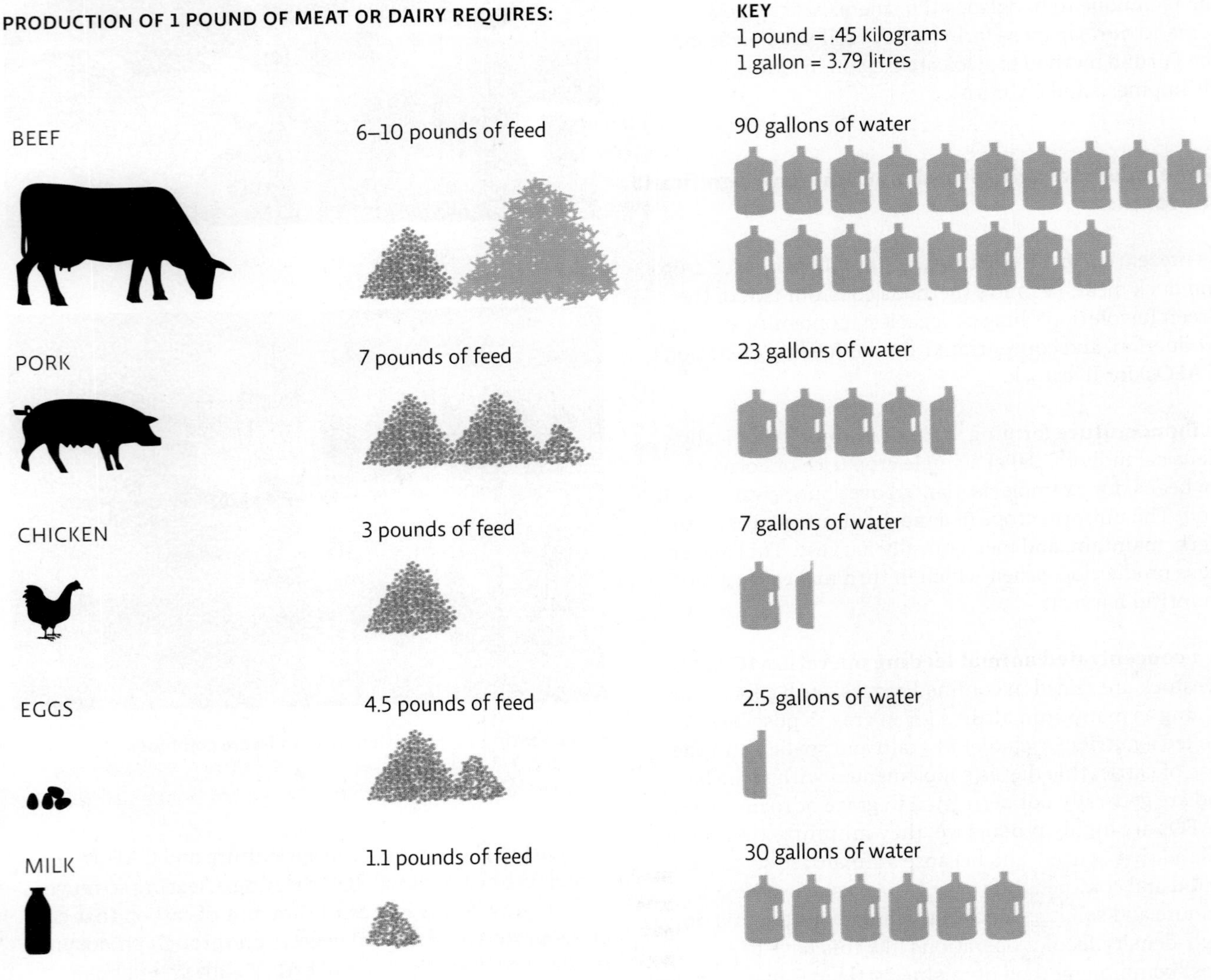
PRODUCTION OF 1 POUND OF MEAT OR DAIRY REQUIRES:
KEY
1 pound = .45 kilograms
1 gallon = 3.79 litres
BEEF
6–10 pounds of feed
90 gallons of water
PORK
7 pounds of feed
23 gallons of water
CHICKEN
3 pounds of feed
7 gallons of water
EGGS
4.5 pounds of feed
2.5 gallons of water
MILK
1.1 pounds of feed
30 gallons of water

Fred Kirshenmann, a distinguished fellow at Iowa State University's Leopold Center for Sustainable Agriculture. "Today, 80% of the state's cultivated land is exclusively corn or soybean. Farming systems that were once supported by complexity and diversity of species have now been replaced by reliance on inputs."

This dependence makes monocrop agriculture both expensive and environmentally taxing. Heavy farm machinery compacts soil and increases its erosion. Use of irrigation can result in soil salinization; as water evaporates from the soil, salts are left behind and become so concentrated that crops can no longer grow.

CAFOs come with their own set of problems: they are environmentally taxing and ethically questionable. While the high grain diets fed to confined animals certainly maximize growth, they are not necessarily good for the animals' health. Such diets may cause liver and stomach disorders, increased susceptibility to infections, and even death. In addition, crowded living conditions (viewed by some as inhumane) make CAFO animals more vulnerable to disease, forcing farmers to use heavy doses of antibiotics to prevent sickness. Such heavy use contributes to antibiotic resistance, which threatens the health of the animals and humans. Resistant microbes that develop in the livestock can make their way into the water supply

↑ Cows at this farm are grass fed and graze using the rotation method. This method is closer to a traditional ecological system than conventional farming.

and spread to other food sources. They can also be passed directly to consumers in the meat itself.

CAFOs have drawn the ire of environmentalists who point out that the grain and soybean used to feed all those livestock would feed far more humans than the meat we harvest from the animals actually does. The reason for this has to do with **feed conversion rates**—how quickly and efficiently any given animal converts the food it eats into body mass (that we then eat). Cattle have a high food conversion rate, meaning it takes a lot of feed to produce a litre of milk and even more to produce a kilogram of meat. It also takes a lot of resources (namely, water and fossil fuel) to produce that feed. Chickens and pigs have slightly better feed conversion rates and thus require slightly fewer resources to grow, but not by much. [INFOGRAPHIC 18.4]

In Canada and the United States, a niche market has developed for "grass-fed" beef, which is produced from animals that are raised on pasturelands as opposed to CAFOs. Grass-fed animals don't gain weight nearly as rapidly as feedlot animals do; they eat less and spend more time moving around, so they don't get quite as fat. But the meat and milk they produce is healthier—it has less saturated fat and more omega-3 fatty acids than CAFO-generated beef. And the animals themselves live better lives—rather than being confined in a tight space, they are free to roam the grasslands and eat the food (namely grass) that they have evolved to process and digest. In fact, unlike CAFO livestock, grass-fed animals can represent a net gain for the human food supply when grazed on land that is unsuitable for human crops; they eat grasses that we can't eat, and turn it into beef and dairy products that we can eat (see Chapter 12).

The ducks on Furuno's farm are like the grass-fed cattle: they are raised in a humane environment, without any undue environmental costs. But whether or not his methods would work as well on the other side of the planet was an open question.

Sustainable agriculture techniques can keep farm productivity high.

Greg and Raquel were well versed in the problems of modern agriculture. Before settling in California to take over the family farm, they had worked as tropical ecologists

---

feed conversion rates How much edible food is produced per unit of feed input.

Infographic **18.5** | **SUSTAINABLE FARMING METHODS**

↓ Many traditional methods are useful for sustainable agriculture because they focus on protecting the soil, the heart of successful farming. Different methods address different issues and are best used to meet a specific need.

CONTOUR FARMING When farming on hilly land, rows are planted along the slope, following the lay of the land, rather than oriented downhill.

REDUCED TILLAGE Planting crops into soil that is minimally tilled reduces soil erosion and water needs (it reduces water evaporation). It also requires less fuel because of less tractor use.

TERRACE FARMING On steep slopes, the land can be levelled into steps. This reduces soil erosion and allows a crop like rice to stay flooded when needed.

CROP ROTATION Planting different crops on a given plot of land every few years helps maintain soil fertility and reduces pest outbreaks since pests from the year before (or their offspring) will not find a suitable food when they emerge in the new season.

STRIP CROPPING Alternating different crops in strips that are several rows wide keeps pest populations low; it is less likely the pests will travel beyond the edge of a strip and they may not find another row of this crop.

COVER CROPS During the off-season, rather than letting a field stand bare, a crop can be planted that will hold the soil in place. Nitrogen-fixing crops that improve the soil, like alfalfa, are often chosen.

Infographic **18.6** | **THE TRADE-OFFS OF ORGANIC FARMING**

FOR	ADVANTAGES	DISADVANTAGES
THE CONSUMER	• No pesticide residue on produce • Foods may contain more antioxidants (the plant's defence against disease). • Better taste • Fresher (because it is not stored and shipped long distances) • Animal products contain no hormones and the animals may be treated more ethically. • Less risk of exposure to antibiotic-resistant bacteria	• Fruits and vegetables may have more blemishes on them. • Shelf life is shorter (not waxed, not picked before ripe). • Usually more expensive • "Greenwashing"—misleading claims are made about the healthiness of organic foods (organic cookies are probably not really healthier than those made with conventional ingredients)

FOR	ADVANTAGES	DISADVANTAGES
THE FARMER/ ENVIRONMENT	• Fewer and less costly inputs • Soil is not degraded and may be enhanced if natural fertilizers and techniques such as cover crops and crop rotation are used. • Beneficial insects are not killed by broad-spectrum pesticides. • More genetic diversity and species diversity makes it less likely that a pest outbreak or other problem will decimate the entire crop. • Less potential for water pollution (no synthetic chemicals)	• More labour intensive • Certification process takes time and is costly • Fewer subsidies available compared to those for industrially grown crops • Crops native to the area or suitable for the climate do best so may not be able to grow all crops in all areas

in Costa Rica, where they learned about traditional, non-industrial farming methods such as terracing, crop rotation, and strip cropping that help protect the soil and keep productivity high without the use of synthetic fertilizers. [INFOGRAPHIC 18.5]

In rice farming, they saw a chance to implement all the concepts and theories they had learned as ecology students, and refined and discussed as scientists, on their own land. "We wanted a farm where success was measured not just in crop yield, but in the overall health of the land," Greg says. "A place where we would count profits, but we would also count the number of Sandhill cranes and California quail we saw populating the area."

They wanted to create a farm that was sustainable and at least partly local (they planned to sell a portion of their rice at local outlets like farmers' markets and food co-ops). **Sustainable agriculture** means depending on farming methods that can be used indefinitely. In addition to looking for sustainably or organically produced food, many consumers are buying food from local farmers; local agriculture supports local economies, provides fresher and thus healthier food to consumers, and minimizes the ecological footprint because food has been shipped over much smaller distances. Of course, sustainable and organic agriculture have their own set of trade-offs: while they might be more environmentally friendly, they are also more expensive and in some cases produce less food per hectare of land. But Greg and Raquel were determined to at least try. [INFOGRAPHIC 18.6]

To achieve their goals, the Massas began by installing a recirculation system to reclaim and reuse irrigation water. They planted native oak trees along field borders to serve as a natural windbreak (windbreaks prevent soil from being carried away by wind erosion), and installed nest boxes for wood ducks, barn owls, and bats so that those wild animals would keep area pests in check. "The idea was to restore as much of the natural biodiversity as

sustainable agriculture Farming methods that do not deplete resources, such as soil and water, faster than they are replaced.

↑ Ducklings in the Massa Organics rice fields

possible," says Greg, "so that we would not need artificial inputs to run the system." They also took their sustainable ideals to the next level and built a straw house to live in—made of 0.6 metre-thick walls of rice bale, coated on both sides with plastic or stucco—that can withstand the unforgiving heat of a Chico summer and maintains a steady temperature almost entirely by itself. The house is fire proof, rodent proof, and as Greg likes to joke, bullet proof, too.

Furuno's method of duck/rice farming would be a good fit for the Massas. Not only would the rice plants rely on natural fertilizers and natural pest control, but the duck eggs and meat produced would be more humanely grown than those produced by factory farms. The ducks would grow up in ponds, not crammed together on slats in a barn with no access to swimming water. They would get to splash around, and express their "duckiness," as Raquel put it.

When the ducks first arrived on the Massa farm, they were just 24-hours-old, cotton ball-sized tufts of yellow feathers. The Massa children cared for them in wooden crates in their barn. But as Raquel soon learned, small ducklings grow mighty fast. In just 2 weeks, they were large enough to turn loose. Furuno had advised stocking about 100 ducks per acre (4047 square metres), but Greg and Raquel did not want to sacrifice that much land for this first attempt, so they fenced off just a quarter-acre (1012 square metres), instead. This amount of space provided plenty of room for their 120 ducks to swim and forage in, but as it turned out, not enough food to support them. "They quickly ate all the weeds in the field," says Greg. They also trampled some of the rice plants in their pond because their section of field was too small. But, even as they ran out of weeds to eat, the baby ducks stayed away from the rice plants. Just as Furuno had insisted they would.

Genetically modified crops may help feed the world.

Though promising, methods like the ones Furuno and the Massas are implementing represent only part of the solution to our emerging food problems. Critics say that by itself, organic farming will never produce as much food as industrial agriculture has, and neither organic nor current industrial practices will be enough to feed 10 billion people. So as population swells, scientists are working on another solution, one that has given rise to its own debates and controversies: genetic engineering.

Genetic engineering forms the basis for what some farmers and scientists like to think of as the **Green Revolution 2.0**; it involves manipulating genes to increase productivity or to make it possible to grow crops in places they normally wouldn't—like marginal land, in drought-plagued regions, or on fields doused in pesticide and herbicide. **Genetically modified organisms (GMOs)** are organisms that have had their genetic information modified in a way that does not occur naturally. This usually involves transferring new genes for desirable characteristics—such as pest resistance, drought resistance, or increased nutrient production—from one species to another, creating a **transgenic organism**. [INFOGRAPHIC 18.7]

Most existing GMO crops are altered to have greater pest resistance (for example, Bt transgenic plants), but scientists are also working on drought-tolerant varieties of plants and have had some success in creating nutritionally enhanced food. The most notable example is golden rice, which was genetically modified to produce extra beta carotene. Deficiencies of beta carotene, which produces vitamin A, is a leading cause of blindness in children.

In Canada, more than 70% of food sold contains GMOs, mainly corn, canola, and soybeans. Most of these crops have an herbicide-tolerant (Ht) gene added so that they can withstand huge doses of herbicides. Scientists are also working to add genetically modified animals to our food supply. AquaBounty has developed a genetically modified Atlantic salmon that grows much faster than normal (thanks to genes from the larger Chinook salmon) and is currently seeking approval from the U.S Food and Drug Administration to bring the fish to market. If approved, this would be the first GMO animal food product sold in the United States. [INFOGRAPHIC 18.8]

In other parts of the world, however, including China and the European Union, GMOs have been met with fierce opposition from consumers who worry about the long-term health impacts and environmental consequences.

One concern is that domestic crops with introduced genes could hybridize (crossbreed) with closely-related wild plants. If a pest-resistance gene became part of weed species genomes, those weeds could grow more aggressively and outcompete other plants, including crops. We already have many herbicide-resistant weeds in North America, though this does not appear to be the result of transfer of genes from GMO crops, but rather from adaptation to a high-pesticide environment. Over 200 weed species have already evolved a herbicide tolerance

Green Revolution 2.0 Focuses on the production of genetically modified organisms (GMOs) to increase crop productivity or create new varieties of crops.
genetically modified organisms (GMOs) Organisms that have had their genetic information modified to give them desirable characteristics such as pest or drought resistance.
transgenic organism An organism that contains genes from another species.

Infographic **18.7** **GENETIC ENGINEERING CAN PRODUCE ORGANISMS WITH USEFUL TRAITS**

↓ Genetic material can be transferred from one organism to another. Plants that receive a gene from the bacterium *Bacillus thuringiensis* (Bt) to produce a toxin that kills pests will produce the toxin themselves, giving the plant protection from many pests.

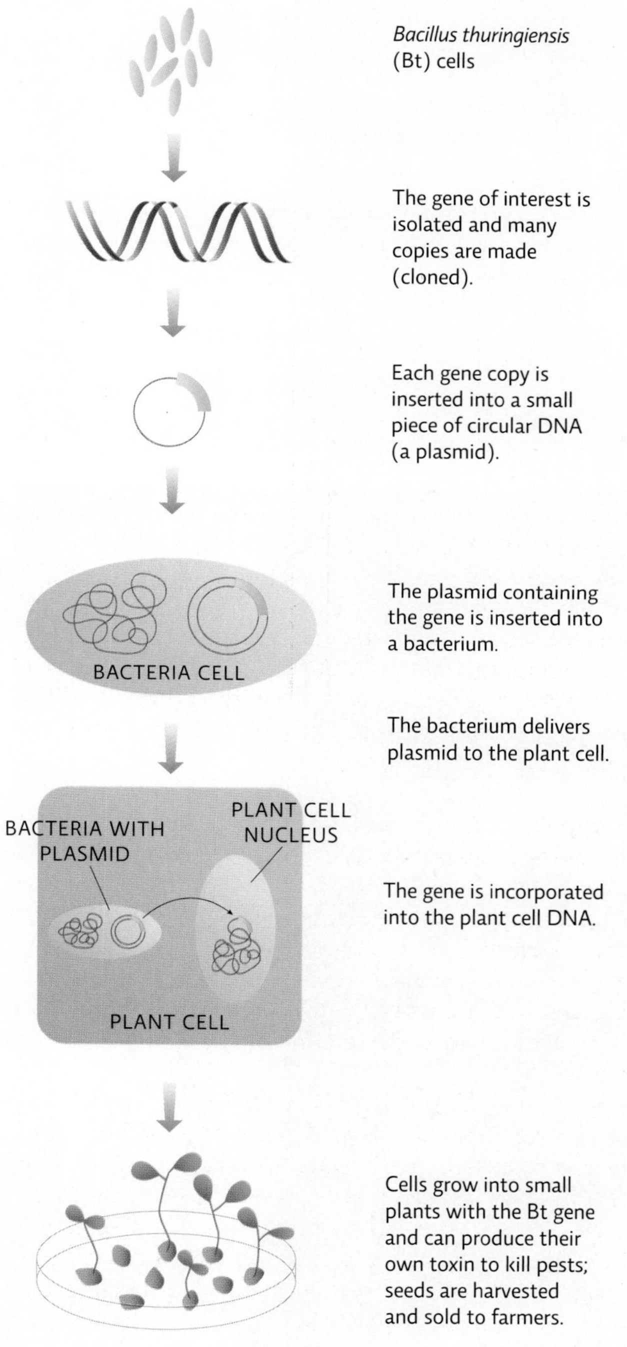
Bacillus thuringiensis (Bt) cells
The gene of interest is isolated and many copies are made (cloned).
Each gene copy is inserted into a small piece of circular DNA (a plasmid).
BACTERIA CELL
The plasmid containing the gene is inserted into a bacterium.
The bacterium delivers plasmid to the plant cell.
BACTERIA WITH PLASMID
PLANT CELL NUCLEUS
PLANT CELL
The gene is incorporated into the plant cell DNA.
Cells grow into small plants with the Bt gene and can produce their own toxin to kill pests; seeds are harvested and sold to farmers.

Infographic **18.8** | **EXAMPLES OF GMOs**

TYPE OF GMO	EXAMPLES
HT (herbicide-tolerant) crops are not killed by the herbicide, so the herbicide can be sprayed on the crop/soil directly, where it will kill weeds but not the crop.	Bromoxynil-tolerant canola and cotton Glyphosate- (Roundup-) tolerant corn, cotton, soybeans Imidazolinone-tolerant wheat
Bt crops contain a gene from *Baccillus thuringiensis,* a naturally occurring bacterium that produces a toxin that kills some pests.	BT corn, potatoes
Nutritionally enhanced food: genes inserted can increase the amount of a particular nutrient or allow the crop to produce a nutrient it would not normally produce.	Golden rice
GM animals are being developed for the food supply.	Salmon that grow faster Pigs that produce more omega-3 fatty acids

↑ Giant ragweed (*Ambrosia trifida*)

(Ht) gene globally. These so-called super weeds, including ragweed and pigweed, can be found in 5 provinces and can tolerate common herbicides such as glyphosate. They take over entire fields, stop combines, and are tough to clear by hand.

Another problem, say opponents, is that GMOs are patented and owned by a few multinational corporations, which makes them more expensive than traditional seeds, and thus a bad option for developing countries. Much of the hunger seen in the world today arises not from lack of food but from an inability of the disempowered to access it. Putting even more of our food supply under the control of a few multinationals could only make this worse.

In the Sacramento Valley, where the Massa farm is located and where virtually all of the U.S. rice crop is grown, farmers are split over the merits of genetic engineering. Some see the advantage of having crops that can better tolerate drought, floods, and disease. "The climate is changing," says one of Greg's neighbours. "We see drier seasons and more extreme weather than we ever have before. And eventually we will need some breed of rice that can endure those changes." But others, including Greg Massa, say that growing genetically engineered rice anywhere, in any quantity, poses a threat to their farms and livelihoods.

Here's why: in 2006, small amounts of experimental strains of genetically engineered rice crept into the U.S.

food supply. Nobody knows how the strain, engineered by the company Bayer to be herbicide resistant, escaped from the Arkansas storage bins where it was being kept. However, when news spread, European retailers pulled all U.S. rice from their shelves, sending rice prices plummeting and threatening the sanctity of the Massa farm and all of its neighbours. "The problem is, even the threat of contamination can kill our businesses," says Greg. "So it's really a danger to grow it anywhere, at least until it wins wider approval. And in the meantime, there are other ways to combat weeds. The duck farming proves that."

At any rate, GMOs are already here, and aren't likely to disappear anytime soon. The bottom line: this technology will almost certainly help address some of our food issues, but as with all new technology, we will have to weigh risks and benefits very carefully with each step forward. And as the global population swells, we will probably need all solutions—crops that have been engineered to resist weeds, and crops grown in more integrated, natural systems, like the duck/rice farms of Japan and California—to feed the world.

For the Massas, duck/rice farming has proven to be the best possible solution to the challenges of modern agriculture. They ended up with duck meat to sell, and though they didn't take any precise measurements of yields during their first trial run, their rice crops did not appear to suffer at all. "We learned a couple of things," Raquel says. "The ducks trampled some of the rice in their pond, which would not have been a problem if we had used a larger section of the field. I also think we used the wrong breed of duck. They were a little too large to move effectively between the dense plants, and they were not active enough in their foraging abilities. These ducks were bred to sit around all day and eat and gain weight quickly for industrial meat production. We are currently researching which breeds to try next."

Still, in the end, the Massas harvested both rice and duck meat. The key to successful duck/rice farming is to harvest the ducks before they get big enough to trample rice plants or strong enough to pluck rice seeds from deep within the mud. The hard part isn't knowing when to do this, but having the resolve for what comes next: killing and eating the ducks. The Massa family struggled, but ultimately felt good about the outcome. "I know the conditions in which they were raised were more humane than 99% of the meat ducks in this country," says Raquel. "They had it good and you can taste that in the finished product."◉

Select references in this chapter:
Hossain, S.T., *et al.* 2005. *Asian Journal of Agriculture and Development*, 2: 79–86.
Reganold, J.P., *et al.* 2010. *PLoS ONE*, 5: e12346.doi:10.1371.
Russell, J.B., and Rychlik, J.L. 2001. *Science*, 292: 1119–1122.

BRING IT HOME

PERSONAL CHOICES THAT HELP

Making food choices for your health and wellness, as well as for the vitality of the environment, requires some thought. Becoming an educated consumer is one way you can reduce your environmental impact, eat a nutrient-rich diet, and contribute to the public dialogue about the safety and sustainability of our food systems.

Individual Steps

→ Support local independent farms by shopping at farmers' markets; buying a share in a community-supported agriculture (CSA) group in your area; and frequenting restaurants that serve food grown locally. For more information, visit the CSA website for your province.

→ Reduce the amount of animal products you eat and try to buy free-range and grass-fed meat whenever possible.

→ Grow your own food. Transportation of food contributes heavily to emissions and fossil fuel usage; planting just a few edibles can greatly decrease your food footprint.

→ Buy organic foods, especially for common produce that typically is heavily treated with chemicals, such as apples, celery, lettuce, strawberries, and peaches; this reduces the chemical residue you ingest and the chemicals released into the environment, and helps provide safer working conditions for people in the agriculture industry. Visit Canadian Organic Growers (www.cog.ca) to find a local organic farm near you.

Group Action

→ Find like-minded people to start a community garden in your area. Community gardens allow apartment dwellers to grow their own food in empty lots or in public spaces. Most are run by municipalities; look for information on your community's website.

Policy Change

→ GMOs offer promise for feeding the world's population, but like all technologies, pose some risks, both known and unknown. Write letters to your newspaper asking for coverage of this issue. Support policies that require the testing, labelling, and traceability of GMO products.

UNDERSTANDING THE ISSUE

CHECK YOUR UNDERSTANDING

1. **The dramatic increases in food production that resulted from the use of high-yield crop varieties coupled with chemical pesticides, irrigation systems, synthetic fertilizers, and modern farming equipment is known as:**
   a. the Green Revolution.
   b. sustainable agriculture.
   c. genetic engineering.
   d. eutrophication.

2. **The ideal pesticide would:**
   a. remain behind on food to continue to kill pests.
   b. allow the development of genetic resistance in the pest.
   c. kill only the target pest.
   d. be persistent in the environment and bioaccumulate in non-target species.

3. **The presence of ducks and azolla in rice cultivation has shown that:**
   a. ducks and other birds must be eliminated from rice fields, as they eat the rice seed before it can grow.
   b. invasive species diminish rice yields.
   c. GM crops such as rice cannot succeed without azolla to provide shade and ducks to provide pollination.
   d. restoring some biodiversity to rice fields reduces pest damage and increases rice yields.

4. **Concentrated animal feeding operations (CAFOs) can contribute to:**
   a. air pollution, because livestock release methane, a potent greenhouse gas.
   b. reducing the impact from overgrazing rangelands as less land is used for feedlot operations.
   c. water contamination, because of the vast amount of feces produced.
   d. All of the above.

5. **Based on feed-conversion ratios, it is more energy efficient to eat chicken than beef because:**
   a. a kilogram of ground beef has fewer calories than a kilogram of ground chicken.
   b. chickens are more efficient than cows at converting the food they are fed into body mass or the food that we eat.
   c. chickens are lower on the food chain than cows.
   d. unlike cows, chickens represent a gain for the human food supply as they eat items we cannot (such as insects) and convert them into meat that we can eat.

6. **Crops that have DNA from entirely unrelated species are referred to as _______ and the process that is used to develop such crops is called _______.**
   a. GMOs; genetic engineering
   b. monocultures; CAFOs
   c. hybrids; sustainable agriculture
   d. HYVs; industrial agriculture

WORK WITH IDEAS

1. Industrial agriculture allows us to grow more food, but many of the technologies used have significant side effects. Discuss three specific examples of how growing our food under such an agricultural system may be harmful to us and to the environment.

2. How is the agricultural system illustrated in the duck/rice farm similar to and/or different from industrial agriculture? Which system do you think is a better way to grow food? Why?

3. Compare and contrast the Green Revolution with the Green Revolution 2.0.

4. What is sustainable agriculture? Give one example each of how we can sustainably raise crops and animals for food. Why is sustainable agriculture a better way to grow food? Provide three reasons.

ANALYZING THE SCIENCE

The data in the following graph come from the Farming Systems Trial (FST) research study conducted by the Rodale Institute in Pennsylvania. This side-by-side comparison of corn and soybeans grown under organic and industrial agricultural systems was started in 1981 and is one of the longest-running studies of its kind.

INTERPRETATION

1. What does this graph show?

2. Calculate the following for conventional and organic systems in both imperial and metric units: profit per unit of yield; energy input per unit of yield; and greenhouse gas emissions per unit of yield. How do the two systems compare for these three parameters?

3. According to the Food and Agricultural Organization of the United Nations, "Organic agriculture has the potential to secure a global food supply, just as conventional agriculture does today, but with reduced environmental impact." How does the data from the FST study support this statement?

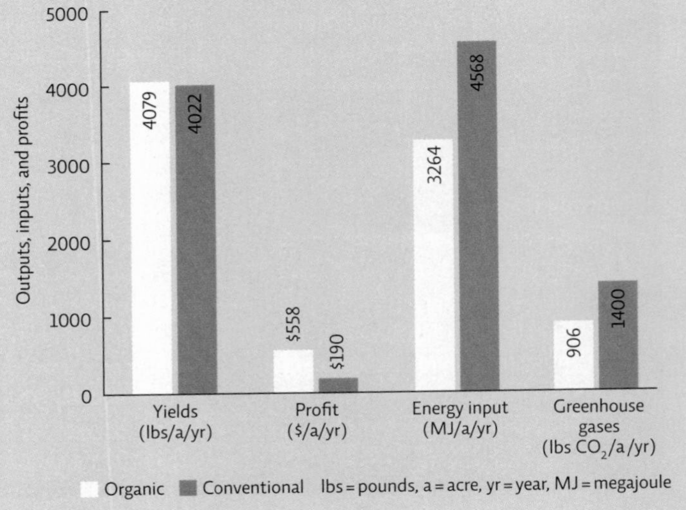
5000
4000
3000
2000
1000
0
Outputs, inputs, and profits
4079
4022
$558
$190
3264
4568
906
1400
Yields (lbs/a/yr)
Profit ($/a/yr)
Energy input (MJ/a/yr)
Greenhouse gases (lbs CO_2/a/yr)
Organic
Conventional
lbs = pounds, a = acre, yr = year, MJ = megajoule

ADVANCE YOUR THINKING

Hint: To answer the following questions, it might be helpful to access the actual Farming Systems Trial report at rodaleinstitute.org/our-work/farming-systems-trial/farming-systems-trial-30-year-report.

4. According to the FST report, even in drought years, the yields for organic corn were approximately 31% greater than for conventional (non-drought-resistant) varieties. At the same time, genetically engineered drought-tolerant varieties had yields that were no more than approximately 13% greater than conventional (non-drought-resistant) varieties. Why might this be the case? Why is this an important finding?

5. The FST report also indicates that crops grown using organic methods produced yields equivalent to conventional crops even though they had more weed competition in their fields. Why might this be the case? What makes this an important finding?

6. According to the FST report, the largest energy input in organic systems was diesel fuel, while the largest for conventional systems was nitrogen fertilizer. How could these energy inputs be changed to make both systems more economically and ecologically sustainable?

EVALUATING NEW INFORMATION

An Internet search for food and agriculture topics brings up a plethora of often conflicting information as well as the names of many organizations with very different agendas. But understanding how to evaluate the validity of the information is necessary in order to make informed decisions about what agricultural system you want to support and the type of food you want available as a consumer.

Consider the question of biotechnology. Is genetically engineered food safe? Is it the way forward in agriculture? The Council for Biotechnology Information says yes, while the Union of Concerned Scientists says no.

Explore the websites for the Council for Biotechnology Information (www.whybiotech.com) and the Union of Concerned Scientists (www.ucsusa.org).

Evaluate the websites and work with the information to answer the following questions:

1. Evaluate the agendas of the two organizations as well as the accuracy of the science behind their positions on GMOs:
   a. Who runs each website? Do the person's/organization's credentials make the information presented on food and agriculture issues, especially GMOs, reliable or unreliable? Explain.
   b. What is the mission of each website? What are the underlying values? How do you know this?
   c. What claims does each website make about the current problems in food production and what the future of agriculture should be, especially with regard to GMOs? Are their claims reasonable? Explain.
   d. How do the websites compare in providing scientific evidence in support of their assessment of agriculture and their position on the role of GMOs? Is the evidence accurate and reliable? Explain.

2. How do the two organizations compare in engaging you as a citizen in agricultural policy and the role of GMOs? Do you have a preference for one approach or the other? Do you think that citizen involvement in policy issues is necessary and effective? Explain your responses.

MAKING CONNECTIONS

A RIGHT TO KNOW? THE BATTLE OVER FOOD LABELLING

Background: Labels on food can include nutritional information, ingredients, and information on how and where the food was produced. Health Canada and the Canadian Food Inspection Agency (CFIA) are the two Canadian government agencies that regulate food labelling. Generally, the CFIA regulates labels that relate to food production (organic standards or the meat and poultry labelling requirements) while Health Canada regulates labels related to health, safety, and nutritional quality. In addition, there are independent nongovernmental organizations that provide labelling services, such as the Marine Stewardship Council for fish certification and the Rainforest Alliance for certification of coffee and chocolate.

For labelling to be effective, there must be consistent, achievable standards, testing and certification services, and enforcement. These can be expensive and when consumers vary significantly on what information they consider important, it is hard to develop a comprehensive food-labelling policy. Advocates of food labelling contend that labels increase clarity for customers, and help improve human health, mitigate environmental hazards, and support certain types of agricultural practices. They would like to see more information on food labels, including data on GM ingredients, irradiated food, the conditions under which livestock was raised, protection for habitat, etc. Opponents of food labelling worry that it will spread unnecessary fear among consumers about certain products and agricultural or food manufacturing practices. They do not believe that it is in the best interests of consumers to invest in the infrastructure that would be necessary to regulate all these labels.

Case: Research the food issues covered in this chapter as well as others (such as those in Chapters 3, 12, and 14) and write a letter to your MP about your position on food labels. In your letter, include the following:

1. An analysis of the pros and cons of labelling versus not labelling food, including a reflection on the values underlying this issue.

2. An evaluation of the different types of labels (those that are currently used as well as some proposed labels) and how useful (or not) you find the information they offer or could offer.

3. Your proposal for a sound food labelling policy, including:
   a. A listing of information that you consider important for consumers to know about the food you eat, with an explanation of why this information is important.
   b. A set of guiding principles that can be applied when considering any new food label, and why these are important.
   c. A discussion of who should be involved in making these decisions about food labels and why they should be involved.

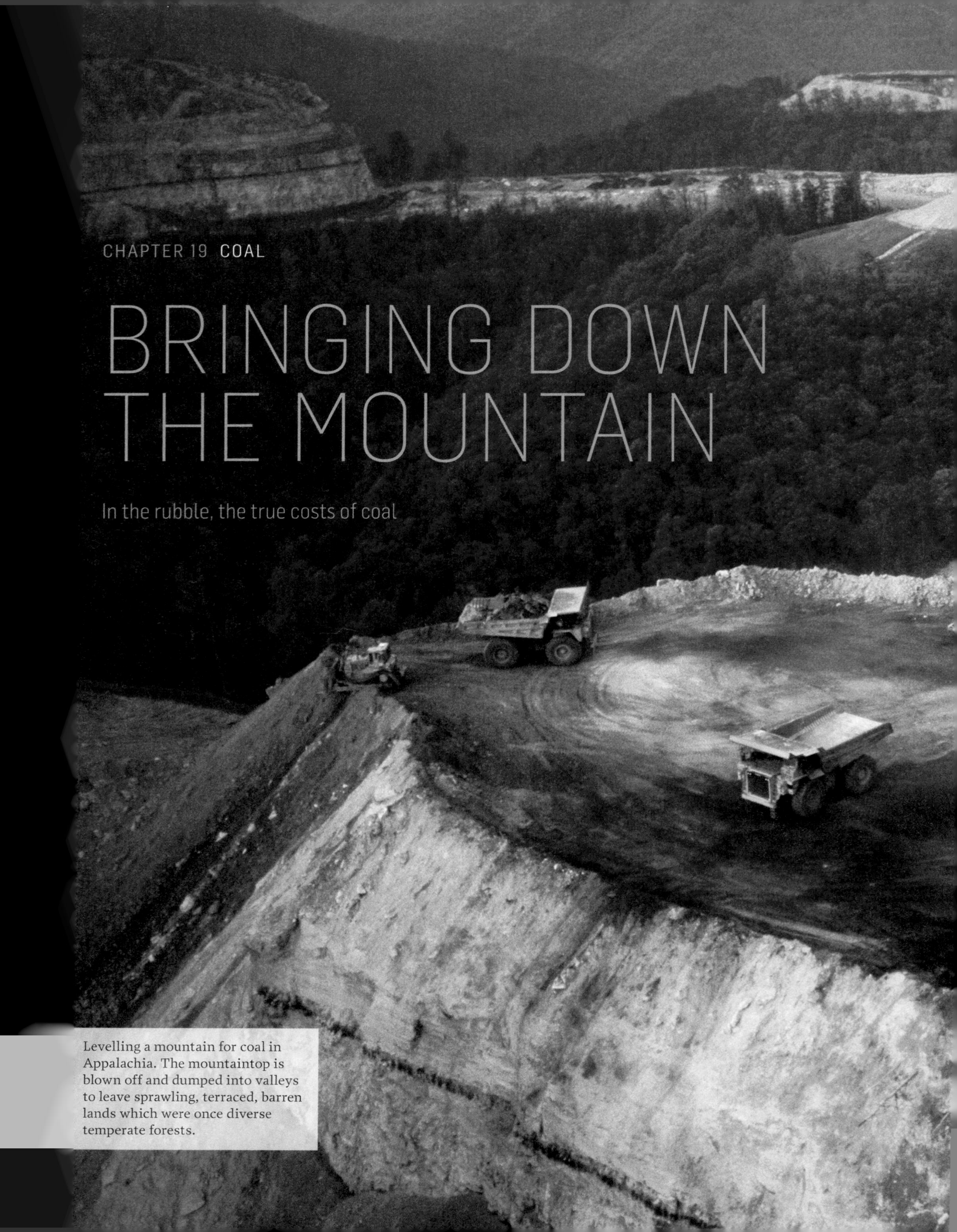

CHAPTER 19 **COAL**

BRINGING DOWN THE MOUNTAIN

In the rubble, the true costs of coal

Levelling a mountain for coal in Appalachia. The mountaintop is blown off and dumped into valleys to leave sprawling, terraced, barren lands which were once diverse temperate forests.

CORE MESSAGE

Human society runs on energy, and coal continues to be a major reliable energy resource. However, coal mining causes irreversible environmental degradation and the by-products of mining and burning coal pose significant health risks. Despite these drawbacks and the availability of renewable, cleaner energy sources, coal's availability and industrial presence keep it a major energy player. Researchers are developing ways to lessen the impact of burning coal, but as long as we continue to use it, the impact of mining will remain.

GUIDING QUESTIONS

After reading this chapter, you should be able to answer the following questions:

→ How important is coal as an energy source and how is it used to generate electricity?

→ What is coal, how is it formed, and what regions of the world have coal deposits that are accessible?

→ What methods are used to mine coal and what are the advantages and disadvantages of each?

→ What are the advantages and drawbacks of burning coal?

→ What new technologies allow us to mine and burn coal with fewer environmental and health problems?

About 300 metres above the foothills of Central Appalachia, near the Kentucky–West Virginia border, a four-seater plane ducks and sways like a tiny boat on an anxious sea. It's windier than expected, and Chuck Nelson, a retired coal miner seated next to the pilot, grips the door in an effort to steady his nerves. The passengers have come to survey the devastation wrought by mountaintop removal—a form of surface mining that involves blasting off up to 120 metres of mountaintop, dumping the rubble into adjacent valleys, and harvesting the thin ribbons of **coal** beneath.

At first, the landscape looks mostly unbroken; mountains made soft and round by eons of erosion roll and dip and rise in every direction, carrying a dense hardwood forest with them to the horizon. But before long, a series of **mountaintop removal** sites come into view. Trucks and heavy equipment crawl like insects across what looks like an apocalyptic moonscape: decapitated peaks and hectares of barren sandstone and shale. Smoke curls up from a brush fire as the side of an existing mountain is cleared for demolition. Orange-and-turquoise-coloured sediment ponds—designed to filter out heavy metal contaminants before they permeate the water downstream—dot the perimeter.

Here and there, a tiny patch of forest clings to some improbably preserved ridge line. "That's where I live," Nelson says, forgetting his air sickness long enough to point out one such patch. "My God, you would never know it was this bad from the ground." The aerial tour has reached Hobet 21, which, at more than 52 square

↓ The controversial Spruce No. 1 coal mine in West Virginia was originally given a permit authorizing it to dump strip mining waste into 11 kilometres of creeks and onto 800 hectares of land. In 2011, the Environmental Protection Agency rejected the permit on the basis of the "irreversible damage" that would be inflicted to the streams, groundwater, and land, including: "the elimination of all fish, killing of birdlife, reduction of habitat value, and risk of human illness."

↑ Seams of coal exposed at a mountaintop removal mining site in Welch, West Virginia

◉WHERE IS APPALACHIA?

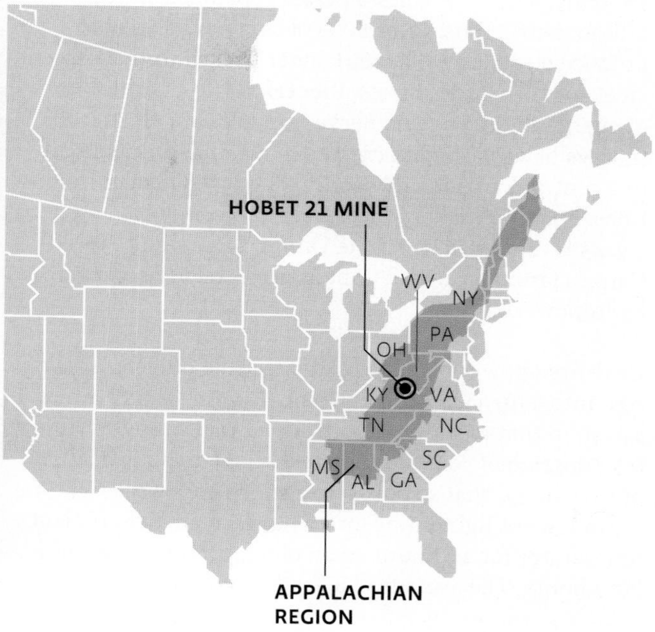
HOBET 21 MINE
WV
NY
PA
OH
KY
VA
TN
NC
SC
MS
AL
GA
APPALACHIAN REGION

kilometres, is the region's largest mining operation. So far, sites like this one have claimed more than 400 000 hectares of forested mountain, across just four states: Kentucky, West Virginia, Virginia, and Tennessee. But there is still more coal to mine. And if the coal companies have their way, tens of thousands of hectares more will be thus obliterated in the coming years.

To stem this tide of destruction, environmental activists have sued the coal industry, the state of West Virginia, and the federal government itself. They argue that mountaintop removal mining destroys biodiversity, pollutes the water beyond recompense, and threatens the health and safety of area residents. And by obliterating the mountains, they say, it also obliterates the culture of Appalachia.

Coal industry reps have countered by decrying the loss of jobs, tax revenue, and business the already impoverished region would suffer if the mines were to close under the weight of too much regulation. They also point out that the culture of Appalachia is as bound to coal mining as it is to the mountains. Both sides count area residents, including miners, among their ranks.

At the heart of the issue is coal itself—the country's dirtiest, and most abundant, energy source—the one most responsible for rising CO_2 levels from electricity production, but also the one the world relies on most heavily and the one that is being consumed most rapidly. As the Appalachian reserves dwindle, debates raging throughout the decimated foothills are reverberating across an energy-addicted nation.

The world depends on coal for most of its electricity production.

Simply put, coal equals energy. **Energy** is defined as the capacity to do work; like all living things, we humans need it, in a biological sense, to survive. But we also need it to run our societies: to heat and cool our homes; operate our cell phones, lamps, and laptops; fuel our cars; and power our industries. Most of our energy comes from **fossil fuels**—nonrenewable carbon-based resources, namely coal, oil, and natural gas—that were formed over millions of years from the remains of dead organisms.

coal Fossil fuel formed when plant material is buried in oxygen-poor conditions and subjected to high heat and pressure over a long time.
mountaintop removal Surface mining technique that uses explosives to blast away the top of a mountain to expose the coal seam underneath; the waste rock and rubble is deposited in a nearby valley.
energy The capacity to do work.
fossil fuels Nonrenewable resources like coal, oil, and natural gas that were formed over millions of years from the remains of dead organisms.

Canada consumes up to 60 million metric tons of coal annually, while the United States consumes almost 1 billion metric tons. Over 80% of this coal is used to produce electricity. The remainder is used to fuel cement, steel, and other industries. **Electricity** is a natural form of energy (lightning and nerve impulses are electrical) that we have learned to create on demand, producing it in a central location and sending it out via transmission lines. Burning coal produces over 40% of global electricity, 45% of electricity in the United States, and 13% in Canada (around 60% of Canadian electricity comes from hydropower).

Coal-fired power plants work by feeding pulverized coal into a furnace to generate heat, which then powers a system that produces electricity. It takes roughly 0.5 kilogram of coal to generate 1 kilowatt-hour (kWh) of electricity; that's enough to run ten 100-watt incandescent light bulbs for an hour, an energy efficient refrigerator for 20 hours, or an older, less efficient one for 7 hours. The average U.S. family of four uses about 11 000 kWh of electricity per year. That comes out to some 5500 kilograms of coal, or 1375 kilograms per person.

So, how does coal stack up against other energy sources? On the one hand, it produces more air pollution than any other fossil fuel. On the other, it is safer to ship, cheaper to extract, and in the United States at least, more abundant by far. In fact, the United States has 10 times more coal than it does oil and natural gas combined; in 2010 alone, Americans mined almost 1 billion metric tons. Most of it came from Wyoming, which leads the nation as a coal producer, and the Appalachian mountains, which follow behind as a close second. While the United States exported some of that yield, the vast majority was used to power U.S. households and businesses. [INFOGRAPHIC 19.1]

In terms of net energy, or **energy return on energy investment (EROEI)**—a metric that allows us to compare the amount of energy we get from any individual

Infographic **19.1** | **ELECTRICITY PRODUCTION FROM COAL**

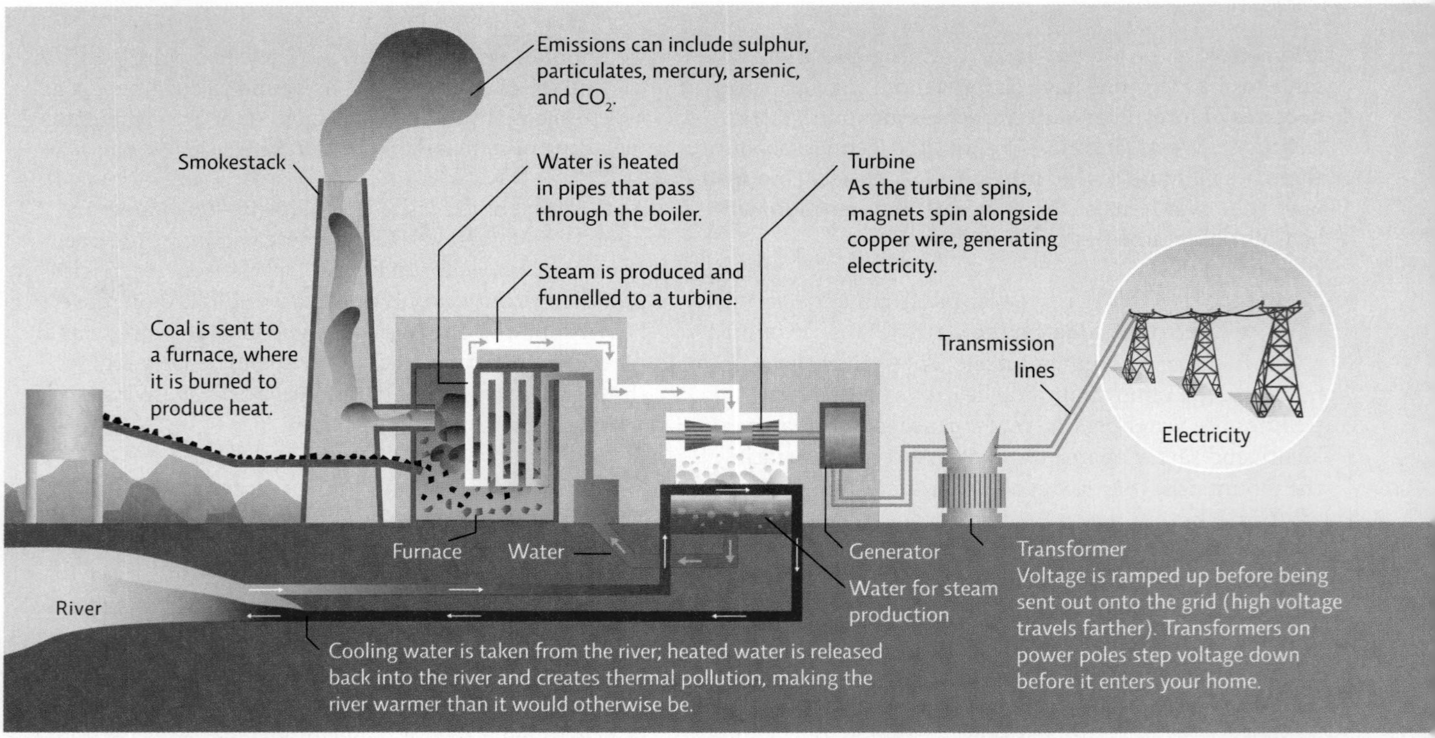
Emissions can include sulphur, particulates, mercury, arsenic, and CO_2.
Smokestack
Water is heated in pipes that pass through the boiler.
Turbine
As the turbine spins, magnets spin alongside copper wire, generating electricity.
Steam is produced and funnelled to a turbine.
Coal is sent to a furnace, where it is burned to produce heat.
Transmission lines
Electricity
Furnace
Water
Generator
Water for steam production
Transformer
Voltage is ramped up before being sent out onto the grid (high voltage travels farther). Transformers on power poles step voltage down before it enters your home.
River
Cooling water is taken from the river; heated water is released back into the river and creates thermal pollution, making the river warmer than it would otherwise be.

↑ The most common way to generate electricity is to heat water to produce steam; the flow of steam turns a turbine inside a generator to produce electricity. This schematic shows a coal-fired plant in Tennessee, which generates 10 billion kilowatt-hours a year by burning 12 700 metric tons of coal a day, supplying electricity to almost 700 000 homes.

↑ A train carrying coal leaves a mountaintop removal mining site and travels through the backyards of homes in Welch, West Virginia.

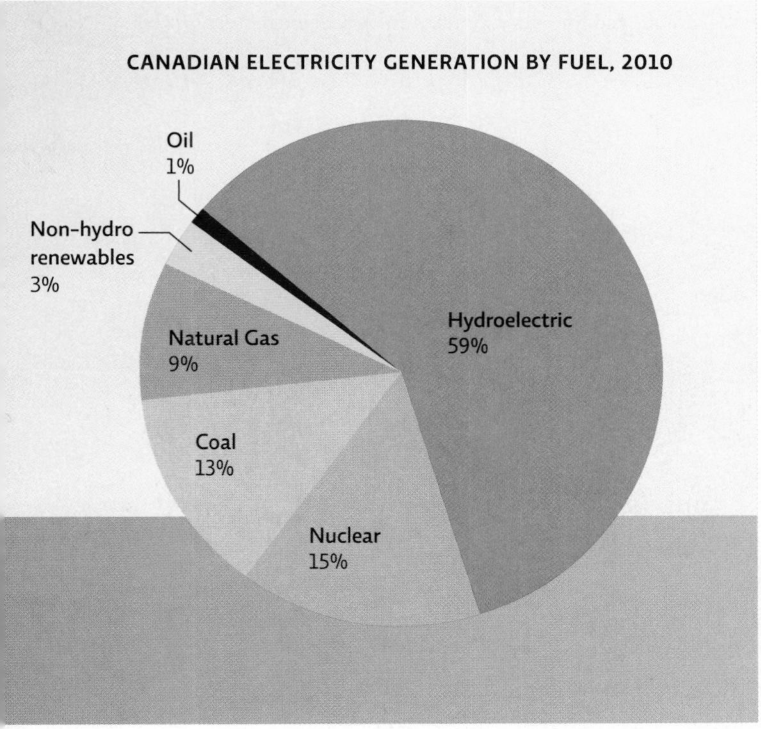
CANADIAN ELECTRICITY GENERATION BY FUEL, 2010
Oil
1%
Non-hydro renewables
3%
Natural Gas
9%
Coal
13%
Nuclear
15%
Hydroelectric
59%

↑ Coal is the main fuel used to produce electricity worldwide but in Canada hydroelectricity dominates. Non-hydro renewables include wind, biomass, solar, and geothermal. Shifting to renewable energy sources and improving energy efficiency and conservation would decrease the role that coal plays in energy production.

source to the amount we must expend to obtain, process, and ship it—coal is neither the best nor the worst. It has an EROEI of about 8:1 (8 units of energy produced for every 1 unit consumed for a net production of 7), compared to 15:1 for oil, 6:1 for nuclear, and 20:1 for wind.

There can be no denying the blessings of coal: this sticky black rock has powered several waves of industrialization—first in Great Britain and the United States, now in China—and in so doing has shaped and reshaped the world as we know it. But as time marches on, the costs of those blessings have become all too apparent: they include an ever-growing list of health impacts—from birth defects to black lung disease—and an equally lengthy roster of environmental costs—not only the destruction of Appalachia, but also the pollution of Earth's atmosphere with CO_2 and other greenhouse gases.

This litany of paradoxes has given rise to a deep national ambivalence in the United States. While U.S. residents are consuming more electricity, and burning more coal, than at any time in history, applications for new coal-fired power plants have been rejected left and right in recent years by determined citizens and local governments. "We're caught in a catch-22," says Scott Eggerud, a forest manager with West Virginia's Department of Environmental Protection. "On one hand, it's like we need the stuff to live; on the other hand, we see that it's kind of killing us." Nowhere is this catch-22 more pronounced than in the foothills of Central and Southern Appalachia.

Coal forms over millions of years.

The Appalachian Mountains were born a few million years before the rise of the dinosaurs, when Greenland, Europe, and North America hovered near the equator as a single giant landmass bathed in a dense tropical swamp. The northwestern bulge of what is now Africa (see Infographic 19.3) pressed into the easternmost edge of North America, pulverizing the colliding continental margins and forcing a colossal mass of land upward, into a mountain range as high as the Himalayas. It was the last and greatest of three violent clashes that joined all the world's land into a single supercontinent, upon which stegosauruses and velociraptors would eventually roam.

The fallout from this cataclysm—the gradual accretion of decomposing swamp vegetation compressed and baked by

electricity The flow of electrons (negatively charged subatomic particles) through a conductive material (such as wire).

energy return on energy investment (EROEI) A measure of the net energy from an energy source (the energy in the source minus the energy required to get it, process it, ship it, and then use it).

Infographic **19.2** | **COAL FORMATION**

↓ Coal is formed over long periods of time as plant matter is buried in an oxygen-poor environment and subjected to high heat and pressure. Places with substantial coal deposits that are retrievable with current technology are called coal reserves.

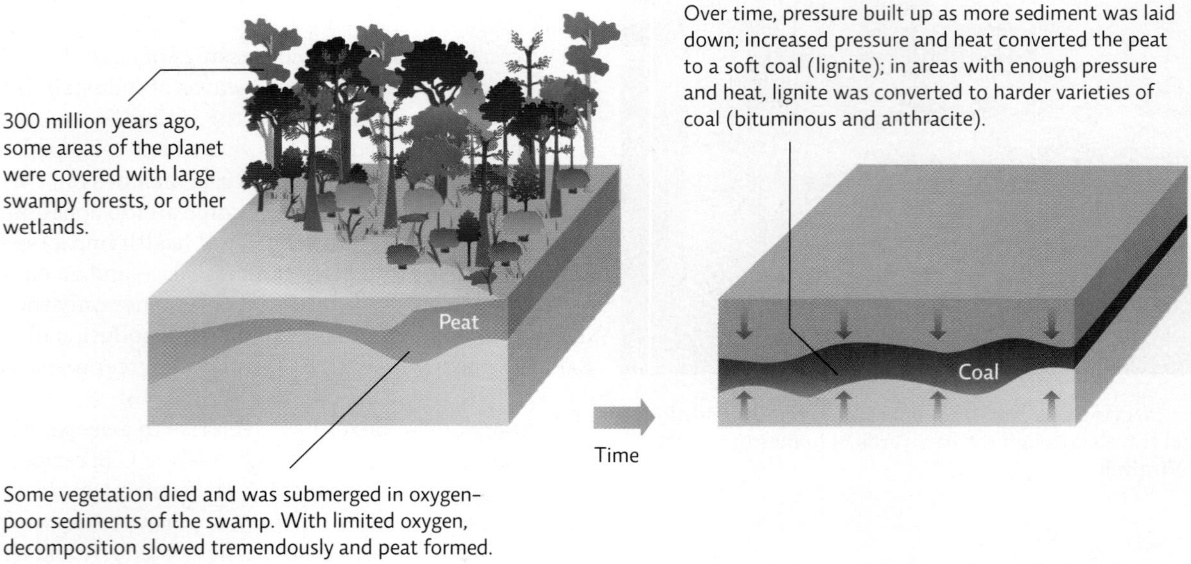
300 million years ago, some areas of the planet were covered with large swampy forests, or other wetlands.
Peat
Some vegetation died and was submerged in oxygen-poor sediments of the swamp. With limited oxygen, decomposition slowed tremendously and peat formed.
Time
Over time, pressure built up as more sediment was laid down; increased pressure and heat converted the peat to a soft coal (lignite); in areas with enough pressure and heat, lignite was converted to harder varieties of coal (bituminous and anthracite).
Coal

↓ Seven metric tons of overburden (the rock, soil, and ecosystem waste) are moved for every metric ton of coal extracted.

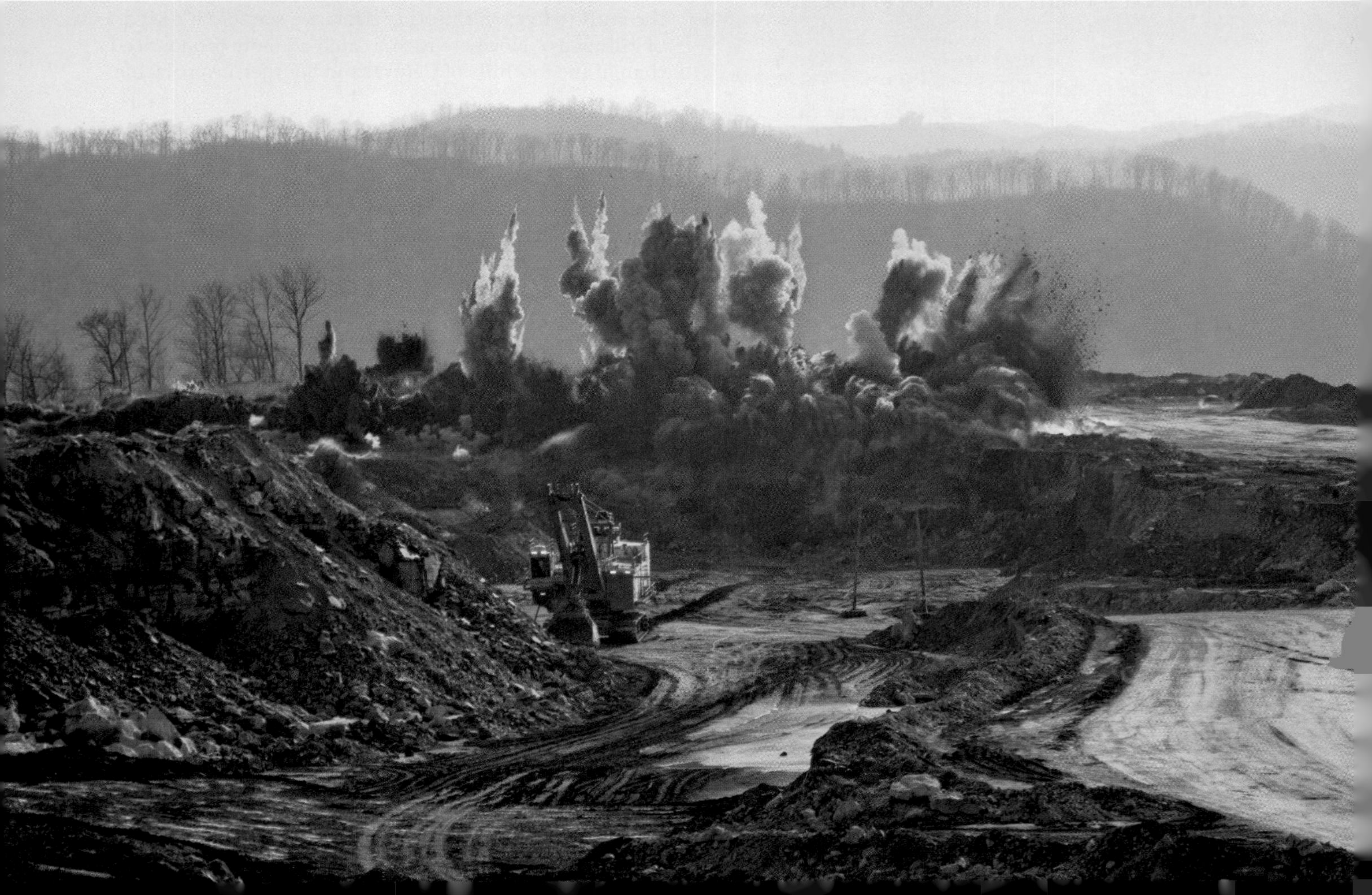

heat and time—established the Appalachian coal beds. As the swamp plants died out, they were buried under a mud so thick it kept oxygen out. Instead of being fully decomposed by bacteria, their remains produced *peat*—a soft mash of partially decayed vegetation. As time passed, and more and more layers of sediment were laid down over the peat, pressure and heat compressed it into the denser rocklike material that we know as coal. [INFOGRAPHIC 19.2]

This same story—tectonic upheaval, followed by deep and rapid burying of organic material, followed by the material's slow compaction into coal—has played out in numerous places across the globe. As a result, coal is found everywhere, though of course, some places have more than others. Europe and Asia hold about 36% of the world's reserves, while the Asian Pacific has about 30%, and North America holds just over 28%. [INFOGRAPHIC 19.3]

The Appalachian beds were once among the most bountiful reserves in the United States; seams as tall as a human adult wound for kilometres through the mountainside and made for easy harvesting. But after 150 or so years of mining, those reserves have dwindled noticeably. At current rates of usage, proven coal reserves (those we know are economically feasible to extract) should last about 120 years—longer if deeper reserves can be accessed. And as the layers of minable coal have grown thinner and harder to reach, the coal industry has become both more sophisticated and more destructive in its approach to extracting the coal.

overburden The rock and soil removed to uncover a mineral deposit during surface mining.

Mining comes with a set of serious trade-offs.

Hobet 21, which has claimed more than 4800 hectares of land, was once the site of several adjacent peaks. To get at the coal beneath those peaks, miners began by clear-cutting the forest above. Next, they drilled holes deep into the side of the mountain (some holes up to 120 metres deep), set dynamite in those holes, and blasted as much as 300 metres of mountain into a mass of rubble known as **overburden**. The miners repeated this process several times, until the layers of coal were exposed. Then, using buckets big enough to hold 20 midsized cars, they scooped that rubble into staircase-shaped mounds that filled in entire neighbouring valleys from the ground up—all told more than 70 million metric tons of overburden

Infographic **19.3** | **MAJOR COAL DEPOSITS OF THE WORLD**

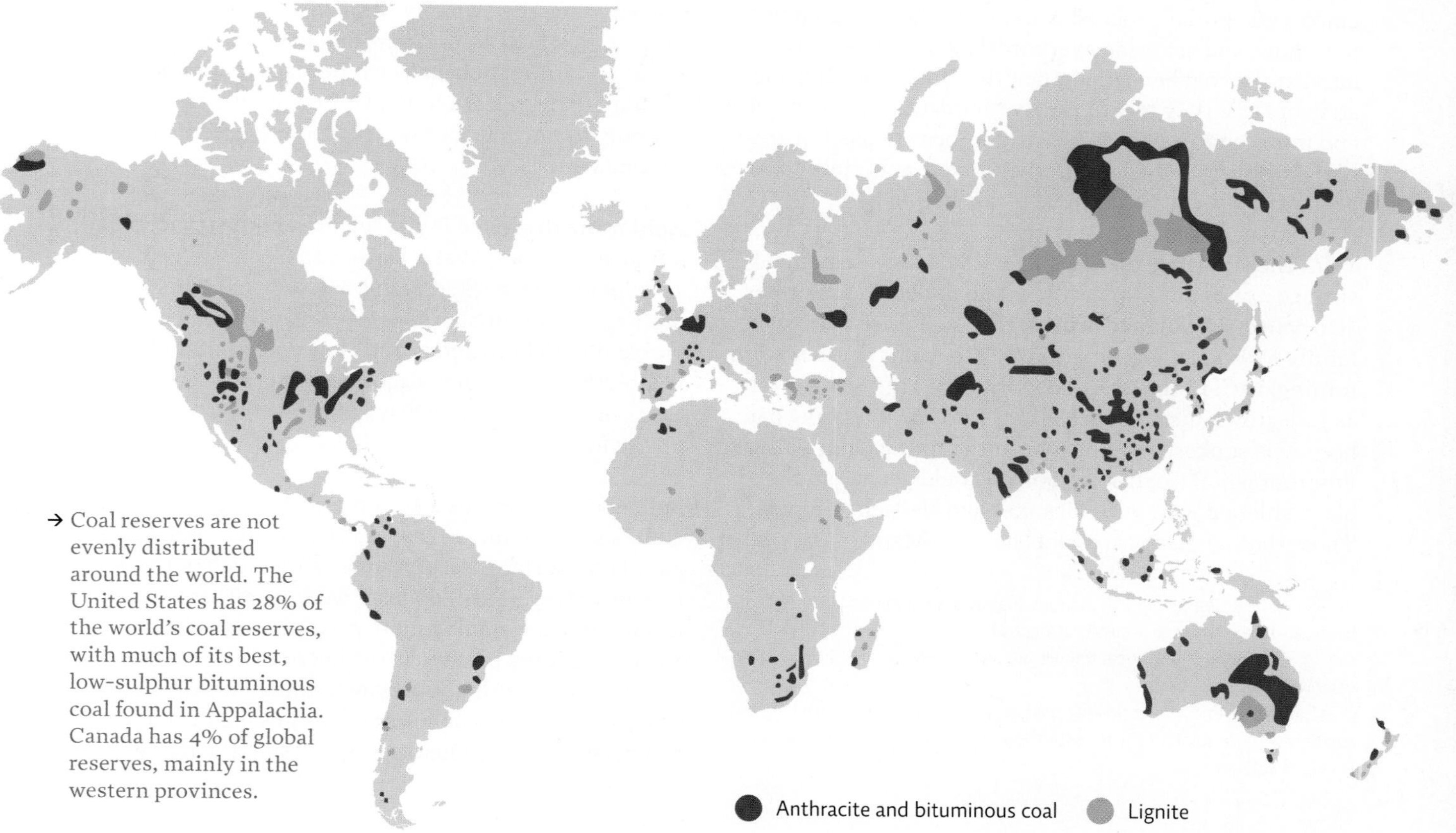
Anthracite and bituminous coal
Lignite

→ Coal reserves are not evenly distributed around the world. The United States has 28% of the world's coal reserves, with much of its best, low-sulphur bituminous coal found in Appalachia. Canada has 4% of global reserves, mainly in the western provinces.

Infographic **19.4** | **MOUNTAINTOP REMOVAL**

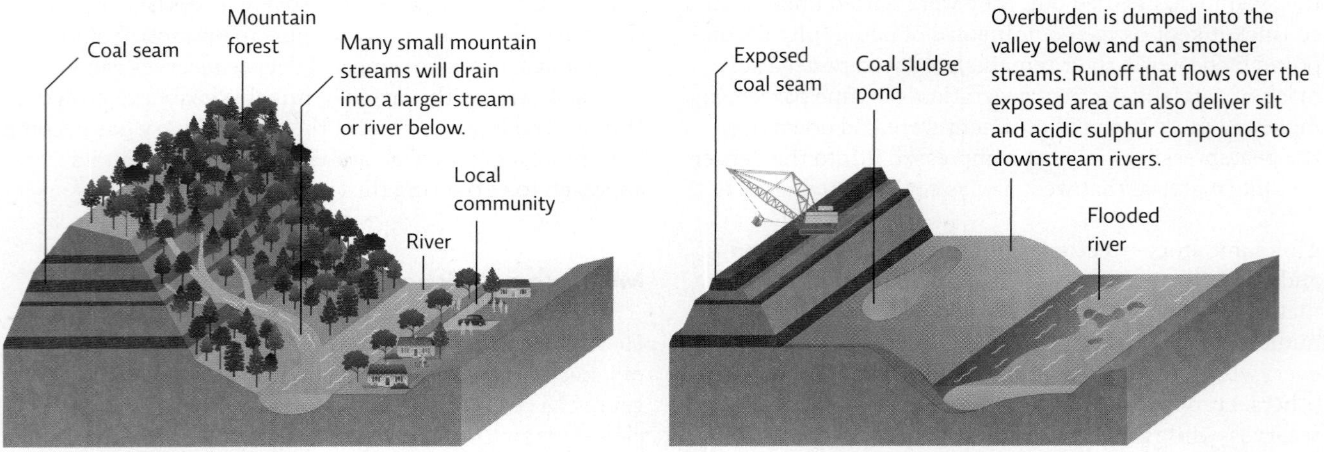
Coal seam
Mountain forest
Many small mountain streams will drain into a larger stream or river below.
Local community
River
Exposed coal seam
Coal sludge pond
Overburden is dumped into the valley below and can smother streams. Runoff that flows over the exposed area can also deliver silt and acidic sulphur compounds to downstream rivers.
Flooded river

↑ Surface mining techniques are used when coal seams are close to the surface. In Appalachia, the forests are first clear-cut and then explosives are used to blast away part of the mountain. Heavy equipment then digs through debris, dumping the overburden (soil and rock) into the nearby valley, burying streams as the valley is filled in. The exposed coal is dug out and some processing is done on site. Coal sludge left over from processing is stored in ponds on the mining site.

each year. The process obliterated the forest habitat, buried countless streams, and permanently reordered the land's natural contours. [INFOGRAPHIC 19.4]

This mountaintop removal mining is just one form of **surface mining**. The other type, known as strip mining, employs a similar process: workers use heavy equipment to remove and set aside overburden so that they can harvest the coal beneath. When they finish mining one strip of land, they return the overburden to the open pit and move on to a new strip. Strip mines are used in areas like Wyoming and Alberta, where the coal is close to the surface and the ground above is fairly level.

With their reliance on explosives and heavy equipment, surface mines are a far cry from the underground mines, also called **subsurface mines,** that sustained the Nelson family for so many generations. "When our daddies were mining, back in the '40s and '50s, the seams were as tall as full-grown men," says Nelson, who since retiring has become a spokesperson for the anti-mining Ohio Valley Environmental Coalition. "So you could get at 'em the old-fashioned way, with pickaxes and sledgehammers." Those days of plenty are gone, he says. Many of the coal seams that remain are too thin to be culled by human hands.

Subsurface mines come with their own challenges. They make up 60% of all coal mines worldwide (50% of those in the United States), but very few have operated in Canada since the Westray Mine disaster in Nova Scotia in 1992, in which an underground methane explosion killed 26 miners. Subsurface mines account for 10% of all methane release in the United States; methane is not only flammable, but is also a potent greenhouse gas [INFOGRAPHIC 19.5].

Acid mine drainage is a problem in both surface and subsurface mines. Water leaches toxic substances such as sulphur from rock, creating acidic pools. This acid drainage contaminates soil and streams and is a major problem with both active and closed mines. Acidic water is directly toxic to many aquatic plants and animals, and alters nutrient cycles in ways that reverberate all the way up the food chain.

But subsurface mines also come with some advantages. Unlike surface mines, they don't disrupt or permanently alter large surface areas. And because much of the work, however risky, is still done by workers rather than by machines, they employ more people. In Appalachia, 100 000 mining jobs were lost between 1980 and 1993 as underground mining gave way to mountaintop removal. The loss of traditional mining jobs has bred yet another controversy. Coal industry reps say that the riskier

surface mining Removing soil and rock that overlays a mineral deposit close to the surface in order to access that deposit.
subsurface mines Sites where tunnels are dug underground to access mineral resources.
acid mine drainage Water flowing past exposed rock in mines, leaching out sulphates. These sulphates react with the water and oxygen to form acids (low-pH solutions).

Infographic **19.5** | **SUBSURFACE MINING**

→ Subsurface mines are used to access many minerals and ores, including coal. Modern mining depends on powerful machinery to drill out tunnels and remove and transport the materials.

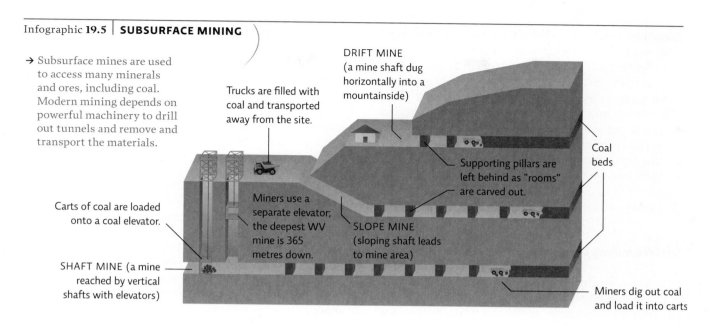

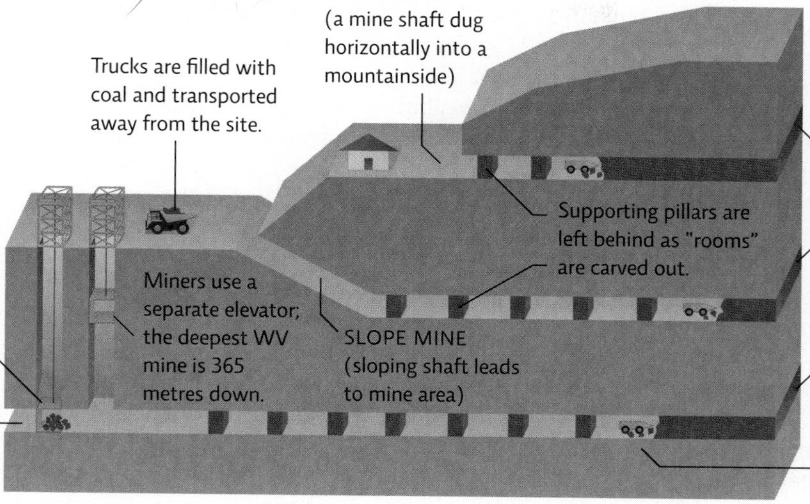
DRIFT MINE
(a mine shaft dug horizontally into a mountainside)
Trucks are filled with coal and transported away from the site.
Coal beds
Supporting pillars are left behind as "rooms" are carved out.
Carts of coal are loaded onto a coal elevator.
Miners use a separate elevator; the deepest WV mine is 365 metres down.
SLOPE MINE
(sloping shaft leads to mine area)
SHAFT MINE (a mine reached by vertical shafts with elevators)
Miners dig out coal and load it into carts

MINING HAZARDS

COAL DUST

Far more miners die from pneumoconiosis (black lung disease) caused by breathing coal dust than from mining accidents. In 2004, the U.S. Centers for Disease Control reported 703 coal miner deaths from pneumoconiosis, compared to 28 accidental deaths.

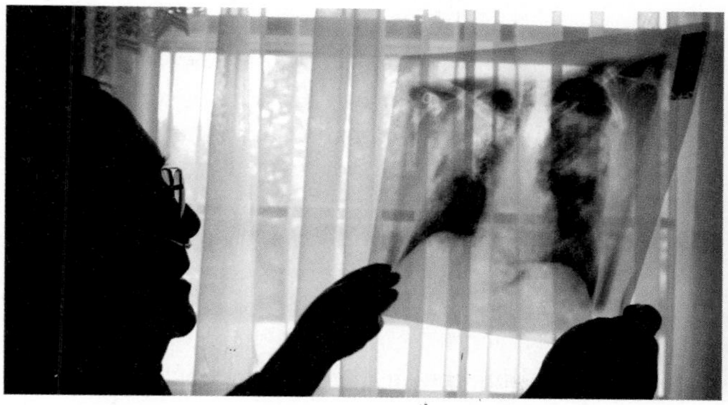

EXPLOSIONS AND MINE COLLAPSE

A coworker sets up miners' gear for coal miners killed or missing in West Virginia's Upper Big Branch Mine explosion in 2010. Methane gas fumes and coal dust are the most common causes of mine explosions; 125 miners died in the Springhill Mine Disaster of 1891 in Nova Scotia, the worst coal-mining accident in Canadian history.

Praying
4 our
Miners

FIRE

Some underground coal mine fires have been burning for hundreds of years.

TOXIC FUMES

A vigil being held for twelve miners who died from carbon monoxide poisoning after an explosion trapped them in the Sago Mine in West Virginia in 2006.

deep-mining jobs are being replaced with higher-paying, safer work—like demolition and heavy equipment operation. But that has not alleviated tension as more jobs are lost than replaced. "It's a double insult," says Tim Landry, a fourth-generation deep miner in West Virginia. "They're not only destroying the land that we love, but they're taking our jobs away, too." In Canada, risky subsurface mining jobs pay around $30/hour, but many mining companies hire foreign workers who are willing to work for lower wages.

But job losses are not the only concern.

Surface mining brings severe environmental impacts.

Bob White is a tiny unincorporated village on the edge of Charleston, West Virginia—just one segment of an endless trickle of trailers and shotgun houses that hug the mountain on each side of long winding valley after long winding valley. Maria Gunnoe, a 40-something waitress, and the daughter, sister, wife, niece, and aunt of coal miners, has lived there, on the same property, all her life. She remembers having free run of the mountain as a child. "When we were kids we used to roam deep in the holler," she says, referring to the mini-valleys that snake through the region's foothills. "We had access to all the resources—food, medicine, water—that these mountains provided."

↑ Maria Gunnoe became an environmentalist after the removal of over 800 hectares as the result of mountaintop mining in her area caused flooding to her home and property, poisoned her well water, and made her daughter sick. Gunnoe received the Goldman Environmental Prize for her organizing efforts in her southern West Virginia community.

Things are different now. Her own children run into big yellow gates and No Trespassing signs wherever they go. The mountains, she says, have been closed off for blasting. And in the past decade, several million metric tons of overburden—an unimaginable mass—have been dumped into the valleys around Bob White.

The upheaval has had a noticeable impact on area residents. For one thing, the loss of forest and the compaction of so much soil has increased both the frequency and severity of flooding (without trees, and with the soil so compressed, the ground can't absorb water). "They've filled in like 15 of the valleys around me, and floods are about 3 times more serious than they ever were before," she says. One 2003 flood nearly swallowed her entire valley—house, barn, family and all.

Floods aren't the only problem. In fact, it's the blasting that most scares Gunnoe. It fills the air with tiny particles of coal dust—easily inhaled and full of toxic substances like mercury and arsenic. Studies show a higher incidence of respiratory illnesses in mining communities. And according to a 2011 study by Melissa Ahern of Washington State University, children in those communities are more likely to suffer a range of serious birth defects, including heart, lung, and central nervous system disorders. "When they were clearing the ridgeline right behind us, we'd get blasted as much as 3 times a day," Gunnoe says. "There were days when we'd have to just stay inside because you couldn't breathe out there. And now, my daughter and I both get nosebleeds all the time."

In addition to filling the air, these toxic substances also permeate the region's rivers, streams, and groundwater. In a 2005 **environmental impact statement** on mountaintop removal mining, the U.S. Environmental Protection Agency (EPA) reported that selenium levels exceed the allowable limits in 87% of streams located downhill from mining operations. Toxic substances were as much as 8 times higher in streams near mined areas than in streams near unmined areas. When tests revealed dangerous levels of selenium in the stream behind Gunnoe's house, she and her neighbours started getting their water from town. "We can't trust the water in our own streams anymore,"

environmental impact statement A document outlining the positive and negative impacts of any action that has the potential to cause environmental damage; used to help decide whether or not that action will be approved.

↑ In 2008, almost 4 billion litres of coal fly ash slurry surged into homes and waterways near Harriman, Tennessee, after breaching a coal ash containment pond. It was the largest fly ash release in U.S. history. Many residents have been permanently displaced.

she says. "It's sad, but this is just not the same place that I grew up in."

On top of all these immediate threats to human health and safety, area residents like Gunnoe worry about long-term damage to the region's natural environment. They are not alone. Around the world, in fact, mining operations are a major cause of environmental degradation. To be sure, geological resources—those like copper, gold, iron ore, and coal, that are buried deep underground and so must be dug up—are essential to the functioning of modern society. But harvesting them unleashes a dangerous mix of toxic substances that are normally sequestered away from functioning ecosystems, and that living things—including humans—are not generally well adapted to handle.

"We can't trust the water in our own streams anymore. It's sad, but this is just not the same place that I grew up in."
—Maria Gunnoe

At surface mines throughout Appalachia, dangerous quantities of iron, aluminum, selenium, and other metals have leached out from the blasted overburden, and may be starting to bioaccumulate and biomagnify up the food chain. Scientists and fishers alike have noted an alarming decline in certain fish that populate area streams and rivers. And recent surveys indicate that biodiversity has decreased in direct proportion to the concentration of such metals in the water. This loss of aquatic life also affects forest life, since many terrestrial animals feed on insects like mayflies and dragonflies that begin their life in the water.

The nutrient cycle in ecosystems surrounding mined areas has also been altered—in some cases, dramatically. Extra sulphates, released from blasted rock, have increased nitrogen and phosphorus availability, and in so doing have led to eutrophication. And as sulphate levels rise, so do populations of sulphate-feeding bacteria. These microbes transform sulphate into hydrogen sulphide, which is toxic to many aquatic plant species.

Meanwhile, overburden has completely buried more than 3200 kilometres of streams, and destroyed an untold range of natural habitats in the process. And destroying habitats invariably threatens local species diversity. Biodiversity in the streams of Appalachia is rivalled only by that in the streams found in rainforests; the diversity of trees is also second only to the tropics. With these ecosystems destroyed on more than 500 mountains in Appalachia, researchers predict the permanent loss of many local plant and animal species.

Infographic **19.6** | **ENVIRONMENTAL AND HEALTH PROBLEMS OF MINING AND BURNING COAL**

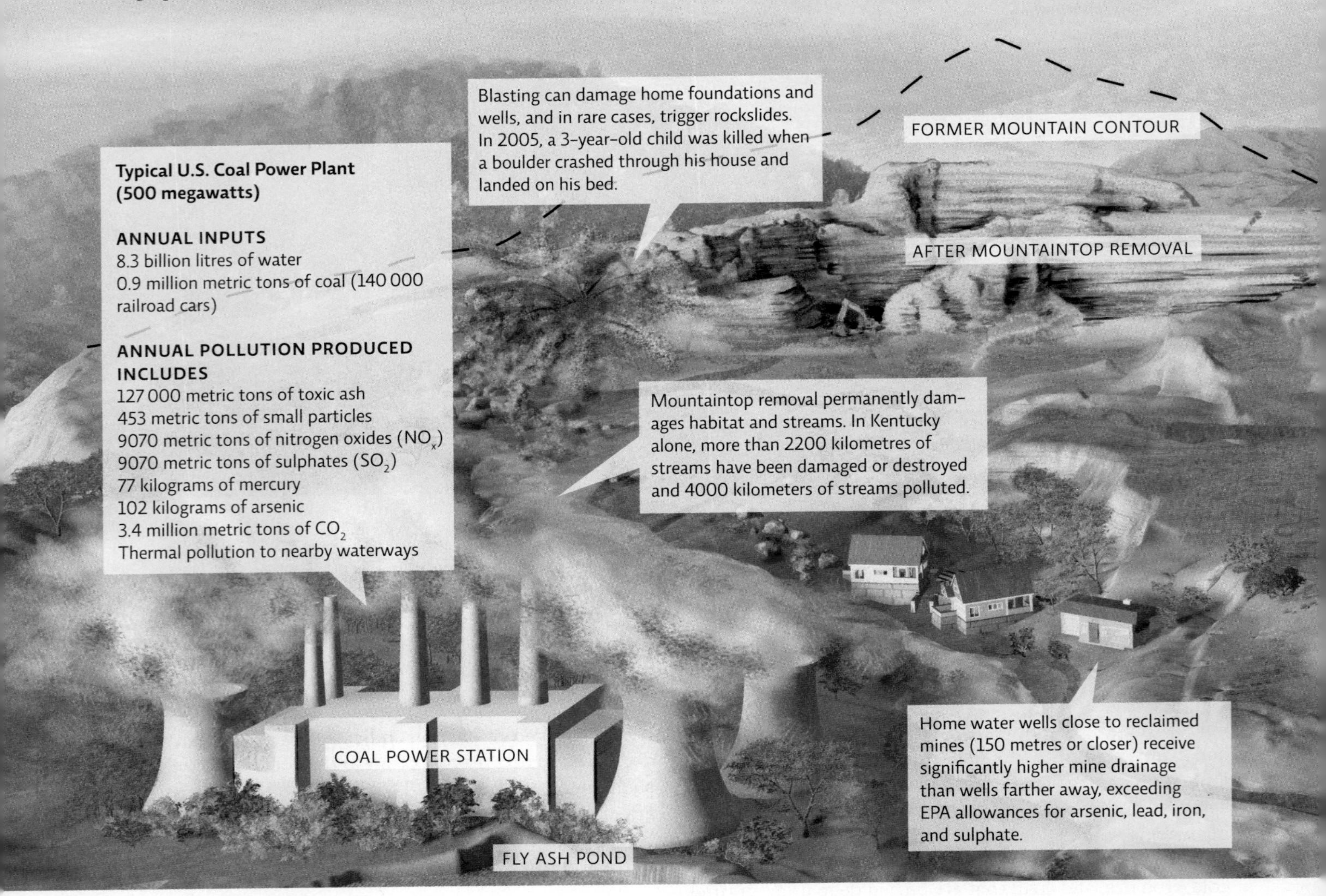

Blasting can damage home foundations and wells, and in rare cases, trigger rockslides. In 2005, a 3-year-old child was killed when a boulder crashed through his house and landed on his bed.
FORMER MOUNTAIN CONTOUR
AFTER MOUNTAINTOP REMOVAL
Typical U.S. Coal Power Plant (500 megawatts)
ANNUAL INPUTS
8.3 billion litres of water
0.9 million metric tons of coal (140 000 railroad cars)
ANNUAL POLLUTION PRODUCED INCLUDES
127 000 metric tons of toxic ash
453 metric tons of small particles
9070 metric tons of nitrogen oxides (NO_x)
9070 metric tons of sulphates (SO_2)
77 kilograms of mercury
102 kilograms of arsenic
3.4 million metric tons of CO_2
Thermal pollution to nearby waterways
Mountaintop removal permanently damages habitat and streams. In Kentucky alone, more than 2200 kilometres of streams have been damaged or destroyed and 4000 kilometers of streams polluted.
COAL POWER STATION
Home water wells close to reclaimed mines (150 metres or closer) receive significantly higher mine drainage than wells farther away, exceeding EPA allowances for arsenic, lead, iron, and sulphate.
FLY ASH POND

Slurry impoundments—reservoirs of thick black sludge that accompany each mining operation—are among the most controversial features of the landscape created by mountaintop removal. They too are a consequence of thinner coal seams. "The thinner seams are messier," says Randal Maggard, a mining supervisor at Argus Energy, a company that has several mining operations throughout Appalachia. "It's 6 inches of coal, 2 inches of rock, 8 inches of coal, 3 inches of rock, and so on. It takes a lot more work to process coal like that." To separate the coal from the rock, miners use a mix of water and magnetite powder known as slurry that coal can float in. Once the sulphur and other impurities have been washed out, the coal is sent for further processing, and the slurry by-product is pumped into artificial holding ponds.

Maggard insists the impoundments are safe. "Before we fill it, we have to do all kinds of drilling to test the bedrock around it," he says. "And then a whole slew of chemical tests on top of that—all to make sure the barrier is impermeable." Still, area residents worry about a breech. With good reason.

In October 2000, the dike holding back the slurry at one of these ponds failed, pouring more than 1 million kilolitres (30 times the Exxon Valdez spill) of toxic sludge into the Big Sandy River of Martin County, Kentucky. The contamination killed all life in some streams and eventually reached the Ohio River, more than 32 kilometres away. The sludge was 1.5 metres deep in some places. The EPA would register it as one of the worst environmental disasters in the eastern United States' history. Eleven years later, sludge can still be found a few centimetres under the river sediments.

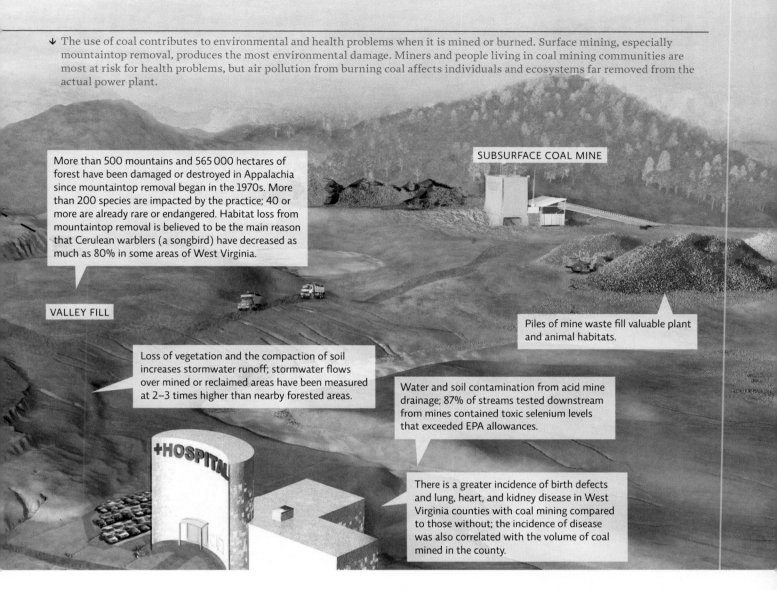

↓ The use of coal contributes to environmental and health problems when it is mined or burned. Surface mining, especially mountaintop removal, produces the most environmental damage. Miners and people living in coal mining communities are most at risk for health problems, but air pollution from burning coal affects individuals and ecosystems far removed from the actual power plant.

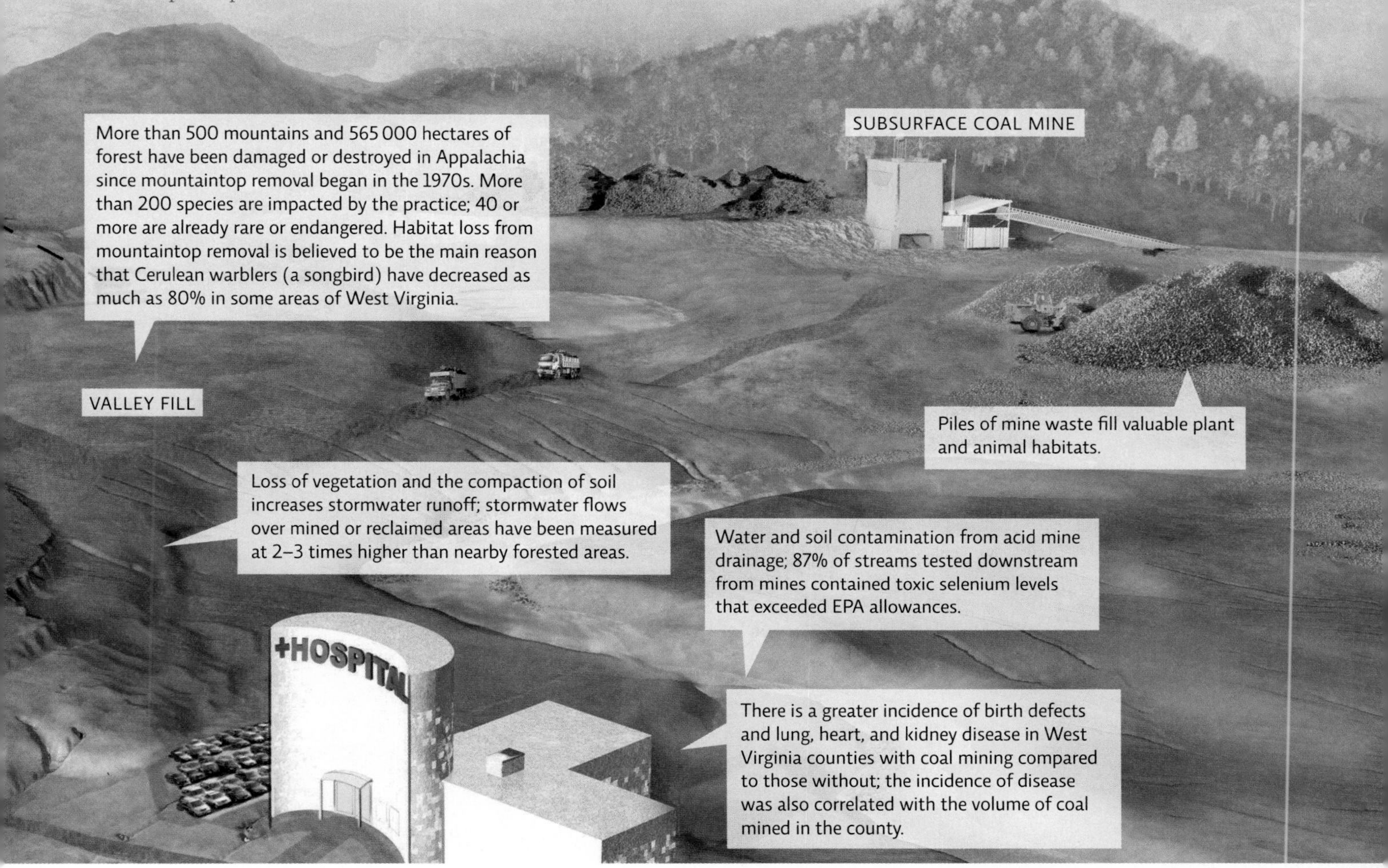
More than 500 mountains and 565 000 hectares of forest have been damaged or destroyed in Appalachia since mountaintop removal began in the 1970s. More than 200 species are impacted by the practice; 40 or more are already rare or endangered. Habitat loss from mountaintop removal is believed to be the main reason that Cerulean warblers (a songbird) have decreased as much as 80% in some areas of West Virginia.
SUBSURFACE COAL MINE
VALLEY FILL
Piles of mine waste fill valuable plant and animal habitats.
Loss of vegetation and the compaction of soil increases stormwater runoff; stormwater flows over mined or reclaimed areas have been measured at 2–3 times higher than nearby forested areas.
Water and soil contamination from acid mine drainage; 87% of streams tested downstream from mines contained toxic selenium levels that exceeded EPA allowances.
+HOSPITAL
There is a greater incidence of birth defects and lung, heart, and kidney disease in West Virginia counties with coal mining compared to those without; the incidence of disease was also correlated with the volume of coal mined in the county.

Can coal's emissions be cleaned up?

Of course it's not just the mining and processing of coal that pollutes the environment.

When coal is burned to produce heat energy, it releases a range of toxic substances that damage the environment and threaten human health: gases (sulphur dioxide, carbon monoxide, nitrogen oxide, and planet-warming carbon dioxide), heavy metals (such as mercury and arsenic), radioactive material (uranium and thorium), and particulate matter (soot) with particles small enough to irritate lung tissue or even enter the bloodstream if inhaled. The EPA's new Mercury and Air Toxic standards of 2011 will reduce emissions of some of the less well-addressed toxic substances by 90% and will save up to $90 billion in human health costs by 2016 (while only costing $9.6 billion to implement).

Coal-fired power plants also generate tons of toxic fly ash—fine ashen particles made up mostly of silica. Some of that ash is diverted to industry, where it's used in concrete production. But the majority of it is buried in hazardous waste landfills or stored in open ponds like the ones used to contain the slurry waste from mining. In 2008, following a heavy rain event, a 15-metre-tall dike from a pond holding coal fly ash from the TVA Kingston Fossil Plant in Tennessee failed, releasing more than 4 billion litres of fly ash into nearby rivers and coating vast expanses of riverside land. The EPA estimates that cleanup will cost $1.2 billion—more when property damage and lawsuit settlements are factored in. Federal regulators have since identified more than 100 other U.S. ash ponds at risk for breach. [INFOGRAPHIC 19.6]

In 2011, public health researchers from Harvard University tallied up the costs of mining and using coal. The analysis

Infographic **19.7** | **CARBON CAPTURE AND SEQUESTRATION (CCS)**

↓ The capture of CO_2 and subsequent storage that prevents it from reentering the atmosphere will greatly decrease coal's contribution to global climate change. A variety of CCS methods are currently in research and development or are being tested as pilot programs.

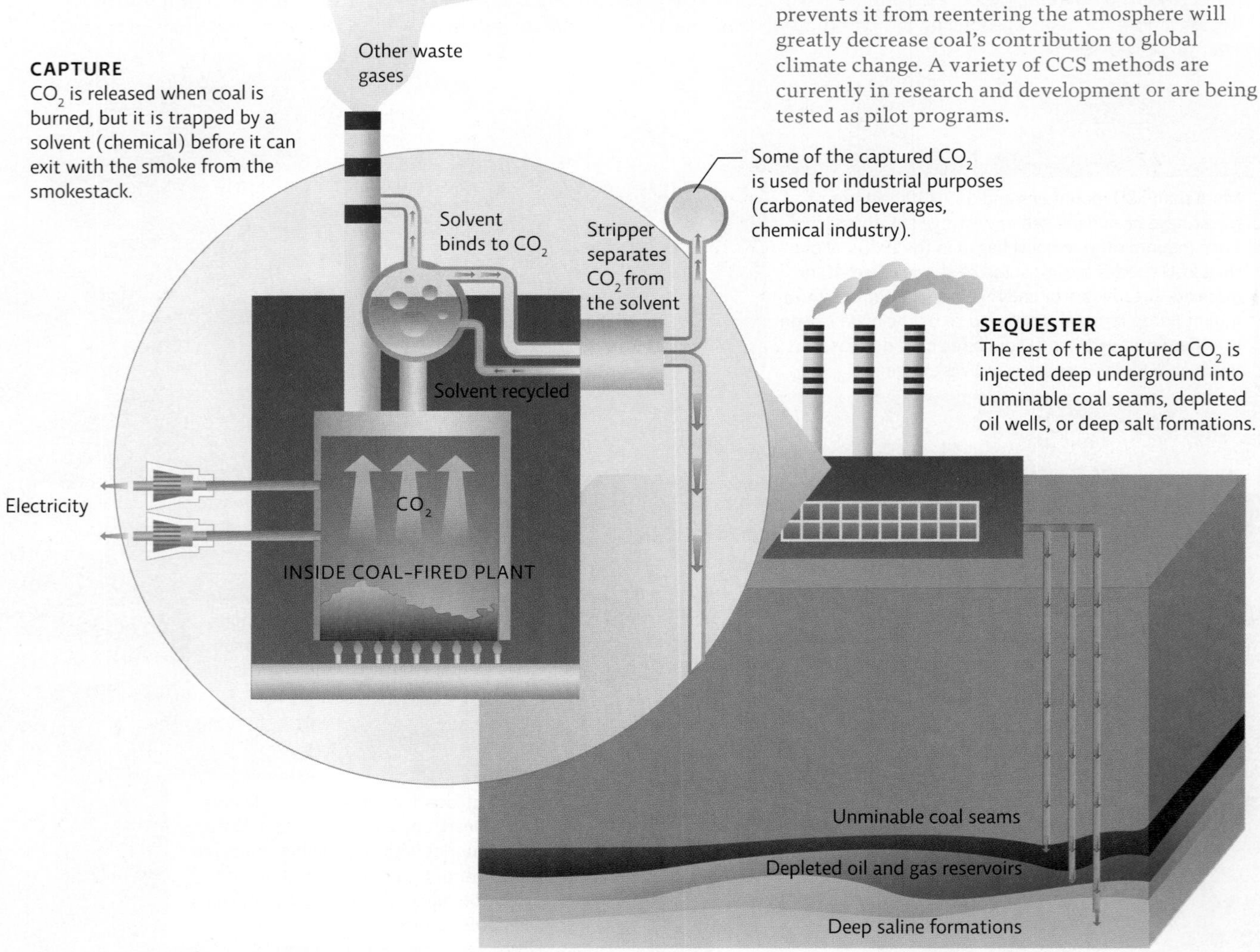
CAPTURE
CO_2 is released when coal is burned, but it is trapped by a solvent (chemical) before it can exit with the smoke from the smokestack.
Other waste gases
Solvent binds to CO_2
Stripper separates CO_2 from the solvent
Some of the captured CO_2 is used for industrial purposes (carbonated beverages, chemical industry).
Solvent recycled
Electricity
CO_2
INSIDE COAL-FIRED PLANT
SEQUESTER
The rest of the captured CO_2 is injected deep underground into unminable coal seams, depleted oil wells, or deep salt formations.
Unminable coal seams
Depleted oil and gas reservoirs
Deep saline formations

took into account many (but not all) of the external costs from mining, shipping, burning, and waste production—that is, the costs that are not currently reflected in the market cost of coal. They estimated that coal costs the U.S. public between \$300 and \$500 billion a year in externalized costs (health, environmental, and property costs). This amounts to an extra \$0.10 to \$0.26 per kWh in external costs—in some cases more than twice what consumers actually pay.

So what can be done? On the one hand, coal provides necessary power. On the other, the processes of mining and then burning it are harming people and the environment as much as they are sustaining lifestyles.

One potential solution is clean coal technology—technology that minimizes the amount of pollution produced by coal. For example, scientists and engineers around the world are working on ways to capture the gases emitted from burning coal. We already have the capacity to capture some emissions like particulate matter and sulphur (see Chapter 21).

The next big challenge is **carbon capture and sequestration (CCS)**—the capture and storage of CO_2 in a way that prevents it from reentering the atmosphere. There are many approaches to CCS; some of them promise to capture more than 90% of CO_2 emissions from coal-fired power plants. But most are still in the research and development stage, and so far, progress has been hindered by the cost of capturing the carbon and by a dearth of good storage options. The extra energy needed to implement CCS must also be considered: estimates for the additional

---

carbon capture and sequestration (CCS) Removing carbon from fuel combustion emissions or other sources and storing it to prevent its release into the atmosphere.

energy needed range from 25% to 40%—this means 25% to 40% more coal that must be mined, processed, transported, and burned. [INFOGRAPHIC 19.7]

Another emerging clean coal technology involves chemically removing some of coal's contaminants before burning it. While it still contains some toxic substances, the final product (a liquid or gaseous fuel, depending on the process used), is cleaner than the original coal. Of course, this process requires energy (thus lowering coal's EROEI) and generates hazardous waste (from the toxic substances found in the coal) that still has to be dealt with.

In fact, as critics are quick to point out, coal can never be truly clean. Each of these "clean coal" technologies produces its own toxic by-products, and none of them eliminates the need to mine coal from deep within Earth's surface.

But if we can't easily avoid using coal, or make it totally clean, we can certainly try to use less of it. One way to accomplish this is to design and use more energy-efficient appliances. Less than 40% of the energy from coal is converted to electricity, but the amount of that energy that goes toward powering appliances depends on how efficient those appliances are. We can also reduce our coal consumption through simple conservation efforts like turning off lights and electronics when we're not using them. Every kilowatt-hour of electricity conserved means roughly 0.45 kilograms less of coal being burned. And a growing number of public utilities now offer "green power" programs in which users can choose to buy electricity that has been generated by solar or wind power instead of coal (for more on energy efficiency and conservation, see Chapter 24).

Reclaiming closed mining sites helps repair the area but can never recreate the original ecosystem.

In some ways, closed mines—those where all the coal has been harvested—look even more alien than active sites like Hobet 21. Instead of the natural sweep of rolling hills, staircase-shaped mounds covered with what looks like lime-green spray paint join one mountain to the next. Atop some of them, gangly young conifers, evenly spaced, strain toward the Sun. The spray paint is really hydroseed, and most of the gangly trees are loblolly pine,

↓ A valley in West Virginia after mining shows none of the original forest, ridges, or streams that were once found there. The grass is grown from "hydroseed," a seed–chemical mixture sprayed onto surfaces to bind the soil to prevent erosion. When hydroseed is applied to rock (often done in highway construction), it grows into a plush lawn of grass that dies after a month or two.

Infographic **19.8** | **MINE SITE RECLAMATION**

↓ U.S. federal law requires that after ceasing operations, the land used for surface coal mines must be restored to close to its original state. However, mining sites never truly or fully recover. After the coal is removed, the area is recontoured to produce a slope and grasses are planted. To replace the mountain streams that were buried by the mining removal process, new channels are constructed to accommodate water flow—but these channels do not support the diverse biological communities that once existed. No matter how much care goes into reclamation, the area will not support a mountain forest community like it once did. However, if done well, a new ecological community may develop.

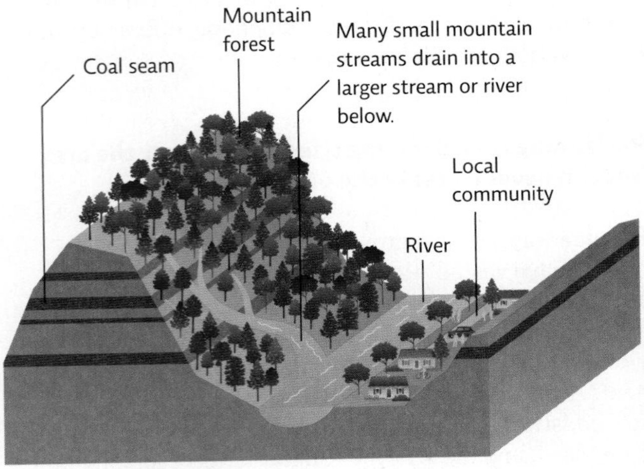
Mountain forest
Many small mountain streams drain into a larger stream or river below.
Coal seam
Local community
River

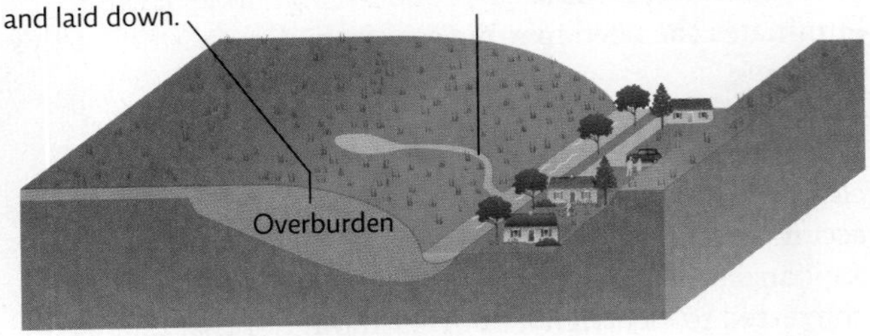
Overburden that was dumped into the valley below is smoothed over to produce a slope. Grasses are planted. Reforestation projects are underway to increase tree plantings in reclaimed areas, but trees do not grow well unless the soil was properly prepared and laid down.
The original mountain streams are gone for good; channels are dug to deliver water down the slope.
Overburden

Reclaimed land may be suitable for other land uses such as agricultural, residential, commercial, or industrial, or can be set aside as a natural area. While there have been some successes on smaller-level reclamation sites, more complex reclamation procedures are required to address mountain sites. Projects such as the Kempton Project in Tucker County, West Virginia, that reclaimed 24 hectares at a cost of $2.3 million, are rare but show that progress is being made. At Kempton, toxic deposits were removed, native trees and grasses were replanted, streams were relocated, and natural wetlands were installed to neutralize acids from two seeps connected to the former mine site. Though not the mountain it once was, the area is considered a success for mountaintop removal reclamation.

a non-native hybrid that foresters are trying to grow in the region.

Such efforts represent the coal industry's attempt to honour the U.S. Surface Mining Control and Reclamation Act, which in 1977 mandated that areas that have been surface mined for coal be "reclaimed" once the mine closes. **Reclamation** requires that the area be returned to a state close to its pre-mining condition.

At some surface mines, at least, reclamation is straightforward (albeit labour-intensive) work: if the mined area was originally fairly flat, the reclamation process includes filling the site with the overburden and contouring the site to match the surrounding land. This relaid rock is then covered with topsoil saved from the original dig. Sometimes, alkaline material such as limestone powder is sprinkled overtop to neutralize acids that have leached into the soil. Vegetation, usually grass, is then planted, leaving other local vegetation to move in on its own. According to the Mineral Information Institute, more than 1 million hectares of coal-mined area have been successfully reclaimed using this strategy.

But reclamation has always been a controversial idea. For one thing, the Appalachian forests were created over eons. In fact, in some ways, they were born of the same calamity that laid the coal beneath the mountains. As the supercontinent drifted, carrying the Appalachian range well north of the equator, a temperate hardwood forest replaced the swamps and grew over time into one of the most biologically diverse ecosystems on the planet. In some areas, a single mountainside may host more tree species than can be found in all of Europe, not to mention songbirds, snails, and salamanders that exist nowhere else on Earth.

For another thing, rebuilding a mountain is considerably more difficult than filling in a strip mine where

reclamation Restoring a damaged natural area to a less damaged state.

the original land was relatively flat. Critics say that so far, there's no evidence that a site as expansive and as thoroughly destroyed as Hobet 21 could ever be truly reclaimed. In one survey, Rutgers ecologist Steven Handel found that trees from neighbouring remnant forests did not readily move in to recolonize mountaintop removal sites, largely because of problems with the soil: at some mines, there was simply not enough of it; at others, it has not been packed densely enough for trees to take root. Those problems have technical solutions, Handel writes, but so far, the seemingly simple act of changing reclamation protocols has been stymied by politics.

And trees are just one facet of reclamation. What about all those streams that were buried? The 1973 U.S. Clean Water Act prohibits the discharge of materials that bury a stream or, if unavoidable, requires mitigation practices that return the stream close enough to its original state such that the overall impact on the stream ecosystem is "non-significant." Both Canada and the United States have federal regulations that require stream restoration. U.S. industry reps argue that they are indeed working to rebuild streams: once the overburden has been reshaped and smoothed over, they dig drainage ditches and line them with stones in a way that resembles a stream or river. But so far, research shows—and most ecologists agree—that such channels don't perform the ecological functions of a stream. "They may look like streams," says ecologist Margaret Palmer, from the University of Maryland. "But form is not function. The channels don't hold water on the same seasonal cycle, or support the same aquatic life, or process contaminants out of the water—all things a natural stream does." [INFOGRAPHIC 19.8]

In Appalachia, the arguments over when and where and how to mine for coal are quickly boiling down to a single intractable question: once it's all gone, how will we clean up the mess we've made? For a story that has played out over geologic time, the question is more immediate than one might think. In West Virginia, coal reserves are expected to last another 50 years, at best. That means no matter what regulations the government imposes, or what methods the coal companies resort to, the day of reckoning will soon be upon us.◉

Select references in this chapter:

Ahern, M., *et al.* 2011. *Environmental Research,* 111: 838–846.

Epstein, P.R., *et al.* 2011. *Annals of The New York Academy Of Sciences: Ecological Economics Review,* 1219: 73–98.

Handel, S.N. 2002. *EIS Technical Study Project for Terrestrial Studies.*

Hendryx, M., and Ahern, M. 2008. *American Journal of Public Health,* 98: 669–671.

Pond, G., *et al.* 2008. *Journal of the North American Benthological Society.* 27: 717–737.

U.S. Environmental Protection Agency. 2005. *Final Programmatic Environmental Impact Statement on Mountaintop Mining/Valley Fills in Appalachia.* Philadelphia, PA: EPA.

Weakland, C.A., and Wood, P.B. 2005. *The Auk,* 122: 497–508.

BRING IT HOME

➲ PERSONAL CHOICES THAT HELP

Although coal is one of the most abundant fossil fuels, its drawbacks are significant. They include CO_2 emissions; the release of air pollutants that cause environmental problems such as acid rain; health problems such as asthma and bronchitis; and massive environmental damage from the mining process. One way to minimize the impact of coal is to reduce consumption of electricity.

Individual Steps

→ Always conserve energy at home and at the workplace.

- Turn off or unplug electronics when not in use.
- Put outside lights on timers or motion detectors so that they only come on when needed.
- Dry clothes outdoors in the sunshine.
- Turn the thermostat up or down a couple of degrees in summer and winter to save energy and money.

Group Action

→ Organize a movie screening of *Coal Country* or *Kilowatt Ours,* which present issues related to coal mining and mountaintop removal from many perspectives.

Policy Change

→ The Appalachian Regional Reforestation Initiative is a great example of how groups, sometimes with very different objectives, can work toward a common goal. Go to arri.osmre.gov to see how this coalition of the coal industry, citizens, and government agencies are working to restore forest habitat on lands used for coal mining.

→ Visit http://content.sierraclub.org/coal/ to find out about events in various states and for the opportunity to weigh in on the decommissioning of outdated coal power plants and the building of new ones in the United States.

UNDERSTANDING THE ISSUE

CHECK YOUR UNDERSTANDING

1. **Which of the following is NOT true about coal?**
   a. It produces more air pollution than other fossil fuels.
   b. It is difficult to ship.
   c. Extraction is relatively cheap.
   d. The United States has an abundant supply of it.

2. **The purpose of mountaintop removal is to:**
   a. promote housing development.
   b. decrease obstacles in airline flight paths.
   c. gain surface access to coal.
   d. promote the regrowth of prairie habitat.

3. **Emissions from a coal-fired power plant include all of the following EXCEPT:**
   a. oxygen.
   b. carbon dioxide.
   c. mercury.
   d. arsenic.

4. **How do miners separate coal from rock?**
   a. By dissolving the rock with acid
   b. By comparing the colours
   c. By checking to see whether the material floats
   d. By looking for bright, reflective surfaces

5. **How long are coal reserves expected to last in West Virginia?**
   a. 50 years
   b. 100 years
   c. 500 years
   d. Forever

6. **Which of the following is an example of reclamation?**
   a. Reworking coal mines to extract more coal
   b. Replanting the site of mountaintop removal with grass and pine trees
   c. Returning land to the people who originally owned it
   d. Filling in a subsurface mine

WORK WITH IDEAS

1. In your own words, describe the process of coal formation. Why is coal a finite resource?

2. What are the pros and cons of the two types of surface mining? Include in your answer a comparison of the extraction process, the energy return on energy investment (including external costs), and the waste-disposal process.

3. Define and describe the process of carbon capture and sequestration. Why is this process necessary? What are the costs and benefits of the process?

4. What are the environmental consequences of surface mining? Use specific examples from the chapter.

5. Describe the process of subsurface mining. What are the costs and benefits? Why do some miners prefer a subsurface mine to a surface mine, while mining companies prefer surface mines?

6. Define reclamation. Explain why, in your own words, reclamation is a controversial idea. Are there examples in which reclamation has succeeded?

ANALYZING THE SCIENCE

The following graphs depict U.S. fatalities in coal mining and other industries.

INTERPRETATION

1. In Graphs A and B, why are both the absolute number of fatalities and the rate of fatalities presented? Which is the fairest measure to use?

2. Based on Graphs A and B, would you rather work in the coal industry or the metal/nonmetal industry? Why?

3 Why do the numbers in Graph B appear to be larger than those in Graph A? For example, compare the numbers for 1968 in both graphs. What conclusion can you make about graphing data in general?

ADVANCE YOUR THINKING

4. The graphs represent fatality data. Do you know what the fatalities are the result of, based on the graph? What do you assume these fatalities are related to? In your opinion, based solely on these numbers, would you risk being a miner, provided the pay and benefits were good?

Graph A

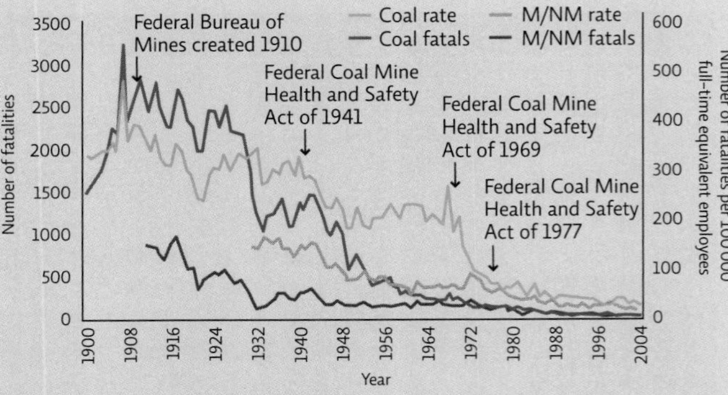
Federal Bureau of Mines created 1910
Coal rate
Coal fatals
M/NM rate
M/NM fatals
Federal Coal Mine Health and Safety Act of 1941
Federal Coal Mine Health and Safety Act of 1969
Federal Coal Mine Health and Safety Act of 1977
Number of fatalities
Number of fatalities per 100 000 full-time equivalent employees
3500
3000
2500
2000
1500
1000
500
0
600
500
400
300
200
100
0
1900
1908
1916
1924
1932
1940
1948
1956
1964
1972
1980
1988
1996
2004
Year

Graph B

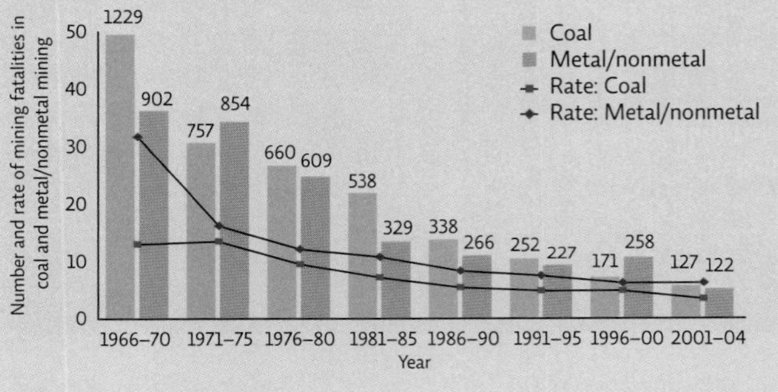
Coal
Metal/nonmetal
Rate: Coal
Rate: Metal/nonmetal
Number and rate of mining fatalities in coal and metal/nonmetal mining
50
40
30
20
10
0
1229
902
757
854
660
609
538
329
338
266
252
227
171
258
127
122
1966–70
1971–75
1976–80
1981–85
1986–90
1991–95
1996–00
2001–04
Year

Graph C

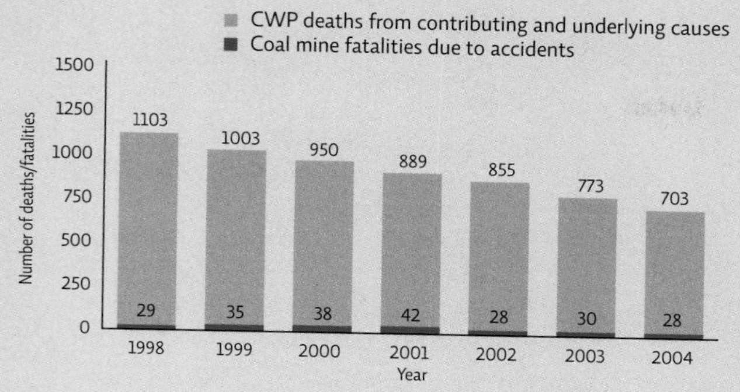

CWP deaths from contributing and underlying causes
Coal mine fatalities due to accidents
1500
1250
1000
750
500
250
0
Number of deaths/fatalities
1103
1003
950
889
855
773
703
29
35
38
42
28
30
28
1998
1999
2000
2001
2002
2003
2004
Year

5. Now, examine Graph C. It depicts the same data as that shown in Graphs A and B, for the years 1998–2004, but shows the deaths (in numbers) due to accidents AND coal workers' pneumoconiosis (CWP, also called black lung disease).
   a. What is one difference in the way these data are reported compared to how they are reported in Graphs A and B?
   b. If your only employment choice were to work in a coal mine, what else would you like to know about the data in Graph C?

6. Based on the information in Graph A, what can you say about fatalities and federal regulations? Do you think it would be safe to dispense with federal regulations and allow mine owner/operators to determine safety standards and practices, now that they know the benefits of safety regulations?

EVALUATING NEW INFORMATION

In the United States, the EPA is responsible, in part, for issuing mining permits, based on the predicted environmental damage from the mines. As with many issues in environmental science, there is often disagreement between government regulators and companies about the extent of environmental damage. Federal regulators argue that some practices are so harmful to the environment that they should be halted; companies argue that halting certain practices will result in severe economic loss.

Go to http://is.gd/ELKPBA. Click on the "Listen to the Story" link, review the photo gallery, and read the story about mountaintop removal.

Evaluate the website and work with the information to answer the following questions:

1. Is this a reliable information source? Does it have a clear and transparent agenda?
   a. Who runs this website? Do this group's credentials make it reliable or unreliable? Explain.
   b. Is the information on the website up to date? Explain.
2. When was this story aired and on what program? What type of information is reported in this story?
3. Now, read an ABC News story about mountaintop removal at http://is.gd/NQMsJH.
   a. Who runs this website? Is it comparable to the first website? Why or why not?
   b. Compare the content of the two stories. Which do you think provides more in-depth coverage of the issue?
   c. Now look for links to other stories about mountaintop removal. Do both sites have links to other websites that you could use to learn more if you wished? Are the links more useful or relevant on one site versus the other?
4. One of the links provided on the ABC site is to Massey Energy, which declined to be interviewed for the story.
   a. Why do you think Massey Energy would decline to be interviewed?
   b. Do a brief Internet search for Massey Energy. What types of stories appear? Do a brief review of the stories. Do these stories give you any more insight into Massey Energy's motives for declining an interview?

MAKING CONNECTIONS

DO WE NEED MORE REGULATIONS?

Background: Coal combustion is a large contributor of greenhouse gases, especially carbon dioxide. Burning coal also releases other compounds into the air, including pollutants such as mercury and sulphur. A partial solution to this problem is to use clean coal technology.

Case: You are an assistant to a provincial/territorial environment minister. Part of your job is to provide the minister with up-to-date information about energy issues. The Canadian Council of Ministers of the Environment (CCME) has been working on developing standards for regulating emissions of mercury (Hg) from coal-fired plants. The companies that already produce power with cleaner fuel (that is, not coal) are in favour of implementing these standards. Companies that primarily use coal are not. You've been asked to research the new standards to help your minister draft a public position statement.

1. As part of your research, find out:
   a. What is the purpose (both immediate and long term) of the new standards?
   b. How long have the new standards been under review and revision?
   c. How will the new standards affect the energy companies in your minister's voting area? (Assume the area is where you are attending school.)
2. Next, find out what opponents and proponents of the new standards predict about the long-term consequences of such standards. Which viewpoint do you think is most similar to those your minister will have? Explain.
3. As a citizen, independent of your role as a ministerial assistant, what is your response to the proponents of the new standards? The opponents? When drafting your response, think about what you know about the history of industry, Environment Canada, Natural Resources Canada, and relevant provincial/territorial agencies, and existing regulations, especially as they relate to coal and current environmental and economic conditions.

The Suncor oil sands operation. The world's largest reserves of oil sands are in Canada and Venezuela. Oil sands could equate to approximately two-thirds of the total global oil resource. There is explosive growth in the oil field areas around Fort McMurray, Alberta.

CHAPTER 20 OIL AND NATURAL GAS

SANDS OF TIME

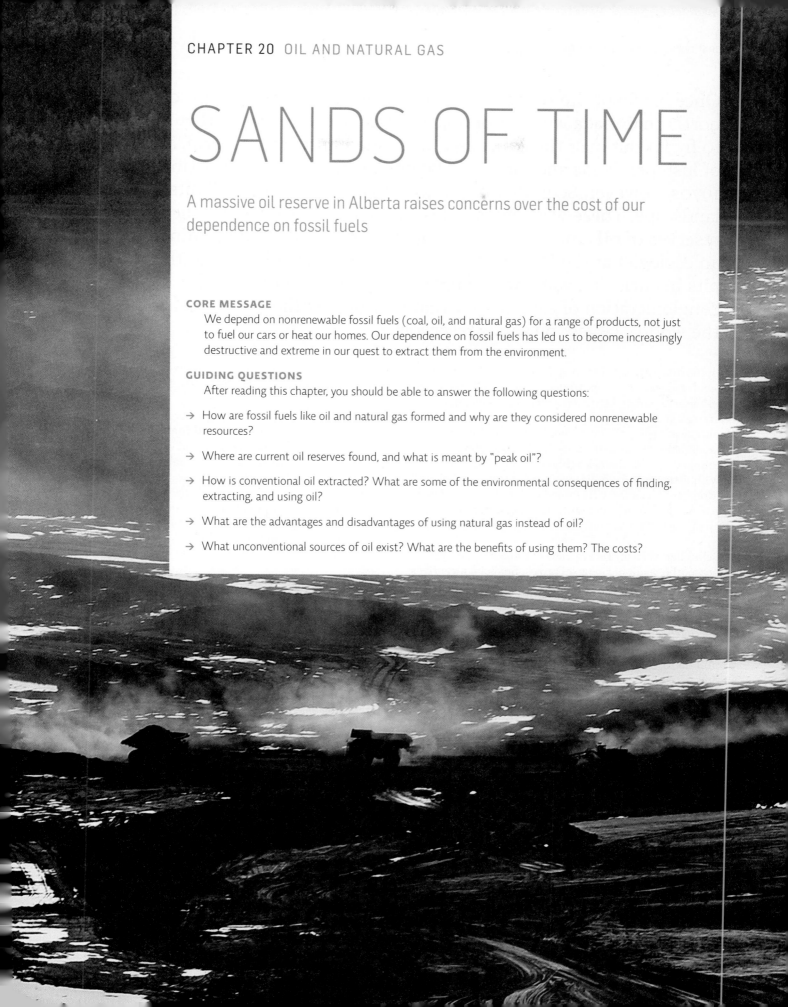

A massive oil reserve in Alberta raises concerns over the cost of our dependence on fossil fuels

CORE MESSAGE

We depend on nonrenewable fossil fuels (coal, oil, and natural gas) for a range of products, not just to fuel our cars or heat our homes. Our dependence on fossil fuels has led us to become increasingly destructive and extreme in our quest to extract them from the environment.

GUIDING QUESTIONS

After reading this chapter, you should be able to answer the following questions:

→ How are fossil fuels like oil and natural gas formed and why are they considered nonrenewable resources?

→ Where are current oil reserves found, and what is meant by "peak oil"?

→ How is conventional oil extracted? What are some of the environmental consequences of finding, extracting, and using oil?

→ What are the advantages and disadvantages of using natural gas instead of oil?

→ What unconventional sources of oil exist? What are the benefits of using them? The costs?

Once his favourite thing to do, David Schindler's fishing trips deep into northern Canada were becoming a source of dread. To get there, he had to fly far out over the Athabasca watershed in central Alberta, an expanse of lush peatlands and forests as far as the eye could see. Starting in the 1970s, however, Schindler began to notice a couple of small pits in the landscape. These were signs industry had begun to tap into the enormous reserves of **oil** embedded in the soil, known as **oil sands**. Schindler, an ecologist at the University of Alberta, would peer down at these pits in curiosity, watching companies try to extract the sticky, dense conglomeration of sand or clay that may hold trillions of barrels of oil—the world's largest oil sands reservoir.

In the mid-1990s, things began to change. The occasional, small excavation pits expanded exponentially, as mining in the area rapidly increased. Once the development started, it didn't stop, Schindler says, and what were once lush bogs and forests quickly became barren, grey landscapes. "Eventually, it looked like somebody had dropped a bomb there, leaving nothing behind but a crater with little spots for buildings, smokestacks, and pipes." Monster trucks carrying mined materials churned up the parched soil; every spot with standing water looked like an oil slick. "It was pretty horrifying." [INFOGRAPHIC 20.1]

Schindler had a bird's-eye view of Canada's energy boom, the result of rising oil prices, financial incentives to invest in oil sands exploration, and technological improvements that made this untapped reserve extractable. By the late 1990s, multiple companies were building massive facilities to mine the oil sands; by 2006, average annual production from these projects exceeded one million barrels per day. And there's no end in sight: production in Canada is set to increase, rising from 2.8 million barrels per day in 2010 to 4.7 million barrels per day in 2025, largely due to increased mining from the oil sands. Much of that oil is exported; indeed, Canada is the largest exporter of crude oil to one of the largest energy users in the world, the United States.

Everywhere, countries are looking for new sources of oil. Oil is one of three principal **fossil fuels** on our planet, along with coal and **natural gas**, and we depend heavily on these sources for energy and to make the everyday products we use (for more on coal, see Chapter 19). Fossil fuels form over millions of years when organisms die and are buried in sediment before they can decompose. Over time, as pressure and heat increase, the buried material goes through a chemical transformation into oil, natural gas, or coal.

Even though they are produced by natural processes, fossil fuels are considered **nonrenewable** because we are quickly using them up in a fraction of the time it takes for them to form. [INFOGRAPHIC 20.2]

Over the decades, we have come to depend heavily on fossil fuels, especially oil. But this effort can cause environmental problems—destroying habitats and blanketing entire regions in spilled oil, such as the 2010 *Deepwater Horizon* explosion that devastated the Gulf of Mexico. This has caused some to openly question whether our unquenchable thirst for oil has gone too far, forcing exploration of new sources of oil such as offshore reserves and oil sands, which are more difficult to extract or process than conventional oil. "Oil sands are the first major commercial foray into these unconventional oils," says Simon Dyer, policy director at the nonprofit Pembina Institute. "As we've finished extracting the easy-to-access oil, technology is unlocking more lower-grade, polluting sources of oil."

Over the years, as he watched the increase in mining of Alberta's oil sands unfold below him with each successive

---

oil Liquid fossil fuel useful as a portable fuel or as a raw material for many industrial products such as plastic and pesticides.

oil sands Geologic formations containing oil in the form of thick, black oil called crude bitumen (also known as *tar sands*).

fossil fuel A nonrenewable natural resource formed millions of years ago from dead plant (coal) or mircroscopic marine-life (oil and natural gas) remains.

natural gas Gaseous fossil fuel composed mainly of simpler hydrocarbon, mostly CH_4 (methane).

nonrenewable resource A resource that is formed more slowly than it is used, or is present in a finite supply.

Infographic **20.1** | **OIL SANDS AND SHALE DEPOSITS IN CANADA AND THE UNITED STATES**

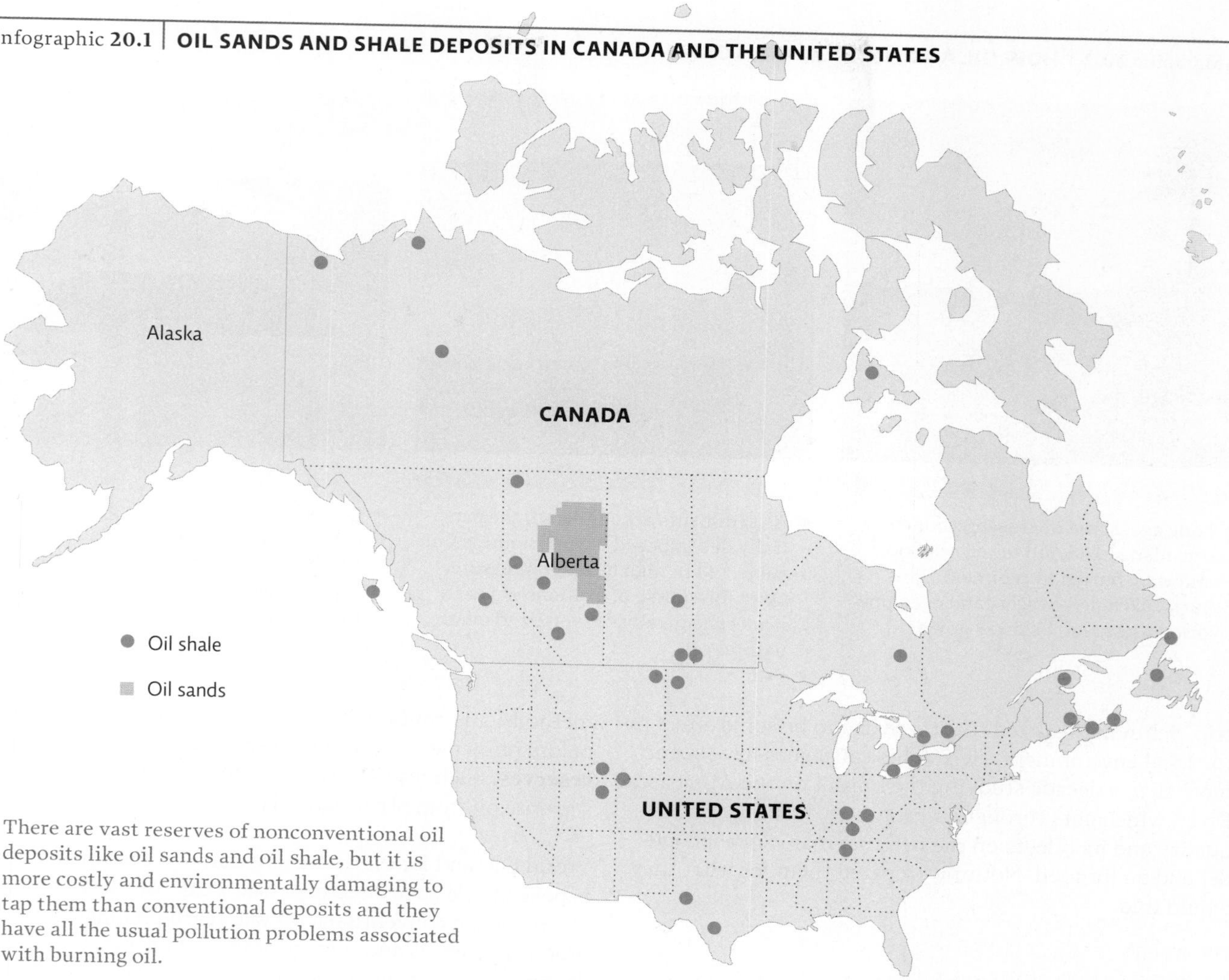
Alaska
CANADA
Alberta
Oil shale
Oil sands
UNITED STATES

↑ There are vast reserves of nonconventional oil deposits like oil sands and oil shale, but it is more costly and environmentally damaging to tap them than conventional deposits and they have all the usual pollution problems associated with burning oil.

↓ Oil deposits exist in a variety of forms. Oil shale is known as "rock that burns" (left). Dense, black bitumen can also be found in sand or clay soils such as the oil sands shown here (right).

Infographic **20.2** | **HOW OIL AND NATURAL GAS FORM**

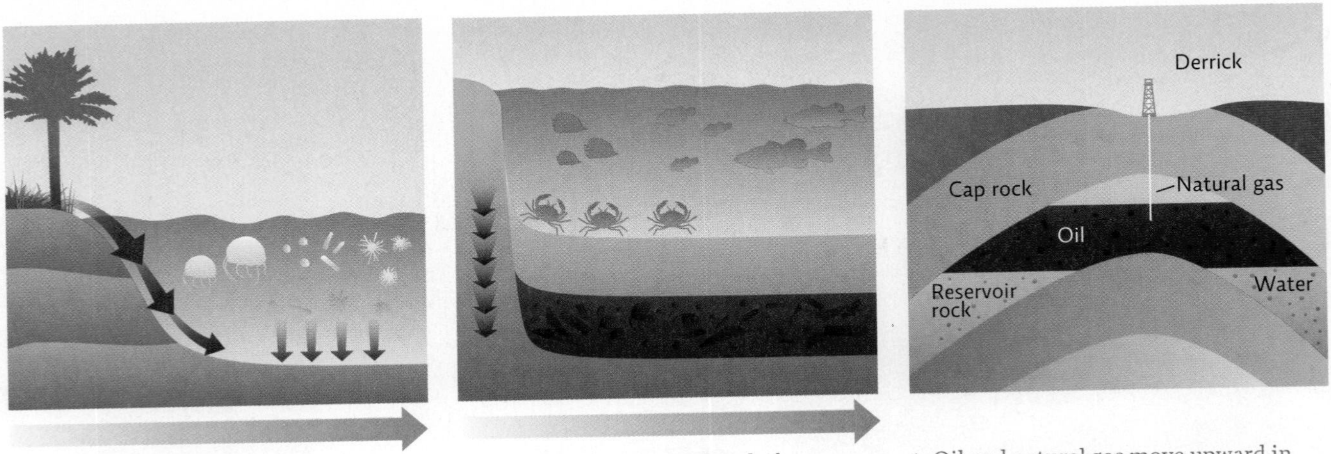
Derrick
Cap rock
Natural gas
Oil
Reservoir rock
Water

↑ Long ago, some microscopic marine organisms died, fell to the sea floor, and were buried in sediment. This burial excluded oxygen, and decomposition was greatly slowed down.

↑ As sediments accumulated, the partially decomposed buried biomass was subjected to high heat and pressure. Over the course of millions of years, it was chemically converted to oil or natural gas.

↑ Oil and natural gas move upward in porous reservoir rock until stopped by a layer of dense cap rock. We tap these deposits by drilling into the porous rock reservoirs.

trip, Schindler realized that there had to be some effect on the local environment. He and his colleagues have spent more than a decade studying the impact on the Athabasca River, which cuts through mining sites, looking at water quality and its effects on the fish that many local people depend on for food. Nothing prepared them for what they would find.

Oil is a valuable resource but it has many drawbacks.

Oil is a liquid fossil fuel made up of hundreds of types of hydrocarbons, organic compounds of hydrogen and carbon. Hydrocarbons take many forms—solid, liquid, or gas—and we have developed methods for extracting them from the depths of Earth. Oil is found some 300 to 9000 metres (0.3-9 kilometres) below the surface, where rigs have to dig deep in order to reach it. **Crude oil** can be found as tiny droplets wedged within the open spaces, or "pores," inside rocks, which are so small you'd need a microscope to see them. Since crude oil from this type of deposit can be pumped out by drilling oil wells, it is considered a **conventional oil reserve**.

There are different types of crude oil deposits. Canada sits on the third-largest proven reserves of oil in the world (only Saudi Arabia and Venezuela have more), 98% of which consist of oil sands. The oil found in oil sands is known as **bitumen**, a heavy, black oil that, in these deposits, is blended with sand, clay, and water. This type of heavy oil cannot be recovered using conventional methods, and contains impurities that must be removed before refinement—making oil sands **unconventional oil reserves**. Such reserves require intensive processing to separate oil from other natural elements.

World demand for oil rises more than 2% each year, and it is possible we have already passed **peak oil**, the moment in time when conventional oil will reach its highest production levels and then steadily and terminally decline. Passing peak oil will have international economic repercussions as the demand for oil outstrips the supply and prices go up. [INFOGRAPHIC 20.3]

Proven reserves are the amount of a fuel that is economically feasible to extract from a deposit using current technology. Conventional oil reserves are not evenly distributed around the planet, leading to political problems between countries that have the oil, like those in the Middle East, and those that do not have enough to meet their own needs, like the United States. Even though

crude oil Liquid fossil fuel that can be extracted from underground deposits by pumping. It is processed into fuels and other products.
conventional oil reserves Liquid fossil fuel deposits that contain freely flowing oil that can be pumped out.
bitumen A thick, sticky oil that may be found in sand or clay deposits.
unconventional oil reserves Recoverable oil deposits that exist in rock, sand, or clay but whose extraction is economically and environmentally costly.
peak oil The moment in time when oil will reach its highest production levels and then steadily and terminally decline.
proven reserves A measure of the amount of a fossil fuel that is economically feasible to extract from a known deposit using current technology.

Canada is a major consumer of fossil fuels, it produces more than it consumes every year. The global distribution problem is compounded by the fact that proven conventional reserves are past peak. At current rates of extraction and use, these proven reserves are expected to last another 40 years (though this number is uncertain because reports of oil reserves tend to be questionable—many nations keep their reserve estimates a secret).

Natural gas reserves are also finite. Russia has the largest amounts of natural gas, with about 25% of the world's conventional reserves; Canada has only 1%, but is the fourth-largest exporter, sending much of it to the United States. Canada itself relies on natural gas for 22% of its energy needs. Total world conventional reserves of natural gas are expected to last 60–100 years at current rates of use. But like oil, natural gas use is increasing at about 2% per year. At this rate, natural gas could run out much sooner than predicted. [INFOGRAPHIC 20.4]

But we are nowhere near "peak oil" for oil sands or shale oil (and, as we shall see, for natural gas). As new technologies and higher energy prices make these unconventional oil sources viable alternatives, they have the potential to change the global energy future—at least in the near term.

Infographic **20.3** | **PEAK OIL FOR CONVENTIONAL RESERVES**

↓ Peak oil represents the point in time when oil will reach its highest production levels and then steadily and terminally decline. It is likely that the world has reached, or will soon reach, peak oil (red arrow). The dip (black arrow) seen in the early 1980s represented an oil glut; conservation efforts during the oil crisis of the 1970s led to reduced demand, but we have since resumed an upward course of oil consumption.

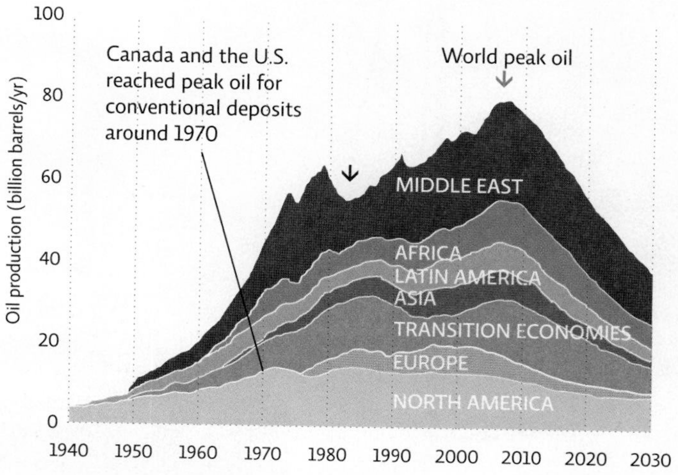
100
80
60
40
20
0
Oil production (billion barrels/yr)
Canada and the U.S. reached peak oil for conventional deposits around 1970
World peak oil
MIDDLE EAST
AFRICA
LATIN AMERICA
ASIA
TRANSITION ECONOMIES
EUROPE
NORTH AMERICA
1940
1950
1960
1970
1980
1990
2000
2010
2020
2030

Infographic **20.4** | **PROVEN OIL AND NATURAL GAS RESERVES**

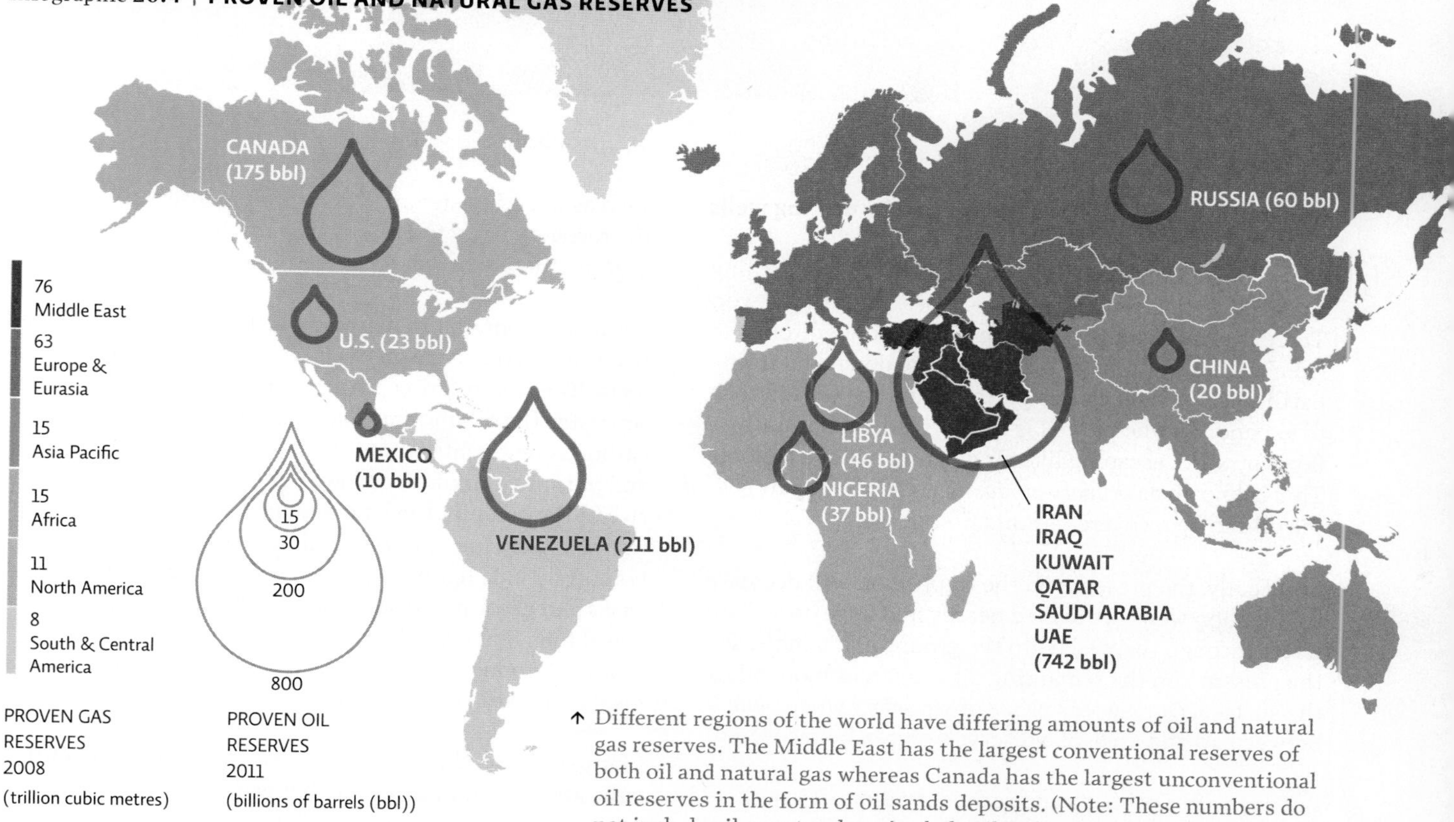
CANADA (175 bbl)
U.S. (23 bbl)
MEXICO (10 bbl)
VENEZUELA (211 bbl)
RUSSIA (60 bbl)
CHINA (20 bbl)
LIBYA (46 bbl)
NIGERIA (37 bbl)
IRAN
IRAQ
KUWAIT
QATAR
SAUDI ARABIA
UAE
(742 bbl)
76
Middle East
63
Europe & Eurasia
15
Asia Pacific
15
Africa
11
North America
8
South & Central America
15
30
200
800
PROVEN GAS RESERVES 2008
(trillion cubic metres)
PROVEN OIL RESERVES 2011
(billions of barrels (bbl))

↑ Different regions of the world have differing amounts of oil and natural gas reserves. The Middle East has the largest conventional reserves of both oil and natural gas whereas Canada has the largest unconventional oil reserves in the form of oil sands deposits. (Note: These numbers do not include oil or natural gas in shale oil and gas reserves.)

Infographic **20.5** | **HOW IT WORKS: EXTRACTING CRUDE OIL**

↓ Conventional oil is obtained by drilling through layers of dense rock to reach the reservoir below. At first, crude oil may easily flow due to the relief of pressure caused by the drill hole. Pumpjacks are used to mechanically pump out more oil when the oil stops flowing freely.

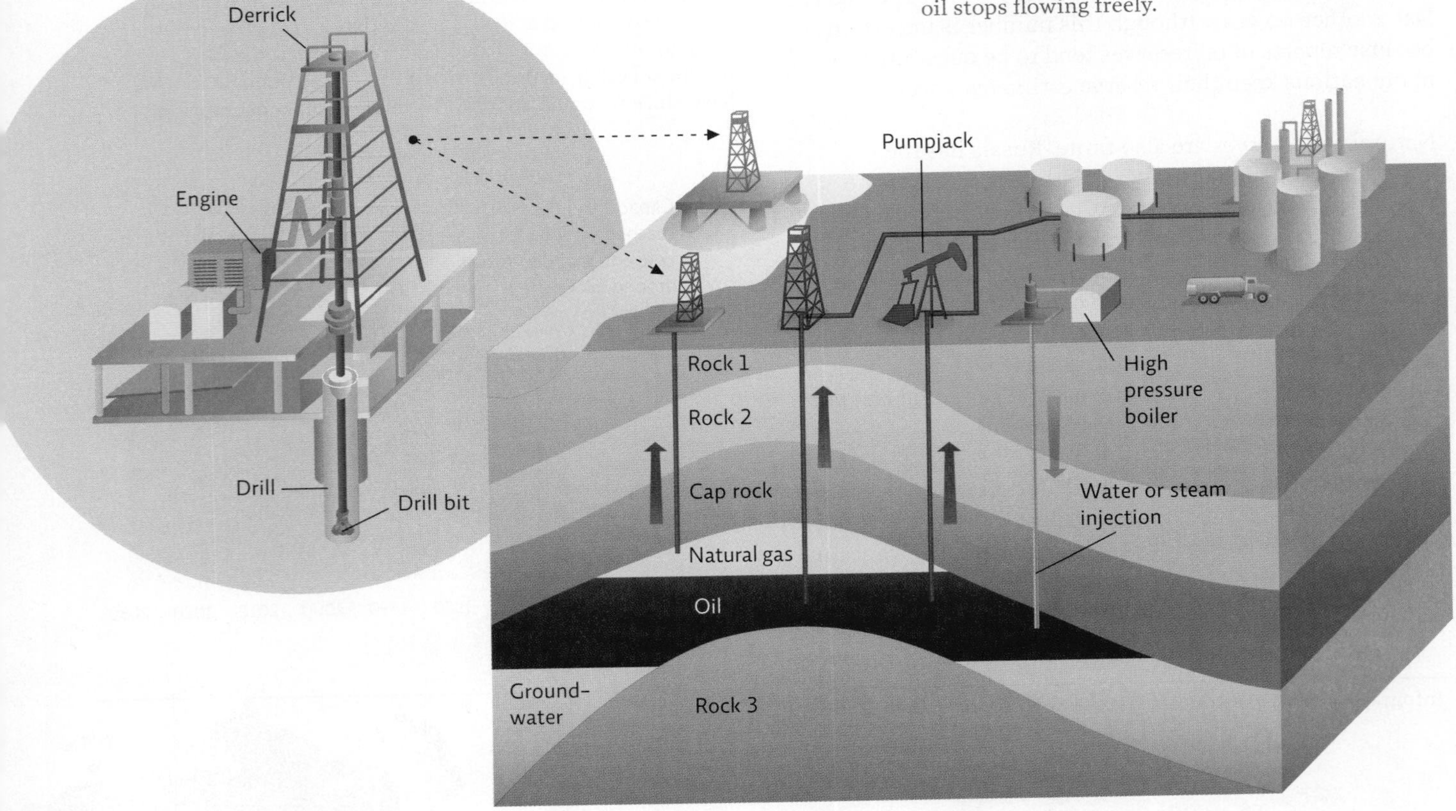

Derrick
Engine
Drill
Drill bit
Pumpjack
Rock 1
Rock 2
Cap rock
Natural gas
Oil
Ground-water
Rock 3
High pressure boiler
Water or steam injection

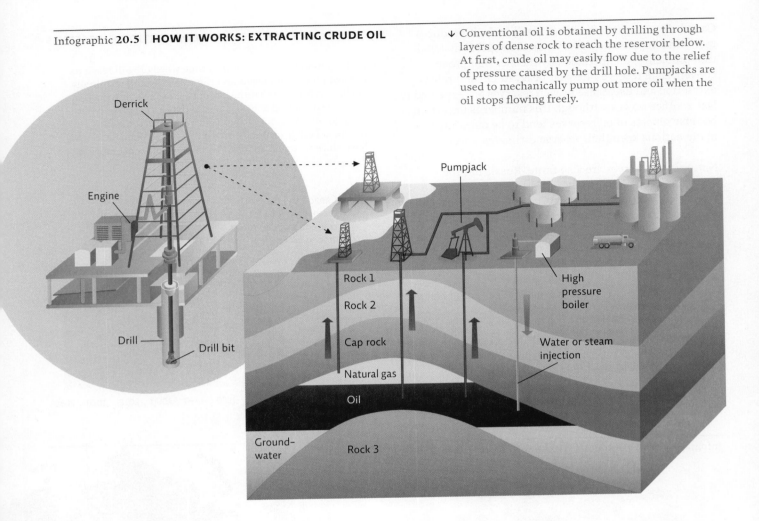

Conventional oil reserves are tapped by drilling wells.

Most conventional crude oil is mined by simply drilling a new oil well, after which nature does much of the work. There is significant pressure on oil deep underground from millions of tons of rocks pressing down and from Earth's heat, which causes gases around the oil and rock to expand. So when a well is first drilled, oil naturally flows upward, escaping like air gushing out of a balloon. This is known as *primary production,* and about 5 to 15% of the oil can be recovered in this first phase.

Eventually, the pressure on the trapped oil will decrease, so injection wells are drilled nearby and huge hoses pump water through the wells into the ground. This increases the pressure on the remaining oil and forces more oil to the surface. During this phase of *secondary production,* between 20 and 40% of a reserve's oil can be recovered.

But even after primary and secondary production, there is a lot of oil left in the ground. *Tertiary production* methods such as injecting steam or CO_2 into the reservoir allow the recovery of up to 60% of the remaining oil but are costly in both money and energy. [INFOGRAPHIC 20.5]

Because it contains so many impurities once it's extracted from the earth, crude oil must be refined into a usable form. The first refining process, called simple distillation, separates the oil into *fractions*, each with different hydrocarbon components. To do this, the crude oil is heated and put into a column, or tower. As the oils boil off and rise in the column, the larger compounds condense back to liquid form and are collected. The first compounds to boil off include lighter liquids, like liquefied petroleum gases and gasoline. Next come jet fuel, kerosene, and diesel fuel. Finally, at temperatures above 600°C, the heaviest products, including grease, wax, asphalt, and residual fuel oil, can be recovered.

Many of the products of distillation, called **petrochemicals**, are used not just as fuel but as raw materials in the production of industrial organic chemicals, pesticides,

Infographic **20.6** | **PROCESSING CRUDE OIL**

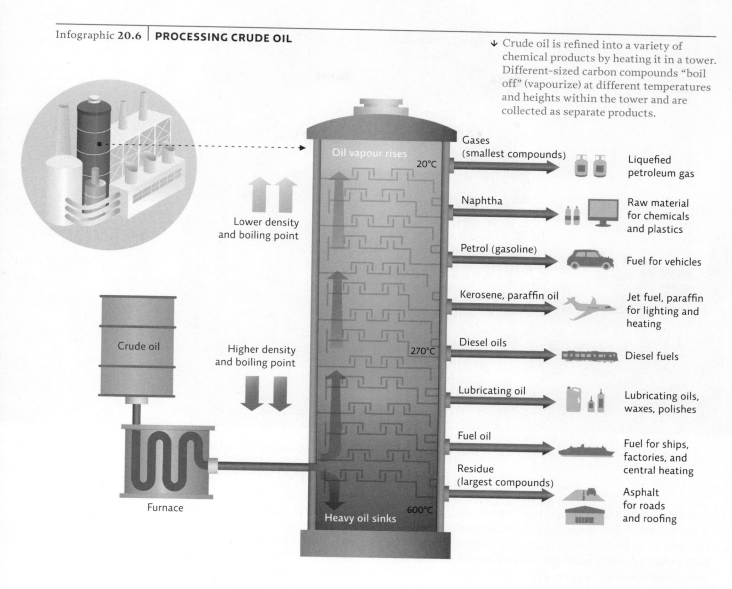

↓ Crude oil is refined into a variety of chemical products by heating it in a tower. Different-sized carbon compounds "boil off" (vapourize) at different temperatures and heights within the tower and are collected as separate products.

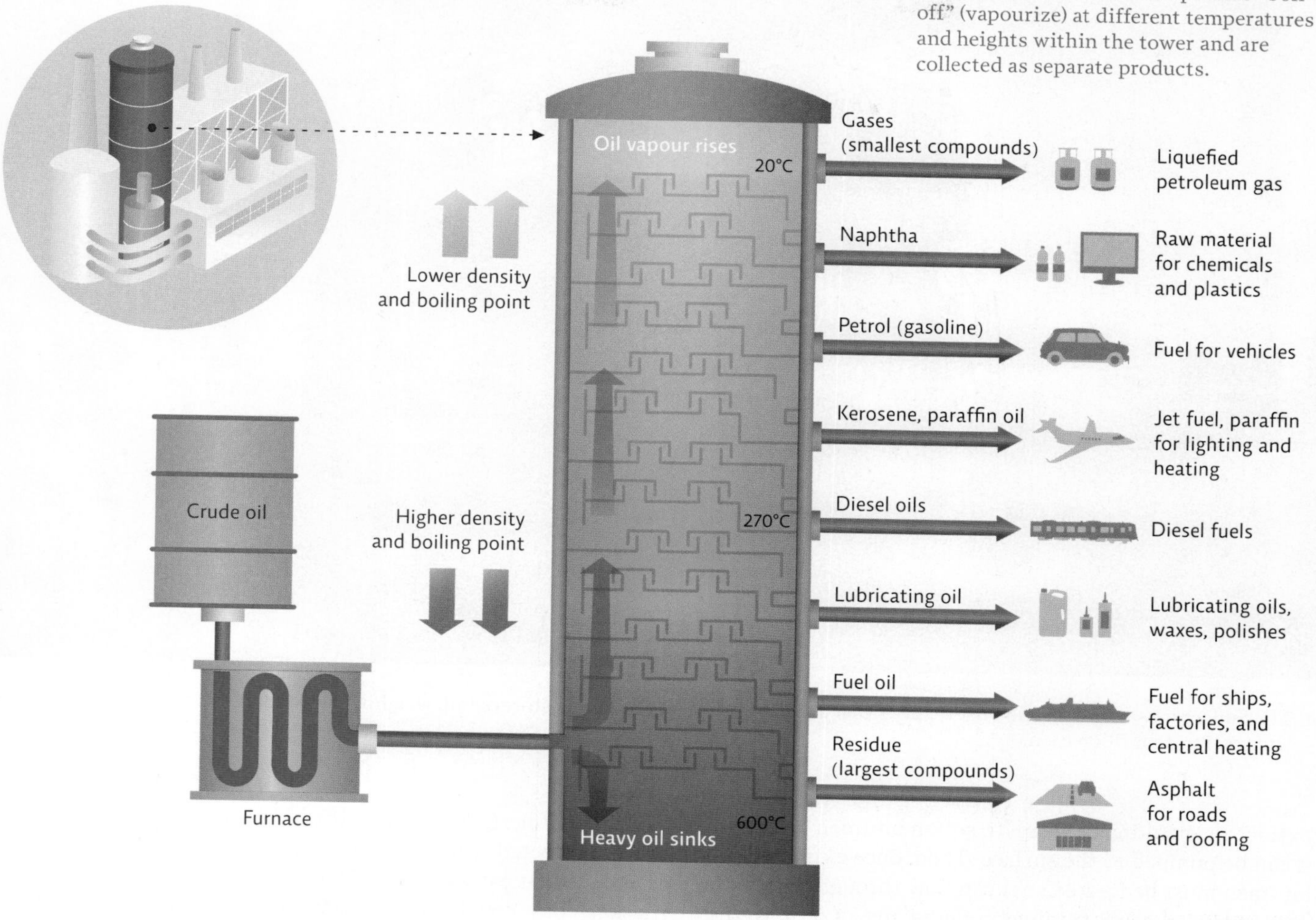

Oil vapour rises
20°C
270°C
600°C
Heavy oil sinks
Lower density and boiling point
Higher density and boiling point
Crude oil
Furnace
Gases (smallest compounds)
Liquefied petroleum gas
Naphtha
Raw material for chemicals and plastics
Petrol (gasoline)
Fuel for vehicles
Kerosene, paraffin oil
Jet fuel, paraffin for lighting and heating
Diesel oils
Diesel fuels
Lubricating oil
Lubricating oils, waxes, polishes
Fuel oil
Fuel for ships, factories, and central heating
Residue (largest compounds)
Asphalt for roads and roofing

plastics, soaps, synthetic fibres, explosives, paints, and medicines. The production of a desktop computer, for example, consumes 10 times the computer's weight in fossil fuels, mostly oil. [INFOGRAPHIC 20.6]

We are increasingly relying on unconventional sources of oil.

Since conventional oil supplies won't last forever, geologists have begun to focus their energy on the unconventional oil reserves.

petrochemicals Distillation products from the processing of crude oil that can have many different uses, including as fuel or as industrial raw materials.
oil shale Compressed sedimentary rocks that contain kerogen, an organic compound that is released as an oil-like liquid when the rock is heated.

The United States possesses the world's largest concentration of **oil shale**, compressed sedimentary rocks that contain kerogen, an organic compound that is released as an oil-like liquid when the rock is heated, a process that is difficult and energy intensive and not yet commercially pursued. Primarily on federal lands, oil shale contains an estimated 1.23 trillion barrels of oil, more than 50 times the country's proven oil reserves, according to the U.S. Bureau of Land Management.

Canada's oil is predominately found in its vast oil sands formations. Some of this oil can be mined on the surface. Flying over the Athabasca watershed, Schindler witnessed bitumen-rich soil being extracted from the surface, then dragged across the landscape by monster trucks and processed at a separate facility. So far, most bitumen has come from surface mining, but an estimated 80% of reserves are situated deep underground. To access these,

CATERPILLAR
159
797B
CAT
797B 159
159

↑ The Suncor oil sands operation uses trucks that are the world's largest: three storeys tall, weighing more than 450 metric tons, and costing $7 million each.

extractors must inject steam to soften bitumen before it can be pumped to the surface. Then, once extracted, bitumen must be treated so it can flow through pipelines. As a result, oil sands mining produces up to 3.6 times the amount of greenhouse gases as conventional oil and gas production—creating the potential for more environmental problems as well, as Schindler would soon discover. [INFOGRAPHIC 20.7]

Oil use comes at a great cost.

Oil extraction is not just financially expensive—it also has environmental consequences at every stage. In order to find sources of oil, companies send seismic waves into the ground that bounce back to reveal the location of possible reserves. But doing so disorients marine wildlife; in 2009, ExxonMobil had to abandon its oil exploration in Madagascar because more than 100 whales had beached themselves, presumably as a result of these seismic exploratory methods.

Drilling can affect wildlife, too. U.S. politicians have long debated the merits of drilling in the Arctic National Wildlife Refuge (ANWR), nearly 70 000 square kilometres of protected wilderness. The U.S. Geological Survey estimates that parts of ANWR located on the northern Alaskan coast could harbour up to 16 billion barrels of crude oil and natural gas reserves. But drilling there—which could begin in the coming decades—would have serious environmental consequences. The coastal area where drilling would take place is home to Arctic foxes, caribou, polar bears, and migratory birds, all of whose habitats could be disturbed by the drilling, affecting migrating populations across Canada and Alaska.

And damage is not limited to the extraction of fossil fuels. Processing and shipping of fossil fuels generates air and water pollution. Pollution generated when fossil fuels are burned is also a major source of air pollution (see Chapter 21) and a main cause of climate change (see Chapter 22). Transporting the oil is another environmental concern and the construction of oil pipelines to carry the oil sands oil to refineries is hotly contested. The proposed Keystone XL pipeline, which would deliver oil from the Alberta oil sands to Texas refineries, is opposed by environmental groups and Canadian and U.S. citizens along its proposed route. An alternative proposal, the Northern

Infographic 20.7 | **GETTING OIL FROM SAND**

SURFACE MINING

EXCAVATION: Shallow deposits of oil sands are recovered with surface mining techniques. The overlying ground (overburden) is removed by huge, multi-storey machines (scoopers) that dig out the oil-laden sands in 90 metric ton scoops. The world's largest trucks, each holding 363 metric tons of sand, then haul away the sand for processing.

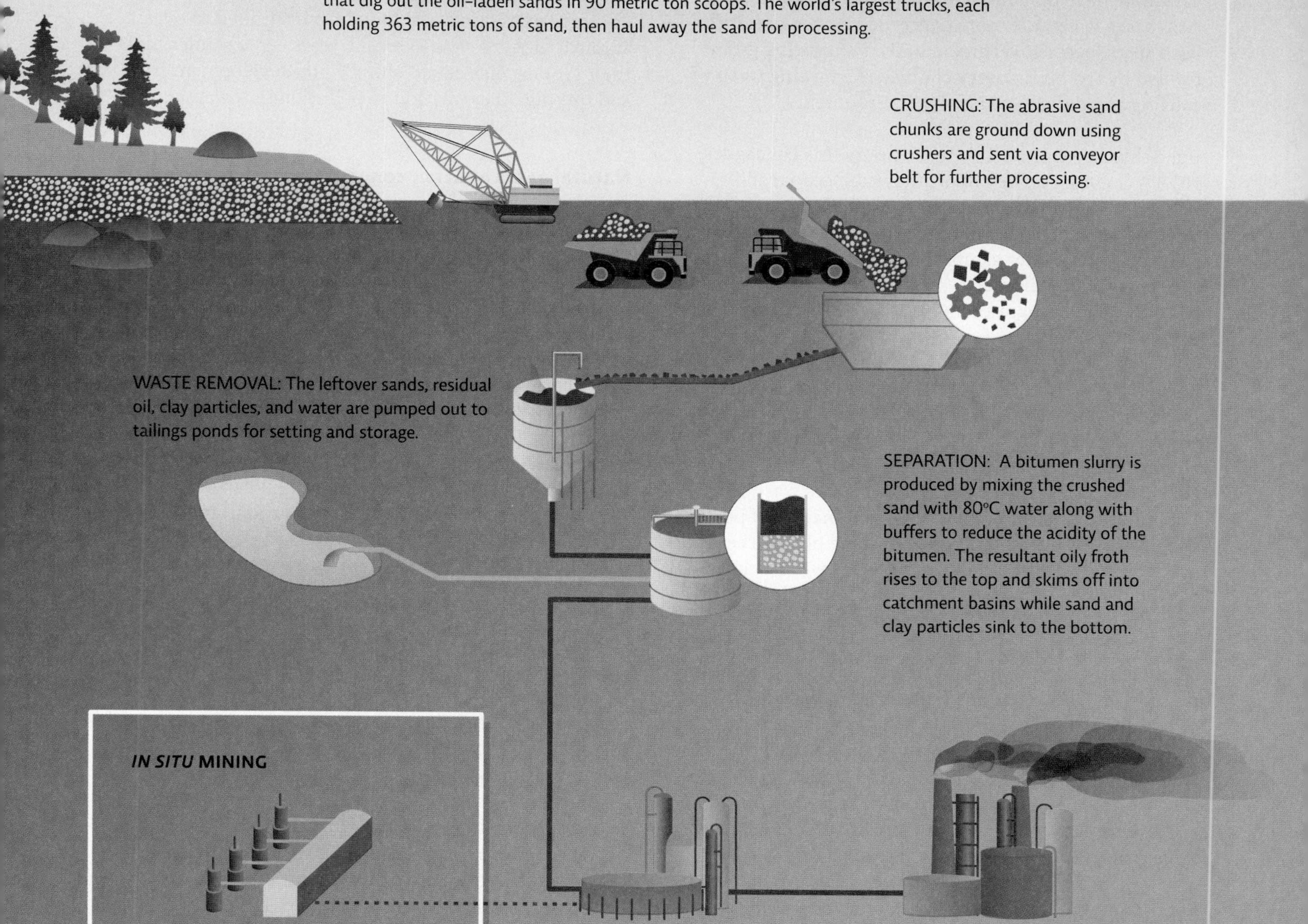

CRUSHING: The abrasive sand chunks are ground down using crushers and sent via conveyor belt for further processing.

WASTE REMOVAL: The leftover sands, residual oil, clay particles, and water are pumped out to tailings ponds for setting and storage.

SEPARATION: A bitumen slurry is produced by mixing the crushed sand with 80°C water along with buffers to reduce the acidity of the bitumen. The resultant oily froth rises to the top and skims off into catchment basins while sand and clay particles sink to the bottom.

***IN SITU* MINING**

STEAM INJECTION: At more than 200 metres underground, most oil sands deposits are too deep for surface mining. To recover the bitumen, pressurized steam is injected into the deposit to melt the hardened bitumen and force it out with the rising steam. This method uses less water than the surface mining processing but requires much more energy to produce and inject the 350°C steam.

COKING AND DILUTION: In the coking tower, some bitumen is heated to break it into shorter chain hydrocarbons to be mixed with the gooey, raw bitumen slurry; this removes excess water and makes for easier pipeline shipping. The coke residue from this heating process (basically a low-grade, dirty form of coal) may be used to power the facility.

UPGRADING: The processed bitumen is converted to crude oil by heating to high temperatures and adding in hydrogen produced from nearby natural gas deposits.

Gateway pipeline, would construct a pipeline to take the oil west through British Columbia for export to Asian markets, also sparking protests among residents in rural and urban areas along the proposed path. [INFOGRAPHIC 20.8]

The more development Schindler saw around the Athabasca River, the more concerned he became. Initially, both the Alberta government and a monitoring group funded by the oil industry claimed the mining wasn't causing any damage to nearby environments.

But Schindler wasn't so sure, so he and his colleagues, NOAA scientist Jeffrey Short (now at Oceana) and Peter Hodson of Queen's University in Ontario, decided to embark on a one-year project to survey the area. They collected water samples from multiple spots along the river and its tributaries in the winter and summer of 2008, then brought them back to the lab to analyze for pollutants.

During the winter, Schindler's then graduate student Erin Kelly led a team of people that took samples of the snowpack near the river, which contained months of precipitation that may have accumulated toxins. Sometimes, collecting samples from such remote areas wasn't easy, Schindler recalls, and relied on a combination of boats, snowmobiles, and helicopters. "In some places, there was so much snow we couldn't tell if the ice was thick enough to land the helicopter, so they would hover the vehicle over the ground and reach down to collect samples."

The scientists found that concentrations of 13 toxic compounds were higher in the tributaries located downstream of oil sands mining than those located upstream, which suggests that the extraction is at least partly responsible. Also of concern were the huge lakes of acidic and toxic wastewater that oil companies store at the mining sites as a by-product of the refining process. The pools are so large that they can be seen from space.

Some contaminants were present at levels 25-50 times higher than those found in areas further from oil sands development. "We knew we would see some contamination, but we didn't know how high it would be," says Schindler, who published his findings in 2009 and 2010.

Schindler and his colleagues found a "bull's-eye" concentration of contamination, highest right around the area where bitumen is mined and processed. This suggests that the industry is playing a significant role. But critics note that small amounts of the compounds are naturally present in the area and since monitoring these compounds only began after the industry moved in, no one knows what the original, natural baseline levels are.

To determine what background or "normal" contamination might be, Joshua Kurek of Queen's University examined sediment cores from area lakes, digging deep enough to examine sediments in the years that preceded oil sands extraction. He found toxins linked to oil sands at much higher levels in recent years (2.5-23 times higher than 1960 levels), establishing a pre-development baseline and linking oil extraction to pollution in the area.

Natural gas is another conventional energy resource.

Natural gas (predominately methane) is an important alternative to oil for some purposes, like generating electricity and heat. Like oil, natural gas reserves are finite. (See Infographic 20.4)

The extracting and processing methods for natural gas and oil are very similar. Natural gas sometimes flows freely up wells due to underground pressure, but most natural gas extraction requires some kind of pumping system to force the gas upward. Natural gas also contains impurities and must be refined before it can be used. However, it is considered the cleanest fossil fuel because it releases much less of the greenhouse gas, carbon dioxide, and other air pollutants than oil or coal does when burned.

As conventional oil and gas reserves shrink, energy companies are looking for alternatives. One increasingly targeted type of fossil fuel reserve is deep natural gas held in shale rock formations. However, some environmental organizations have raised concerns about the safety of the extraction process used to access these reserves. For instance, to help release natural gas from the ground after drilling, fluids containing toxic chemicals are injected underground at high pressure in a process known as hydraulic fracturing or "**fracking**." An estimated 30 to 70% of this liquid resurfaces with the natural gas, bringing the toxic chemicals along with it. These chemicals have the potential to contaminate ground and drinking water, soil, and air. Methane itself may also contaminate groundwater near these wells. Though areas with methane deposits may normally have some methane in the groundwater, fracking may increase the amount. [INFOGRAPHIC 20.9]

---

fracking (hydraulic fracturing) The extraction of oil, or more commonly natural gas, from rock (typically shale) via the propagation of fractures using pressurized fluid.

Infographic **20.8** | **ENVIRONMENTAL COSTS OF OIL**

↓ Oil has environmental costs at every stage—from exploration for reserves, to extraction, processing, transporting, and burning. If these, often externalized, costs were added to the price of our fuel and oil-based goods, the price of the goods would be much higher.

WILDLIFE DISRUPTION

Sonic exploration for oil caused whales to beach near Busselton, Australia.

HABITAT LOSS

An open-pit oil sands mine in Alberta where boreal forest once stood.

EXPLOSIONS

The *Deepwater Horizon* burns.

AIR POLLUTION

Burning oil is a major source of air pollution in Mexico City.

OIL SPILL

A severely oiled pelican being rescued in Louisiana. The bird lived.

HABITAT FRAGMENTATION

In Alberta, 80% of cutlines used for seismic exploration collect water and become canals, fragmenting habitat.

Infographic **20.9** | **FRACKING**

↓ Natural gas or oil held tightly in rock can be accessed via hydraulic fracturing. Concerns about water pollution make this a contentious resource in populated areas.

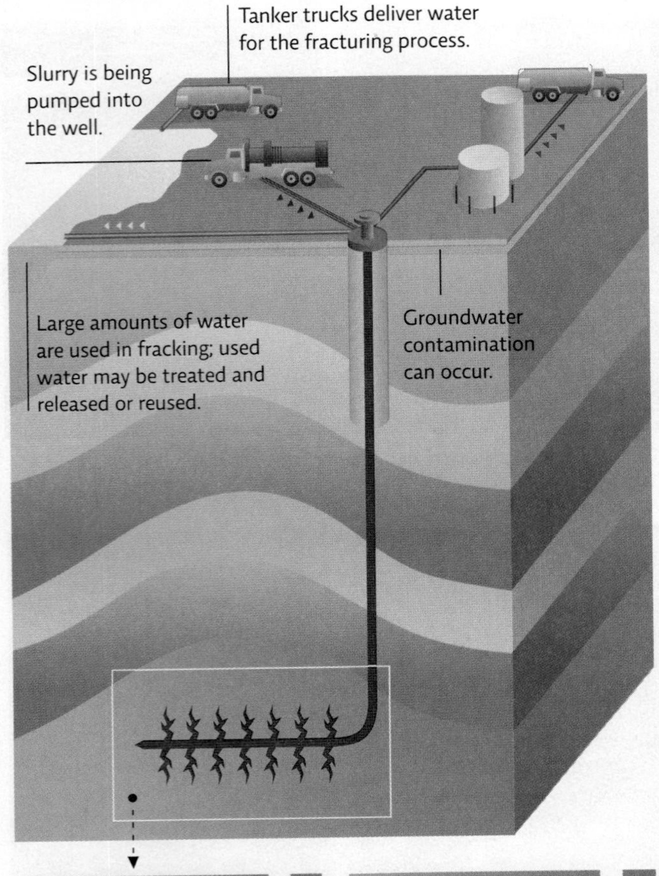
Tanker trucks deliver water for the fracturing process.
Slurry is being pumped into the well.
Large amounts of water are used in fracking; used water may be treated and released or reused.
Groundwater contamination can occur.

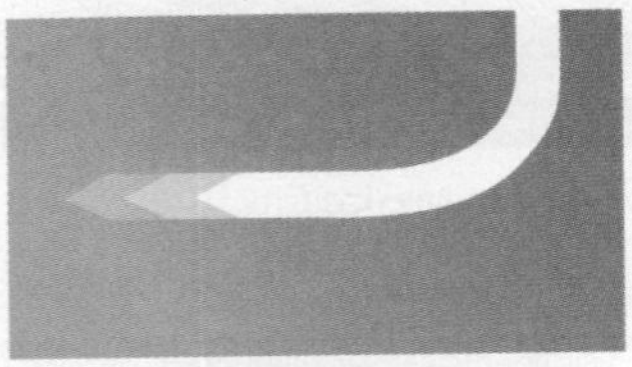

1. For deep deposits, a well is drilled down to gas-bearing rock and then extended horizontally.

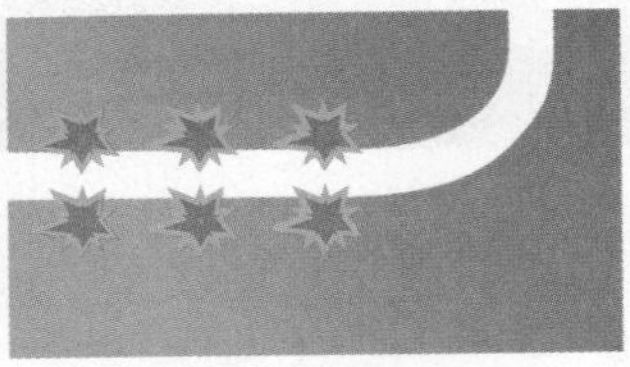

2. Holes are blasted into the rock using explosive charges, to create large fractures.

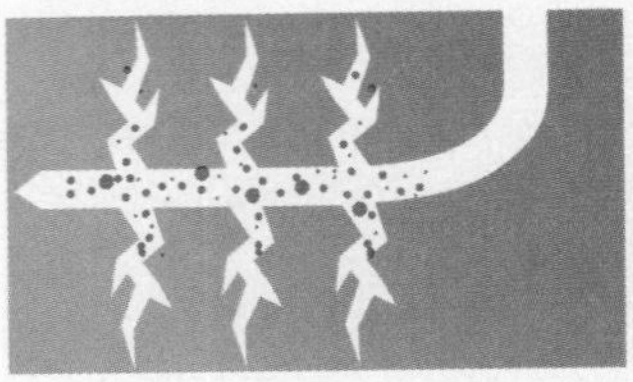

3. A slurry of sand, water, and chemicals is pumped into the fractures, to further crack open the rock, hold the fractures open, and make the gas more readily extractable.

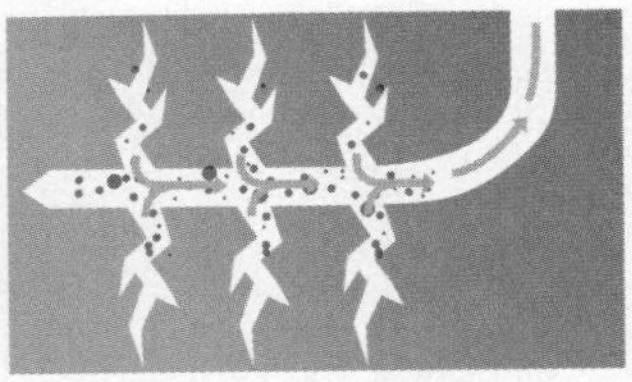

4. Natural gas can now flow through the more permeable, fractured rock and be extracted.

Oil consumption drives extraction.

Not all countries use energy equally. The United States, for instance, uses more energy than any other country—its overall primary energy use tripled from 1949 to 2009, according to the U.S. Energy Information Administration. Canada is one of the top users of energy, as well. Transportation accounts for a large part of this energy; a relatively small population spread over a large area uses a lot of fossil fuel to get around. Canadians also use a lot of energy for heating, a result of our cold climate.

Much of this energy comes from fossil fuels—coal, oil, and natural gas, with crude oil as the dominant source.

Canada is consistently ranked as one of the top energy producers each year. Oil sands and other sources of heavy oil are estimated to contain around 20 trillion barrels of oil in place, and much more may still be undetected. More than 140 000 square kilometres contain oil sands; as of 2007, the Canadian Association of Petroleum Producers estimates that more than 170 billion barrels of proven reserves still remain. Another oil resource, a non-kerogen type of oil found in oil shale known as *tight oil*, is also being accessed via fracking, adding to Canada's overall production and reserves.

In other countries, the energy balance isn't as favourable: the United States uses much more oil than it produces. In 2009, it produced only 11% of the world's oil, but consumed 22% of it. Most of the rest was imported from Canada, Venezuela, and Saudi Arabia. This, however, may be changing. The United States is also applying fracking technology to tap its own tight oil deposits in North Dakota and Texas. Whether this will be enough to reduce its imports from Canada remains to be seen.

At the same time, the United States is trying to reduce its consumption of oil, according to a report published in 2009 by IHS Cambridge Energy Research Associates. The report predicts that U.S. oil use peaked in 2005 and is now in decline, partly due to improvements in vehicle fuel efficiency. But in developing countries such as China and India—the world's second and fourth largest consumers of oil, respectively, with Japan in third—oil consumption is still rapidly increasing. This makes it more profitable for Canada and other countries to extract as much oil as possible from the earth, regardless of the consequences.

Our economy and lifestyle are dependent upon reliable access to affordable energy, or **energy security**. Yet there

energy security Having access to enough reliable and affordable energy sources to meet one's needs.

↑ Fisherman Ed Cooper, 76, shows his former fishing spot on the Athabasca River, which he claims is now too polluted to fish due to its proximity to oil sands mines and extraction facilities near the town of Fort MacKay, Alberta.

are many reasons that our energy supplies might become unreliable or unaffordable. These include dwindling supplies, increasing demand, dependence on energy imports, competition from other countries, and a cartel or monopoly increasing prices or decreasing oil supplies.

As supplies dwindle and demand continues to increase, easily extracted oil becomes less available, forcing companies to use more powerful—and potentially more dangerous—technologies. But will the risks always be worth the benefits?

Schindler, for one, has seen the effects of the pollution at the Athabasca River first-hand. One of the contaminants that he and his team consistently found in high amounts was mercury. This toxic heavy metal is emitted into the atmosphere during the processing of bitumen into oil, then falls back down onto land and water surfaces as part of dry particles and in precipitation. One form of mercury, methylmercury, is particularly problematic because it readily accumulates in organisms, and biomagnifies at higher levels of food chains. This becomes a major problem for nearby residents who rely on fish for food. Ingesting dangerous quantities of mercury can affect all systems of the body, especially the nervous system, and is particularly dangerous for pregnant women and children. (See Chapter 3 for more on bioaccumulation and biomagnification of toxins like mercury.)

Other pollutants are known cancer-causers, and Athabasca fish have a high incidence of malformations such as spinal deformities, tumours, and problems with their eyes and mouth. "I've seen fish with one eye much bigger than the other, or a fish with two tails," says Schindler. "I've got a collection of photos."

For some people living in the region, simply cutting out fish isn't always an option, explains Schindler. Many are poor, and there are few livestock farms in northern latitudes, making beef or pork very expensive. The only protein alternatives are often canned meats, such as Spam.

Clearly, pollution from oil mining has many ripple effects. According to Frances Beinecke, president of the nonprofit Natural Resources Defense Council, "as long as we keep the driving addiction and the oil addiction that we have we will be jeopardizing really fertile, fragile ecosystems all around the country—onshore and offshore."

The future of oil sands mining remains unclear.

When Schindler and his colleagues first published their analysis of environmental contamination from oil sands mining, the reaction was mixed.

↑ The 1300-kilometre Trans-Alaska Pipeline delivers oil across Alaska to the southern port of Valdez. An even longer pipeline has been proposed to take oil from Canada's oil sands to Texas.

Schindler received some unpleasant letters from members of the oil industry, for instance, accusing him of exaggerating the state of the science to undermine the development of oil sands mining.

But other scientists were eager to confirm the data. Environment Canada, for instance, took additional snow samples and examined contamination in lake sediments from areas near the oil plants, and obtained similar findings. Panels of experts reviewed the research and upheld the results. Other groups evaluated the industry's monitoring program, which initially claimed the development was not causing any additional harm, and concluded their analysis was substandard, says Schindler.

"If you are confident in your science, you know there will be people who try to follow up on it," says Schindler. "And if you did good science, they will find the same results. Eventually, you are going to be vindicated."

Schindler's scientific rigour is "beyond reproach," says Dyer. His 2010 analyses of the environmental effects of oil sands mining appeared in one of the most prestigious peer-reviewed journals, *Proceedings of the National Academy of Sciences*. No industry monitoring group can say the same, he adds. "It shouldn't have taken this long for people to see what was going on."

Today, the Canadian Association of Petroleum Producers pledges on its website to implement new techniques and practices that mitigate the impact of development on the region's air, land, and water, such as operating facilities that maintain regional objectives for air quality.

The Canadian government is also discussing creation of a new system to monitor the environmental effects of oil sands mining, which would be run by outside experts. In the meantime, development has rapidly increased. In 2011, Canada produced nearly 3.7 million barrels of oil per day, more than half of which came from the oil sands. By 2035, our annual average could reach 6.6 million barrels per day. In 2009, oil sands already represented 6.5% of Canada's total greenhouse gas emissions; "at a time when we need to be reducing emissions, it's the fastest growing source of greenhouse gases in Canada," says Dyer.

Even without the environmental concerns, the path to expanded exploration of the oil sands may not be smooth, predicts Schindler. The United States is Canada's primary oil customer, making both countries dependent on each other. Even though Canada produces more energy than it consumes, it still imports oil from the United States, partly due to geography that makes it difficult to transport oil extracted from western provinces across the country to the more populated eastern provinces. But the United States has said it wants to be more energy self-sufficient, and rely less on imports. This will force Canada to find new customers for its oil, such as China, and develop the infrastructure to transport it.

So how do we increase energy security and reduce our dependence on fossil fuels so that we can avoid using dangerous extraction methods? There are many possibilities, but the main methods include reducing imports, exploiting local energy sources, developing alternative energy sources, and reducing energy needs through increased conservation and energy efficiency. Iceland, for instance, is increasing its energy security by focusing on **energy independence**, such that it will meet all of its energy needs without imports by the year 2050.

"Here in Canada, the debate over oil sands is often framed as very black or white—do we pursue it or not? But the truth is somewhere in the middle," says Dyer. "Current development is going too fast, and monitoring standards need to be higher. But there may be a responsible level of oil sands production, and that's what we've got to determine."◉

energy independence Meeting all of one's energy needs without importing any energy.

Select references in this chapter:
Kelly, E., *et al.* 2009. *Proceedings of the National Academy of Sciences*, 106:22346–22351.
Kelly, E., *et al.* 2010. *Proceedings of the National Academy of Sciences*, 107: 16178–16183.
Kurek, J., *et al.* 2013. *Proceedings of the National Academy of Sciences* [Online], www.pnas.org/cgi/doi/10.1073/pnas.1217675110.

BRING IT HOME

PERSONAL CHOICES THAT HELP

Environmental and health concerns that surround our acquisition and use of fossil fuels continue to grow as we dig deeper and use more extreme methods to obtain oil and natural gas, highlighting our dependence on nonrenewable energy resources. By decreasing our fossil fuel use, we can reduce the pressure on oil companies to pursue sources of oil that have a greater potential for environmental damage, such as deep-water drilling or bitumen deposits like those in the oil sands and shales.

Individual Steps

→ Minimize your fuel use when driving by planning ahead to condense shopping trips and errands and to reduce total distance driven and by parking as far as you can from your entrance and getting some exercise instead of wasting gas. Not driving? See where you can safely walk, bike, or use public transportation.

→ Reduce your use of disposable plastics like water bottles and single-serve food containers, and always recycle plastics when possible.

→ Turn your thermostat down in the winter to reduce your energy use.

Group Action

→ Organize a carpool system in your community to reduce the number of single-passenger car trips.

→ Organize a screening of the documentary *Tipping Point: The Age of the Oil Sands.*

Policy Change

→ Work with your school's administration to encourage public transportation and increased bike usage on your campus.

→ Visit www.pembina.org/oil-sands and read about the Pembina Institute's position on oil sands. Voice your own opinion to your local or national government officials.

UNDERSTANDING THE ISSUE

CHECK YOUR UNDERSTANDING

1. What is the correct definition of fossil fuel "proven reserves"?
a. Storage areas for emergency supplies of fossil fuels
b. Areas where there are known stores of untapped fossil fuels held in reserve for the future
c. Known sources of untapped fossil fuels that are expected to be both economically and technologically recoverable
d. Sources of fossil fuels that have not yet been discovered, but which geologists have predicted to exist

2. To the best of our knowledge, how many years will we be able to rely on conventional oil reserves as a fuel source?
a. 10 years
b. 20 years
c. 40 years
d. 80 years

3. Which of the following describes the process of fracking?
a. Piping a mixture of water, sand, and chemicals deep underground to force natural gas to the surface
b. Blasting cracks, or fractures, in Earth's surface to search for natural gas
c. Drilling for oil under the ocean's surface
d. Surface mining processes that separate oil from sand or clay sediments

4. Examine Infographic 20.6. According to the data, which of the following has the highest boiling point?
a. Jet fuel
b. Gasoline
c. Naphtha
d. Fuel oil

5. How much more proven oil is located in the Middle East than in Canada?
a. 37 times as much
b. 77 times as much
c. 200 times as much
d. 4 times as much

6. Why do analysts think that oil and gasoline will never be inexpensive again?
a. Oil supplies are controlled by political enemies of developed nations.
b. The world has passed peak oil production.
c. Steps that prevent oil spills from ever happening again will keep the price of oil high.
d. As oil companies spend money on research and development of alternative energy sources, those costs will be passed on to consumers through higher fuel costs.

WORK WITH IDEAS

1. Describe the process of conventional oil extraction. Include in your description the following terms: primary production, secondary production, tertiary production, pumpjack, cap rock, and oil reservoir.

2. Compare and contrast the pursuit of conventional oil reserves and nonconventional oil sands reserves. What are the advantages and disadvantages of each?

3. In what ways might natural gas be a better fossil fuel option than oil? What are the drawbacks to natural gas recovery and use?

4. If oil and natural gas are formed from once-living organisms, why are they considered to be nonrenewable resources?

5. Should Canada continue to develop its nonconventional oil and gas resources? If so, should additional environmental protection be required? Support your answer.

6. In addition to driving less and/or driving more fuel-efficient cars, what can you do to reduce your overall oil consumption?

ANALYZING THE SCIENCE

The following graph shows the costs of extracting oil in different parts of the world.

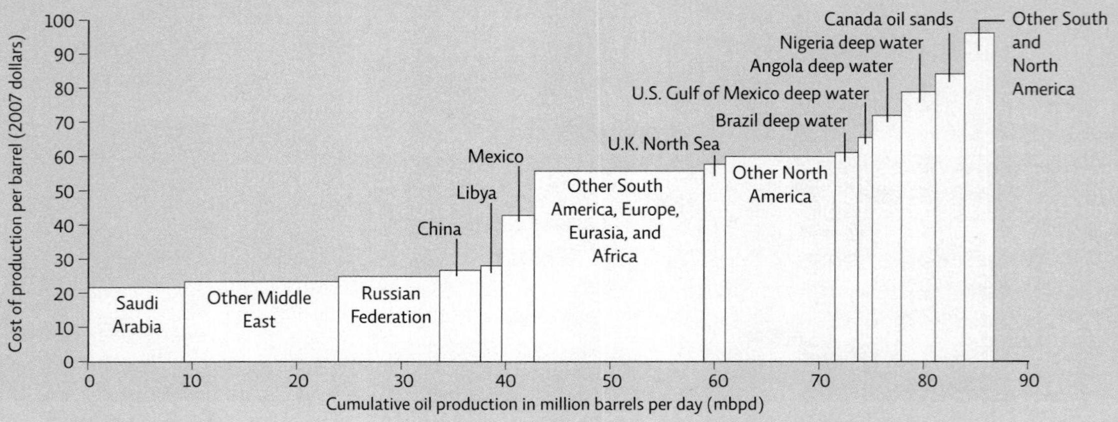
Cost of production per barrel (2007 dollars)
0
10
20
30
40
50
60
70
80
90
100
Cumulative oil production in million barrels per day (mbpd)
Saudi Arabia
Other Middle East
Russian Federation
China
Libya
Mexico
Other South America, Europe, Eurasia, and Africa
U.K. North Sea
Other North America
Brazil deep water
U.S. Gulf of Mexico deep water
Angola deep water
Nigeria deep water
Canada oil sands
Other South and North America

INTERPRETATION

1. Approximately how much oil is produced in "Other North America" compared to Saudi Arabia and "Other Middle East"? If you were the president of ExxonMobil, where would you suggest drilling for oil, given the information in this graph?
2. Why is the bar for "Other Middle East" short and wide while the bar for "Canada oil sands" is tall and narrow? Why is it cheaper to extract oil in Saudi Arabia than in Canada?
3. Calculate the cost per day to produce oil in Saudi Arabia and compare it to the cost of oil from the Canadian oil sands. (Hint: Estimate the height of each bar to get the cost to produce a barrel; then estimate the width of each bar to get the total number of barrels in a given region. Divide the cost by the total number of barrels to get the cost to produce oil in that region.)

ADVANCE YOUR THINKING

4. The graph lists only the economic costs associated with oil production. What specifically are the other costs (including environmental, social, and economic)? How do we decide when the costs outweigh the benefits of oil extraction and production?
5. In your opinion, when do the environmental costs (externalities) outweigh the benefits (status quo) of a fossil fuel economy?
6. How did we come to rely so heavily on fossil fuels, and what are the consequences of our dependence? Do you think we will be able to switch to an alternative fuel source before we run out of fossil fuels?

EVALUATING NEW INFORMATION

Oil from the Canadian oil sands is transported via pipelines. The proposed Northern Gateway pipeline would carry oil from Alberta, through British Columbia to the port city of Kitimat for access to Asian markets. This route takes the pipeline through rural and urban areas as well as First Nations' lands and faces opposition by those concerned about environmental risks of the pipeline and those opposed to extracting oil sands oil in general. Proponents point to the economic benefits that come with the pipeline.

Go to the following websites and read the articles about this issue: www.pipeupagainstenbridge.ca/ and www.northerngateway.ca/.

Evaluate the websites and work with the information to answer the following questions:

1. Are these reliable information sources? Do they have a clear and transparent agenda?
   a. Who runs these websites? Do the organizations' credentials make them reliable or unreliable? Explain.
   b. Do you suspect any bias on the part of these organizations? What might that bias be and how could you verify their major claims?

2. Briefly summarize what you believe are the most important pros and cons of the Northern Gateway pipeline. Based on your reading, do you support this pipeline project? Justify your answer.

3. What else would you like to know about this project? Do an Internet search to find more information. List at least two more sources, including their urls. In addition, answer Questions 1a and 1b for each source. Based on the new information you have gathered, has your opinion about the project changed? Why or why not?

MAKING CONNECTIONS

FRACKING IN CANADA

Background: Western Canada has the biggest natural gas shale fracking operations in North America, with several major sites in British Columbia and Alberta supported in part by government financial incentives such as reduced royalties and assistance in road building and pipeline costs. The regulation of shale gas production on private land is largely controlled by each provincial or territorial government. In early 2011, Quebec Environment Minister Pierre Arcand announced a halt to all shale gas development in that province while the process and its impacts on the nearby St. Lawrence River are more thoroughly studied.

Case: You are a resident of Quebec, living in a community near a proposed natural gas drilling site. Shale gas development could bring a big economic boost to your area but the environmental concerns about fracking have alarmed some community members, who oppose it. You have been asked to research what is known about the environmental impacts of fracking and present the information at a public consultation meeting, so members of council can make an informed decision about what is right for the community.

1. Be sure to include the following in your explanation:
   a. A description of the fracking process and where it is currently being used on a regional and local scale
   b. Possibilities for economic improvement and job creation related to the project.
   c. General environmental concerns related to fracking.

2. Prior to your move to Quebec, you lived in Alberta and became familiar with the much publicized case of Jessica Ernst and her contaminated well water, contamination that occurred only after drilling began near her home. You decide to present a detailed case study about Jessica Ernst's problems. Be sure to include the following in your presentation:
   a. A detailed report of Ernst's case, including her credentials, the particular problems she faced, and the testing record of her well water
   b. Responses of oil companies to the people affected by fracking
   c. Responses from Alberta Environment and Water and other government agencies to both the oil companies and the people affected by fracking

3. In your summation to the council, provide your opinion of fracking and whether or not it should be permitted in your community.
   a. What is your recommendation to the council? Why?
   b. What other types of information might the council members want before they make a decision? Where would you suggest they go to find the answers to their questions?

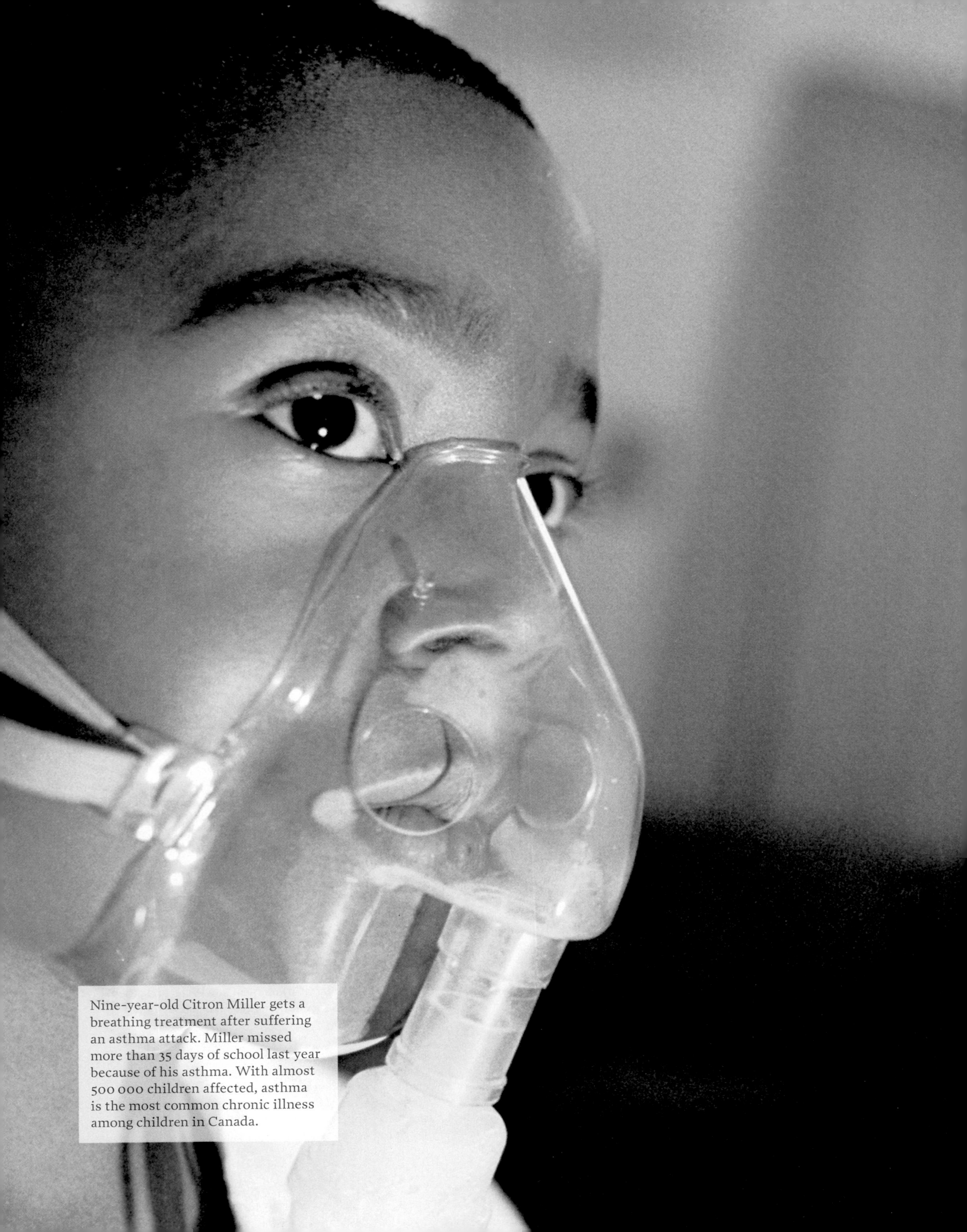

Nine-year-old Citron Miller gets a breathing treatment after suffering an asthma attack. Miller missed more than 35 days of school last year because of his asthma. With almost 500 000 children affected, asthma is the most common chronic illness among children in Canada.

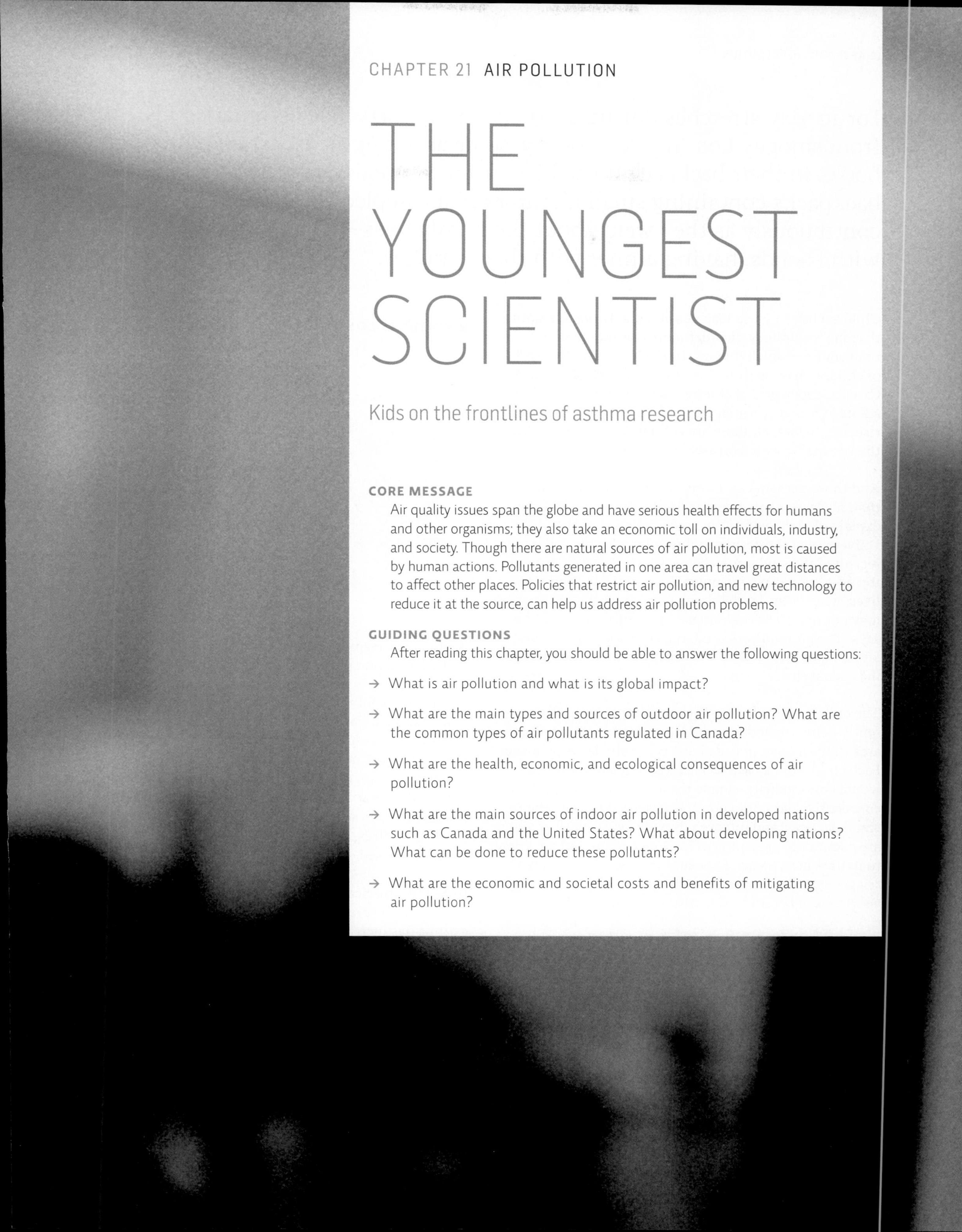

THE YOUNGEST SCIENTIST

Kids on the frontlines of asthma research

CORE MESSAGE

Air quality issues span the globe and have serious health effects for humans and other organisms; they also take an economic toll on individuals, industry, and society. Though there are natural sources of air pollution, most is caused by human actions. Pollutants generated in one area can travel great distances to affect other places. Policies that restrict air pollution, and new technology to reduce it at the source, can help us address air pollution problems.

GUIDING QUESTIONS

After reading this chapter, you should be able to answer the following questions:

→ What is air pollution and what is its global impact?

→ What are the main types and sources of outdoor air pollution? What are the common types of air pollutants regulated in Canada?

→ What are the health, economic, and ecological consequences of air pollution?

→ What are the main sources of indoor air pollution in developed nations such as Canada and the United States? What about developing nations? What can be done to reduce these pollutants?

→ What are the economic and societal costs and benefits of mitigating air pollution?

For 10-day stretches during 2003 and 2004, forty-five asthmatic kids from smoggy Los Angeles County, some as young as 9, carried more than books in their backpacks. The kids, from two regions of the county, wore backpacks containing small monitors that sampled the air around them continuously as they went about their daily lives—going to class, playing with friends, having dinner with their families.

Children have a lot to teach us about **asthma**. It's not just that the respiratory disorder more commonly afflicts kids than adults—which it does. (In the United States, 10% of children have asthma, compared to 7.7% of adults. In Canada, about 15% of children have asthma, as do 8.5% of adults.) Or that research on children with asthma is on the rise. No, in fact, children are actually conducting some of the seminal research on asthma themselves.

And those personal air monitors are far more accurate than measurements taken at local monitoring stations, explains Ralph Delfino, an epidemiologist at the University of California, Irvine, who with his colleagues, recruited the students to help collect the data. "A monitoring station can be many miles from where the subject lives, where they go to school, etc.—so that measurement may not represent their actual exposure very well," he says. The air monitors used in Defino's study detected levels of harmful pollutants in the air surrounding each individual child.

The children selected for the experiment were all currently being treated for mild to moderate asthma and the area they all lived in had significant vehicle air pollution. Each child wore a backpack containing a monitor that would continuously sample the air around the child for 10 consecutive days; the monitor measured the amount of small particles and nitrogen dioxide (NO_2), both of which are commonly present in vehicle emissions and are known to irritate lung tissue. Exposure to these pollutants can trigger asthma symptoms such as wheezing, coughing, or shortness of breath in sensitive individuals.

Ten times a day, the children exhaled into a special bag that assessed their breath for nitric oxide (NO), a chemical marker of airway inflammation—a telltale symptom of asthma.

When the 10 days were over, Delfino compared the types of pollutants to which the kids were exposed with their nitric oxide levels at similar points in time. The goal was to paint a picture of the types and amounts of pollutants that exacerbate asthma.

◉WHERE IS LOS ANGELES, CALIFORNIA?

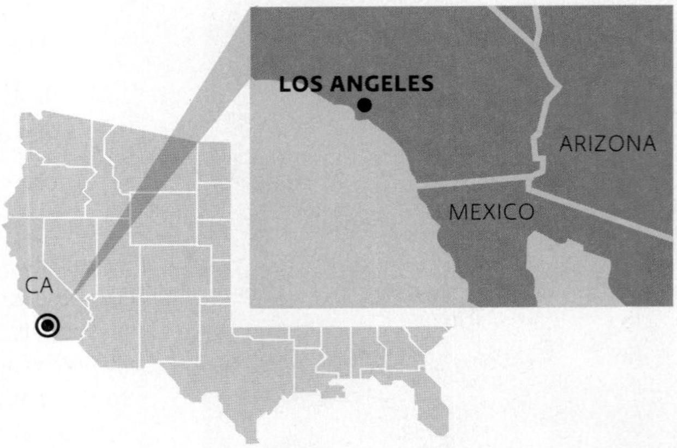
LOS ANGELES
ARIZONA
MEXICO
CA

Asthma is a respiratory ailment marked by inflammation and constriction of the narrow airways of the lungs. Delfino's work is important in part because asthma is one of the most common chronic childhood diseases in the United States, Canada, and other developed nations and a major cause of childhood disability. The United Kingdom has the highest incidence of asthma, with more than 15% of its population diagnosed in 2006 (compared to 14.1% in Canada and 10.9% in the United States). Developing nations are also seeing a rise in asthma, especially in urban centres. In all areas, asthma rates are likely underdiagnosed.

In North America, asthma is the leading cause of school absences. The prevalence of childhood and adolescent asthma has increased by 25-75% per decade during the period since 1960. Currently, 35.5 million people in the United States and Canada are diagnosed with asthma. Because there are more adults than children, overall more adults have asthma, but a larger percentage of children have the disease.

asthma A chronic inflammatory respiratory disorder characterized by "attacks" during which the airways narrow, making it hard to breathe; can be fatal.

The Los Angeles skyline is obscured by smog.

Researchers believe that **air pollution**—contaminants, either from natural sources or human activities, that cause health or environmental problems—may play a key role in the recent asthma spike. In addition, air pollution has been linked to cancer, infection, and other respiratory diseases—and it not only harms us, but also plants, animals, and even our buildings, bridges, and statues.

Outdoor air pollution has been under examination for many years.

As far back as the 1930s, scientists recognized a link between outdoor air pollution and human illness. In 1930, for instance, 63 people died and 1000 were sickened in Belgium when a temperature inversion—a situation that occurs when the temperature is higher in upper regions of the atmosphere than in the lower, causing pollutants to become trapped near Earth's surface—led to a sudden spike in lower atmospheric sulphur levels. And 1952 was the year of the famous Great Smog in London, UK, when acid aerosols trapped in the lower atmosphere in London killed 4000 people.

In many urban areas and in developed nations today, much of the air pollution comes from vehicle exhaust and industry emissions, including emissions from coal-fired power plants. While less-developed nations do have outdoor pollution, their biggest problem is often indoor air pollution from small particles released through burning solid fuels such as charcoal, wood, or animal waste. However, only in the last 25 years have scientists really been able to tease out the link between air pollution and asthma. [INFOGRAPHIC 21.1]

Infographic **21.1** | **AIR POLLUTION IS A WORLDWIDE PROBLEM**

↓ The World Health Organization (WHO) recognizes air pollution as a major threat to human health. Outdoor air pollution in urban areas is responsible for at least 1.3 million deaths annually. Indoor air pollution may be responsible for 2 million or more premature deaths per year, mostly in developing nations; about half of these deaths involve children under the age of 5.

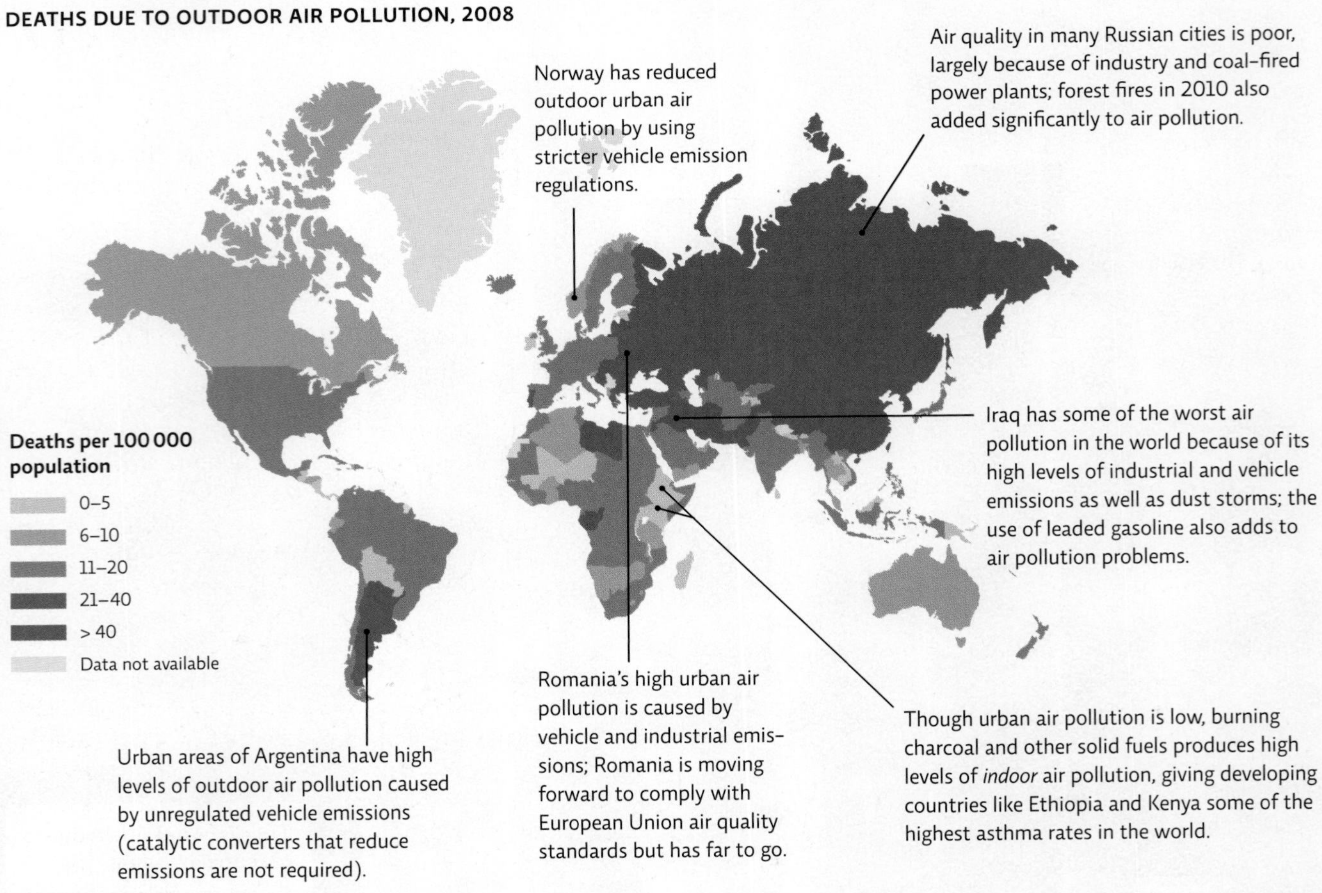
DEATHS DUE TO OUTDOOR AIR POLLUTION, 2008
Norway has reduced outdoor urban air pollution by using stricter vehicle emission regulations.
Air quality in many Russian cities is poor, largely because of industry and coal-fired power plants; forest fires in 2010 also added significantly to air pollution.
Deaths per 100 000 population
0–5
6–10
11–20
21–40
> 40
Data not available
Iraq has some of the worst air pollution in the world because of its high levels of industrial and vehicle emissions as well as dust storms; the use of leaded gasoline also adds to air pollution problems.
Urban areas of Argentina have high levels of outdoor air pollution caused by unregulated vehicle emissions (catalytic converters that reduce emissions are not required).
Romania's high urban air pollution is caused by vehicle and industrial emissions; Romania is moving forward to comply with European Union air quality standards but has far to go.
Though urban air pollution is low, burning charcoal and other solid fuels produces high levels of *indoor* air pollution, giving developing countries like Ethiopia and Kenya some of the highest asthma rates in the world.

↑ Controlled burns of agriculture fields, like this one near Madoc, Ontario, help clear land for more planting but release particulate matter (small particles) into the air, contributing to respiratory distress in sensitive individuals.

Outdoor air pollution includes chemicals and small particles in the atmosphere that can be either natural in origin—arising from sandstorms, volcanic eruptions, or wildfires—or come from humans, such as pollution released from factories and vehicles during the combustion of fossil fuels, or from burning any biomass such as wood, crop waste, or garbage. Of these anthropogenic sources, **primary air pollutants** are pollutants released directly from both mobile sources (such as cars) and stationary sources (such as industrial plants). In addition, some primary air pollutants react with one another or with other chemicals in the air to form **secondary air pollutants**. For example, **ground-level ozone** forms when some of the pollutants released during fossil fuel combustion react with atmospheric oxygen in the presence of sunlight. Acid rain is another common secondary pollutant.

air pollution Any material added to the atmosphere (naturally or by humans) that harms living organisms, affects the climate, or impacts structures.
primary air pollutants Air pollutants released directly from both mobile sources (such as cars) and stationary sources (such as industrial and power plants).
secondary air pollutants Air pollutants formed when primary air pollutants react with one another or with other chemicals in the air.
ground-level ozone A secondary pollutant that forms when some of the pollutants released during fossil fuel combustion react with atmospheric oxygen in the presence of sunlight.

Pollution can also move from the troposphere up into the stratosphere, a much thinner layer of the atmosphere that extends from 11 to 50 kilometres above Earth. This region contains the "ozone layer," an area with high ozone (O_3) concentrations. As we saw in Chapter 2, ozone in the stratosphere is significant because it serves as Earth's sunscreen, blocking some of the dangerous ultraviolet (UV) radiation from the sun. Air pollution can have grave impacts on this layer; for instance, chlorofluorocarbons (CFCs)—compounds that contain carbon, chlorine, and fluorine—can travel up into the stratosphere and destroy ozone.

Don't confuse ground-level ozone with stratospheric ozone depletion. These are two very different problems though they deal with the same molecule—O_3. Ozone in the stratosphere is a good thing—but you don't want to breathe it in as it can directly damage the sensitive tissue of the lungs. Even plants are damaged by the corrosive action of ozone.

Infographic **21.2** | **OUTDOOR AIR POLLUTION**

↓ There are many sources of outdoor air pollution, both natural and anthropogenic. These sources release primary pollutants, some of which may be converted to different chemicals (secondary pollutants). Prevailing winds transport pollution that reaches the upper troposphere or stratosphere around the globe—no area is immune to air pollution. Agricultural and industrial pollutants have been found in Arctic and Antarctic air, delivered by these prevailing winds.

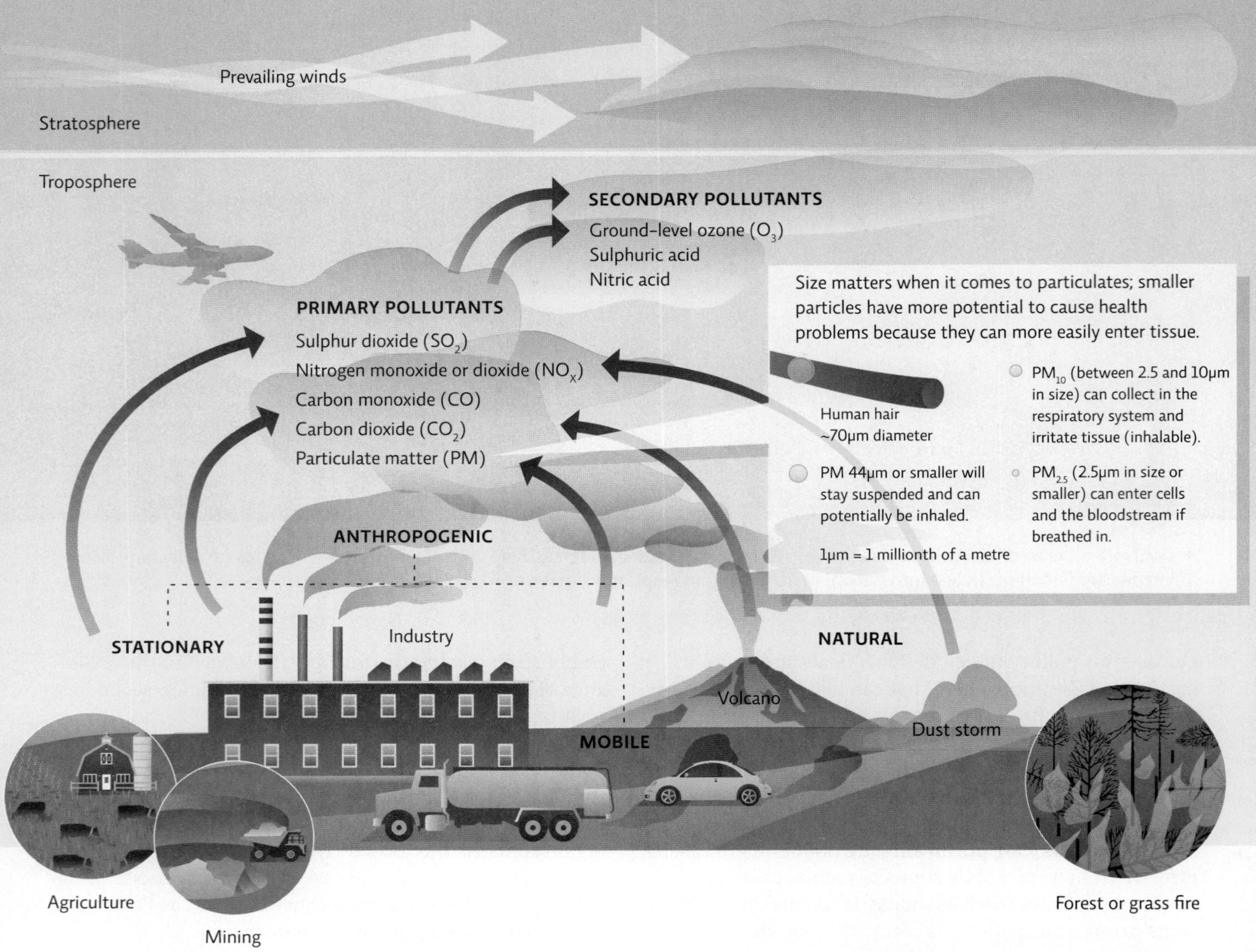
Prevailing winds
Stratosphere
Troposphere
SECONDARY POLLUTANTS
Ground-level ozone (O_3)
Sulphuric acid
Nitric acid
PRIMARY POLLUTANTS
Sulphur dioxide (SO_2)
Nitrogen monoxide or dioxide (NO_X)
Carbon monoxide (CO)
Carbon dioxide (CO_2)
Particulate matter (PM)
Size matters when it comes to particulates; smaller particles have more potential to cause health problems because they can more easily enter tissue.
Human hair
~70μm diameter
PM_{10} (between 2.5 and 10μm in size) can collect in the respiratory system and irritate tissue (inhalable).
PM 44μm or smaller will stay suspended and can potentially be inhaled.
$PM_{2.5}$ (2.5μm in size or smaller) can enter cells and the bloodstream if breathed in.
1μm = 1 millionth of a metre
ANTHROPOGENIC
STATIONARY
Industry
NATURAL
Volcano
Dust storm
MOBILE
Agriculture
Mining
Forest or grass fire

Outdoor air pollution, in all forms, is one of the most dramatic contributors to asthma. When the U.S. Environmental Protection Agency (EPA) started regulating pollution back in 1971, the agency did not know the extent to which air pollution could affect health; in fact, the EPA administrator noted at the time that the agency's clean air regulations were "based on investigations conducted at the outer limits of our capability to measure connections between levels of pollution and effects on man." Nevertheless, in 1971, the EPA set standards for the most common but problematic pollutants. Five of these were chemical air pollutants: sulphur oxides, carbon monoxide, nitrogen oxides, ground-level ozone, and lead; standards were also set for **particulate matter (PM)**—particles or droplets small enough to remain aloft in the air for long periods of time. [INFOGRAPHIC 21.2]

These six "criteria pollutants" are still considered the most problematic and health-threatening pollutants in the United States. A 1993 study published by Harvard University researchers helped establish the link between air pollution and impaired health. A follow-up study estimated that particulate pollution—which includes soot, ash, dust, smoke, pollen, and small, suspended droplets (aerosols)—accounts for 75 000 premature deaths per year in the United States. [INFOGRAPHIC 21.3]

Although all particulates reduce visibility, it is the smallest particles—those with a diameter less than 2.5 micrometres (μm), about 1/40th the diameter of a human hair—that aggravate asthma and other chronic lung diseases, and increase one's risk for death.

Outdoor air pollution has many sources.

Where does outdoor air pollution come from? The burning of fossil fuels commercially, industrially, and residentially contributes heavily to outdoor air pollution. Coal and oil burning releases emissions that can produce **smog**—a term coined by combining the words "smoke" and "fog"—which is a hazy air pollution that contains a variety of pollutants, including sulphur dioxide (SO_2), nitrogen oxides (NO and NO_2—together expressed as NO_X), particulates, and ground-level ozone.

Factories, incinerators, and mining operations also release pollutants. Industrial pollution releases what is called **point source pollution**, in that it is possible to identify the pollution's exact point of entry into the environment. Hypothetically, point source pollution is easier to monitor and regulate than is **non-point source pollution**—pollution from dispersed or mobile sources like vehicles and lawn mowers.

Agriculture is yet another source of non-point source outdoor pollution. Toxic pesticides sprayed on crops can become airborne and drift as far as 30 kilometres; confined animal feeding operations produce significant odour problems and particulate pollution; and animal waste contributes to global warming by releasing the greenhouse gas methane.

In his study in Los Angeles County, Delfino found that particles contained in diesel fuel exhaust were among the worst asthma culprits. In addition, particulate levels were higher in Riverside, one of two regions he tested; the researchers concluded that Riverside had more pollution because it was downwind of the main urban areas in L.A.

In addition to the six criteria pollutants, the EPA also recognizes 187 hazardous air pollutants that can have adverse effects on human health, even in small doses. These toxic substances may cause cancer, developmental defects, or may damage the central nervous system or other body tissues. They include *volatile organic compounds* (VOCs),

particulate matter (PM) Particles or droplets small enough to remain aloft in the air for long periods of time.

smog Hazy air pollution that contains a variety of pollutants including sulphur dioxide, nitrogen oxides, tropospheric ozone, and particulates.

point source pollution Pollution that enters the air from a readily identifiable source such as a smokestack.

non-point source pollution Pollution that enters the air from dispersed or mobile sources.

Infographic **21.3** | **THE HARVARD SIX CITIES STUDY LINKED AIR POLLUTION TO HEALTH PROBLEMS**

→ To see if there was a link between health and small particulate pollution ($PM_{2.5}$; that is, particulate matter smaller than 2.5 micrometres) researchers at Harvard University compared death rates and $PM_{2.5}$ levels in six U.S. cities. The data reveal that as particulate pollution increases, so does the mortality (death) rate. $PM_{2.5}$ pollution and deaths went down in every city between the two sampling periods evaluated—likely a result of stricter pollution limits in the 1990s.

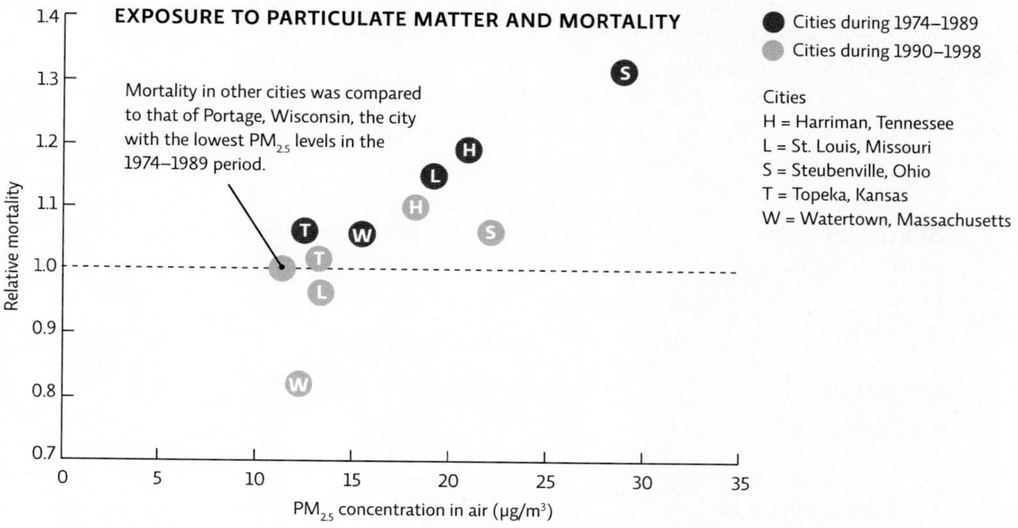
EXPOSURE TO PARTICULATE MATTER AND MORTALITY
Mortality in other cities was compared to that of Portage, Wisconsin, the city with the lowest PM2.5 levels in the 1974–1989 period.
Cities during 1974–1989
Cities during 1990–1998
Cities
H = Harriman, Tennessee
L = St. Louis, Missouri
S = Steubenville, Ohio
T = Topeka, Kansas
W = Watertown, Massachusetts
Relative mortality
1.4
1.3
1.2
1.1
1.0
0.9
0.8
0.7
0
5
10
15
20
25
30
35
PM2.5 concentration in air (μg/m3)

Infographic **21.4** | **SOURCES AND EFFECTS OF AIR POLLUTANTS**

CRITERIA AIR POLLUTANT	SOURCE	HEALTH/ENVIRONMENTAL EFFECTS
CARBON MONOXIDE (CO) From incomplete combustion of any carbon-based fuel (and most combustion is incomplete)	Vehicles, forest fires, volcanoes	Interferes with red blood cells' ability to carry oxygen; causes headaches and can lead to asphyxia (death)
SULPHUR DIOXIDE (SO_2) From burning fossil fuels, which all contain sulphur (e.g. coal)	Industry, volcanoes , dust	Respiratory irritant; harms plant tissue and can be converted to sulphuric acid which damages plants, aquatic creatures, and concrete structures
NITROGEN OXIDES (NO_x = NO AND NO_2) From the reaction of nitrogen in fuel or air with oxygen at high temperatures (usually during combustion of a fuel)	Vehicles, industry, nitrification by soil and aquatic bacteria	Respiratory irritant; can increase susceptibility to infection; can be converted to nitric acid, which damages plants, aquatic organisms, and even concrete structures; overfertilizes ecosystems; can cause eutrophication (see Chapter 16)
GROUND-LEVEL OZONE (O_3) Formed from reactions between NO_x and VOC with oxygen in the presence of sunlight	Vehicles are the main source of NO_x; VOCs can be released from manufactured products or be released directly by industry. Some are also released by trees and other plants.	Respiratory irritant that reduces overall lung function; can reduce photosynthesis in plants
PARTICULATE MATTER (PM) Tiny airborne particles or droplets, smaller than 44 micrometres. The smaller the particle, the more dangerous it is for tissue.	Released during the combustion of any fuel or activity that produces dust; also produced by forest fires, dust storms, and even sea spray	Respiratory irritant; can reduce respiratory and cardiovascular function; reduces visibility; particles can end up in aquatic or terrestrial ecosystem supplying nutrients or acids that can harm organisms that live there
LEAD (Pb) Additive to gasoline, paint, and other solvents; mostly phased out of the Canadian gas supply (99.8% is lead-free), but still used in some aircraft and in competition vehicles (racecars)	Lead-based paint in older homes and from other countries; leaded gasoline; soil erosion and volcanoes	Damages nervous, excretory, immune, reproductive, and cardiovascular systems; can accumulate in soils and in the tissues of organisms and can biomagnify up a food chain (see Chapter 3)

OTHER AIR POLLUTANTS	SOURCE	HEALTH/ENVIRONMENTAL EFFECTS
VOLATILE ORGANIC COMPOUNDS (VOCs) Organic molecules (hydrocarbons) that easily evaporate	Solvents, paints, glues, and other organic chemicals; plants naturally release VOCs	Those from human sources can be directly toxic or disruptive to living organisms, including humans; contribute to ground-level ozone formation
MERCURY (Hg) Naturally occurring element	Burning coal; mining and smelting operations; forest fires and volcanoes	A major neurotoxin that can disrupt development in embryos and young children; can bioaccumulate in individuals and biomagnify up the food chain
CARBON DIOXIDE (CO_2)	Burning carbon-based fuels such as fossil fuels; forest fires and normal decomposition	Nontoxic so no health effects at normal levels of exposure; greenhouse gas that contributes to climate change, affecting ecosystems worldwide

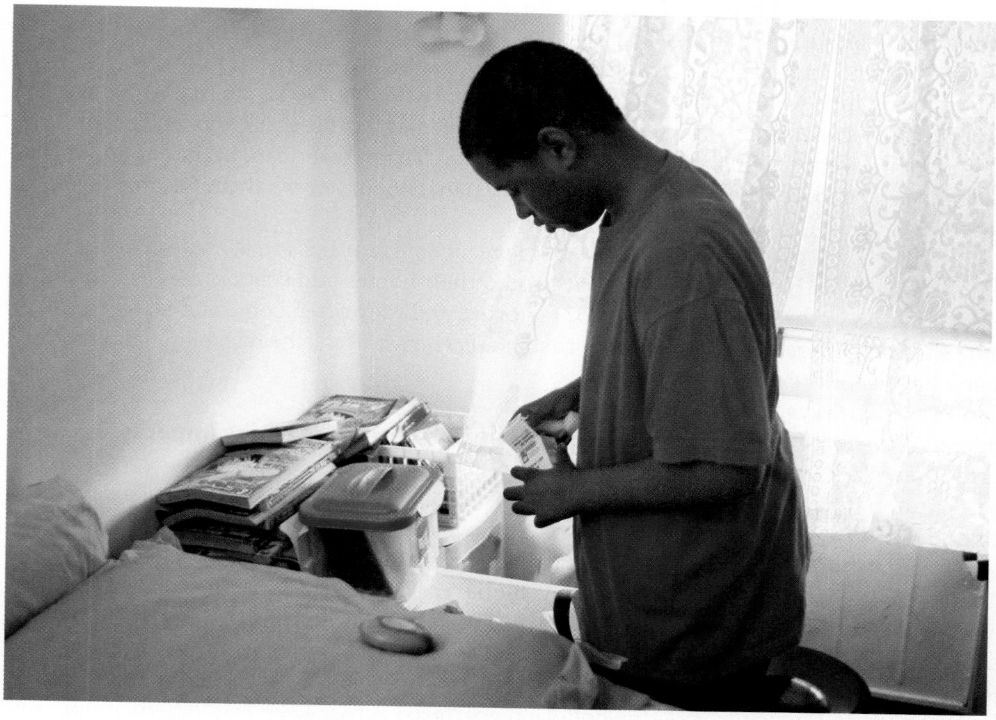

← Derrick Reliford, 14, in his Bronx, New York, home. At 10 years old, he participated in the New York University study to measure how much pollution Bronx children were exposed to. The researchers found that students in the South Bronx were twice as likely to attend a school near a highway as were children in other parts of the city. The South Bronx is home to some of the highest asthma hospitalization rates for children in New York City.

a variety of chemicals that readily evaporate but don't dissolve in water. VOCs are released by natural sources such as wetlands, and household products including paint, carpets, and cleaners; the main outdoor source is fossil fuel combustion. Another chemical class of concern is *polycyclic aromatic hydrocarbons* (PAHs), which are released as by-products from the combustion of wood, tobacco, coal, or diesel. And in 2007, the EPA ruled that greenhouse gases are air pollutants, giving the agency the authority to regulate carbon dioxide (CO_2) emissions. The Canadian government does not officially acknowledge CO_2 as being toxic, but it nonetheless was added to the CEPA Schedule (list) of Toxic Substances in 2005, so that it can be regulated. [INFOGRAPHIC 21.4]

The air we breathe affects our lungs, especially those of children.

If anyone was born to study air pollution, it was Kari Nadeau. Growing up near smoggy Newark, New Jersey, she suffered as a child from terrible asthma and allergies, which she always suspected were related to the pollution surrounding her. Her mom was a public health school nurse and her dad worked for the EPA. "I always had these questions lingering about how much the environment affects people with asthma or allergies," she recalls. Was air pollution a culprit in her respiratory woes? Her instincts told her yes.

And now her research does, too. An assistant professor of pediatric immunology & allergy at Stanford University, Nadeau recently uncovered something surprising: when she looked at the blood collected from kids who came to her clinic from Fresno, California—kids who frequently had terrible asthma—she saw that they had different-looking immune systems than other California kids. Specifically, their immune system's "peacekeeping" functions didn't work as well in keeping asthma-producing inflammation at bay. This observation piqued her curiosity: "I thought, what's different in Fresno? So I went on the Internet and searched and saw that Fresno is the second-most polluted city in the country," she says.

Nadeau is now collaborating with scientists in Fresno to study the link between immune system regulatory cells, asthma, and air pollution. To understand the differences among kids from Fresno versus other areas in California, she collected blood from 71 asthmatic children who had spent their entire lives in Fresno, 30 healthy children from Fresno, 40 asthmatic kids from less-polluted Palo Alto, and 40 healthy kids from Palo Alto. She found that all the kids who grew up in Fresno had far higher levels of common pollutants in their blood than did the Palo Alto kids—and the higher their pollutant levels were, the more likely they were to have asthma. Based on her research, Nadeau thinks that air pollutants stifle the activity of genes responsible for maintaining normal immune system function—and that by doing so, they increase the risk for asthma and allergies.

Nadeau's work at Stanford suggests that PAHs are key to understanding the link between air pollution and asthma. "Levels of PAHs and the severity of the asthma were both directly related to the function of immune system cells in our patients," she says, and levels of PAHs were 7 times higher in Fresno, where asthma tended to be more common and severe, than in Palo Alto.

Other studies also support the link between asthma symptoms and air pollution. More children in the South Bronx are hospitalized for asthma than anywhere else in New York State, and since many Bronx children live or attend schools adjacent to congested highways, Bronx congressman José Serrano wondered if the two factors might be related. In 2002, he asked New York University environmental scientist George Thurston if he would be willing to conduct a study to find out. "We thought about it for a nanosecond, and then said, 'sure,'" Thurston recalls. In a study reminiscent of Delfino's, Thurston recruited 40 South Bronx fifth-graders to tote wheeled backpacks containing personal air monitors for a month while rating their respiratory symptoms 3 times a day. "You rolled it, so it wasn't really that heavy," Derrick Reliford, one of the students in the study told the *New York Times*. "They were the rock stars of the class—everybody wanted to help them with the backpacks," Thurston recalls.

The children came from four different schools, two of which were close to a highway and two of which were not. Thurston found that, sure enough, the children who went to schools or lived closer to highways were exposed to more air pollution—in particular, diesel fuel exhaust—and they also had more severe respiratory symptoms.

Low-income or minority areas often have some of the worst air. This raises questions of **environmental justice**—the concept that access to a clean, healthy environment is a basic human right. Polluting industries such as power plants or incinerators are often placed in areas where residents have less ability to fight for their rights—less money, less education, little or no voice in local government. In some cases, even when socioeconomic status is accounted for, minority communities still face more exposure to pollution than average, an example of **environmental racism**. A study conducted in Southern California by Brown University researcher Rachel Morello-Fosche found that one's risk for developing cancer from exposure to polluted air increased as income decreased. And in general, cancer risk was higher for minorities (Asian, African American, Latino) than for the majority (Caucasians) no matter what the income level.

Children of low-income families are at particular risk: as in the Bronx, their homes and schools are near major roads or factories, and they often come and go to school during rush hour when traffic is heaviest and smog forms.

Travelling pollution has far-reaching impacts.

One of the most problematic characteristics of air pollution is that it moves. Air pollution produced in one city can end up harming humans and other species halfway around the globe. For example, a large proportion of air pollution in Southern Ontario originates in the Ohio Valley, where it is released by tall smokestacks of coal-burning electrical power plants. Prevailing winds bring the pollution northeast across Lake Erie. There, it not only pollutes the air but also produces **acid deposition**—sulphur and nitrogen emissions that react with oxygen and water to form acids that can fall back to ground as acid rain or snow. Acidification of soil due to acid deposition can change the soil chemistry and mobilize toxic metals such as aluminum, hindering plants' ability to take up water. Acids leach nutrients from the soil, too, reducing the amount of calcium, magnesium, and potassium available to plants in topsoil. Taken together, these impacts can decrease plant growth, weaken plants so they are more vulnerable to disease or pests, and even kill them. This acid deposition remains a problem in the northeastern United States and eastern Canada. (See Chapter 13 for more on the pH scale and acidification). [INFOGRAPHIC 21.5]

Proof that air pollution can travel long distances can be found in the far north. Prevailing air currents pick up pollutants from the western United States, conveying them all the way to Lake Laberge in the Yukon, where the moisture condenses, forms clouds, and falls on the lake as rain or snow. Thus, even the most isolated regions on Earth are vulnerable to the effects of air pollution, because atmospheric and hydrologic circulation moves chemical and particulate pollutants around the globe.

Acid rain in eastern Canada, pollution in Lake Laberge, and stratospheric ozone depletion are **transboundary pollution** problems because regions that suffer from the pollution are not necessarily the ones that released the pollutants. This means that even if an area does not

---

environmental justice The concept that access to a clean, healthy environment is a basic human right.
environmental racism Occurs when minority communities face more exposure to pollution than average for the region.
acid deposition Precipitation that contains sulphuric or nitric acid; dry particles may also fall and become acidified once they mix with water.
transboundary pollution Pollution that is produced in one area but falls in a different area, possibly a different nation.

Infographic **21.5** | **ACID DEPOSITION**

↓ Burning fossil fuels releases sulphur and nitrogen oxides. These compounds react in the atmosphere to form acids. Acid rain, snow, fog, and even dry particles can fall to Earth as acid deposition with the potential to alter the pH of lakes and soil, damaging plant and animal life.

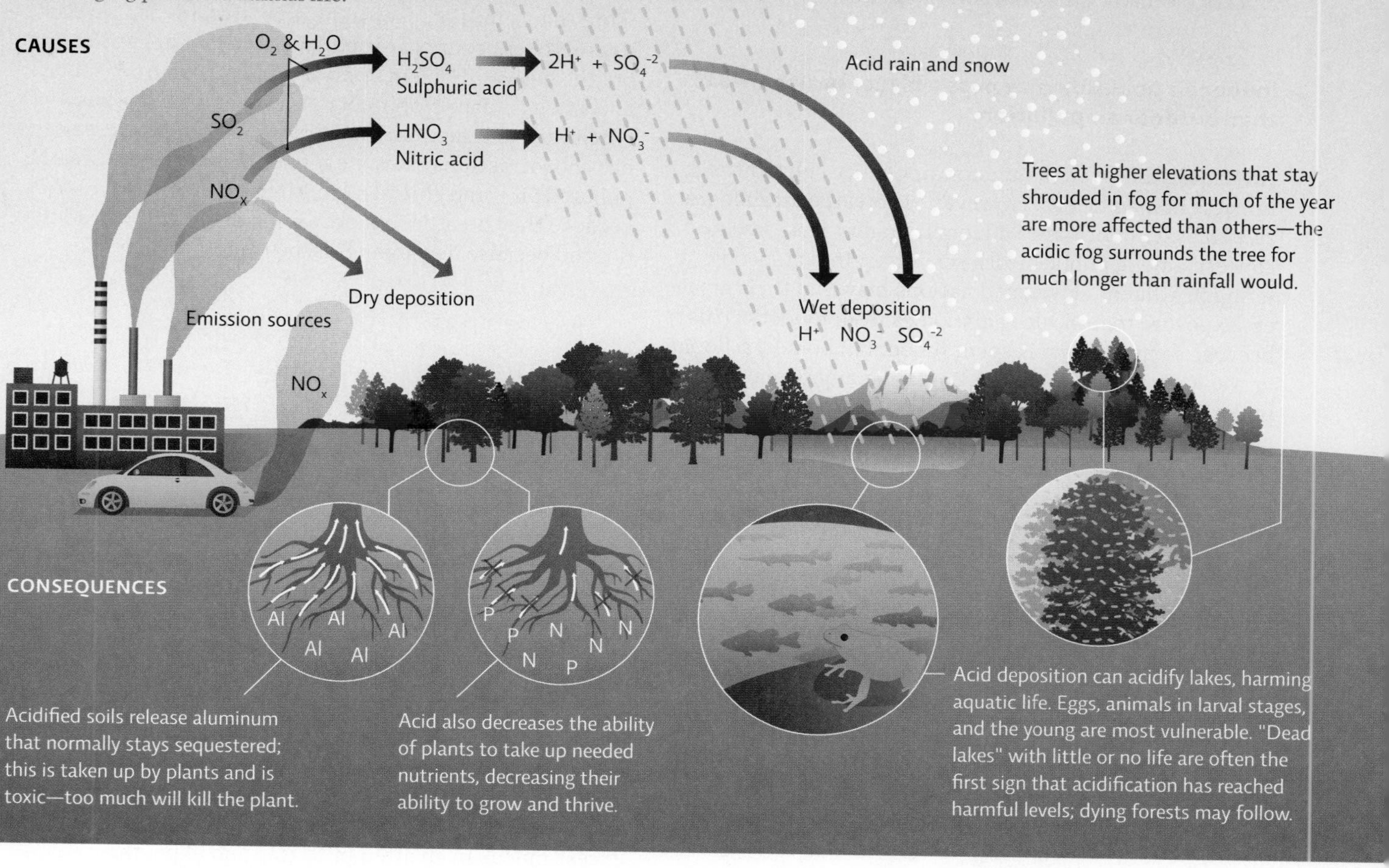
CAUSES
O_2 & H_2O
SO_2
NO_x
H_2SO_4
Sulphuric acid
HNO_3
Nitric acid
$2H^+ + SO_4^{-2}$
$H^+ + NO_3^-$
Acid rain and snow
Dry deposition
Emission sources
NO_x
Wet deposition
H^+ NO_3^- SO_4^{-2}
Trees at higher elevations that stay shrouded in fog for much of the year are more affected than others—the acidic fog surrounds the tree for much longer than rainfall would.
CONSEQUENCES
Al Al Al Al Al
P P N N N N P
Acidified soils release aluminum that normally stays sequestered; this is taken up by plants and is toxic—too much will kill the plant.
Acid also decreases the ability of plants to take up needed nutrients, decreasing their ability to grow and thrive.
Acid deposition can acidify lakes, harming aquatic life. Eggs, animals in larval stages, and the young are most vulnerable. "Dead lakes" with little or no life are often the first sign that acidification has reached harmful levels; dying forests may follow.

ACID DEPOSITION HAS DECREASED

↓ Restrictions imposed by the U.S. Clean Air Act, the Canadian Environmental Protection Act, the Canada-U.S. Air Quality Agreement, and the Geneva Convention on Long Range Trans-boundary Air Pollution have all helped to decrease acid deposition in the United States and Canada. Smokestack scrubbers remove sulphur from coal burning, reducing the SO_2 released. Emission-control technologies on vehicles, such as the catalytic converter, convert dangerous combustion by-products to safer emissions (such as converting NO_x to N_2). This reduces, but doesn't eliminate, these dangerous emissions.

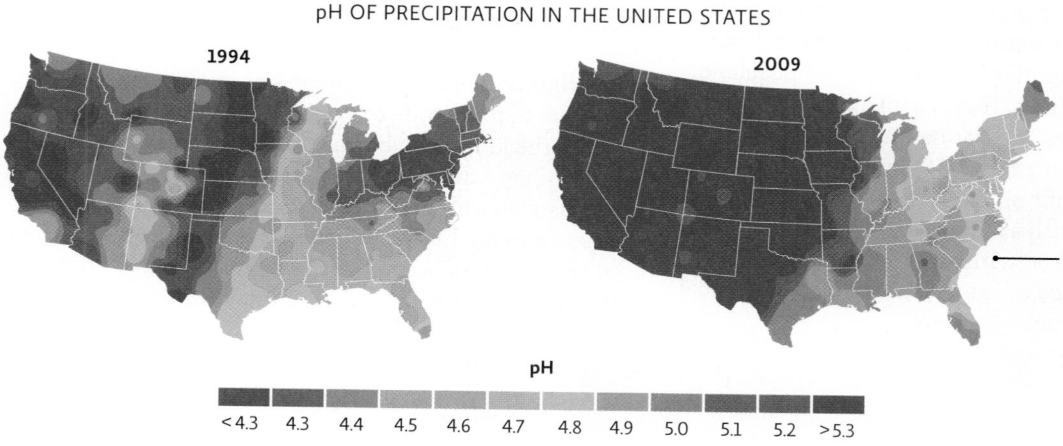
pH OF PRECIPITATION IN THE UNITED STATES
1994
2009
pH
<4.3 4.3 4.4 4.5 4.6 4.7 4.8 4.9 5.0 5.1 5.2 >5.3

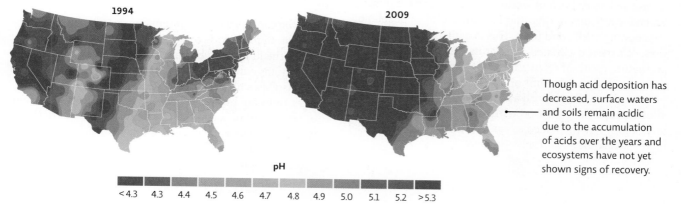

Though acid deposition has decreased, surface waters and soils remain acidic due to the accumulation of acids over the years and ecosystems have not yet shown signs of recovery.

produce pollution itself, its air may still be toxic. With Environment Canada's Air Quality Health Index available online, people can search for up-to-date air quality reports for many Canadian communities.

Indoor air pollution may pose a bigger health threat than outdoor air pollution.

As Nadeau discovered, outdoor air quality substantially impacts human health. However, we breathe air indoors as well as outdoors, and indoor air quality is a growing concern among public health scientists. In fact, people living in affluent, developed nations may find their greatest exposure to unhealthy air comes from indoors. This is because so much time is spent indoors in homes, schools, or the workplace; these areas contain many potential air pollution sources. For instance, cigarette smoke causes significant health problems, including eye, nose, and mucous membrane irritation; lung damage, which can exacerbate or cause asthma; and lung cancer. Items in our home, like paint, cleaners, and furniture, release VOCs, which can also cause health problems.

Outdoor pollutants can also find their way into our buildings. Radon is a naturally occurring radioactive gas produced from the decay of uranium in rock. It can seep through the foundations of homes and accumulate in basements; exposure to radon can cause lung cancer. Not every area has the type of rock that produces radon, but buildings constructed over areas where soil or groundwater is contaminated with VOCs, such as areas with underground chemical storage tanks, also present an infiltration risk. [INFOGRAPHIC 21.6]

In developing countries, where many people cook and heat with open fires, smoke and soot from burning wood, charcoal, dung, or crop waste are major sources of indoor pollution. Kirk Smith, a professor of environmental health at the University of California, Berkeley, has found that indoor fires increase the risk of pneumonia, tuberculosis, chronic bronchitis, lung cancer, cataracts, and low birth weight in babies born of women who are exposed during pregnancy. "Considering that half the world's households are cooking with solid fuels, this is a big problem," Smith says. "Current estimates are that indoor fires cause the premature death of 1.5 to 2 million women and children per year." Many nonprofit organizations are stepping up to meet this problem with a simple, $50 solar cooker that allows people to cook food without building a fire. This technology has the added advantage of not depleting local biomass resources for fuel.

Air pollution is responsible for myriad health and environmental problems.

The World Health Organization (WHO) estimates that more than 3 million people die prematurely each year as a result of exposure to air pollution. Respiratory ailments are the biggest problem because particulates from soot and smog damage respiratory tissue and increase susceptibility to infection. In addition, as Nadeau's research has shown, asthma rates are higher in people who breathe polluted air—and children are particularly sensitive because they breathe in more air for their size than adults do, and because developing tissue is more vulnerable.

> More than 3 million people die prematurely each year as a result of exposure to air pollution.

Delfino's team of researchers followed up their first study with a second in 2007 that included 53 students with asthma (aged 9–18) with air monitors strapped to their backs. The students also had to breathe into detectors that recorded how much air they were able to blow out from their lungs at once—a measure of lung function. The results suggested that high levels of particulate pollution actually decrease lung function.

Lungs are particularly vulnerable to these small particulates because they get so much exposure (we breathe all the time) and the tissue itself is delicate. Irritants like particles, dust, and pollen can cause the lungs to produce excess mucus in an attempt to trap and expel the irritant. The lining of the airways can become inflamed and swell; in people with asthma, the irritation may trigger muscle contractions which close off the airway completely. Particles smaller than 2.5 μm can actually penetrate cells of the lungs or enter the bloodstream, where they are delivered to other cells of the body. If these particles come from the combustion of fossil fuels or other industrial sources, they may contain toxic substances, leading to problems associated with toxic exposure.

Since the cardiovascular system depends on the respiratory system to provide oxygen for the body, anything that impairs the lungs also harms the cardiovascular system, which might explain the fact that people living in polluted areas also have higher rates of heart attacks and strokes. Cancer rates are higher in people exposed to air pollution, too—exposure to secondhand smoke and radon are the leading environmental causes of lung cancer, and exposure to smog and vehicle emissions is linked to increased risk for lung and breast cancer.

Infographic **21.6** | **SOURCES OF INDOOR AIR POLLUTION**

↓ For most people, the greatest exposure to air pollution comes from being indoors. There are many sources of air pollution in a home or other building, as these structures tend to trap pollutants, keeping concentrations high. One can reduce exposure by avoiding or limiting the use of carpets, upholstered items, and furniture made with toxic glue and formaldehyde. Safer cleaners and low-VOC paints are readily available. Simple behaviours like taking off your shoes before entering the house and using a vacuum equipped with a HEPA filter will also help—especially important if you have indoor pets. Good ventilation and properly working heating and air conditioning units help keep indoor air pollutants at bay.

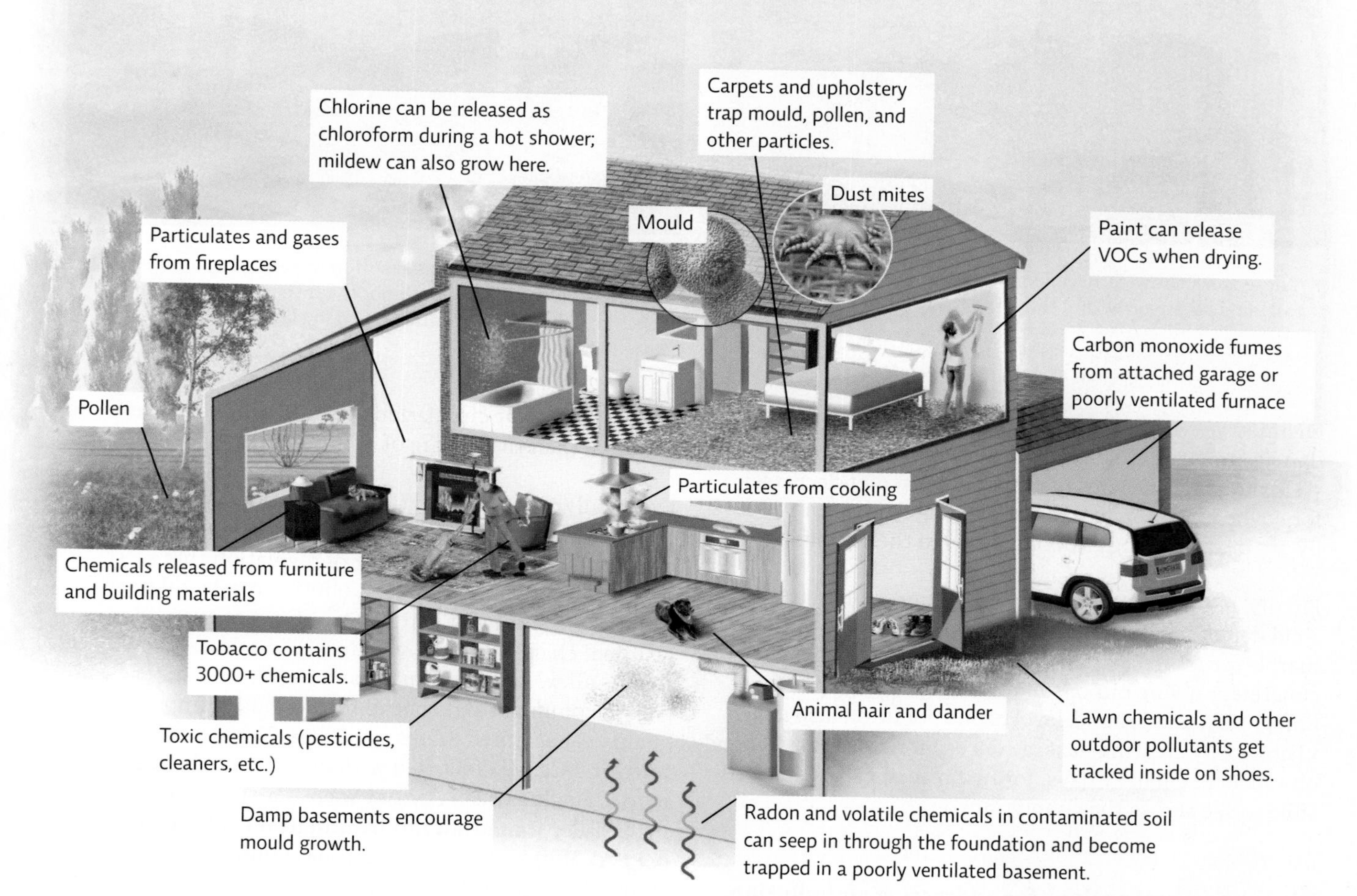
Chlorine can be released as chloroform during a hot shower; mildew can also grow here.
Carpets and upholstery trap mould, pollen, and other particles.
Dust mites
Mould
Particulates and gases from fireplaces
Paint can release VOCs when drying.
Carbon monoxide fumes from attached garage or poorly ventilated furnace
Pollen
Particulates from cooking
Chemicals released from furniture and building materials
Tobacco contains 3000+ chemicals.
Animal hair and dander
Lawn chemicals and other outdoor pollutants get tracked inside on shoes.
Toxic chemicals (pesticides, cleaners, etc.)
Damp basements encourage mould growth.
Radon and volatile chemicals in contaminated soil can seep in through the foundation and become trapped in a poorly ventilated basement.

Again, researchers have found that pollution causes babies to be born prematurely and with low birth weight. In addition, air pollution increases the risk that a pregnant woman will suffer from preeclampsia, a form of pregnancy-related hypertension. "We have learned that PAHs end up in the placenta and fetus," explains Beate Ritz, an epidemiologist at the University of California, Los Angeles, School of Public Health.

But humans aren't the only creatures suffering the ill effects of air pollution. Many animals suffer the same respiratory distress as humans—all lung tissue is very vulnerable to air pollution. Invertebrates as a group, especially aquatic ones, seem to be more directly impacted by air pollution (toxic effects or reproductive declines) than are vertebrates. But aquatic vertebrates like fish and amphibians are certainly feeling severe impacts. Declines in North Atlantic salmon populations have been linked to air pollution–induced water acidity. The higher acidity causes aluminum to build up in water—aluminum is toxic to fish, especially juvenile salmon.

Plant tissues are also vulnerable to pollutants like smog and ozone, as they cause direct damage to sensitive cell membranes. Exposure can damage a leaf's ability to photosynthesize, preventing healthy growth and compromising its survival. Lichens are particularly vulnerable to air pollution such as nitrogen emissions, and their decline is seen as a warning of the potential for damage to forests and crops. Together with changes in soil

↑ These trees show damage from the ozone pumped out of the tall vertical pipes (white) that ring the area.

↑ Exposure to acid rain has resulted in yellowing and loss of needles, decreasing overall photosynthesis and stunting the growth of these conifers at high elevations in the Austrian Alps.

chemistry—which can hinder plant growth—pollution damage to plant crops will ultimately cause global crop yields to fall: estimates of crop yields predict decreases for soybean, wheat, and maize ranging from 4% to 26% by 2030, amounting to dollar losses in the billions.

Finally, pollution damages buildings and monuments. Acid deposition literally eats away at limestone and marble structures; it can etch glass and damage steel and concrete, causing billions of dollars of damage per year. Smog, SO_4, and ground-level ozone pollution also lower visibility by creating haze, a concern for areas that depend on tourism. On hazy days, for instance, it can be impossible to see across the Grand Canyon.

We have several options for addressing air pollution.

Since air pollution often travels to areas that do not produce significant amounts of pollution themselves, regulating air pollution is a particular challenge. How does one country regulate pollution that travels through the atmosphere from another country?

In developed countries, the original approach to dealing with air pollution from human activities was to spread it out—*the solution to pollution is dilution*. Factories, power plants, and other point sources built tall smoke stacks to send emissions high into the atmosphere so that they wouldn't pool at the site of production. The idea was that if dispersed, the amount of pollution in any one area would be too low to cause a problem. But this approach simply doesn't work—industry releases too much pollution, and air circulation patterns cause some areas to get more than their share of pollution.

Eventually it became clear that regulation would be necessary. The typical approach in the 1970s was **command and control** regulation, a type of regulation that sets national limits on how much pollution can be released into the environment and imposes fines or even brings criminal charges against violators who release more than is allowed. Both Canada and the United States have passed command and control regulations to improve air quality. The United States passed its Clean Air Act in 1963, and Canada followed with our **Clean Air Act** in 1970. Canada's Clean Air Act was later subsumed under the broader **Canadian Environmental Protection Act (CEPA)** of 1999. These acts set maximum amounts, or air quality standards, for emissions of key pollutants or the presence of key pollutants in ambient air. Both Environment Canada and the provinces are responsible for monitoring air quality as well as for developing, implementing, and enforcing approved compliance plans. As a result of the Clean Air Act and related legislation, Canada has seen major reductions in common air pollutants. Removing lead from gasoline, for instance, reduced lead air pollution by 98% from 1970 levels. Sulphur pollution has also been significantly reduced.

A new Clean Air Act was adopted in Canada in 2006, containing measures to fight smog and greenhouse gas pollution. Under this Act, 2003 emissions of greenhouse gases are targeted to be cut by 45 to 65% by the year 2050, vehicle fuel consumption will be reduced, and there are targets for reducing ozone and smog by 2025. The

↑ Statue in Trafalgar Square, in London, England, shows erosion that exceeds normal weathering and is likely due to acid rain.

effectiveness of this Act has been challenged by opposition parties and the public, who have asserted that the Act does little to prevent climate change and more must be done. Canada and the United States have both been criticized globally for their weak positions on controlling greenhouse gases.

The fact that our air is cleaner today than it was in the 1960s—even with a larger population and more industry—is evidence that regulations can be effective. Still, these improvements are costly to industry and farmers, and that cost is usually passed on to consumers. For this reason, many individuals and groups oppose such policies, charging that the restrictions are excessive or that the government goes too far in trying to regulate emissions. Some environmentalists worry that if we weaken or dismantle the environmental legislation that protects our air and water (and by extension, our health and ecosystems), we face the return of a highly contaminated environment, compromised health, and diminished ecosystem function and services.

In addition to command and control regulation, there are other ways to curb pollution. One example is **green taxes** on environmentally undesirable actions, such as an extra tax on less fuel-efficient vehicles. **Tax credits**, reductions in the amount of tax one pays in exchange for environmentally beneficial actions, fall on the other side of the spectrum. Tax credits encourage consumers to pursue options that might be more expensive than conventional options (such as the purchase of a hybrid automobile); as more people buy the products, the industries that make them can scale up and bring prices down.

Governments also offer **subsidies**, free money or resources intended to promote environmentally friendly

command and control Regulations that set an upper allowable limit of pollution release which is enforced with fines and/or incarceration.

Clean Air Act First passed in 1970, this Act now falls under the broader Canadian Environmental Protection Act, which sets standards for dangerous air pollutants. The more recent, and different, Clean Air Act of 2006 additionally addresses smog and greenhouse gases.

Canadian Environmental Protection Act (CEPA) A broad Act passed in 1999 to set standards aimed at preventing pollution and protecting the environment and human health.

green tax Tax (fee paid to government) assessed on environmentally undesirable activities.

tax credit A reduction in the tax one has to pay in exchange for some desirable action.

subsidies Free government money or resources intended to promote desired activities.

Infographic 21.7 | **APPROACHES TO REDUCING AIR POLLUTION**

↓ We have many approaches that can be used to lessen air pollution, including technology to reduce emissions before a fuel is burned (see Infographic 19.7) and technology to capture emissions after a fuel is burned, as shown below. Some policy tools, such as cap-and-trade, encourage the use of "best available control technologies," that is, the current technology that releases the fewest amount of pollutants. There are economic costs to implementing these changes, but some benefits include new jobs, a competitive advantage for industries that can successfully reduce emissions, and a healthier society and ecosystem.

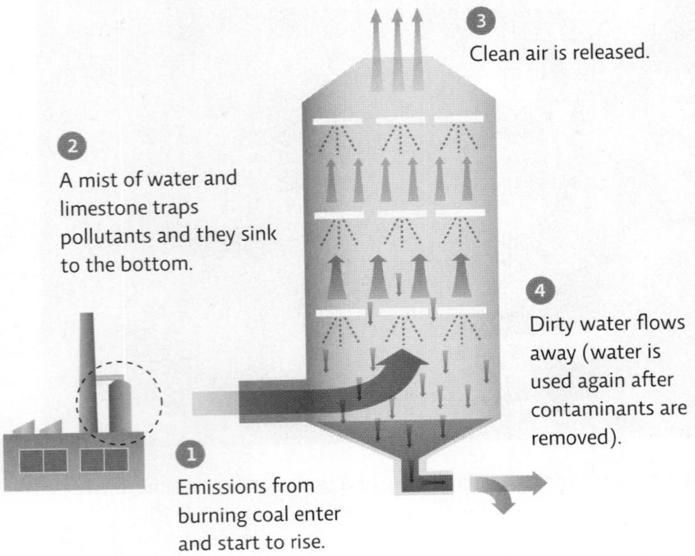
TECHNOLOGY: SMOKESTACK SCRUBBER
3 Clean air is released.
2 A mist of water and limestone traps pollutants and they sink to the bottom.
4 Dirty water flows away (water is used again after contaminants are removed).
1 Emissions from burning coal enter and start to rise.

↑ A variety of methods are available to capture harmful pollutants before they leave the smokestack. Smokestack scrubbers send the emissions though a mist of water and limestone to trap contaminants and prevent their release.

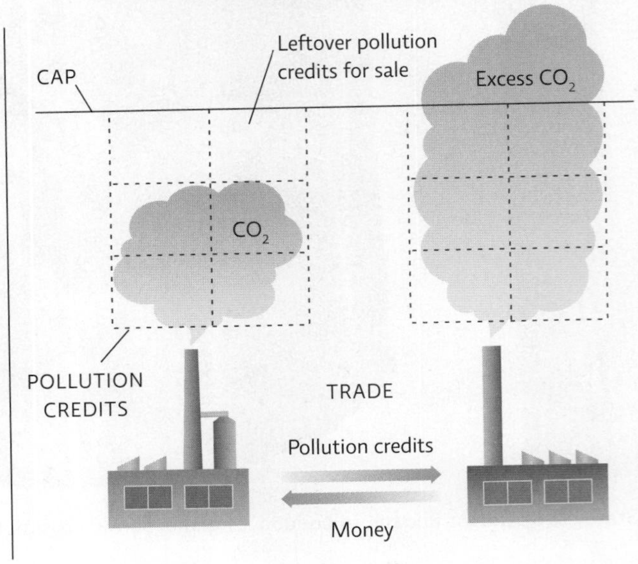
POLICY: CAP-AND-TRADE
CAP
Leftover pollution credits for sale
Excess CO_2
CO_2
POLLUTION CREDITS
TRADE
Pollution credits
Money

↑ In cap-and-trade programs, a cap (upper limit) is set for the total pollution that can be released in an area and pollution credits are issued to each producer that identify how much pollution each can release. If a producer implements changes that reduce pollution emissions below their total credit, they can sell their leftover credits to another producer who exceeds their allotment.

activities. And with **cap-and-trade**, also called *permit trading*, a government or regulatory agency sets an upper limit on emissions for a pollutant on a nationwide or regional level, and then gives or sells permits to polluting industries. Users that reduce their pollution emissions below what their permit allows can sell their remaining credits to other users who exceed their allotments. Over time, pollution levels can be reduced as the cap—or limit—is lowered. A cap-and-trade program successfully reduced sulphur pollution from coal-fired power plants in the United States in the 1990s. A downside to cap-and-trade programs is that pollution can become concentrated in areas where industries choose to buy additional permits rather than reduce emissions.

Technology can also play a big role in improving air quality. To improve indoor air quality we can install air filters and better ventilation systems. To curb pollution emissions from manufacturing, end-of-pipe solutions like scrubbers, filters, electrostatic precipitators, and catalytic converters trap pollutants before they are released. In addition, technology can inspire cleaner methods for extracting energy out of fossil fuels, as is the case with "clean coal" technologies described in Chapter 19. [INFOGRAPHIC 21.7]

---

cap-and-trade Regulations that set upper limits for pollution release. Producers are issued permits that allow them to release a portion of that amount; if they release less, they can sell their remaining allotment to others who did not reduce their emissions enough.

Mitigating or preventing air pollution costs money, but it is money well spent because it prevents far greater losses down the line—especially in terms of human health. According to a 2003 study published in the *Journal of Allergy and Clinical Immunology*, the average annual cost of care for an asthma patient in the United States is $4912, with 65% of that going to medications, hospital admissions, and nonemergency doctor visits. The remaining 35% goes to indirect costs like lost time at work. Ultimately, asthma costs Canada and the United States billions of dollars each year. Nadeau, Delfino, and others hope their research helps policymakers realize just how useful curbing pollution can be. "We're talking about the air we breathe," Thurston says. "There's nothing more communal than that."◉

POWERED BY CLEAN
2190
marta
ARTS CTR STA
GET YOUR HOME EXPOSED!
gamlshon

↑ Mass transit options that decrease the number of cars on the road will reduce air pollution. Buses that run on compressed natural gas emit fewer emissions overall but the particulates they release are very small so, while better than a traditional diesel bus, they are still not pollution free.

Select references in this chapter:

Cisternas, M., *et al.* 2003. *Journal of Allergy and Clinical Immunology*, 111: 1212–1218.

Delfino, R.J., *et al.* 2006. *Environmental Health Perspectives*, 114: 1736–1743.

Delfino, R.J., *et al.* 2008. *Environmental Health Perspectives*, 116: 550–558.

Dockery, D., *et al.* 1993. *New England Journal of Medicine*, 329: 1753–1759.

Laden, F., *et al.* 2006. *American Journal of Respiratory and Critical Care Medicine*, 173: 667–672.

Morello-Frosch, R., *et al.* 2002. *Environmental Health Perspectives*. 110: 149–154.

Nadeau, K., *et al.* 2010. *Journal of Allergy and Clinical Immunology*. 126: 845–852.

Spira-Cohen, A., *et al.* 2011. *Environmental Health Perspectives*. 119: 559–565.

BRING IT HOME

➲ PERSONAL CHOICES THAT HELP

Individuals can have an effect on air quality by researching the threats to their area, making appropriate behaviour changes, and supporting legislation that limits the production of air pollutants.

Individual Steps

→ Reduce your exposure to indoor air pollution by reducing your use of harsh cleaning products, synthetic air fresheners, vinyl products, and oil-based candles.

→ Avoid outdoor exercise during poor air quality days. Go to http://www.ec.gc.ca/cas-aqhi/default.asp?lang=En&n=450C1129-1 to find the air quality health index in your community or in another community in your province or territory.

→ Buy a radon detector and carbon monoxide detector for your house to keep you and your family safe.

Group Action

→ Organize a car-free day at your school, community, or workplace, to reduce emissions from vehicles.

→ Work with community leaders and businesses to sponsor a "free public transit" day.

Policy Change

→ If your community does not have public transit, ask community leaders to investigate bringing it to your area.

→ There are many groups working to improve our air quality. Find one in your region and see what issues they are addressing (see www.cleanair.ca).

UNDERSTANDING THE ISSUE

CHECK YOUR UNDERSTANDING

1. In the preindustrial era, what was the primary source of air pollution in cities?
   a. Volcanic eruptions
   b. Sandstorms
   c. Wood smoke
   d. Pollen

2. Given the relationship between asthma and air pollution, where would you raise a family to decrease the risk of asthma?
   a. In an area with low VOCs but moderate to high particulates
   b. In an urban area
   c. Away from major highway systems
   d. In a valley where most people use wood to heat their homes

3. Air pollution that results when chemicals in the atmosphere react to form a new pollutant is called:
   a. primary pollution.
   b. secondary pollution.
   c. point source pollution.
   d. particulate pollution.

4. According to the chapter, over 35.5 million people in the United States and Canada have been diagnosed with asthma. What is the estimated economic cost related to caring for the U.S. patients?
   a. $1000 per patient per year
   b. Billions of dollars per year
   c. About $1 million
   d. It is impossible to estimate the cost, due to unknowns such as lost time at work.

5. Which of the following is NOT a financial incentive to encourage less pollution?
   a. Green taxes
   b. Cap-and-trade
   c. Tax breaks
   d. Green subsidies

6. What pollutant did Canada add to the CEPA Schedule of Toxic Substances so that it can be regulated, despite denying that it is toxic?
   a. Ground-level ozone
   b. Sulphur
   c. Lead
   d. Carbon dioxide

WORK WITH IDEAS

1. Using the information presented in Infographic 21.6, explain why indoor air pollution is a cause of growing concern, especially with regard to health problems.

2. Compare and contrast point source and non-point source pollution. Provide examples of common pollutants from each source. Explain why industrial pollutants are found in even the most remote places on Earth.

3. List the six criteria air pollutants regulated by the EPA. How have the levels of the pollutants in the atmosphere changed since the Clean Air Act/CEPA were implemented? Based on your answer, would you say that the Clean Air Act has been successful at reducing air pollution?

4. Describe the policy of cap-and-trade. What are some downsides to it? Why do some members of industry dislike this policy as a means of dealing with increasing levels of CO_2?

5. Describe the types of problems that air pollution causes in ecosystems and human health. What is the overall impact in terms of worldwide mortality?

6. Where do you personally receive the most exposure to air pollution? What can you do to lessen its impact?

ANALYZING THE SCIENCE

The graph on the following page indicates levels of ground-level ozone and particulate matter that exceeded national reference levels (the level above which health or ecosystem problems occur) in areas and cities in Canada.

INTERPRETATION

1. What does the *y* axis represent? Choose a city and describe the data for that city.

2. The legend states that $PM_{2.5}$ data is given for 4 years. PM stands for "particulate matter." The number 2.5 represents the size of the particulate. From a health standpoint, why are the $PM_{2.5}$ values reported?

3. How many years of data are graphed for ozone? For particulate matter? Does the difference in the amount of time over which the data have been collected make a difference in your interpretation of these data?

ADVANCE YOUR THINKING

4. Which city would be the worst for your health, based on its levels of pollution? Why?

5. Based on the type of pollution present, what can you predict about the causes of pollution in Sault Ste. Marie versus Montréal? (Hint: Find the two cities on a map and compare the size of each city, their weather, and their primary industries.)

Comparison of the average percent of time smog pollutant reference levels were exceeded in select Canadian cities and the Fraser Valley

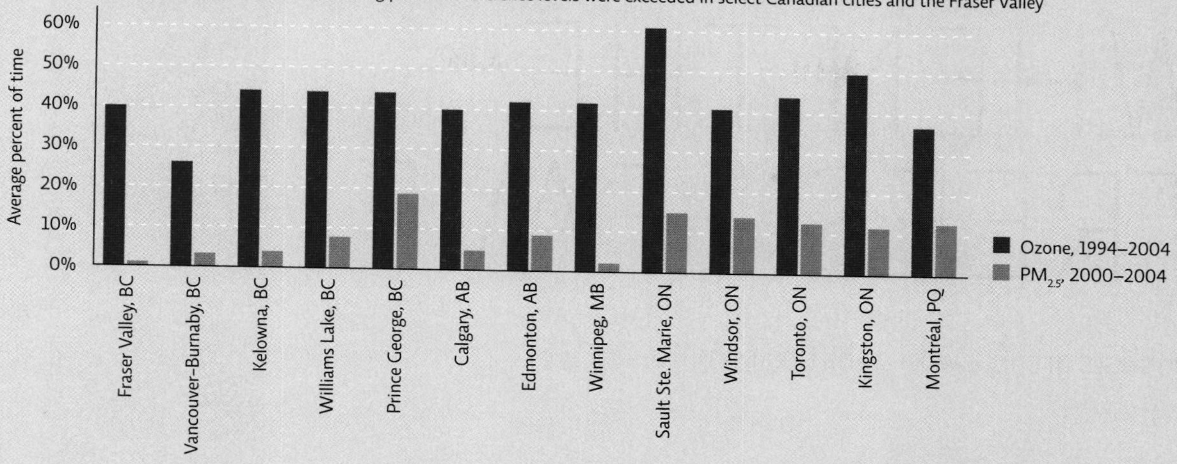
60%
50%
40%
30%
20%
10%
0%
Average percent of time
Fraser Valley, BC
Vancouver-Burnaby, BC
Kelowna, BC
Williams Lake, BC
Prince George, BC
Calgary, AB
Edmonton, AB
Winnipeg, MB
Sault Ste. Marie, ON
Windsor, ON
Toronto, ON
Kingston, ON
Montréal, PQ
Ozone, 1994–2004
$PM_{2.5}$, 2000–2004

EVALUATING NEW INFORMATION

The EPA is tasked with regulating pollutants in the United States. As part of this process, the agency collects and records data for many pollutants, but not all of them. The federal EPA is assisted in this endeavour by state EPAs. However, it is impossible to collect air quality data about every locality in the United States, so most data is collected in and around cities.

Go to www.stateoftheair.org. Choose a state and look at the data for any county. Record these data. Then, across the top bar, click on "Key Findings" and read about how the grades were calculated for each county. Finally, click on "Health Risks" and read about the specific health risks associated with both ozone and particulate matter.

Evaluate the website and work with the information to answer the following questions:

1. Is this a reliable information source? Does the organization have a clear and transparent agenda?
   a. Who runs the website? Do this organization's credentials make it reliable or unreliable? Explain.
2. What grade did the area you chose receive for both ozone and particulates?
   a. Based on what you read about how the grade was determined, do you feel the grading system is too lax or too strict?
3. Based on what you know about the levels of pollution in your own area in Canada and the effects of ozone and particulates on human health, do you believe that the regulations of the Clean Air Act / CEPA should be loosened, tightened, or remain the same? Should more areas be monitored, or is it sufficient to monitor cities? Why?

MAKING CONNECTIONS

LUNG DISEASE: WHERE YOU LIVE MAKES A DIFFERENCE

Background: Ground-level ozone and suspended particulate matter are two of the most common air pollutants. They are also causes of health problems, especially respiratory illnesses such as asthma. One of the primary sources of ground-level ozone is vehicles. Particulate matter comes from the combustion of organic matter, including wood, fossil fuels, and tobacco.

Case: You are a parent of two children, both of whom have asthma. You are also the primary caregiver for your aging grandmother, who suffers from chronic lung disease. You and your spouse have both been offered jobs in Phoenix, Arizona, and Hamilton, Ontario. The jobs in each location are equivalent, you have family in both locations, and both locations offer the benefits of large cities. Your decision has come down to which city would be the healthiest for your children and grandmother.

1. Prepare a PowerPoint presentation about each city for your family. In it, include the following:
   a. Pictures of each city
   b. Typical weather for each month of the year, including average ambient air temperature, rainfall, and average number of sunny days
   c. Geography of the area, including altitude, nearby mountain ranges, and bodies of water
   d. Population of each city
   e. Average number of days with air pollution warnings
   f. Typical industries, especially those that would contribute to air pollution
   g. Any natural sources of air pollution
   h. Average incidence of asthma or other lung problems in each city
2. At the end of your presentation, recommend one of the cities to your family, using the evidence you gathered to support your recommendation. Provide the sources you used so that your family can evaluate the thoroughness and reliability of your research.

WHEN THE TREES LEAVE

Scientists grapple with a shifting climate

CORE MESSAGE

One of the biggest environmental problems facing humanity today is climate change. Evidence overwhelmingly points to the fact that climate is changing and that human activity is predominantly responsible for the changes. Climate change is impacting species, ecosystems, and the health and well-being of people around the globe, with more changes to come. Science can help us evaluate the changes that are happening, investigate causes, and also provide information to help make sound policy for dealing with a changing climate.

GUIDING QUESTIONS

After reading this chapter, you should be able to answer the following questions:

→ What is the difference between climate and weather? Why is a change of a few degrees in average global temperatures more concerning than day-to-day weather changes of a few degrees?

→ What is the evidence that climate change is currently occurring? How do scientists determine past and present temperatures and CO_2 concentrations?

→ What natural and anthropogenic factors affect climate, and which are implicated in the climate change we are experiencing now? How might positive feedback loops affect climate?

→ What are the projections for future warming and what are the potential effects of further warming?

→ What actions can we take to respond to a world with a changing climate?

Flames engulf the sky and the ancient forest of the North Woods of Minnesota in a forest fire that occurred unusually early in the season.

Lee Frelich was examining a 700-year-old cedar tree when he first noticed the smoke curling up into the sky over Minnesota's great North Woods. Within an hour, his entire view was filled, so Frelich, an ecologist with the University of Minnesota, and his companions—a photographer and a journalist who had cajoled him into taking them on a tour of the iconic boreal forest—trekked up to the north end of Ham Lake, away from the calamity. They would remain trapped there, amid dense, ancient stands of spruce and fir, for 3 full days, as fire claimed over 30 000 hectares around them.

The fire's magnitude was not surprising. Neither, really, was the fact that the flames laid bare a patch of forest that had not burned since 1801.

What was surprising was the timing. It was the first weekend in May—unusually early in the year for such a tremendous fire, especially given that thick ice had just melted off the lake a few days before. Frelich's hiking companions, who knew that such fires tended to come in late summer, were surprised. Frelich wasn't. To him it was just one more not-so-subtle reminder that **climate change** was rapidly throwing this ancient landscape into flux.

Other reminders have become commonplace: earlier springs, later winters, some tree species dying off, others popping up in unexpected places. Frelich and his colleagues worry that if current projections hold true, the forests themselves could vanish—converted by stress and time into scrubland or savannah.

◉WHERE ARE THE NORTH WOODS OF MINNESOTA?

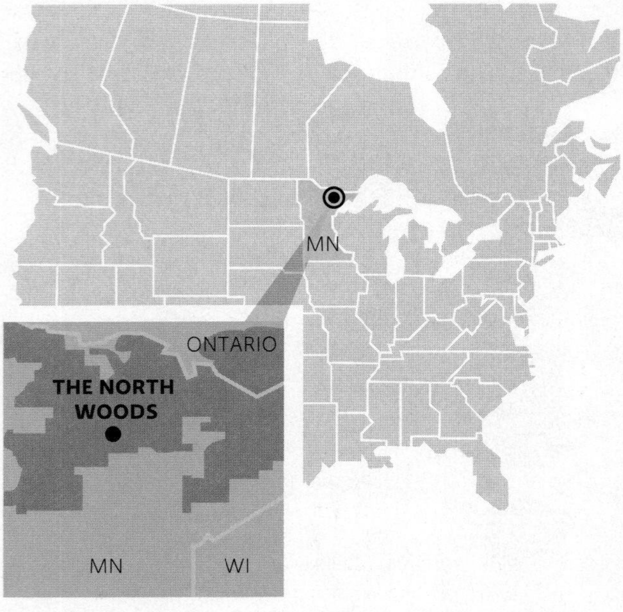
MN
ONTARIO
THE NORTH WOODS
MN
WI

Minnesota is not alone. In fact, the great North Woods are but one example of a whole planet in distress. In Africa and the American West, prairies are giving way to deserts (see Chapter 12). In tropical seas around the world, coral reefs are dying off (see Chapter 13). And everywhere, biodiversity is being threatened on a scale not seen since the last mass extinction 65 million years ago (see Chapter 9). Scientists say that all of these changes are occurring as global climate warms with unprecedented speed. What that might mean for the future of our planet is something they are still trying to figure out.

Climate is not the same thing as weather.

Weather refers to the meteorological conditions in a given place on a given day, whereas **climate** refers to long-term patterns or trends. Put another way, the actual temperature on any given day is the weather, while the range of expected values, based on location and time of year, is the climate. We use what we know about a region's climate to predict the weather: seasonal shifts come at about the same time each year, and winter lows and summer highs generally hover close to expected norms.

Weather can and does vary—sometimes considerably—from one day to the next. But no single weather event—no individual storm, flood, drought, or wildfire (not even one as colossal and ill-timed as the one that trapped Frelich and his friends)—can be attributed to global warming. Global warming is the province of climate. It refers to a rapidly shifting range of temperatures that scientists have measured in myriad locations around the world—not just a few warmer days here and there, but higher high temperatures, more and longer heat waves, earlier springs, and later winters.

↑ University of Minnesota ecologist Lee Frelich surveys the damage to the burnt forest immediately following the Ham Lake fire.

To be sure, climate change is not a new phenomenon. In fact, Minnesota's climate has been in flux for thousands of years. Fossil plant and pollen records show that, after the North American ice cap retreated—some 10 000 years ago—the climate warmed so dramatically that tree species' ranges shifted northward at a rate of 50 kilometres per century. Pines, oaks, and other deciduous species replaced the spruce trees that had covered most of the region. As summers became warmer, water levels fell, prairie plants took root, and birches and pine moved north. Then, about 6000 years ago, the climate cooled a bit, and trees began migrating south and west once again. (Such shifts are sometimes called "tree migration" but the more accurate term is "tree range migration," since they are really changes in a species' range.)

Scientists say things are changing much more quickly now. Instead of thousands of years, organisms that inhabit a given region—not only the boreal forest around Ham Lake, but ecosystems everywhere—might have just a few decades to adapt or migrate as climate change makes their current homes uninhabitable. In Canada the impacts of climate change will be seen even more quickly in the north, where sea ice, glaciers, and permafrost are already melting.

Evidence of global climate change abounds.

By all accounts, the forests of Northern Minnesota are places of uncommon majesty. Moose and deer meander through an ocean of trees—some of them close to 1000 years old—that grow out of colossal, hundreds-of-metres-high granite hills whose outcrops reflect the colours of day: pale, soft pink when the Sun rises, and deep, sombre rouge when it sets. The serenity belies an unsettling truth: this forest, which has existed for more than 3000 years—since the early days of the Roman Empire—and even inspired the U.S. Wilderness Act of 1964, could vanish within the next century.

To Frelich's well-trained eye, the signs are obvious. He has spent his entire adult life trekking through this ancient landscape, observing it and cataloguing the changes—centimetre by centimetre, leaf by leaf. So it's no surprise that when he looks on the placid landscape,

climate change Alteration in the long-term patterns and statistical averages of meteorological events.
weather The meteorological conditions in a given place on a given day.
climate Long-term patterns or trends of meteorological conditions.

he sees a catastrophe unfolding. In one patch of forest, scrawny, young birch trees bud several weeks ahead of schedule. In another, adult birches are dying off rapidly, leaving a graveyard of bony white trunks. "A long growing season is not good for this tree," Frelich says, "because it goes hand in hand with warmer soil, which the paper birch doesn't tolerate well." Meanwhile, red maple, a temperate species that grows as far south as Louisiana, but is far less common up north, is thriving in northeast Minnesota's Sea Gull Lake area. "When you have red maples growing as much as 4 feet in a single year," says Frelich, "you are not talking about a boreal climate anymore."

Regional climate data correlates well with the changes Frelich sees. In the past two decades, springtime has come ever earlier to the region—a week or two sooner than the historical average, according to the Minnesota Department of Natural Resource's climatology office. Nine of the state's 20 warmest years have been recorded since 1981. And for the first time in recorded history, Minnesota logged three mild winters in a row, each with record highs: 1997, 1998, and 1999.

Those data correspond to larger global trends. Overall, 2000 to 2010 is the warmest decade on record, since climatologists started keeping records back in 1850. The seven warmest years occurred between 1998 and 2010, with 2005 and 2010 tying for warmest individual year. According to the U.S. National Oceanic and Atmospheric Administration (NOAA), the global land average temperature increased by 0.66°C in 2010, compared to the 20th-century average. The year 2012 was a record-breaker in North America with U.S temperatures 1.8°C higher than normal. Even more telling, 2012 represented the 36th year in a row that exceeded the long-term average—the sign of a shift in climate rather than just natural variability of weather. In higher latitudes, it's even worse: temperatures have increased by 5°C or more. Sea surface temperatures have also increased, by almost 3°C in some places. In Canada, 2012 ranked as the 4th warmest year on record, with areas such as Windsor and Quebec City experiencing record-breaking spring and summer temperatures. The winter of 2011/2012 was 3.6°C higher than the baseline average from 1961-1990, and this was only the 3rd warmest winter. The winter of 2009/2010 holds the record so far, at 4.1°C above baseline.

A few degrees might not seem like much. But even such seemingly small changes in climate can have tremendous impacts on weather, and thus on natural ecosystems and human societies. [INFOGRAPHIC 22.1]

Infographic **22.1** | **A CHANGE IN AVERAGE TEMPERATURE: WHY DO ONLY A FEW DEGREES MATTER?**

↓ A shift of only a few degrees in the average global temperature will likely result in more frequent and extreme heat waves. Compared to the mid-20th-century average, we have increased about 1.0°C. To put this in perspective, at the end of the last ice age 10 000 years ago, Earth's average temperature was only about 3.0°C colder. At that time there was an ice sheet 1.6 kilometres thick as far south in North America as Chicago.

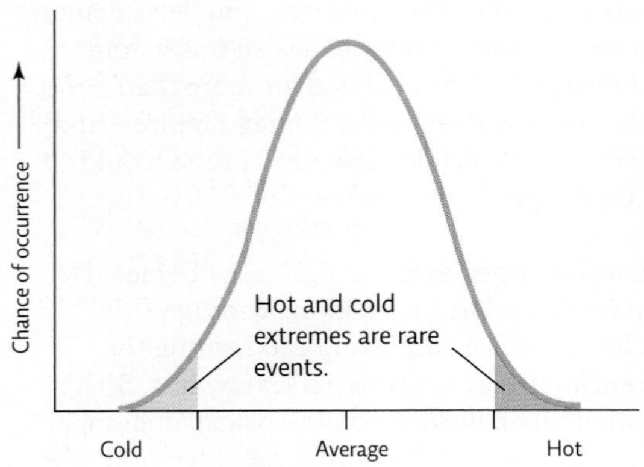
Chance of occurrence
Hot and cold extremes are rare events.
Cold
Average
Hot

In any stable climate time period (decade, century, millennium), there is some climate variability. In some years, the average temperature is hotter or colder than others but, on average, extremely cold and hot years occur infrequently. Most often, temperatures fall somewhere in the middle.

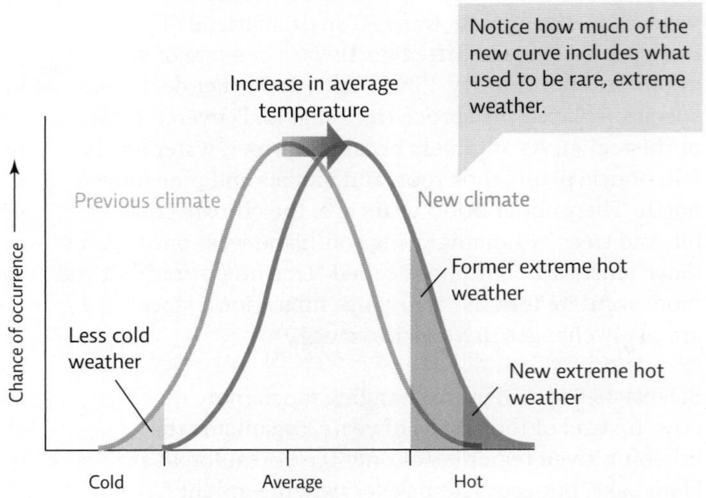
Notice how much of the new curve includes what used to be rare, extreme weather.
Increase in average temperature
Previous climate
New climate
Chance of occurrence
Former extreme hot weather
Less cold weather
New extreme hot weather
Cold
Average
Hot

An increase of a few degrees in the "average temperature" shifts the entire curve to the right. This means more years that have the "extreme" hot weather; it also means the affected area will likely set new records for extreme heat.

↑ Icebergs 60 metres tall, formerly part of the Greenland Ice Sheet, float into the North Atlantic Ocean. Because this ice used to be on land, it contributed to sea level rise when it fell into the ocean.

In fact, we are already seeing some big effects, especially in the Arctic, which is particularly vulnerable to climate change because warming that occurs there causes ice to melt, which triggers additional warming. Warming in the Arctic is also affecting weather around the world. A recent study shows that climate change is altering the northern hemisphere's polar jet stream, slowing it down and making it "wavier," meaning the peaks and troughs of the jet stream are extending farther north and south as the jet stream moves from west to east. This brings colder than normal air farther south and can slow down the movement of weather systems, sometimes causing them to stall and dump excessive amounts of snow or rain in one area. Essentially, cold Arctic air is reaching farther south than normal, more often than normal, and for longer periods of time, contributing to extreme winter weather events such as those seen in North America and Europe in recent years.

So how quickly is the Arctic warming? In 2011, researchers at Cambridge University evaluated the weather patterns of Ellesmere Island, Nunavut, in Canada's Arctic, and found that spring and summer temperatures were 11–16°C higher than previous years—making the climate similar to a boreal climate 1600 to 2250 kilometres farther south. But as large as that increase was, it was not entirely surprising. In 2007 and some subsequent years, researchers have measured record losses of sea ice (ice that floats on top of the ocean) in the region, including the record low that occurred in September 2012. Based on the current rate of melting, the entire Arctic could be completely ice free in the summers come 2035.

And it's not just Arctic ice that's melting. In fact, glaciers on the other side of the planet are dwindling just as fast as those in the Arctic, if not faster. Overall, sea ice is decreasing about 3–4% per decade. In Antarctica, the Larsen B Ice Shelf—a colossal sheet of sea ice, more than 210 metres thick and about half the size of Prince Edward Island—collapsed in 2002 in just a couple of weeks. This stunning spectacle, which took climatologists by surprise, was repeated in 2008, when the Wilkins Ice Shelf (also in Antarctica) collapsed, again in the space of 2 weeks.

And as ice on land melts, sea levels are rising—by an average of 10 to 20 centimetres during the last century. About half of this rise is due to land-based ice melt and the other half to thermal expansion—the expansion of water molecules as they heat up. So far, rising sea levels have displaced hundreds of thousands of coastal-dwelling people around the world. [INFOGRAPHIC 22.2]

Infographic 22.2 | **EVIDENCE FOR CLIMATE CHANGE**

↓ We know that a variety of factors can alter global temperature, but what is the physical evidence that temperatures have actually increased and that the climate is changing? In other words, what do we predict we would see if warming were occurring and what do we actually see when we test those predictions?

WARMER TEMPERATURES

If climate is indeed warming, we expect to see warmer global temperatures, on average, than in the recent past (more temperature anomalies in the direction of warming).

TEMPERATURE ANOMALIES, AUGUST 2011 (COMPARED TO 1971–2000 BASELINE)

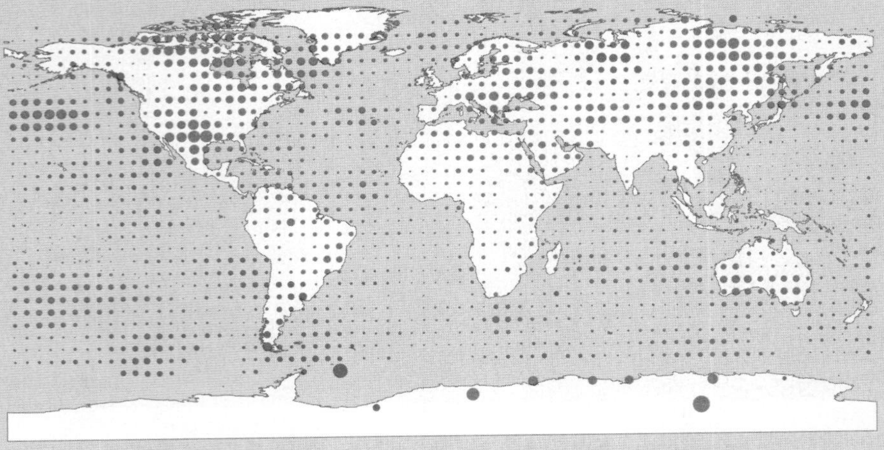

Colder than normal
Warmer than normal
-5°C
-4°C
-3°C
-2°C
-1°C
0°C
1°C
2°C
3°C
4°C
5°C
Temperature

NOAA records show that August 2011 was the second-warmest August, behind 1998, since records began in 1880. This was the 142nd consecutive month that the monthly global land temperature has been above the long-term average.

ANNUAL GLOBAL AVERAGE TEMPERATURE (LAND AND OCEAN)

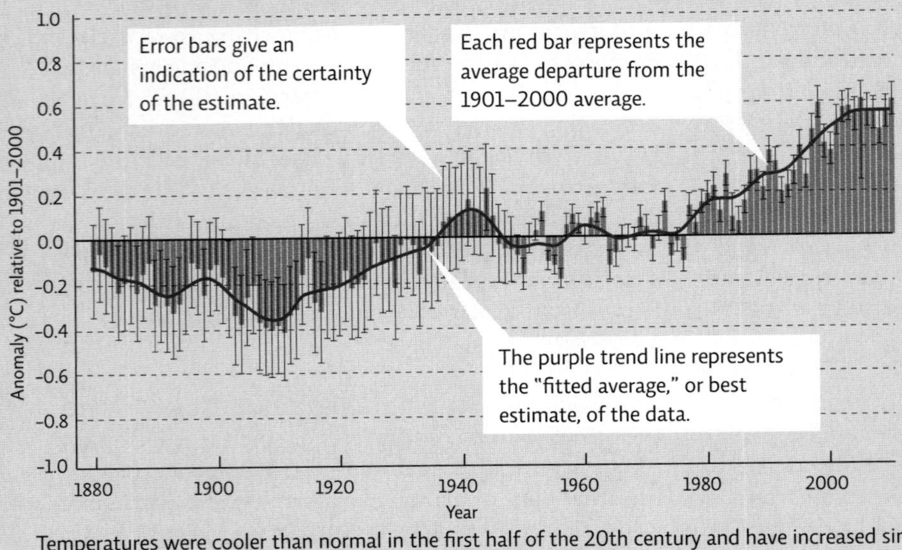

Error bars give an indication of the certainty of the estimate.
Each red bar represents the average departure from the 1901–2000 average.
The purple trend line represents the "fitted average," or best estimate, of the data.
Anomaly (°C) relative to 1901–2000
1.0
0.8
0.6
0.4
0.2
0.0
-0.2
-0.4
-0.6
-0.8
-1.0
1880
1900
1920
1940
1960
1980
2000
Year

Temperatures were cooler than normal in the first half of the 20th century and have increased since the low seen in 1910. Since the 1980s, temperatures have increasingly exceeded the 20th-century average. The recent plateau may be caused by the release of more air pollution particles that reflect away some incoming solar radiation.

WARMER TEMPERATURES SHOULD LEAD TO

MELTING ICE
If temperatures are warming, we would expect to see more ice melt.

CUMULATIVE LOSS OF GLACIER MASS

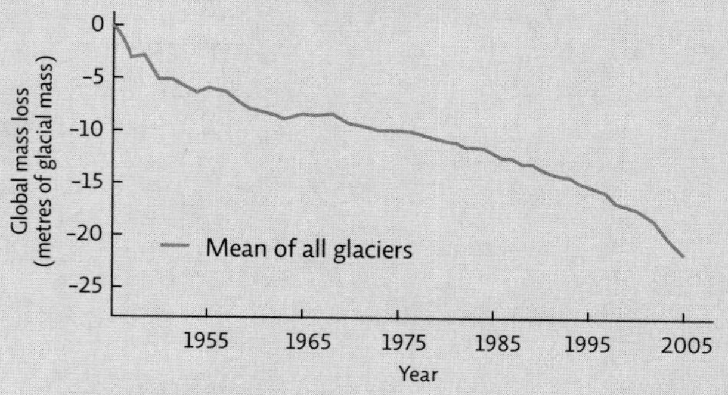
0
-5
-10
-15
-20
-25
Global mass loss
(metres of glacial mass)
Mean of all glaciers
1955
1965
1975
1985
1995
2005
Year

Since the middle of the 20th-century, glaciers around the world have a net loss of ice.

SEA LEVEL RISE
We would also expect to see an increase in sea level as land-based ice melts and as warmer seawater expands.

SEA LEVEL CHANGE OVER TIME

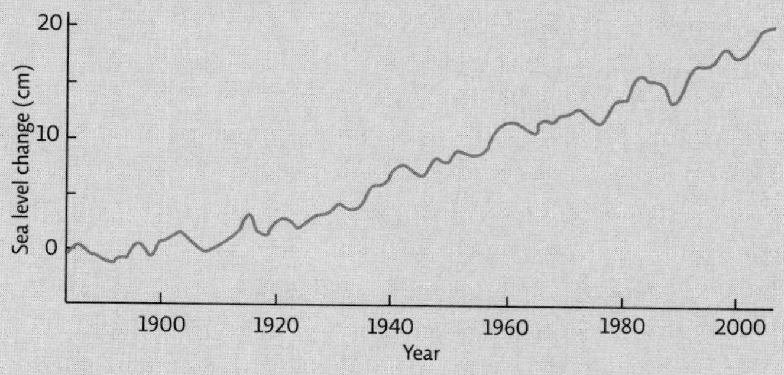
20
10
0
Sea level change (cm)
1900
1920
1940
1960
1980
2000
Year

Sea level has indeed risen 10–20 cm over the last century.

WEATHER EXTREMES
A warmer climate causes more water to evaporate from Earth's surfaces and is expected to produce more extreme weather (heat waves, tropical storms) but also should produce unusual weather as the atmosphere redistributes this heat and moisture around the planet. Areas close to large bodies of water are expected to be wetter (more rain or snow), whereas inner continental areas are expected to be drier as warmer temperatures increase the loss of water from soil, but not enough to fall back down as rain.

PRECIPITATION CHANGES Some areas received more precipitation than normal in 2010, as indicated by the size of the circles; others have received less. The departure from normal has increased since 2000, when the U.S. National Climatic Data Center first started collecting these data—in other words, the circles have gotten bigger each year since 2000.

PRECIPITATION ANOMALIES, JAN 2010 TO DEC 2010
(COMPARED TO A 1961–1990 BASELINE)

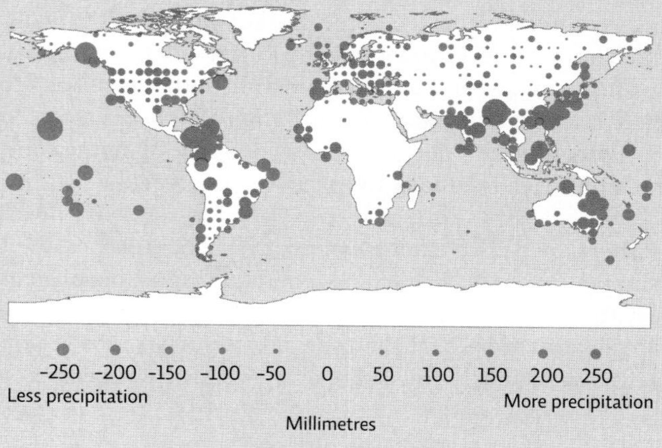
-250 -200 -150 -100 -50 0 50 100 150 200 250
Less precipitation
More precipitation
Millimetres

INCREASE IN STORMS The intensity of Atlantic hurricanes is measured using the Power Dissipation Index, which takes into account maximum wind speeds and duration of the storm; higher values indicate more intense storms. The increase seen since 1970 tracks the observed changes in sea surface temperature closely.

HURRICANE INTENSITY IN THE ATLANTIC OCEAN
(YEARLY AVERAGE)

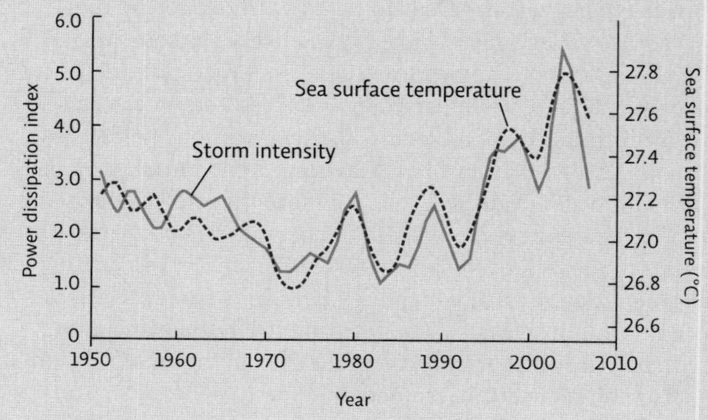
6.0
5.0
4.0
3.0
2.0
1.0
0
Power dissipation index
Sea surface temperature
Storm intensity
27.8
27.6
27.4
27.2
27.0
26.8
26.6
Sea surface temperature (°C)
1950
1960
1970
1980
1990
2000
2010
Year

Infographic **22.3** | **THE GREENHOUSE EFFECT**

↓ Life on Earth depends on the ability of greenhouse gases in the atmosphere to trap heat and warm the planet. More greenhouse gases, however, mean more trapped heat and a warmer planet (an enhanced greenhouse effect).

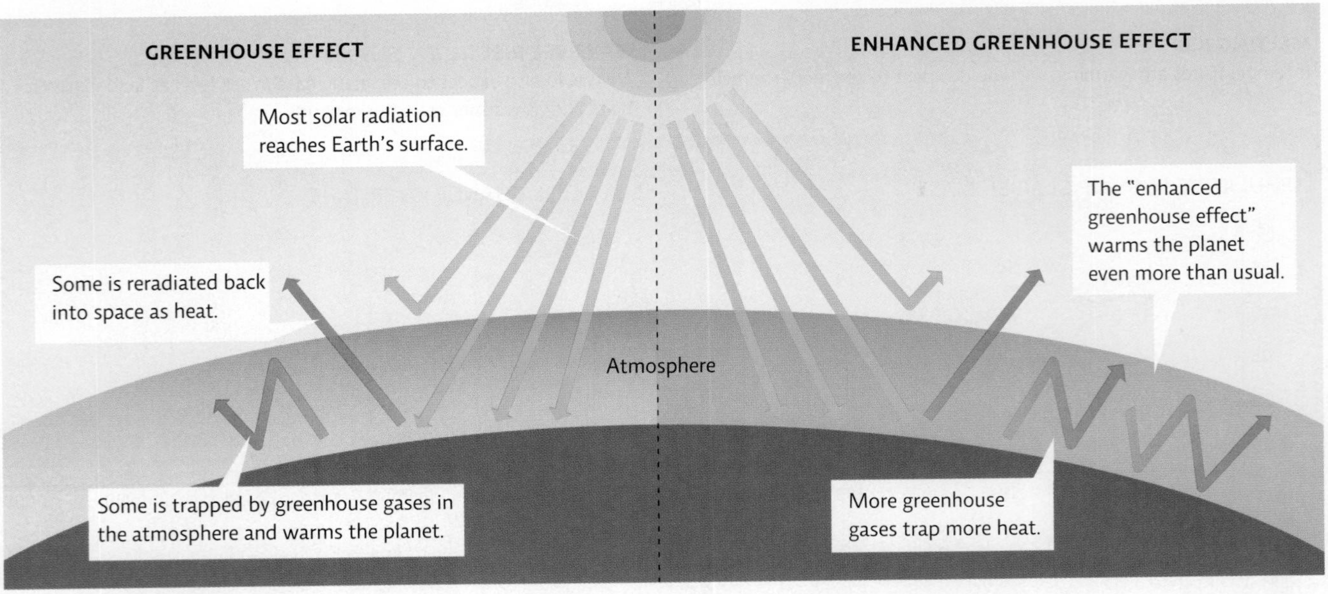
GREENHOUSE EFFECT
ENHANCED GREENHOUSE EFFECT
Most solar radiation reaches Earth's surface.
The "enhanced greenhouse effect" warms the planet even more than usual.
Some is reradiated back into space as heat.
Atmosphere
Some is trapped by greenhouse gases in the atmosphere and warms the planet.
More greenhouse gases trap more heat.

↓ Carbon dioxide (CO_2) is the greenhouse gas released by human actions that currently has the biggest impact on climate. Because different greenhouse gases have different abilities to trap heat, their heat-trapping capacity is expressed as CO_2 equivalents—the amount of CO_2 that would produce the same warming. For example, since a molecule of methane (CH_4) traps 25 times as much heat as CO_2, one methane molecule is equivalent to 25 CO_2 molecules. Much less methane is released in tonnage compared to CO_2, but because each molecule of methane is so much more potent a greenhouse gas than CO_2, it too has a big impact on climate.

GREENHOUSE GASES: GLOBAL EMISSIONS (IN CO_2 EQUIVALENTS), 2004

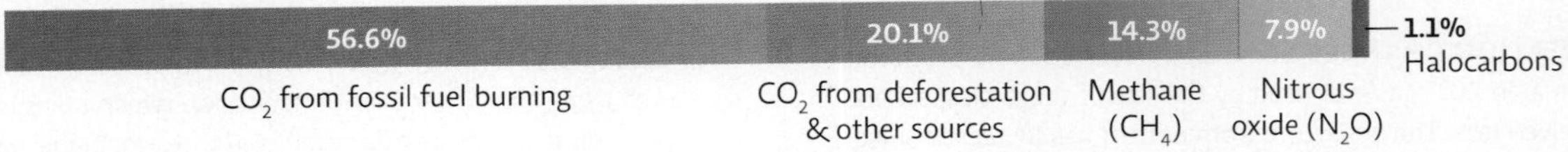
56.6%
20.1%
14.3%
7.9%
1.1%
Halocarbons
CO2 from fossil fuel burning
CO2 from deforestation & other sources
Methane (CH4)
Nitrous oxide (N2O)

A variety of factors affect climate.

So why is all this happening?

In the past few decades, scientists have discovered that the level of certain gases in Earth's atmosphere is on the rise. These gases, called **greenhouse gases**—which include carbon dioxide, methane, and nitrous oxide—trap heat and help warm Earth, in a process known as the **greenhouse effect**. Normally, the greenhouse effect is a good thing—without it, the average temperature on Earth would be around -15°C—that's more than 30° colder than the planet's current average temperature! But starting in the 1980s, scientists began to see evidence of an enhanced greenhouse effect, which they believe is from the release of greenhouse gases from burning fossil fuels and other industrial and agricultural practices—in other words, from human activities. [INFOGRAPHIC 22.3]

Greenhouse gases are just one type of **radiative forcer**, or factor that can affect global climate. Another kind of forcer that plays a role in present warming trends is **albedo**, the ability of a surface to reflect away solar radiation. Light-coloured surfaces, like glaciers and meadows, have a high albedo—they reflect sunlight, and thus heat, away from the planet's surface. Darker surfaces, like water and dark asphalt, have low albedo—they absorb sunlight, and heat along with it, and then reradiate that heat back to the atmosphere. As surfaces with high albedo are replaced by those with low albedo, not only does the planet warm, but a **positive feedback loop** can be triggered.

Glaciers are a good example of positive feedback: as temperatures rise, glaciers melt and ice (high albedo) gives way to water (low albedo). Because this new watery surface absorbs more heat than the old icy surface, the region warms even faster—replacing even more ice with water,

Infographic **22.4** | **ALBEDO CHANGES CAN INCREASE WARMING VIA POSITIVE FEEDBACK**

↓ Albedo is a measure of the reflectivity of a surface. The lighter-coloured the surface, the higher the albedo. Unreflected (absorbed) light is reradiated as heat, so surfaces with a low albedo release more heat to the atmosphere than high albedo surfaces.

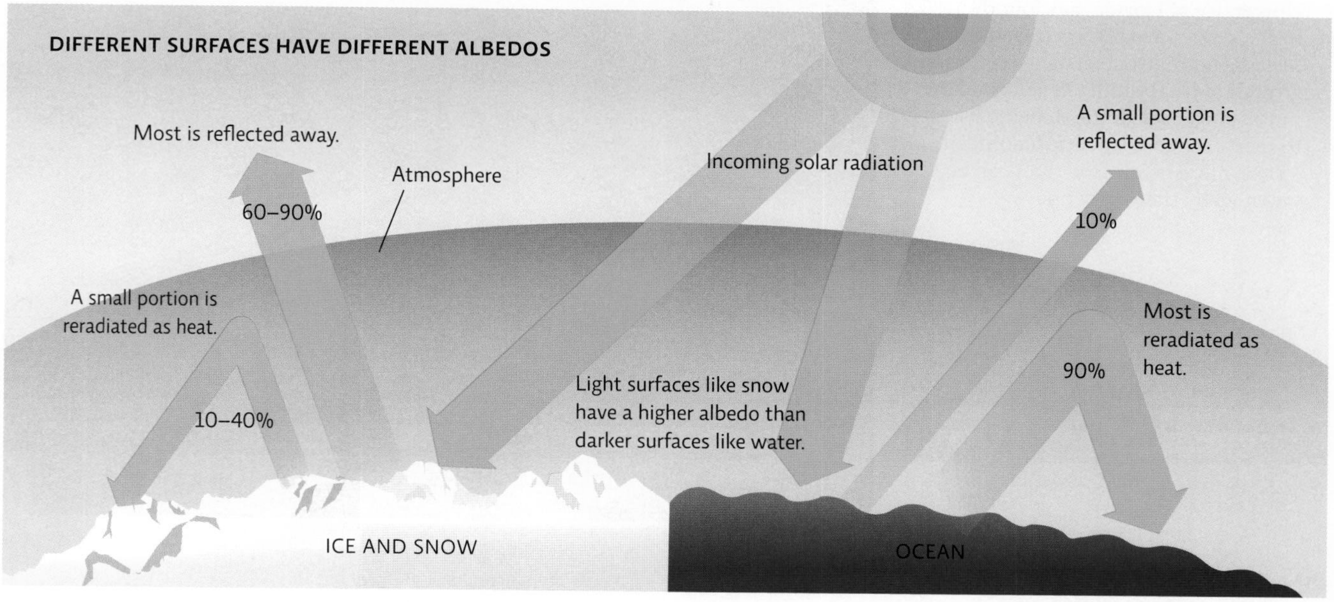
DIFFERENT SURFACES HAVE DIFFERENT ALBEDOS
Most is reflected away.
Atmosphere
60–90%
Incoming solar radiation
A small portion is reflected away.
10%
A small portion is reradiated as heat.
10–40%
Light surfaces like snow have a higher albedo than darker surfaces like water.
90%
Most is reradiated as heat.
ICE AND SNOW
OCEAN

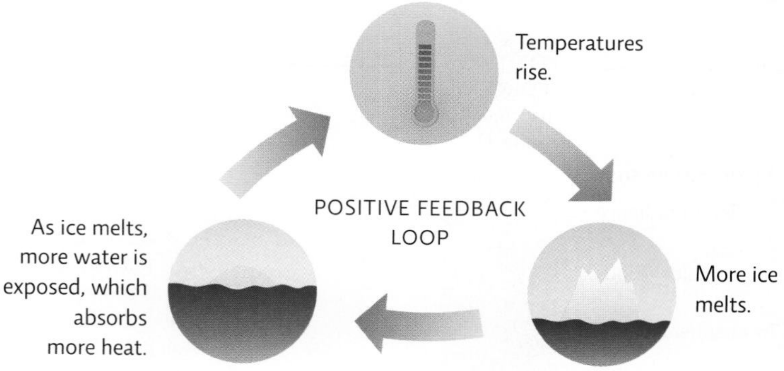
Temperatures rise.
POSITIVE FEEDBACK LOOP
As ice melts, more water is exposed, which absorbs more heat.
More ice melts.

← When sea ice melts, it uncovers water, a darker surface with a lower albedo. This activates a positive feedback loop: as the exposed water absorbs sunlight and releases more heat into the atmosphere, more ice melts and more water is exposed, which then absorbs more sunlight, releasing even more heat, and so on. Positive feedback loops represent changes that trigger additional change in the same direction (warming that triggers even more warming); they do not indicate a "positive" or beneficial event.

and so on. Canada may lose 1/5th of its ice sheet by 2100. The Canadian ice sheet is the 3rd largest in the world, after Antarctica and Greenland, and this melting alone could raise sea levels 3.5 centimetres. [INFOGRAPHIC 22.4]

There are other positive feedback loops, too. In the Arctic, for example, the upper levels of permafrost (land that normally remains frozen year-round) are melting during the summer months. When these areas thaw, they release stored carbon, adding more greenhouse gases to the atmosphere. These gases warm the area further, causing even more permafrost melt. This is very significant in Canada, where half of the landmass is permafrost.

There are also other natural forcers, including clouds. Some have a high albedo and thus work to cool the planet; others trap reradiated heat from the planet's surface and thus have a warming effect. If warming temperatures cause the formation of more of the high-albedo clouds, this could trigger cooling—a **negative feedback loop**. Right now, clouds have a net cooling effect; whether this

greenhouse gases Molecules in the atmosphere that absorb heat and reradiate it back to Earth.
greenhouse effect The warming of the planet that results when heat is trapped by Earth's atmosphere.
radiative forcer Anything that alters the balance of incoming solar radiation relative to the amount of heat that escapes out into space.
albedo The ability of a surface to reflect away solar radiation.
positive feedback loop Changes caused by an initial event that then accentuate that original event (for example, a warming trend gets even warmer).
negative feedback loop Changes caused by an initial event that trigger events that then reverse the response (for example, warming leads to events that eventually result in cooling).

→ Climate change research in Alaska monitors CO_2 release from thawing permafrost, along with tundra growth. The project, led by Ted Schuur of University of Florida, has found that in the short term, CO_2 released from melting permafrost leads to increased growth of local tundra vegetation. But in the longer term, the thawing leads to increased atmospheric loading of CO_2 as more is released than can be taken up by the vegetation.

Infographic 22.5 | **CLIMATE FORCERS**

→ A variety of factors can warm or cool the planet. Positive forcers have a net effect of warming; negative forcers cool the climate. Greenhouse gases trap heat in the atmosphere and warm it, while aerosols like sulphate emissions cool it.

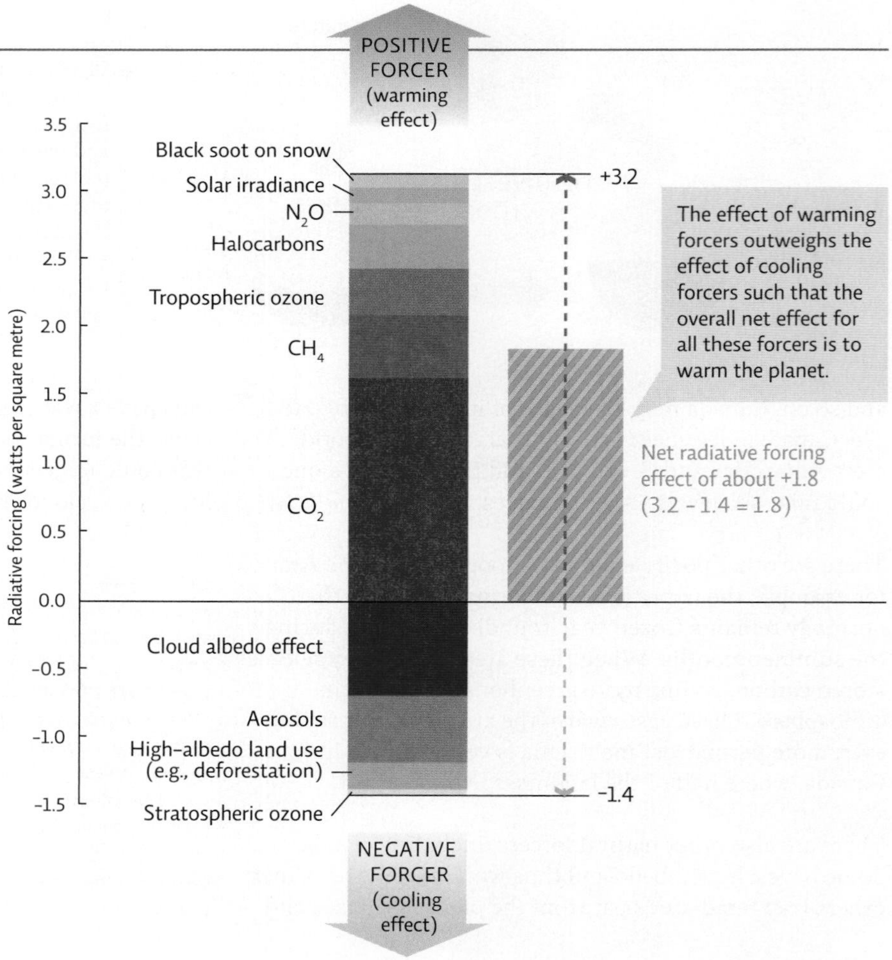
POSITIVE FORCER (warming effect)
Radiative forcing (watts per square metre)
3.5
3.0
2.5
2.0
1.5
1.0
0.5
0.0
-0.5
-1.0
-1.5
Black soot on snow
Solar irradiance
N₂O
Halocarbons
Tropospheric ozone
CH₄
CO₂
Cloud albedo effect
Aerosols
High-albedo land use (e.g., deforestation)
Stratospheric ozone
+3.2
-1.4
The effect of warming forcers outweighs the effect of cooling forcers such that the overall net effect for all these forcers is to warm the planet.
Net radiative forcing effect of about +1.8 (3.2 - 1.4 = 1.8)
NEGATIVE FORCER (cooling effect)

trend will continue in the future remains to be seen. [INFOGRAPHIC 22.5]

Volcanic eruptions and changes in solar irradiance (like sunspot cycles) are also considered natural forcers, as both have been known to impact climate in the past, though only over short time frames and not as severely as greenhouse gases. Scientists also believe that **Milankovitch cycles** (predictable long-term cycles of Earth's position relative to the Sun) played an important role in earlier climate change events. [INFOGRAPHIC 22.6]

To figure out how much any given forcer or feedback loop is contributing to current warming, or to guess at what future climate might look like based on what we are seeing now, climate scientists must do more than monitor current atmospheric conditions; they must also gather data (on temperature, CO_2 levels, etc.) from the distant past. They do this by studying a wide variety of clues that have been left behind—ice and sediment cores, tree rings, coral reef growth layers, even fossilized mud at the bottom of lakes and rivers (see Chapter 1). These various sources have consistently corroborated one another: ice cores paint the same picture as tree rings, and coral reef layers confirm what pollen sediments tell us. This consistency enables us to trust their overall story. [INFOGRAPHIC 22.7]

Climate scientists have used this wealth of historical data to develop what they call *climate models*—computer programs that allow them to make future climate projections by plugging in all the current values (for temperature, CO_2, global air circulation patterns, etc.). These models are used to see how altering the value of certain parameters (say, increasing the amount of atmospheric CO_2) might impact future climate. It may seem ironic that climatologists can predict what the climate will be like 100 years from now, when meteorologists often have a hard time getting the weekly weather forecast right. But scientists say that climate is actually easier to predict than weather. Climate refers to general trends, while weather is much more specific; it is more like a close-up view rather than a view from afar.

Milankovitch cycles Predictable variations in Earth's position in space relative to the Sun that affect climate.

Infographic **22.6** | **MILANKOVITCH CYCLES HELP EXPLAIN PAST CLIMATE CHANGE**

↓ Warm periods and ice ages of the past can be attributed in part to Earth's position in space relative to the Sun. Earth has three different cycles that can each have an impact on climate. The current warming we are experiencing cannot be explained by any of these cycles—Earth is currently not in a part of any cycle in which it would have greater warming.

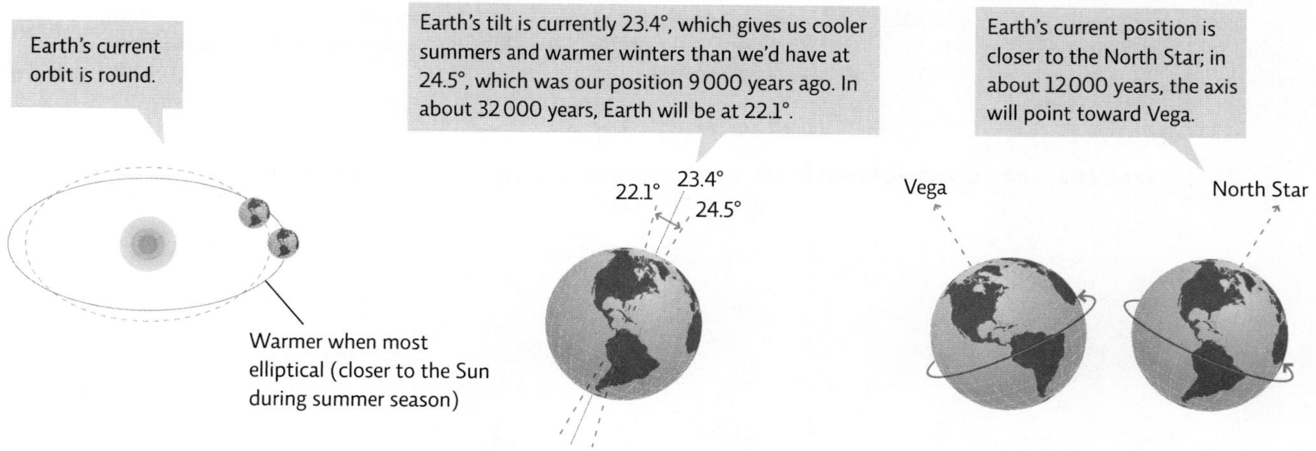
Earth's current orbit is round.
Warmer when most elliptical (closer to the Sun during summer season)
Earth's tilt is currently 23.4°, which gives us cooler summers and warmer winters than we'd have at 24.5°, which was our position 9 000 years ago. In about 32 000 years, Earth will be at 22.1°.
22.1°
23.4°
24.5°
Earth's current position is closer to the North Star; in about 12 000 years, the axis will point toward Vega.
Vega
North Star

ORBITAL ECCENTRICITY The shape of Earth's orbit around the Sun varies over a 100 000-year cycle from mostly round to more elliptical.

AXIAL TILT The angle of Earth's tilt as it spins on its axis changes in a 41 000-year cycle. The greater the angle, the greater the extremes between seasons (hotter summers and colder winters).

AXIAL PRECESSION Earth "wobbles" on its axis, changing not the angle but the direction the axis points in a 20 000-year cycle. This changes the orientation of Earth to the Sun and affects the severity of the seasons. When Earth is tilted toward Vega, it is also tilted toward the Sun during summer, making summers hotter in the northern hemisphere.

Infographic 22.7 | **DIRECT AND INDIRECT MEASUREMENTS OF TEMPERATURE AND CO_2**

↓ We use a variety of methods to measure temperature and CO_2 levels. Current temperatures are monitored directly using instruments such as thermometers and satellites; a variety of analytical equipment is used to measure CO_2 levels in the air. Charles Keeling began taking precise CO_2 measurements at the Mauna Loa Observatory, Hawaii, in 1958, a location chosen because the air there received no local pollution to compromise the data. We can also estimate values from the past indirectly by evaluating physical evidence such as sediment and ice cores, and tree-ring data. Ice cores are particularly helpful; scientists can date the sections by counting the annual layers. They then measure atmospheric gases like CO_2 from the bubbles in the ice core sample to determine levels at the time the ice was laid down; measuring the ratio of 2 isotopes of oxygen, O-16 and O-18, gives a very accurate estimate of temperature at the time.

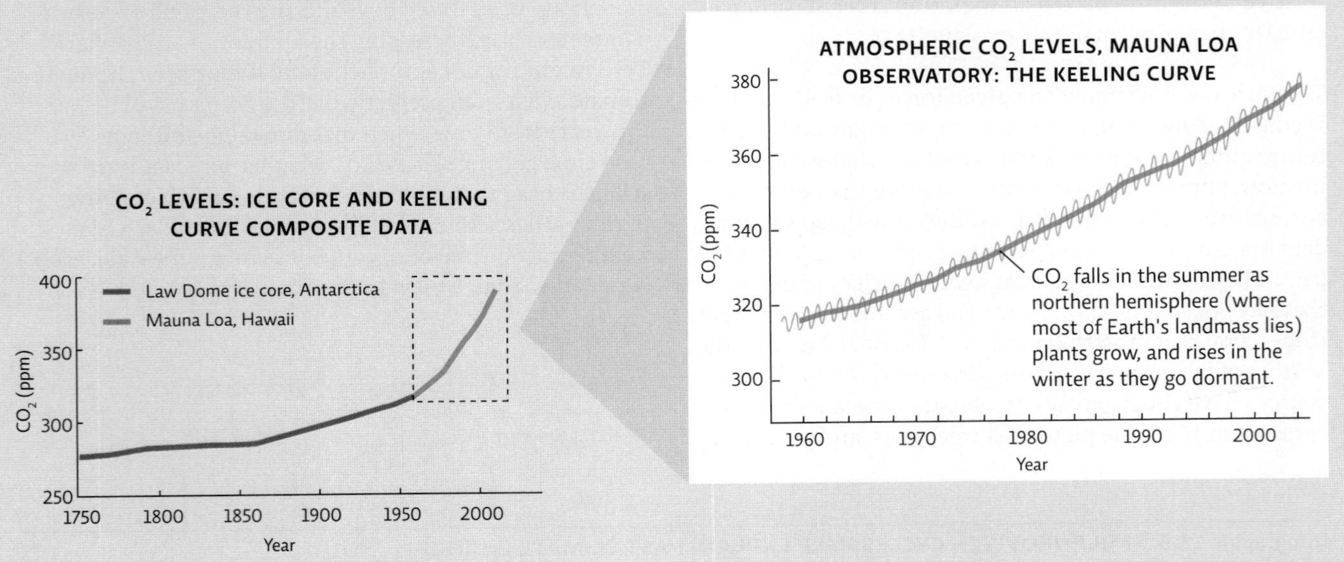

CO2 LEVELS: ICE CORE AND KEELING CURVE COMPOSITE DATA
Law Dome ice core, Antarctica
Mauna Loa, Hawaii
CO2 (ppm)
400
350
300
250
1750
1800
1850
1900
1950
2000
Year
ATMOSPHERIC CO2 LEVELS, MAUNA LOA OBSERVATORY: THE KEELING CURVE
380
360
340
320
300
CO2 (ppm)
CO2 falls in the summer as northern hemisphere (where most of Earth's landmass lies) plants grow, and rises in the winter as they go dormant.
1960
1970
1980
1990
2000
Year

↓ A comparison of historic CO_2 levels and temperatures, as determined from the Antarctic Vostok ice core, show that the two parameters have been closely aligned over the past 400 000 years. It turns out that the relationship between CO_2 and temperature is one of cause and effect. In the far past, natural events such as differences in Earth's orbit triggered warming, resulting in the release of more CO_2 (an effect), which then caused even more warming (a cause). Today, humans are the source of much of the extra CO_2 (and other greenhouse gases) being released. In this case, the CO_2 release is preceding the warming—it is the initial cause. The bottom line is that no matter the reason for the release of extra greenhouse gases such as CO_2, temperatures change as CO_2 increases and decreases.

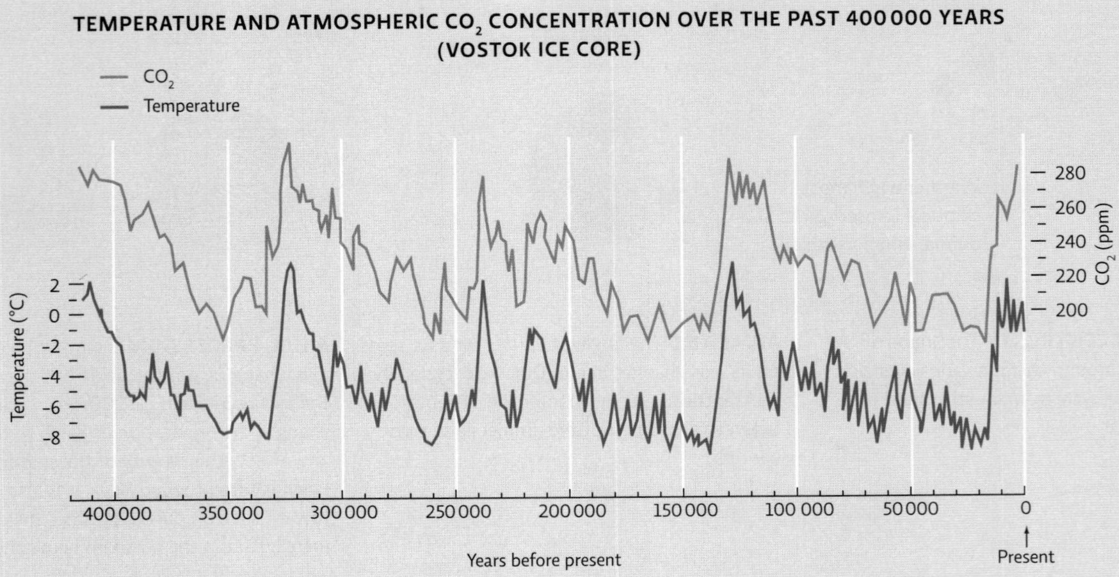

TEMPERATURE AND ATMOSPHERIC CO2 CONCENTRATION OVER THE PAST 400 000 YEARS (VOSTOK ICE CORE)
CO2
Temperature
Temperature (°C)
2
0
-2
-4
-6
-8
280
260
240
220
200
CO2 (ppm)
400 000
350 000
300 000
250 000
200 000
150 000
100 000
50 000
0
Years before present
Present

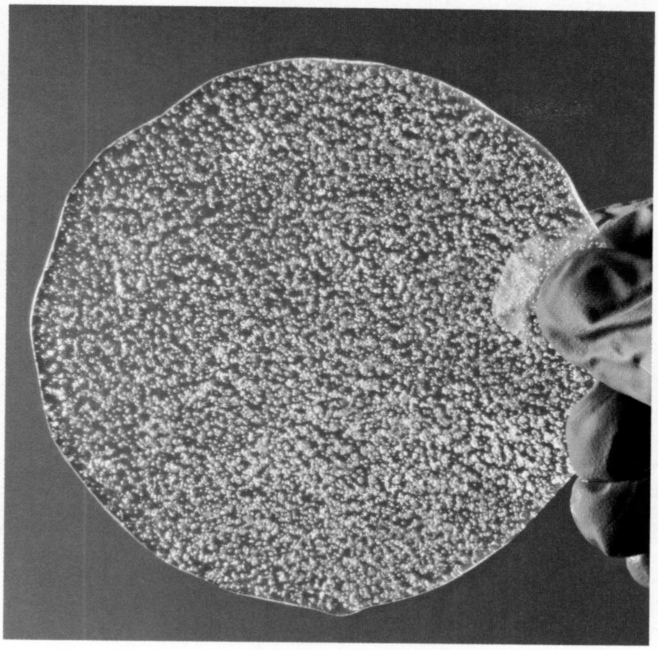

↑ A researcher holds a thin slice of ice from an ice core extracted from Antarctica. The core contains trapped air bubbles that can reveal information about the atmosphere and temperatures of the past.

Based on all the clues they have gathered and analyzed, scientists agree that all the natural forcers combined are not enough to account for the rapid climate change that is currently underway. Only when we consider both natural and **anthropogenic** (related to human actions) forcers together do current trends make sense. [INFOGRAPHIC 22.8]

Current climate change has both human and natural causes.

The **Intergovernmental Panel on Climate Change (IPCC)** is an international scientific body, established by the United Nations and the World Meteorological Organization in 1988, that is made up of thousands of scientists from around the world. They evaluate all the climate science that is published through peer review and compile it into a cohesive series of publications that explain what is understood about the current state of the climate; they also make recommendations that may inform government action.

anthropogenic Caused by or related to human action.
Intergovernmental Panel on Climate Change (IPCC) An international group of scientists who evaluate scientific studies related to any aspect of climate change to give thorough and objective assessment of the data.

Infographic 22.8 | **WHAT'S CAUSING THE WARMING?**

↓ Climate scientists use computer models (multiple mathematical equations) that take into account the major factors that are known to have affected past climates in order to see what might be responsible for recent warming. Data about natural and anthropogenic factors can be fed into the computer model separately and then together to see which circumstances match up with the warming that has been observed.

COMPUTER MODELS' RECONSTRUCTION OF PAST TEMPERATURES

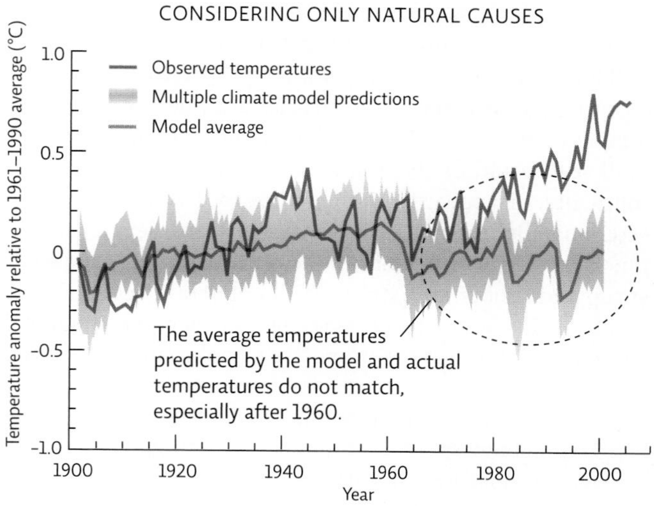
CONSIDERING ONLY NATURAL CAUSES
Temperature anomaly relative to 1961–1990 average (°C)
1.0
0.5
0
-0.5
-1.0
Observed temperatures
Multiple climate model predictions
Model average
The average temperatures predicted by the model and actual temperatures do not match, especially after 1960.
1900
1920
1940
1960
1980
2000
Year

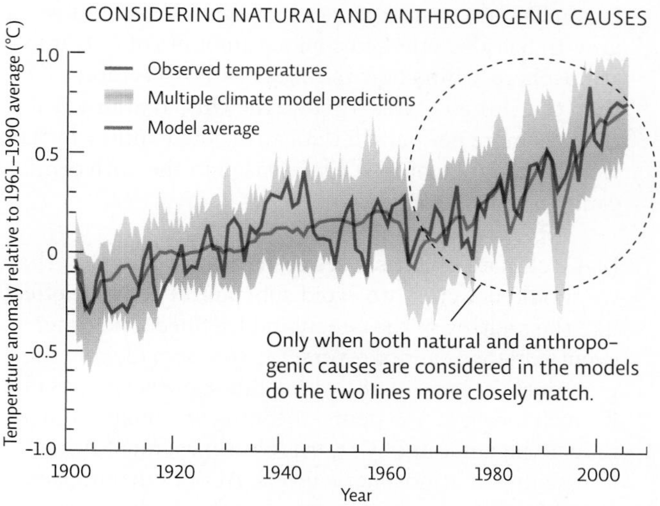
CONSIDERING NATURAL AND ANTHROPOGENIC CAUSES
Temperature anomaly relative to 1961–1990 average (°C)
1.0
0.5
0
-0.5
-1.0
Observed temperatures
Multiple climate model predictions
Model average
Only when both natural and anthropogenic causes are considered in the models do the two lines more closely match.
1900
1920
1940
1960
1980
2000
Year

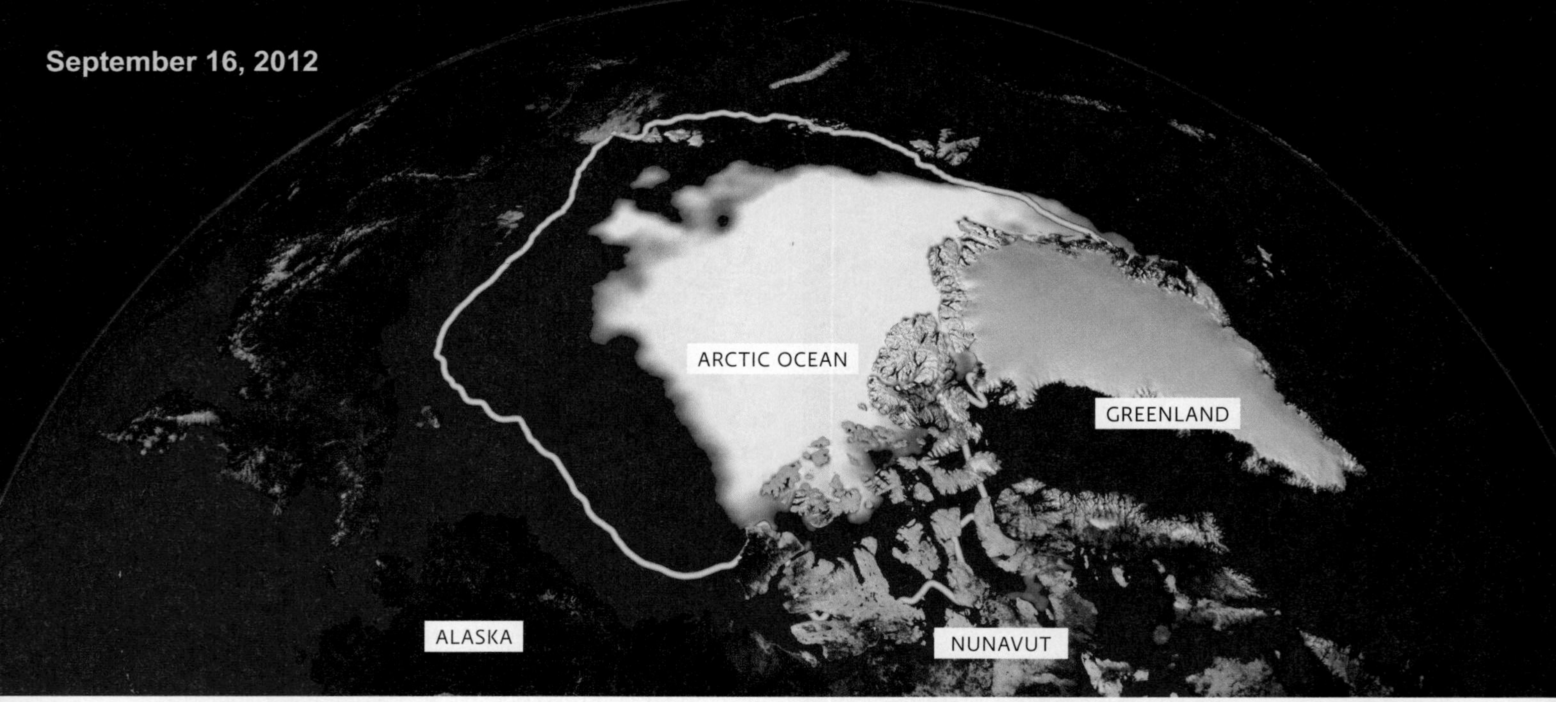
September 16, 2012
ARCTIC OCEAN
GREENLAND
ALASKA
NUNAVUT

↑ NASA satellite data reveals that 2012's sea ice extent in the Arctic Ocean reached a new record low on September 16, 2012, as shown here; this was far smaller than the 30-year average (in yellow). This new record extent of 3.41 million square kilometres was 760 000 square kilometres less than the previous record set in 2007.

Here's what everyone agrees on so far: Earth's atmosphere is changing dramatically and with alarming speed. Analysis of CO_2 trapped in ice cores reveals that current levels are higher than at any time in the last 800 000 years and are increasing steadily. In 2011, the atmospheric concentration of CO_2 was 390 ppm (parts per million). This is considerably higher than the preindustrial level of about 280 ppm—a level that had been maintained for millennia.

The majority of this change is due to human activities, especially the burning of fossil fuels, and the consequent release of greenhouse gases like CO_2 into the atmosphere. For most of modern history, the United States has been the biggest emitter of CO_2. In 2007, however, China took the lead: the proliferation of coal-fired power plants that has fuelled the country's recent economic and industrial growth has also released copious amounts of CO_2 into the atmosphere. China now releases nearly 30% more CO_2 than the United States (though the United States still releases more per person than any other country in the world, and much of the CO_2 released in the 20th century came from U.S. sources).

Many climate scientists set the upper limit of CO_2 that we should not cross (to avoid substantial negative effects, like the melting of the Greenland Ice Sheet and other events that we cannot reverse) at 560 ppm (double the preindustrial level of 280 ppm). But some scientists place it much lower, at 350 ppm—meaning we must not just reduce the amount of CO_2 we release, we must bring down current atmospheric levels. At our current pace (rapid fossil fuel consumption combined with sluggish efforts to curb our emissions) we will probably surpass that 560 threshold before the end of this century. We have already set into motion a chain of events that are dramatically changing the face of the planet, such as glacier and permafrost melt, ocean acidification, and loss of carbon sinks and vital habitats. Exceeding current levels of CO_2 will only kick these and other events into an even higher gear as we warm the planet further.

But while the causes of climate change are scientifically well established, the future is still riddled with uncertainty. On the whole, we are likely in for more variable weather—not just warmer weather. While some regions are already enduring heat extremes, other areas will have colder winters as weather patterns shift.

One of the biggest uncertainties is the effect that changes in mean annual temperature and precipitation will have on the world's forests. Some species may be able to migrate as temperatures rise and their optimum temperature range shifts northward, but which ones? Will spruce and fir species, which thrive in colder environments, disappear from the landscape? Will iconic and economically important trees like sugar maple and jack pine survive in sufficient numbers to sustain Canada's syrup and timber industries? Will important ecological relationships be fractured as some species move or adapt while others (perhaps prey species or pollinators) do not?

Somewhere, buried in the reams of data that scientists like Frelich have spent decades accumulating, lie clues to answering those very questions.

Some tree species are already migrating north; that doesn't mean they will survive.

Like Frelich, Chris Woodall has spent his entire adult life studying the great forests—both temperate and boreal—that stretch from the northeastern United States well into Canada. Unlike Frelich, Woodall spends most of his time in front of a computer screen, crunching numbers. He works for the U.S. Department of Agriculture's forest inventory program, which maintains roughly 100 000 permanent plots throughout the region—segments of forest where the USDA monitors a host of variables, from temperature and precipitation to tree growth and sapling density. It's a tremendous database, and for the past decade, scientists have used it to develop computerized models of how a warmer climate might change forests in the future. But until recently, no one had looked at whether or not they were already changing.

Woodall's logic was simple: mature trees tell you where the current range is. Seedlings tell you where that range will be in the future. By comparing the ratio of seedlings to mature trees, one should be able to say whether or not any given species is on the move. Using the most recent data collected from 30 states and some 66 000 inventory plots, Woodall compared tree-seedling densities to forest biomass for more than two dozen tree species.

He was astounded by what he found. For most of those species—spruce, jack pine, sugar maple, and several others—the mean location for seedlings proved to be significantly further north than their associated mature trees. "It's like if they had a gravity centre, it would be pulling them northward."

Whether they will actually survive in new locations is another story. While studies of the average locations for trees and saplings show northward migration, subsequent studies show contraction, or a loss of trees, at the northern edges. "The obvious question is, 'Well, how does that reconcile with the findings on mean, that they are moving north?'" Woodall says. "The answer is that it's like buffalo rushing off a cliff. As climate warms, you've got all these species gravitating to the northernmost edges of their traditional ranges. But that doesn't mean they will survive there long term."

Back in the North Woods, Frelich is working to understand why. So far, he's identified several forces that seem to be working in concert. "Warming triggers a whole cascade of events," he says. "Factors that make tree ranges shift northward, and factors that prevent those trees from thriving in their new, more northerly habitats."

One of the biggest factors, he says, is deer, which have proliferated like mad in recent years. "In a warmer climate," Frelich says, "you'd expect the maple to advance in the understory, so that as the spruce die off, the maple are ready to take over. Likewise in the south: as maple move northward in response to warming, oak should move in to fill the void." But it turns out that deer like maple much more than they like spruce, and oak even more than maple. "So in places where the deer population is very high, the trees are having a hard time adapting to climate change because the deer are eating up all the early migrators."

> "It's like buffalo rushing off a cliff. As climate warms, you've got all these species gravitating to the northernmost edges of their traditional ranges."
> —Chris Woodall

And as climate warms, other stresses abound: snowpack melts earlier, causing more severe water deficits in summer, right when trees need extra water to survive. The whole landscape dries out, creating conditions that favour intense fires and stressing trees so much that they become easy prey for beetle infestation. Pine beetles are a natural part of the life cycle in western forests, but the current outbreak, underway for more than a decade, is unlike anything seen before. "It used to get down to 40 below, every couple years, and that would keep things in check by killing the beetles off," says Frelich. "But that isn't happening anymore." Mountain pine beetle infestation has impacted 18.1 million hectares of British Columbia's forests.

There are many other indicators that species are responding to climate change. In fact, that they are responding is evidence itself that climate is changing. Because communities are complex assemblages of many species, there are concerns that important community connections will become uncoupled as species respond in different ways to climate change. [INFOGRAPHIC 22.9]

Climate change has environmental, health, and economic consequences.

The Boundary Waters Canoe Area surrounding Ham Lake—where Frelich and his colleagues were trapped—is the most heavily used chunk of the U.S. National Wilderness Preservation System. The boreal forest, its

Infographic 22.9 | **SPECIES' RESPONSES** ↓ Species have evolved to live and thrive in certain habitats. If the climate is changing enough to alter ecosystems, we expect to see species responding by changing where they live or the timing of important temperature-dependent biological events. Species' responses such as shifting ranges or earlier blooming and hatching may be the best evidence that climate is actually changing—it is unlikely that these temperature-dependent events would change in this way if the planet were not getting warmer.

SOME SPECIES ARE MOVING TO HIGHER ALTITUDES
A 2008 study of the elevation distribution of 171 forest plants looked at where plants were found as well as the optimum elevation for growth. Data were compared for plants from two different time periods: 1905–1985 and 1986–2005. On average, plants shifted their range to higher elevations 29 metres per decade.

SOME IMPORTANT COMMUNITY CONNECTIONS ARE UNCOUPLED
Caterpillars, an important food source that birds feed their young, hatch based on temperature cues; the caterpillars are hatching 15 days sooner than they did in 1985. Pied flycatchers, birds that migrate north to their breeding grounds based on day-length cues (and thus have not changed their migration timing), are arriving too late to take advantage of peak caterpillar hatching.

FLOWERING PLANTS ARE BLOOMING EARLIER
Since many biological events are linked to the arrival of spring, we would expect to see earlier blooming, leaf out, and reproduction in species sensitive to temperature cues. One study showed that in Alberta, aspen, an early spring bloomer, blooms 2 weeks earlier now than normal. Other plants that normally bloom later in spring bloomed 0–6 days earlier.

lakes, hiking trails, and breathtaking wildlife entice some 200 000 visitors every year, providing roughly 18 000 tourism jobs that pay a total of $240 million in wages. Global warming and shifting tree ranges threaten all of that.

"Everyone's worried about losing the forests," Frelich says. "Resort owners, people who own cabins up there, local outfitters that rent camping gear—and the tourists themselves. If the forests burn too much, or if they descend into savannah, or can no longer support the iconic wildlife—like moose, lynx, and boreal owls—that people have come to expect, tourism will dry up. Because no one will want to vacation there."

And it's not just the tourism industry that will suffer in a warmer North Woods. The region as a whole supports about 1000 forestry and logging jobs, which pay about $50 million in wages. On top of that, some individual species have become industries unto themselves—for example, the sugar maple.

Maple syrup production is heavily dependent on climate. The flow of sweet, sticky tree sap that eventually covers your pancakes is governed by changes in air pressure, which are in turn governed by changes in air temperature. When the temperature drops below freezing, the tree acts as a giant suction system, pulling the sap out of its branches, down into its roots. When the temperature rises above freezing, this action is reversed; the pressure gradient forces sap up from the roots, through the branches and out of any holes—including ones that syrup makers have drilled for taps.

Traditionally, climate in northeastern North America—from Minnesota to Maine, to southern Ontario, Quebec, and New Brunswick—has provided the optimal freeze-thaw patterns for this process, which syrup makers call "sugaring." But in recent years, the transition from winter to spring has accelerated, leaving fewer freeze-thaw cycles and less sap overall. Warmer daytime temperatures in Canada have increased the number of freeze-thaw cycles, and some experts say that Canada is already in the middle

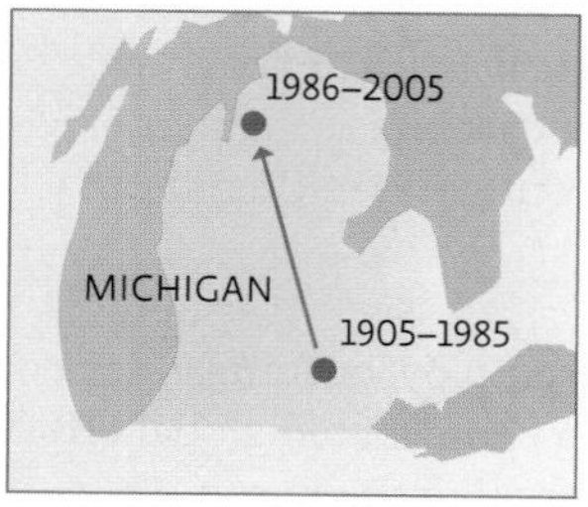
1986–2005
MICHIGAN
1905–1985

SOME SPECIES ARE MOVING TO HIGHER LATITUDES
Several studies report a northward shift in the range of several, but not all, species of birds that have been studied. For example, the blue-gray gnatcatcher has extended its breeding range more than 300 kilometres northward since the 1970s.

The sugar maple was one of 11 out of 15 species that showed an average northern shift in range of 21 kilometres. Most seedlings are found in the northern-most regions, a sign that this species is migrating north.

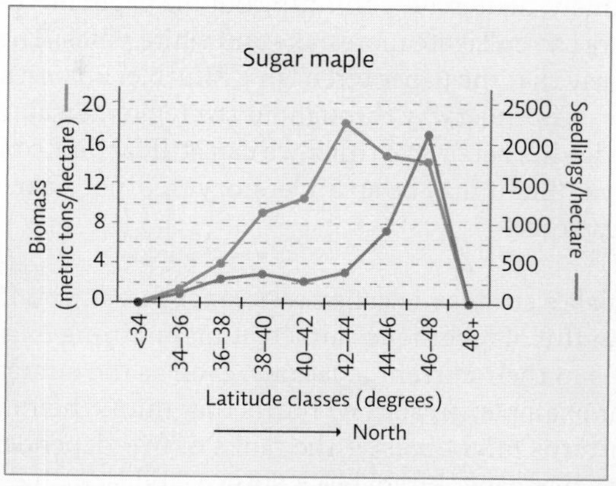
Sugar maple
Biomass (metric tons/hectare)
0
4
8
12
16
20
Seedlings/hectare
0
500
1000
1500
2000
2500
<34
34–36
36–38
38–40
40–42
42–44
44–46
46–48
48+
Latitude classes (degrees)
North

SOME SPECIES' POPULATIONS ARE EXPANDING
Many species of insect pests are on the increase, resulting in the decimation of forests around the world. Forests across the northern hemisphere are affected by several species of bark beetles. The pest populations normally die back in the winter, helping to keep them in check and allowing the trees to recover. Winter temperatures no longer get cold enough to kill the beetles in some areas, allowing the beetles to thrive year-round. The U.S. Forest Service estimates that more than 1.6 million hectares have been affected in the western United States.

of a syrup boom. "If current trends continue," Woodall says, "it's not impossible that the entire industry could one day be lost to Canada."

For Woodall, the stakes are both more basic and more terrifying than the loss of any given industry: societies that don't protect their forests fail, he says. And it's easy to see why. "Forests stabilize soil and clear water of pollutants," he says. "In fact, the vast majority of Americans drink water that comes from a forested watershed. That means trees are as crucial to our survival as the water we drink." The loss of forests is also another positive feedback loop that threatens to exacerbate climate change: forests are an important carbon sink—they store trillions of metric tons of CO_2 in their plants and soils. When they are burned, or die and decompose, much of that carbon is released into the atmosphere (see Chapter 11 for more on the ecosystem services of forests).

To be sure, some people and places will benefit from climate change. In Greenland, for example, warmer temperatures have enabled farmers to grow a wider variety of crops than they have been able to in the past. Warmer weather during the summer months has also opened the Northwest Passage—a long sought-after shipping route north of Canada through the Arctic Ocean—which would significantly reduce the transport time for ships that otherwise have to take the southern route through the Panama Canal. High-latitude land in Canada and Siberia will likely become warmer and more habitable—lessening the incidences of cold-related health problems and deaths.

But other areas will suffer more harm than good. Coastal flooding is already affecting low-lying areas, from Bangladesh to New Orleans. Extreme weather has already begun to claim both valuable crops and human lives. And infectious tropical diseases have begun to migrate north of their traditional range (dengue fever, for example, has made its way from Africa and South America into Texas and Florida).

Confronting climate change is challenging.

Jack Rajala's timber company owns over 14 000 hectares of commercial forest land in Itasca County, Minnesota. In recent years, he's started doing things differently—namely, deliberately thinning out his paper birches in an effort to cultivate more oaks and white pines. That's not to say that the paper birch isn't valuable. But with massive die-offs underway throughout the region, Rajala needs to hedge his bets. "We think we can still facilitate birch," he says. "But it may be an understory tree, not a canopy tree anymore."

Rajala's strategy is called resistance forestry; it includes a handful of techniques aimed at maintaining existing species in their current locations, even as the climate shifts. For example, prescribed burns that mimic historic fire patterns might bolster the ranks of fire-dependent species like paper birch, black spruce, and jack pine, allowing them to spread over a wider area and enhancing their genetic diversity. Planting seeds instead of saplings also helps species hold their ground; natural selection favours the hardiest field-grown seedlings, and so may yield a population better able to survive environmental stresses.

Scientists refer to such efforts, which are intended to minimize the extent or impact of climate change, as **mitigation**.

Mitigation includes any attempt to seriously curb the amount of CO_2 we are releasing into the atmosphere—either by using carbon capture techniques to remove the greenhouse gas from our air and sequester it underground, or by consuming fewer fossil fuels to begin with. In 2004, Princeton University researchers Stephen Pacala and Robert Socolow proposed a "stabilization wedge" strategy—a step-by-step implementation of currently available technology; each step could prevent the release of 1 billion metric tons of carbon. Any 8 of the 15 steps, or "wedges," would stabilize CO_2 in the atmosphere at close to 525 ppm in the next 50 years. [INFOGRAPHIC 22.10]

On a national or global scale, mitigation efforts can be facilitated in a variety of ways, such as command and control regulations that limit greenhouse gas release; tax breaks; green taxes (in this case, **carbon taxes**); and market-driven programs such as carbon cap-and-trade (see Chapter 21). Financial incentives that encourage the development and use of noncarbon fuels and more energy-efficient technology is critical (see Chapters 23 and 24).

No matter which strategies we employ, curbing greenhouse gas emissions will take a coordinated global effort, meaning that world superpowers like the European Union, the United States, and China will have to cooperate with the developing nations of the world. So far, efforts have been fraught with obstacles and lack of cooperation. In 1997, an international treaty called the Kyoto Protocol was ratified by every UN nation except the United States. The treaty set different but specific targets of CO_2 release reduction for various nations; the United States objected because the protocol set much higher reduction requirements for developed countries than it did for developing nations. Notoriously, Canada withdrew from the Kyoto Protocol in 2011; no other country has ever backed out of Kyoto. The federal environment minister justified the withdrawal by asserting that since neither China nor the United States, the two biggest emitters, were covered by the protocol, it would not work, and that since Canada had failed to meet its Kyoto CO_2 emission targets thus far, it would have to back out to avoid paying the resulting fines.

Additional criticisms of Kyoto underscore the trouble with confronting such a global problem. Some of those who opposed Kyoto said that it went way too far in curbing greenhouse gas emission; they argued that placing any kind of limit on CO_2 would hurt the economy because it would force industries to spend money updating their infrastructure, limit the amount of work they could do, and place them at a disadvantage compared with countries who had lower reduction targets under the treaty.

Other critics said that Kyoto did not go far enough; given the overwhelming evidence, these critics felt we needed to set much higher reduction targets to make any dent in the problem. They also felt that setting concrete, legally binding reduction targets in developed nations would actually stimulate the economy because it would force companies in those nations to develop new, cleaner, more efficient technologies that other countries would then buy.

The conflicting views on Kyoto illustrate why climate change is considered a *wicked problem*— one that is resistant to resolution because it is fraught with complexity, change, incomplete information, and lack of sociopolitical acceptance. Climate change is a classic example of this, and it challenges the bedrock of modern civilization: energy use. Some argue that applying the **precautionary principle** now could help avoid, or at least lessen, some of the most serious outcomes of a changing global climate.

---

mitigation Preventative efforts intended to minimize the extent or impact of a problem such as climate change.

carbon taxes Governmental fees imposed on activities that release CO_2 into the atmosphere, usually on fossil fuel use.

precautionary principle Acting in a way that leaves a safety margin when the data is uncertain or severe consequences are possible.

Infographic **22.10** | **OUR CHOICES WILL AFFECT HOW MUCH THE PLANET WARMS WHICH, IN TURN, WILL INFLUENCE THE SEVERITY OF THE IMPACTS**

↓ The IPCC has several climate change response scenarios that predict how much the climate will change (average temperature increase) based on how the human population responds. These scenarios take into account how quickly fossil fuel use is decreased, that is, how cooperative nations are in sharing technologies and supporting global initiatives to address resource consumption and fossil fuel use.

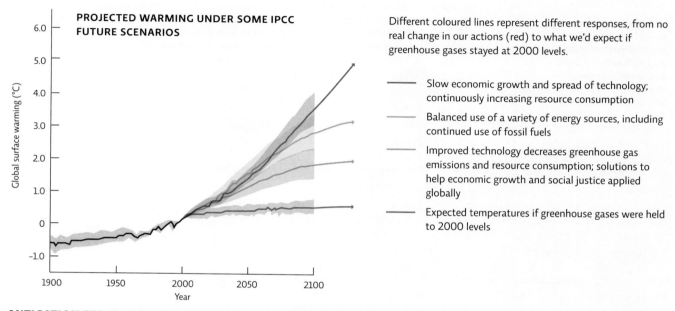

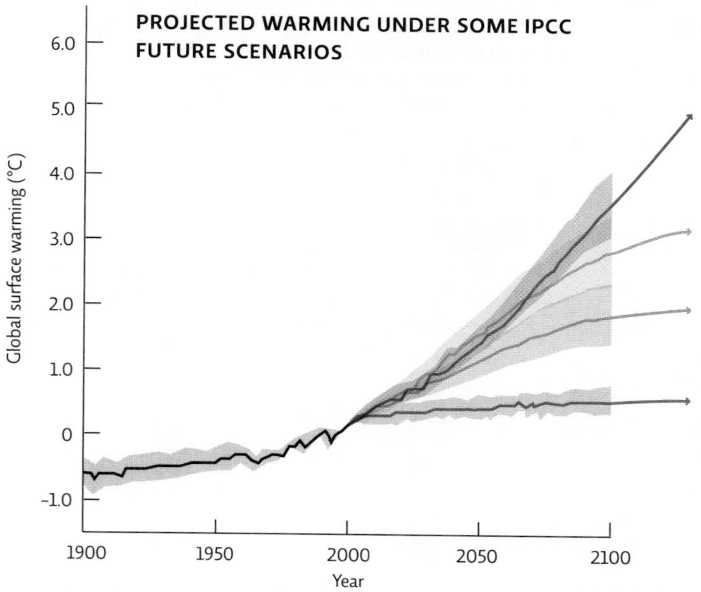
PROJECTED WARMING UNDER SOME IPCC FUTURE SCENARIOS
Global surface warming (°C)
6.0
5.0
4.0
3.0
2.0
1.0
0
-1.0
1900
1950
2000
2050
2100
Year

Different coloured lines represent different responses, from no real change in our actions (red) to what we'd expect if greenhouse gases stayed at 2000 levels.

- Slow economic growth and spread of technology; continuously increasing resource consumption
- Balanced use of a variety of energy sources, including continued use of fossil fuels
- Improved technology decreases greenhouse gas emissions and resource consumption; solutions to help economic growth and social justice applied globally
- Expected temperatures if greenhouse gases were held to 2000 levels

MITIGATION STRATEGIES HELP REDUCE FACTORS THAT LEAD TO CLIMATE CHANGE

↓ We can take steps to curb climate change by reducing emissions of greenhouse gases and by making better resource and land-use decisions. This will lessen the eventual peak warming we might experience. Pacala and Socolow estimate that employing any 8 of 15 potential stabilization wedges, which each reduce CO_2 emissions by 1 gigaton (1 billion metric tons) per year over the next 50 years, would allow the atmosphere to stabilize close to 500 ppm.

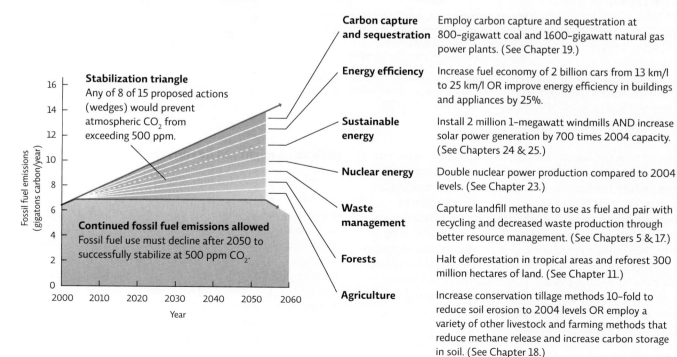

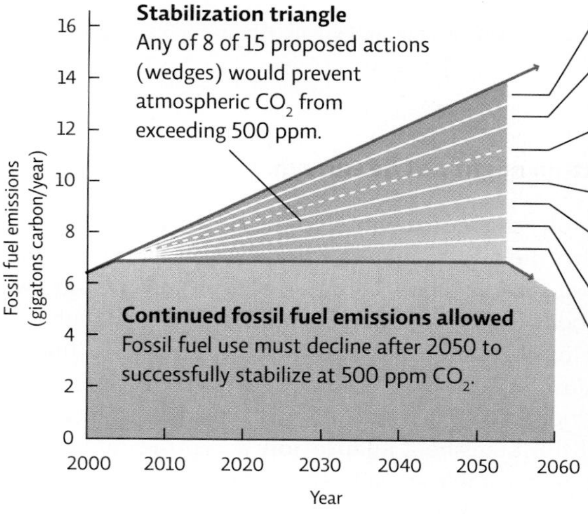
Stabilization triangle
Any of 8 of 15 proposed actions (wedges) would prevent atmospheric CO_2 from exceeding 500 ppm.
Continued fossil fuel emissions allowed
Fossil fuel use must decline after 2050 to successfully stabilize at 500 ppm CO_2.
Fossil fuel emissions (gigatons carbon/year)
16
14
12
10
8
6
4
2
0
2000
2010
2020
2030
2040
2050
2060
Year

Carbon capture and sequestration — Employ carbon capture and sequestration at 800-gigawatt coal and 1600-gigawatt natural gas power plants. (See Chapter 19.)

Energy efficiency — Increase fuel economy of 2 billion cars from 13 km/l to 25 km/l OR improve energy efficiency in buildings and appliances by 25%.

Sustainable energy — Install 2 million 1-megawatt windmills AND increase solar power generation by 700 times 2004 capacity. (See Chapters 24 & 25.)

Nuclear energy — Double nuclear power production compared to 2004 levels. (See Chapter 23.)

Waste management — Capture landfill methane to use as fuel and pair with recycling and decreased waste production through better resource management. (See Chapters 5 & 17.)

Forests — Halt deforestation in tropical areas and reforest 300 million hectares of land. (See Chapter 11.)

Agriculture — Increase conservation tillage methods 10-fold to reduce soil erosion to 2004 levels OR employ a variety of other livestock and farming methods that reduce methane release and increase carbon storage in soil. (See Chapter 18.)

Infographic **22.11** | **CLIMATE CHANGE IMPACTS ON CANADA**

↓ Climate change is already impacting Canada and additional impacts are expected in the future as more warming is experienced. Annual economic costs of these impacts are estimated to be \$5 billion by 2020 and as much as \$43 billion by 2050. One extremely important impact will be the loss of carbon storage capacity, especially in boreal forests and wetlands, which will lead to greater atmospheric methane and CO_2.

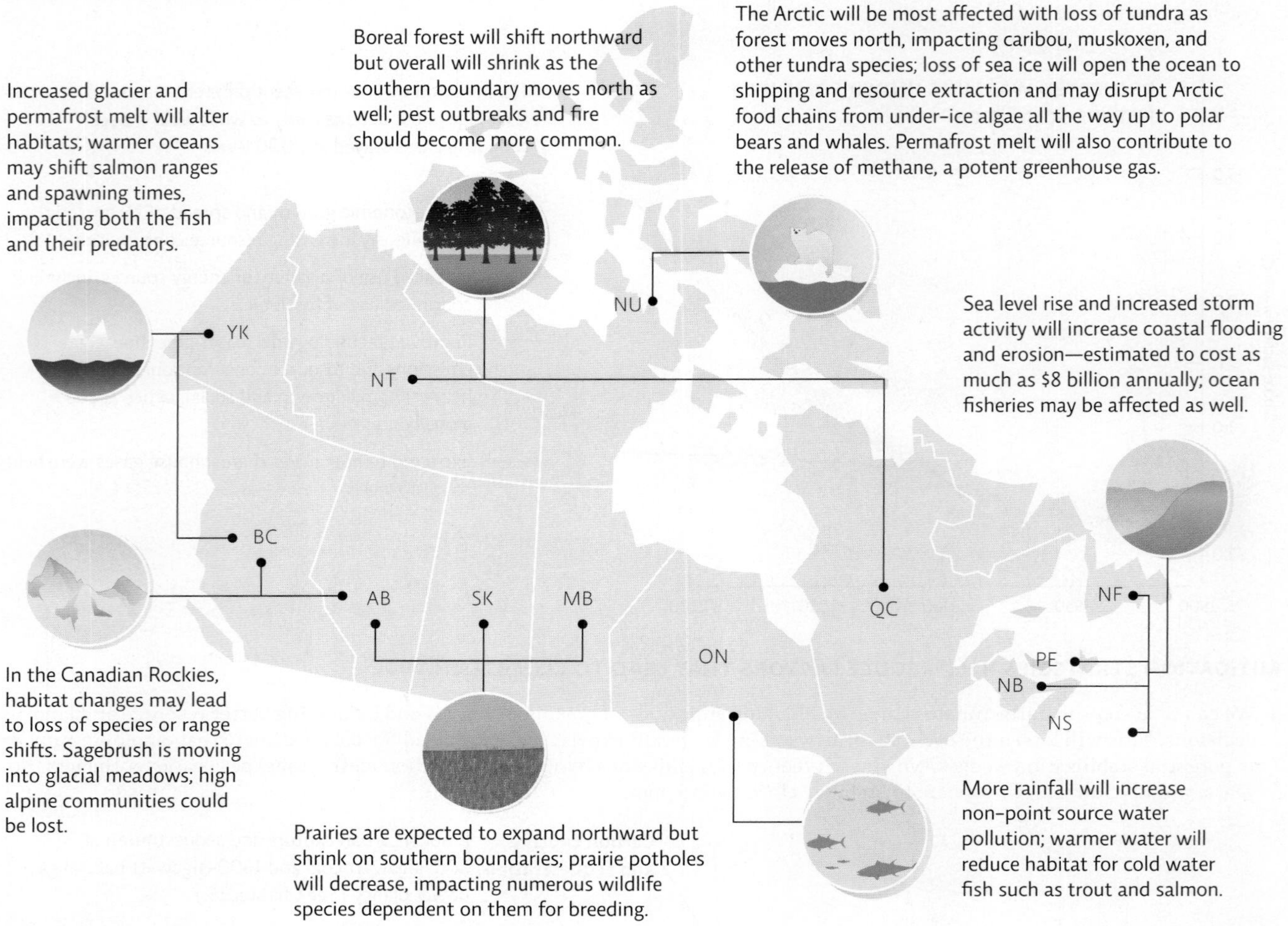
Increased glacier and permafrost melt will alter habitats; warmer oceans may shift salmon ranges and spawning times, impacting both the fish and their predators.
Boreal forest will shift northward but overall will shrink as the southern boundary moves north as well; pest outbreaks and fire should become more common.
The Arctic will be most affected with loss of tundra as forest moves north, impacting caribou, muskoxen, and other tundra species; loss of sea ice will open the ocean to shipping and resource extraction and may disrupt Arctic food chains from under-ice algae all the way up to polar bears and whales. Permafrost melt will also contribute to the release of methane, a potent greenhouse gas.
Sea level rise and increased storm activity will increase coastal flooding and erosion—estimated to cost as much as \$8 billion annually; ocean fisheries may be affected as well.
YK
NU
NT
BC
AB
SK
MB
QC
ON
NF
PE
NB
NS
In the Canadian Rockies, habitat changes may lead to loss of species or range shifts. Sagebrush is moving into glacial meadows; high alpine communities could be lost.
Prairies are expected to expand northward but shrink on southern boundaries; prairie potholes will decrease, impacting numerous wildlife species dependent on them for breeding.
More rainfall will increase non-point source water pollution; warmer water will reduce habitat for cold water fish such as trout and salmon.

Canada's large cities are also bracing for health impacts of poorer air, heat waves, and infectious disease outbreaks. Annual costs for these added health problems is estimated to be in the billions of dollars by 2050.

At any rate, the Kyoto Protocol expired on December 31, 2012; so far, despite annual meetings, the international community has had no success in drafting a replacement treaty. The 2011 UN Climate Change Conference in Durban, South Africa, resulted in a legally binding agreement to adopt a future international climate treaty by 2015, but failed to produce any actual treaty. Some nations (including Germany and the United Kingdom) have made their own progress toward CO_2 reductions, but other nations (including Canada, China, and the United States) have increased CO_2 release since 1997.

Mitigation might not be enough.

Meanwhile, change is coming to the North Woods. And as birches die off and maples try to expand their territory, as moose falter and pine beetles thrive—those who know the woods best say that mitigation will not be enough. "Resisting climate change at this point is like paddling upstream," says Frelich. "It might buy us some time, but it's not going to save the day." So, he says, we need to start thinking about **adaptation**: accepting climate change as inevitable and adjusting as best we can. For human societies at large, this means taking steps to ensure a sufficient water supply in areas where freshwater supplies may dry up; it means planting different crops or shoring up coastlines against rising sea levels; it means preparing for heat waves and cold spells and outbreaks of infectious diseases. Canada is now focusing its efforts and funding on adaptation rather than mitigation, taking the positive outlook that new opportunities will arise with climate change. [INFOGRAPHIC 22.11]

In the North Woods, it might mean facilitation—moving tree species to entirely new ranges where they don't currently grow, based on the notion that the speed of climate change will make it impossible for natural tree migratory processes, such as seed dispersal, to occur. "The idea is that if we want the forest to adapt, we will have to help it along," says Frelich.

Facilitation has no shortage of critics, many of whom say such tinkering is both dangerous and unnecessary. "Facilitation is my nightmare," says John Almindinger, a forest ecologist with Minnesota's Department of Natural Resources. "That we'll start to believe we're smart enough to figure out how to move things. It's sheer hubris." Besides, he says, many if not most tree species seem to be moving just fine on their own, along traditional forest migration routes. So far, the U.S. Forest Service agrees; the agency does not allow such bold interventions as planting pines inside the wilderness.

Still, some skeptics are coming around to the idea. "We've changed the landscape through development and agriculture," says Peter Reich, a colleague of Frelich at the University of Minnesota. "And we've changed the climate, too, with fossil fuel consumption. So we might now need to change the way we manage wild lands to compensate."

On this much, everyone seems to agree: if northern Minnesota is to remain fully forested in the coming century, something will have to be done. "We see it already," says Rajala. "The impact of climate change will be too big to just let nature take its course."◉

Select references in this chapter:
Beaubien, E., and Hamann, A. 2011. *BioScience*, 61: 514–524.
Both, C., *et al.* 2009. *Journal of Animal Ecology*, 78: 73–83.
Environmental Protection Agency. 2010. *Climate Change Indicators in the United States.*
Francis, J. A., and Vavrus, S. J. 2012. *Geophysical Research Letters*, 39(6): L06801, doi:10.1029/2012GL051000
Frelich, L., and Reich, P. 2009. *Natural Areas Journal*, 29: 385–393.
Frelich, L., and Reich, P. 2010. *Frontiers in Ecology and the Environment*, 8: 371–378.
Hitch, A., and Leberg, P. 2007. *Conservation Biology*, 21: 534–539.
Lenoir, J., *et al.* 2008. *Science*, 320: 1768–1771.
NOAA National Climatic Data Center. 2010. *State of the Climate: Global Analysis for Annual 2010.* http://www.ncdc.noaa.gov/sotc/global/2010/13.
Pacala, S., and Socolow, R. 2004. *Science*, 305: 968–972.
Woodall, C.W., *et al.* 2009. *Forest Ecology and Management*, 257: 1434–1444.

adaptation Efforts intended to help deal with a problem that exists, such as climate change.

BRING IT HOME

PERSONAL CHOICES THAT HELP

The effects of climate change are already being felt by humans, other species, and ecosystems around the globe. By changing our actions, we can decrease the greenhouse gases we produce and show policy-makers that citizens are interested in preventing global climate change.

Individual Steps

→ Do your part to reduce carbon emissions by conserving energy. Walk or ride a bike instead of driving a car. Share a ride with a coworker rather than driving alone. Negotiate with your employer to telecommute. Live close to where you work or go to school. Reduce your heating and cooling energy use, and always turn off electronics and lights when not in use.

→ If your utility company offers renewable energy, buy it.

→ Reduce the carbon footprint of your food by decreasing the amount of feedlot-produced meat you eat. Buy your food as locally as possible to reduce energy used in transportation.

→ Avoid air travel, but if you must, buy carbon offset credits if they are available.

Group Action

→ Volunteer to help build a zero-energy Habitat for Humanity home.

→ Organize a community lecture on climate change with a local university expert or meteorologist as the speaker.

→ Organize an event at your school or community to raise awareness about global climate change and ways to prevent it. Go to 350.org to join a current campaign and for other program ideas.

Policy Change

→ Consider writing, calling, or visiting your member of parliament to tell him or her to support funding for research and development of clean and renewable sources of energy. In addition, insist that your MP support the funding of science, especially those efforts to understand and confront climate change.

UNDERSTANDING THE ISSUE

CHECK YOUR UNDERSTANDING

1. Which of the following best describes radiative forcers?
   a. Factors that can alter climate
   b. Events that increase the likelihood of a future change in the same direction as past changes
   c. Long-term patterns in meteorological conditions
   d. Natural factors, such as volcanoes, that affect the amount of sunlight that reaches the surface of Earth

2. Current atmospheric CO_2 levels are at 390 parts per million (ppm). What is the upper limit of CO_2 predicted by most scientists that we can live with and still avoid substantial negative effects?
   a. 400 ppm
   b. 560 ppm
   c. 680 ppm
   d. 750 ppm

3. Scientists have documented that climate change is a normal process. What is the PRIMARY reason that the current warming trend is so alarming to those same scientists?
   a. It raises the possibility of another ice age.
   b. We have no means of stopping it.
   c. It is solely due to human activities.
   d. It is occurring at a rapid rate.

4. According to some, the current increase in average global temperature is due to the "wobble" effect. If this were true, which of the following would also be true?
   a. The axis of Earth would be shifted to point toward the star Vega, resulting in hotter summers in the northern hemisphere.
   b. Earth would be rotating around the Sun in an ellipse as opposed to a circle.
   c. Earth would be tilted on its axis at an angle that maximizes the differences between summertime highs and winter lows (temperatures).
   d. The Sun would be rotating around Earth in an ellipse as opposed to a circle.

5. Which of the following is NOT a mitigation strategy for climate change?
   a. Green taxes
   b. Carbon cap-and-trade
   c. Carbon-sequestration techniques
   d. Planting different crops

6. Which of the following factors tend to increase the albedo effect and contribute to cooling?
   a. Melting glaciers
   b. Sulphate aerosols
   c. Clear, sunny days
   d. Increasing CO_2 levels

WORK WITH IDEAS

1. Compare and contrast the major radiative forcers, including both those that are natural and those that are humanmade. Overall, which forcers are currently having the most effect on global climate? Are they natural or produced by human actions?

2. What is the IPCC? Why is it important?

3. Describe greenhouse gases. What is their function? What human actions have led to an increase in the amount of greenhouse gases in the atmosphere? What has been the result?

4. Describe the types of problems that global climate change will cause for Canada. Of the impacts given in Infographic 22.11, which do you feel is likely to cause the biggest problem? Why?

5. In the winter of 2010, the northeastern part of the United States had several large snowstorms that resulted in record high snowfall amounts. How does this weather fit in with the notion of global climate change?

6. Describe the process of facilitation and its goal. Why do some scientists claim that facilitation is a bad idea?

ANALYZING THE SCIENCE

The graphs on the following page show 2 of the 15 northern species evaluated in Chris Woodall's study of tree-range migration mentioned in this chapter. The total standing biomass and the total number of seedlings of each species are shown at different latitudes within the study area.

INTERPRETATION

1. What does the purple line on each graph represent? What does the blue line represent?

2. There are 111 kilometres between two adjacent latitude lines. How far apart are the latitude classes shown here? How many kilometres wide is the study site (from latitude 34° to 48°)?

3. Look at each graph to determine at which latitude class each species shows the most biomass per hectare. Which tree species has more standing biomass at its peak: balsam poplar or American basswood? How can you tell?

ADVANCE YOUR THINKING

4. Is either of these tree populations exhibiting a range migration shift? Present the evidence for your conclusions.

5. Of the two species shown here, which would you expect to see at higher altitudes on a mountainside and which at lower altitudes? Assuming both species could migrate, what do you predict will happen to the populations of these two species if the climate warms a little? If it warms a lot? Explain your answers.

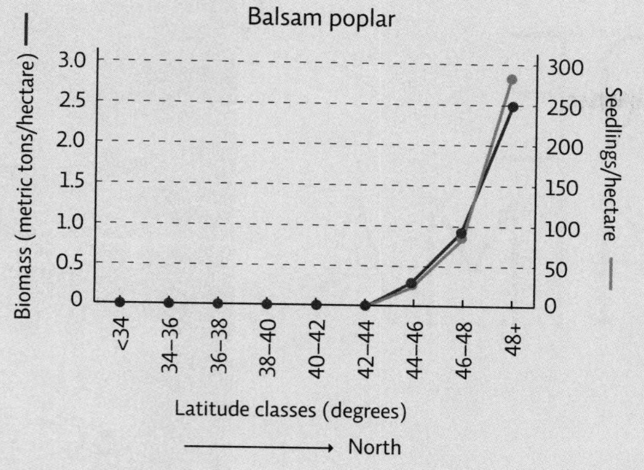
Balsam poplar
Biomass (metric tons/hectare)
Seedlings/hectare
3.0
2.5
2.0
1.5
1.0
0.5
0
300
250
200
150
100
50
0
<34
34–36
36–38
38–40
40–42
42–44
44–46
46–48
48+
Latitude classes (degrees)
North

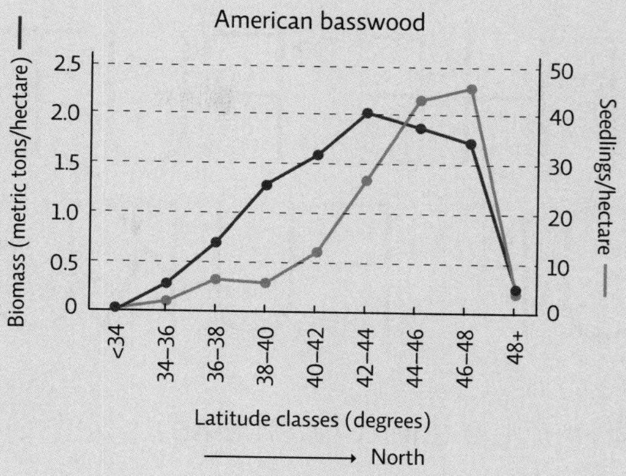
American basswood
Biomass (metric tons/hectare)
Seedlings/hectare
2.5
2.0
1.5
1.0
0.5
0
50
40
30
20
10
0
<34
34–36
36–38
38–40
40–42
42–44
44–46
46–48
48+
Latitude classes (degrees)
North

EVALUATING NEW INFORMATION

Among scientists, there is broad consensus (97%) that climate change is significantly caused by human activity. Yet in a recent poll conducted by the BBC (news.bbc.co.uk/2/hi/science/nature/8500443.stm), 74% of the public are either not convinced or deny that humans are causing global warming. Members of the public get their information from a variety of media sources and information posted on the Internet. How can there be such a large disconnect between scientists and the public?

Go to the Global Warming Hoax page (www.globalwarminghoax.com/news.php?extend.133).

Evaluate the website and work with the information to answer the following questions:

1. Is this a reliable information source? Does it have a clear and transparent agenda?
   a. Who runs the website? Do this person's/group's credentials make the site reliable or unreliable? Explain.
   b. What is the primary message of the website? What evidence is it providing in the short article on Antarctic sea ice?
   c. Do you have any questions about the data presented? If so, what are they?

Now go to the Skeptical Science website (www.skepticalscience.com). Click on "Most Used Climate Myths" on the left menu.

2. Is this a reliable information source? Does it have a clear and transparent agenda?
   a. Who runs the website? Do this person's/group's credentials make the site reliable or unreliable? Explain.
   b. What is the primary message of this website? What types of evidence does it provide to support its message?
   c. Click on the "Antarctica is gaining ice" link. Read the article carefully and compare the main point of the article to the article on the Global Warming Hoax site. Based on this article, what are the fundamental flaws of the first article?
   d. Which website do you find more credible? Why?

MAKING CONNECTIONS

IS CLIMATE CHANGE THE BIGGEST CHALLENGE OF OUR TIME?

Background: Many people feel that climate change is the biggest challenge of our generation. While skeptics remain and large portions of the public are unconvinced, the evidence supporting a changing global climate continues to accumulate. Why aren't people convinced by this evidence? Susan Hassol, a climate change communicator, provides a number of explanations. These include:

- The way climate change is presented in the media as a story with two sides.
- A general lack of science literacy among the public.
- A lack of understanding of the scientific process by the public.
- A fundamental misunderstanding between scientists and laypeople involving the use of language—essentially a communication issue.
- A psychological barrier that results when people are presented with something that frightens them.

Case: As a member of your campus environmental organization, you have been asked to participate in a workshop for teachers about the importance of science in education. Your presentation is limited to 30 minutes. Using one of the five preceding explanations as a starting point, and using global climate change as an example, create a presentation that will convince audience members that science is a necessary part of education, especially if we want to solve problems such as climate change.

Make sure that your presentation includes information from experts in the field, presented in both text form and as graphs and figures, and uses language that the average layperson will understand.

THE FUTURE OF FUKUSHIMA

Can nuclear energy overcome its bad rep?

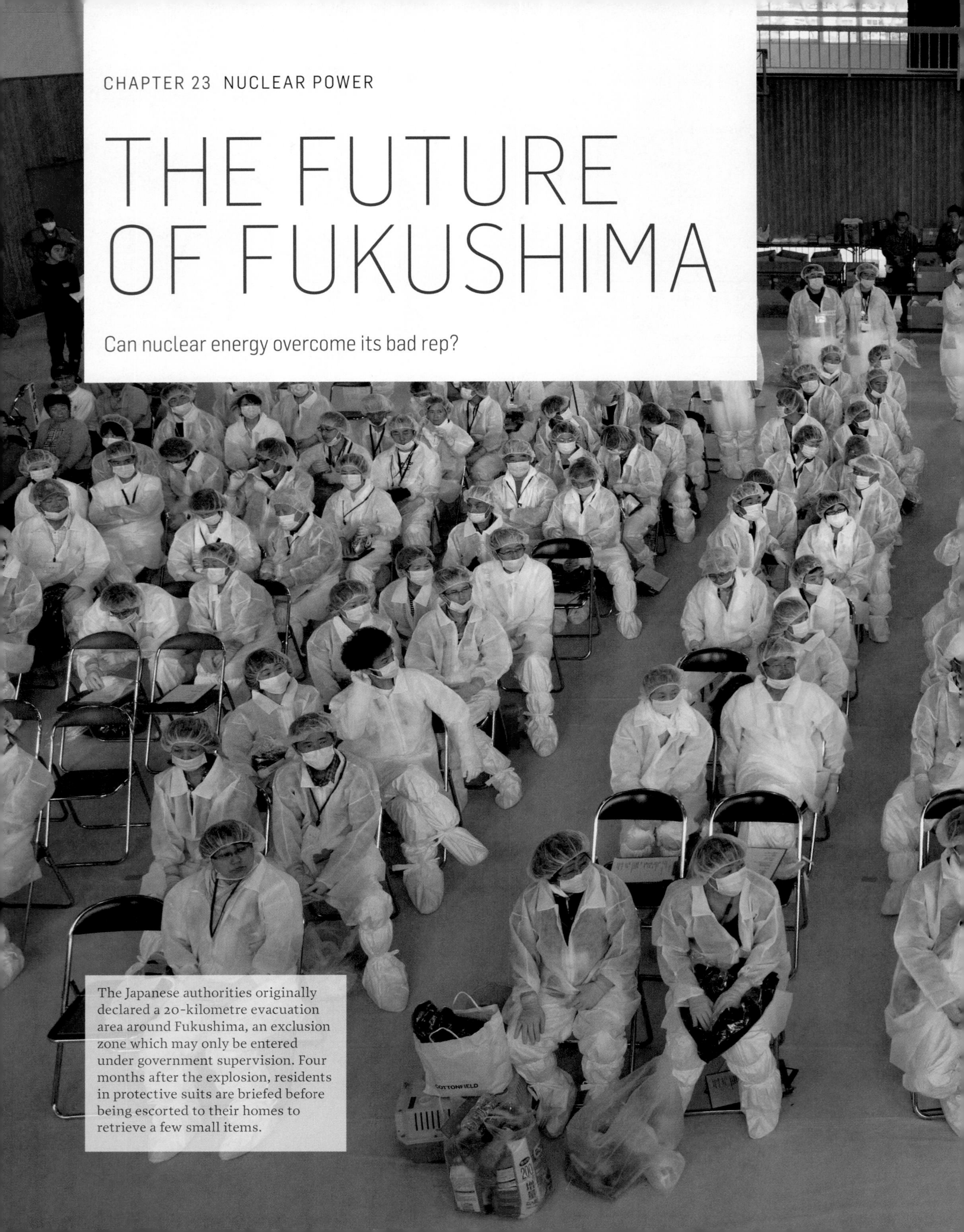
COTTONFIELD

The Japanese authorities originally declared a 20-kilometre evacuation area around Fukushima, an exclusion zone which may only be entered under government supervision. Four months after the explosion, residents in protective suits are briefed before being escorted to their homes to retrieve a few small items.

CORE MESSAGE

Nuclear energy can create tremendous amounts of power, and with concerns over fossil fuel supplies and climate change, nuclear energy has the potential to be an increasingly important part of the world's energy future. However, there are serious safety concerns with nuclear power, including vulnerability to natural disasters, radioactive waste disposal, and potential for weapons production.

GUIDING QUESTIONS

After reading this chapter, you should be able to answer the following questions:

→ What are radioactive isotopes and why are they important for nuclear power?

→ How do radioactive isotopes decay and what types of radiation are produced? How dangerous is this radiation?

→ How is nuclear energy harnessed to generate electricity in a fission reactor?

→ What problems are associated with the production of nuclear fuel and the disposal of nuclear waste?

→ What are the advantages and disadvantages of nuclear power?

↑ The crippled Fukushima Daiichi Nuclear Power Station 10 months after the disaster.

→ Satellite image of the damaged Fukushima Daiichi Nuclear Power Station on March 12, 2011.

The Fukushima Daiichi Nuclear Power Station is a maze of steel and concrete perched right on Japan's Pacific coast, just 240 kilometres north of Tokyo; its six nuclear reactors supply some 4.7 GW (1 gigawatt = 1 billion watts) of electric power to the country, making it one of the largest nuclear power plants in the world. On March 11, 2011, when a magnitude 9.0 earthquake struck 130 kilometres north of the plant, there were more than 6000 workers inside. The quake caused a power outage, and in the darkness, chaos ensued: men and women groped desperately for ground that would not stabilize beneath their hands and feet, shouting in panic as steel and concrete collided around them. When the shaking stopped, emergency lights came on, revealing a cloud of dust. But that was only the beginning of the disaster.

The earthquake had erupted beneath the ocean floor, triggering a tsunami that would arrive at the plant in two distinct waves. The first wave was not big enough to breach the 10-metre-high concrete wall that had been built between the plant and the sea. But the second wave, a fearsome mass of water that came 8 minutes later, was. At four storeys tall, it bulldozed a string of protective barriers, sent buses and cars and trucks careening into pipes and levers and control panels, and eventually settled, in deep black pools, around the reactors themselves.

They were the strongest earthquake and largest tsunami in the country's long memory. Together, they would claim some 20 000 lives along a 400-kilometre stretch of coast (roughly equal to the driving distance between Toronto and Ottawa). But as the ground steadied and the water subsided, the world's attention would quickly turn to a third disaster, even more precarious and potentially deadly than the first two: the risk of nuclear meltdown at Daiichi.

◉WHERE IS FUKUSHIMA, JAPAN?

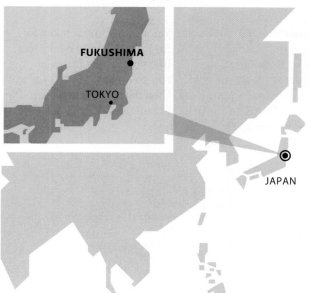

FUKUSHIMA
TOKYO
JAPAN

It's no surprise that the story of nuclear power pivots on calamity. Ever since its potential was first demonstrated, humankind has scurried relentlessly between two competing goals—the desire to harness nuclear energy for our own ends, and the impulse to protect ourselves from its destructive capacity. When the ground trembled beneath Fukushima, concerns over greenhouse gases and global warming had been pushing much of the world—including Canada and the United States—toward the former. As the people of Japan scrambled to respond, the world watched closely.

Nuclear power harnesses the heat released in nuclear reactions to produce electricity.

In some ways, electricity generated using **nuclear energy** is no different than other forms of thermoelectric power (those that use heat to produce electricity). Just like power plants that run on oil or coal, nuclear plants use heat to boil water and produce steam, which is then used to generate electricity. The difference, really, is in where that heat comes from; coal and oil plants create it by burning fossil fuels. At a nuclear power plant, heat is produced

nuclear energy Energy released when an atom is split (fission) or combines with another to form a new atom (fusion); can be tapped to generate electricity.

Infographic **23.1** | **ATOMS AND ISOTOPES**

→ All matter is made up of atoms. Each atom is made up of subatomic particles: *protons* and *neutrons* in the nucleus (centre) of the atom, make up the mass of the atom. Orbiting around the nucleus are much smaller particles called *electrons*. Different combinations of these subatomic particles produce specific *elements*—a chemical substance made up of only one kind of atom. The number of protons, the *atomic number*, is unique to each element. For example, any atom with only 2 protons is an atom of the element helium. The sum of the number of protons and neutrons gives an element its *mass number*. Helium, shown here, is an element that has 2 protons, 2 neutrons, and 2 electrons.

THE ATOM

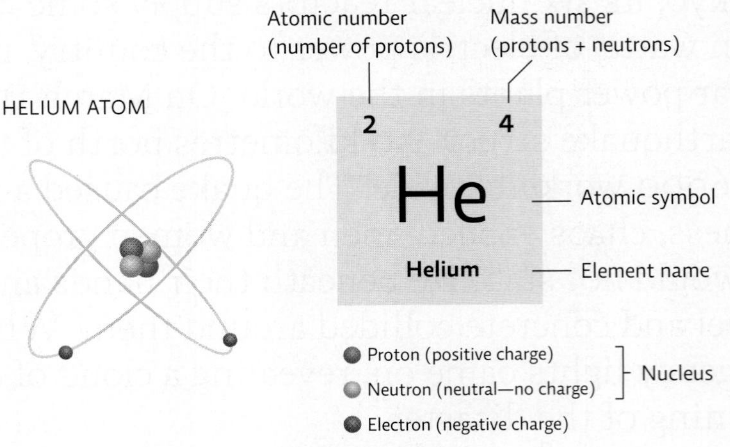
HELIUM ATOM
Atomic number
(number of protons)
Mass number
(protons + neutrons)
2
4
He
Atomic symbol
Helium
Element name
Proton (positive charge)
Neutron (neutral—no charge)
Nucleus
Electron (negative charge)

→ *Isotopes* are atoms that have the same atomic number (number of protons) but a different number of neutrons, and thus a different mass number. An atom with 92 protons is uranium; it can have 146 neutrons, and is called U-238 (92 protons + 146 neutrons = 238). Another uranium isotope, U-235, has 143 neutrons (92 protons + 143 neutrons = 235).

ISOTOPES

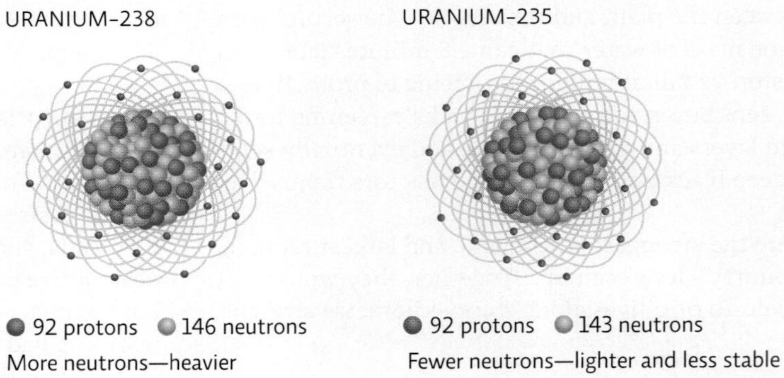
URANIUM-238
URANIUM-235
92 protons
146 neutrons
More neutrons—heavier
92 protons
143 neutrons
Fewer neutrons—lighter and less stable

through a controlled nuclear reaction—usually a **nuclear fission** reaction.

Nuclear fission reactions are those that result in the splitting of an atom. Their main ingredient is a special type of atom known as a *radioactive isotope*. Some elements can exist in two or more forms—each form has the same number of protons and electrons but a different number of neutrons, and hence, a different atomic mass; these different versions of the atom are called **isotopes**. [INFOGRAPHIC 23.1]

Most isotopes are stable, meaning they do not spontaneously lose protons or neutrons. But some are **radioactive**—they emit subatomic particles and heat energy (radiation) in a process known as radioactive decay. Radioactive decay is measured in half-lives. An isotope's **radioactive half-life** is the amount of time it takes for half of the radioactive material in question to decay to a new form. So after one half-life, 50% of the material will decay; in the next half-life, 50% of what's left (or 25% of the original amount) will then decay. After ten half-lives, just 0.1% of the original radioactive material is left. [INFOGRAPHIC 23.2]

nuclear fission Nuclear reaction that occurs when a neutron strikes the nucleus of an atom and breaks it into two or more parts.
isotopes Atoms that have different numbers of neutrons in their nucleus but the same number of protons.
radioactive Atoms that spontaneously emit subatomic particles and/or energy.
radioactive half-life The time it takes for half of the radioactive isotopes in a sample to decay to a new form.

Infographic **23.2** | **RADIOACTIVE DECAY**

↓ The rate of decay for a given radioactive isotope is predictable and expressed as a *half-life*—the amount of time it takes for half of the original radioactive material (*parent*) to decay to the new *daughter* material (a new isotope, or even a new atom if protons are lost). Radioactive isotopes and their daughter radioactive isotopes continue to decay until they form a stable isotope that no longer loses particles. For instance, U-238 decays initially to thorium-234, which itself will decay over time. The entire decay sequence of U-238 includes progression through at least 13 isotopes (each step with its own half-life that ranges from milliseconds to thousands of years) until the final isotope decays into lead-206, a stable atom.

RADIOACTIVE DECAY

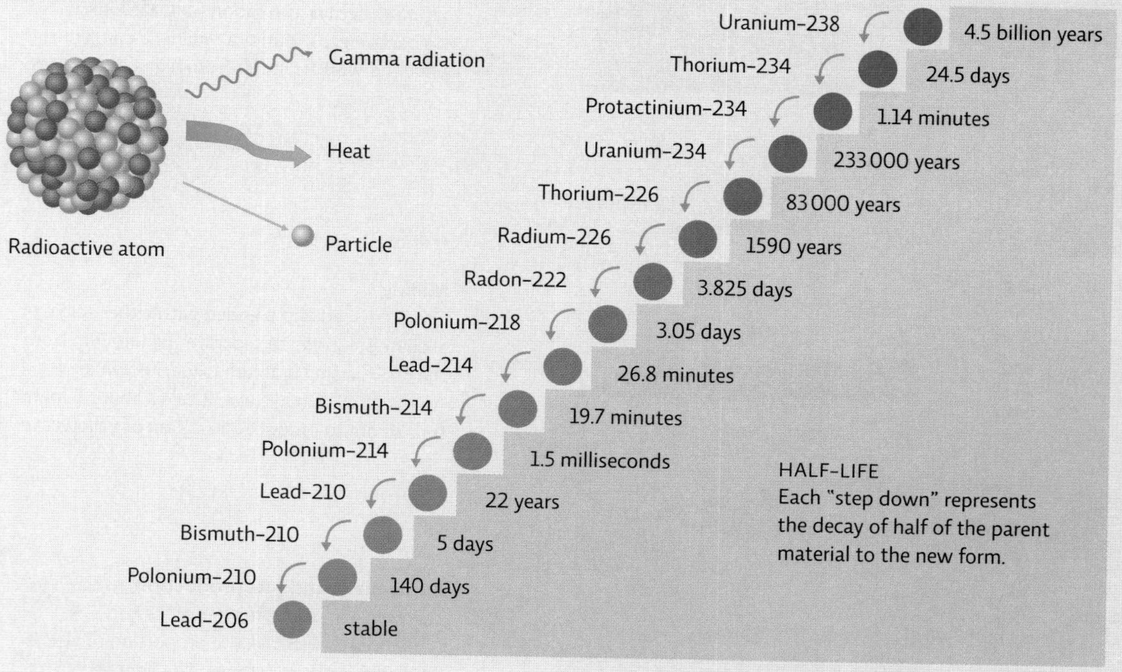

Gamma radiation
Heat
Particle
Radioactive atom
Uranium-238
4.5 billion years
Thorium-234
24.5 days
Protactinium-234
1.14 minutes
Uranium-234
233 000 years
Thorium-226
83 000 years
Radium-226
1590 years
Radon-222
3.825 days
Polonium-218
3.05 days
Lead-214
26.8 minutes
Bismuth-214
19.7 minutes
Polonium-214
1.5 milliseconds
Lead-210
22 years
Bismuth-210
5 days
Polonium-210
140 days
Lead-206
stable
HALF-LIFE
Each "step down" represents the decay of half of the parent material to the new form.

RADIOACTIVE HALF-LIFE

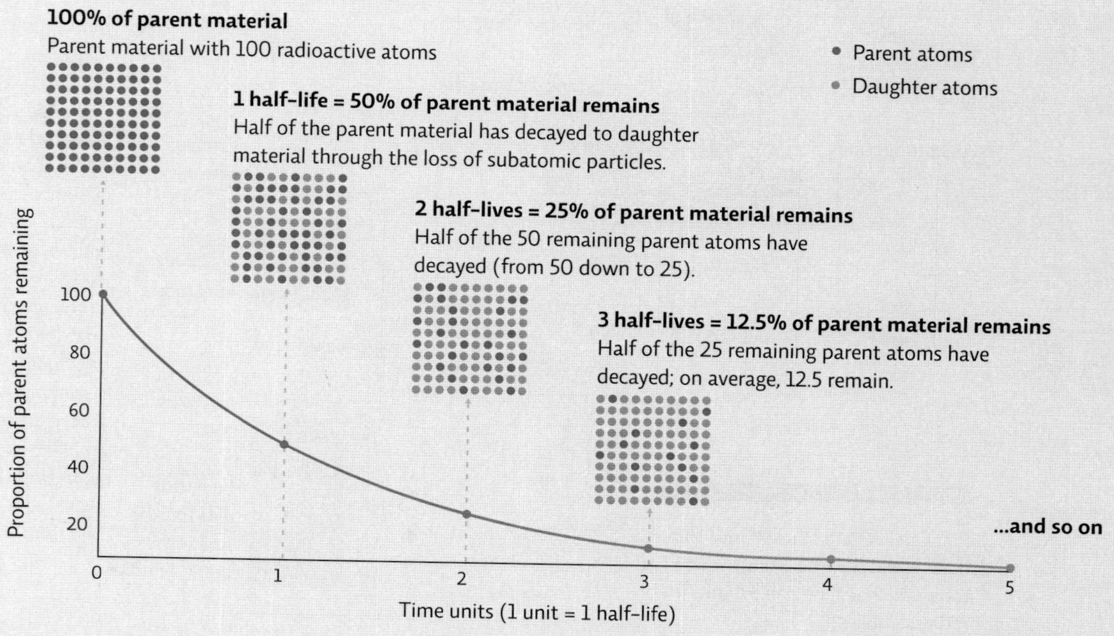

100% of parent material
Parent material with 100 radioactive atoms
Parent atoms
Daughter atoms
1 half-life = 50% of parent material remains
Half of the parent material has decayed to daughter material through the loss of subatomic particles.
2 half-lives = 25% of parent material remains
Half of the 50 remaining parent atoms have decayed (from 50 down to 25).
3 half-lives = 12.5% of parent material remains
Half of the 25 remaining parent atoms have decayed; on average, 12.5 remain.
...and so on
Proportion of parent atoms remaining
100
80
60
40
20
0
1
2
3
4
5
Time units (1 unit = 1 half-life)

Infographic 23.3 | **NUCLEAR FUEL PRODUCTION**

↓ Uranium ore (rock that contains uranium) is mined and goes through many stages of processing to produce fuel suitable for a nuclear reactor. The process creates hazardous waste at every step.

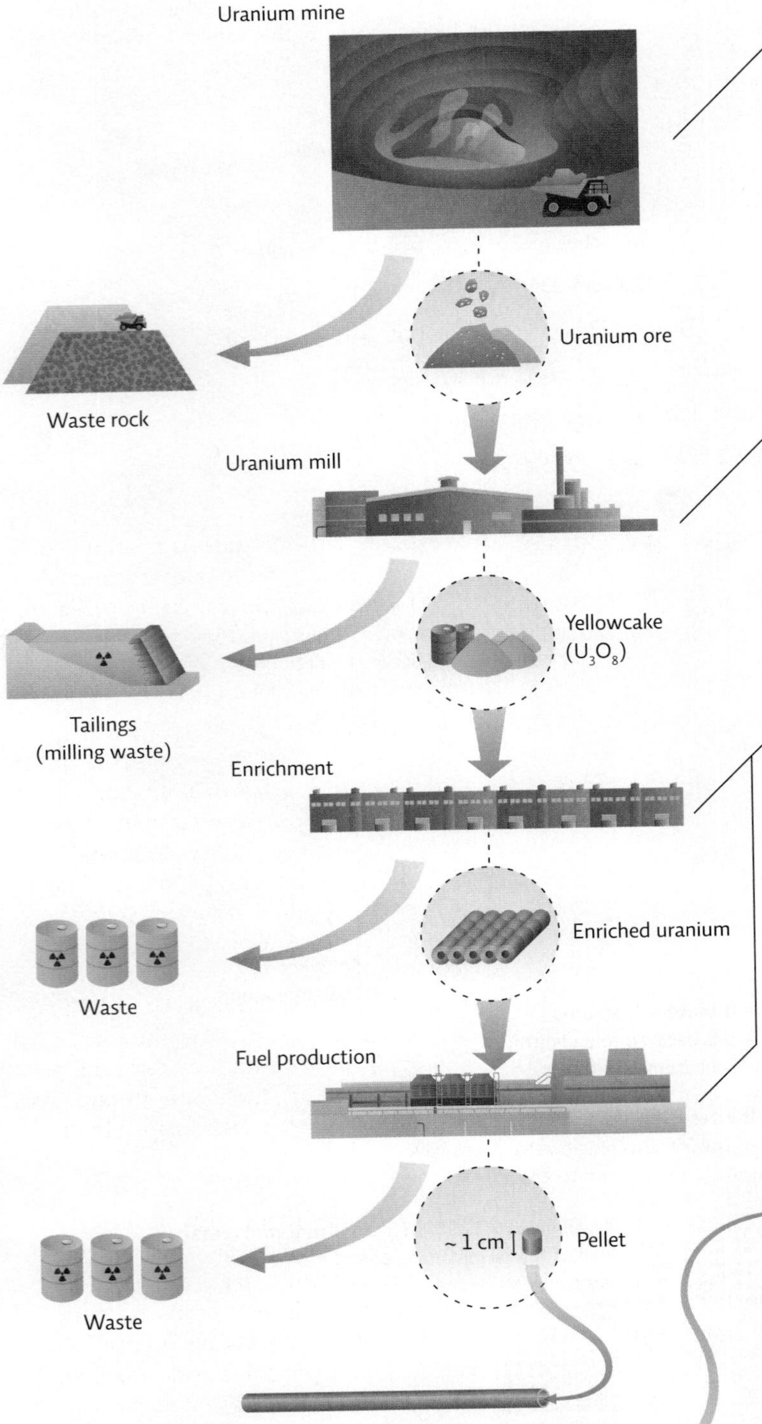
Uranium mine
Uranium ore
Waste rock
Uranium mill
Yellowcake (U_3O_8)
Tailings (milling waste)
Enrichment
Enriched uranium
Waste
Fuel production
~1 cm
Pellet
Waste

Small pellets are stacked into thin, hollow metal 3.5 to 5 metre-long fuel rods, which will be packed into fuel rod assemblies.

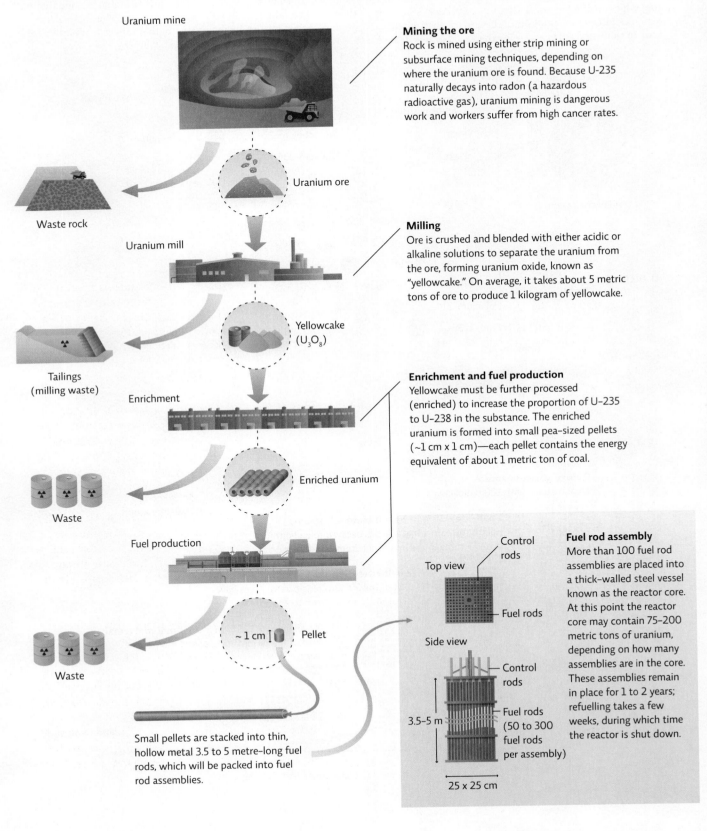

Mining the ore
Rock is mined using either strip mining or subsurface mining techniques, depending on where the uranium ore is found. Because U-235 naturally decays into radon (a hazardous radioactive gas), uranium mining is dangerous work and workers suffer from high cancer rates.

Milling
Ore is crushed and blended with either acidic or alkaline solutions to separate the uranium from the ore, forming uranium oxide, known as "yellowcake." On average, it takes about 5 metric tons of ore to produce 1 kilogram of yellowcake.

Enrichment and fuel production
Yellowcake must be further processed (enriched) to increase the proportion of U-235 to U-238 in the substance. The enriched uranium is formed into small pea-sized pellets (~1 cm x 1 cm)—each pellet contains the energy equivalent of about 1 metric ton of coal.

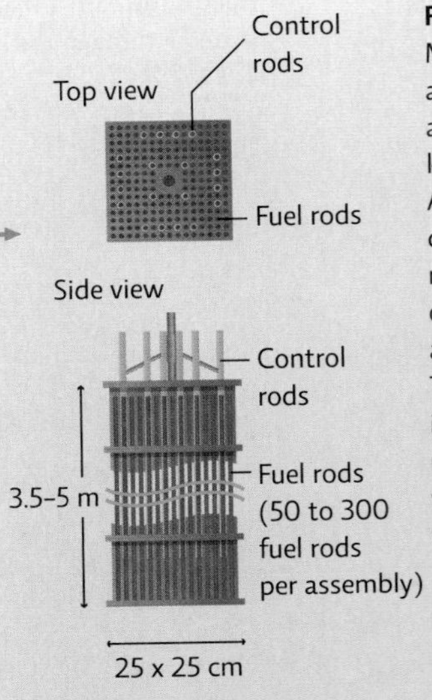
Top view
Control rods
Fuel rods
Side view
Control rods
Fuel rods (50 to 300 fuel rods per assembly)
3.5–5 m
25 x 25 cm

Fuel rod assembly
More than 100 fuel rod assemblies are placed into a thick-walled steel vessel known as the reactor core. At this point the reactor core may contain 75–200 metric tons of uranium, depending on how many assemblies are in the core. These assemblies remain in place for 1 to 2 years; refuelling takes a few weeks, during which time the reactor is shut down.

Infographic **23.4** | **NUCLEAR FISSION REACTION**

↓ Fission, or the breaking apart of atoms, begins when an atom like U-235 is bombarded with a neutron—this breaks the atom into other smaller atoms and releases free neutrons, which in turn hit other U-235 atoms, causing them to split and release neutrons, and so on. The reaction in the fuel assembly is controlled by the insertion of control rods of nonfissionable material which absorb some of the free neutrons.

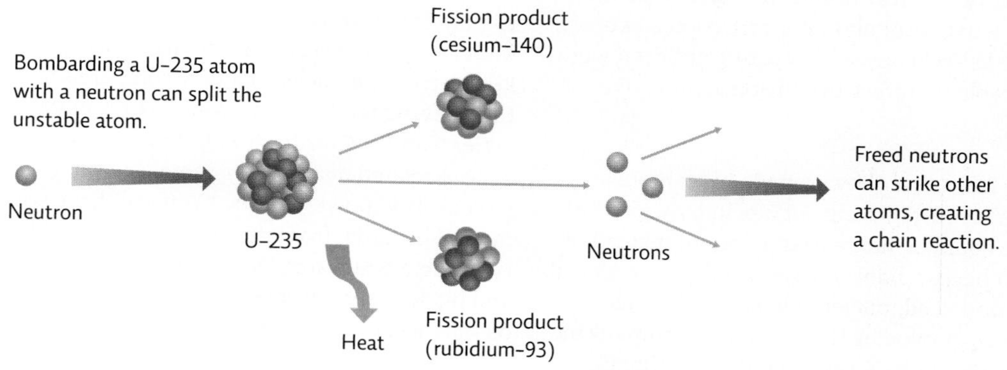
Fission product
(cesium-140)
Bombarding a U-235 atom
with a neutron can split the
unstable atom.
Neutron
U-235
Heat
Fission product
(rubidium-93)
Neutrons
Freed neutrons
can strike other
atoms, creating
a chain reaction.

Most nuclear reactors use uranium, which has several isotopes. U-238 is the most stable and is the most abundant form of uranium—it makes up roughly 99% of Earth's total supply. But it's U-235, the most reactive form, that is mined from Earth, processed into nuclear fuel (which, like all mining work, involves both safety and environmental hazards), and packed into the **fuel rods** that are used in facilities like Fukushima. [INFOGRAPHIC 23.3]

A nuclear fission chain reaction begins when U-235 in the fuel rods is deliberately bombarded with a neutron. This bombardment makes the nucleus unstable, causing it to split into a variety of two or more smaller atoms and releasing two or three additional neutrons in the process. These newly released neutrons then hit other U-235 atoms, causing them to split and release even more neutrons, and so on. [INFOGRAPHIC 23.4]

Unlike the type of nuclear reaction at work in a nuclear bomb, which uses much more radioactive material , setting off a chain reaction that is almost instantaneous, the reactions at nuclear power plants are highly controlled. **Control rods**—made of materials such as boron or graphite that absorb neutrons—are placed in the fuel rod assembly between the fuel rods to control the speed of the reaction. They can be added (to slow down or stop the reaction) or removed (to make it go faster).

fuel rods Hollow metal cylinders filled with uranium fuel pellets for use in fission reactors.

control rods Rods that absorb neutrons and slow the fission chain reaction.

Even controlled, this chain reaction releases a tremendous amount of heat—10 million times more than is released by burning fossil fuels like coal or oil. The heat is used to boil water, which produces steam, which turns turbines that create electricity.

All types of thermoelectric power take a lot of water—that's why power plants are sited near rivers and oceans; but at the moment, nuclear power requires the most—on average, around 2500 litres per megawatt hour (MWh), compared with 1900 litres per MWh for coal, and 605 litres per MWh for natural gas. (A MWh is the production of 1 megawatt—1 million watts—over an hour's time.) That means a typical 1000-MW nuclear reactor requires roughly 2 500 000 litres of water per minute to flow through the cooling system. Though most of this water (more than 95%) is returned to the source (river or ocean), there are still problems with the release of warmer-than-normal water back into the environment, as well as damage to aquatic life that gets trapped in or against intake filters.

The reason for all that water is simple: with nuclear energy, water is needed not only to produce steam but also to keep spent fuel rods cool and to prevent the reactor from overheating (remember, heat is produced from radioactive decay of fission by-products, which continues even after the reactor is shut down and fission stops; so spent fuel needs constant cooling). Without water to cool them, fuel rods can melt, releasing large amounts of radioactivity; the fuel rod metal casing can also get hot enough to react with steam in a way that produces highly explosive hydrogen gas.

Nuclear energy has a troubled history.

Nuclear energy is the most concentrated source of energy on Earth. Its fearsome power was first demonstrated in 1945, when the U.S. military dropped atomic bombs over the Japanese cities of Hiroshima and Nagasaki. The bombs brought an end to World War II, but they also wreaked havoc on an entire nation of civilians: radiation sickness, cancers that killed slowly and which young children were especially vulnerable to, infertility in some, and birth defects in others.

In 1953, U.S. President Eisenhower made his famous "Atoms for Peace" speech, laying out a plan by which this destructive force could be harnessed for good: instead of building bombs, cheap, reliable energy would be produced. In the years that followed, nuclear physics indeed gave rise to a litany of technologies that benefitted humankind, from radiocarbon dating to X-rays to radiation therapy for cancer.

But while there are now 432 nuclear power plants around the world, nuclear energy itself remains mired in controversy.

Proponents argue that uranium ore (uranium-containing rock) is both more abundant and produces a more efficient fuel than any fossil fuel. For example, 1.0 kilgram of uranium produces the same amount of energy as about 100 000 kilograms of coal. And the operating costs, per kilowatt-hour, for nuclear power are comparable to those of coal.

Of course, as critics are quick to point out, that estimate does not factor in the great expense of building, maintaining, and then decommissioning nuclear plants. In Canada, it costs just under $4 billion to build one nuclear reactor (over 5 times the cost of a coal power plant), and from $270 million to $1.8 billion to decommission one (most nuclear reactors have an expected lifespan of 40-60 years, after which they need to be disassembled and the radioactive components stored and guarded).

One thing both sides agree on is that nuclear energy is a cleaner way to produce electricity. The processes of generating it emit much less CO_2 than the analogous processes for any fossil fuel, and virtually none of the other problematic combustion by-products, like sulphur dioxide, nitrogen oxides, and particulate matter. According to the Canadian Nuclear Association, switching from fossil fuels to nuclear energy would be the single most effective way to reduce greenhouse gas emissions in Canada. Existing Canadian nuclear power plants prevent the emission of about 90 million metric tons of greenhouse gases annually compared to coal power. Worldwide, they prevent close to 2.5 billion metric tons of CO_2 from entering the atmosphere.

And despite some persistent fears, research suggests that living near a nuclear power plant is actually safer than living near a coal-fired one. A 2011 study found no increased risk of birth defects for those living within 10 kilometres of a nuclear facility compared to the risk for those living farther away. Meanwhile, epidemiologist Javier García-Pérez has found that in Spain, the number of cancer-related deaths does increase as one moves closer to coal-fired power plants. "You still have some environmental hazards, from mining uranium and from radioactive waste and water," says Charles Powers, a professor and nuclear energy scientist at Vanderbilt University. "But on balance, nuclear is far cleaner than any fossil fuel."

Still, the debates over safety remain unresolved. Proponents point out that, considering the number of existing plants, and the length of time they have been operating, accidents have been exceedingly few and far between. But opponents say that such safety claims ignore two key points: the vulnerability of nuclear power plants to natural disasters (which, as we will see, can lead to nuclear meltdown and the release of radioactive material into the environment), and the potential for nuclear fuel to be stolen and weaponized. The radioactive waste produced by nuclear power plants is another huge safety issue: it's extremely dangerous; there's a lot of it, and we have yet to come up with a plan for disposing of it safely.

Public opinion has been divided over nuclear energy since the 1980s, after two infamous nuclear accidents made global headlines. The first was a partial meltdown at the Three Mile Island plant near Middletown, Pennsylvania, in 1979 (due to an electrical failure followed by a flurry of operator errors). The second was a full nuclear meltdown at the Chernobyl reactor in what is now the Ukraine, in the spring of 1986.

The Three Mile Island incident did not result in any deaths or major public health problems. The Chernobyl meltdown was considerably more severe. An explosion triggered by a test that went awry sent tremendous amounts of radiation wafting over much of western Russia and Europe. More than a fifth of the surrounding farmland remains unusable to this day, and the World Health Organization estimates that the radiation will ultimately be responsible for some 4000 deaths when all is said and done. That figure does not include cancer-related deaths, which range from 60 000 (according to one European report) to nearly 1 million (according to a Russian report).

> The radioactive waste produced by nuclear power plants is another huge safety issue: it's extremely dangerous; there's a lot of it, and we have yet to come up with a plan for disposing of it safely.

For their part, the Japanese were terrified of nuclear power after the bombings of Hiroshima and Nagasaki. (The popular Godzilla movies were actually based on a fictional reptile that had been mutated by a nuclear reaction and was coming to exact his revenge on humankind!) But as their country entered its own era of industrialization and economic growth, they were forced to overcome those fears. "Japan had no other natural energy source," says Frank N. von Hippel, nuclear physicist and arms control expert at Princeton University. "Nuclear was their only ticket to becoming a world superpower."

Nuclear accidents can be devastating.

On the day of the quake, each of the three operating reactors at the Fukushima plant held about 25 000 fuel rods, each one about 4 metres long and filled with pellets of enriched uranium. The power outage had not only plunged the plant into darkness, but also stopped the normal delivery of water to those reactors.

As the world watched, workers at the plant tried everything they could think of to get cold water on the hot fuel. They tried to bring fire trucks and emergency power vehicles in, but the quake and tsunami had rendered the roads impassable. In one desperate attempt, some workers even searched the parking lot for vehicles that might have survived the tsunami with their batteries intact. But it was all to no avail.

Not only were the rods in danger of melting, and of producing hydrogen gas, but without a steady supply of coolant, they were rapidly boiling away all the existing water, thereby creating huge steam pressure in the reactor. To prevent an explosion, the steam would have to be released through a vent. But the design of the Fukushima reactors made this an especially tricky feat.

↓ Police guard a checkpoint at the edge of the exclusion zone leading to the town of Minami Soma, just north of Daiichi. The sign reads "Keep Out."

立入禁止
立入禁止
災害対策基本法により
南相馬市
車両通行止

Infographic 23.5 | **NUCLEAR REACTORS**

↓ The most common type of nuclear power plant is a pressurized water reactor (PWR), with 265 in operation around the world. The reactor at Fukushima is a boiling water reactor (BWR), one of 94 in the world. Both designs use fuel assemblies with control rods and water as a cooling and steam source. Temperatures are kept "down" to about 1400°C, well below the melting temperature of the uranium fuel (2800°C) and the metal casing of the fuel rods (2200°C). If the reaction is not kept cool with circulating water, a meltdown can occur. Even if "shut down" by inserting all the control rods to absorb the neutrons and stop the chain reaction, heat will still be produced by the natural decay of the isotopes (they are not being split by bombardment; they are simply spontaneously losing particles).

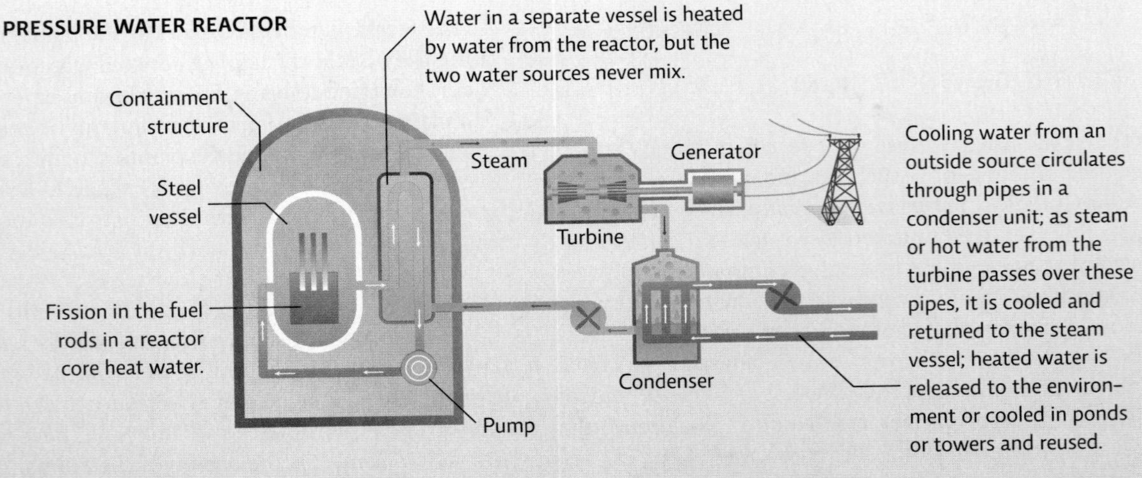
PRESSURE WATER REACTOR
Water in a separate vessel is heated by water from the reactor, but the two water sources never mix.
Containment structure
Steel vessel
Fission in the fuel rods in a reactor core heat water.
Steam
Generator
Turbine
Condenser
Pump
Cooling water from an outside source circulates through pipes in a condenser unit; as steam or hot water from the turbine passes over these pipes, it is cooled and returned to the steam vessel; heated water is released to the environment or cooled in ponds or towers and reused.

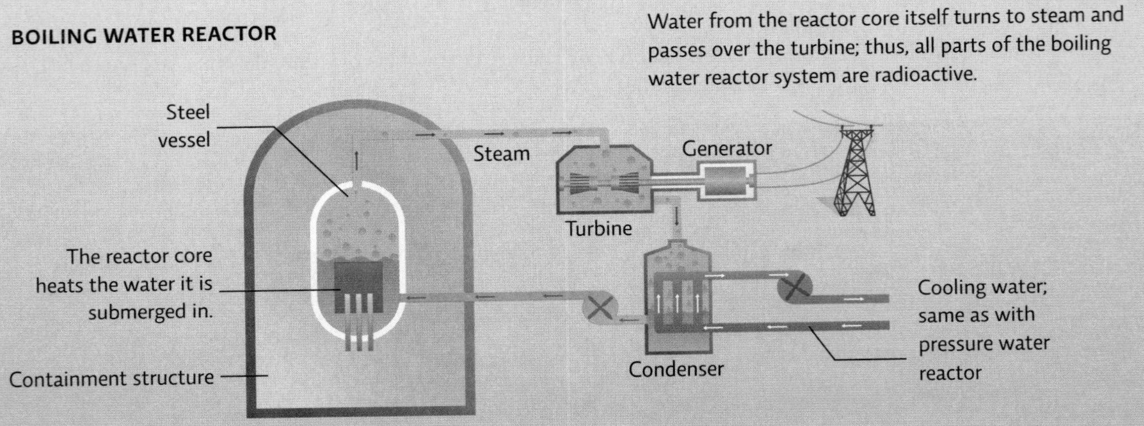
BOILING WATER REACTOR
Water from the reactor core itself turns to steam and passes over the turbine; thus, all parts of the boiling water reactor system are radioactive.
Steel vessel
Steam
Generator
Turbine
The reactor core heats the water it is submerged in.
Condenser
Cooling water; same as with pressure water reactor
Containment structure

There are several different types of fission reactors. The most common type worldwide is a *pressurized water reactor* (PWR), where the steam that turns the turbine is not exposed to radiation: nuclear fission in the core heats water under pressure (like a pressure cooker); that water (which is exposed to radiation) is then piped through, and thus heats, a separate container of water (which is not exposed to radiation); it is the steam from this radiation-free water that turns the turbine. The reactors at Fukushima, however, were *boiling water reactors* (BWR); BWRs produce steam in the reactor core itself. This means that both the steam and the turbine become radioactive in the process. [INFOGRAPHIC 23.5]

It also meant that opening the vent would be akin to pumping radiation straight into the air. Still, not opening it would almost certainly be worse: if any one of the reactors exploded, it would release much, much more radiation than the steam from the vent. "It was this horrible double-edged sword," says von Hippel. "Exactly the kind of situation that we always feared with the BWRs—a lose-lose."

On the morning of March 12th, six workers gathered in the plant's main control room, where each of them swallowed several iodine pills; by loading their thyroid glands with iodine beforehand, they hoped to prevent their bodies from absorbing radioactive iodine from the steam that would pour out of the vent. They donned thick, heavy firefighter uniforms, along with oxygen tanks and masks, in an effort to shield themselves from alpha and beta radiation. Both alpha and beta radiation come from particles (alpha particles are basically a helium nucleus; beta particles are essentially electrons). Neither type can penetrate the skin very well. But if they enter the body through ingestion, both kinds can linger for years, causing organ damage and cancer.

None of the equipment would protect the workers from gamma radiation, which is much more energetic than alpha or beta particles and can easily penetrate walls, skin, and other surfaces. The only way for workers to avoid over-exposure to gamma rays would be to get in and out as quickly as possible. Dosiometers—mini radiation detectors attached to each fire suit—would let them know, with loud, alarmlike beeps, when they had exceeded the legal limits of exposure. [INFOGRAPHIC 23.6]

As the sun climbed high over the ravaged coastline, the team of six headed into the first reactor to search in darkness for the vent. The plan was to work in teams of two, and to relay their efforts so that no one worker was left hovering over the vent for too long. It took the first team 11 minutes to find the manual gate valve, crank it open a quarter of the way, and retreat. With the vent open partway, the radiation levels climbed so fast that the second team could not even reach it; in just 6 minutes, one of them absorbed a dose of radiation that surpassed the legal limit allowed for 5 years. The third team did not even try.

That afternoon, the building housing reactor No. 1 exploded, sending concrete and steel debris flying through the air and setting off a chaotic, and seemingly irreversible, chain reaction: radiation levels around the plant climbed exponentially, deterring efforts to vent the other two reactors. In the days that followed, both of them also exploded.

The generation of nuclear waste is a particularly difficult problem to address.

The reactors weren't the only problem. Experts around the world were particularly concerned about Fukushima's radioactive waste.

To understand why, it helps to know a little about radioactive waste. In general, there are two kinds: low-level radioactive waste and high-level radioactive waste.

Infographic **23.6** | **RADIOACTIVE ISOTOPES CAN RELEASE ONE OR MORE OF THREE DIFFERENT KINDS OF RADIATION**

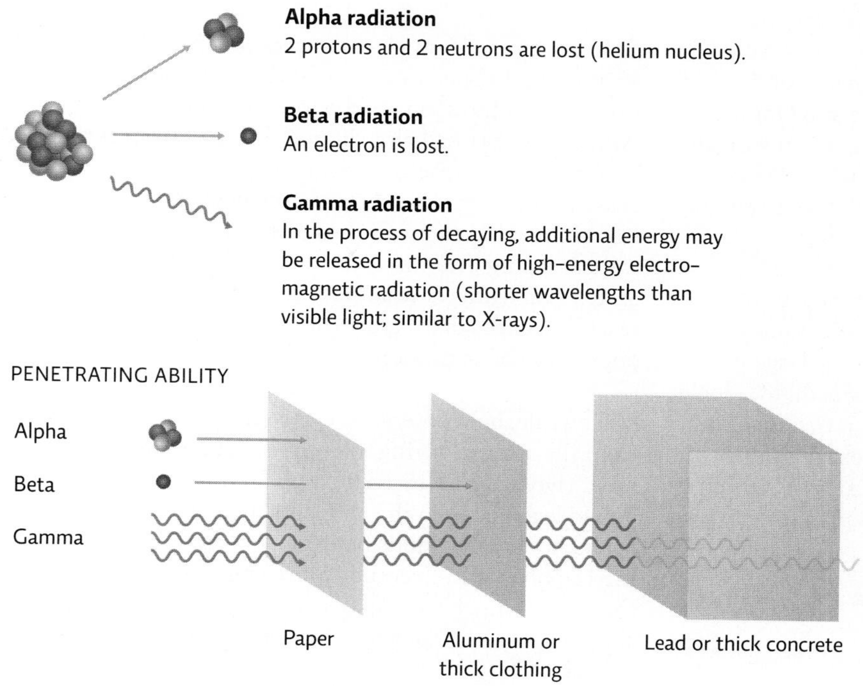
Alpha radiation
2 protons and 2 neutrons are lost (helium nucleus).
Beta radiation
An electron is lost.
Gamma radiation
In the process of decaying, additional energy may be released in the form of high-energy electromagnetic radiation (shorter wavelengths than visible light; similar to X-rays).
PENETRATING ABILITY
Alpha
Beta
Gamma
Paper
Aluminum or thick clothing
Lead or thick concrete

← The different types of radiation have differing abilities to penetrate surfaces. Alpha particles don't travel far and can't even penetrate paper; they do not penetrate skin but can be harmful if inhaled or ingested in food or water. The much smaller beta particles can penetrate the upper layers of the skin but are easily stopped by a thin sheet of aluminum or very heavy clothing. High-energy gamma rays can penetrate the deepest but can be stopped by thick or very dense material such as 20 centimetres of concrete or 2.5 centimetres of lead. Exposure to radiation, especially gamma rays, can damage a wide variety of body organs, leading to radiation sickness; it can also cause mutations that lead to cancer or cause birth defects.

Low-level radioactive waste (LLRW) is material that has low amounts of radiation relative to its volume, and can usually be safely buried. This includes clothing, gloves, tools, etc., that have been exposed to radioactive material. In 2010, Canada produced over 5000 cubic metres of LLRW and had accumulated over 2.3 million cubic metres.

High-level radioactive waste (HLRW) is another story. As its name suggests, it's more reactive than LLRW; in fact, because so many radioactive by-products are created in fission reactions, HLRW is actually much more radioactive than the original fuel rods. In 2010, Canada produced almost 300 cubic metres of HLRW and had more than 9000 cubic metres of the stuff in storage. HLRW includes nuclear fuel rods, other fuel forms, and some liquids.

So far, we have no safe, reliable, long-term storage plan for this waste. Spent fuel rods are stored onsite in steel-lined pools, where at least some isotopes—those with short half-lives—can decay to safe levels. Isotopes with longer half-lives require more time to reach this point. Because the isotopes produced in the fission reaction or by the decay of these isotopes are all mixed together, the waste has to be stored for as long as it takes the longest half-life material to decay to safe levels—in some cases that's more than 2 million years.

Radioactive waste in Canada is stored in concrete or steel containers and pools at numerous sites across the country, none of which are permanent. Most radioactive waste is in interim storage at the nuclear power and radioisotope production reactor sites where they were generated. Some HLRW goes to Atomic Energy of Canada Limited (AECL) research sites in Manitoba and Ontario, or to facilities in the United States. All Canadian storage sites, including those belonging to the AECL, will have to be decommissioned or remediated in the future to comply with Canada's Nuclear Fuel Waste Act.

Permanent storage awaits development of long-term waste management facilities. The Canadian strategy is to create a "deep geological repository" in the largely impermeable, granitic precambrian (or "Canadian") shield, which covers most of eastern and northern Canada. This strategy has been criticized by environmental groups as being an "out-of-sight, out-of-mind" solution; however, it is the only plan we have thus far. For this plan to work, we would also need to develop plants to properly containerize the waste before long-term storage. Even if a long-term storage option is approved and constructed, we still have the problem of how to safely transport nuclear waste across the country, a dilemma not likely to be resolved in the near future. [INFOGRAPHIC 23.7]

Meanwhile, back at Fukushima, radioactive waste posed yet another deadly threat. Each of the six reactors had its own swimming pool-like container where used radioactive fuel was stored. The pools were full of years' worth of waste, and they relied exclusively on water to prevent overheating. If just one of those pools ran dry, the consequences would be catastrophic: by some estimates, a worker standing next to a dry pool could receive a fatal dose of radiation in just 16 seconds.

On March 13th, just as experts around the world were contemplating such an event, a fire broke out around the spent fuel pool near reactor No. 4.

Responding to a nuclear accident is difficult and dangerous work.

By March 14th, three days after the earthquake and tsunami, all but a handful of workers had evacuated. The world media would dub the cohort that stayed behind "The Fukushima 50." The radiation levels in the main control room of reactor No. 2 had climbed so high that those remaining workers had to rotate out at regular intervals to avoid being poisoned with radiation.

Around the world, nuclear experts worried aloud over what might happen next: would there be another explosion? Would the spent-fuel pools run completely dry? Would fuel that had already melted into a heap at the bottom of some reactors now melt through their steel vessels and react with the concrete below? And how far would the wind carry all the radioactive vapours that were escaping into the atmosphere? Experts in the United States said it was at least possible that the vapours could reach the edges of Tokyo—the world's largest city, with a population of 35 million.

Meanwhile, helicopters were scooping up buckets of water from the sea and trying to dump them on the reactors. First, they were deterred by high radiation levels. They bolted lead plates to the choppers' bottoms and tried again, but strong winds blew most of the water askew of the Daiichi plant. Eventually, fire engines made their way through. Using hoses designed for jet-fuel fires, they were able to get some water to the places where it was most needed.

---

low-level radioactive waste (LLRW) Material that has a low level of radiation for its volume.
high-level radioactive waste (HLRW) Spent fuel rods or nuclear weapons production waste that is still highly radioactive.

Infographic **23.7** | **RADIOACTIVE WASTE**

↓ Radioactive waste does not just come from nuclear power plants, but is also generated by industry, the medical field, research laboratories, and weapons production. All of these users produce low-level radioactive waste (LLRW); nuclear power and weapons production is responsible for almost all of the high-level radioactive waste (HLRW). Though much of this material is highly dangerous and remains so for centuries, we currently have no safe way to dispose of it.

HLRW is a by-product of nuclear fission. Canada lacks a long-term storage facility for HLRW; most spent fuel rods are currently stored onsite at nuclear power plants in steel-lined pools. Some have been moved out of the pools and into dry casks. European and Asian nations are looking into the possibility of joint disposal facilities, though little progress has been made.

← Technicians load highly enriched uranium (HEU) fuel assemblies into casks at the Institute of Nuclear Physics in Kazakhstan.

LLRW includes contaminated items such as clothing, filters, gloves, and other items exposed to radiation. Short-half-life LLRW can be stored until it is no longer radioactive and then disposed of as regular trash. LLRW with a longer half-life is stored in casks. In Canada, most LLRW remains at the reactor site where it was generated, and is stored in steel or concrete containers, or in concrete or steel-lined pools. This is true for both active and decommissioned reactor sites.

← Above-ground casks hold longer half-life LLRW (also known as intermediate-level radioactive waste) at Chalk River Nuclear Laboratory in Ontario. These casks can be easily monitored and the material retrieved for disposal once it has decayed to safe levels.

MILL TAILINGS Mining and crushing uranium ore produces small particle (sandlike consistency) waste. It contains low levels of radioactive isotopes with long half-lives, such as radium, thorium, and uranium. Piles of mill tailings are stored near milling facilities and covered with clay and rock to prevent the release of radioactive material into the atmosphere—by law, they must remain covered for at least 200 years.

← Mine tailings are stored in a large open pit at the Ranger Uranium mine in Australia; rainwater fills the pit during the rainy season and there are concerns that the water could overflow the dam in heavy rain years.

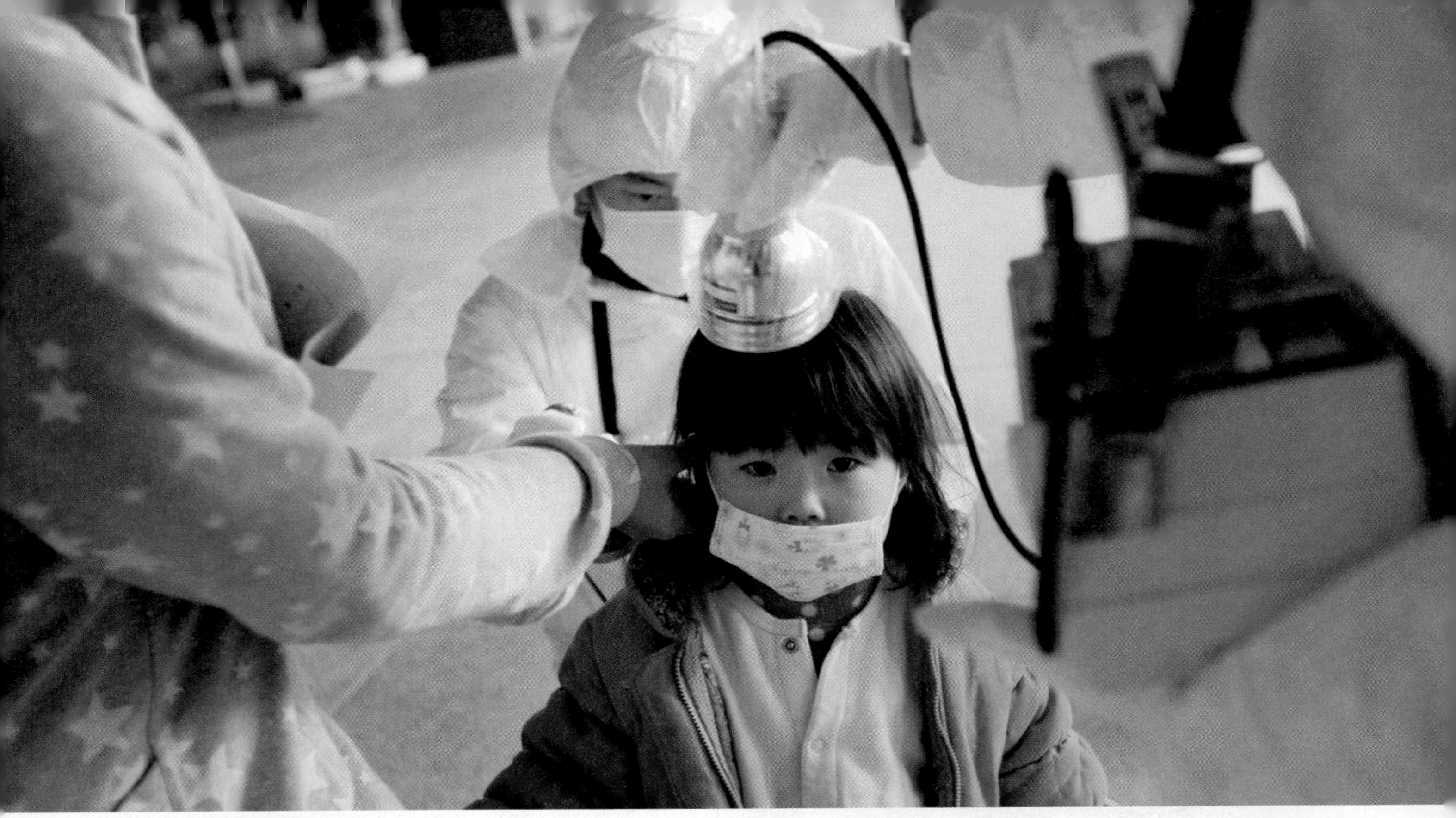

↑ Displaced people who were evacuated from Minamisoma, Futaba, and other towns located near the Fukushima Daiichi Nuclear Power Plant are checked for traces of radiation before they are permitted to enter a sports facility in Fukushima City, where about 1200 evacuees found temporary shelter.

The saving move, however, was already in the works by then. "Even as buildings exploded," the *New Yorker* reported, "some of the workers had hooked up a train of fire trucks capable of generating enough pressure to inject water directly into the fuel cores. The drenching continued around the clock, and it went on for months. The process was ungainly, and it produced millions of litres of radioactive waste that will be dangerous to store, but it probably did more than any other measure to avert a far worse disaster."

The impacts of nuclear accidents can be far reaching.

On March 16th, members of the U.S. Nuclear Regulatory Commission (NRC) determined that anyone more than 80 kilometres from the plant was probably safe from atmospheric radiation emitted by the disaster, based on computer models they had developed.

The finding did not quell public fears. In fact, long after the likely dispersal of radiation had been mapped and made widely known, people in many communities were fiercely divided about whether or not to evacuate. Local newspapers published daily radiation levels alongside weather reports, and average people sorted as best they could through reams of dense, technical information—not all of it reliable—as they tried to make decisions about the health and safety of their families.

For months, it seemed, one new shocking discovery or scandalous revelation followed the next: the Japanese government discovered stores selling beef, spinach, and other food containing small amounts of radiation and scrambled to recall those products. Authorities warned parents not to give local milk to their children because children's cells grow faster and that makes them more vulnerable to the effects of radiation. Distrustful of government reports, and frustrated by the lack of consistent information, average citizens began collecting and testing their own soil and water—sometimes with disturbing results. On October 14, for example, the *New York Times* reported that citizens groups had found 20-odd spots in and around Tokyo that were contaminated with potentially dangerous levels of radioactive cesium.

The impact on the natural environment has been more difficult to determine. "It depends so much on luck," says Powers. "Even if everyone involved in the response does everything perfectly, we're still at the mercy of the wind." At first, the wind at Fukushima Daiichi blew steadily out to sea. But eventually it turned inland, to the northwest, carrying all those radioactive vapours with it. Rain and snow captured some particles, laying them deep in the

↑ The Gösgen Nuclear Plant is located within the town of Däniken, Switzerland. Steam rising from a cooling tower greets children as they walk home from school.

mountains, forests, and streams around Fukushima. It will take decades before we know the full effect they will have on the ecosystems they have become a part of. But 5 months after the quake, Japan's central government acknowledged that the area within 3 kilometres of the plant will likely be uninhabitable for decades.

Will nuclear power play a role in future energy?

The March 2011 tsunami would become the most expensive natural disaster in human history, with losses estimated at $300 billion. In its wake, countries around the world began rethinking their plans to expand their own nuclear energy programs. Germany resolved to phase out all of its nuclear power plants by 2022. Austria, Italy, and Switzerland also reconsidered. While Canadians were horrified by the tragedy in Japan, there was little backlash against nuclear power in Canada. Government agencies reassured the public that all of our reactors are located in low-earthquake-risk regions and are built to meet "seismic standards." However, most reactors are also located in highly-populated areas, such as the Durham region east of Toronto. Critics asserted that, while a nuclear emergency might be unlikely, if one did occur, millions of people would be at risk. Proponents, meanwhile, held fast to their contention that nuclear power could be used safely and would be an essential, unavoidable part of weaning the planet from its fossil fuel dependence.

Meanwhile, back in Japan, just 2 months after the quake and tsunami, the prime minister responded to public pressure and global criticism by calling for a temporary shutdown of one nuclear plant in the country's centre, where scientists have estimated an 87% chance of a big quake sometime in the next three decades. Other plants that were closed for routine maintenance were told to hold off on reopening, so that, by summer's end, less than a third of Japan's 54 reactors were still running.

The course change was short-lived. Once factory owners began warning that such drastic power cuts would quickly lead to a recession, many of the deactivated power plants sprung back to life by order of the same prime minister who had closed them. Ultimately, the country's leaders would agree to close some of its oldest plants but leave most—36 out of 54, by one estimate—up and running for the foreseeable future. One official told the *New Yorker* that anything else would be "idealistic but very unrealistic." Speaking at an October 2011 forum sponsored by the American Association for the Advancement of Science, Gregory Jaczko, chairman of the NRC, predicted that the impact of Fukushima on the long-term development of nuclear power will be minimal.

Infographic **23.8** | **NUCLEAR POWER: TRADE-OFFS**

↓ What role will nuclear power play in the future? Like all of our energy options, nuclear power has advantages and disadvantages which must be weighed when making this decision.

ADVANTAGES	DISADVANTAGES
Operating costs are comparable to those of a fossil fuel power plant; the technology is available now.	Nuclear power plants are much more expensive to build (just under \$4 billion to build a fission reactor), maintain, and decommission (\$270 million to \$1.8 billion to decommission after a 40- to 60-year lifespan) than fossil fuel power plants.
No CO_2 is released during operation, so nuclear power does not contribute to climate change (though some CO_2 is released during mining and processing).	Mining and processing ore for nuclear fuel produces hazardous waste; surface mining damages the environment and can pollute air and water.
Dependable amounts of electricity can be produced—no worries about nighttime or cloudy days (solar power) or windless days (wind power).	Radioactive waste is very hazardous and we still have no long-term plan for dealing with the waste that is produced from the fission reaction.
Power production can be increased or decreased to meet demand (up to the capacity of the facility).	Shipping waste (by truck or rail) is also a safety concern and vehemently opposed by those who live on the transport route.
Uranium supplies should last 80 years but other isotopes can also be used; some reactor designs actually *produce* fuel during the reaction process.	Though accidents are rare in the history of nuclear power, they can have extremely serious consequences and long-term impacts when they occur (Chernobyl and Fukushima).
Fuels used for nuclear reactors are energy rich—one small uranium pellet contains about the same amount of energy as 1 metric ton of coal.	Some methods of nuclear power production produce radioisotopes that could be used in nuclear weapons production; facilities for the processing of fuel could hide weapons programs.

Less than a year after the fires had cooled, experts would agree that the Daiichi disaster was something of a draw. "People who support nuclear energy can't argue anymore that the risk is nonexistent," says Powers. "But at the same time, neither can opponents say that a meltdown would automatically result in a zillion immediate casualties."

In a paper published before the disaster at Fukushima, British nuclear engineers Robin Grimes and William Nuttall argued that nuclear power could enjoy a "renaissance" and increase its contribution to electricity production in the future, if only we improved our nuclear technology. They pointed to new "third generation" BWR and PWR reactor designs (and others) available now, and other technologies ("fourth generation") in development that should be safer and more efficient, producing more electricity with less fuel. Some designs also produce less waste material that could be used to make nuclear weapons, reducing concerns about weapons proliferation.

The third-generation designs in use today are safer, according to safety assessments, and include more fail-safe responses (technical responses that automatically kick in if a problem occurs). But of course, the 2011 earthquake and tsunami have illustrated the need to reevaluate the potential damage of natural disasters. The renaissance would also require better waste-handling options, including the reprocessing of spent fuel (which would decrease the waste produced and increase overall efficiency).

Given the problems associated with fossil fuels, it is likely that nuclear power will continue to have a place in our energy mix. But in making choices about how to pursue our energy future, we must consider the costs and benefits of all our energy options. A "benefit analysis" must evaluate how well a particular energy source meets our energy needs. A "cost analysis" must consider not just the monetary cost of getting kilowatts delivered to our homes; it must also include the environmental and social costs associated with every step of the energy source's life, from acquisition to production to delivery. In addition, as mountaintop removal, oil spills, climate change, and Fukushima demonstrate, there is also a risk assessment that must take place. We must ask two very crucial questions: how risky is the venture (an assessment we can do with at least some degree of accuracy) and how much risk are we willing to take? [INFOGRAPHIC 23.8]

"There is definitely a certain weighing that has to take place," says Powers. "Global warming on one hand, nuclear accidents like Fukushima on the other."

The future of nuclear energy is uncertain.

On August 23, 2011, a magnitude 5.8 earthquake erupted beneath the state of Virginia, causing the North Anna Nuclear Power Plant there to tremble with much greater force than its reactors were designed to withstand: Dry casks, each weighing more than 90 metric tons and filled with spent fuel rods, shifted several centimetres. It was the region's largest quake in more than a century, and the first time such a calamity had struck an American nuclear power station. Just 5 days later, when a category 5 hurricane by the name of Irene struck the East Coast, workers at three other nuclear power plants noticed that emergency sirens had failed to function properly. And at one plant—the Indian Point Power Plant, the one closest to Manhattan—a discharge canal carrying (nonradioactive) water from the cooling system overflowed due to the high river levels.

On September 9th, the NRC staff suggested ordering power plants to review their ability to survive quakes and floods "without unnecessary delay."◉

Select references in this chapter:
García-Pérez, J., *et al.* 2009. *Science of the Total Environment*, 407: 2593–2602.
Grimes, R.W. and Nuttall, W.J. 2010. *Science*, 329: 799–803.
Queißer-Luft, A., *et al.* 2011. *Radiation and Environmental Biophysics*, 50: 313–323.

BRING IT HOME

PERSONAL CHOICES THAT HELP

Nuclear energy has been rebranded as "green energy" because it does not emit greenhouse gases. Technology has improved the safety of nuclear facilities; however, there are still safety issues and valid concerns over the long-term storage of nuclear waste. In addition, cost, national security, and uranium supply problems make nuclear power a complicated energy solution.

Individual Steps

→ Use the My Community link on the Canadian Nuclear Safety Commission website (nuclearsafety.gc.ca/eng) to find out if you have nuclear reactors where you live.

Group Action

As you would expect, the policies endorsed by a particular group will depend on their overall view of nuclear energy. To see two examples, check out the following sites and see how they compare.

→ Visit the Canadian Nuclear Association (www.cna.ca), which is a pronuclear organization, for information about the economic and environmental benefits of nuclear energy.

→ Visit the Nuclear Energy Information Service (neis.org), which is a nonprofit antinuclear organization committed to a nuclear-free future.

Policy Change

→ What is your opinion about the necessity of nuclear energy in Canada? The Canadian Nuclear Safety Commission's Reading Room (nuclearsafety.gc.ca/eng/readingroom/index.cfm) maintains news releases, fact sheets, hearing and meeting documents, and reports regarding nuclear energy in Canada. This resource can help you understand current policy, funding issues, and upcoming legislation.

Vermonter to Vermonter
VPIRG

UNDERSTANDING THE ISSUE

CHECK YOUR UNDERSTANDING

1. **Atoms that have the same number of protons and electrons but different numbers of neutrons are called:**
   a. radioactive.
   b. ions.
   c. isotopes.
   d. subatomic.

2. **After 4 half-lives, about how much radioactive parent material is left?**
   a. 25%
   b. 40%
   c. 0.5%
   d. 6%

3. **What is the expected lifespan of a nuclear reactor?**
   a. 40–60 years
   b. 100–150 years
   c. about 200 years
   d. less than 20 years

4. **For what problem related to nuclear power do we not yet have a workable solution?**
   a. Finding a substitute for uranium once it is used up
   b. Safe disposal of HLRW
   c. Producing steam that is not radioactive to turn the turbine
   d. Controlling the fission reaction

5. **Nuclear fission is a reaction that:**
   a. splits an atom, releasing energy.
   b. combines two or more atoms, producing energy.
   c. results in large explosions.
   d. is required to make atoms radioactive.

6. **There are different types of radiation. Which one is most energetic and can therefore penetrate many surfaces, including skin?**
   a. alpha radiation
   b. beta radiation
   c. gamma radiation
   d. particle radiation

WORK WITH IDEAS

1. Describe the process of nuclear fission. Explain how it can lead to a chain reaction in a fission reactor.

2. What are the pros and cons of generating energy from a nuclear source versus a fossil fuel source (coal)? Include in your answer a comparison of the extraction, energy-generating, and waste disposal processes. Based SOLELY on these comparisons, which of these is a better resource to use if we want to minimize environmental degradation?

3. From an economic standpoint, which type of electricity production (nuclear or fossil fuel) is less expensive? Explain. Remember to include both internal and external costs in your answer.

4. Using Infographic 23.4, describe the steps used in mining and processing uranium to the point where it is packed into fuel rods.

5. Suppose a natural disaster that compromises the integrity of a nuclear power plant had occurred in Canada. What might engineers in Canada be able to do immediately that engineers at the Fukushima Daiichi plant in Japan were reluctant to do, based on the differences in design of the plants?

6. In a nuclear fuel assembly, control rods are used to regulate the nuclear reaction. But even if all the control rods are inserted into the fuel assembly, heat is still produced. Explain why.

ANALYZING THE SCIENCE

The following graph depicts the number of accidents that occur in a variety of industries in the United States.

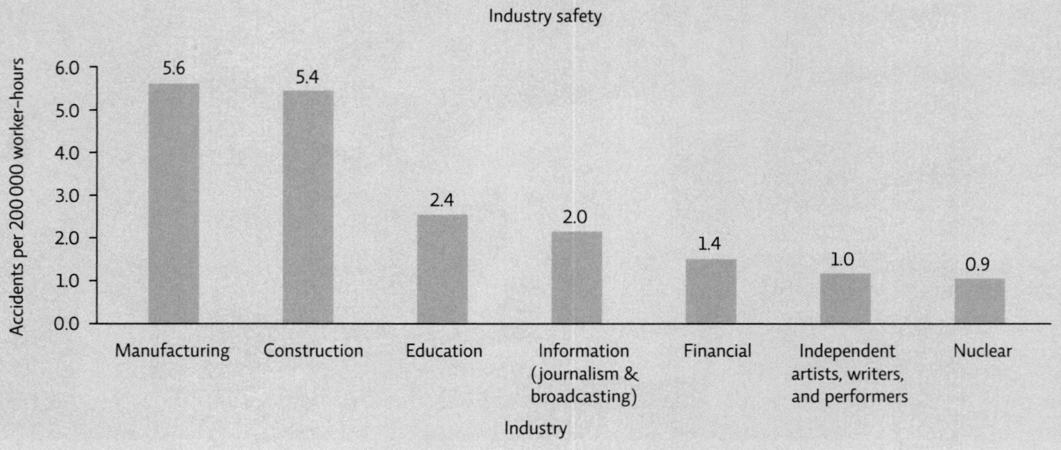
Industry safety
Accidents per 200 000 worker-hours
6.0
5.0
4.0
3.0
2.0
1.0
0.0
5.6
5.4
2.4
2.0
1.4
1.0
0.9
Manufacturing
Construction
Education
Information (journalism & broadcasting)
Financial
Independent artists, writers, and performers
Nuclear
Industry

INTERPRETATION

1. What does the height of each bar represent?
2. Based on the graph, which two industries have the highest accident rate?
3. Based on the graph, which two industries have the lowest accident rate?

ADVANCE YOUR THINKING

4. Is there bias in this graph? How would you decide? Think about and answer the following questions:
   a. Which industries are represented? Are they comparable?
   b. Do the data for nuclear workers include accidents that occur as a result of mining, processing, and production?
   c. Which of these industries do you think has the strictest occupational health and safety regulations? regulations? Might that account for differences in the number of accidents?
5. If you had access to safety data from all industries, which industries would you show in the graph to obtain a fair comparison to the nuclear industry?

EVALUATING NEW INFORMATION

The Canadian Nuclear Safety Commission (CNSC) is tasked with regulating nuclear power in Canada to "protect the health, safety, and security of Canadians and the environment." The members of the commission review applications for licences according to regulatory requirements, make recommendations to the Commission, and enforce compliance with the Nuclear Safety and Control Act, regulations, and any licence conditions imposed by the Commission (http://nuclearsafety.gc.ca/eng/about/index.cfm). The CNSC maintains a detailed website with information about nuclear power plants, the CNSC's regulatory framework, and licence applications.

Go the CNSC's Nuclear Facilities in Canadian Communities page (http://nuclearsafety.gc.ca/eng/mycommunity/facilities/index.cfm). Find your province or territory on the list.

Evaluate the website and work with the information to answer the following questions:

1. Is this a reliable information source? Does the organization have a clear and transparent agenda?
   a. Who runs the website? Do this person's/group's credentials make the information reliable or unreliable?
   b. Is the information on the website up to date? When was the website last updated? Explain.
2. How many nuclear facilities (active or being decommissioned) are in Canada? What types of nuclear sites are listed here?
3. For your province or territory (or another province or territory if yours is not listed here), click on each link for the nuclear sites listed and answer the following questions:
   a. For each nuclear site in this province or territory, describe what the site is, what it contains (in terms of nuclear material), the environmental issues associated with the site (if any), and the plan of action to correct the issues.
   b. If there are nuclear power reactors in this province or territory, write down information about their age, intended use, current status, and specific locations. Find the location on a map and record the closest large body of water.
4. Consider the information you have gathered, and do a basic risk assessment analysis for one of the sites.
   a. Do you think the facility is a danger to the environment? To human health? Why or why not?
   b. Is the facility located in an area that may have natural disasters? If so, what are they? If a natural disaster were to occur, predict some outcomes.

MAKING CONNECTIONS

WHAT DO WE DO WITH NUCLEAR WASTE PRODUCTS?

Background: One of the problems with nuclear material is that the radiation generated is harmful to life AND it persists in the environment for a long time—from hundreds of thousands to millions of years. To date, we have no long-term solution to the problem of what to do with much of this material. Currently, material is stored in what are described as "temporary" facilities (see Infographic 23.7), although the reality is that they are de facto permanent facilities.

Case: You are member of a citizen's group seeking to prevent the transportation of nuclear waste on the highway that goes through your town. Before joining the group, you were unaware that nuclear waste was being transported by truck to disposal sites. You must research the issue and make a presentation to the members of the group. (The CNSC website mentioned above is a good starting place.) Make sure that you include the following:

1. What type of nuclear waste is transported across the country?
2. Where are the waste storage facilities for low-level nuclear waste? For waste incidental to reprocessing?
3. How is nuclear waste transported? What regulations are in place to provide safe transportation?
   a. Based on your research, what recommendations would you make to your group concerning the current transportation of nuclear waste? For example, is it relatively safe? Have there been releases of radiation to the environment because of transportation? What type of regulations could your locality enact to enhance safety while the material is being transported through your region?
   b. In the event that a long-term disposal site for spent nuclear fuel is approved, how would you suggest that the material be transported to the site from the nuclear power plants across Canada?

CHAPTER 24 SUN, WIND, AND WATER ENERGY

FUELED BY THE SUN

A tiny island makes big strides with renewable energy

Playing soccer under the shade of wind turbines on Samsø Island, Denmark, the first island in the world with 100% renewable energy. Collectively, Samsø's land-based turbines produce about 26 million kilowatt-hours a year, enough to meet the island's demands for electricity.

CORE MESSAGE

In order to become a sustainable society, we need to transition to reliable, renewable energy sources with acceptable environmental and social impacts. No single energy source can replace fossil fuels. Instead, a variety of methods, selected to meet the needs of the population, and availability of local energy sources, will help communities shift to sustainable energy use. Fortunately, we have many good options already at our disposal, with other new methods currently in research and development.

GUIDING QUESTIONS

After reading this chapter, you should be able to answer the following questions:

→ What are the characteristics of a sustainable energy source and what role does renewable energy play in terms of global energy production?

→ How do wind and solar power technologies capture energy? What are the advantages and disadvantages of wind and solar power and how does each compare to fossil fuels in terms of true costs?

→ In what ways can we harness geothermal energy and the power of water and what are some of the trade-offs associated with each resource?

→ What combination of actions did Samsø take to become an energy-positive island and what is the take-home message to other communities who might want to reduce their use of fossil fuels?

→ What roles do conservation and energy efficiency play in helping us meet our energy needs sustainably?

On a typical cold, misty January day in Denmark in 2003, many of the 4100 residents of a small island gathered together at the beach. Everyone, including the mayor, strained their eyes to see the faint outline of several structures, each over 30 storeys tall, located more than 3 kilometres offshore.

Nestled in the crook of Denmark's mainland, Samsø is home to a small, windswept community of Danish farmers known for their sweet strawberries and tender early potatoes. It is a quiet and serene place. Yet it has been the site of a dramatic revolution—a community transformation that made headlines around the world.

The transformation began on that cold day in 2003. Finally, the mayor pushed a button, and the offshore structures slowly creaked to life. Through the grey mist and rain, people could gradually see the massive blades begin to rotate, converting the power of wind into energy. It was a landmark day in the island of Samsø's ambitious attempt to become the greenest, cleanest, and most energy-independent place on Earth.

"That was a very big moment," recalls Søren Hermansen, a Samsø resident who was key in getting the community behind the project. "Nobody really thought it would happen when we started."

◉WHERE IS SAMSØ?

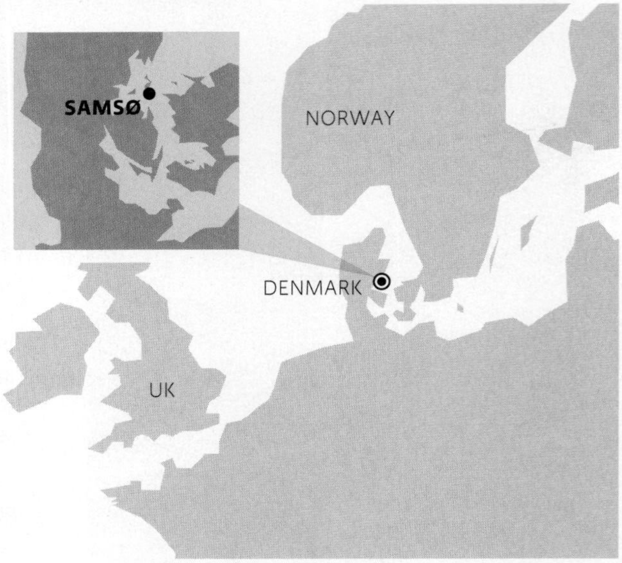
SAMSØ
NORWAY
DENMARK
UK

Reliance on renewable energy is a characteristic of a sustainable ecosystem and society.

Samsø used to be just like most communities on Earth: fully dependent on fossil fuels like coal, oil, and gas. As recently as 1998, Samsø imported all of its energy resources from the mainland: tankers hauled oil into its ports and electricity generated from burning coal was imported via cables.

But in 1997, the government of Denmark decided it was important to promote the idea of **renewable energy**, energy from sources that are replenished over short time scales or that are perpetually available. To do so, they announced a competition: which local area or island could become self-sufficient on renewable energy? The contest invited applicants to describe which renewable resources were available in their community and how they would be used to replace fossil fuels.

The purpose of the contest was to put communities on track to be more sustainable. To qualify as a **sustainable energy** source, the energy must be renewable, with a low enough environmental impact that it can be used for the long term. As a result, any source of energy that causes environmental damage when it is captured or produced, such as generating waste or pollution, is not ideal. For example, experts continue to debate the value of "clean coal"—this type of coal does burn more cleanly than typical coal, but causes environmental damage when it is extracted and requires a lot of energy to process into a cleaner fuel—while still generating toxic waste (for more, see Chapter 19).

Even though Samsø is small—about 26 kilometres long and only 7 kilometres at its widest—it has plenty of the natural resources that are becoming increasingly important sources of sustainable energy. A small firm put together an application for the contest, and the little island won.

↑ Samsø's offshore wind turbines provide much of the electricity used by the island. Since 2005, Samsø has produced more electricity than it uses and exports the extra to mainland Denmark and beyond to neighbouring countries.

To become sustainable, Samsø would harness the power of one of its most plentiful natural resources.

A major part of Samsø's transformation was the installation of 11 onshore and 10 offshore wind turbines designed to harness the power of **wind energy**, or energy contained in the motion of air across Earth's surface. There is no shortage of wind on Samsø, and the powerful breezes turn huge blades (up to 40 metres in length) that are connected to a generator, converting mechanical energy into electricity.

Nine of the onshore turbines on Samsø were purchased collectively by groups of farmers who bought shares in their construction. "People on the island are personally invested in this," says Bernd Garbers, a German engineer who lives on the island and is a consultant for the firm that won the energy contest, Samsø Energy Academy. The machines produce 75% of the electricity used by the islanders.

Samsø uses about 500 billion kilojoules (kJ) of energy each year. A barrel of oil produces 6.15 million kJ, so

renewable energy Energy from sources that are replenished over short time scales or that are perpetually available.
sustainable energy Energy from sources that are renewable and have a low environmental impact.
wind energy Energy contained in the motion of air across Earth's surface.

Infographic **24.1** | **RENEWABLE ENERGY USE**

↓ According to the U.S. Energy Information Administration, as of 2010, renewable sources of energy contributed 3.46 trillion kilowatt-hours (kWh) of total energy production worldwide (about 18% of the total). That amount is projected to more than double to 7.97 trillion kWh by 2035—this will be 22.7% of the total at that time. Of renewable sources, biomass, mostly fuels like wood, charcoal, and animal waste, makes up the largest proportion, followed by energy produced by hydroelectric dams.

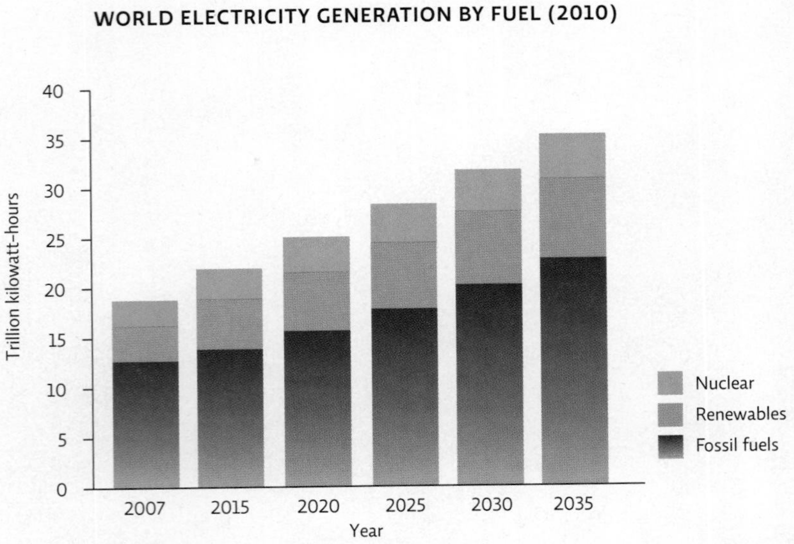
WORLD ELECTRICITY GENERATION BY FUEL (2010)
Trillion kilowatt-hours
40
35
30
25
20
15
10
5
0
2007
2015
2020
2025
2030
2035
Year
Nuclear
Renewables
Fossil fuels

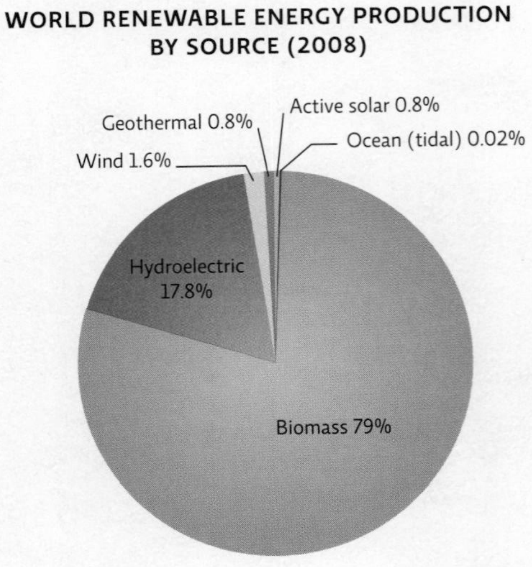
WORLD RENEWABLE ENERGY PRODUCTION BY SOURCE (2008)
Active solar 0.8%
Geothermal 0.8%
Ocean (tidal) 0.02%
Wind 1.6%
Hydroelectric 17.8%
Biomass 79%

Samsø uses 81 300 *barrels of oil equivalents (BOE)* annually (that's 500 billion divided by 6.15 million). The entire human population uses somewhere between 50 and 70 billion BOE per year, and only about 18% of that energy comes from renewable resources. Canada fulfills most of its energy needs through renewable energy, mainly through hydroelectric power. [INFOGRAPHIC 24.1]

Ideally, Samsø residents would have swapped their vehicles for cars that run on hydrogen or electricity, but those technologies were too expensive and not efficient enough to use on the island. Since Samsø would continue to rely on conventional fuels for transportation, engineers decided to install large offshore wind turbines to produce an equivalent amount of clean energy, offsetting uses by motor vehicles and boats. Islanders are working toward a gradual transition to electric vehicles and those powered by biofuels (see Chapter 25). The island's small size helps make all this feasible since daily "commutes" are short and kilmetres travelled per person per day are lower than in most industrialized nations.

Far off the coast, the offshore turbines are especially efficient at generating energy because wind conditions are better at sea. In some especially windy locations, people have built strategically placed *wind farms*, which can contain dozens of turbines. In 2010, the world's largest was located in Texas, the leading producer of wind energy in the United States. It has 627 wind turbines that generate enough electricity—781.5 megawatts (MW)—to power 230 000 homes per year. Larger wind farms are under construction, with one in Oregon slated to produce 845 MW. Although wind power is on the rise in Canada, the largest wind farm to date, in Gros Morne, Québec, only generates 211 MW annually. [INFOGRAPHIC 24.2]

But wind energy isn't perfect. First, even for a blustery location like Samsø, wind is intermittent—it stops and starts irregularly, not producing a steady stream of power. And wind turbines are not cheap. Each onshore turbine costs the Samsø islanders the equivalent of $1 million; offshore turbines rang up at $5 million apiece. Beyond cost, wind turbines can create noise, and some people see them as an eyesore. They can also have an impact on the local environment, threatening birds and bats, who are unable to nest near turbines or are killed by rotating turbine blades. But switching energy reliance from fossil fuels to wind and other alternatives might have a net benefit to bird and bat populations. Wind farms globally cause 0.3 to 0.4 bird fatalities per GWh while fossil

Infographic **24.2** | **WIND TURBINES GENERATE ELECTRICITY**

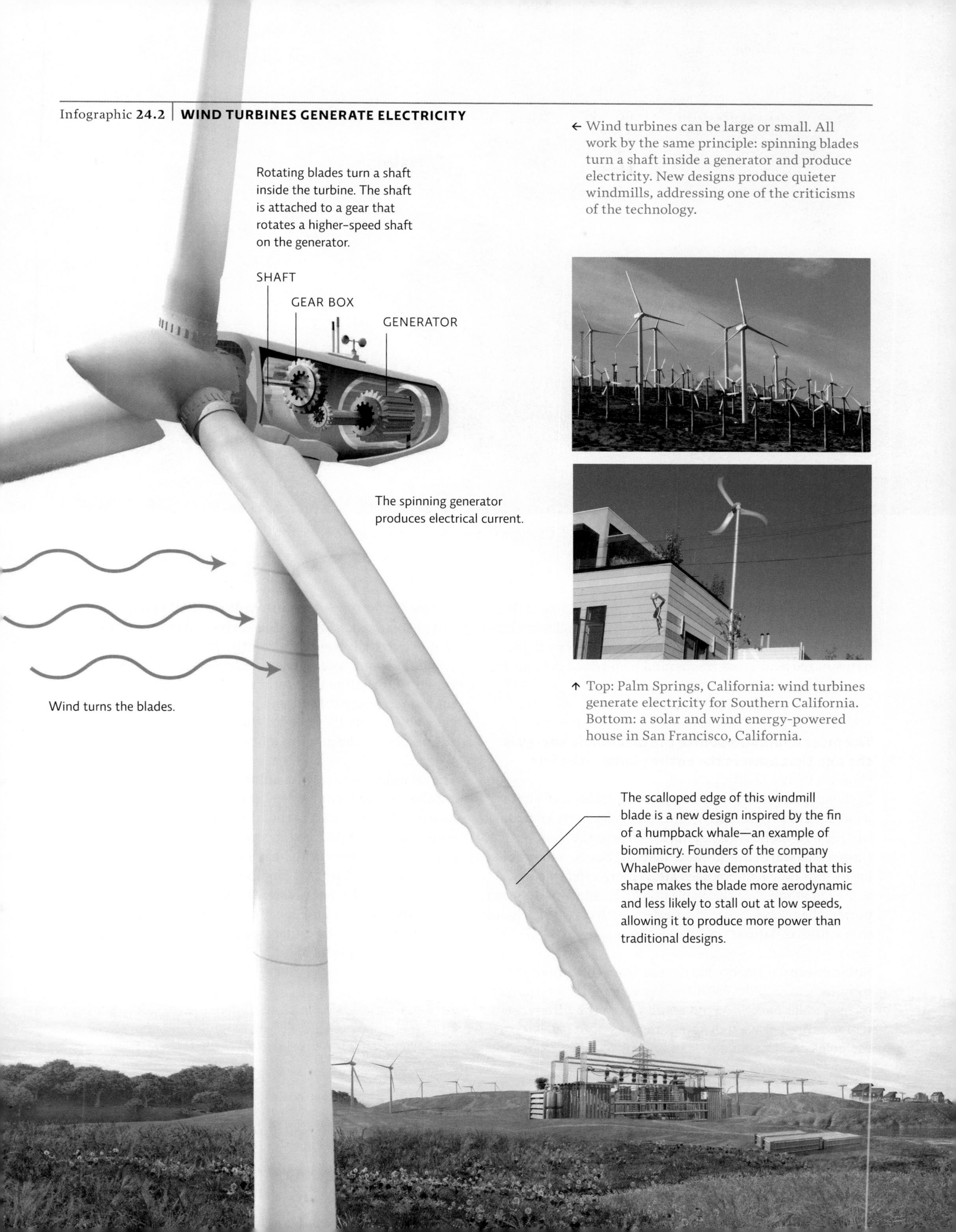
Rotating blades turn a shaft inside the turbine. The shaft is attached to a gear that rotates a higher-speed shaft on the generator.
SHAFT
GEAR BOX
GENERATOR
The spinning generator produces electrical current.
Wind turns the blades.
The scalloped edge of this windmill blade is a new design inspired by the fin of a humpback whale—an example of biomimicry. Founders of the company WhalePower have demonstrated that this shape makes the blade more aerodynamic and less likely to stall out at low speeds, allowing it to produce more power than traditional designs.

← Wind turbines can be large or small. All work by the same principle: spinning blades turn a shaft inside a generator and produce electricity. New designs produce quieter windmills, addressing one of the criticisms of the technology.

↑ Top: Palm Springs, California: wind turbines generate electricity for Southern California. Bottom: a solar and wind energy-powered house in San Francisco, California.

↑ A solar heating plant near Marup, Samsø Island. Changes in Samsø energy use mean that instead of importing electricity, the island exports it.

fuel-powered electricity generation causes 5.2 per GWh. Still, to decrease the risk from windmills, engineers have tried to avoid placing them in known migratory flight paths or close to areas frequented by birds of prey such as eagles.

The most abundant source of sustainable energy is the one that powers the entire planet—the Sun.

Each year, Earth receives a staggering amount of energy from the Sun, more than 4 million exajoules (1 exajoule is 10^{18}, or a million trillion, joules). We use the Sun for many things—to warm our homes, heat our pools, and provide light—but new technologies allow more effective use, and even storage, of the Sun's power. In addition to wind power, the islanders on Samsø decided to tap directly into this amazing natural resource.

Solar energy is energy harnessed from the Sun in the form of heat or light. Wind power is actually an indirect form of solar energy. Wind results from the difference in temperature between different regions of Earth, such as the poles and the equator, causing air to move from cooler regions to warmer regions.

Solar energy can be used in two ways, through active or passive technologies. Around the countryside in Samsø, homes are dotted with **photovoltaic (PV) cells**, also called solar cells. PV cells are **active solar technologies** that convert solar energy directly into electricity. If just 4% of the world's deserts were covered in PV cells, it would supply all of the world's electricity needs. When the Samsø renewable energy project first began, locals were able to buy PV cells for their homes at a low price subsidized by the government.

The islanders also rely on **solar thermal systems**, another active technology that captures solar energy for heating. At the north of the island, rows and rows of solar collectors in a field face the sky, absorbing the Sun's rays and using that energy to heat a massive tank of water to 71°C. The hot water is then piped into 178 homes in the area for use in heating systems. Another 200 homes on the island, ones farther away from district heating plants, have individual solar collectors to capture solar energy for heat. [INFOGRAPHIC 24.3]

solar energy Energy harnessed from the Sun in the form of heat or light.
photovoltaic (PV) cells Also called solar cells, PV cells convert solar energy directly into electricity.
active solar technologies The use of mechanical equipment to capture, convert, and sometimes concentrate solar energy into a more usable form.
solar thermal system An active technology that captures solar energy for heating.

↑ At least 450 residents of Samsø own shares in the onshore turbines, and a roughly equal number own shares in those offshore. Residents are proud of their accomplishments in creating the first island in the world operating on 100% renewable energy.

Infographic **24.3** | **SAMSØ: THE ENERGY POSITIVE ISLAND**

↓ Samsø is pursuing several methods to help it produce enough renewable energy to meet all its needs.

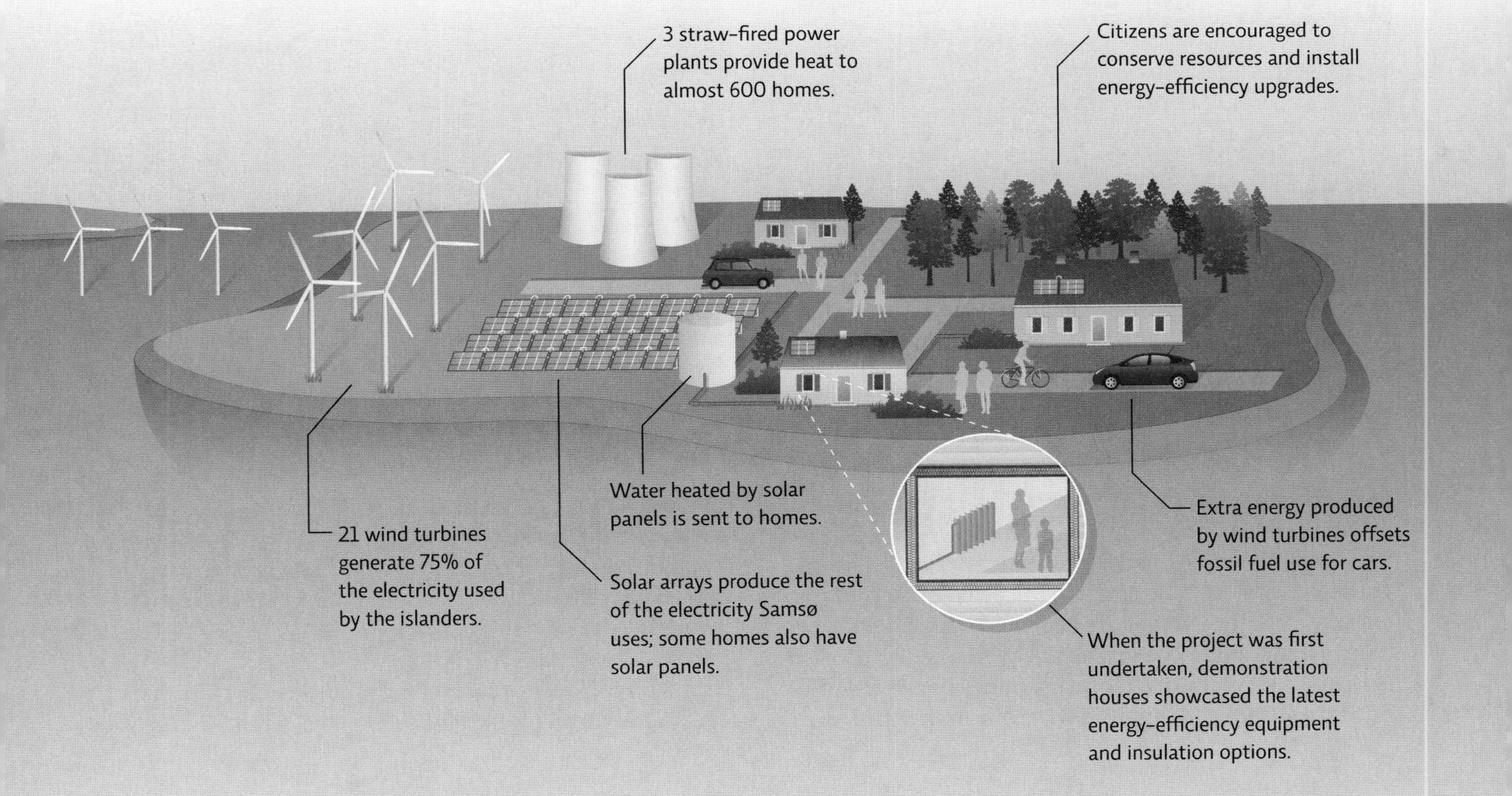

3 straw-fired power plants provide heat to almost 600 homes.
Citizens are encouraged to conserve resources and install energy-efficiency upgrades.
Water heated by solar panels is sent to homes.
21 wind turbines generate 75% of the electricity used by the islanders.
Solar arrays produce the rest of the electricity Samsø uses; some homes also have solar panels.
Extra energy produced by wind turbines offsets fossil fuel use for cars.
When the project was first undertaken, demonstration houses showcased the latest energy-efficiency equipment and insulation options.

Infographic **24.4** | **SOLAR ENERGY TECHNOLOGIES TAKE MANY FORMS**

↓ There are many ways to capture and use solar energy. Passive solar homes are constructed in a way that maximizes solar heating potential.

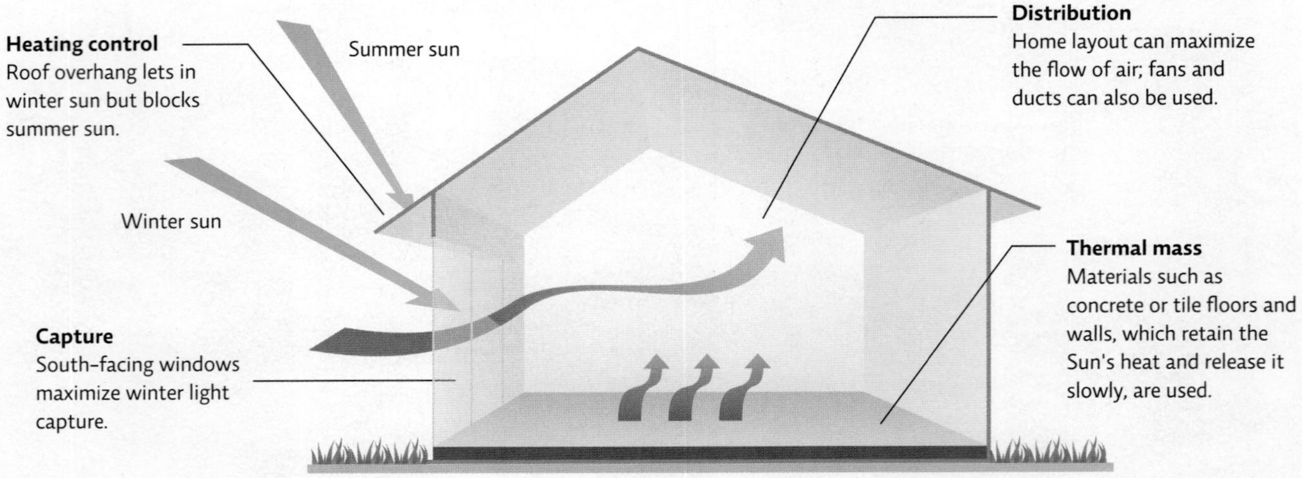
Heating control
Roof overhang lets in winter sun but blocks summer sun.
Summer sun
Winter sun
Distribution
Home layout can maximize the flow of air; fans and ducts can also be used.
Thermal mass
Materials such as concrete or tile floors and walls, which retain the Sun's heat and release it slowly, are used.
Capture
South-facing windows maximize winter light capture.

↓ Active solar technologies capture and convert solar energy for another use such as heating water or producing electricity.

↑ Solar hot-water heaters are the solar technology with the quickest payback for the average homeowner. Sunlight heats a fluid in pipes which then heats water in a tank to be used in the home.

↑ Photovoltaic panels capture sunlight and convert it to electricity that can be used by the homeowner or fed back to the grid and sold to the local utility.

Less expensive alternatives to PV cells or solar thermal systems are **passive solar technologies**. A greenhouse is a simple example of such a system: it captures heat without any electronic or mechanical assistance. Many energy-conscious homes are designed with passive energy in mind, incorporating strategically oriented windows to maximize sunlight in a room and dark-coloured walls or floors to absorb that light and heat the home.

Because solar energy is conceptually simple, safe, and clean, with no noise or moving parts, it is the most popular member of the renewable energy club; today, thousands of buildings around the world are powered by PV cells. [INFOGRAPHIC 24.4]

But like wind energy, solar energy is plagued by intermittency and start-up costs. PV cells are becoming cheaper every day, but it can still take a home or business owner many years to recoup the cost of installation. Intermittency is a bigger problem. Sunlight is only available for half of each day—even less in places like Anchorage, Alaska, which see only a few hours of pale sunlight in the winter. Because of this, it is unlikely that any community will rely solely on solar power for its energy needs.

This is a hallmark of a sustainable energy future—because sustainable energy sources have their own strengths and weaknesses, no single source will likely meet the needs of any particular community, certainly not those of the entire world. But together, each can provide its own contribution. For example, solar is productive in daylight hours, whereas wind tends to blow harder at night; thus these two renewable sources complement each other in most places.

But in some regions, it makes more sense to tap other sources of renewable energy.

Energy that causes volcanoes to erupt and warms hot springs can also heat our homes.

Unlike Samsø, some communities are fortunate enough to be located near sources of **geothermal energy**. A tremendous amount of heat is produced deep in Earth from radioactive decay of isotopes; temperature increases with depth and the Earth's core may be 5000°C. It is the same heat that bubbles hot springs and causes geysers to erupt from the ground at Yellowstone National Park in Wyoming. Within the last century, technological advances have enabled us to tap into these vast resources for heat and electricity. "There has been a huge surge in the development of geothermal resources," says Wilfred Elders, co-chief scientist of an ambitious geothermal project known as the Iceland Deep Drilling Project (IDDP). "It is one of the most viable and economically attractive sources of renewable energy."

There are a wide variety of geothermal systems currently in use. **Geothermal heat pumps** (also called ground-source heat pumps) are used in more than half a million homes around the world—not to generate electricity, but to reduce our use of it. If you have ever been in a cave, you have probably been struck by the cool, constant temperatures there—always around 12.5°C, whether it's freezing outside or in the middle of a heat wave. No matter what the temperature above, underground remains a constant 12.5°C. Engineers bury fluid-filled pipes, bringing the fluid to that temperature. They then pump the fluid into homes, essentially providing people with year-round 12.5°C temperatures. In the winter, the home only needs to be warmed up from 12.5°C, rather than the outside colder temperatures, and in the summer, this system provides natural cooling. Such pumps are fairly expensive to install, yet have lower monthly energy bills than conventional heating and cooling systems. Areas with more extreme climates save enough money in monthly bills to offset the cost of installation—a metric known as **payback time**—in as little as 5 years. [INFOGRAPHIC 24.5]

> Because sustainable energy sources have their own strengths and weaknesses, no single source will likely meet the needs of any particular community.

On the other end of the spectrum are **geothermal power plants**. Generators above geothermal wells use steam released from hydrothermal reservoirs (hot springs) to spin turbines, producing electricity. Hydrothermal reservoirs are areas in Earth's crust that hold heat energy in the form of water, either as steam or liquid. Geysers in Northern California power the world's largest geothermal electrical system, generating more than 1000 MW of energy—enough for a city as large as San Francisco. Geothermal power plants are reliable and efficient, but their potential is entirely dependent on location. Because drilling is expensive, only sites with enough heat to generate significant electricity are considered for development. These tend to be hot zones, locations like Iceland and the western United States, where the tectonic plates in Earth's crust pull away from or rub against each other, allowing the hot magma to flow upward through cracks in the rock to reach areas closer to the surface.

This type of renewable energy is becoming more popular: between 2005 and 2010, use of geothermal power worldwide increased by 20%, according to the International Geothermal Association, and is expected to grow even further through 2015.

Harnessing such a powerful heat source can be difficult, however. In early 2007, an earthquake of magnitude 3.4

---

passive solar technologies The capture of solar energy (heat or light) without any electronic or mechanical assistance.
geothermal energy The heat stored underground, contained in either rocks or fluids.
geothermal heat pump A system that actively moves heat from the underground into a house to warm it or removes heat from a house to cool it.
payback time The amount of time it would take to save enough money in operation costs to pay for the equipment.
geothermal power plants Power plants that use the heat of hydrothermal reservoirs to produce steam and turn turbines to generate electricity.

on the Richter scale shook the town of Basel, Switzerland. The quake was attributed to a geothermal mining project in the area, and the project was halted. The Iceland drilling project fell into serious problems when the drill repeatedly got stuck 2 kilometres deep into a volcanic crater, leading to months of jammed drill bits and broken pipes. The ultimate goal of the IDDP: drill into a high-temperature geothermal system (the Krafla volcano, in this case), penetrate twice as deep as conventional geothermal wells (which are usually less than 3000 metres deep), and reach *supercritical fluids*, substances with temperatures above 370°C that are confined by such intense pressure they exist in a limbo state between liquid and gas. The resulting superheated steam could generate 10 times the electricity of most geothermal wells.

Power can be generated from moving water—but at potentially significant environmental and social costs.

The island of Samsø is shaped curiously like a violin. It is also surrounded by one of the most significant sources of renewable energy worldwide: water.

Humans have harnessed the power of falling water for thousands of years, ever since early civilizations used watermills to grind grain into flour. Today, energy produced from moving water—known as **hydropower**—supplies more electricity to the population than any other single renewable resource. Approximately 60% of electricity in Canada and 20% around the world is generated

Infographic **24.5** | **GEOTHERMAL ENERGY CAN BE HARNESSED IN A VARIETY OF WAYS**

↓ The high temperatures found underground in some regions can be tapped to generate electricity. Geothermal heat can also be piped directly to communities to provide heat or hot water. Geothemal energy can also be tapped using ground-source heat pumps, reducing the cost to heat and cool individual buildings.

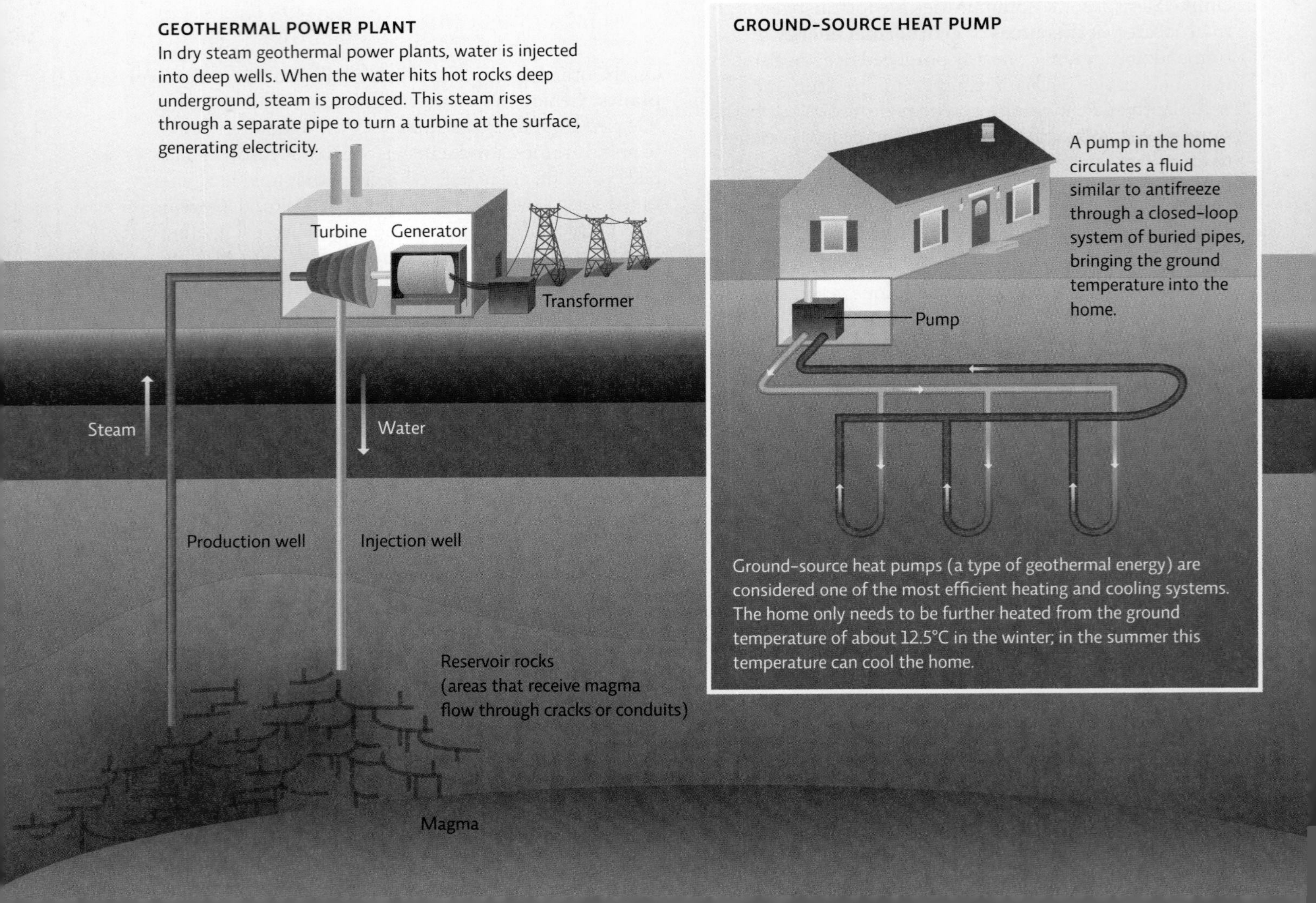
GEOTHERMAL POWER PLANT
In dry steam geothermal power plants, water is injected into deep wells. When the water hits hot rocks deep underground, steam is produced. This steam rises through a separate pipe to turn a turbine at the surface, generating electricity.
Turbine
Generator
Transformer
Steam
Water
Production well
Injection well
Reservoir rocks (areas that receive magma flow through cracks or conduits)
Magma
GROUND-SOURCE HEAT PUMP
A pump in the home circulates a fluid similar to antifreeze through a closed-loop system of buried pipes, bringing the ground temperature into the home.
Pump
Ground-source heat pumps (a type of geothermal energy) are considered one of the most efficient heating and cooling systems. The home only needs to be further heated from the ground temperature of about 12.5°C in the winter; in the summer this temperature can cool the home.

↑ The geothermal heat produced from the Reykjanes Power Plant in Iceland is mostly used to heat freshwater, which heats 89% of the houses in Iceland.

↑ Installation of a geothermal heating system in a residence. Closed-loop tubing can be installed horizontally in shallow trenches or vertically in much deeper wells. Here, a horizontal loop is being installed.

from hydropower. Large-scale hydroelectric power plants at giant dams are the source of most of that power.

Like wind energy, hydropower is an indirect form of solar energy. (See Chapter 15 for more information about the water cycle.) Hydropower is abundant, clean, and does not typically produce greenhouse gases.

There are a variety of ways to harness the energy of moving water—from ocean waves and the tides, to capturing energy released from variations in ocean temperatures—but the most common way to generate hydropower is with dams. The Grand Coulee Dam on the Columbia River in Washington State is one of the largest concrete structures on the planet and the third-largest dam. The more than 9 million cubic metres of concrete that make up the dam could form a 1.2-metre-wide sidewalk wrapped twice around the equator. The dam is the largest electrical power producer in the United States, with a total generating capacity of 6809 MW. The reservoir created by the dam, an artificial lake that pools behind the structure (called an impoundment), stretches some 240 kilometres back to the Washington/British Columbia border.

The buildup of water in the reservoir creates an enormous amount of pressure; water diverted from the top of the dam flows through long pipes, called penstocks, to turbines below. As the water rushes past the curved blades of the turbines, they spin and generate electricity. Three power plants at the Grand Coulee contain 24 generators, which produce enough electricity to power two cities the size of Vancouver.

But hydropower from large dams is far from an ideal resource. It wasn't a real option on Samsø, for instance, because the island lacks high mountain peaks whose runoff would feed large rivers. Importantly, hydroelectric systems are also responsible for the loss of major habitats and the displacement of tens of millions of people around the globe. It is the most debated and contentious of any renewable resource.

For nearby Aboriginal communities, the Grand Coulee was a humanmade disaster that sacrificed not only land, but a staple of their economy and culture—salmon. Salmon are migratory fish: they hatch in the freshwater tributaries of the Columbia, migrate downstream into the Pacific Ocean to spend most of their lives in salt water, then return to the freshwater environment of their origin to spawn. But the Grand Coulee Dam blocked the way; salmon and trout runs upstream of the dam were completely wiped out.

British Columbia, Manitoba, Newfoundland, Labrador, Quebec, and the Yukon produce over 90% of their electricity from moving water. In British Columbia's mountainous terrain, there are dams on several major waterways and many of the same problems arise there. These dams create water impoundments that flood upstream valleys and reduce flow downstream. Flooded areas are no longer useful as hunting grounds for local First Nations communities, and fish are unable to migrate upriver to spawn, reducing fish populations, especially for trout and salmon. In the flatter terrain of northern

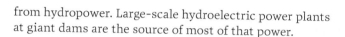

hydropower The energy produced from moving water.

DANGER
HIGH WATER FLOWS
OCCUR WITHOUT WARNING

↑ A run-of-the-river hydroelectricity station. In such stations, the river still flows freely and is not impounded to produce a lake.

Québec, tall dams are uncommon, but there are several diversion and dam mega-projects, such as the massive James Bay project, which generates over 10 000 MW annually. These northern projects have come at a cost to local Aboriginal communities who have lost traditional hunting grounds to extensive flooding. Flooding of organic soils leads to an increase in mercury bioavailability to the food chain. In many reservoirs, mercury levels in fish have risen, making their use as an alternative source of food and income a risky choice. Greenhouse gas emissions also increase temporarily in new reservoirs, although the long-term impacts of forest or peatland flooding on carbon cycling are not yet fully understood.

Not all hydropower systems have the same impact as a giant dam. A *run-of-the-river hydroelectric* system, often a low stone or concrete wall, doesn't block the water; it merely directs some of the flowing water past a turbine and thus generates electricity from the natural flow and elevation drop of a river, which is less disruptive to the river ecosystem. But this system is only suitable for rivers with dependable flow rates year-round. Such systems are useful in remote areas where electricity can be costly, and they are good candidates to supplement solar power systems.

The true cost of various energy technologies can be difficult to estimate.

One recurring theme with renewable energy technologies is that they are more expensive than fossil fuel methods. Even though sunlight, wind, and water are free, constructing and installing the solar cells, wind turbines, and hydroelectric systems needed to harness their energy are not.

But perhaps we should not be asking why renewables are so expensive, but instead asking why fossil fuels are so cheap. We need to consider *all* the costs—environmental, social, and economic—of these technologies to fairly compare them. Meaning, even if the costs to construct and install a technology are high, those costs are offset by the lower environmental costs of the sustainable technology. This makes comparisons between the energy technologies difficult. Various studies have tried, and so far agree on several points: fossil fuels, led by coal, have the highest external (environmental) costs. Purely from an economic standpoint, solar energy is the most expensive form of energy. But when external costs are added to production costs to estimate a true cost per kilowatt-hour, wind leads the way as the least expensive method.

It was this logic that helped convince the people of Samsø to personally invest in the ambitious project to convert to 100% sustainable energy. Such a project would not be possible everywhere, says Hermansen. Indeed, he wasn't even sure at first it would succeed, because it needed the buy-in of every one of the island's 4000-plus residents—literally. This is the only way the project would work, he and others believed—if it was owned by the people. For years, Hermansen and others had to go practically door to door, trying to convince everyone to become part owners in the island's sustainable energy infrastructure. "If we didn't own these wind turbines and other sustainable energy projects on Samsø, a big power company would," says Hermansen. "Eventually, people will get their money back."

At first, the islanders were skeptical. "People were a little scared in the beginning. They were afraid the changes would disturb the landscape and affect tourism," says Garbers. But attitudes slowly changed, thanks largely to Hermansen's efforts.

Hermansen was born on Samsø and grew up there on a family farm. He was an environmental studies teacher with an enthusiasm for all things renewable when he heard about the contest, and it wasn't long before he became the project's first employee. Through extensive community meetings and seminars, he rallied the islanders to the cause. "We are lucky to have him," says Garbers. "If it had been some guy from the mainland promoting the

---

conservation Efforts that reduce waste and increase efficient use of resources.

↑ The Grand Coulee Dam, located on the Columbia River in central Washington, is the largest concrete structure in the United States. In addition to producing up to 6.5 million kilowatts of power, the dam irrigates more than 200 000 hectares of farmland and provides wildlife and recreation areas. However, its construction eradicated salmon runs and inundated Aboriginal villages and ancient burial sites.

project, maybe the islanders wouldn't have believed him. But Søren was local." This constellation of forces—a small, close-knit community open to the project, community members who personally invested in the venture, led by a local individual with a passion for sustainability—helped Samsø achieve its goal.

By now, most of Samsø's residents have invested, starting at around $600 for 1 share in a wind turbine. The logic, says Hermansen, is that all residents share the profits—in a good wind year, the island sells its electricity, and people who purchased 10 shares receive 10 shares' worth of profits. He estimates that each Samsø resident has invested an average of $20,000 over the last 10 years in revolutionizing electricity on the island.

Meeting our energy needs with renewable sources becomes more likely when we pair them with energy conservation measures.

Even though the people of Samsø saw the importance of investing in sustainable energy, they realized that one of the best ways to help achieve energy independence would be to simply use less of it. As energy advisors like to say, *the greenest kilowatt is the one you never use*. Luckily, there are lots of ways to reduce electricity use right now. [INFOGRAPHIC 24.6]

For instance, residents were encouraged to focus on energy efficiency, relying on technology that uses less energy for a given output. In Samsø's cold climate, one of the first conservation efforts revolved around heating homes, mainly by transitioning away from electrical heaters. Energy home audits pinpointed other steps individuals could take, such as using straw, wood chips, and other sources of biomass for heat. Gas stoves were replaced by more efficient electric stoves; people began heating water using electricity generated by the wind turbines or solar panels. Residents were encouraged to apply for grants toward upgrades that made their homes more energy efficient. Similar programs exist in Canada. (See Chapter 26 for more ways consumers can cut energy use.)

Another basic approach to using less energy relies on **conservation**—making choices that result in less energy

Infographic 24.6 | **SAVING ENERGY**

↓ Energy conservation is about wise use: using less and using energy more efficiently. Many of the steps below can be taken immediately, with no investment; others require money and time to implement. Whatever you can do will not only reduce your energy use, it will reduce your energy bill and the pollution generated from producing that energy.

No-cost ways to save energy:

- Turn off your computer and monitor when not in use.
- Plug home electronics, such as TVs and DVD players, into power strips; turn the power strips off when the equipment is not in use.
- Lower the thermostat on your hot-water heater to 50°C; turn it off if you will be away for several days.
- In the winter, open curtains on south-facing windows to let in heat and light during the day; close them at night to reduce heat loss.
- Be sure to close the fireplace damper when not in use to prevent loss of heat.
- Take short showers instead of baths to reduce hot water use.
- Wash only full loads of dishes and clothes; wash clothes with cold water.

Steps that cost money but have quick payback times:

- The first step is to have an energy assessment performed to identify where your home is energy inefficient (check online for home audit programs and incentives in your province).
- Install attic and wall insulation to meet recommendations for your area.
- Weatherstrip and caulk doors and window frames; insulate electrical outlets on outside walls with inexpensive foam cutouts. Check heating ducts for leaks and seal if needed.
- Insulate hot-water pipes and the hot-water heater, especially if the hot-water heater is located in an unheated area of your home.
- Install a programmable thermostat—you can program it to automatically adjust the temperature when you are not at home or are sleeping.
- Look for the Energy Star and Energuide labels on home appliances and products, and for R-2000 builiding standards in new homes.
- Upgrade to a front-loading washing machine that uses less energy (and less water—which also saves energy).

More expensive options with payback times of greater than 5 years:

- Replace older windows with double-pane windows suitable for your area—a variety of low emissivity (low-e) window products restrict loss of heating or cooling while allowing in plenty of light.
- Replace the heating, ventilation, and air cooling (HVAC) system with an energy-efficient model; consider a ground-source heat pump.
- Install an on-demand (tankless) hot water heater; this saves energy because water is not constantly being heated, only to cool back down, requiring that it be heated again.

consumption. Much of conservation involves changing people's behavior; since lighting typically accounts for about 20% of the average home's electric bill, simply turning off the lights when you leave a room can make a big difference. Other options include studying in a room with ample natural light instead of a dark corner that requires artificial light, and lighting your home with compact fluorescent lights (CFLs) or LEDs (light-emitting diodes), which consume less electricity. [INFOGRAPHIC 24.7]

In just a few years, the island of Samsø has transformed itself, capturing the attention of countries around the world, impressed by how a small farming community became the most energy efficient place on Earth. By 2005, the island was producing more energy from renewable resources than it was using. This transition has the added benefit of cleaner air—Samsø has reduced its nitrogen and sulphur air pollution by 71% and 41%, respectively, and its CO_2 emissions have fallen to below zero, thanks to the fact that the windmills offset more CO_2 than their vehicles emit.

Much of that success stems from community members working together to solve their own local problems, says Hermansen. As he told *Time* magazine in 2008, "People say: 'Think globally and act locally.' But I say you have to think locally and act locally, and the rest will take care of itself."◉

Select references in this chapter:
Holm, A. *et al.* 2010. *Geothermal Energy International Market Update.* Geothermal Energy Association.
Sovacool, B. K. 2009. *Energy Policy*, 37, Issue 6: 2241–2248.

Infographic **24.7** | **ENERGY EFFICIENCY**

Improved technologies can reduce our use of energy and other resources by wasting less. For example, we generally use light bulbs because we want light. But about 95% of the electricity that powers an incandescent bulb produces heat – only 5% goes to light. CFL and LED lights produce much more light than heat and thus need less electricity to produce the same amount of light. Products that bear the Energy Star® label are independently certified to meet or exceed high energy efficiency standards.

ENERGY STAR LIGHTING
Compact fluorescent lightbulbs (CFLs) use 75% less energy and last 10 times longer than traditional incandescent bulbs. These bulbs contain a very small amount of mercury (so are considered hazardous waste) but if used widely would reduce the amount of mercury released to the environment by reducing the amount of coal burned to generate electricity (the major source of mercury contamination).

Light-emitting diodes (LED bulbs) use 80–90% less energy and last 25 times longer than traditional incandescent bulbs and contain no mercury.

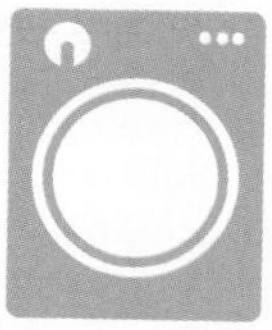

ENERGY STAR CLOTHES WASHER
Uses about half the energy of and 35–50% less water than a traditional washer

ENERGY STAR DISHWASHER
At least 25% more energy efficient than the minimum Canadian standard; sensors determine correct water temperature and cycle duration for complete cleaning.

ENERGY STAR REFRIGERATOR
Better insulation and high-efficiency compressors make new Energy Star refrigerators much more efficient than traditional models, using 25–30% less energy; in 10 years or less, the Energy Star refrigerator will pay for itself.

BRING IT HOME

PERSONAL CHOICES THAT HELP

A key component to developing a sustainable society is the use of renewable energies. Renewable energy decreases harmful impacts of mining waste associated with nonrenewable energy, plus reduces the amount of air and water pollution produced when the fuel is processed and burned. The efficiency and availability of renewable energy is rapidly increasing because of the development of new trends and technologies.

Individual Steps

→ Contact your energy provider to see if you can purchase a percentage of your energy from a renewable source.

→ Regardless of your home's energy source, make sure you are using energy efficiently. Review Infographic 24.6 and visit oee.nrcan.gc.ca/home for more ideas.

→ If you have an iPhone, download the PVme app to see how many solar panels you would need to meet your household energy needs.

→ A major barrier to solar energy for many people is the cost. For information on tax incentives, rebates, and other programs that make using renewable energy easier, check out oee.nrcan.gc.ca/corporate/statistics/neud/dpa/policy_e/programs.cfm or look for resources in your province.

Group Action

→ Invest in providing solar energy to lower-income communities; such initiatives are available in several provinces.

Policy Change

→ Contact your MPs and MPPs and ask them to support or sponsor legislation that provides financial incentives for the purchase of renewable technologies.

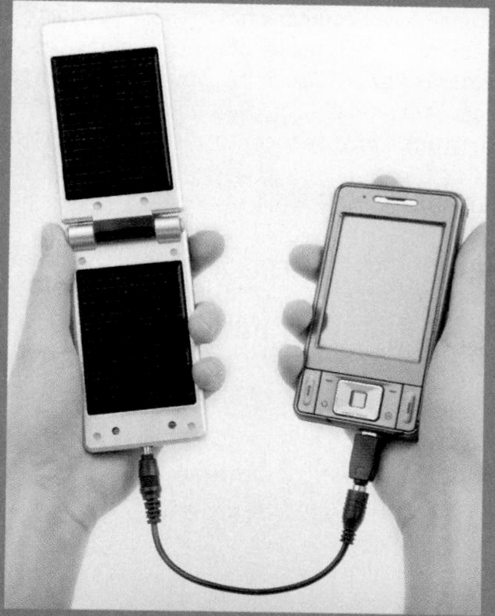

UNDERSTANDING THE ISSUE

CHECK YOUR UNDERSTANDING

1. Which of the following is a source of renewable energy?
a. Natural gas
b. Wind
c. Oil
d. Coal

2. As a member of the city council in a small town in New Brunswick, you must vote on a new source of energy for your growing community. What should be the first step in determining where the energy will come from?
a. Rely on historical practices and build a new hydroelectric power plant.
b. Pass a new law that requires all buildings to have solar panels installed.
c. Conduct an assessment of available energy sources, renewable and nonrenewable.
d. Provide citizens a tax subsidy for improving the energy efficiency of their homes.

3. Which of the following renewable energy sources contributes the largest proportion of energy worldwide?
a. Wind
b. Water
c. Solar
d. Biomass

4. Which of the following is considered one of the most efficient cooling and heating systems?
a. Ground-source heat pumps
b. Passive solar
c. Solar roofing tiles
d. Electric baseboard heat

5. What type of energy might you recommend to someone who lives in a remote area with lots of small streams, a rainy season, and a dry season?
a. Geothermal heat pump
b. Large-scale hydroelectric
c. Solar and run-of-the-river hydropower
d. Traditional power plant

6. In some parts of the world, women (and children) walk hours in each direction to collect wood as fuel for heating homes and preparing meals. The long walk is necessary because there is no closer source of wood; it has all been harvested. Which of the following technologies do you think would help ease the burden on women AND benefit the environment?
a Passive solar heating
b. Solar roofing tiles
c. Solar cookers
d. Solar hot-water heaters

WORK WITH IDEAS

1. Describe the characteristics of the five sustainable energy sources presented here. Why are all sources not practical for every location?

2. Compare and contrast the costs and benefits of energy generated from wind, water, sun, geothermal heat, and biological matter.

3. Explain the process of harvesting geothermal energy from supercritical fluids as if you are talking to someone unfamiliar with the method. Anticipate in your description questions the person might have, such as "Why can't we do that here?"

4. Describe the various types of hydropower. In your description, include information about scale, cost, availability, and environmental impact.

ANALYZING THE SCIENCE

A regional power company has proposed building a wind farm in your community. There is plenty of wind, so many people in the community think this is a good idea. Some, however, have cited the environmental impacts of windmills, especially bird and bat mortality. Examine the following graphs and use them to answer the questions that follow.

INTERPRETATION

1. In one or two sentences, summarize the information provided in each graph or table.

2. For each graph or table, provide a title and a legend. What is the take-home message for each graph or table?

ADVANCE YOUR THINKING

3. Based on the data and where you live, what do you think the impact of a wind farm in your area would be on bird and bat mortality? If a wind farm were installed, what recommendations could you make to reduce bat mortality? Make sure to discuss your answer in relation to each of the graphs/tables.

Graph 1

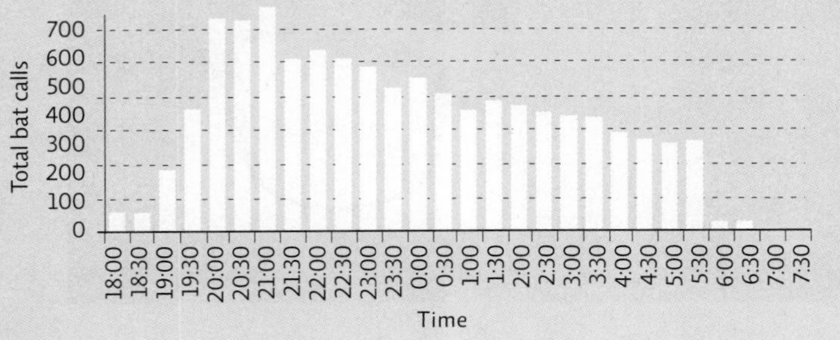
Total bat calls
700
600
500
400
300
200
100
0
18:00
18:30
19:00
19:30
20:00
20:30
21:00
21:30
22:00
22:30
23:00
23:30
0:00
0:30
1:00
1:30
2:00
2:30
3:00
3:30
4:00
4:30
5:00
5:30
6:00
6:30
7:00
7:30
Time

Graph 2

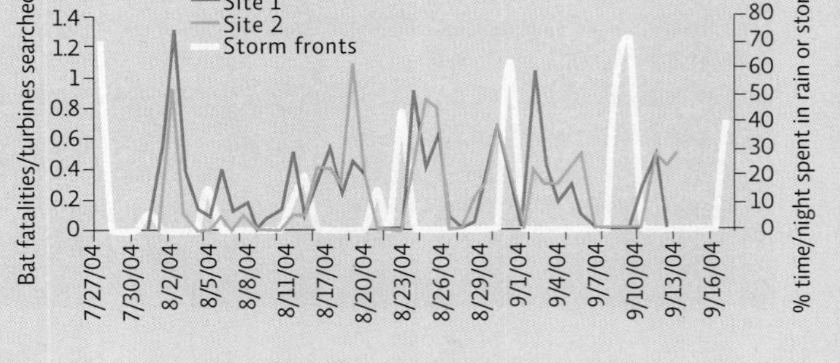
Site 1
Site 2
Storm fronts
Bat fatalities/turbines searched
1.4
1.2
1
0.8
0.6
0.4
0.2
0
7/27/04
7/30/04
8/2/04
8/5/04
8/8/04
8/11/04
8/17/04
8/20/04
8/23/04
8/26/04
8/29/04
9/1/04
9/4/04
9/7/04
9/10/04
9/13/04
9/16/04
% time/night spent in rain or storm
80
70
60
50
40
30
20
10
0

Graph 3

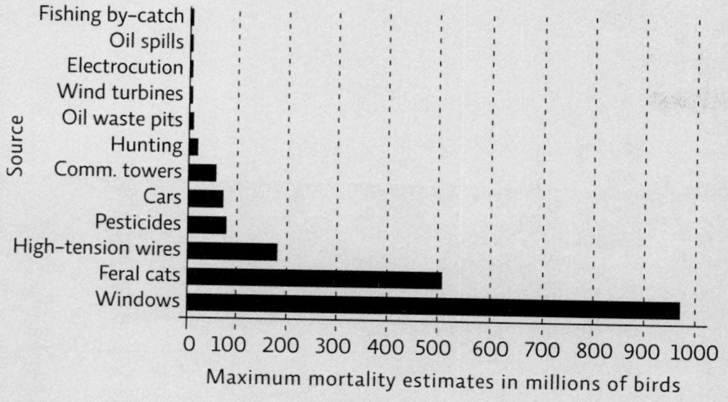
Fishing by-catch
Oil spills
Electrocution
Wind turbines
Oil waste pits
Hunting
Comm. towers
Cars
Pesticides
High-tension wires
Feral cats
Windows
Source
0 100 200 300 400 500 600 700 800 900 1000
Maximum mortality estimates in millions of birds

Table 1

Date	Buffalo Ridge, MN	Vansycle, OR	Buffalo Mtn., TN	Stateline OR/WA	Foote Creek Rm, WY	Total bat carcasses	%
May 1–15	0	0	0	–	0	0	0.0
May 16–31	1	0	0	–	1	2	0.4
June 1–15	0	0	0	–	1	1	0.2
June 16–30	3	0	0	–	2	5	0.9
July 1–15	9	0	9	0	2	13	2.8
July 16–31	88	0	0	0	26	110	22.2
Aug 1–15	127	0	10	0	19	151	28.2
Aug 16–31	75	4	0	11	33	128	23.9
Sep 1–15	52	4	8	0	21	81	13.1
Sep 16–30	4	2	–	10	0	20	3.7
Oct 1–15	1	0	0	8	2	11	2.1
Oct 16–31	2	0	0	0	0	2	0.4
Nov 1–15	0	0	0	1	0	1	0.2

EVALUATING NEW INFORMATION

Renewable energy resources have become a priority around the world as we grapple with the cost and limited supply of fossil fuels, not to mention the effect on the environment resulting from their use. How do we know which energy source is appropriate to invest in, with both our time and money? Do we have to give up our current standard of living if we switch to an alternative energy source?

Go to Natural Resources Canada's energy website (www.nrcan.gc.ca/energy/home). Review this main page and follow links on energy sources and renewable energy. Take note of the tables that indicate what energy sources are used where in Canada.

Evaluate the website and work with the information to answer the following questions:

1. Is this a reliable information source? Does it have a clear and transparent agenda?
   a. Who runs this website? Do this organization's credentials make it reliable/unreliable? Explain.
   b. Who are the authors? What are their credentials? Do they have the scientific background and expertise to lend credibility to the website?
2. In your own words, briefly summarize what you believe are the most important pros and cons of alternative energy resources based on your readings. Do any of these also apply to traditional fossil fuels? Would you be willing to invest your money in research and development of alternative energy sources? Why or why not?
3. What fuel provides most of the energy in your region? Were you aware of this, or are you surprised?
4. How does your community differ from the nation in terms of both fuel sources and emissions?
5. Based on what you know about the weather and geography in your area, what types of alternative renewable fuel sources would you recommend? If you need help with local weather statistics, you can do an Internet search for your region.

MAKING CONNECTIONS

A NEW ENERGY MODEL FOR CENTRAL AMERICA?

Background: The Central American Arc is a chain of volcanoes extending throughout Central America. It is caused by a subduction zone, and massive eruptions have occurred in the past. The countries that the volcanoes are in are all developing nations that are trying to increase their standard of living, which includes providing electricity to all citizens. However, these countries lack fossil fuel resources, so they must import them. In addition, they lack the infrastructure to provide electricity in remote areas.

Case: You are a member of a philanthropic organization that provides expert advice and start-up funds to localities (large and small) in Central America to improve energy facilities and infrastructure. The government of Nicaragua has approached your organization for advice about which types of energy to invest in. Before your organization can make reliable recommendations, you need to gather information about possible sources of energy.

1. Research possible energy fuel sources for Nicaragua. As you conduct your research, answer the following questions:
   a. What are the typical weather patterns?
   b. What is the topography of the country like?
   c. Where do people live?
   d. What types of fuel are available?
   e. What types of fuels can the country afford, now and in the future?
2. Answer these same questions for the country of Iceland.
3. Compare your results for the two countries. Based on your research and comparisons, write a brief report to your organization that details your findings. In your report, make sure to include answers to the following questions:
   a. What recommendations should your organization make to the Nicaraguan government for both short-term and long-term solutions to the problem of energy resources?
   b. Do your recommendations apply to the country as a whole, or do you have region-specific recommendations?

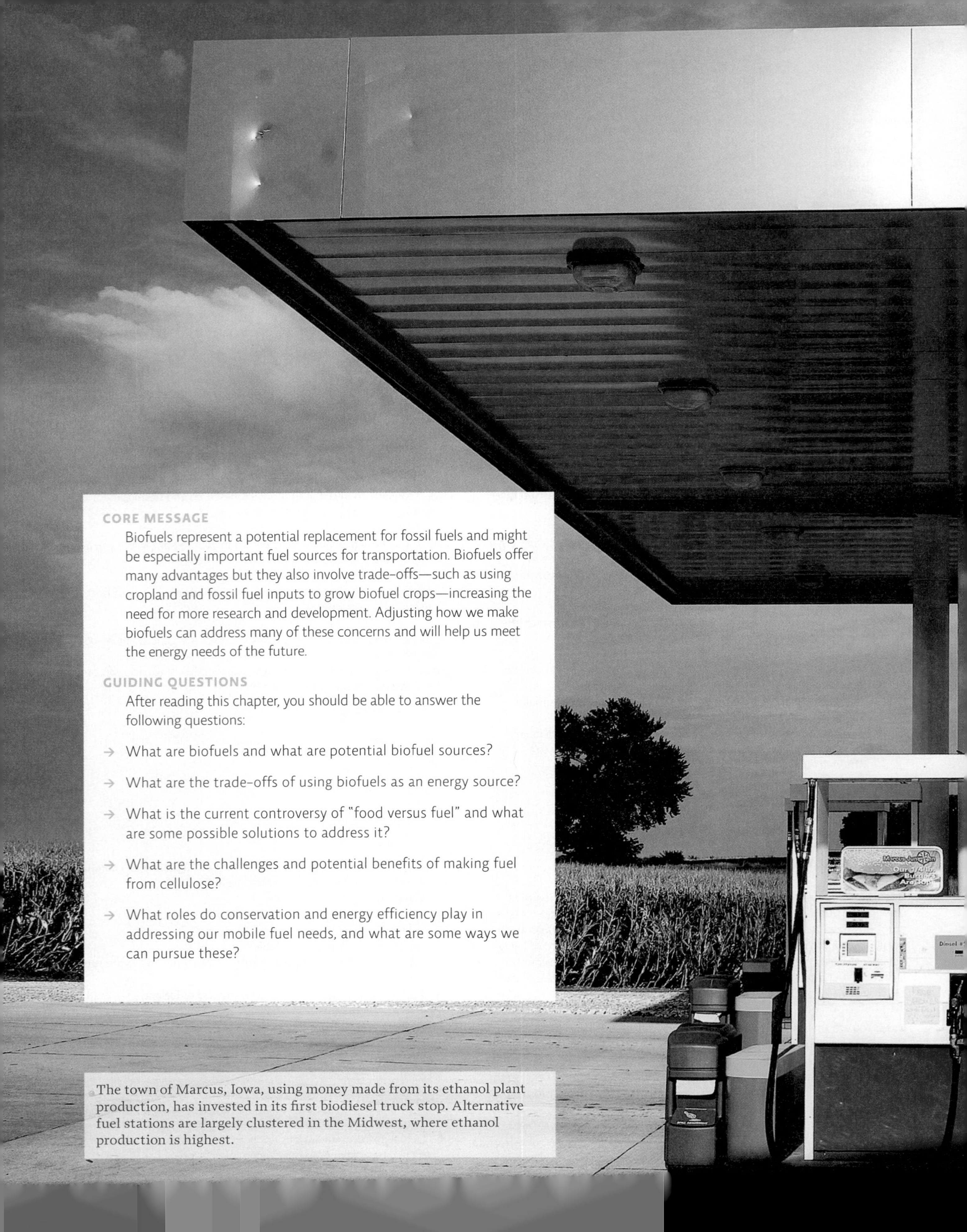

CORE MESSAGE

Biofuels represent a potential replacement for fossil fuels and might be especially important fuel sources for transportation. Biofuels offer many advantages but they also involve trade-offs—such as using cropland and fossil fuel inputs to grow biofuel crops—increasing the need for more research and development. Adjusting how we make biofuels can address many of these concerns and will help us meet the energy needs of the future.

GUIDING QUESTIONS

After reading this chapter, you should be able to answer the following questions:

→ What are biofuels and what are potential biofuel sources?

→ What are the trade-offs of using biofuels as an energy source?

→ What is the current controversy of "food versus fuel" and what are some possible solutions to address it?

→ What are the challenges and potential benefits of making fuel from cellulose?

→ What roles do conservation and energy efficiency play in addressing our mobile fuel needs, and what are some ways we can pursue these?

The town of Marcus, Iowa, using money made from its ethanol plant production, has invested in its first biodiesel truck stop. Alternative fuel stations are largely clustered in the Midwest, where ethanol production is highest.

CHAPTER 25 **BIOFUELS**

GAS FROM GRASS

Will an ordinary prairie grass become the next biofuel?

Bio Diesel
10%

↑ The Cedar Creek Ecosystem Science Reserve in Minnesota has established more than 1100 long-term experimental plots and 2300 permanent observational plots distributed across 22 old fields.

→ University of Minnesota ecologist David Tilman (right) and the Nature Conservancy's Joe Fargione (left) study how clearing land for crops releases CO_2.

The rain stopped, and the plants began to die. In 1987, a massive drought struck Minnesota. At the Cedar Creek Ecosystem Science Reserve, David Tilman helplessly watched the grasslands he had been cultivating for more than 5 years—part of a project to test how different amounts of nitrogen and other resources affect growth—wither away.

◉WHERE IS THE CEDAR CREEK RESERVE?

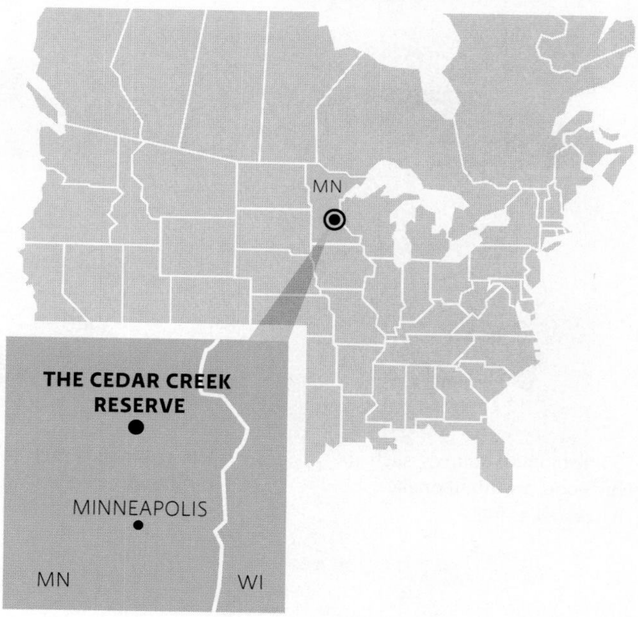
MN
THE CEDAR CREEK RESERVE
MINNEAPOLIS
MN
WI

But while some areas thinned out and became barren, others kept the auburn and green of healthy prairie grasslands. For some reason, certain plots were less affected by the drought. Tilman, a bow tie-sporting ecologist at the University of Minnesota, decided to find out why. Little did he know, his simple plan to study grass in Minnesota would unexpectedly lead him to a new and potentially better type of **biofuel**—a substance that provides the energy needed to power engines—made from a mixture of prairie grasses.

"We weren't thinking about biofuels at all," recalls Tilman. "When you have a new idea, you start imagining what might happen, but you never know what will really happen until you try the experiment."

Biofuels are a potentially important alternative to fossil fuels.

The term "biofuels" makes headlines daily, but the concept is not new—people have been deriving energy from **biomass** for millennia, for instance through burning wood, peat, corn stalks, and other crop residue for heat and cooking.

Currently, the world relies primarily on fossil fuels as the main source of energy—but burning them has had a profoundly negative impact on our environment (see Chapters 19–22 for more). And with increasing demand for energy, continued dependence on these nonrenewable fuels is no longer a viable option. Communities are experimenting with ways to produce electricity without fossil fuels (see Chapter 24) and with sustainable alternatives to power motor vehicles.

Biofuels are derived from material from living or recently living organisms (biomass) or their by-products. Fuels derived from biomass are considered renewable since the raw materials can be naturally replenished relatively quickly. Biofuels also have the advantage of being locally produced and thus they reduce a nation's dependence on other countries for energy.

But biofuels are not a simple solution to our energy problems. For one, their raw materials or **feedstocks**—crops, animal waste, and wood products—require resources to grow. Furthermore, to function as a fuel, these materials must be converted into another form, a process that uses energy.

Today, there is a massive effort to render that process more sustainable. Biofuels currently provide about 4% of the energy used in the United States and Canada (more than any other type of renewable resource other than hydroelectric). According to the U.S. Department of Energy, global biofuel supply and demand is expected to grow more than sixfold, from 45 billion to 315 billion litres over the next two decades.

But, as Tilman was about to learn, replacing fossil fuels—especially petroleum—is no easy task.

---

biofuels Solids, liquids, or gases that produce energy from biological material.
biomass Material from living or recently living organisms or their by-products.
feedstock Biomass sources used to make biofuels.

Infographic 25.1 **BIOFUEL SOURCES**

↓ Biomass (material of biological origin) can be burned directly or converted to other forms of fuel, such as bioethanol or biodiesel. Some crops are specifically grown just for fuel production (fuel crops) but waste material can also be used. As long as the feedstocks (biomass sources used in production) are grown and harvested sustainably, biofuels are a sustainable resource.

DIRECT BIOMASS ENERGY

A variety of materials, such as wood, dried manure, and crop waste, can be burned directly. This is the most common energy source in less-developed regions of the world but it can lead to problems such as deforestation and air pollution.

Some biomass sources, such as firewood, are intentionally harvested as fuel.

Waste biomass such as cornstalks can be collected and burned; manure can be dried and then burned.

Composite "briquettes" are made from compressing flammable material such as paper, grass, forest, and agricultural waste chips along with shredded plastic (which increases flammability) and are a promising alternative to charcoal.

Biofuels can come from unexpected sources.

Back in Minnesota, Tilman set about testing to see why half his grasslands seemed resistant to drought. After trying "a whole variety of things," Tilman and his team came to a simple conclusion: the plants growing in the most diverse areas—those with up to 20 species of grasses and flowers—were the most able to weather periods of drought. Tilman was baffled. Why should plants competing with each other be more successful during a drought?

He and his team of ecologists decided to study the question empirically. They spread out under a bright blue sky at the Cedar Creek Reserve in Minnesota and planted a smorgasbord of seeds among 152 plots, each plot roughly the size of half a tennis court. The plants began to sprout: gold bunches of junegrass, stately spires of dark blue lupine, bright yellow tufts of goldenrod, and tall, spiky western wheatgrass. Some plots had only 1 species, some 2, others up to 16. Each plant was a **perennial**, meaning it would grow back every year, differing from **annual** plants such as corn and soybeans, which must be replanted each year. The team used no fertilizer and only watered plots in the first few weeks after planting.

One of the species was switchgrass, a common North American perennial that grows thick and tall (up to 3.5 metres), with roots penetrating 3 metres into the soil. As Tilman watched his plants grow, he was not the only person thinking about switchgrass. The crop was rapidly becoming a household name for another reason—as a potential source of biofuels.

Often, biofuels are derived from **fuel crops**, those specifically grown to make biofuels. For instance, **bioethanol** can be derived from crops such as corn and sugarcane during a process of fermentation and distillation, similar to that used to make alcoholic beverages. **Biodiesel**, a different kind of biofuel produced from oils, is most often made from high-oil crops like soybeans. [INFOGRAPHIC 25.1]

INDIRECT BIOMASS ENERGY

Biomass can also be converted to other forms and then burned. This produces more energy-rich fuels that are less bulky than the original biomass; these fuels also burn more cleanly (less particulate matter) than the original forms and can fuel engines that currently run on gasoline or diesel fuel.

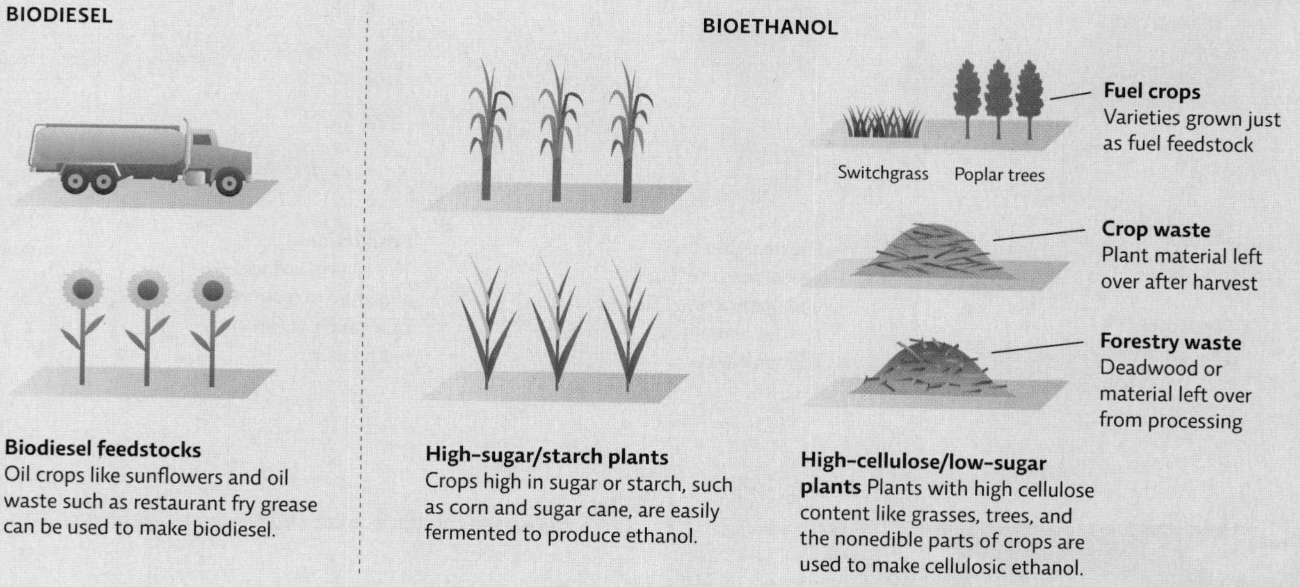
BIODIESEL
BIOETHANOL
Fuel crops
Varieties grown just as fuel feedstock
Switchgrass
Poplar trees
Crop waste
Plant material left over after harvest
Forestry waste
Deadwood or material left over from processing
Biodiesel feedstocks
Oil crops like sunflowers and oil waste such as restaurant fry grease can be used to make biodiesel.
High-sugar/starch plants
Crops high in sugar or starch, such as corn and sugar cane, are easily fermented to produce ethanol.
High-cellulose/low-sugar plants Plants with high cellulose content like grasses, trees, and the nonedible parts of crops are used to make cellulosic ethanol.

Today, ethanol is primarily used as an additive, mixed with gasoline to produce a cleaner burning fuel. It's also more corrosive than gasoline, so mixtures at the gas pump typically contain no more than 10% ethanol (E10 fuels). Some so-called "flex fuel" vehicles are equipped to use up to 85% ethanol mixtures, and thousands of these vehicles are on the road today. Because ethanol has only about two-thirds of the energy found in gasoline, however, it is not suitable for high-demand engines like those in airplanes and large trucks. Biodiesel, however, is more energy-rich than ethanol, and can be used in these applications; it can also be used directly in diesel engines, with little or no modification.

There's another convenient source of biofuels—*biowaste*, often in the form of organic leftovers such as crop residues, garbage, or manure. For instance, even though ethanol usually comes from corn and sugarcane, it can also be produced from the stalks, husks, and leaves of plants left over after a crop has been harvested. Any kind of organic material has the potential to be converted to liquid biofuels such as ethanol, or to a gaseous product similar to natural gas known simply as biogas.

The idea of converting waste to energy is appealing—it has the dual benefit of both producing energy and dealing with waste at the same time. For instance, communities are struggling to deal with methane released from decomposing landfill waste, which acts as a potent greenhouse gas, each molecule trapping 25 times more heat in the atmosphere than one of carbon dioxide. Some communities are now working to capture that gas for

---

perennial Plants that live for more than a year, growing and producing seed year after year.
annual Plants that live for a year, produce seed, and then die.
fuel crops Crops specifically grown to be used to produce biofuels.
bioethanol An alcohol fuel made from crops like corn and sugarcane in a process of fermentation and distillation.
biodiesel A liquid fuel made from vegetable oil, animal fats, or waste oil that can be used directly in a diesel internal combustion engine.

Infographic **25.2** | **WASTE TO ENERGY**

↓ Garbage and agricultural, industrial, and food waste can be used to create a variety of biofuels. Biodiesel can be produced from leftover used restaurant fryer oil, or any oil waste product, such as waste from slaughterhouses. It also has the advantage of being a rather simple production process suitable for large or small scale. Small operations like the Vancouver Biodiesel Co-op in British Columbia, serve the local community by picking up waste oil from local restaurants, converting it to biodiesel, and then selling it to local users.

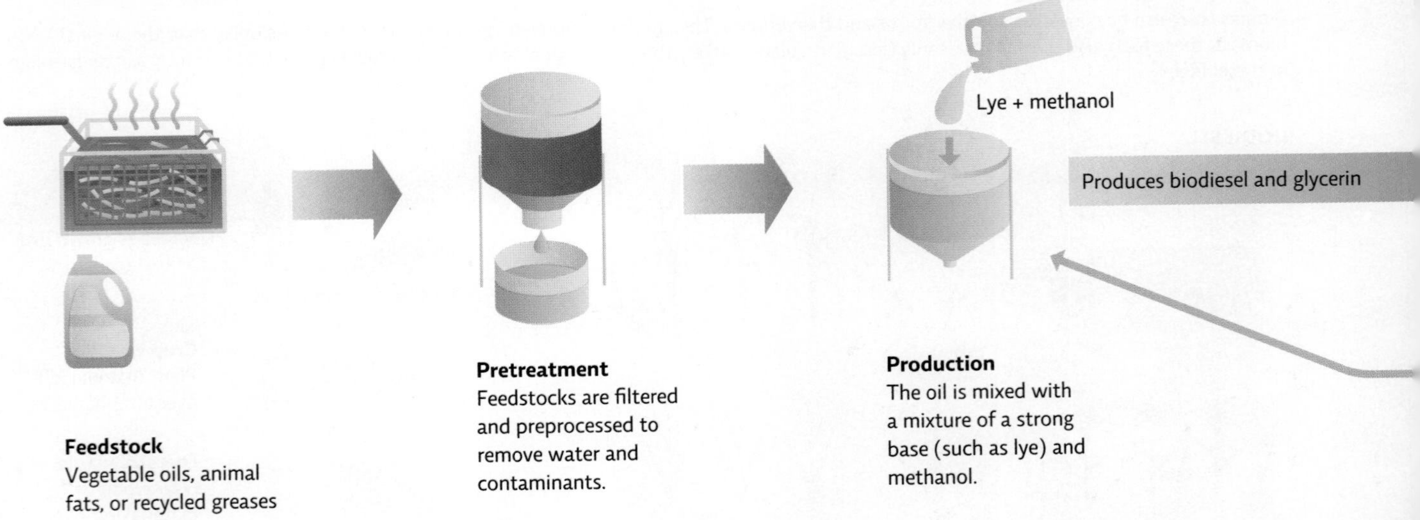
Lye + methanol
Produces biodiesel and glycerin
Feedstock
Vegetable oils, animal fats, or recycled greases
Pretreatment
Feedstocks are filtered and preprocessed to remove water and contaminants.
Production
The oil is mixed with a mixture of a strong base (such as lye) and methanol.

↑ Collection of used vegetable oil from a restaurant to recycle into biodiesel fuel.

energy use. Of 2300 municipal solid waste landfills in the United States, more than 450 have landfill gas projects. Canada has 64 landfill gas recovery facilities.

Biodiesel can also easily be made from waste such as animal fats or waste oil. Indeed, a new industry is emerging around the disposal of used fryer oil from restaurants; once a liability that restaurants paid to have hauled away, this oil can be picked up free of charge by biodiesel entrepreneurs who turn it into a fuel that can be sold to local diesel users.

At the moment, biodiesel comprises only a small portion of the biofuels produced, and cannot be used in gasoline engines. But biodiesel buses and recycling trucks are on the road all around North America. Currently, biodiesel is the only biofuel to be certified by the EPA, having passed the safety tests required by the Clean Air Act.

Biodiesel can also be produced on a small scale. Across North America, communities are taking matters into their own hands and forming small biodiesel cooperatives that produce relatively small amounts of biofuel for personal use. In Vancouver, British Columbia, the Vancouver Biodiesel Co-op provides its 300 plus members with high-quality, certified biodiesel from local-source supplier Lower Mainland Biodiesel. The Co-op tracks its contribution to CO_2 emission reduction in real time, which is already over 180 967 kilograms. [INFOGRAPHIC 25.2]

People have spent the last 100 years trying to use biomass to power engines: Henry Ford's first Model T, rolled out in 1908, was designed to use ethanol as fuel. But it wasn't until the second half of the 20th century that biofuels became big business—and more recently, a hotbed of controversy.

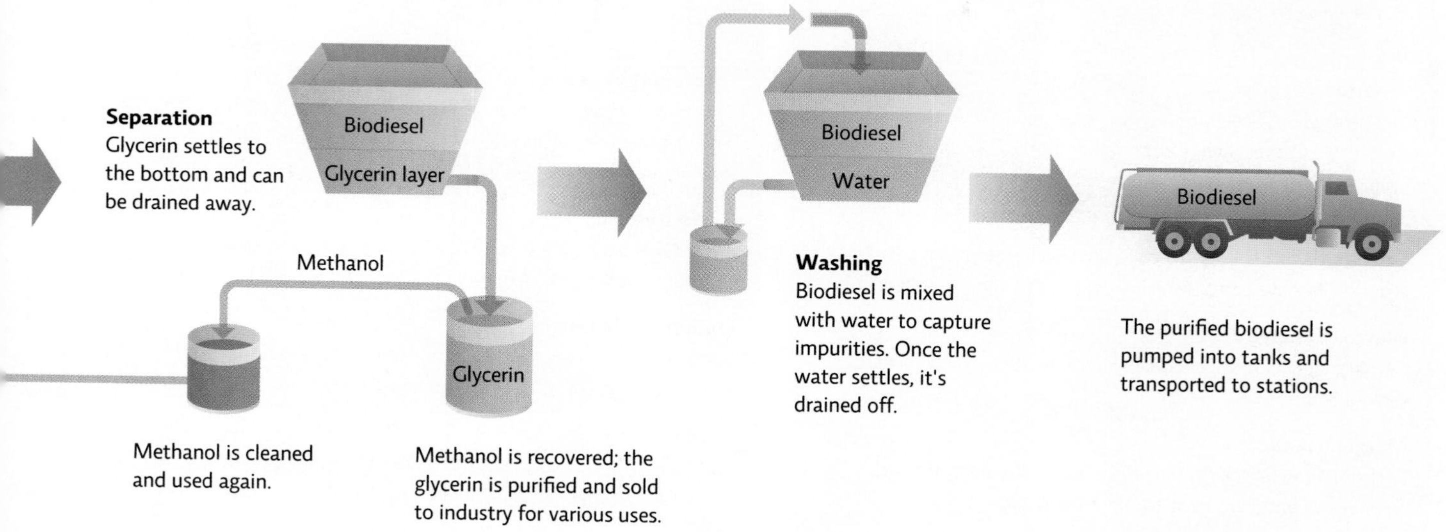
Separation
Glycerin settles to the bottom and can be drained away.
Biodiesel
Glycerin layer
Methanol
Glycerin
Methanol is cleaned and used again.
Methanol is recovered; the glycerin is purified and sold to industry for various uses.
Biodiesel
Water
Washing
Biodiesel is mixed with water to capture impurities. Once the water settles, it's drained off.
Biodiesel
The purified biodiesel is pumped into tanks and transported to stations.

Turning grass into gas is less environmentally friendly than it sounds.

Despite the attraction of putting waste to work, today most biofuels are made from crops grown specifically for energy use. Biodiesel, for instance, typically stems from crops with a high oil content, such as soybeans or canola (rapeseed), and these are the top two biodiesel crops in Canada. For decades, sugar cane has served as the primary source of ethanol in Brazil. But in North America, corn is the ethanol fuel crop of choice.

In the mid-1970s, fuel shortages revived public interest in alternatives to fill cars, fuel stoves, and light homes. As the world's leading producer of corn, the United States had the infrastructure in place to create corn-based ethanol production facilities, which began to pop up in rural farming towns all over the midwestern United States. Between 1999 and 2009, the total number of operating ethanol plants in the United States tripled, growing from 50 to more than 150. The Energy Policy Act, passed by the U.S. Congress in 2005, required that the country boost its biofuel production to 7.5 billion gallons (28 billion litres) by 2012. Only 4 years later, in 2009, the country had exceeded that goal, topping 11 billion gallons (40 billion litres). Today, nearly 50% of America's gasoline contains some amount of ethanol, according to the American Petroleum Institute. In Canada, as of 2011, all gasoline has been required to have at least 5% renewable content, and as of 2013, all diesel must have at least 2% renewable content.

In the beginning, biofuels were welcomed as an ideal green solution, reducing our dependence on foreign oil and saving the planet along the way. But slowly, that sparkling image began to fade.

Over time, experts realized that the initial attempts to create biofuels weren't much more environmentally friendly than the fossil fuels they were intended to replace. Corn is one of the most energy-intensive crops to grow and harvest, requiring energy inputs like fertilizers and pesticides (made from fossil fuel), as well as fuel needed for the operation of farm machinery; these early corn ethanol projects used as much (or almost as much) energy as they produced. Water was also a concern: one study estimated that a car running on ethanol uses the equivalent of 120 litres of water for every kilometre (compared to 2 to 3 litres of water used to produce a litre of gasoline).

Another worry is that many natural ecosystems are being converted to farmland for biofuels, endangering native species that live there, such as orangutans in the

Infographic **25.3** | **BIOFUEL TRADE-OFFS**

ADVANTAGES	DISADVANTAGES
Renewable Can be replaced over short time scales	**Energy content** May not be as high as fossil fuels so more biofuel must be burned to produce the same energy
Versatile Can be used in a variety of ways or converted to fuels that can run our existing machinery; especially important as fuel for mobile vehicles	**Total energy used** If traditional monoculture crops like corn or soybeans are used that require high fossil fuel inputs, the overall energy return of the biofuel is diminished.
Local Can be grown or collected locally, based on what grows best or is available in a given area, reducing the need to import energy	**Overharvesting** If too much biomass is taken, it can reduce forest and grassland contributions to ecosystem services.
Waste to energy Garbage, agricultural, industrial, or food waste can be a feedstock for biofuels.	**Equipment redesign** Some equipment may need to be retooled or redesigned to use biofuels (i.e., vehicles that use ethanol fuel mixtures greater than 10%).
Marginal land Fuel crops can be chosen that will grow on marginal land, which frees up the best land for food production.	**Food issues** Can take over good cropland and reduce the amount of food grown, reducing food supplies; this can drive up the price of food
Pollution Biofuels burn more cleanly and are less toxic than petroleum fuels.	**Pollution** Burning biomass directly (wood, manure, etc.) produces high particulate matter pollution.
Carbon neutral or negative If grown with minimal or no fossil fuel inputs, the carbon released is equal to or less than that taken out of the atmosphere.	**Water** Biofuel crops, especially monoculture annuals like corn and soybeans, consume more water per kWh than is used for traditional fossil fuels (nontraditional oil shale and sands use more water than biofuels).

rainforests of Malaysia and African elephants in Ethiopia. These biofuel crops also displace food crops—farmers who might normally grow corn for food or animal feed are growing it for fuel. This switch drove up the price of corn in 2007, and caused riots in Mexico, Haiti, and other nations around the world where corn is a staple food item. And clearing land for biofuel crops releases carbon stored in soil and plants, adding greenhouse gases to the atmosphere as well as disrupting all the benefits those ecosystems provide, such as habitat for a variety of organisms. In addition, the use of crop waste to produce biofuels has its drawbacks—namely the removal of material that normally would have been plowed back into the soil to decompose and add back nutrients. [INFOGRAPHIC 25.3]

Advances in recent years, mostly in increased production per hectare, now make corn a better option than it used to be. Even though early attempts to make ethanol from corn often used more energy than they produced, the U.S. Department of Energy recently estimated corn's EROEI (energy return on energy investment) at 1.38, meaning it produced 0.38 units of energy for every unit of energy invested. (By comparison, the EROEI of coal is >80.) But even if the energy return could be further improved, should we be using prime farmland to produce a fuel crop? Is there a way to grow biofuel feedstocks on marginal lands? Along with many other researchers, Tilman was discovering some interesting insights on just how to do that.

Tilman's experiments showed the importance of biodiversity.

Every August for 10 years, as part of their ongoing quest to understand why plants growing in the most diverse areas could more easily weather periods of drought, Tilman's team spread out and harvested small strips of vegetation from each plot. They dried and weighed the plants, determining how much biomass—in this case, plant material—had grown. The results floored them.

↑ Corn grain is processed at Archer Daniels Midland (ADM) ethanol and corn syrup production plant in Decatur, Illinois. Long known as a food company, ADM has invested millions of dollars in biofuel projects.

"Frankly, we never imagined the effects of diversity would be as large as they turned out to be," says Tilman. The biomass analysis showed that plots with 16 species of plants were an average of 238% more productive than fields planted with only one species (monoculture). In a way, this was not a surprise—ecologists have found that plants often grow better together than by themselves because their differences can complement each other. For example, one species may shed its leaves in the late spring, depositing valuable nutrients into the soil that another species takes up in the summer. Ecological communities with high species diversity such as these arose over time as individuals competed for resources. This led to resource partitioning—each species uses particular resources and is very efficient at accessing and using those resources. Species diversity also leads to mutualistic and commensal relationships, all of which maximize energy uptake and nutrient flow in the community, increasing overall biomass (see Chapter 8).

cellulosic ethanol Bioethanol made by breaking down cellulose in plants; a difficult process that has yet to be scaled up to meet large production goals.
carbon sequestration The storage of carbon in a form that does not readily release the carbon to the atmosphere or water.

Tilman kept thinking about the results—specifically, about how they could model biofuel crops after what nature had already figured out: biodiverse ecosystems that are naturally adapted to their environment are the most productive.

Ethanol is easily made from the starch in corn, but it can also be derived from cellulose, a compound that forms in the cell walls of all plants and makes up the bulk of the fibrous structure of a plant like switchgrass. Like with other forms of biofuels, people had been trying to produce fuel from cellulosic material such as wood for decades.

Recently, **cellulosic ethanol** that can be produced from crops such as switchgrass has become a popular alternative to corn ethanol. For starters, it has the potential to circumvent the food controversy, since it doesn't have to occupy land that could support food crops. Additionally, switchgrass is resistant to many pests and plant diseases and requires little fertilization, so it consumes less energy to produce. Even more attractive to alternative energy aficionados, switchgrass, like other plants, absorbs CO_2 from the atmosphere and stores carbon in its roots, a process called **carbon sequestration**. By working with a perennial like switchgrass instead of an annual like corn, much of the plant (and its carbon) would remain behind

Infographic **25.4** | **LIHD CROPS OFFER ADVANTAGES OVER TRADITIONAL MONOCULTURE BIOFUEL CROPS**

↓ A life-cycle analysis by David Tilman and colleagues shows that low-input, high-diversity (LIHD) grassland plants offer a higher net energy return and release fewer greenhouse gases than do the traditional monoculture fuel crops like corn and soybeans.

ENERGY RETURN ON ENERGY INVESTMENT COMPARISON

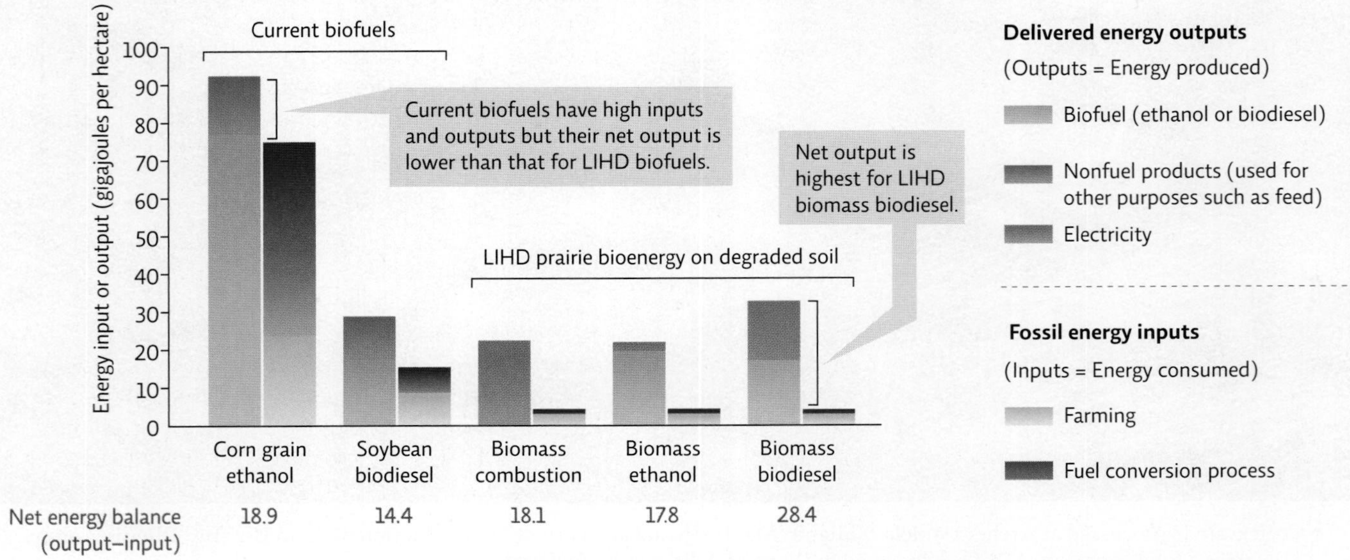
Energy input or output (gigajoules per hectare)
100
90
80
70
60
50
40
30
20
10
0
Current biofuels
Current biofuels have high inputs and outputs but their net output is lower than that for LIHD biofuels.
Net output is highest for LIHD biomass biodiesel.
LIHD prairie bioenergy on degraded soil
Corn grain ethanol
Soybean biodiesel
Biomass combustion
Biomass ethanol
Biomass biodiesel
Net energy balance (output–input)
18.9
14.4
18.1
17.8
28.4
Delivered energy outputs
(Outputs = Energy produced)
Biofuel (ethanol or biodiesel)
Nonfuel products (used for other purposes such as feed)
Electricity
Fossil energy inputs
(Inputs = Energy consumed)
Farming
Fuel conversion process

↓ Compared to gasoline, all biofuels produced less greenhouse gas emissions in an EROEI analysis. In particular, fuels from LIHD grasses grown on marginal land produced significantly lower greenhouse gas emissions than higher-input conventional biofuel crops (corn and soybeans).

REDUCTION OF GREENHOUSE GAS (GHG) EMISSIONS BY BIOFUELS

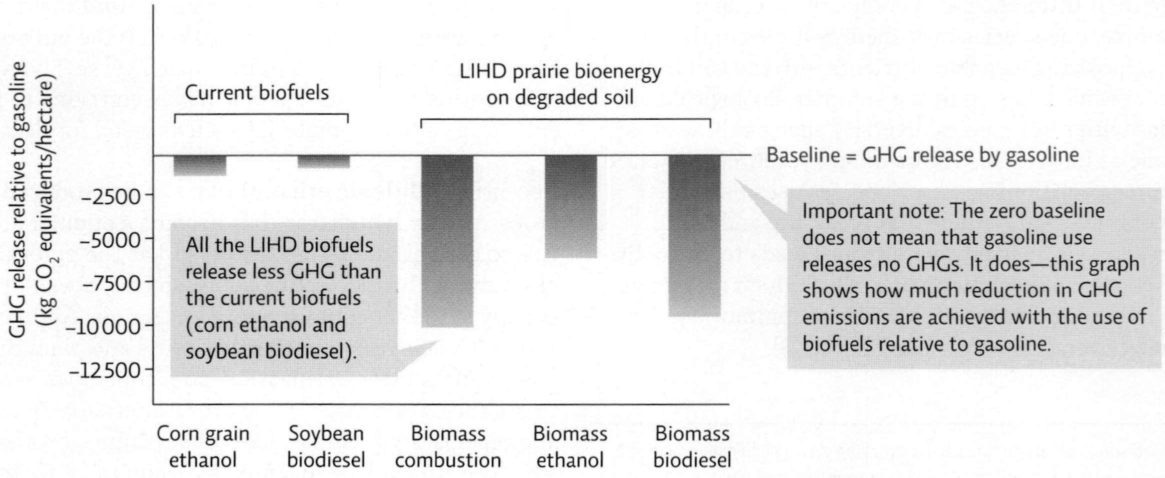
GHG release relative to gasoline (kg CO2 equivalents/hectare)
-2500
-5000
-7500
-10000
-12500
Current biofuels
LIHD prairie bioenergy on degraded soil
Baseline = GHG release by gasoline
All the LIHD biofuels release less GHG than the current biofuels (corn ethanol and soybean biodiesel).
Important note: The zero baseline does not mean that gasoline use releases no GHGs. It does—this graph shows how much reduction in GHG emissions are achieved with the use of biofuels relative to gasoline.
Corn grain ethanol
Soybean biodiesel
Biomass combustion
Biomass ethanol
Biomass biodiesel

in the deep root system, ready to grow again immediately after harvesting.

However, people who mentioned switchgrass as a source of biofuels "were always talking about monocultures," Tilman remembers, and growing switchgrass as a crop by itself. Tilman realized that by growing switchgrass in high-diversity mixtures of many plant species, a plot could produce a lot more biomass with the same or smaller amounts of irrigation, fertilizer, and management. That meant more biofuel for fewer inputs.

"We kept saying, 'we gotta tell people about this!'" Tilman recalls. To do so, he hired Jason Hill, then an assistant professor of biology at St. Olaf College in Minnesota. Hill collected Tilman's biomass data for each plot over the 10 years, as well as the recorded levels of carbon in the soil and roots of the plants, and analyzed the net energy that could be produced from the grasslands. He compared those data to the amounts of energy harvested from corn and soybeans. The results were overwhelming: the 16-species plots were not only capable of producing more bioenergy than monocultures, they sequestered 14 times more carbon in the soil than the amount released by growing and processing the crop into a biofuel. In other words, they made the grasslands carbon negative—they stored more carbon than was released to grow or use the fuel. Finally, Hill and Tilman noted that diverse grasses can be grown on neglected agricultural lands and in marginal soil, therefore not displacing food production or destroying habitats. The two researchers published their results in a 2006 issue of *Science*, demonstrating for the first time the feasibility and potential of biofuels from low-input, high-diversity grasslands, called LIHD biofuels. [INFOGRAPHIC 25.4]

> "When you have a new idea, you start imagining what might happen, but you never know what will really happen until you try the experiment."
> —David Tilman

The media went crazy: "Wild flowers could provide a solution to global warming," hailed a Scottish newspaper; "It sounds too good to be true," wrote a journalist at *Discovery News*. Unfortunately, that's exactly what some other scientists thought of Tilman's conclusions—too good to be true.

Criticism came swiftly, and from close to home. In the building next door to Tilman's at the University of Minnesota, members of the United States Department of Agriculture (USDA) and professors in the University's agronomy department began to discuss their concerns about Tilman's paper. Michael Russelle, a USDA soil scientist and adjunct professor at the university, suspected that Tilman's group had overestimated the carbon sequestration of the grasslands because of the way they managed the plots—by burning the plots at the end of the season after estimating productivity, they potentially left behind more carbon than harvesting would have done. Because of this, Russelle felt that the energy accounting was "misleading."

"Many issues have rapid political implications, and this is one of those. In this case, we felt that the paper wasn't warranted by the data, so we just felt it had to be challenged," recalls Russelle. He approached Tilman and Hill to inform them that he and colleagues were planning to write a rebuttal (an official commentary in response to Tilman's paper). Tilman and Hill were surprised but accommodating. "It's good to have a public dialogue over a controversial paper," says Hill. "You just don't necessarily expect it to come from the building next to you," he adds with a laugh.

The three men met for coffee, but by the end of the meeting, they remained divided. "They didn't convince me," says Russelle. The dissenters published their comment in *Science*, but Tilman and Hill didn't back down, and published a response, backing up their results and study design with other published research. "We essentially said we appreciate your concerns, but we do not think they changed the outcome of the conclusions at all," says Hill.

Today, the two groups still maintain opposing viewpoints, but both recognize the importance of the dialogue. "Science is a continual process of putting forth a hypothesis, testing it, evaluating the results, and moving forward," says Russelle.

There is another rising biofuel star: algae.

In January of 2009, Continental Airlines ran a test flight of a Boeing 737 powered by a blend of normal fuel and a new type of biofuel manufactured from algae.

Like Tillman, algae researcher Stephen Mayfield didn't set out to discover a new type of biofuel. But after spending decades studying and tweaking the genetics

↑ Researcher Emma Valdez and a coworker monitor algae cultures at Sapphire Energy's facility in San Diego, California. Dubbed "green crude," oil from the algae can be processed in existing refineries into jet fuel, gasoline, or diesel.

of algae, he and his colleagues at the University of California, San Diego, had an epiphany in 2006. At that time, Mayfield was watching TV footage of the glaciers shrinking from climate change, and paying more for gas every time he fuelled up. During a meeting to discuss an ongoing research project to engineer algae to produce drugs such as antibiotics, a new idea arose. "We realized, 'wait a minute, if we can do that, we can produce fuel.'"

Like tiny factories, algae use photosynthesis to convert CO_2 and sunlight into sugars and then convert some of the sugars into oil, which can be harvested and converted to biodiesel. Algae are now making the same kind of headlines as LIHD biofuels, and for good reason.

Algae have some clear advantages over other biofuel sources. Like switchgrass, algae won't compete with food. They grow very quickly and can be raised on salty water, so they don't use up freshwater. Algae are also predicted to generate 30 times more oil per hectare than other plants used for biodiesel, such as soybean, oil palms, and rapeseed.

But algae can be finicky—it is difficult to maintain the ideal conditions to maximize oil production—and the facilities are expensive. Plus, unlike corn and other land crops, people haven't been harvesting algae on a large scale for generations, so scientists have to devise new infrastructure. Even so, algae-based biofuels are starting to look like big business. There are hundreds of U.S. companies now invested in turning algae oil into fuel, according to the U.S.'s National Renewable Energy Laboratory. One company Mayfield cofounded, Sapphire Energy, recently opened the commercial test phase of its "green crude" algae farm - 120 hectares of algae ponds in a New Mexico facility. While there are over 35 algal biofuel companies in the United States, Canada only has 3 to date.

Those investments may one day pay off. After Continental Airlines analyzed data from the test flight, they found that the algae biofuel blend resulted in a 1.1% increase in fuel efficiency over regular jet fuel and didn't affect the jet's ability to perform flight manoeuvres such as a mid-flight engine shutdown and restart. [INFOGRAPHIC 25.5]

There are many reasons why biofuels have not solved our dependence on fossil fuels.

Despite the hopes of Tilman and others, cellulosic ethanol is not likely to arrive at a pump anytime soon. For one, it's not easy to get the energy out of the plants: while starch from corn is soft, cellulose in grasses is complex and durable, not easily degraded. "There's a reason we make houses and furniture out of cellulosic material such as wood," says Mayfield. "It doesn't like to spontaneously turn into something else." Several attempts use high temperatures, acid, enzymes, and yeast to tear apart the strong bonds—a slow, expensive, and energy-intensive process that has proven difficult to scale up. [INFOGRAPHIC 25.6]

There is currently no large-scale commercial facility online for converting cellulose-rich biomass to ethanol, whether from switchgrass monocultures or LIHD grasslands, but there are some smaller prototype facilities, and more are under construction around the world. One began operation in Vonore, Tennessee, in 2010 as part of the $70.5 million Biofuels Initiative of the University of Tennessee. The goal is to create a template for the logistics of growing, handling, storing, and transporting the thousands of bales of switchgrass that would be processed daily at a commercial facility. "It's the holy grail," says Michael Palmer, a professor of botany at Oklahoma State University who tracks biofuel research. "Millions and millions of dollars from the government

Infographic **25.5** | **BIOFUELS FROM ALGAE**

↓ Algae can be grown as a feedstock for biofuels. Biodiesel is the most common product, but the sugars in algae can also be extracted to produce ethanol, while the solids can even be harvested to produce a high-protein animal feed. Self-contained operations can grow lots of algae in a small space (in tubes or vats) or in outdoor algae ponds—which take up more land area. A New Zealand company, Aquaflow, even plans to harvest wild algae from coastal areas with algal blooms.

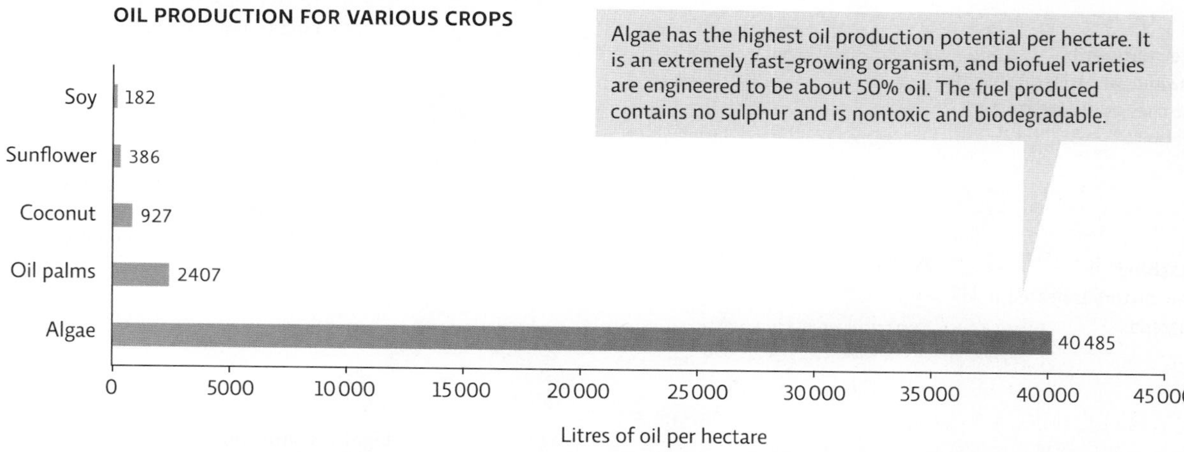
OIL PRODUCTION FOR VARIOUS CROPS
Algae has the highest oil production potential per hectare. It is an extremely fast-growing organism, and biofuel varieties are engineered to be about 50% oil. The fuel produced contains no sulphur and is nontoxic and biodegradable.
Soy
182
Sunflower
386
Coconut
927
Oil palms
2407
Algae
40 485
0
5000
10 000
15 000
20 000
25 000
30 000
35 000
40 000
45 000
Litres of oil per hectare

↑ Researchers test different types of algae to see which produces the best oil to use for biodiesel formation.

↑ Algae can be grown in enclosed, clear containers or, like here, in open ponds. Warm, sunny areas give the greatest yields.

Infographic **25.6** | **BIOETHANOL PRODUCTION**

↓ Ethanol can be produced from the biological fermentation of any plant material. Food sources like grains and sugarcane are easily broken down by yeast (fermentation), whereas plant material high in cellulose is more challenging to break down and requires stronger chemical reactions. Research to improve the breakdown of cellulose and make the process more efficient is underway.

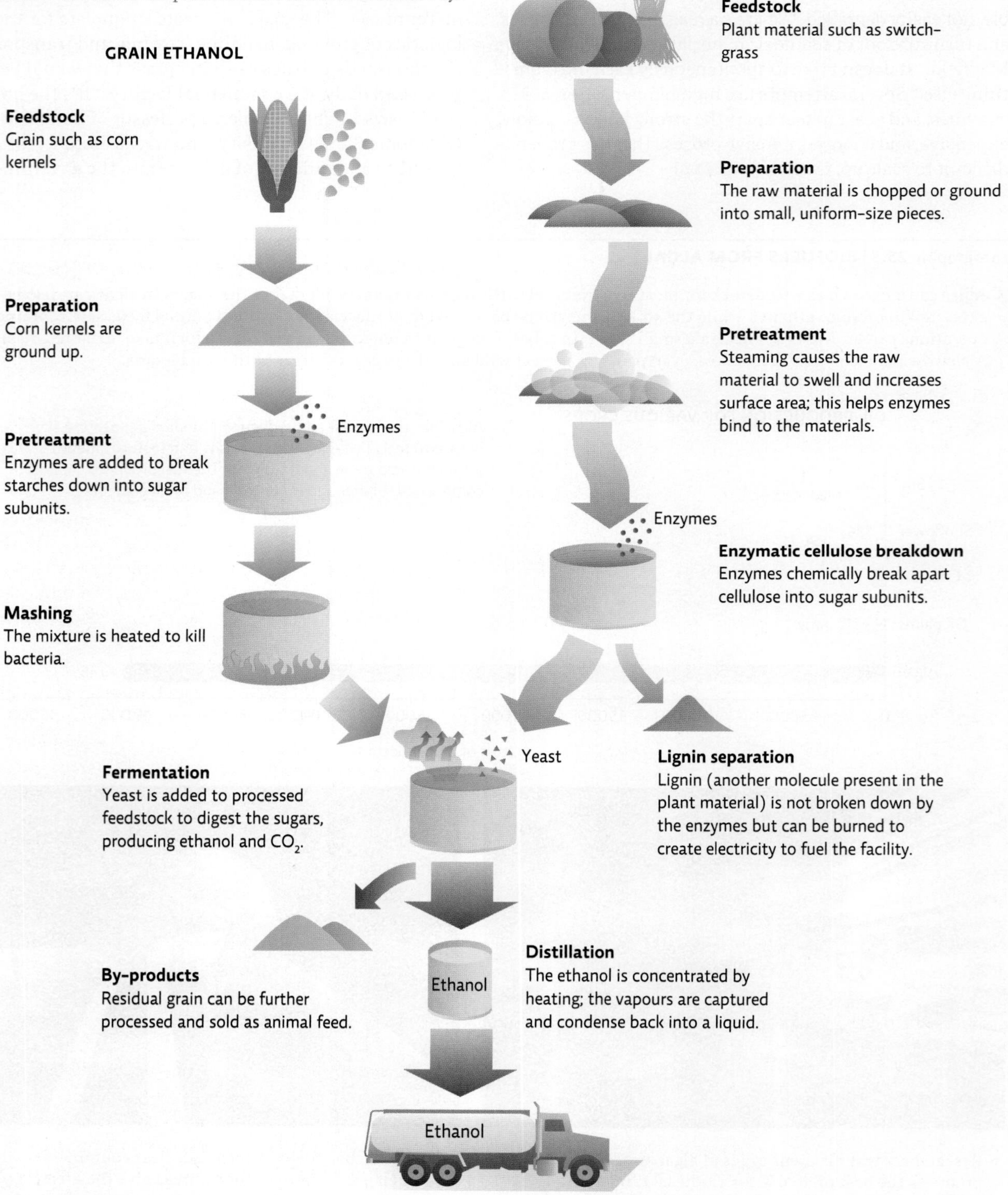
CELLULOSIC ETHANOL
Feedstock
Plant material such as switch-grass
GRAIN ETHANOL
Feedstock
Grain such as corn kernels
Preparation
The raw material is chopped or ground into small, uniform-size pieces.
Preparation
Corn kernels are ground up.
Pretreatment
Steaming causes the raw material to swell and increases surface area; this helps enzymes bind to the materials.
Enzymes
Pretreatment
Enzymes are added to break starches down into sugar subunits.
Enzymes
Enzymatic cellulose breakdown
Enzymes chemically break apart cellulose into sugar subunits.
Mashing
The mixture is heated to kill bacteria.
Yeast
Lignin separation
Lignin (another molecule present in the plant material) is not broken down by the enzymes but can be burned to create electricity to fuel the facility.
Fermentation
Yeast is added to processed feedstock to digest the sugars, producing ethanol and CO_2.
Distillation
The ethanol is concentrated by heating; the vapours are captured and condense back into a liquid.
By-products
Residual grain can be further processed and sold as animal feed.
Ethanol
Ethanol

← Virginia Tech professor, John Fike, kneels in a thick stand of switchgrass. Dr. Fike's research focuses on the productivity and environmental suitability of different biofuel crops.

and private sectors are being pumped into the technology of cellulosic conversion." Verenium Biofuels Corporation, a Massachusetts-based firm that has proprietary rights for special enzymes used in the breakdown of cellulose, recently sold its business and demonstration facility in Louisiana to BP for $98 million. The goal is to produce low-cost cellulosic ethanol from grasses, wood chips, agricultural waste, or even garbage.

Algae, too, will take several years before it can be scaled up, predicts Mayfield. It's relatively easy to convert algae to diesel fuel, he says, but building the infrastructure to do this on a large scale—such as massive ponds—delays progress. He predicts that algae will become widely used as a biofuel source in the United States before switch-grass, however, simply because it has a bigger customer: the U.S. military, which can pay a premium for renewable diesel to fly its jets. Furthermore, there already is a major source of ethanol—corn. "I think algae are more likely to be commercially deployed in the next several years."

Multiple solutions will be needed to help replace fossil fuels.

Like all strategies to replace fossil fuels, biofuels will never have just one golden ticket winner. Instead, a patchwork of feedstocks grown and processed across the United States and Canada will help provide our countries' fuel. The good news is that we have a wide variety of options, which enables us to focus on the types of biofuels that work best for different needs or in different regions.

Whatever the future, one thing is sure—to be truly environmentally friendly, we may simply have to stop using so much fuel. This means increasing energy efficiency to decrease waste, and basic conservation—simply using less. The energy efficiency of vehicles can be improved with lighter materials, more aerodynamic shapes, more energy-efficient engines, and hybrid technologies that pair electric motors with internal combustion engines. Totally electric vehicles would lessen the demand for liquid fuels. Drivers can also increase the fuel efficiency of whatever vehicle they drive by adopting better driving habits and of course, by driving less overall. [INFOGRAPHIC 25.7]

In 2007, the U.S. Congress passed the Energy Independence and Security Act, which requires the United States to produce 36 billion gallons (135 billion litres) of renewable and alternative fuels per year by 2022 (up from 11 billion gallons [41 billion litres] produced in 2009)—equivalent to about one-quarter of the entire country's yearly use of gasoline.

The United States is behind schedule—the economic downturn and obstacles to scaling up cellulosic ethanol have stalled progress in recent years. The goal for 2010 was 250 million gallons (945 million litres) of cellulosic ethanol; that year, the United States produced only 25 million gallons (94.5 million litres)—off by a factor of 10. Uncertainties about whether tax credits and subsidies will remain available make cellulosic ethanol a risky investment, and many venture capitalists have pulled out of projects. In Canada, many projects have been put on hold or slowed considerably due to similar delays in policy and incentive development.

And yet, biofuel projects are moving ahead. Many U.S. states offer rebates or tax exemptions on fuels for commercial vehicles that operate on biodiesel blends or E-85 fuels, and financial incentives like tax credits or grants are offered to biodiesel producers or distributors. In August of 2011, the U.S. federal government launched a $510 million program, in the form of matching funds, to

Infographic 25.7 | ENERGY EFFICIENCY AND CONSERVATION ARE PART OF THE SOLUTION

↓ Choosing the vehicle that meets your needs but that is also fuel efficient is an important part of the solution to our energy problems. Go to the Office of Energy Consumption's website (http://oee.nrcan.gc.ca/transportation/tools/fuelratings/ratings-search.cfm) to compare different makes and models of cars for fuel efficiency. Fuel efficiency will also save you money—a car that is 4.25 km/L more efficient than another will save the driver almost $1,500 in fuel costs per year (assumes 24 000 kilometres driving per year with gasoline at $1.15/L). All the major (and some smaller) car companies have hybrid and/or electric vehicles on the road or in planning. Though hybrids are generally more fuel efficient than other cars, many traditional (and less expensive) models get 15 to 20 km/L (5-6.7 L/100 km).

PLUG-IN HYBRID VEHICLES

140
160
100
PS
KEY
TEST CAR
EV

Gasoline-hybrid vehicles have a battery system, charged when the car is braking or coasting, that helps run the motor. The car goes from electric to gasoline mode seamlessly. Plug-in hybrids can be plugged into a household 120V AC outlet to add charge to the battery and extend the battery range.

ELECTRIC VEHICLES

A public charging station in Israel. A charging station infrastructure will be a vital part of a future that includes electric vehicles. Public charging stations are springing up across Canada. Owners can visit the website of the nonprofit Electric Mobility Canada to locate the nearest EV charging station.

↓ Driving habits that maximize fuel efficiency will make our biofuels go farther.

HOW CAN YOU INCREASE YOUR FUEL EFFICIENCY?

DRIVE MORE EFFICIENTLY

Don't be an aggressive driver
Avoid "jackrabbit" quick starts; avoid speeding up to a stop.

Don't speed
The majority of cars are most fuel efficient between 50 and 80 kilometres per hour; when possible, try to drive in your top gear (cruising gear).

Use cruise control
By keeping a more constant speed, fuel efficiency increases.

Avoid idling
Turn off your car if you will be parked, even briefly. Starting the car back up takes less fuel than idling for more than a few seconds. Hybrid cars that automatically turn off the gas engine save idling gas waste.

TAKE CARE OF THE CAR

Tune it up
Keep your vehicle in tune to maximize fuel efficiency—don't skip maintenance visits.

Keep tires properly inflated
Under- or overinflated tires decrease fuel efficiency and make for unsafe driving; check tire inflation regularly, especially when the seasons change—air pressure increases in warm weather and decreases when it gets cold.

Unload the trunk
Avoid carrying excess weight in the vehicle—the heavier the load, the lower the fuel efficiency.

PLAN YOUR OUTINGS

Combine trips and plan your route
A cold car is less fuel efficient; combining trips while driving a warm car—even starting and stopping several times as you drive around town—will save gas; plan your route to avoid backtracking.

Commuting
Drive the most fuel-efficient car available when commuting; consider working from home when possible and carpool with coworkers if you can; using high-occupancy vehicle (HOV) lanes can improve your fuel efficiency and get you to work more quickly.

help finance biofuel projects directed toward military transportation applications.

Biofuels are also growing in popularity around the world. Brazil, a world leader in ethanol biofuel, has no light vehicles that run on 100% gasoline, and Brazilian car manufactures have developed vehicles that run on ethanol blends or 100% ethanol. China and India have national policies mandating the increase of the ethanol component of gasoline. In addition, China, India, and other developing countries are supporting biodiesel production from locally grown crops such as the succulent plant jatropha. Meanwhile, China is pursuing the development of technologies to produce biofuels from municipal garbage. Smaller-scale programs are also underway in many nations around the world.

Despite ongoing controversies and setbacks, the future of biofuels looks bright.

Working with Nature Conservancy scientist Joseph Fargione, Tilman and Hill set about to determine exactly how much CO_2 is released by clearing land for crop biofuels, which involves burning vegetation and plowing carbon-rich soil. The results were shocking: the team found it could take decades—or even centuries—of biofuel production to make up for the amount of CO_2 released during land conversion, what they called the **carbon debt**. For instance, it will take 93 years to make up for the carbon debt resulting from digging up central grasslands in the United States to produce corn ethanol. To repay the carbon debt for clearing peatland rainforest for palm biodiesel in Indonesia and Malaysia: 423 years. The only biofuel with zero carbon debt? Bioethanol from prairie biomass grown on marginal cropland, making this crop a strong choice in many Canadian regions.

The results solidified Tilman's hopes for a new type of biofuel that circumvents the problems of corn ethanol, such as fuel from switchgrass or algae. "Biofuels, if used properly, can help us balance our need for food, energy, and a habitable and sustainable environment," Tilman wrote in a 2007 opinion piece in the *Washington Post*. "We have the knowledge and the technology to start solving these problems."◉

carbon debt The cumulative reduction of the carbon dioxide absorption or storage capacity of ecosystems, e.g. by converting land from a natural ecosystem to agricultural uses.

Select references in this chapter:
Fargione, J., *et al.* 2008. *Science*, 319: 1235–1238.
Russelle, M.P., *et al.* 2007. *Science*, 316: 1567b.
Tilman, D., *et al.* 2006. *Nature*, 441: 629–632.
Tilman, D., *et al.* 2006. *Science*, 314: 1598–1600.
Tilman, D., *et al.* 2007. *Science*, 316: 1567c.

BRING IT HOME

PERSONAL CHOICES THAT HELP

Biofuels represent a potential replacement for fossil fuels and might be an especially important fuel for transportation. Despite the promise of biofuel use, there are also trade-offs such as the fossil fuel inputs needed to grow fuel crops. Developing new biofuel technologies may help us meet the energy needs of the future.

Powered By Pure Plant Oil
www.bloomingfutures.com

Individual Steps

→ Visit ecogeek.org to find information on experiments and new advances in biofuel technology.

→ Review Infographic 25.7 and take steps to make your vehicle and driving more fuel efficient; make the investment in a hybrid or electric vehicle if you can.

Group Action

→ Host a movie night to watch *FUEL*, an award-winning film that looks at the history of biofuels as well as possible solutions for the future.

→ Join a biofuel co-op in your area that provides alternatives to power your vehicle.

Policy Change

→ Use the Office of Energy Efficiency site (http://oee.nrcan.gc.ca/transportation/alternative-fuels/780) to learn about sustainable biofuel options and stay current on biofuel programs.

UNDERSTANDING THE ISSUE

CHECK YOUR UNDERSTANDING

1. Which of the following best describes a fuel crop?
a. Crops that are specifically grown to make biofuels
b. Waste biomass from food crops that is converted to fuel
c. Waste oil from restaurants used to make biodiesel
d. Manure patties remaining in a field grazed by herbivores

2. What is a LIHD biofuel?
a. Biofuel with a high-density lipid content
b. Biofuel produced from waste oil, with a low impact on the environment
c. Biofuel produced from algae grown in high density in fertilizer-rich liquid
d. Biofuel produced from grasslands with high species diversity, grown with few inputs

3. Which of the following produces the most oil per hectare of crop?
a. Soybeans
b. Algae
c. Sunflowers
d. Oil palms

4. How do plants sequester carbon?
a. During photosynthesis, carbon is stored in the leaves of plants.
b. Plants take in atmospheric CO_2 and store carbon in their roots.
c. Plants convert CO_2 into oxygen, replacing CO_2 in the atmosphere.
d. Sugar is converted to CO_2 and released during respiration.

5. Compared to gasoline, which of the following produces the smallest amount of greenhouse gases?
a. Corn grain ethanol
b. Soybean biodiesel
c. LIHD biodiesel
d. LIHD ethanol

6. Direct sources of biomass energy include wood, corn stalks, grass bales, and _______.
a. sugar cane
b. used restaurant oil
c. dried manure
d. biodiesel

WORK WITH IDEAS

1. Describe the environmental issues associated with biofuels. Include issues specifically related to the use of corn.
2. You are an investor looking to get into the biofuels market, and you plan to choose one source for generating biofuel. Based on the information in this chapter, which source would you choose and why?
3. List and describe the current sources of fuel (both direct and indirect) that come from biomass. Include in your description the sources of biofuel and the feasibility of producing large quantities of biofuel from these sources (to the best of our knowledge at the present time).
4. Compare and contrast corn and switchgrass as sources of biofuels. What are the benefits of each? What are the costs?
5. Describe the process of biodiesel production using leftover restaurant oil. Is this possible on a small scale?
6. Explain, in your own words, why biofuels (especially cellulose and algae) have not replaced fossil fuels.

ANALYZING THE SCIENCE

The following graph shows the total environmental impact of biofuels compared to greenhouse gas emissions. Each data point is plotted relative to the emissions and impact of gasoline; fuels that fall in the white area of the graph are considered to be better choices.

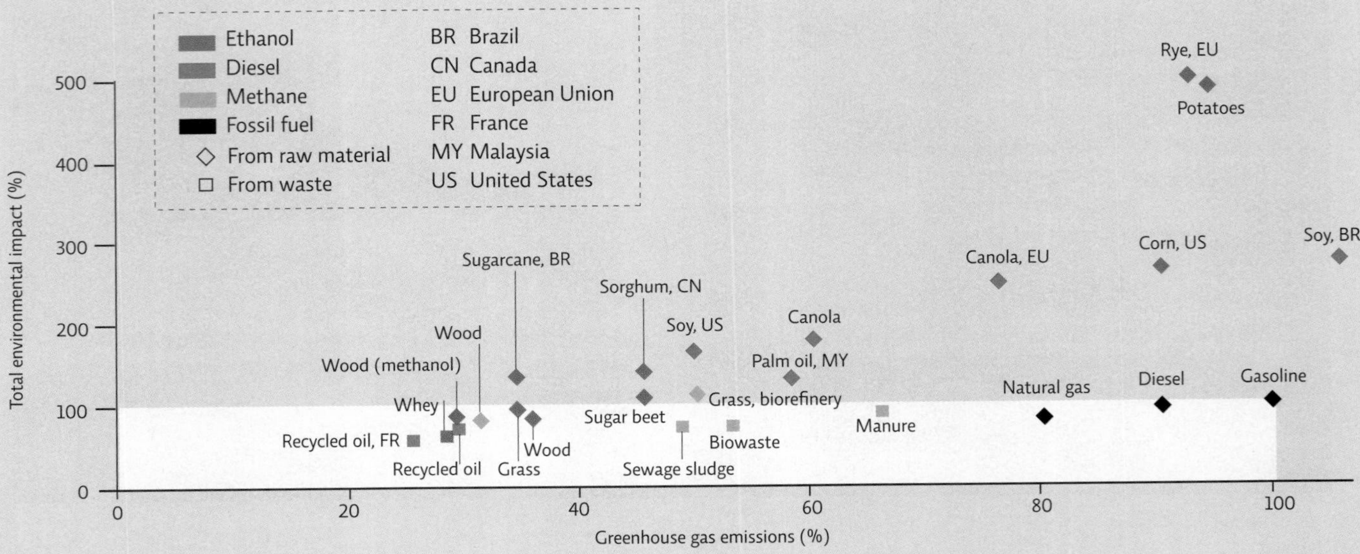
Ethanol
Diesel
Methane
Fossil fuel
From raw material
From waste
BR Brazil
CN Canada
EU European Union
FR France
MY Malaysia
US United States
Total environmental impact (%)
Greenhouse gas emissions (%)
0
100
200
300
400
500
20
40
60
80
Rye, EU
Potatoes
Soy, BR
Corn, US
Canola, EU
Sugarcane, BR
Sorghum, CN
Soy, US
Canola
Wood
Palm oil, MY
Wood (methanol)
Grass, biorefinery
Whey
Sugar beet
Manure
Natural gas
Diesel
Gasoline
Recycled oil, FR
Wood
Biowaste
Recycled oil
Grass
Sewage sludge

INTERPRETATION

1. What does each axis of this graph represent?

2. What types of fuel are being compared, and what are their original sources?

3. Locate each of the following on the graph: U.S. corn ethanol, Brazilian (BR) sugarcane ethanol, Brazilian soy diesel, Malaysian (MY) palm oil diesel, and gasoline. Note that the first four are the most economically important biofuel sources.
   a. Rank all five in terms of their total environmental impact, from highest to lowest. Which has the highest environmental impact and which has the lowest? Assuming that all were equally abundant, which would be the most environmentally friendly choice, based on total environmental impact?

ADVANCE YOUR THINKING

4. What kinds of things does the measure "total environmental impact" likely include? (Hint: Remember to look at Infographic 25.3.)

5. Why is the environmental impact for Brazilian soy diesel high?

6. Assuming that the scientists who evaluated these data are correct, what type of fuel contributes the least amount of greenhouse gases and has the lowest total environmental impact? Why do you think its impact is low relative to other fuel sources?

7. According to the graph, which biofuel would make the most *environmental* sense to use in North America, given our current standard of living and the resources we have access to?

EVALUATING NEW INFORMATION

The Canadian Renewable Fuels Association (CRFA) website provides information about biodiesel and ethanol and their uses. Go to that site (greenfuels.org) and explore it by opening each of the main links at the top of the page.

Evaluate the website and work with the information to answer the following questions:

1. What type of information does the site provide under each link?

2. Is this a reliable information source? Does it have a clear and transparent agenda?
   a. Who runs the website? Do this person's/organization's credentials make the site reliable or unreliable?
   b. Is the information on the website up to date? When was the last time the website was updated? Explain.

3. Go to the "Resource Centre" via the main link at the top of the page, and review the materials provided on ethanol and biodiesel.
   a. Explore these resources. What are the environmental benefits of ethanol and biodiesel according to the CRFA?
   b. What are the economic benefits of ethanol and biodiesel noted?
   c. What evidence is given in support of these claims? In your opinion, is this evidence sufficient? Explain.
   d. If you drove a vehicle that could use biodiesel, would you use it? Why or why not?

MAKING CONNECTIONS

THE FUTURE OF ENERGY

Background: Environmental science is complex because there are no "magic bullets" that solve environmental problems. Each potential response comes with trade-offs. A major goal of environmental science is to identify sustainable choices that reduce our impact in a manner that is socially, culturally, and economically acceptable. Such choices occur in every sphere of society and are everyone's responsibility.

Case: As an environmental science student, you have learned about the pros and cons of sources of energy from fossil fuels to renewable energy. You have been assigned to draft a proposal for future energy policy on university campuses in your province or territory that will be sent to the provincial or territorial governing body of the higher education system.

1. Be sure to include the following in your proposal:
   a. The number and size of public institutions in your province or territory.
   b. The types of energy that are currently being used on each campus. For example, what type of fuel does the local bus system use? What is the source for electricity production in your provice or territory? Do any of the schools have mechanisms in place to either reduce energy consumption or to generate their own power?

2. Based on your research, what recommendations would you make to the governing body to decrease the quantity of energy consumption and to increase the production of sustainable energy on campuses in your province or territory? In your recommendations, include options for the institutions as a whole and for individuals in those institutions. Remember to consider the location of each institution in your province or territory, and assume there will be money to fund such an initiative.

3. Conclude your report with a section on benefits to the institution and to members of the institution (faculty, staff, and students). In addition to becoming more energy efficient, how will this initiative help your campus?

CARSON

Clay Garden, built on the site of a burned-down home by a resident across the street, provides urban farming opportunities for local residents. In the background are the Webster Morrisania public housing projects, which provide homes for some of the poorest people in the Bronx.

CHAPTER 26 URBANIZATION AND SUSTAINABLE COMMUNITIES

THE GHETTO GOES GREEN

In the Bronx, building a better backyard

CORE MESSAGE

Cities can be both an environmental blessing and a curse. Using green strategies to plan or retrofit cities can benefit citizens, business, and the environment—not to mention reduce environmentally related health problems and degradation of natural resources.

GUIDING QUESTIONS

After reading this chapter, you should be able to answer the following questions:

→ What is the pattern of global urbanization and megacity growth in recent decades?

→ What are the trade-offs associated with cities or urban areas?

→ What is environmental justice? How does urban flight contribute to and result from environmental justice problems?

→ What environmental problems does suburban sprawl generate?

→ How can we create cities that are environmentally sustainable and promote a good quality of life for the residents?

As she walked her dog Xena—a scruffy puppy she had found tied to a tree in her South Bronx neighbourhood—Majora Carter considered her options. The 32-year-old aspiring filmmaker had moved back home to save money while she attended graduate school. Initially, she had wanted as little to do with the decaying neighbourhood as possible. But then she'd gotten involved in a local artists' group and taken work at a community development centre. Now, a colleague at the city parks department was offering her a $10,000 grant to come up with a waterfront development project for her neighbourhood. Carter was balking.

At the moment, she and her neighbours were busy fighting a mammoth waste facility that the city was trying to move from Staten Island to the East River waterfront. With 30 transfer stations in the South Bronx, their tiny parcel of New York already handled 40% of the entire city's commercial waste, not to mention having four power plants, two sludge processing plants, and the largest food distribution centre in the world. All told, some 60 000 diesel trucks passed through the neighbourhood every week. In exchange for this burden, area residents boasted the highest asthma and obesity rates in the country, along with some of the poorest air quality. Another waste facility would only make matters worse. Consumed with this battle, Carter wasn't sure she had the time or energy to take on a development project.

Besides, the idea of developing waterfront property in the South Bronx seemed a bit naive to her. Like most of her neighbours, Carter had lived in the neighbourhood most of her life; she knew full well how inaccessible the surrounding river was to residents.

◉WHERE IS THE BRONX, NEW YORK?

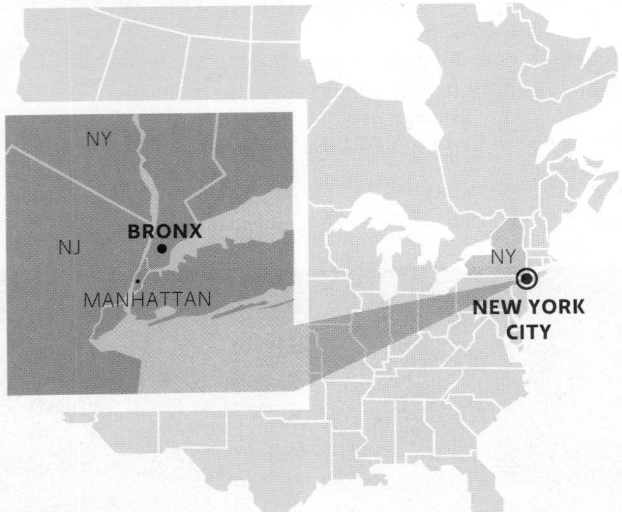
NY
NJ
BRONX
MANHATTAN
NY
NEW YORK CITY

The waterfront—all of it—had long been claimed by industry. There was simply nothing left to develop, she thought, as she and Xena made their way along their usual route—past the transfer station, roaring with diesel-powered, garbage-filled 18-wheelers, along a winding string of garages filled with auto glass shops, metal work, and produce shipments.

And then Xena began pulling her toward an abandoned lot, one they had passed a million times without bothering to notice. After a futile effort to resist, Carter allowed herself to be led down a garbage-strewn path, through a ramble of towering weeds. There at the end, sparkling in the early morning light, was the East River. Carter stood in awe. How many other forgotten patches of waterfront were there, she wondered? Maybe the river wasn't so inaccessible after all.

More people live in cities than ever before.

For the first time in human history, more than half the world's population lives in **urban areas**—densely populated regions that include both cities and the suburbs that invariably surround them. In Canada the proportion is even higher: 80% of Canadians are urban dwellers. **Urbanization**, the migration of people to large cities, is happening around the world at an unprecedented rate. As global population swells, rural lands are morphing into urban and suburban ones, and ordinary cities are growing into *megacities*—those with at least 10 million residents. North America has only 3 megacities—Mexico City and New York City (approximately 20 million each) and Los Angeles (approximately 13 million). Canada's largest city, Toronto, has fewer than 6 million inhabitants. [INFOGRAPHIC 26.1]

urban areas Densely populated regions that include cities and the suburbs that surround them.
urbanization The migration of people to large cities; sometimes also defined as the growth of urban areas.

Concrete Plant Park, a 3-hectare park on the site of a former concrete batch mix plant, is one of the milestones in the creation of the Bronx River Greenway, an environmental effort to transform the Bronx River. New York City officials and local activists re-established salt marshes on the riverbank once strewn with trash and tires and opened it up to recreation and water activities.

Infographic **26.1** | **URBANIZATION AND THE GROWTH OF MEGACITIES**

→ The world's population is becoming more urban. The growth of megacities, those with at least 10 million people, has increased dramatically over the last half century. In 1950, there were only two megacities: New York City with 12.3 million, and Tokyo, Japan, with 11.3 million. In 2011, there were 25 megacities, with Tokyo's population of more than 37 million making it the largest city in the world.

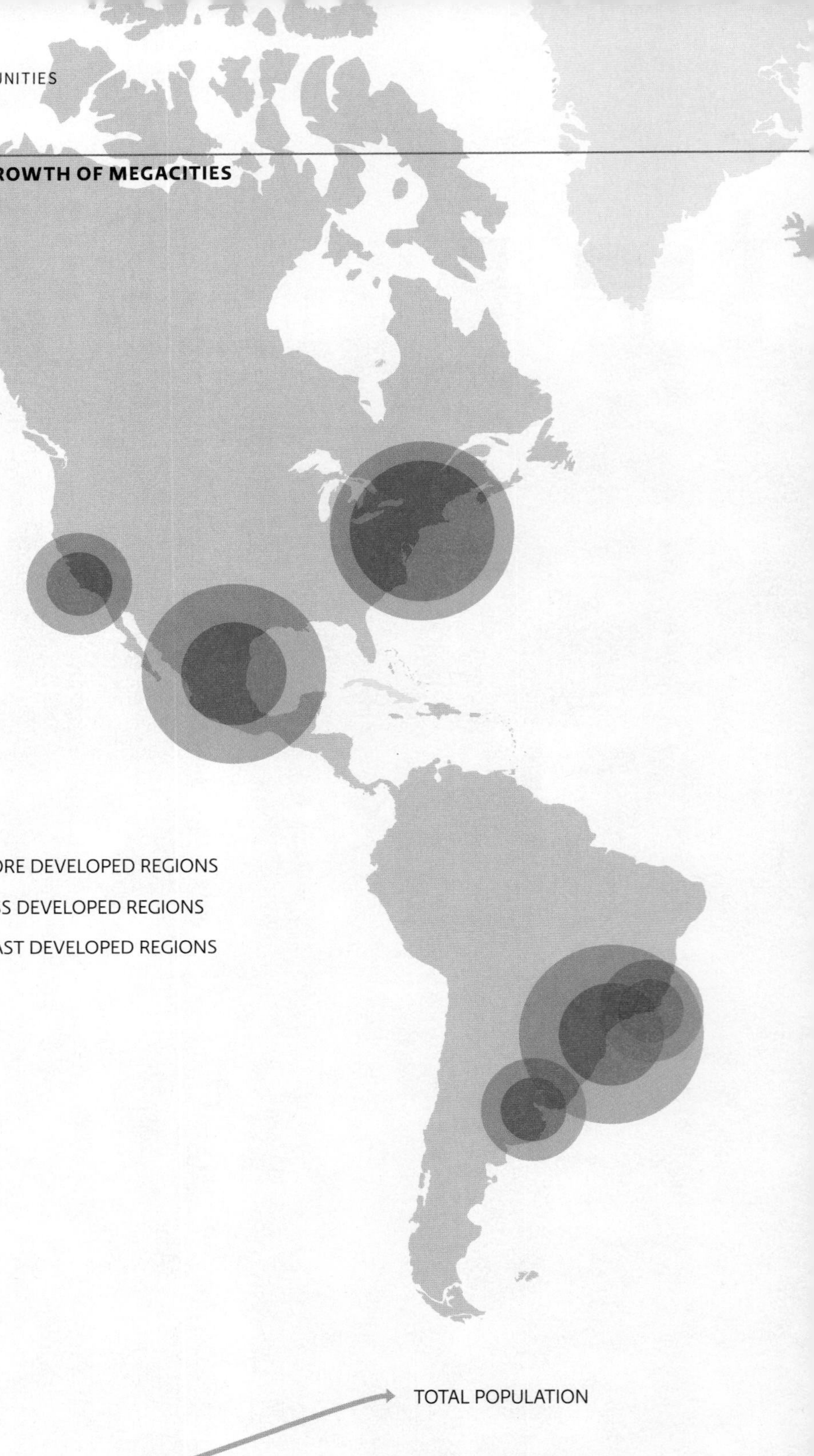

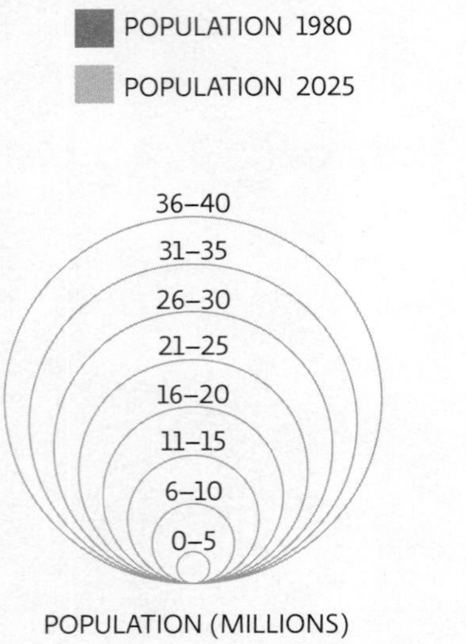

POPULATION 1980
POPULATION 2025
MORE DEVELOPED REGIONS
LESS DEVELOPED REGIONS
LEAST DEVELOPED REGIONS
36–40
31–35
26–30
21–25
16–20
11–15
6–10
0–5
POPULATION (MILLIONS)

→ World population is continuing to increase, though at a slower rate than in the 20th century.

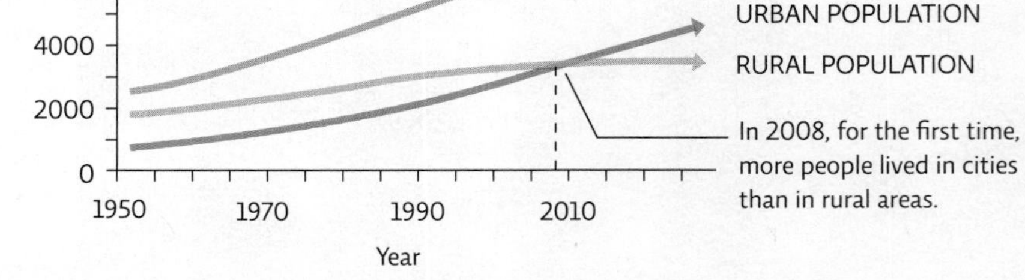

Millions
8000
6000
4000
2000
0
1950
1970
1990
2010
Year
TOTAL POPULATION
URBAN POPULATION
RURAL POPULATION
In 2008, for the first time, more people lived in cities than in rural areas.

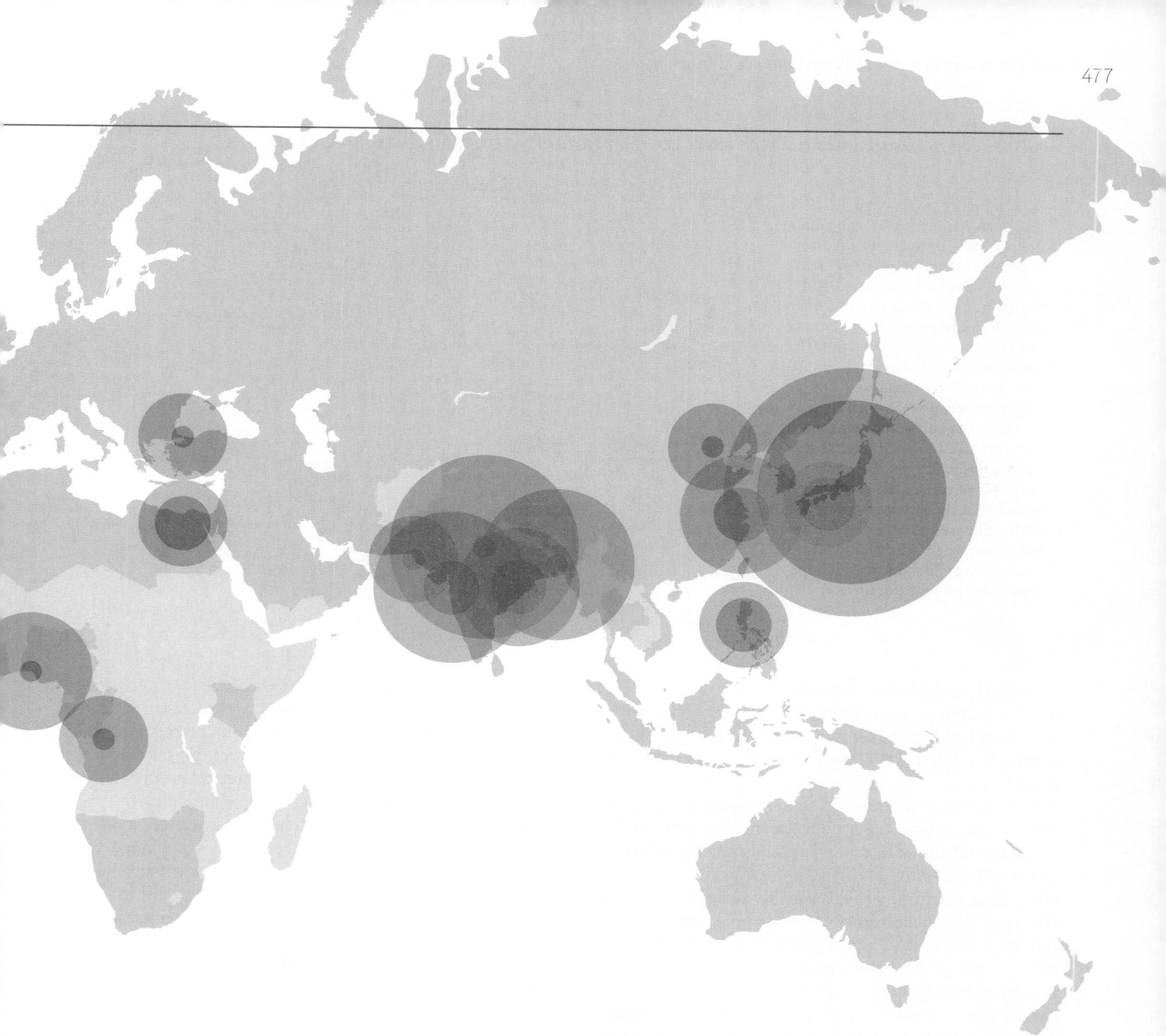

→ The United Nations predicts that by 2025, there will be 37 cities with populations over 10 million. Most of those cities will be in Asia; no Canadian city will be among them. Toronto, Canada's largest urban area, will have a projected population of almost 7 million people, ranking it as the 64th largest urban population in 2025.

POPULATION OF THE 10 METROPOLITAN AREAS PROJECTED TO HAVE THE LARGEST POPULATIONS IN 2025

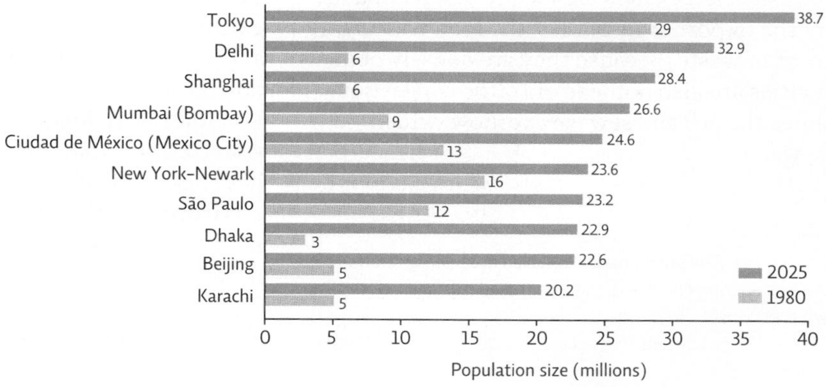
Tokyo 38.7 29
Delhi 32.9 6
Shanghai 28.4 6
Mumbai (Bombay) 26.6 9
Ciudad de México (Mexico City) 24.6 13
New York-Newark 23.6 16
São Paulo 23.2 12
Dhaka 22.9 3
Beijing 22.6 5
Karachi 20.2 5
2025
1980
0 5 10 15 20 25 30 35 40
Population size (millions)

Infographic **26.2** | **MANY URBAN AREAS HAVE LOWER PER CAPITA FOOTPRINTS THAN AVERAGE**

→ Due to higher population densities, less personal vehicle travel, smaller homes, and efficiencies of scale, people living in large urban areas typically have a lower ecological footprint than those in suburban areas (but not necessarily lower than rural areas). A 2009 study by geographer David Dodman found that, of the cities analyzed, the only two urban areas with higher footprints than the national average were in China—Shanghai and Beijing (only Shanghai is shown here). However, a look at the data shows that their footprints were not high compared to other large cities but rather, the national footprint in China is very low due to the large rural, low-income population.

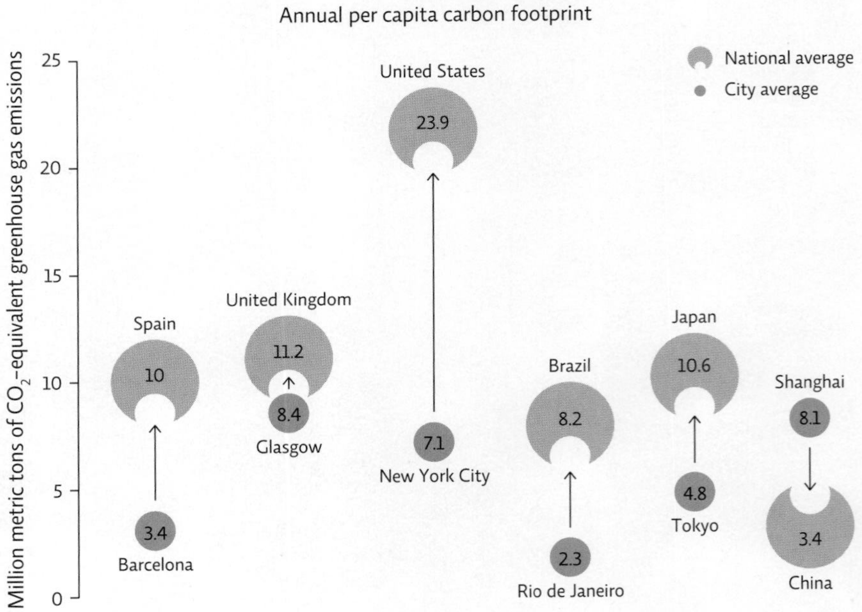
Annual per capita carbon footprint
National average
City average
Million metric tons of CO_2-equivalent greenhouse gas emissions
25
20
15
10
5
0
Spain
10
3.4
Barcelona
United Kingdom
11.2
8.4
Glasgow
United States
23.9
7.1
New York City
Brazil
8.2
2.3
Rio de Janeiro
Japan
10.6
4.8
Tokyo
Shanghai
8.1
3.4
China

To be sure, cities bring some obvious advantages to their inhabitants: more job opportunities, better access to education and health care, and more cultural amenities, to name a few. But as far as the environment is concerned, urbanization is both a blessing and a curse. On the plus side, concentrating people in smaller areas (building up rather than out) can reduce development of outlying agricultural land and wild spaces and thus protect existing farms and ecosystems. Higher population densities also make some environmentally friendly practices more cost effective. For example, it's easier to implement recycling and mass transit programs in cities because there are more people to share the cost of these services. Living in smaller homes that are closer to needed amenities and having access to mass transit also decreases the energy use—and carbon footprint—of urban dwellers, compared to those who live in suburban areas. [INFOGRAPHIC 26.2]

On the minus side, large cities are locally unsustainable: they require the import of resources like food and energy and the export of waste. Because they are densely populated, most cities are also hotbeds of traffic congestion (which pollutes the air) and sewage overflow (which pollutes the water).

urban heat island effect The phenomenon in which urban areas are warmer than the surrounding countryside due to pavement, dark surfaces, closed-in spaces, and high energy use.

environmental justice The concept that access to a clean, healthy environment is a basic human right.

Another problem stems from the way cities are designed and built—namely, the replacement of vegetation with pavement. Plants absorb water, filter air, and regulate area temperatures; pavement and concrete do not. In fact, the blacktop that covers most cities prevents rainwater from being absorbed into the ground, which in turn diminishes groundwater supplies and can lead to flooding (see Chapter 16). Cities also require an abundance of energy. This trifecta—too few plants, too much pavement, and high energy use—conspires to trap the solar heat that is absorbed and reflected by buildings, making most cities warmer than their surrounding countrysides. This phenomenon is known as the **urban heat island effect**. [INFOGRAPHIC 26.3]

All urban dwellers are vulnerable to the health effects associated with pollutants. But in most cities, the pros and cons of city living are unevenly realized. For example, Toronto is a wealthy, populous city, but most of the cultural amenities, top-notch health-care facilities, and job opportunities are concentrated in the central downtown area, while poorer communities such as South Riverdale were historically chosen as the sites of industrial facilities that pollute the area. This imbalance has spawned a whole new area of activism known as **environmental justice**, based on the idea that no community should be saddled with more environmental burdens and less environmental benefits than any other.

The movement is particularly relevant in the most impoverished cities in the world. Mumbai, the largest

Infographic **26.3** | **TRADE-OFFS OF URBANIZATION**

ADVANTAGES	DISADVANTAGES
Lower impact per person due to smaller homes and less travelling.	**Concentrated wastes** that have to be transported away
Higher energy efficiency in stacked housing than in freestanding buildings.	**Urban heat island effect** increases energy needs and can have health consequences.
More transportation options lessen the need for personal vehicles; closer proximity to destinations make walking and mass transit viable options.	**Dependence on food and resource inputs** from outside the city
Zoning ordinances are easier to implement.	**Disease and violence** may be higher in concentrated inner-city areas.
More job opportunities; local collaboration from a diverse community fosters innovation and ingenuity.	**Traffic congestion** and its associated air pollution
More services for citizens, including more educational and cultural opportunities and better health-care options.	**Less green space** leads to stormwater problems.

↓ A volunteer gardener at Finca Del Sur, a garden in the South Bronx, tends the corn stalks while a passenger train goes by in the background. The garden was created on an empty plot of land bordered by a highway exit ramp and a commuter train line.

Infographic **26.4** | **URBAN FLIGHT CONTRIBUTES TO SUBURBAN SPRAWL**

MONTRÉAL CMA (CENSUS METROPOLITAN AREA)

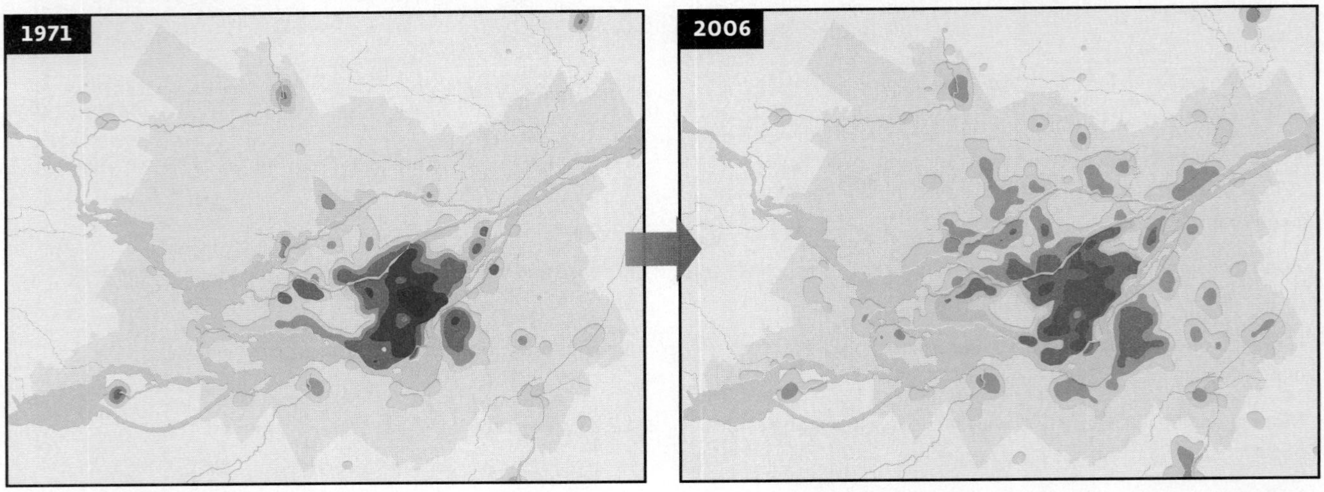
1971
2006

POPULATION DENSITY (PERSONS/KM²)

12 000 and over | 6000 to 11 999 | 3000 to 5999 | 1500 to 2999 | 400 to 1499

↑ Urban flight, the movement of people out of inner-city areas, is often driven by the decay of urban areas ("flight from blight") and the influx of lower-income groups who are often minorities.

city in India, has nearly 20 million people. Almost 7 million of these are slum dwellers who live in horrid, overcrowded conditions—as many as 45 000 people per hectare—without adequate sanitation or running water. One report showed only one toilet per 1440 residents. Globally, more than 1 billion of the world's people live in slums, mostly in large cities in developing countries.

With 90% of future population growth predicted to occur in large cities, urban planners are desperately searching for ways to create cities where the basic needs of residents are met and where environmental benefits outweigh environmental costs. The story of how the South Bronx waterfront was lost and then reclaimed provides important lessons about how to do this.

Suburban sprawl consumes open space and wastes resources.

In the late 1940s when Carter's father first bought the house Carter would grow up in, the South Bronx was a mostly European-descended white working class suburb of Manhattan. But, as more Hispanic and Black Americans moved to the area, seeking better opportunities, whites moved to nearby commuter towns. The process of people leaving a city centre for surrounding areas, originally made possible by the automobile and later by mass transit, is known as **urban flight**. While in many cities—especially those in developing nations—immigration exceeds emigration, urban flight today is triggered by a variety of forces, including overcrowding, noise and air pollution, the high cost of city living, and, in some cases, racial tensions. [INFOGRAPHIC 26.4]

No matter what the cause, urban flight results in **suburban sprawl**—a slow conversion of rural areas into suburban and exurban ones. **Exurbs** are more sparsely populated towns beyond the immediate suburbs whose residents also commute into the city for work.

As its name suggests, suburban sprawl tends to spread out over long corridors in an unplanned and often inefficient

urban flight The process of people leaving an inner-city area to live in surrounding areas.
suburban sprawl Low-population-density developments that are built outside of a city.
exurbs Towns beyond the immediate suburbs whose residents commute into the city for work.

Infographic **26.5** | **SUBURBAN SPRAWL**

↓ Suburban development often leads to suburban sprawl—low population density in developments that appear outside of a city. Homes typically get larger the farther they are from the city and residents have a larger ecological footprint (larger homes and more time spent driving). The suburbs now have their own suburbs—the exurbs, which are commuter towns that are beyond the traditional suburbs but whose residents still commute into the city, often an hour or more each way. Both suburbs and exurbs often displace farmland and wildlands.

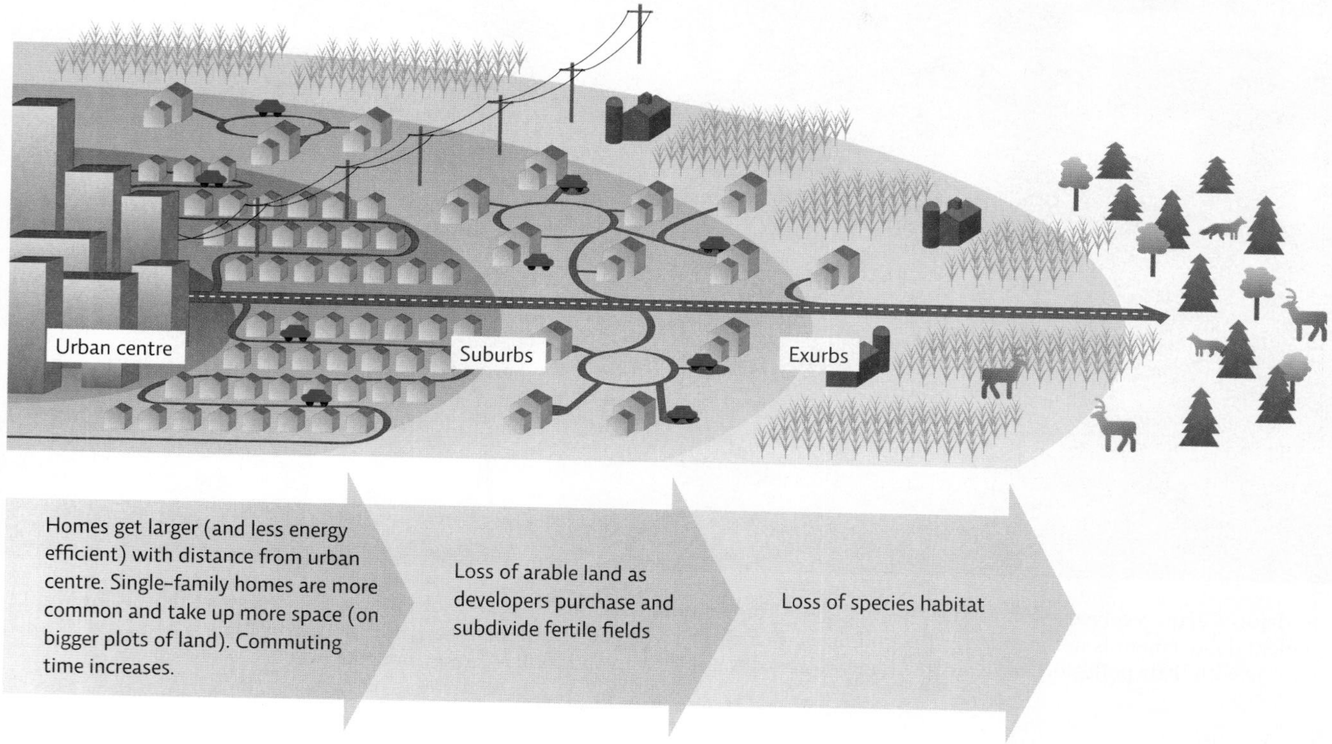
Urban centre
Suburbs
Exurbs
Homes get larger (and less energy efficient) with distance from urban centre. Single-family homes are more common and take up more space (on bigger plots of land). Commuting time increases.
Loss of arable land as developers purchase and subdivide fertile fields
Loss of species habitat

manner. By covering ever-greater swaths of terrain with concrete and pavement, sprawl reduces the amount of land available for farming, wildlife, and ecosystem services. Because of the haphazard way in which they develop, the resulting communities are heavily dependent on driving. Unlike cities, which are densely populated and can accommodate mass transit systems, suburbs and exurbs force residents to drive almost everywhere they need to go. And because suburban homes are typically larger than urban ones (and exurban homes are often even larger than suburban ones), they tend to have a greater ecological footprint. [INFOGRAPHIC 26.5]

Urban planners of today are well aware of these perils and often work to mitigate them. But in the 1960s, prolific U.S. urban planner Robert Moses was all too happy to accommodate urban flight and the sprawl that came with it. It was Moses who commissioned the Cross Bronx Expressway, a highway that enabled commuters from suburban Westchester County in the north to completely bypass the Bronx as they travelled in and out of Manhattan each day. But the expressway also displaced 600 000 Bronx residents and further segregated the ailing borough from the rest of New York City. "The South Bronx was utterly cut off," says Marta Rodriguez, a lifelong Bronx resident and a colleague of Carter's. "We didn't stand a chance."

To make matters worse, in the Bronx and elsewhere in the 1960s, urban flight in New York City led to redlining—the process whereby banks rule certain sections of a city off limits to any type of investment. Bronx landlords quickly discovered that if their neighbourhood was redlined, torching their buildings and collecting the insurance money would yield greater profits than renting or selling. By the 1970s, the burning of tenement houses had spiralled out of control, so much so that at one Yankees game, an addled sportscaster famously declared, "Ladies and gentlemen, the Bronx is burning." And as the shopping centres and apartment buildings were shuttered or burned, other industries took their place—namely, the garbage disposal operations and auto parts manufacturers that had been shunned by wealthier enclaves. Before long, the South Bronx had been transformed into an industrial wasteland.

↑ Majora Carter received a MacArthur "genius" grant for her work with Sustainable South Bronx. She stresses that the environmental movement is not just one of the middle-class majority who can afford to buy organic food, drive hybrid cars, and live in areas with little pollution. Low-income families also deserve a clean and healthy environment.

In most cities, including those in the United States and Canada, *zoning laws*—laws that restrict the type of development allowed in a given area—create a buffer between commercial and residential areas so that factories aren't wedged between houses. But in the Bronx, such laws were routinely ignored. Without other options, area residents were forced to accept factories and warehouses built on the ashes of apartment complexes. And because these new industrial neighbours preferentially hired commuters from outside the Bronx, unemployment rates skyrocketed—along with crime, poverty, and asthma.

It turned out that the patch of waterfront Carter and Xena had stumbled upon was a relic of the Moses-era highway expansion, sitting as it did beneath the Sheridan Expressway, a stretch of highway originally meant to cut across the entire northeast Bronx. The Sheridan was abandoned when planners realized it would run through the Bronx Zoo, a popular tourist destination. But by then the damage was done. Hunts Point—Carter's neighbourhood—had been isolated and the surrounding waterfront condemned to wasteland.

Carter knew that replacing the abandoned lot with a park would be a big first step toward righting some of the wrongs that her community had endured. The trees and plants would trap pollutants from the air, preventing them from infiltrating people's lungs. The grass and soil would absorb rainwater so that it could no longer carry trash and detritus from the streets into the river. And claiming even a small patch of waterfront for themselves would give Carter and her neighbours a sense of ownership, not to mention a connection to nature and a place to stretch their legs.

Indeed, studies by urban planning expert Reid Ewing and others have shown that parks improve both the physical and psychological health of people who live near them. And cities themselves—cities that, like the Bronx, were once plagued by drug trafficking and rampant gun violence—have shown that more green space can also mean less crime. In Bogotá, Colombia, in the late 1990s, for example, a particularly environmentally conscious mayor noticed that while his city was designed to accommodate heavy automobile traffic, the vast majority of his electorate did not drive. So he narrowed municipal thoroughfares from five lanes to three, expanded bike lanes and pedestrian walkways, and established a string of parks and public plazas throughout the city. The result? People stopped littering. Crimes rates dropped. And slowly but surely, city residents reclaimed their streets.

Bogotá was not so different from the South Bronx, Carter thought. If that city could go green on a third-world budget, surely she and her colleagues could raise enough money to do the same. Starting with the $10,000 seed grant from the city parks department, they leveraged a small fortune in additional grants, donations, and private investment, until they had finalized plans to build a $3 million park, complete with gardens, grassy knolls, and East River kayaking. Hunts Point Riverside Park—the spot that Carter stumbled upon—would be the borough's first waterfront park in more than 60 years. But that was just the beginning.

Carter knew that replacing the abandoned lot with a park would be a big first step toward righting some of the wrongs that her community had endured.

Environmental justice requires engaged citizens.

Energized by their successful riverside park project, Carter and her neighbours formed a nonprofit called the Sustainable South Bronx (SSBx). The group immediately set its sights on an even grander vision: they would create a greenbelt around the entire community—2.4 kilometres of waterfront greenway, 5 hectares of new waterfront open space, and 14 kilometres of green streets (those with landscaped medians)—all connected by an interlinking system of bike and pedestrian pathways that stretched from the Hunts Point Riverside Park, around the South Bronx's winding edges, all the way to the existing 162 hectare park on Randall's Island. They would also disassemble the Sheridan Expressway and turn it into over 11 hectares of additional parkland, some of which they would designate as *conservation easements*—tracts of land that the city would sign a legally binding agreement not to develop.

It was an ambitious agenda indeed—an expensive one, too—and would require the support of administrators and elected officials from the Bronx to Manhattan to the state capital in Albany. "There is a big fear that environmental justice is fiscally irresponsible," says Carter. "People running the city think 'how can we spend money on parks when we're coming up short on schools, and clinics, and job training, and health care?' What they don't realize is that parks can actually help with those things, too." Parks not only increase community pride, but also create green jobs and improve health.

Convincing community members of these benefits would prove as difficult as convincing legislators. Getting them to come out and oppose a landfill was one thing; area residents knew all too well what another trash heap would do to their neighbourhood. But getting them to support a park? They had more pressing concerns. Theirs was the poorest congressional district in the city; at that moment, more than 20% of residents were unemployed. And their neighbourhood hadn't had a waterfront park in more than 60 years, let alone an entire greenway. Why bother now? "We'd ask people, 'What would you like to see in your neighbourhood?' and they really didn't have an answer," Rodriguez says. "They'd never been asked that question before. It was as if having parks was too far in the future for them." In fact, the SSBx vision folded readily into a

↑ Manitoba Hydro Place in Winnipeg is one of the world's most environmentally friendly office buildings. Its tall solar chimney improves ventilation by enhancing the movement of air through the facility.

growing movement aimed at making cities more environmentally friendly and socially equitable. **New Urbanism**, as the movement is called, maintains that cities (both now and in the future) have the capacity to reduce our per-person ecological footprint, even as they improve the quality of life for people, provided they are designed properly. New examples of this type of community are already cropping up—one of the biggest in North America is McKenzie Towne in southeast Calgary, designed to provide plenty of green spaces, public transportation, and a variety of businesses within walking distance that limit the need for a car.

On top of the carbon savings, the consensus emerging from environmental scientists, sociologists, and economists is that the future lies in cities—that's where most people will live and perhaps is where most people should live. Cities promote interaction among a diverse group of people. This promotes the exchange of ideas and lessens cultural and economic barriers.

The future depends on making large cities sustainable.

Sustainable cities are those where the environmental pros outweigh the cons—where sprawl is minimized, walkability is maximized, and the needs of inhabitants are met locally. In recent years, urban planners have come up with a wide range of strategies for accomplishing these goals. To achieve self-sufficiency, for example, a sustainable city might maintain a mixture of open and agricultural land along its outskirts. Such land could provide a large part of the local food, fibre, and fuel crops, along with recreational opportunities and ecological services. Waste and recycling facilities could also be located nearby, along with other enterprises aimed at producing resources needed by area residents. To stave off sprawl, the same city might establish *urban growth boundaries*—outer city limits beyond which major development would be prohibited. Keeping any outward growth that does occur as close to mass transit as possible minimizes the impacts of transportation, just as building "up" (a parking garage) rather than "out" (an expansive parking lot), minimizes the amount of land used. To encourage more walking and less driving, zoning laws might allow for mixed land uses, where residential areas are located reasonably close to commercial and light industrial ones.

Of course, building an ideal city from scratch is easy compared with the task of overhauling an existing city, especially when that city is as densely populated and ever-expanding as New York or Toronto. Upgrading decaying infrastructure like roads, public places, and sewage and water lines can be more expensive than new construction and the process is disruptive to residents. Urban retrofits are certainly possible but sometimes their very success raises property values to the point that the original residents can no longer afford to live in their own neighbourhoods. Even so, there are plenty of ways that cities can push themselves into the environmental plus column. For example, **infill development**—the development of empty lots within a city—can significantly reduce suburban sprawl. And even the most car-friendly of cities has a range of options for reducing traffic congestion and the air pollution that comes with it: reliable public transportation, car sharing programs that allow residents to use cars when needed for a monthly fee, and sidewalks and overhead passageways that allow pedestrians to safely cross busy roads. The cumulative effect of strategies like these, which help create walkable communities with lower ecological footprints, is known as **smart growth**. [INFOGRAPHIC 26.6]

Persuading people to support smart growth, as Carter and her colleagues soon discovered, was a matter of showing them that the benefits could be economic as well as environmental. "You need to show them what we call the triple bottom line," says James Chase, vice-president of SSBx (and Carter's husband), referring to the economic, social, and environmental impacts of any decision (see Chapters 1 and 5). "Developers, government, and residents all need some tangible, positive return." A major park project would surely be a boon for all three. Developers would be guaranteed millions in waterfront development contracts. Residents could look forward to cleaner air and water, a prettier neighbourhood, and better health as a result. The government would save a bundle in health-care costs. The greenbelt would also spur the local economy—such a vast stretch of public space would attract street vendors, food stands, bicycle shops, and sporting goods stores.

It would also require a green workforce. Some of the undeveloped property that Carter and her neighbours hoped to convert into parkland was contaminated with hazardous waste. These sites are called *brownfields*, and they require a special type of cleanup, or remediation, before they can be developed.

The surrounding wetlands, suffering from decades of neglect, would also need to be restored. And maintaining the new trees, plants, and parks they hoped to create

---

New Urbanism A movement that promotes the creation of compact, mixed-use communities with all of the amenities for day-to-day living close by and accessible.

infill development The development of empty lots within a city.

smart growth Strategies that help create walkable communities with lower ecological footprints.

Infographic **26.6** | **SUSTAINABLE CITIES AND SMART GROWTH**

↓ Smart growth can be applied to large cities or to smaller communities. It employs strategies that make efficient use of land to create pleasant livable communities with a lower ecological footprint than current suburban areas.

Take advantage of compact building design and incorporate environmentally friendly technologies.
Create a range of housing opportunities and choices.
Renovate and develop existing communities (rather than building outside the city).
Foster distinctive, attractive communities with a strong sense of place.
Al's Diner
Encourage community and stakeholder collaboration; make development decisions that are fair and cost effective.
Mix land uses to place residential and commercial areas together.
COMMUNITY MEETING
All Welcome!
Provide urban green space; preserve farmland and critical environmental areas.
Create walkable and bike-friendly neighbourhoods.
Provide a variety of "clean" transportation choices into and around the city.
050
NATURAL GAS BUS
1127
CNG

Infographic **26.7** | **GREEN BUILDING**

↓ There are many steps that can be taken to build or retrofit a building so that it has less environmental impact and is a healthier environment for those who live, work, or go to school there. The nonprofit group, Canada Green Building Council, certifies buildings through its LEED program (Leadership in Energy and Environmental Design). Buildings receive a standard, silver, gold, or platinum rating, based on a variety of criteria that include energy efficiency, sustainable building material use, and innovative design.

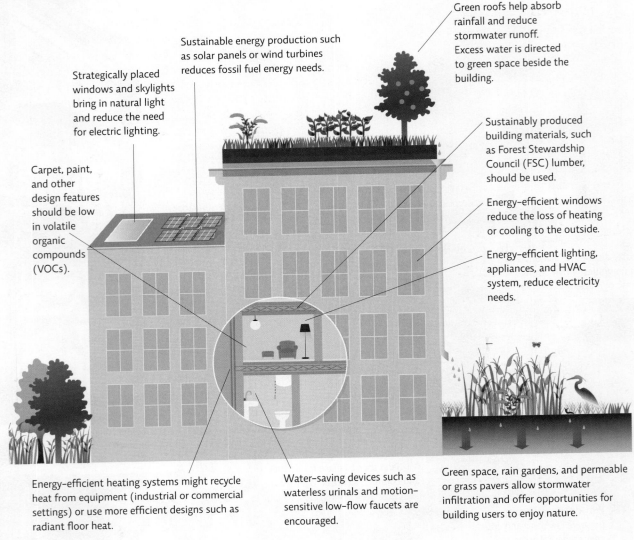

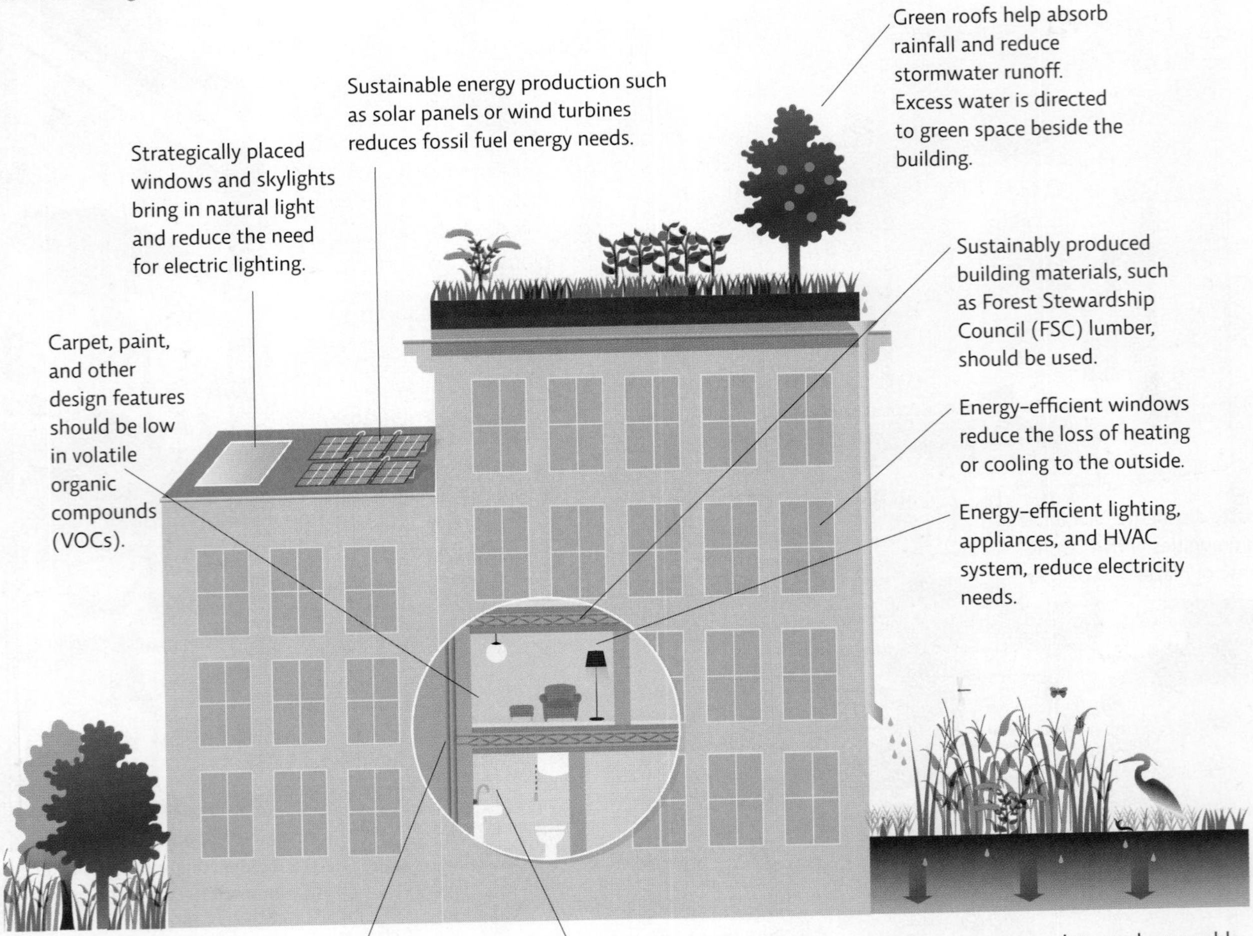
Green roofs help absorb rainfall and reduce stormwater runoff. Excess water is directed to green space beside the building.
Sustainable energy production such as solar panels or wind turbines reduces fossil fuel energy needs.
Strategically placed windows and skylights bring in natural light and reduce the need for electric lighting.
Carpet, paint, and other design features should be low in volatile organic compounds (VOCs).
Sustainably produced building materials, such as Forest Stewardship Council (FSC) lumber, should be used.
Energy-efficient windows reduce the loss of heating or cooling to the outside.
Energy-efficient lighting, appliances, and HVAC system, reduce electricity needs.
Energy-efficient heating systems might recycle heat from equipment (industrial or commercial settings) or use more efficient designs such as radiant floor heat.
Water-saving devices such as waterless urinals and motion-sensitive low-flow faucets are encouraged.
Green space, rain gardens, and permeable or grass pavers allow stormwater infiltration and offer opportunities for building users to enjoy nature.

would require a workforce trained in urban forestry. Anxious to claim these emerging professions—all of which promised job security and living wages—for their own community, Carter and her neighbours launched Bronx Environmental Stewardship Training—a green job training program that teaches South Bronx residents the principles of urban forestry, brownfield remediation, and wetland restoration. Program participants—many of them ex-convicts and high school dropouts facing prison time—also learn how to install solar panels and retrofit older buildings to make them energy efficient. So far, 82% of the participants have found jobs in the green economy and 15% have gone on to college.

The green economy includes a movement known as **green building**, that is, the construction of buildings that are better for the environment and the health of those

---

green building Construction and operational designs that promote resource and energy efficiency, and provide a better environment for occupants.

LEED (Leadership in Energy and Environmental Design) A third-party certification program administered by the nonprofit Green Building Council that awards a rating (standard, silver, gold, or platinum) to buildings that include environmentally sound design features.

who use them. The best green buildings are awarded a **LEED (Leadership in Energy and Environmental Design)** certification, which is internationally recognized. The Bronx Library Center is a silver-certified LEED building. It earned the silver certification by recycling 90% of the waste materials created during the construction of the building, using architectural design and efficient heating and cooling systems to save 20% of energy costs, and using sustainably grown wood in 80% of the construction lumber.

Carter and her team also launched Smart Roofs, a green-roof and green-wall installation company. Green roofs are one type of rain garden—an area seeded with plants suited to local temperature and rainfall conditions (see Chapter 16 for more on rain gardens). A 2004 study by the New York City Department of Design and Construction found that consumers could save more than $5 million in annual cooling costs if green roofs were installed on just 5% of the city's buildings. According to a study by Columbia University, the same amount of green roofing could achieve an annual reduction of 318 000 metric tons of greenhouse gases. And Riverkeep, an environmental nonprofit, found that green roofs can retain over 3 cubic metres of stormwater for every $1,000 of investment—easing pressure on the city's overburdened sewer systems and mitigating water pollution from storm runoff. Green roofs also lessen the urban heat island effect. [INFOGRAPHIC 26.7]

Green roofs require factory-made artificial soil (ordinary topsoil is too heavy and can clog drainage systems). And to prevent leaks, a specially designed membrane system must be installed. Producing the artificial soil, creating and installing the membranes, and planting the right mix of species to trap water requires labour—skilled labour. "Once we figured out the employment factor, we had a win-win-win," says Chase. "There's all these jobs—good jobs—that are going to take off in the next decade, but that not a lot of people know how to do right now. It was a clear opportunity for us." To help the company take off, SSBx secured tax credits from the state legislature. Building owners who install green roofs on at least 50% of their available rooftop now receive a 1-year property tax credit of up to $100,000. Carter offered up her own roof as the first test case.

By the time the Hunts Point Riverside Park opened, dozens of cities across North America—from Vancouver to Miami—had taken up the mantle of sustainability and smart growth.◉

Select references in this chapter:
Beckett, K., and Godoy, A. 2010. *Urban Studies*, 47: 277–301.
Dodman, D. 2009. *Environment and Urbanization*, 21: 185–201.
Ewing, R., *et al.* 2003. *American Journal of Health Promotion*, 18: 47–57.
United Nations Population Division. 2008. *An Overview Of Urbanization, Internal Migration, Population Distribution And Development In The World.* New York: United Nations.

BRING IT HOME

PERSONAL CHOICES THAT HELP

A sustainable community is one that promotes economic and environmental health and social equity. It is one in which the health and well-being of all citizens are considered, while those citizens help implement and maintain the community.

Individual Steps

→ Research products before you purchase them to understand the impact of your consumption choices (www.goodguide.com). Investigate and support sustainable businesses in your area (www.greendirection.ca).

→ If you have a balcony or yard, plant flowers, vegetables, or trees.

→ Support local businesses by shopping and dining close to home.

Group Action

→ Join neighbourhood cleanup days. If you can't find one, organize one.

→ Start a petition to get more bike lanes in your city. Ride public transit more often.

→ Find out how colleges and universities are working toward sustainable practices at www.AASHE.org.

Policy Change

→ Attend a meeting of your city or regional council and ask members to look into smart growth opportunities.

→ See how well you can plan for a sustainable community. Play the new PC strategy game Fate of the World and see how policies you put in place impact global climate change, rainforest preservation, and resource use.

UNDERSTANDING THE ISSUE

CHECK YOUR UNDERSTANDING

1. In 2025, where will most of the world's megacities be found?
a. North America
b. Europe
c. Africa
d. Asia

2. In response to redlining in the 1960s, landlords in the Bronx:
a. increased rents, forcing working-class people to move out of the borough.
b. subdivided apartments into smaller units to increase the number of renters.
c. burned their buildings to collect insurance money.
d. constructed walkways for tenants to access the nearest subway.

3. A city that promotes smart growth:
a. encourages development at the city edges.
b. provides tax incentives for people who own more than one car.
c. allows vacant lots to accumulate in the city for a more open look.
d. mixes land uses to place residential and commercial areas together.

4. Which of the following is an example of environmental injustice?
a. Locating industries away from where people live
b. Building garbage dumps in high-poverty, low-income areas
c. People chaining themselves to trees to prevent the trees from being cut down
d. Preventing the construction of a dam to save an endangered species of fish

5. What unexpected impact did Bogotá experience as a result of building more green space?
a. Decline of mass transit use
b. Lower crime rates
c. Lower property values
d. More athletic injuries

6. Which of the following describes a brownfield?
a. An area that requires cleanup of pollution and hazardous waste before it can be developed
b. A field of dormant grasses in the fall
c. An area where treated sewage is spread to dry and create compost
d. In a city, a vacant lot where buildings once stood

WORK WITH IDEAS

1. Explain the triple bottom line, using the Bronx waterfront restoration as an example.

2. Describe a green roof system. What are the costs and benefits? Why might green roofs be especially beneficial in cities?

3. You are taking a road trip with friends, driving from Montreal to Calgary on the TransCanada Highway. Along the way, your friends begin to notice that the temperatures are more bearable driving through the country than the city. Explain why this happens.

4. Describe a LEED-certified building. What criteria are used to evaluate a building for LEED certification?

ANALYZING THE SCIENCE

MEASURED VERSUS EXPECTED EUIS IN LEED-CERTIFIED BUILDINGS

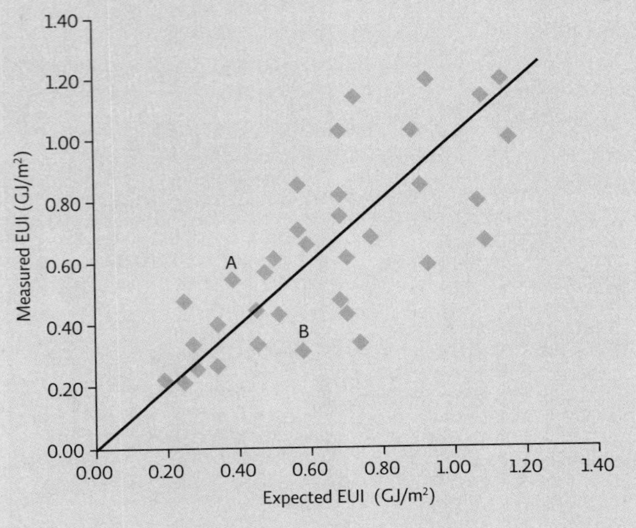

1.40
1.20
1.00
0.80
0.60
0.40
0.20
0.00
0.00
0.20
0.40
0.60
0.80
1.00
1.20
1.40
Measured EUI (GJ/m²)
Expected EUI (GJ/m²)
A
B

One common measurement of how well a building performs with regard to energy usage is its Energy Use Intensity (EUI). EUI is expressed as energy used (in gigajoules) relative to a building's size so that structures of different sizes can be compared. This graph plots data for 36 LEED-certified office buildings, comparing their actual performances to expected performances based on their designs. (Each data point represents a different building's EUI.) The line shown is not a trend line for the actual data; it is a line that bisects the graph at a 45° angle, allowing us to see how closely the expected values match the observed values.

INTERPRETATION

1. Look at the points labelled A and B. For each, what was the expected EUI predicted by the design plan? What was the actual measured EUI for each of these buildings?

2. Which of these two buildings, A or B, is performing better than expected?

3. Give the coordinate points for the one building that performed exactly as expected.
4. What does the line in the scatter plot tell you about those buildings above or below it?

ADVANCE YOUR THINKING

5. About how well do LEED-certified buildings meet or exceed their predicted EUI?
6. The average EUI for non-LEED office buildings is 2.19 GJ/m^2. How do LEED-certified buildings, even those that do not perform as well as expected, compare to this average?

EVALUATING NEW INFORMATION

More and more people around the world live in cities, and the number of megacities is increasing. There are costs and benefits to living in cities, and in an effort to increase the benefits, there is a growing movement toward "greening" cities. Environmental and social scientists have published a number of studies documenting the effects of greener cities on environmental and human health.

Go to the website Smart Growth Canada Network (www.smartgrowth.ca). Read the introduction to the site.

Evaluate the website and work with the information to answer the following questions:

1. Is this a reliable information source? Does it have a clear and transparent agenda?
   a. Who runs this website? Do the organization's credentials make it reliable/unreliable? Explain.
   b. Who are the authors? What are their credentials? Do they have the scientific background and expertise to lend credibility to the website?

On the menu bar, click on the "Useful Research" link. Then, choose one of the reports to examine.

2. What kind of information is provided in the report? Where is the information drawn from?
3. What are the major conclusions made in the report? Do you agree with them and how do these conclusions confirm or contrast with what is happening in your city?
4. Having read this report, do you feel that Smart Growth is the best model to use for urban development? Why or why not?
5. Do some online research to find out what critics of Smart Growth are saying about its effectiveness. What are the major criticisms of Smart Growth? Who are the main critics, and do you feel that they have the credentials to make these criticisms? Do you agree with these criticisms? Why or why not?

MAKING CONNECTIONS

BUILDING GREEN

Background: Poor air quality, excessive runoff, and lack of green space are all major problems in urban areas. One solution that helps address all three issues is green roofs.

Case: You are a resident of Toronto, one of the most urbanized areas of Canada. Smog from both urban and industrial sources impacts air quality, and residents have asked the city to investigate whether Toronto would benefit from green roof technology. As a member of the consulting firm that the city has hired to assess this possibility, your job is to investigate the pros and cons of green roofs in Toronto and to make recommendations for how and where to install them to have the best outcomes. Your three-part report must also address the triple bottom line.

1. In Part 1, summarize the history of green roofs in Toronto, briefly outlining:
   a. the current area covered by green roofs in Toronto (known as the *areal extent*).
   b. where these green roofs are located and who has installed them.
   c. the cost of installing these green roofs.
2. In Part 2, summarize the potential for green roofs in Toronto, noting:
   a. the different green roof types that are available, and which is best in this area.
   b. the potential areal extent that could be converted to green roofs.
   c. target companies and organizations that could be approached to implement green roof technology.
   d. the cost of installing a typical green roof.
3. In Part 3, present a triple bottom line analysis of green roof technology in Toronto, making sure to include all the pros and cons for each of the three areas.

GLOSSARY

A

abiotic The nonliving components of an ecosystem, such as rainfall and mineral composition of the soil. (Chapter 6)

acid deposition Precipitation that contains sulphuric or nitric acid; dry particles may also fall and become acidified once they mix with water. (Chapter 21)

acidification The lowering of the pH of a solution. (Chapter 13)

acid mine drainage Water flowing past exposed rock in mines, leaching out sulphates. These sulphates react with the water and oxygen to form acids (low-pH solutions). (Chapter 19)

active solar technologies The use of mechanical equipment to capture, convert, and sometimes concentrate solar energy into a more usable form. (Chapter 24)

adaptation A trait that helps an individual survive or reproduce (Chapter 10); efforts intended to help us deal with a problem that exists, such as climate change. (Chapter 22)

adaptive management Plan that allows room for altering strategies as new information comes in or the situation itself changes. (Chapter 2)

additive effects When exposure to two or more chemicals has an effect equivalent to the sum of their individual effects. (Chapter 3)

age structure The part of a population pyramid that shows what percentage of the population is distributed into various age groups of males and females. (Chapter 4)

age structure diagram A graphic that displays the size of various age groups, with males shown on one side of the graphic and females on the other. (Chapter 4)

air pollution Any material added to the atmosphere (naturally or by humans) that harms living organisms, affects the climate, or impacts structures. (Chapter 21)

albedo The ability of a surface to reflect away solar radiation. (Chapter 22)

annual Plants that live for a year, produce seed, and then die. (Chapter 25)

antagonistic effects When exposure to two or more chemicals has a smaller effect than the sum of their individual effects normally would. (Chapter 3)

anthropocentric worldview A human-centred view that assigns intrinsic value only to humans. (Chapter 1)

anthropogenic Caused by or related to human action. (Chapters 1 and 22)

anthropogenic climate change Alterations to climate resulting from human impact. (Chapter 10)

applied science Research whose findings are used to help solve practical problems. (Chapter 1)

aquaculture Fish-farming; the rearing of aquatic species in tanks, ponds, or ocean net pens. (Chapter 14)

aquifer An underground, permeable region of soil or rock that is saturated with water. (Chapter 15)

artificial selection When humans decide which individuals breed and which do not in an attempt to produce a population with desired traits. (Chapter 10)

asthma A chronic inflammatory respiratory disorder characterized by "attacks" during which the airways narrow, making it hard to breathe; can be fatal. (Chapter 21)

atmosphere Blanket of gases that surrounds Earth and other planets. (Chapter 2)

B

background rate of extinction The average rate of extinction that occurred before the appearance of humans or occurs outside of mass extinction events. (Chapter 10)

benthic macroinvertebrates Easy-to-see (not microscopic) arthropods such as insects and crayfish that live on the stream bottom. (Chapter 16)

bioaccumulation The buildup of substances in the tissues of an organism over the course of its lifetime. (Chapter 3)

biocapacity The ability of land or aquatic ecosystems to produce resources and assimilate our waste. (Chapter 5)

biocentric worldview A life-centred approach that views all life as having intrinsic value, regardless of its usefulness to humans. (Chapter 1)

biodegradable Capable of being broken down by living organisms. (Chapter 17)

biodiesel A liquid fuel made from vegetable oil, animal fats, or waste oil that can be used directly in a diesel internal combustion engine. (Chapter 25)

biodiversity The variety of life on Earth; it includes species, genetic, and ecological diversity. (Chapters 1 and 9)

biodiversity hotspot An area that contains a large number of endemic but threatened species. (Chapter 9)

bioethanol An alcohol fuel made from crops like corn and sugarcane in a process of fermentation and distillation. (Chapter 25)

biofuels Solids, liquids, or gases that produce energy from biological material. (Chapter 25)

biological assessment Sampling an area to see what lives there as a tool to determine how healthy the area is. (Chapter 16)

biomagnification The increased levels of substances in the tissue of predatory (higher trophic level) animals that have consumed organisms that contain bioaccumulated toxic substances. (Chapter 3)

biomass The sum of all organic material—plant and animal matter—that makes up an ecosystem, including material from living or recently living organisms or their by-products. (Chapters 6 and 25)

biome One of many distinctive types of ecosystems determined by climate and identified by the predominant vegetation and organisms that have adapted to live there. (Chapter 6)

biomimicry Using nature as a model to inspire sustainable solutions to environmental problems. (Chapter 5)

biosphere The sum total of all of Earth's ecosystems. (Chapter 6)

biotic The living (organic) components of an ecosystem, such as the plants and animals and their waste (dead leaves, feces). (Chapter 6)

biotic potential (*r*) The maximum rate at which the population can grow due to births if each member of the population survives and reproduces. (Chapter 7)

birth rate The number of births per 1000 individuals per year. (Chapter 7)

bitumen A thick, sticky oil that may be found in sand or clay deposits. (Chapter 20)

boom-and-bust cycles Fluctuations in population size that produce a very large population followed by a crash that lowers the population size drastically, followed again by an increase to a large size and a subsequent crash. (Chapter 7)

boreal forests Coniferous forests found at high latitudes and altitudes characterized by low temperatures and low annual precipitation. (Chapter 11)

bottleneck effect When population size is drastically reduced, leading to the loss of some genetic variants, and resulting in a less diverse population. (Chapter 10)

bycatch Non-target species that become trapped in fishing nets and are usually discarded. Some methods, like trawling, have very high bycatch levels, and discards often exceed the actual target species catch. (Chapter 14)

C

Canadian Environmental Protection Act (CEPA) A broad act passed in 1999 to set standards aimed at preventing pollution and protecting the environment and human health. (Chapter 21)

canopy Upper layer of a forest formed where the crowns (tops) of the majority of the tallest trees meet. (Chapter 11)

cap-and-trade Regulations that set upper limits for pollution release. Producers are issued permits that allow them to release a portion of that amount; if they release less, they can sell their remaining allotment to others who did not reduce their emissions enough. (Chapter 21)

carbon capture and sequestration (CCS) Removing carbon from fuel combustion emissions or other sources and storing it to prevent its release into the atmosphere. (Chapter 19)

carbon cycle Movement of carbon through biotic and abiotic parts of an ecosystem. Carbon cycles via photosynthesis and cellular respiration as well as in and out of other reservoirs such as the oceans, soil, rock, and atmosphere. It is also released by human actions such as fossil fuel burning. (Chapter 6)

carbon debt The cumulative reduction of the carbon dioxide absorption or storage capacity of ecosystems, e.g., by converting land from a natural ecosystem to agricultural uses. (Chapter 25)

carbon sequestration The storage of carbon in a form that does not readily release the carbon to the atmosphere or water. (Chapter 25)

carbon sinks Places such as forests, ocean sediments, and soil, where accumulated carbon does not readily re-enter the carbon cycle. (Chapter 11)

carbon taxes Governmental fees imposed on activities that release CO_2 into the atmosphere, usually on fossil fuel use. (Chapter 22)

carrying capacity (*K*) The population size that a particular environment can support indefinitely without long-term damage to the environment (Chapters 1 and 7); the population size that an area can support for the long term; it depends on resource availability and the rate of per capita resource use by the population. (Chapter 4)

cause-and-effect relationship An association between two variables that identifies one (the effect) occurring as a result of or in response to the other (the cause). (Chapter 2)

cellular respiration The process in which all organisms break down sugar to release its energy, using oxygen and giving off CO_2 as a waste product. (Chapter 6)

cellulosic ethanol Bioethanol made by breaking down cellulose in plants; a difficult process that has yet to be scaled up to meet large production goals. (Chapter 25)

Chemicals Management Plan Joint Health and Environment Canada program to assess and research chemical substances and improve product safety. (Chapter 3)

Clean Air Act First passed in 1970, this act now falls under the broader Canadian Environmental Protection Act, which sets standards for dangerous air pollutants. The more recent, and different, Clean Air Act of 2006 additionally addresses smog and greenhouse gases. (Chapter 21)

climate Long-term patterns or trends of meteorological conditions. (Chapter 22)

climate change Alteration in the long-term patterns and statistical averages of meteorological events. (Chapter 22)

climax community The end stage of ecological succession in which the conditions created by the climax species are suitable for the plants that created them so they can persist as long as their environment remains unchanged. (Chapter 8)

climax species Species that move into an area at later stages of ecological succession. (Chapter 8)

closed-loop system A production system in which the product is returned to the resource stream when consumers are finished with it, or is disposed of in such a way that nature can decompose it. (Chapter 5)

clumped distribution Individuals are found in groups or patches within the habitat. (Chapter 7)

coal Fossil fuel formed when plant material is buried in oxygen-poor conditions and subjected to high heat and pressure over a long time. (Chapter 19)

coevolution Two species each provide the selective pressure that determines which traits are favoured by natural selection in the other. (Chapter 10)

coliform bacteria Bacteria often found in the intestinal tract of animals; monitored for fecal contamination of water. (Chapter 15)

collapsed fishery A fishery in which annual catches fall below 10% of their historic high; stocks can no longer support a fishery. (Chapter 14)

command and control Regulations that set an upper allowable limit of pollution release which is enforced with fines and/or incarceration. (Chapter 21)

commensalism A symbiotic relationship between individuals of two species in which one benefits from the presence of the other but the other is unaffected. (Chapter 8)

community ecology The study of all the populations (plants, animals, and other species) living and interacting in an area. (Chapter 8)

competition Species interaction in which individuals are vying for limited resources. (Chapter 8)

composting Providing good conditions for the decomposition of biodegradable waste, producing a soil-like mulch. (Chapter 17)

concentrated animal feeding operation (CAFO) Meat or dairy animals being reared in confined spaces, maximizing the number of animals that can be grown in a small area. (Chapter 18)

condensation Conversion of water from a gaseous state (water vapour) to a liquid state. (Chapter 15)

conservation Efforts that reduce waste and increase efficient use of resources. (Chapter 24)

conservation reserve program Farmers and ranchers are paid to keep damaged land out of production to promote recovery. (Chapter 12)

consumer An organism that eats other organisms to gain energy and nutrients; includes animals, fungi, and most bacteria. (Chapters 6 and 8)

control group The group in an experimental study that the test group's results are compared to; ideally, the control group will differ from the test group in only one way. (Chapter 2)

control rods Rods that absorb neutrons and slow the fission chain reaction. (Chapter 23)

conventional oil reserves Liquid fossil fuel deposits that contain freely flowing oil that can be pumped out. (Chapter 20)

coral reef Large underwater structures formed by colonies of tiny animals (coral) that produce calcium carbonate exoskeletons that over time build up; found in shallow, warm, tropical seas. (Chapter 13)

core species Species that prefer core areas of a habitat—areas deep within the habitat, away from the edge. (Chapters 8 and 10)

correlation Two things occur together—but it doesn't necessarily mean that one caused the other. (Chapter 2)

cradle-to-cradle Management of a resource that considers the impact of its use at every stage from raw material extraction to final disposal or recycling. (Chapter 5)

critical thinking Skills that enable individuals to logically assess information, reflect on that information, and reach their own conclusions. (Chapter 3)

crude birth rate The number of offspring per 1000 individuals per year. (Chapter 4)

crude death rate The number of deaths per 1000 individuals per year. (Chapter 4)

crude oil Liquid fossil fuel that can be extracted from underground deposits by pumping. It is processed into fuels and other products. (Chapter 20)

cultural eutrophication Eutrophication specifically caused by human activities (most eutrophication is caused by humans). (Chapter 16)

D

dam Structure that blocks the flow of water in a river or stream. (Chapter 15)

death rate The number of deaths per 1000 individuals per year. (Chapter 7)

debt-for-nature swap A wealthy nation forgives the debt of a developing nation in return for a pledge to protect natural areas in that developing nation. (Chapter 9)

decomposers Organisms such as bacteria and fungi that break organic matter all the way down to constituent atoms or molecules in a form that plants can take back up. (Chapter 8)

deforestation Net loss of trees in a forested area. (Chapter 11)

demographic factors Population characteristics such as birth rate or life expectancy that influence how a population changes in size and composition. (Chapter 4)

demographic transition Theoretical model that describes the expected drop in once-high population growth rates as economic conditions improve the quality of life in a population. (Chapter 4)

density dependent Factors, such as predation or disease, whose impact on the population increases as population size goes up. (Chapter 7)

density independent Factors, such as a storm or an avalanche, whose impact on the population is not related to population size. (Chapter 7)

dependent variable The variable in an experiment that is evaluated to see if it changes due to the conditions of the experiment. (Chapter 2)

depleted fisheries The fish population is well below historic levels and the population's reproductive capacity is low, meaning that recovery will be slow, if at all. (Chapter 14)

desalination The removal of salt and minerals from seawater to make it suitable for consumption. (Chapter 15)

desertification The process that transforms once-fertile land into desert. (Chapter 12)

detritivores Consumers (including worms, insects, crabs, etc.) who eat dead organic material. (Chapter 8)

developed country A country that has a moderate to high standard of living on average and an established market economy. (Chapter 4)

developing country A country that has a lower standard of living than a developed country, and has a weak economy; may have high poverty. (Chapter 4)

discounting future value Giving more weight to short-term benefits and costs than to long-term ones. (Chapter 5)

dissolved oxygen (DO) The amount of oxygen in the water. (Chapter 16)

dose-response curve A graph of the effects of a substance at different concentrations or levels of exposures. (Chapter 3)

E

ecocentric worldview A system-centred view that values intact ecosystems, not just the individual parts. (Chapter 1)

eco-industrial parks Industrial parks in which industries are physically positioned near each other for "waste-to-feed" exchanges (the waste of one becomes the raw material for another). (Chapter 17)

ecolabelling Providing information about how a product is made and where it comes from; allows consumers to make more sustainable choices and support sustainable products and the businesses that produce them. (Chapter 5)

ecological diversity The variety within an ecosystem's structure, including many communities, habitats, niches, and trophic levels. (Chapter 9)

ecological economics Branch of economics that considers the long-term impact of our choices on people and the environment. (Chapter 5)

ecological footprint The land area needed to provide the resources for, and assimilate the waste of, a person or population. (Chapters 1 and 5)

ecological succession Progressive replacement of plant (and then animal) species in a community over time due to the changing conditions that the plants themselves create (more soil, shade, etc.). (Chapter 8)

economics The social science that deals with the production, distribution, and consumption of goods and services. (Chapter 5)

ecosystem All of the organisms in a given area plus the physical environment in which, and with which, they interact. (Chapters 6 and 8)

ecosystem services Essential ecological processes that make life on Earth possible (Chapter 5); benefits that are important to all life, including humans, provided by functional ecosystems; includes such things as nutrient cycles, air and water purification, and ecosystem goods, such as food and fuel. (Chapter 9)

ecotones Regions of distinctly different physical areas that serve as boundaries between different communities. (Chapter 8)

ecotourism Low-impact travel to natural areas that contributes to the protection of the environment and respects the local people. (Chapter 11)

edge effects The different physical makeup of the ecotone that creates different conditions that either attract or repel certain species (for instance, it is drier, warmer, and more open at the edge of a forest and field than it is further in the forest). (Chapter 8)

edge species Species that prefer to live close to the edges of two different habitats (ecotone areas). (Chapters 8 and 10)

effluent Wastewater discharged into the environment. (Chapter 15)

electricity The flow of electrons (negatively charged subatomic particles) through a conductive material (such as wire). (Chapter 19)

emergent The region where a tree that is taller than the canopy trees rises above the canopy layer. (Chapter 11)

emigration The movement of people out of a given population. (Chapter 4)

empirical evidence Information gathered via observation of physical phenomena. (Chapter 2)

empirical science A scientific approach that investigates the natural world through systematic observation and experimentation. (Chapter 1)

endangered A species that faces a very high risk of extinction in the immediate future. (Chapter 10)

endemic Describes a species that is native to a particular area and is not naturally found elsewhere. (Chapters 9 and 10)

endocrine disruptor A substance that interferes with the endocrine system, typically by mimicking a hormone or preventing a hormone from having an effect. (Chapter 3) energy The capacity to do work. (Chapter 19)

energy flow The one-way passage of energy through an ecosystem. (Chapter 6)

energy independence Meeting all of one's energy needs without importing any energy. (Chapter 20)

energy return on energy investment (EROEI) A measure of the net energy from an energy source (the energy in the source minus the energy required to get it, process it, ship it, and then use it). (Chapter 19)

energy security Having access to enough reliable and affordable energy sources to meet one's needs. (Chapter 20)

environment The biological and physical surroundings in which any given living organism exists. (Chapter 1)

Environment Canada A Canadian federal government department responsible for environmental protection and weather monitoring. (Chapter 2)

environmental ethic The personal philosophy that influences how a person interacts with his or her natural environment and thus affects how one responds to environmental problems. (Chapter 1)

environmental impact assessment An evaluation of the positive and negative impacts of a proposed environmental action, including alternative actions that could be pursued. (Chapter 7)

environmental impact statement A document outlining the positive and negative impacts of any action that has the potential to cause environmental damage; used to help decide whether or not that action will be approved. (Chapter 19)

environmental justice The concept that access to a clean, healthy environment is a basic human right. (Chapters 21 and 26)

environmental literacy A basic understanding of how ecosystems function and of the impact of our choices on the environment. (Chapter 1)

environmental racism Occurs when minority communities face more exposure to pollution than average for the region. (Chapter 21)

environmental science An interdisciplinary field of research that draws on the natural and social sciences and the humanities in order to understand the natural world and our relationship to it. (Chapter 1)

epidemiologist A scientist who studies the causes and patterns of disease in human populations. (Chapter 3)

estuary Region where rivers empty into the ocean. (Chapter 13)

eutrophication Nutrient enrichment of water bodies, which typically leads to algal overgrowth and oxygen depletion, and which can occur naturally or via human activities. (Chapter 16)

evaporation The conversion of water from a liquid state to a gaseous state. (Chapter 15)

evolution Differences in the gene frequencies within a population from one generation to the next. (Chapter 10)

e-waste Unwanted computers and other electronic devices such as discarded televisions and cellphones. (Chapter 17)

exclusive economic zones (EEZs) Zones that extend 200 nautical miles (370 kilometres) from the coastline of any given nation, where that nation has exclusive rights over marine resources, including fish. (Chapter 14)

experimental study Research that manipulates a variable in a test group and compares the response to that of a control group that was not exposed to the same variable. (Chapter 2)

exponential growth Population size becomes progressively larger each breeding cycle; produces a J curve when plotted over time. (Chapter 7)

external costs Costs that are associated with a product or service, but are not taken into account when a price is assigned to that product or service. (Chapter 5)

extinct/extinction The complete loss of a species from an area; may be local (gone from a specific area) or global (gone throughout the world). (Chapter 10)

extirpated Describes a species that is locally extinct in one or more areas but still has some individual members in other areas. (Chapter 9)

extirpation Local extinction of a species (Chapter 7); local extinction in one or more areas, though some individuals exist in other areas. (Chapter 10)

exurbs Towns beyond the immediate suburbs whose residents commute into the city for work. (Chapter 26)

F

falsifiable An idea or a prediction that can be proved wrong by evidence. (Chapter 2)

feed conversion rates How much edible food is produced per unit of feed input. (Chapter 18)

feedstock Biomass sources used to make biofuels. (Chapter 25)

fisheries The industry devoted to commercial fishing or the places where fish are caught, harvested, processed, and sold. (Chapter 14)

Fisheries Act The federal law that regulates fishing and the harvesting of marine plant life, and protects marine habitats. (Chapter 14)

Fisheries and Oceans Canada The federal agency that protects oceans and manages fisheries. (Chapter 14)

food chain A simple, linear path starting with a plant (or other photosynthetic organism) that identifies what each organism in the path eats. (Chapter 8)

food miles The distance a food travels from its site of production to the consumer. (Chapter 18)

food web A linkage of all the food chains together that shows the many connections in the community. (Chapter 8)

forest floor The lowest level of the forest, containing herbaceous plants, fungi, leaf litter, and soil. (Chapter 11)

fossil fuel A nonrenewable resource like coal, oil, and natural gas that was formed over millions of years from the remains of dead organisms (Chapter 19); a nonrenewable natural resource formed millions of years ago from dead plant (coal) or microscopic marine-life (oil and natural gas) remains. (Chapter 20)

fossil record The total collection of fossils (remains, impressions, traces of ancient organisms) found on Earth. (Chapter 10)

founder effect When a small group with only a subset of the larger population's genetic diversity becomes isolated and it evolves into a different population, missing some of the traits of the original. (Chapter 10)

fracking (hydraulic fracturing) The extraction of oil, or more commonly natural gas, from rock (typically shale) via the propagation of fractures using pressurized fluid. (Chapter 20)

freshwater Water that has few dissolved ions such as salt. (Chapter 15)

fuel crops Crops specifically grown to be used to produce biofuels. (Chapter 25)

fuel rods Hollow metal cylinders filled with uranium fuel pellets for use in fission reactors. (Chapter 23)

G

gendercide The systematic killing of a specific gender (male or female). (Chapter 4)

gene frequencies The assortment and abundance of particular variants of genes relative to each other within a population. (Chapter 10)

genes Stretches of DNA, the cell's hereditary material, that each direct the production of a particular protein and influence an individual's traits. (Chapter 10)

genetic diversity The heritable variation among individuals of a single population or within the species as a whole. (Chapters 9 and 10)

genetic drift The change in gene frequencies of a population over time due to random mating that results in the loss of some gene variants. (Chapter 10)

genetically modified organism (GMO) Organism that has had its genetic information modified to give it desirable characteristics such as pest or drought resistance. (Chapter 18)

geothermal energy The heat stored underground, contained in either rocks or fluids. (Chapter 24)

geothermal heat pump A system that actively moves heat from the underground into a house to warm it or removes heat from a house to cool it. (Chapter 24)

geothermal power plants Power plants that use the heat of hydrothermal reservoirs to produce steam and turn turbines to generate electricity. (Chapter 24)

grasslands A biome that is predominately grasses, due to low rainfall, grazing animals, and/or fire. (Chapter 12)

green building Construction and operational designs that promote resource and energy efficiency, and provide a better environment for occupants. (Chapter 26)

green business Doing business in a way that is good for people and the environment. (Chapter 5)

Green Revolution Plant-breeding program in the mid-1900s that dramatically increased crop yields and led the way for mechanized, large-scale agriculture. (Chapter 18)

Green Revolution 2.0 Focuses on the production of genetically modified organisms (GMOs) to increase crop productivity or create new varieties of crops. (Chapter 18)

green tax Tax (fee paid to government) assessed on environmentally undesirable activities. (Chapter 21)

greenhouse effect The warming of the planet that results when heat is trapped by Earth's atmosphere. (Chapter 22)

greenhouse gases Molecules in the atmosphere that absorb heat and reradiate it back to Earth. (Chapter 22)

ground-level ozone A secondary pollutant that forms when some of the pollutants released during fossil fuel combustion react with atmospheric oxygen in the presence of sunlight. (Chapter 21)

groundwater Water found underground in aquifers. (Chapter 15)

gross primary productivity A measure of the total amount of energy captured via photosynthesis and transferred to organic molecules in an ecosystem. (Chapter 8)

growth rate The percent increase of population size over time; affected by births, deaths, and the number of people moving into or out of a regional population. (Chapter 4)

H

habitat The physical environment in which individuals of a particular species can be found. (Chapters 6 and 8)

habitat destruction Altering a natural area in a way that makes it uninhabitable for the species living there. (Chapter 10)

habitat fragmentation Destruction of part of an area that separates suitable habitat patches from one another; patches that are too small may be unusable for some species. (Chapter 10)

hazardous waste Waste that is toxic, flammable, corrosive, explosive, or radioactive. (Chapter 17)

Health Canada The federal agency responsible for protecting Canadians' health. (Chapter 3)

hectare (ha) Metric unit of measure for area; 1 ha = 2.5 acres (ac). (Chapter 11)

herbivore An animal that feeds on plants. (Chapter 12)

high-level radioactive waste (HLRW) Spent fuel rods or nuclear weapons production waste that is still highly radioactive. (Chapter 23) hormone A chemical released by organisms that directs cellular activity and produces changes in how their bodies function. (Chapter 3)

hydropower The energy produced from moving water. (Chapter 24)

hypothesis A possible explanation for what we have observed that is based on some previous knowledge. (Chapter 2)

hypoxia A situation in which the level of oxygen in the water is inadequate to support life. (Chapter 16)

I

immigration The movement of people into a given population. (Chapter 4)

in vitro study Research that studies the effects of experimental treatment cells in culture dishes rather than in intact organisms. (Chapter 3)

in vivo study Research that studies the effects of an experimental treatment in intact organisms. (Chapter 3)

incinerators Facilities that burn trash at high temperatures. (Chapter 17)

independent variable The variable in an experiment that the researcher manipulates or changes to see if it produces an effect. (Chapter 2)

indicator species The species that are particularly vulnerable to ecosystem perturbations, and that, when we monitor them, can give us advance warning of a problem. (Chapter 8)

infant mortality rate The number of infants who die in their first year of life per every thousand live births in that year. (Chapter 4)

inferences Conclusions we draw based on observations. (Chapter 2)

infill development The development of empty lots within a city. (Chapter 26)

infiltration The process of water soaking into the ground. (Chapter 15)

information literacy The ability to evaluate the quality of information. (Chapter 3)

instrumental value The value or worth of an object, organism, or species based on its usefulness to humans. (Chapters 1 and 9)

Intergovernmental Panel on Climate Change (IPCC) An international group of scientists who evaluate scientific studies related to any aspect of climate change to give thorough and objective assessment of the data. (Chapter 22)

internal costs Those costs—such as raw materials, manufacturing costs, labour, taxes, utilities, insurance, and rent —that are accounted for when a product or service is evaluated for pricing. (Chapter 5)

intrinsic value The value or worth of an object, organism, or species based on its mere existence; it has an inherent right to exist. (Chapters 1 and 9)

invasive species A non-native species (a species outside of its range) whose introduction causes or is likely to cause economic or environmental harm or harm to human health. (Chapter 10)

IPAT model An equation (I = P x A x T) that measures human impact (I), based on three factors: population (P), affluence (A), and technology (T). (Chapter 5)

isotopes Atoms that have different numbers of neutrons in their nucleus but the same number of protons. (Chapter 23)

K

keystone species A species that impacts its community more than its mere abundance would predict, often altering ecosystem structure. (Chapter 8)

K-selected species Species that have a low biotic potential and that share characteristics such as long lifespan, late maturity, and low fecundity; generally show logistic population growth. (Chapter 7)

L

law of conservation of matter Matter can neither be created nor destroyed; it only changes form. (Chapter 17)

LD_{50} (lethal dose 50%) The dose of a substance that would kill 50% of the test population. (Chapter 3)

leachate Water that carries dissolved substances (often contaminated) that can percolate through soil. (Chapter 17)

LEED (Leadership in Energy and Environmental Design) A third-party certification program administered by the nonprofit Green Building Council that awards a rating (standard, silver, gold, or platinum) to buildings that include environmentally sound design features. (Chapter 26)

life expectancy The number of years an individual is expected to live. (Chapter 4)

limiting factor The critical resource whose supply determines the population size of a given species in a given biome. (Chapter 6)

logical fallacies Arguments which attempt to sway the reader without using evidence. (Chapter 3)

logistic growth The kind of growth in which population size increases rapidly at first but then slows down as the population becomes larger; produces an S curve when plotted over time. (Chapter 7)

low-level radioactive waste (LLRW) Material that has a low level of radiation for its volume. (Chapter 23)

M

marine protected areas (MPAs) Discrete regions of ocean that are legally protected from various forms of human exploitation. (Chapter 14)

marine reserves Restricted areas where all fishing is prohibited and absolutely no human disturbance is allowed. (Chapter 14)

maximum sustainable yield (MSY) Harvesting as much as is sustainably possible for the greatest economic benefit (Chapter 11); the amount that can be harvested without decreasing the yield in future years. (Chapter 14)

Milankovitch cycles Predictable variations in Earth's position in space relative to the Sun that affect climate. (Chapter 22)

minimum viable population The smallest number of individuals that would still allow a population to be able to persist or grow, ensuring long-term survival. (Chapter 7)

mitigation Preventative efforts intended to minimize the extent or impact of a problem such as climate change. (Chapter 22)

monoculture Farming method in which one variety of one crop is planted, typically in rows over huge swaths of land, with large inputs of fertilizer, pesticides, and water. (Chapter 18)

Montreal Protocol International treaty that laid out plans to phase out ozone-depleting chemicals like CFC. (Chapter 2)

mountaintop removal Surface mining technique that uses explosives to blast away the top of a mountain to expose the coal seam underneath; the waste rock and rubble are deposited in a nearby valley. (Chapter 19)

municipal solid waste (MSW) Everyday garbage or trash (solid waste) produced by individuals or small businesses. (Chapter 17)

mutualism A symbiotic relationship between individuals of two species in which both parties benefit. (Chapter 8)

N

National Forest Strategy Canada's plan to incorporate economic, social, and ecological principles into forest management. (Chapter 11)

natural capital The wealth of resources on Earth. (Chapter 5)

natural gas Gaseous fossil fuel composed mainly of a simpler hydrocarbon, mostly CH_4 (methane). (Chapter 20)

natural interest Readily produced resources that we could use and still leave enough natural capital behind to replace what we took. (Chapter 5)

natural selection The process by which organisms best adapted to the environment (the fittest) survive to reproduce, leaving more offspring than less well-adapted individuals. (Chapter 10)

negative feedback Reduction or reversal of an effect by its own influence on the process giving rise to it (e.g., changes brought on by warming lead to cooling). (Chapter 13)

negative feedback loop Changes caused by an initial event that trigger events that then reverse the response (for example, warming leads to events that eventually result in cooling). (Chapter 22)

net primary productivity (NPP) A measure of the amount of energy captured via photosynthesis and stored in photosynthetic organisms. (Chapter 8)

New Urbanism A movement that promotes the creation of compact, mixed-use communities with all of the amenities for day-to-day living close by and accessible. (Chapter 26)

niche The role a species plays in its community, including how it gets its energy and nutrients, what habitat requirements it has, and what other species and parts of the ecosystem it interacts with. (Chapters 6 and 8)

nitrogen cycle Continuous series of natural processes by which nitrogen passes from the air to the soil, to organisms, and then returns back to the air or soil through decomposition or denitrification. (Chapter 6)

nitrogen fixation Conversion of atmospheric nitrogen into a biologically usable form, carried out by bacteria found in soil or via lightning. (Chapter 6)

nondegradable Incapable of being broken down under normal conditions. (Chapter 17)

non-point source pollution Runoff that enters the water from overland flow and can come from any area in the watershed (Chapter 16); pollution that enters the air from dispersed or mobile sources or enters the water from overland flow. (Chapter 21)

nonrenewable resource A resource that is formed more slowly than it is used, or is present in a finite supply. (Chapter 20)

nonrenewable resources Resources whose supply is finite or not replenished in a timely fashion. (Chapter 1)

nuclear energy Energy released when an atom is split (fission) or combines with another to form a new atom (fusion); can be tapped to generate electricity. (Chapter 23)

nuclear fission Nuclear reaction that occurs when a neutron strikes the nucleus of an atom and breaks it into two or more parts. (Chapter 23)

nutrient cycles Movement of life's essential chemicals or nutrients through an ecosystem. (Chapter 6)

O

observational study Research that gathers data in a real-world setting without intentionally manipulating any variable. (Chapter 2)

observations Information detected with the senses—or with equipment that extends our senses. (Chapter 2)

oil Liquid fossil fuel useful as a portable fuel or as a raw material for many industrial products such as plastic and pesticides. (Chapter 20)

oil sands Geologic formations containing oil in the form of thick, black oil called crude bitumen (also known as *tar sands*). (Chapter 20)

oil shale Compressed sedimentary rocks that contain kerogen, an organic compound that is released as an oil-like liquid when the rock is heated. (Chapter 20) open dumps Places where trash, both hazardous and nonhazardous, is simply piled up. (Chapter 17)

organic agriculture Farming that does not use synthetic fertilizer, pesticides, or other chemical additives like hormones (for animal rearing). (Chapter 18)

overburden The rock and soil removed to uncover a mineral deposit during surface mining. (Chapter 19)

overexploited fisheries More fish are taken than is sustainable in the long run, leading to population declines. (Chapter 14)

overgrazing Too many herbivores feeding in an area, eating the plants faster than they can regrow. (Chapter 12)

overharvesting Human activity that removes more of a resource than can be replaced in the same time frame, such as taking too many individuals from a population. (Chapter 10)

overpopulation More people living in an area than its natural and human resources can support. (Chapter 4)

ozone Molecule with 3 oxygen atoms that absorbs UV radiation in the stratosphere. (Chapter 2)

P

parasitism A symbiotic relationship between individuals of two species in which one benefits and the other is negatively affected (a form of predation). (Chapter 8)

particulate matter (PM) Particles or droplets small enough to remain aloft in the air for long periods of time. (Chapter 21)

passive solar technologies The capture of solar energy (heat or light) without any electronic or mechanical assistance. (Chapter 24)

pastoralists Individuals who herd and care for livestock as a way of life. (Chapter 12)

payback time The amount of time it would take to save enough money in operation costs to pay for the equipment. (Chapter 24)

peak oil The moment in time when oil will reach its highest production levels and then steadily and terminally decline. (Chapter 20)

peer review A process whereby researchers submit a report of their work to a group of outside experts who evaluate the study's design and results of the study to determine whether it is of high-enough quality to publish. (Chapters 2 and 3)

perennial Plants that live for more than a year, growing and producing seed year after year. (Chapter 25)

persistence The ability of a substance to remain in its original form; often expressed as the length of time it takes a substance to break down in the environment. (Chapter 3)

petrochemicals Distillation products from the processing of crude oil that can have many different uses, including as fuel or as industrial raw materials. (Chapter 20)

phosphorus cycle Series of natural processes by which the nutrient phosphorus moves from rock to soil or water, to living organisms, and back to the soil. (Chapter 6)

photovoltaic (PV) cells Also called solar cells, PV cells convert solar energy directly into electricity. (Chapter 24)

pioneer species Plant species that move into an area during early stages of succession; these are often *r* species and may be annuals, species that live one year, leave behind seeds, and then die. (Chapter 8)

point source pollution Pollution that can be traced back to discrete sources such as wastewater treatment plants or industrial sites (Chapter 16); pollution that enters the air from a readily identifiable source such as a smokestack. (Chapter 21)

policy A formalized plan that addresses a desired outcome or goal. (Chapter 2)

pollution Hazardous or objectionable substances that are released into the environment; also includes noise and light. (Chapter 10)

pollution standards Allowable levels of a pollutant that can be released over a certain time period. (Chapter 16)

population All the individuals of a species that live in the same geographic area and are able to interact and interbreed. (Chapter 7)

population density The number of individuals per unit area. (Chapters 4 and 7)

population distribution The location and spacing of individuals within their range. (Chapter 7)

population dynamics The changes over time of population size and composition. (Chapter 7)

population growth rate The change in population size over time (births minus deaths over a specific time period). (Chapter 7)

population momentum The tendency of a young population to continue to grow even after birth rates drop to "replacement rates"—2 children per couple. (Chapter 4)

positive feedback Changes caused by an initial event accentuate that original event (e.g., changes brought on by warming lead to even more warming). (Chapter 13)

positive feedback loop Changes caused by an initial event that then accentuate that original event (for example, a warming trend gets even warmer). (Chapter 22)

potable Water clean enough for consumption. (Chapter 15)

precautionary principle A principle that encourages acting in a way that leaves a margin of safety when there is a potential for serious harm but uncertainty about the form or magnitude of that harm (Chapters 2 and 3); acting in a way that leaves a safety margin when the data is uncertain or severe consequences are possible. (Chapter 22)

precipitation Rain, snow, sleet, or any form of water falling from the atmosphere. (Chapter 15)

prediction A statement that identifies what is expected to happen in a given situation. (Chapter 2)

primary air pollutants Air pollutants released directly from both mobile sources (such as cars) and stationary sources (such as industrial and power plants). (Chapter 21)

primary sources Sources that present new and original data or information, including novel scientific experiments or observations and first-hand accounts of any given event. (Chapter 3)

primary succession Ecological succession that occurs in an area where no ecosystem existed before (for example, on bare rock with no soil). (Chapter 8)

producer An organism that converts solar energy to chemical energy via photosynthesis (Chapter 6); a photosynthetic organism that captures solar energy directly and uses it to produce its own food (sugar). (Chapter 8)

proven reserves A measure of the amount of a fossil fuel that is economically feasible to extract from a known deposit using current technology. (Chapter 20)

R

radiative forcer Anything that alters the balance of incoming solar radiation relative to the amount of heat that escapes out into space. (Chapter 22)

radioactive Atoms that spontaneously emit subatomic particles and/or energy. (Chapter 23)

radioactive half-life The time it takes for half of the radioactive isotopes in a sample to decay to a new form. (Chapter 23)

rain garden Runoff area that is planted with water-tolerant plants to slow runoff and promote infiltration. (Chapter 16)

random distribution Individuals are spread out over the environment irregularly with no discernible pattern. (Chapter 7)

range of tolerance The range, within upper and lower limits, of a limiting factor that allows a species to survive and reproduce. (Chapter 6)

rangeland Grassland used for grazing of livestock. (Chapter 12)

receptor A structure on or inside a cell that binds a particular molecule, such as a hormone, thus allowing the molecule to affect the cell. (Chapter 3)

reclamation Restoring a damaged natural area to a less damaged state. (Chapter 19)

recycle The fourth of the waste-reduction "4 Rs": return items for reprocessing to make new products. (Chapter 17)

reduce The second of the waste-reduction "4 Rs": make choices that allow you to use less of a resource by, for instance, purchasing durable goods that will last or can be repaired. (Chapter 17)

refuse The first of the waste-reduction "4 Rs": choose NOT to use or buy a product if you can do without it. (Chapter 17)

renewable energy Energy that comes from an infinitely available or easily replenished source (Chapter 1); energy from sources that are replenished over short time scales or that are perpetually available. (Chapter 24)

replacement fertility rate The rate at which children must be born to replace those dying in the population. (Chapter 4)

reproductive strategies How quickly a population can potentially increase, reflecting the biology of the species (lifespan, fecundity, maturity rate, etc.). (Chapter 7)

reservoir (or sink) Abiotic or biotic component of the environment that serves as a storage place for cycling nutrients (Chapter 6); artificial lake formed when a river is impounded by a dam. (Chapter 15)

resilience The ability of an ecosystem to recover when it is damaged or perturbed. (Chapter 8)

resource partitioning When different species use different parts or aspects of a resource, rather than competing directly for exactly the same resource. (Chapter 8)

restoration ecology The science that deals with the repair of damaged or disturbed ecosystems. (Chapter 8)

reuse The third of the waste-reduction "4 Rs": use a product more than once for its original purpose or another purpose. (Chapter 17)

riparian areas The land area adjacent to a body of water that is affected by the water's presence (for example, water-tolerant plants grow there) and that affects the water itself (for example, provides shade). (Chapter 16)

risk assessment Weighing the risks and benefits of a particular action in order to decide how to proceed. (Chapter 3)

rotational grazing Moving animals from one pasture to the next in a predetermined sequence to prevent overgrazing. (Chapter 12)

r-selected species Species that have a high biotic potential and that share other characteristics such as short lifespan, early maturity, and high fecundity. (Chapter 7)

S

saltwater intrusion The inflow of ocean (salt) water into a freshwater aquifer that happens when an aquifer has lost some of its freshwater stores. (Chapter 15)

sanitary landfills Disposal sites that seal in trash at the top and bottom to prevent its release into the atmosphere; the sites are lined on the bottom, and trash is dumped in and covered with soil daily. (Chapter 17)

science A body of knowledge (facts and explanations) about the natural world, and the process used to get that knowledge. (Chapter 2)

scientific method Procedure scientists use to empirically test a hypothesis. (Chapter 2)

secondary air pollutants Air pollutants formed when primary air pollutants react with one another or with other chemicals in the air. (Chapter 21)

secondary sources Sources that present and interpret information from primary sources. Secondary sources include newspapers, magazines, books, and most information from the Internet. (Chapter 3)

secondary succession Ecological succession that occurs in an ecosystem that has been disturbed; occurs more quickly than primary succession because soil is present. (Chapter 8)

selective pressure A nonrandom influence affecting who survives or reproduces. (Chapter 10)

sex ratio The relative number of males to females in a population; calculated by dividing the number of males by the number of females. (Chapter 4)

sliding reinforcer Actions that are beneficial at first but that change conditions such that their benefit declines over time. (Chapter 1)

smart growth Strategies that help create walkable communities with lower ecological footprints. (Chapter 26)

smog Hazy air pollution that contains a variety of pollutants including sulphur dioxide, nitrogen oxides, tropospheric ozone, and particulates. (Chapter 21)

social traps Decisions by individuals or groups that seem good at the time and produce a short-term benefit, but that hurt society in the long run. (Chapter 1)

soil erosion The removal of soil by wind and water that exceeds the soil's natural replacement. (Chapter 12)

solar energy Energy harnessed from the Sun in the form of heat or light. (Chapter 24)

solar thermal system An active technology that captures solar energy for heating. (Chapter 24)

solubility The ability of a substance to dissolve in a liquid or gas. (Chapter 3)

species A group of plants or animals that have a high degree of similarity and can generally only interbreed among themselves. (Chapter 6)

Species at Risk Act (SARA) The primary law under which biodiversity is protected in Canada. (Chapter 9)

species diversity The variety of species in an area, includes how many are present (richness), and their abundance relative to each other (evenness). (Chapters 8 and 9)

species evenness The relative abundance of each species in a community. (Chapter 8)

species richness The total number of different species in a community. (Chapter 8)

statistics The mathematical evaluation of experimental data to determine how likely it is that any difference observed is due to the variable being tested. (Chapter 2)

stormwater runoff Water from precipitation that flows over the surface of the land. (Chapters 11 and 16)

stratosphere Region of the atmosphere that starts at the top of the troposphere and extends up to about 50 kilometres; contains the ozone layer. (Chapter 2)

subsidies Free government money or resources intended to promote desired activities. (Chapter 21)

subsurface mines Sites where tunnels are dug underground to access mineral resources. (Chapter 19)

suburban sprawl Low-population-density developments that are built outside of a city. (Chapter 26)

surface mining Removing soil and rock that overlays a mineral deposit close to the surface in order to access that deposit. (Chapter 19)

surface water Any body of water found above ground such as oceans, rivers, and lakes. (Chapter 15)

sustainable A method of using resources in such a way that we can continue to use them indefinitely; capable of being continued without degrading the environment. (Chapters 1 and 5)

sustainable agriculture Farming methods that do not deplete resources, such as soil and water, faster than they are replaced. (Chapter 18)

sustainable development Development that meets present needs without compromising the ability of future generations to do the same. (Chapters 1 and 5)

sustainable energy Energy from sources that are renewable and have a low environmental impact. (Chapter 24)

sustainable fishery A fishery that ensures that fish stocks are maintained at healthy levels, the ecosystem is fully functional, and fishing activity does not threaten biological diversity. (Chapter 14)

sustainable forest management A forest management approach that blends ecosystem conservation with economic and social purposes. (Chapter 11)

sustainable grazing Practices that allow animals to graze in a way that keeps pastures healthy and allows grasses to recover. (Chapter 12)

symbiosis A close biological or ecological relationship between two species. (Chapter 8)

synergistic effects When exposure to two or more chemicals has a greater effect than the sum of their individual effects normally would. (Chapter 3)

T

tar sands Geologic formations containing oil in the form of thick, black oil called crude bitumen (also known as *oil sands*). (Chapter 20)

tax credit A reduction in the tax one has to pay in exchange for some desirable action. (Chapter 21)

temperate forests Found in areas with four seasons and a moderate climate, receive 750–1500 millimetres of precipitation per year, and may include conifers and/or hardwood deciduous trees (lose their leaves in the winter). (Chapter 11)

tertiary sources Sources that present and interpret information from secondary sources. (Chapter 3)

test group The group in an experimental study that is manipulated somehow such that it differs from the control group in only one way. (Chapter 2)

testable A possible explanation that generates predictions for which empirical evidence can be collected to verify or refute the hypothesis. (Chapter 2) theory A widely accepted explanation of a natural phenomenon that has been extensively and rigorously tested scientifically. (Chapter 2)

threatened A species that is likely to become endangered in the near future. (Chapter 10)

time delay Actions that produce a benefit today set into motion events that cause problems later on. (Chapter 1)

total allowable catch (TAC) The maximum amount (weight or numbers of fish or shellfish) of a particular species that can be harvested per year or fishing season in a given area; meant to prevent overfishing. (Chapter 14)

total fertility rate (TFR) The number of children the average woman has in her lifetime. (Chapter 4)

toxicologist A scientist who studies the specific properties of any given potentially toxic substances. (Chapter 3)

toxic substances Substances that cause damage to living organisms through immediate or long-term exposure. (Chapter 3)

trade-offs The imperfect and sometimes problematic responses that we must at times choose between when addressing complex problems. (Chapter 1)

tragedy of the commons The tendency of an individual to abuse commonly held resources in order to maximize his or her own personal interest. (Chapter 1)

transboundary pollution Pollution that is produced in one area but falls in a different area, possibly a different nation. (Chapter 21)

transgenic organism An organism that contains genes from another species. (Chapter 18)

transpiration The loss of water vapour from plants. (Chapter 15)

triple-bottom line The combination of the environmental, social, and economic impacts of our choices. (Chapters 1 and 5)

trophic levels Feeding levels in a food chain. (Chapter 8)

tropical forests Found in equatorial areas with warm temperatures year-round and high rainfall; some have distinct wet and dry seasons but none has a winter season. (Chapter 11)

troposphere Region of the atmosphere that starts at ground level and extends upward about 11 kilometres. (Chapter 2)

true cost The sum of both external and internal costs of a good or service. (Chapter 5)

U

ultraviolet (UV) radiation Short-wavelength electromagnetic energy emitted by the Sun. (Chapter 2)

unconventional oil reserves Recoverable oil deposits that exist in rock, sand, or clay but whose extraction is economically and environmentally costly. (Chapter 20)

understory The smaller trees, shrubs, and saplings that live in the shade of the forest canopy. (Chapter 11)

uniform distribution Individuals are spaced evenly, perhaps due to territorial behaviour or mechanisms for suppressing the growth of nearby individuals. (Chapter 7)

urban areas Densely populated regions that include cities and the suburbs that surround them. (Chapter 26)

urban flight The process of people leaving an inner-city area to live in surrounding areas. (Chapter 26)

urban heat island effect The phenomenon in which urban areas are warmer than the surrounding countryside due to pavement, dark surfaces, closed-in spaces, and high energy use. (Chapter 26)

urbanization The migration of people to large cities; sometimes also defined as the growth of urban areas. (Chapter 26)

W

waste Any material that humans deem to be unwanted. (Chapter 17)

wastewater Used and contaminated water that is released after use by households, industry, or agriculture. (Chapter 15)

watershed The land area surrounding a body of water over which water such as rain could flow and potentially enter that body of water. (Chapter 16)

water cycle The movement of water from gaseous to liquid states through various water compartments such as surface waters, soil, and living organisms. (Chapter 15)

water monitoring Collection of water samples from different parts of a body of water (surface and deeper water), particularly for comparison with what would be expected in healthy systems. (Chapter 16)

water pollution The addition of anything that might degrade the quality of the water. (Chapter 16)

water scarcity Not having access to enough clean water. (Chapter 15)

water table The uppermost water level of the saturated zone of an aquifer. (Chapter 15)

weather The meteorological conditions in a given place on a given day. (Chapter 22)

wetland An ecosystem that is permanently or seasonally flooded. (Chapter 15)

wind energy Energy contained in the motion of air across Earth's surface. (Chapter 24)

worldviews The window through which one views one's world and existence. (Chapter 1)

Z

zero population growth The absence of population growth; occurs when birth rates equal death rates. (Chapter 4)

APPENDIX 1
BASIC MATH SKILLS

Math skills are needed to evaluate data and even to understand much of the information in news reports. Here we present a review of some basic skills that will be useful in this class and in other science classes.

AVERAGES (MEANS)

To calculate an average, add all the numbers in the data set and divide by the number of numbers.

Example: The sum of these numbers is 10 547; 10 547 / 8 = 1318.375. You could round this off for an average or mean of 1318.

Data set

1004	766
2349	988
456	1203
1882	1899

WORK WITH AVERAGES

Problem 1: If the following numbers were your grades on exams, what would be your exam average?
84, 73, 93, 95, 79, 86

PERCENTAGES/FREQUENCIES

To **convert a fraction to a percentage**, divide the numerator (top number) by the denominator (bottom number) and multiply by 100. Example: To express the fraction $^{2}/_{5}$ as a decimal, divide 2 by 5, which equals 0.4. Multiply this by 100 for your answer: 40%.
To **convert a decimal to a percentage**, multiply by 100; a shortcut for this is simply to move the decimal over two places to the right.
Example: $0.08 \times 100 = 8\%$

WORK WITH PERCENTAGES

Problem 2: If 8 out of 32 frogs in a pond have deformities, what percentage of frogs have deformities?

Problem 3: In a pond, 25% of the frogs have leg deformities. If there are 100 frogs in the pond, how many have deformities?

Problem 4: If there are 68 frogs in the pond and 25% have deformities, how many have deformities? (First, make an estimate based on your answer to Problem 3—will it be a higher or lower number? This will help you decide if the answer you calculate is reasonable or whether you might need to recalculate.)

SCIENTIFIC NOTATION

In science, we often use very large or very small numbers. To make these easier to present, scientists use scientific notation, which multiplies a number (called the coefficient) by 10 (the base) raised to a given power (the exponent). If the coefficient is 1, we can leave it off and simply show the base and exponent (e.g. $1 \times 10^2 = 10^2$). The exponent tells us how many orders of magnitude larger or smaller to make the number. In other words, the exponent is telling us how many zeros the number will have: $10^2 = 100$; $10^3 = 1000$, and so on. Negative exponents represent decimals; for example: $10^{-2} = 0.01$; $10^{-3} = 0.001$, and so on.

Here is a simple shorthand way to evaluate numbers given in scientific notation. Move the decimal place to the right if 10 has a positive exponent, and to the left if the exponent is negative. The number of spaces the decimal place is moved is equal to the exponent. For example, 10^2 tells us to move the decimal place 2 spaces to the right; 10^{-2} means we move it 2 spaces to the left.

By convention, we always designate the coefficient as a whole number (2) or a decimal, with the decimal point at the "10" position (2.3). In other words we would write 2.3×10^5, not 23×10^4. Both are technically correct but the first is the preferred format.

Examples:
$2 \times 10^6 = 2\,000\,000$
$2.36 \times 10^5 = 236\,000$
$4.99 \times 10^{-4} = 0.000499$

Some typical values you might run across include:
10^6 = 1 million
10^9 = 1 billion
10^{12} = 1 trillion

MEASUREMENTS AND UNITS OF MEASURE

There are many handy conversion calculators on the Internet, but it is still useful to have a general idea of how large various units of measure are and how metric, English, Imperial, and U.S. systems of measurement compare.

LENGTH

Metric
1 kilometre (km) = 1000 metres (10^3)
1 metre (m) = 100 centimetres
1 centimetre (cm) = 10 millimetres
1 millimetre (mm) = 0.000001 metres (10^{-6})
1 micrometre (μm) = 0.000000001 metres (10^{-9})
1 nanometre (nm) = 0.000000000001 metres (10^{-12})

English/Imperial/U.S.
1 mile (mi) = 5280 feet
1 yard (yd) = 36 inches (in) or 3 feet
1 foot (ft) = 12 inches

Conversions
1 km = 0.621 mi
1 m = 39.4 in
1 cm = 0.394 in

1 mi = 1.609 km
1 yd = 0.914 m
1 in = 2.54 cm

MASS

Metric
1 metric ton (mt or tonne) = 1000 kilograms
1 kilogram (kg) = 1000 grams
1 gram (g) = 1000 milligrams
1 milligram (mg) = 0.001 grams (10^{-3})

English/Imperial/U.S.
1 U.S. ton (t) = 2000 pounds (lbs)
1 pound (lb) = 16 ounces (oz)

Conversions
1 mt = 2200 lb
1 kg = 2.2 lb
1 lb = 454 g
1 lb = 0.454 kg
1 g = 0.035 oz

VOLUME

Metric
1 litre (L) = 1000 millilitres
1 millilitre (ml) = 0.001 litres

English/Imperial/U.S.
1 gallon (gal) = 4 quarts
1 quart (qt) = 2 pints or 4 cups
1 pint (pt) = 16 fluid oz

Conversions
1 L = 0.265 gal or 1.06 qt
1 gal = 3.79 L

AREA

Metric
1 hectare (ha) = 10 000 square m

English/Imperial/U.S.
1 acre (ac) = 4840 square yards (yd^2)

Conversions
1 ha = 2.47 ac
1 ac = 0.405 ha

CONCENTRATIONS

Metric
1 part per million (ppm) = 1 mg/L
1 part per billion (ppb) = 1 μg/L
1 part per trillion (ppt) = 1 ng/L

TEMPERATURE CONVERSIONS

Celsius (°C) to Fahrenheit (°F): $°F = (°C \times {}^{5}/_{9}) + 32$
Fahrenheit (°F) to Celsius (°C): $°C = (°F - 32) \times {}^{5}/_{9}$

In general:
1°C = 1.8°F
1°F = 0.56°C

Answers to problems:
1. 85 **2.** $^{8}/_{32}$ = 0.25 = 25% **3.** 100 × 0.25 = 25 **4.** 68 × 0.25 = 17

APPENDIX 2
DATA-HANDLING AND GRAPHING SKILLS

This tutorial offers a quick look at the basics of working with data and graphing.

Scientists gather data to learn about the natural world. Data can be organized into graphs, which are "pictures" or visual representations of the data. Because they can condense and organize large amounts of information, graphs are often easier to interpret than a simple list of numbers. They show relationships between two or more variables that help us determine whether the variables are correlated in any way and allow us to look for trends or patterns that might emerge. To be effective, graphs should be constructed according to conventions, and must be accurately plotted and properly labelled. Certain types of graphs are more suitable than others to show particular types of data, so it is important to choose the correct graph for your data.

The following sections describe variables found in graphs; data tables; and the types of graphs commonly used in environmental science.

VARIABLES

The **independent variable** is the parameter the experimenter manipulates—it could be whether or not a group is exposed to a treatment (given a medicine, exposed to a particular wavelength of light), is part of a distinct group (trees at specified distances from a stream), or is a group followed over a period of time (monitored daily, yearly, etc.). If you were setting up a data table in which to record the data your experiment would produce, you would be able to fill in the values for the independent variable *before beginning* the actual experiment.

The **dependent variable** is the response being measured in the experiment—the responding variable. The experiment is being conducted to see if this variable is "dependent on" the independent variable. In other words, when you change the independent variable, does the dependent variable change as a result? If you were setting up a data table in which to record data, you would be able to include a column heading for the dependent variable, but you would not be able to enter the values until the experiment was complete. There may be more than one dependent variable being tested.

DATA TABLES

Data tables have a conventional format. The independent variable is shown in the left-hand column and the data for the dependent variable or variables are shown in columns to the right of that. To be useful, the data table needs to have a descriptive title (what data are we looking at?) and the units of measure must be included.

ANNUAL ATLANTIC HERRING CATCH

Year	Herring catch (1000 metric tons)
1965	731
1970	580
1975	382
1980	270
1985	180
1990	150
1995	45
2000	20
2005	26

Independent variable (Year)

Dependent variable (Herring catch)

Units of measure are given—here we multiply each value by 1000 to find the number of metric tons of catch that year; using the 1000-fold conversion in the units allows us to use numbers that are easier to interpret.

TYPES OF GRAPHS

A. LINE GRAPHS

In science, researchers often test the effect of one variable on another. Line graphs are used when the independent variable is represented by a numerical sequence (e.g., 1, 2, 3) rather than discrete categories (e.g., red, yellow, blue). The dependent variable is always a numerical sequence.

Steps to producing a line graph

1. **Determine the *x*-axis and the *y*-axis.** The independent variable is usually shown on the *x*-axis (the horizontal axis). In a line graph, this variable is one that changes in a predictable, numerical sequence, such as the passing of time, increasing concentrations of a solution being tested, habitat distance from the seashore, etc. The data being collected (the response being observed) represent the dependent variable, which is shown on the *y*-axis (the vertical axis). The axes require a descriptive title that indicates exactly what each axis represents, along with units of measure, if needed.

2. **Set up the axes.** Set up each axis so that the largest data value for that variable is close to the end of the axis, leaving as much of the available space for your graph as possible. Aim for 5 to 15 "ticks" (the small dividing marks) on any given axis—don't overload it with 50 tiny ticks or have so few that it is hard to place data points. It is essential to evenly space the ticks on a given axis, keeping the increments the same numerical size. On the sample graph, all the *x*-axis ticks are 10 years apart; all the *y*-axis ticks are 100 units apart. The increments will depend on your particular data; however, be sure they are presented in the same size. Give the graph a descriptive title.

3. **Plot the points.** Plot each point by finding the *x*-value on the *x*-axis and moving up until you reach the *y*-value position across from the *y*-axis. If you are graphing more than one set of data, draw the data points as different shapes or colours and provide a legend to identify each data set.

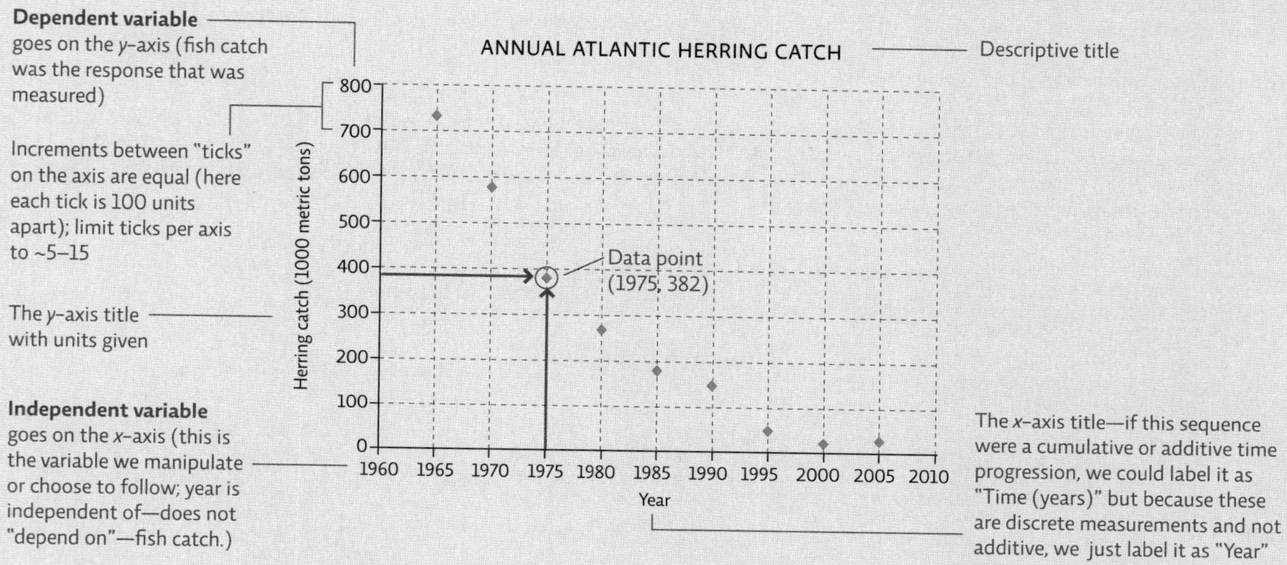

Dependent variable goes on the *y*-axis (fish catch was the response that was measured)
Increments between "ticks" on the axis are equal (here each tick is 100 units apart); limit ticks per axis to ~5–15
The *y*-axis title with units given
Independent variable goes on the *x*-axis (this is the variable we manipulate or choose to follow; year is independent of—does not "depend on"—fish catch.)
ANNUAL ATLANTIC HERRING CATCH
Descriptive title
Herring catch (1000 metric tons)
800
700
600
500
400
300
200
100
0
Data point (1975, 382)
1960 1965 1970 1975 1980 1985 1990 1995 2000 2005 2010
Year
The *x*-axis title—if this sequence were a cumulative or additive time progression, we could label it as "Time (years)" but because these are discrete measurements and not additive, we just label it as "Year"

4. **Draw the line.** Once data points are in place, you can draw a line through the data that highlights the trend in that data, but don't simply connect the dots unless they all line up exactly. Step back and visualize what kind of trend the data are showing and draw a line that approximates that trend. These trend lines can be mathematically determined but can also be fairly accurately estimated by simply drawing in a line that goes through the centre of the data—about as many points will be above as below the line. You can draw a straight line or you may elect to draw a curve to accommodate shifts in the trend.

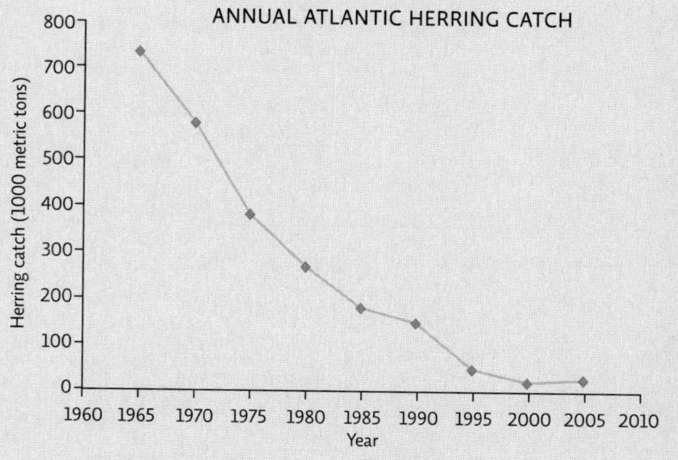

ANNUAL ATLANTIC HERRING CATCH
Herring catch (1000 metric tons)
800
700
600
500
400
300
200
100
0
1960 1965 1970 1975 1980 1985 1990 1995 2000 2005 2010
Year

"Connecting the dots" like this implies that each data point is perfectly accurate and that this exactly represents the relationship between the two variables.

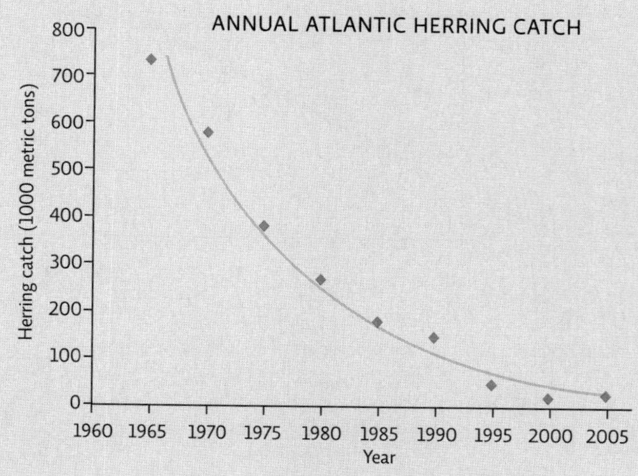

ANNUAL ATLANTIC HERRING CATCH
Herring catch (1000 metric tons)
800
700
600
500
400
300
200
100
0
1960 1965 1970 1975 1980 1985 1990 1995 2000 2005
Year

Drawing in a "trend line" that floats through the cloud of points is a more accurate estimation of the actual relationship seen between the two variables. We could draw a straight trend line for this data, but since it seems that the rate of decline lessens as times goes by, the curve seen here may better represent the relationship.

Interpreting the data

Once the graph is made, we can evaluate the data and draw conclusions. The first step is to simply describe the relationship seen—this is a statement of the *results* (observations). Here we see that between 1965 and 1980, herring catches dropped off dramatically and thereafter continued to drop, but more slowly. Now that we understand the relationship between the two variables, we can draw *conclusions*—make some inferences: What might have caused this relationship? What else may be true because this relationship exists? We could infer from these data that the herring population size also decreased in this time frame. If we know that the same number of fishers were fishing for herring the same number of days each year, the slower decline after 1980 might represent the fact that the fish are more difficult to catch because the population size is smaller. We might also conclude that it has not been as profitable to fish commercially for herring since 1980 as it was in the 1960s and 1970s. These last three statements are conclusions (inferences) based on the results of the study (observations); they are not observations themselves.

Interpolation and extrapolation (projections)

Plotting line graphs also allows us to estimate values of *y* or *x* within the range of our data set, values that we did not actually measure (*interpolation*). We can also create a projection of data points beyond our data set (*extrapolation*) by extending the line. This assumes the same trend will hold at higher or lower *x*-axis values, which may or may not be true; therefore, extrapolations are not likely to be as accurate as interpolations. Extrapolations, also called projections, are often shown as dashed lines.

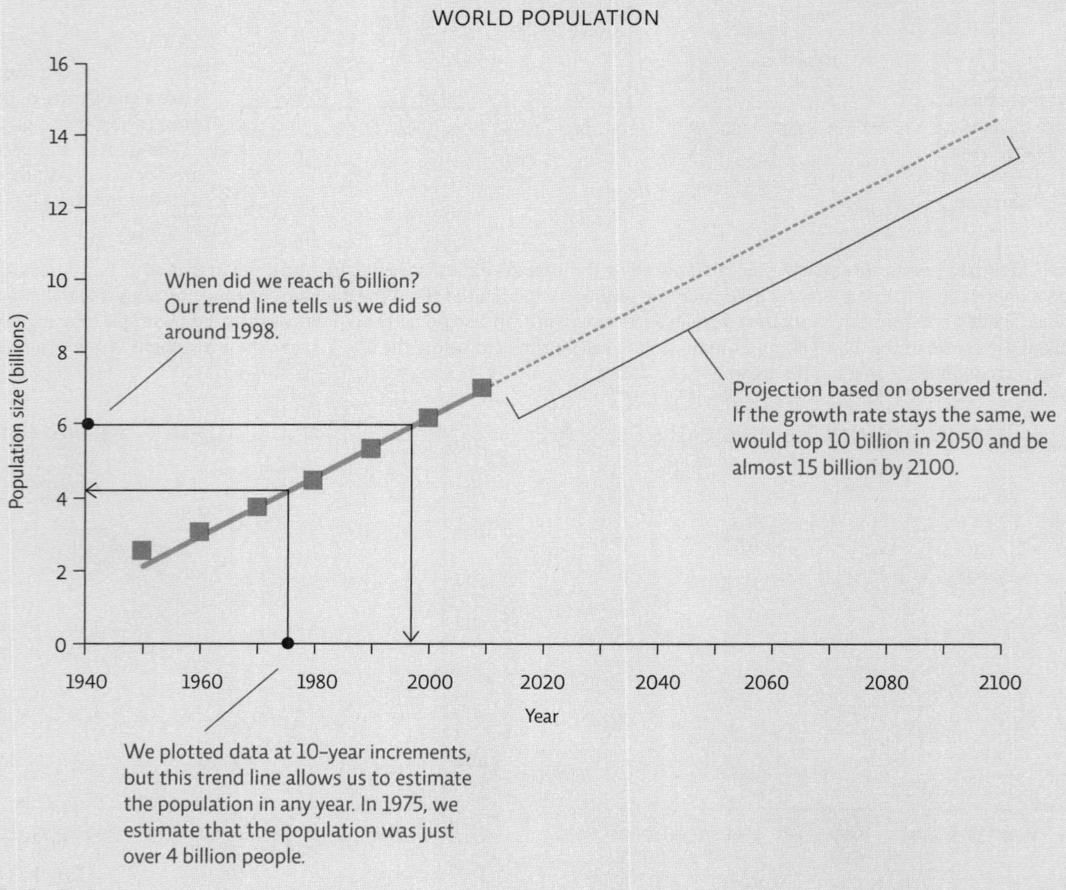
WORLD POPULATION
Population size (billions)
16
14
12
10
8
6
4
2
0
1940
1960
1980
2000
2020
2040
2060
2080
2100
Year
When did we reach 6 billion? Our trend line tells us we did so around 1998.
Projection based on observed trend. If the growth rate stays the same, we would top 10 billion in 2050 and be almost 15 billion by 2100.
We plotted data at 10-year increments, but this trend line allows us to estimate the population in any year. In 1975, we estimate that the population was just over 4 billion people.

B. SCATTER PLOTS

Scatter plots (with or without a trend line) are used when any *x*-value could have multiple *y*-values. For instance, in the second graph shown here, data were collected from various countries. Girls in four of the countries surveyed receive, on average, 4 years of schooling; therefore, there are 4 data points over the *x*-axis value of 4. But each of those countries had different *y*-values (total fertility rate). Here it would make no sense to "connect the dots." The resulting line would be impossible to follow.

It is more appropriate to construct a line that passes through the cloud of points and shows the "trend," just as we did with the line graph. Data points can be entered into a computer graphing program to calculate a "best fit" line, but you can also estimate the path yourself. To pick the best fit line, draw a line (straight or curved) that passes centrally through the cloud of data points, with about as many points above as below the line—the occasional point far away from the others won't impact the line significantly. When completed, your line may or may not be straight, but it should not connect the dots.

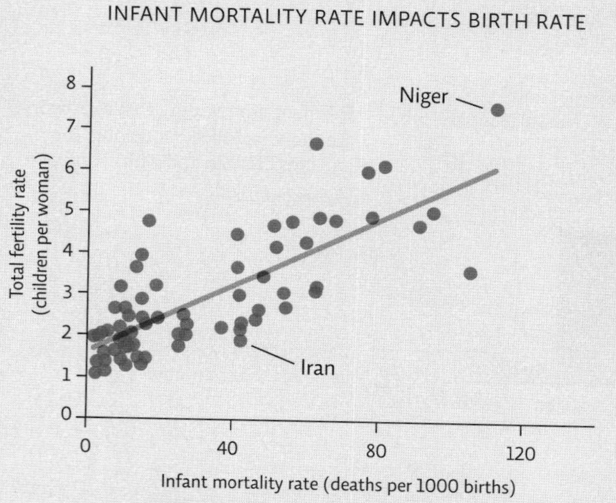

INFANT MORTALITY RATE IMPACTS BIRTH RATE
Niger
Iran
Total fertility rate (children per woman)
0
1
2
3
4
5
6
7
8
0
40
80
120
Infant mortality rate (deaths per 1000 births)

A *positive correlation*—as one variable increases, so does the other—an upward-sloping line

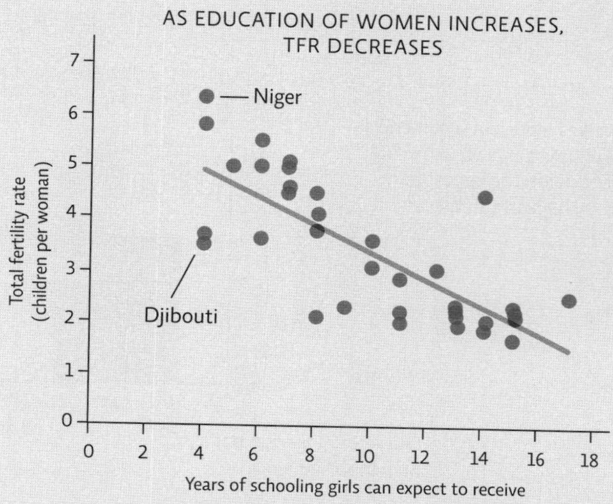

AS EDUCATION OF WOMEN INCREASES, TFR DECREASES
Niger
Djibouti
Total fertility rate (children per woman)
0
1
2
3
4
5
6
7
0
2
4
6
8
10
12
14
16
18
Years of schooling girls can expect to receive

A *negative correlation*—as one variable increases, the other decreases—a downward-sloping line

C. PIE CHARTS

Pie charts are useful when the groups represented by the independent variable are all discrete categories (e.g., red, yellow, blue) rather than a numerical sequence (e.g., 1, 2, 3). In addition, the categories also represent all the subsets of a whole—all the category values add up to 100%. In other words, you have the entire pie! Data values and/or category titles can be shown either inside each "slice" or outside the pie. The data could also be shown as a bar graph (see Section D on the next page), but showing it as a pie chart instead allows one to more easily compare the size of each group to the other groups and to the whole.

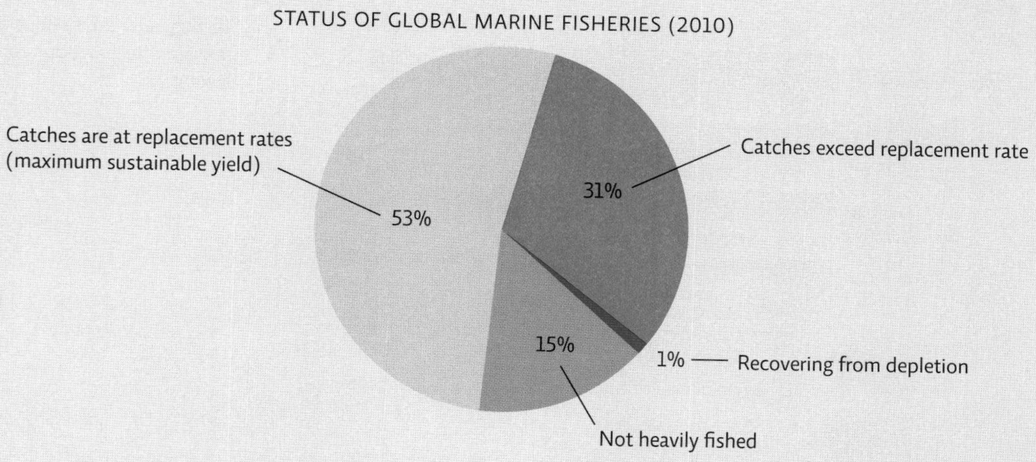

STATUS OF GLOBAL MARINE FISHERIES (2010)
Catches are at replacement rates (maximum sustainable yield)
53%
31%
Catches exceed replacement rate
15%
1%
Recovering from depletion
Not heavily fished

D. BAR GRAPHS

Bar graphs are appropriate in some cases. As with a pie chart, the key consideration is whether the independent variable (the variable you are manipulating in your experiment) is part of a numerical sequence (in which case, a line graph or scatter plot would be used) or represents discrete, separate groups. When you have discrete, separate groups, a bar graph can be used—it would make no sense to connect the data from one group to the next in a line. As with a line graph or scatter plot, the independent variable usually goes on the *x*-axis and the dependent variable is shown on the *y*-axis.

For example, researchers examined the stomach contents of birds from two different colonies. "Colony" is the independent variable because the researchers chose to see if where a bird lived would affect the type of food it ate. The volume of each food type ingested is the dependent variable—it is the data that the researcher set out to find and it may change according to colony location.

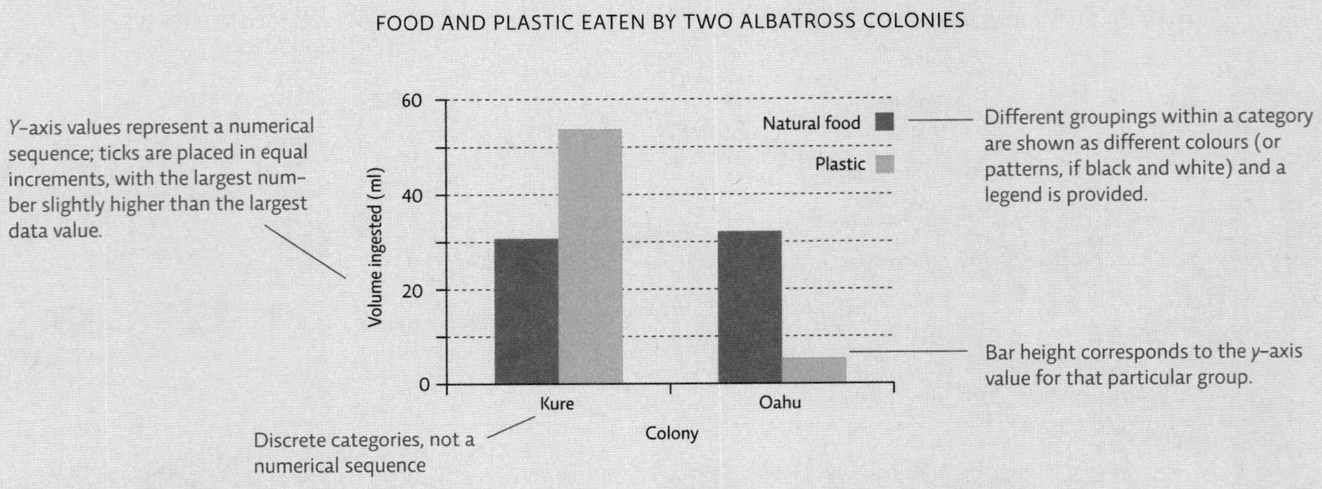
FOOD AND PLASTIC EATEN BY TWO ALBATROSS COLONIES
Y-axis values represent a numerical sequence; ticks are placed in equal increments, with the largest number slightly higher than the largest data value.
Volume ingested (ml)
60
40
20
0
Natural food
Plastic
Kure
Oahu
Colony
Different groupings within a category are shown as different colours (or patterns, if black and white) and a legend is provided.
Bar height corresponds to the y-axis value for that particular group.
Discrete categories, not a numerical sequence

Sometimes it is easier to place the independent variable on the *y*-axis if the labels themselves are long. This prevents the need to place labels sideways, making them harder to read.

The graph below, which shows the percent of women who would like to use birth control but have no access to it (unmet need), displays the independent variable (region) on the *y*-axis and the dependent variable (percent with unmet need) on the *x*-axis.

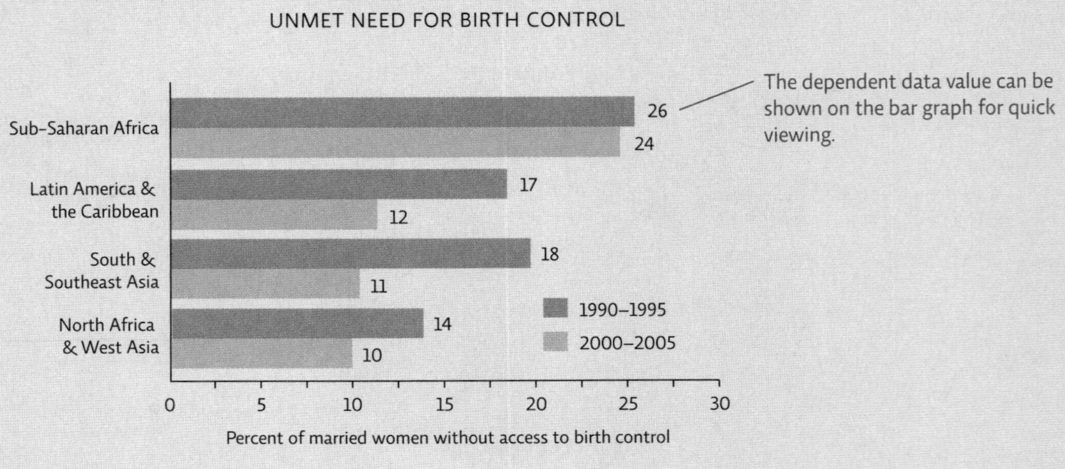
UNMET NEED FOR BIRTH CONTROL
Sub-Saharan Africa
26
24
Latin America & the Caribbean
17
12
South & Southeast Asia
18
11
North Africa & West Asia
14
10
1990–1995
2000–2005
0
5
10
15
20
25
30
Percent of married women without access to birth control
The dependent data value can be shown on the bar graph for quick viewing.

E. AREA GRAPHS

Another useful graph that allows us to view the relative proportion of all the groups being compared is an area graph. It is used when we have a line graph (the independent variable on the *x*-axis is a numerical sequence) showing multiple lines. Each data set (line) is part of a larger group—here we show total fish catch broken down by type of fish. Each line is graphed "on top" of the other, and the space between the lines is filled in with a different colour. This is useful because at any given *x*-axis point (say, the year 1968) we can see what the total fish catch was as well as how much each type of fish contributed to the total catch. The width of the "ribbon" for each fish type at that point represents its *y*-axis value—in this case its catch in 1000 metric tons. (The *y*-axis value opposite the ribbon represents the total for all groups.)

FISH CATCH BY COMMERCIAL GROUP: NEWFOUNDLAND-LABRADOR SHELF

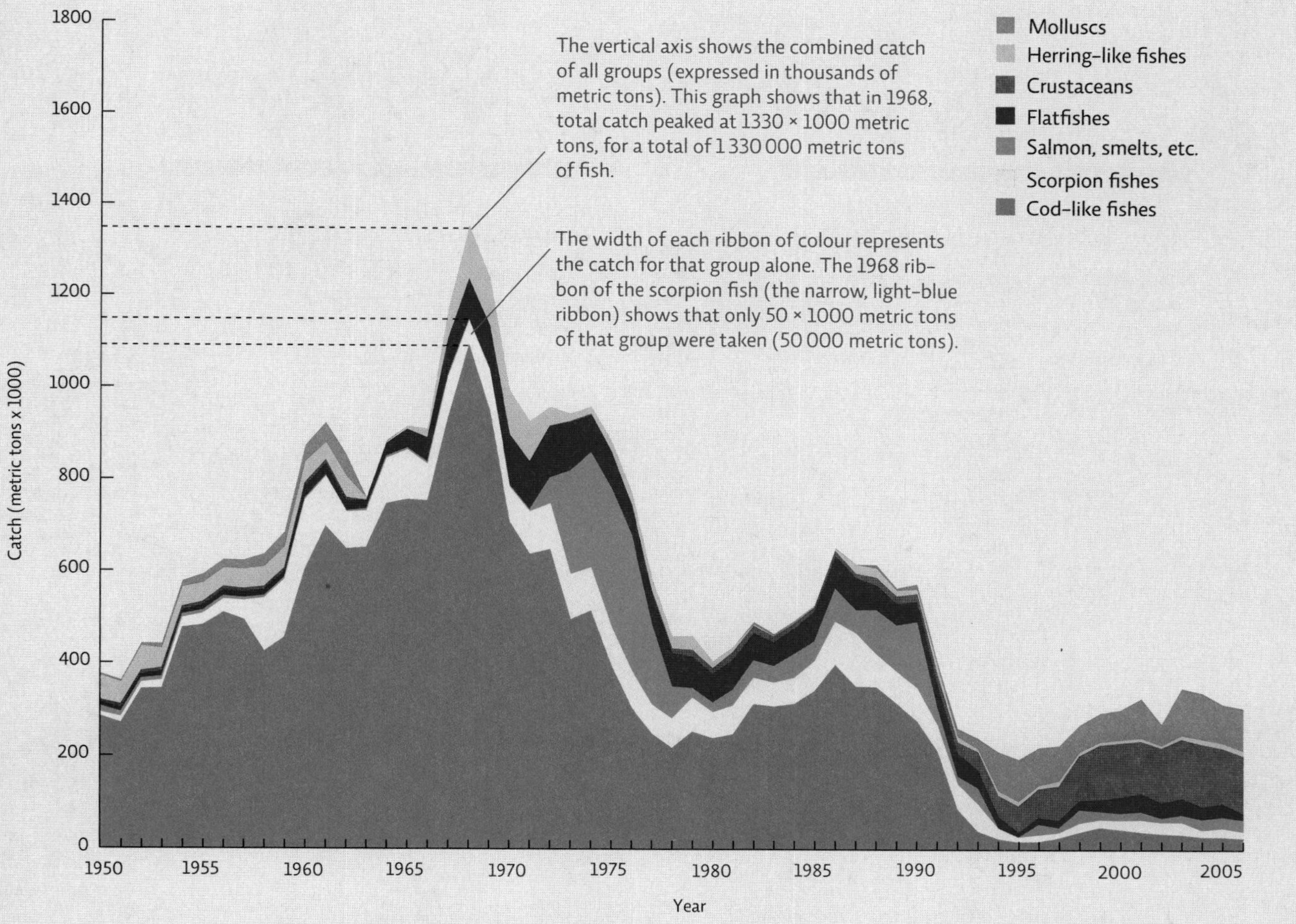
The vertical axis shows the combined catch of all groups (expressed in thousands of metric tons). This graph shows that in 1968, total catch peaked at 1330 × 1000 metric tons, for a total of 1 330 000 metric tons of fish.
The width of each ribbon of colour represents the catch for that group alone. The 1968 ribbon of the scorpion fish (the narrow, light-blue ribbon) shows that only 50 × 1000 metric tons of that group were taken (50 000 metric tons).
Molluscs
Herring-like fishes
Crustaceans
Flatfishes
Salmon, smelts, etc.
Scorpion fishes
Cod-like fishes
Catch (metric tons x 1000)
1800
1600
1400
1200
1000
800
600
400
200
0
1950
1955
1960
1965
1970
1975
1980
1985
1990
1995
2000
2005
Year

APPENDIX 3
STATISTICAL ANALYSIS

DESCRIPTIVE STATISTICS

In science it is not enough to simply collect data and graph it in order to draw conclusions. Suppose we see a difference between data collected for different groups. How different must the data sets be in order to conclude that the groups are different from each other? And how can we determine whether the treatment we applied—say, growing plants with a new fertilizer—really affected growth? We turn to statistical analysis.

Let's look at an example. We are growing two sets of plants, identical in every way except that one is grown without any fertilizer (the control group) and the other is grown with fertilizer (the test group). To draw conclusions, examine the values in the data table.

We begin with descriptive statistics—what are the characteristics of our data set? We calculate useful statistical values for each data set such as the *mean* (the average), the *range* (the highest value minus the lowest value), and the *sample size* (the number of subjects in each group). We might also calculate other values (with the help of any number of readily available online or calculator-based programs) that help describe the data set, such as *standard deviation* (the average amount of variation of each data value from the mean) and *standard error* (a measure that gives us an idea of how accurate our calculated mean really is, based on the standard deviation). Standard error bars are often shown with data as ± values (for example, 14.9 ± 1.4) or as error bars on a graph.

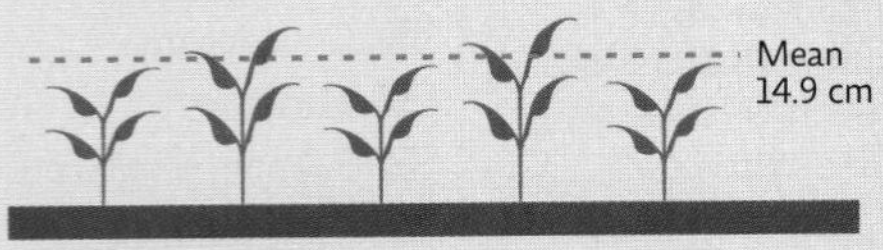
Mean
14.9 cm

CONTROL GROUP—
GROWN WITHOUT FERTILIZER

Sample size: 10
Range: 15
Mean: 14.9
Std deviation: 4.3
Std error: 1.4

Height (cm)
5
11
14
15
15
16
17
18
18
20

Mean
19 cm

TEST GROUP—
GROWN WITH FERTILIZER

Sample size: 10
Range: 4
Mean: 19
Std deviation: 1.2
Std error: 0.38

Height (cm)
17
17
18
18
19
20
20
20
20
21

AVERAGE HEIGHT OF PLANTS GROWN WITH AND WITHOUT FERTILIZER

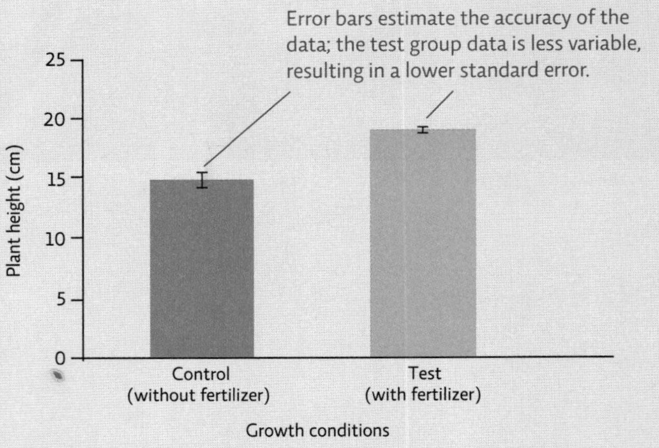
Error bars estimate the accuracy of the data; the test group data is less variable, resulting in a lower standard error.
Plant height (cm)
0
5
10
15
20
25
Control
(without fertilizer)
Test
(with fertilizer)
Growth conditions

INFERENTIAL STATISTICS

We can take our analysis further and evaluate the data with an inferential statistical test to determine how likely it is that the data we obtained from the two groups in our experiment actually represent different responses or whether our two groups are both just subsets of a single, larger group. The statistical test gives us a *p-value*—a number that tells us how much overlap there is between the data sets. In science, we generally require that there be no more than a 5% overlap between the two data sets. If the high end of one set (the control group here) overlaps just a little with the low end of the other data set (our test group), and this overlap is no more than 5%, we can conclude that the two groups most likely represent two distinct populations, a result of the treatment we applied—in this case, the addition of fertilizer. If the overlap had been more than 5% (a p-value > 0.05), we would not have sufficient evidence to conclude that they were indeed two different groups, but instead we would say they were likely to be a single group that varied widely.

The data showed this:

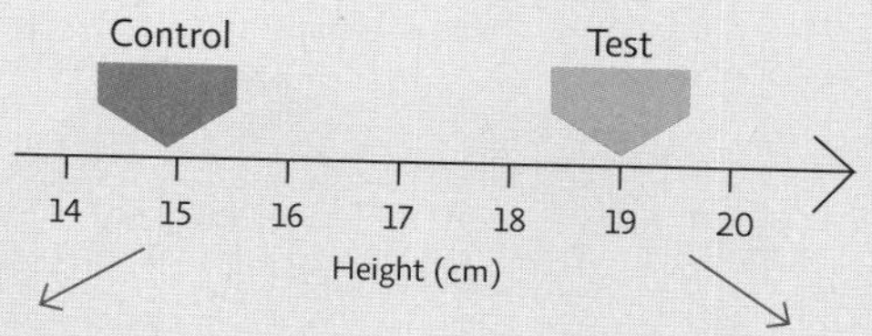
Control
Test
14
15
16
17
18
19
20
Height (cm)

The average height of the plants in the two groups is different—but are they different enough to be considered two different populations?

Is it this?

...or this?

THE DATA POINTS COULD ALL BE PART OF ONE POPULATION

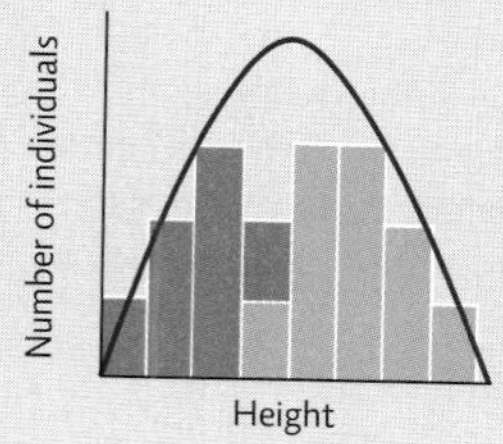
Number of individuals
Height

If the two groups are really part of one population, we might have inadvertently put faster-growing individuals in the test group and/or slower-growing ones in the control group. If that is true, retesting or using a larger sample size should produce some short control and some tall test plants.

THE DATA POINTS COULD REPRESENT 2 SEPARATE POPULATIONS

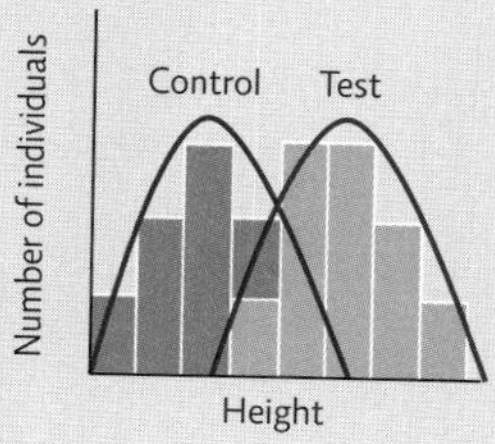
Number of individuals
Control
Test
Height

If the two groups really did respond differently to the treatment (fertilizer use), retesting or using a larger sample size should still produce this same trend, with most test plants being larger than control plants.

For this data set, a *t-test* (a simple statistical test) yields a p-value of 0.035—our data sets overlap 3.5%. Therefore, we can conclude that the two groups are different at the 0.05 level. Because our experimental design eliminated other variables that might have affected growth (the only difference between the two groups was whether plants received fertilizer), it is reasonable to conclude that the greater growth was caused by the fertilizer. As you read about studies and evaluate the authors' conclusions, look for the p-value given with the analysis of the data—if the calculated p-value is larger than 0.05, the author will probably conclude that there is not sufficient evidence to conclude that the variable tested had an effect.

OTHER FACTORS AFFECT RESULTS

VARIABILITY

Data sets with a lot of variability are less likely to show significant differences, even if the means are different—there is more of a chance that the two data sets overlap so much that they must be considered part of the same population.

LITTLE VARIABILITY

LOTS OF VARIABILITY

SAMPLE SIZE IS IMPORTANT

Small sample sizes are less reliable because, due to sampling error, we may have inadvertently sampled mostly unusual subjects; this would give us an incorrect picture of the entire population.

Mean

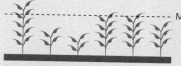

This smaller sample overestimates the mean.

SMALLER CONTROL GROUP

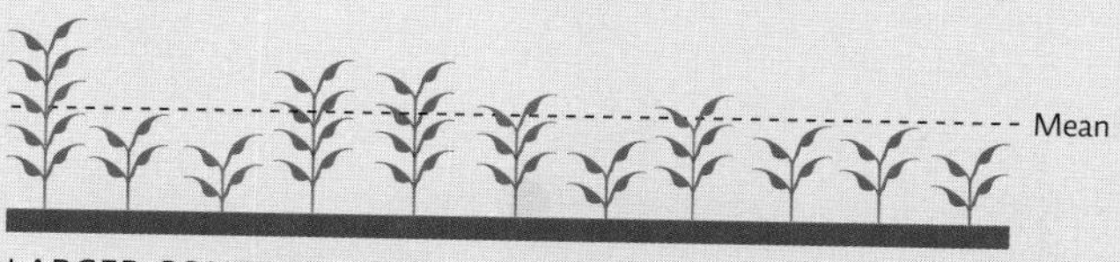
Mean

LARGER CONTROL GROUP

This larger sample suggests that larger plants are not the norm.

APPENDIX 4
GEOLOGY

EARTH IS A DYNAMIC PLANET THAT IS CONSTANTLY CHANGING

Earth is composed of discrete layers; mineral and fossil fuel deposits are found in the layers of Earth's crust. Powerful geologic forces are constantly but slowly rearranging the face of Earth.

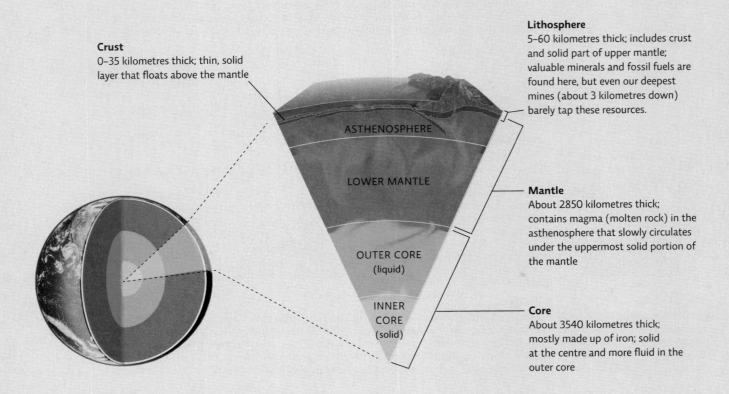

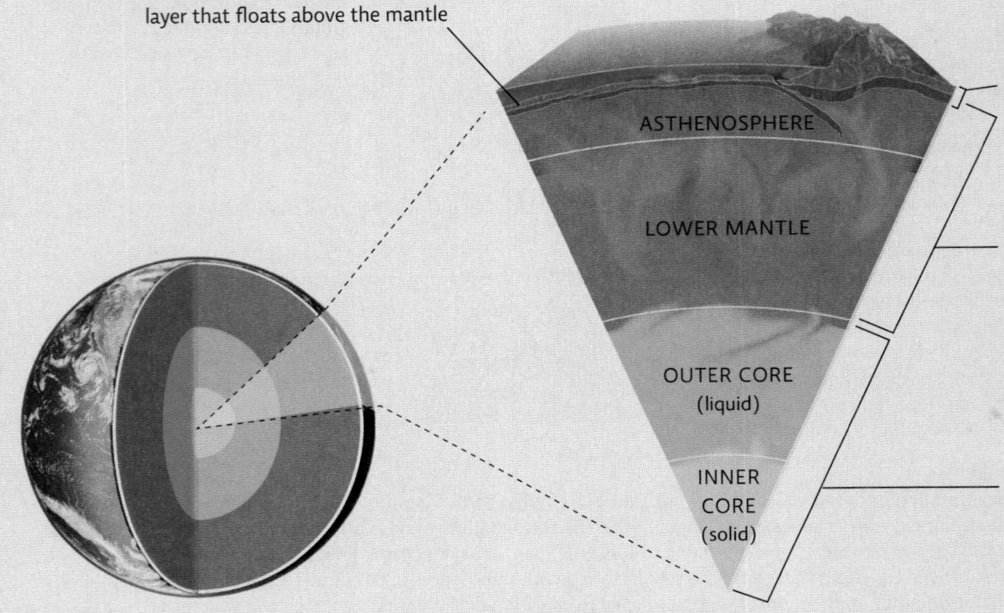
Crust
0–35 kilometres thick; thin, solid layer that floats above the mantle
Lithosphere
5–60 kilometres thick; includes crust and solid part of upper mantle; valuable minerals and fossil fuels are found here, but even our deepest mines (about 3 kilometres down) barely tap these resources.
ASTHENOSPHERE
LOWER MANTLE
OUTER CORE
(liquid)
INNER
CORE
(solid)
Mantle
About 2850 kilometres thick; contains magma (molten rock) in the asthenosphere that slowly circulates under the uppermost solid portion of the mantle
Core
About 3540 kilometres thick; mostly made up of iron; solid at the centre and more fluid in the outer core

EARTH'S CRUST

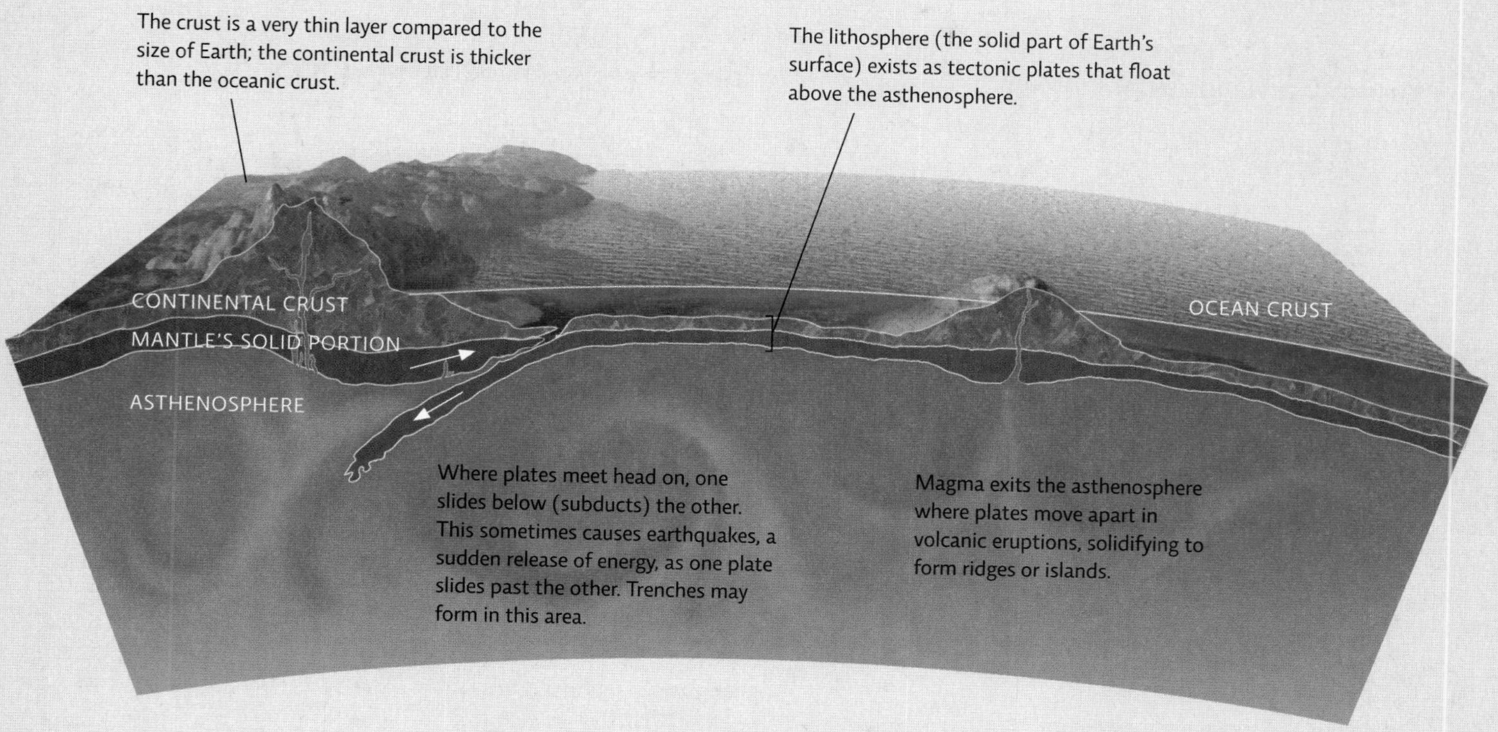
The crust is a very thin layer compared to the size of Earth; the continental crust is thicker than the oceanic crust.
The lithosphere (the solid part of Earth's surface) exists as tectonic plates that float above the asthenosphere.
CONTINENTAL CRUST
MANTLE'S SOLID PORTION
ASTHENOSPHERE
OCEAN CRUST
Where plates meet head on, one slides below (subducts) the other. This sometimes causes earthquakes, a sudden release of energy, as one plate slides past the other. Trenches may form in this area.
Magma exits the asthenosphere where plates move apart in volcanic eruptions, solidifying to form ridges or islands.

THE ROCK CYCLE CONSTANTLY FORMS AND REFORMS ROCKS IN EARTH'S CRUST

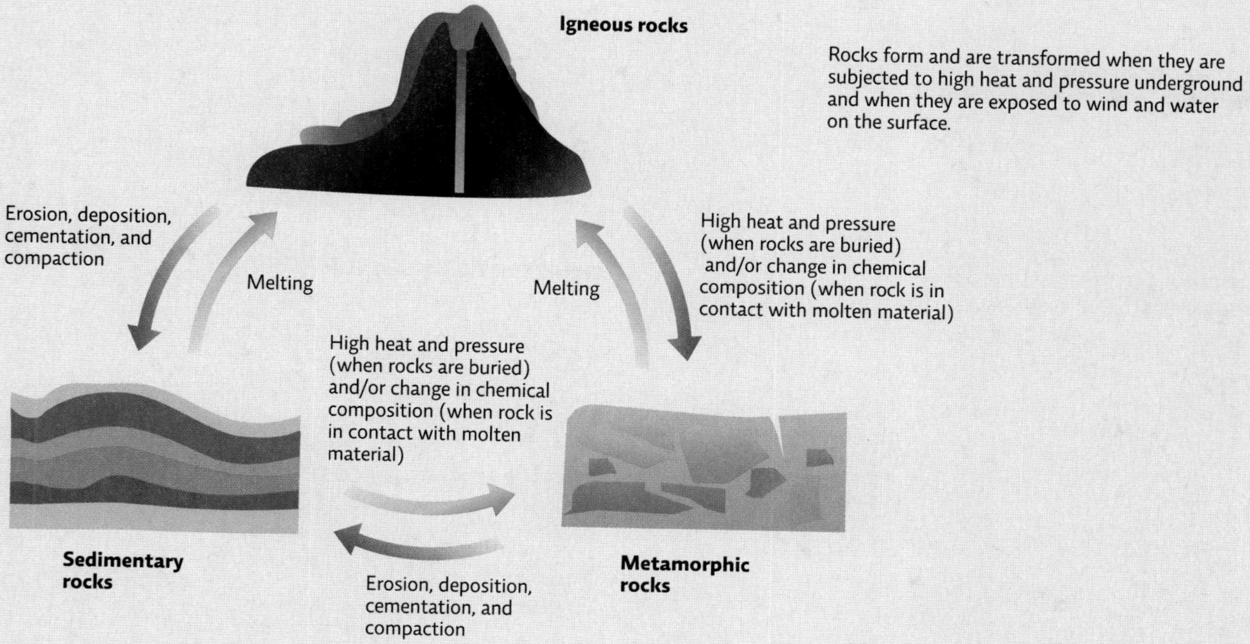
Igneous rocks
Rocks form and are transformed when they are subjected to high heat and pressure underground and when they are exposed to wind and water on the surface.
Erosion, deposition, cementation, and compaction
Melting
Melting
High heat and pressure (when rocks are buried) and/or change in chemical composition (when rock is in contact with molten material)
High heat and pressure (when rocks are buried) and/or change in chemical composition (when rock is in contact with molten material)
Sedimentary rocks
Metamorphic rocks
Erosion, deposition, cementation, and compaction

APPENDIX 5
ANSWERS TO *CHECK YOUR UNDERSTANDING* QUESTIONS

CHAPTER 1 ENVIRONMENTAL LITERACY

1. b
2. a
3. b
4. c
5. d
6. a

CHAPTER 2 SCIENCE LITERACY AND THE PROCESS OF SCIENCE

1. a
2. c
3. b
4. d
5. c
6. c

CHAPTER 3 INFORMATION LITERACY

1. a
2. d
3. a
4. c
5. b
6. d

CHAPTER 4 HUMAN POPULATIONS

1. a
2. c
3. c
4. b
5. d
6. b

CHAPTER 5 ECOLOGICAL ECONOMICS AND CONSUMPTION

1. b
2. a
3. d
4. d
5. c

CHAPTER 6 ECOSYSTEMS AND NUTRIENT CYCLING

1. d
2. b
3. a
4. c
5. a
6. b

CHAPTER 7 POPULATION ECOLOGY

1. d
2. d
3. a
4. c
5. d
6. b

CHAPTER 8 COMMUNITY ECOLOGY

1. d
2. c
3. a
4. b
5. c
6. b

CHAPTER 9 BIODIVERSITY

1. a
2. b
3. c
4. b
5. b
6. c

CHAPTER 10 EVOLUTION AND EXTINCTION

1. c
2. d
3. a
4. a
5. a
6. c

CHAPTER 11 FORESTS

1. b
2. d
3. a
4. c
5. c
6. b

CHAPTER 12 GRASSLANDS

1. a
2. b
3. b
4. a
5. b
6. a

CHAPTER 13 MARINE ECOSYSTEMS

1. c
2. a
3. d
4. a
5. b
6. b

CHAPTER 14 FISHERIES AND AQUACULTURE

1. c
2. b
3. b
4. d
5. c
6. d

CHAPTER 15 FRESHWATER RESOURCES

1. d
2. a
3. b
4. c
5. d
6. a

CHAPTER 16 WATER POLLUTION

1. d
2. b
3. c
4. a
5. a
6. c

CHAPTER 17 SOLID WASTE

1. c
2. a
3. b
4. c
5. a
6. d

CHAPTER 18 AGRICULTURE

1. a
2. c
3. d
4. d
5. b
6. a

CHAPTER 19 COAL

1. b
2. c
3. a
4. c
5. a
6. b

CHAPTER 20 OIL AND NATURAL GAS

1. c
2. c
3. a
4. d
5. d
6. b

CHAPTER 21 AIR POLLUTION.

1. c
2. c
3. b
4. b
5. b
6. d

CHAPTER 22 CLIMATE CHANGE

1. a
2. b
3. d
4. a
5. d
6. b

CHAPTER 23 NUCLEAR POWER

1. c
2. d
3. a
4. b
5. a
6. c

CHAPTER 24 SUN, WIND, AND WATER ENERGY

1. b
2. c
3. d
4. a
5. c
6. c

CHAPTER 25 BIOFUELS

1. a
2. d
3. b
4. b
5. c
6. c

CHAPTER 26 URBANIZATION AND SUSTAINABLE COMMUNITIES

1. d
2. c
3. d
4. b
5. b
6. a

CREDITS/SOURCES

INFOGRAPHIC SOURCES AND REFERENCES

CHAPTER 1:

IG 1.7 Adapted from UN Human Development Indices, 2008 (http://is.gd/7UqClm)

CHAPTER 2:

IG 2.6 Adapted from UN Vital Ozone Graphics, 2007 (http://is.gd/vMwgpH)
IG 2.7 Satellite images, Source: NASA Graphs adapted from ozone graph from "Refrigerant Reclaim Australia," 2009 Annual Report (https://www.refrigerantreclaim.com.au/AR06/) and (http://is.gd/eW5qb1)

CHAPTER 3:

IG 3.3 Adapted from:
Nagel, S. C., *et al.* (1997). *Environmental Health Perspectives*. 105: 70–76.
Ishido, M. & J. Suzuki (2010). *Journal of Health Science*. 56: 175–181.
Lang, I. A., *et al.* (2008). *Journal of the American Medical Association*. 300: 1303–1310.

CHAPTER 4:

IG 4.1 Data from *UN World Population Prospects, the 2010 Revision*.
IG 4.2 Bar graph adapted from Population Reference Bureau, *2010 World Population Data Sheet*; map adapted from World Resources Institute (http://is.gd/t2A8lx)
IG 4.3 Table from Population Reference Bureau, *2010 World Population Data Sheet*; graph adapted from the United Nations Department of Economic and Social Affairs
IG 4.5 Data from U.S. Census Bureau
IG 4.6 Infant mortality adapted from CIA Factbook, 2011 estimates; Desired Fertility adapted from Pritchett, L. (1994). *Population and Development Review*. 20: 1–55; Education adapted from Pew Research Center, *The Future of the Global Muslim-Population*, 2011; Family planning adapted from Sedgh, G., *et al.* (2007). *Women With an Unmet Need for Contraception in Developing Countries and Their Reasons for Not Using a Method*, Guttmacher Institute.
IG 4.7 Adapted from Population Action International
IG 4.8 Adapted from *The Ecological Footprint Atlas 2010*

CHAPTER 5:

IG 5.1 Adapted from Costanza, R., *et al.* (1997). *Nature*. 387:253–260.
IG 5.2 Data from WWF *Living Planet Report, 2012* and *Footprint Interactive Graph* (http://wwf.panda.org/about_our_earth/all_publications/living_planet_report/living_planet_report_graphics/footprint_interactive/)

CHAPTER 6:

IG 6.4 Map adapted from http://is.gd/Kh0ni4
Biome graph adapted from R. Ricklefs, (2000) *The Economy of Nature*, W.H. Freeman, New York

CHAPTER 7:

IG 7.6 Predator-prey graph adapted from Krebs, C. J. (2010). *Proceedings of the Royal Society B*. 278 (1705):481–489.

CHAPTER 8:

IG 8.1 Map adapted from Natural Resources Canada (http://cfs.nrcan.gc.ca/pages/125)

CHAPTER 9:

IG 9.2 Data from *Number of Living Species in Australia and the World Report*, 2009
IG 9.5 Adapted from Conservation International (PDF: http://is.gd/PsXTqP)

CHAPTER 10:

IG 10.3 Map adapted from Savidge, J.A. (1987) *Ecology*, 68: 660–668

CHAPTER 11:

IG 11.1 Map adapted from http://is.gd/Kh0ni4
IG 11.3 Adapted from Costanza, R., *et al.* (1997). *Nature*. 387:253–260.
IG 11.4 Map adapted from UN Food and Agriculture Organization (PDF: http://is.gd/ma67Ps)

CHAPTER 12:

IG 12.1 Map adapted from http://is.gd/Kh0ni4
IG 12.3 Source: U.S. Department of Agriculture (http://is.gd/pFRKaX)

CHAPTER 13:

IG 13.1 Graph adapted from NOAA (http://is.gd/pAuD5o); maps adapted from Cao, L. & K. Caldeira. (2010) Climatic Change.99, 1–2.
IG 13.3 Map and graphs adapted from UN Environmental Programme (http://is.gd/VI5oz2)

CHAPTER 14:

IG 14.1 Map adapted from Aotearoa (http://en.wikipedia.org/wiki/Atlantic_cod); graph adapted from UN Millennium Ecosystem Assessment (http://is.gd/25KrJl)
IG 14.3 Graph and fish catch adapted from The Sea Around Us Project
IG 14.4 Data from UN Food and Agriculture Organization
IG 14.5 Map adapted from the Sea Around Us Project (www.seaaroundus.org/eez/124.aspx) and Fisheries and Oceans Canada (www.dfo-mpo.gc.ca/oceans/marineareas-zonesmarines/mpa-zpm/index-eng.htm)

CHAPTER 15:

IG 15.3 Map adapted from UN Food and Agriculture Organization (http://is.gd/1TgsbR); graph adapted from the

Population Reference Bureau; Domestic Water Use Graph adapted from Hoekstra, A. Y. and A. K. Chapagain (2007). *Water Resources Management*, 21:35–48.
IG 15.4 Adapted from World Water Assessment Programme (WWAP); gallons of water per food or product from *National Geographic* (http://is.gd/MCeBGJ)
IG 15.7 Pie chart adapted from University of Georgia Cooperative Extension; data for water use per 1,000 kwh from Institute of Electrical and Electronics Engineers

CHAPTER 16:

IG 16.5 Based on the protocol for the *Save Our Steams* water monitoring program of the Izaak Walton League of America.

CHAPTER 17:

IG 17.1 Line graph adapted from UN (2012) *Vital Waste Graphics 3* (http://www.grida.no/publications/vg/waste3/); Data for pie chart from Environment Canada (2012) *Human Activity and the Environment: Solid Waste Report*
IG 17.3 Source: EPA
IG 17.5 From: Young, L.C., *et al.* (2009). PLoS ONE 4(10): e7623

CHAPTER 18:

IG 18.3 From: Furuno, Takao. *The Power of the Duck*. Tasmania: Takari Publications, 2001
IG 18.4 Adapted from Smil, V. 2001 *Feeding the world: A challenge for the twenty-first century*. Cambridge, MA: MIT Press.

CHAPTER 19:

IG 19.1 Source: Tennessee Valley Authority; pie chart data from Natural Resources Canada (www.nrcan.gc.ca/energy/sources/electricity/1387)
IG 19.3 Map adapted from Britannica Encyclopedia (http://is.gd/MUgevK)
IG 19.5 Adapted from Kentucky Geological Survey (http://is.gd/nyyEbh)
IG 19.7 Adapted from World Coal Association

CHAPTER 20:

IG 20.1 Data for map from U.S. Energy Information Agency (http://www.eia.gov/analysis/studies/usshalegas/) and Alberta Geological Survey (http://www.ags.gov.ab.ca/energy/oilsands/)
IG 20.3 Adapted from Energy Watch Group (February 2008) *Crude Oil – The Supply Outlook, Revised Edition*
IG 20.4 Source: U.S. Energy Information Administration, *Oil & Gas Journal*
IG 20.7 Adapted from *Scientific American* (http://www.scientificamerican.com/slideshow.cfm?id=how-to-turn-tar-sands-into-oil-slideshow)
IG 20.9 Adapted from Technovelgy (http://is.gd/LerVhk)

CHAPTER 21:

IG 21.1 Source: the World Health Organization
IG 21.2 Source: EPA
IG 21.3 Adapted from Laden, Francine, *et al.*, (2006). *American Journal of Respiratory and Critical Care Medicine*. 173:667–672.
IG 21.4 Adapted from PhysicalGeography.net; maps adapted from National Atmospheric Deposition Program
IG 21.5 Source: EPA
IG 21.6 Smokestack scrubber adapted from Encyclopedia Britannica; cap-and-trade adapted from the *Washington Post*

CHAPTER 22:

IG 22.1 Adapted from IPCC, 4th Assessment Report, Working Group 1 Report: *The Physical Science Basis* and Environment Canada, Climate Trends and Variations Bulletin
IG 22.2 Temperature anomalies source: National Climatic Data Center/NOAA; annual global average temperature (land and ocean) source: National Climatic Data Center/NOAA; Cumulative loss of glacier ice source: United Nations Environmental Programme ; sea level change over time source: National Climatic Data Center/NOAA; precipitation changes from: National Climatic Data Center/NOAA; Increase in storms from: EPA
IG 22.3 Bar graph adapted from Intergovernmental Panel on Climate Change, 4th Assessment Report, Working Group 1 Report: *The Physical Science Basis*
IG 22.5 Adapted from Intergovernmental Panel on Climate Change, 4th Assessment Report, Working Group 1 Report: *The Physical Science Basis*
IG 22.7 Upper graphs source: adapted from NASA; lower graph source: adapted from United States Global Change Research Program
IG 22.8 Adapted from Intergovernmental Panel on Climate Change, 4th Assessment Report, Working Group 1 Report: *The Physical Science Basis*
IG 22.10 Top graph adapted from Intergovernmental Panel on Climate Change, 4th Assessment Report, Working Group I Report: *The Physical Science Basis*; bottom graph source: Pacala, S. & R. Socolow (2004). *Science*, 305:968–972.
IG 22.11 Adapted from Environment Canada and the National Round Table on the Environment and the Economy, 2012 report: *Reality Check: The State of Climate Progress in Canada* (http://collectionscanada.gc.ca/webarchives2/20130322165457/http://nrtee-trnee.ca/wp-content/uploads/2012/06/reality-check-report-eng.pdf), and NASA News Report: *Climate Change May Bring Big Ecosystem Changes* (http://www.jpl.nasa.gov/news/news.php?release=2011-387)

CHAPTER 23:

IG 23.1 Adapted from aboutnuclear.org
IG 23.2 Decay chain: adapted from Department of Energy; bottom graph adapted from GeoKansas
IG 23.3 Flow chart adapted from World Information Service On Energy (WISE) Uranium Project; Fuel assembly adapted from climateandfuel.com
IG 23.4 Adapted from atomicarchive.com
IG 23.5 Adapted from Nuclear Regulatory Commission and the Union of Concerned Scientists
IG 23.6 Image adapted from Stannered (http://is.gd/PV4jfL)

CHAPTER 24:

IG 24.1 Bar graph source: U.S. Energy Information Administration, *International*

Energy Statistics Database; pie chart source: U.S. Energy Information Administration
IG 24.2 Adapted from US. Department of the Interior, *Wind Energy Development Programmatic Environmental Impact Statement*
IG 24.3 Adapted from *Power and Energy* (http://is.gd/5B34FF)
IG 24.4 Source: U.S. Department of Energy
IG 24.5 Source: U.S. Department of Energy (http://is.gd/WECJHW); ground source heat pump adapted from Tennessee Valley Authority
IG 24.7 Sources: Natural Resources Canada and U.S. Department of Energy

CHAPTER 25:
IG 25.3 Adapted from Tilman, D., *et al* (2006) *Science* 314:1598-1600.
IG 25.5 Adapted from oilgae.com (http://is.gd/yQDsh0)
IG 25.6 Adapted from alternate-energy-sources.com
IG 25.7 Sources: Natural Resources Canada (Office of Energy Efficiency) and U.S. Department of Energy

CHAPTER 26:
IG 26.1 Map source: UN Population Division; CIA World Factbook; Urban vs. rural line graph adapted from United Nations, Population Division; megacities bar graph adapted from United Nations, *World Urbanization Prospects, the 2011 Revision*
IG 26.2 Adapted from *National Geographic*
IG 26.4 Adapted from Stats Canada (http://www12.statcan.gc.ca/census-recensement/2006/as-sa/97-550/tables-tableaux-notes-eng.cfm#maps)
IG 26.6 Source: Smart Growth BC (http://www.smartgrowth.bc.ca/Default.aspx?tabid=133)

ANALYZING THE SCIENCE DATA SOURCES

CHAPTER 1:
Modified from Brown, L.R., "World on the Edge—Food and Agriculture Data—Livestock and Fish." pp. 7, 15. Earth Policy Institute. http://www.earthpolicy.org/datacenter/pdf/book_wote_livestock.pdf.

CHAPTER 2:
Modified from http://undsci.berkeley.edu/article/ozone_depletion_01. Figure 28 - A Plot of Chlorine Monoxide and Ozone Concentrations from Ozone Depletion: Uncovering the hidden hazard of hairspray. Originally from: Anderson, J.G., *et al.* 1989. Ozone destruction by chlorine radicals within the Antarctic vortex: the spatial and temporal evolution of ClO–O_3 anticorrelation based on in situ ER-2 data. *Journal of Geophysical Research* 94:11465–11479.

CHAPTER 3:
Adapted from http://pollutioninpeople.org/results/report/chapter-3/metals_1 (Figure 3: Mercury levels in participant hair; http://pollutioninpeople.org/results/report/chapter-6/ddt_pcb_2) (Figure 7: DDT exposure was measured as the breakdown product p,p'DDE in blood serum).

CHAPTER 4:
Adapted from United Nations, Department of Economic and Social Affairs, Population Division (2011): World Population Prospects: The 2010 Revision. New York. (Updated: 15 April 2011).
http://esa.un.org/unpd/wpp/Analytical-Figures/htm/fig_1.htm (Graph A); (Updated: 5 July 2011).
http://esa.un.org/unpd/wpp/Analytical-Figures/htm/fig_13.htm (Graph B).

CHAPTER 5:
Modified from Brown, L.R., "Learning from China: Why the Existing Economic Model Will Fail." Earth Policy Institute. 8 Sep 2011. http://www.earthpolicy.org/data_highlights/2011/highlights18.

CHAPTER 6:
Modified from: http://www.globalchange.umich.edu/globalchange1/current/lectures/kling/rainforest/rainforest_table.html. Original data from: J. Terborgh 1992. Diversity and the tropical rain forest. Scientific American Library, W. H. Freeman, New York, xii + 242 pages.

CHAPTER 7:
Modified from "Kaibab Plateau Deer Population: 1907–1940" http://www.hhh.umn.edu/centers/stpp/pdf/KaibabPlateauExercise.pdf (Redrawn from Leopold, A)

CHAPTER 8:
Adapted from Kayranli B., *et al* (2010) "Carbon Storage and Fluxes within Freshwater Wetlands: a Critical Review," *Wetlands* 30:111–124.

CHAPTER 9:
Modified from Cincotta, R.P., *et al* (2000) *Nature.* VOL 404. 27

CHAPTER 10:
Adapted from "OVERPOPULATION: A Key Factor in Species Extinction." Center for Biological Diversity. http://www.biologicaldiversity.org/campaigns/overpopulation/.

CHAPTER 11:
Adapted from "FAO Forestry Paper 163: Global Forest Resources Assessment 2010: Main report." Food and Agriculture Organization of the United Nations, Rome 2010. http://www.fao.org/docrep/013/i1757e/i1757e00.htm Chapter 6 - Protective functions of forest resources.

CHAPTER 12:
Adapted from Heidenreich, B. "What are Global Temperate Grasslands Worth? A Case for their Protection." Figure 1, Habitat Conversion and Protection in the World's 13 Terrestrial Biomes. The World Temperate Grasslands Conservation Initiative, International Union for Conservation of Nature. July 2009. http://www.iucn.org/about/union/commissions/wcpa/wcpa_puball/wcpa_pubsubject/wcpa_grasslandspub/?4266/What-are-Global-Temperate-Grasslands-worth-A-case-for-their-protection

Originally from: Hoekstra J.M., *et al*, 2005. "Confronting a biome crisis: global disparities of habitat loss and protection." *Ecology Letters*. 8:23 – 29.

CHAPTER 13:
Modified from Burke, L., *et al*, "Reefs at Risk Revisited." February 2011. Page 42, Table 4.1 http://www.wri.org/publication/reefs-at-risk-revisited.

CHAPTER 14:
Modified from Talberth, J., *et al*, "The Ecological Fishprint of Nations: Measuring Humanity's Impact on Marine Ecosystems." http://www.wwf.dk/dk/Service/Bibliotek/Hav+og+fiskeri/Rapporter+mv./FishprintofNations
2006. Page 8, Figure 3.

CHAPTER 15:
Adapted from Figure "DALYs attributable to water, sanitation and hygiene (diarrhea), 2004." World Health Organization. Public Health Information and Geographic Information Systems (GIS). 2011. http://gamapserver.who.int/mapLibrary/Files/Maps/Global_wsh_daly_2004.png

CHAPTER 16:
Adapted from Collins, S. J. and Russell, R. W. (2009) "Toxicity of road salt to Nova Scotia amphibians." *Environmental Pollution*. 157(1):320-324.

CHAPTER 17:
Adapted from "Plastic Debris in the World's Oceans." Table 2.1 Number and Percentage of Marine Species Worldwide with Documented Entanglement and Ingestion Records. Greenpeace. www.unep.org/regionalseas/marinelitter/.../plastic_ocean_report.pdf

CHAPTER 18:
Modified from Rodale Institute. "The Farming Systems Trial." Figure "Comparison of FST Organic and Conventional Systems." http://www.rodaleinstitute.org/fst30years

CHAPTER 19:
Modified from National Institute for Occupational Safety and Health, 2007 World Report. http://www.cdc.gov/niosh/programs/mining/risks.html (Graphs A and B)
Table 2-1 as reported by the National Center for Health Statistics, CDC. http://blogs.wvgazette.com/coaltattoo/files/2010/10/blacklungdeathchart1.jpg (Graph C)

CHAPTER 20:
Adapted from Murphy, D., *Business Insider*. http://www.businessinsider.com/does-peak-oil-even-matter-2010-12. Figure 1. Estimates of the cost of production for oil production from various locations. Data from Cera.[4]

CHAPTER 21:
Modified from Environment Canada, National Air Pollution Surveillance (NAPS) Network, Ottawa, 2004. http://www.ecoinfo.org/env_ind/region/smog/smog_e.cfm.

CHAPTER 22:
Adapted from Woodall, C. W., *et al.* (2009). "An indicator of tree migration in forests of the eastern United States." *Forest Ecology and Management*, 257: 1434–1444.

CHAPTER 23:
Modified from Clean Energy Insight. http://www.cleanenergyinsight.org/wp-content/uploads/2009/08/comparingindustrysafety_graph.jpg. U.S. Bureau of Labor Statistics, 2007.

CHAPTER 24:
Adapted from Sibley, David. "Causes of Bird Mortality." November 18th, 2010. Sibley Guides: Identification of North American birds and trees. http://www.sibleyguides.com/conservation/causes-of-bird-mortality/. (Graph 1)
Adapted from Erickson, W., *et al.* (2001). "Avian Collisions with Wind Turbines: A summary of existing studies and comparisons to other sources of Avian Collision Mortality in the United States." West Inc., prepared for the National Wind Coordinating Committee. http://www.duke.edu/web/nicholas/bio217/ptb4/batdata.html (Table 1)
Adapted from Arnett, E., *et al.* (2005). "Relationships Between Bats and Wind Turbines in Pennsylvania and West Virginia: An assessment of fatality search protocols, patterns of fatality, and behavioral interactions with wind turbines." Report prepared for Bats and Wind Energy Cooperative. (Graphs 2 and 3)

CHAPTER 25:
Modified from: http://jcwinnie.biz/wordpress/imageSnag/GHG_from_various_transportation_fuels.png
Originally from Scharlemann, J. P. W. and Laurance, W. F. (2008) *Science*. 43-44.

CHAPTER 26:
Graph adapted from New Buildings Institute (2008) *Energy Performance of LEED® for New Construction Buildings: Final Report* (http://newbuildings.org/sites/default/files/Energy_Performance_of_LEED-NC_Buildings-Final_3-4-08b.pdf); other data from the EPA (http://www.energystar.gov/index.cfm?fuseaction=buildingcontest.eui).

PHOTOS

COVER:
Rolf Hicker/All Canada Photos/Corbis

FRONT MATTER:
p. II Courtesy Linnea Avallone/Concordiasi Team; p. IV Mike Grandmaison/Corbis; p. VII Andreas Strauss/Getty Images; p. VIII Ken Canning/E+/Getty Images; p. X Benjamin Rusnak/Zuma Press/Newscom; p. XII Justin Lambert/Getty Images; p. XV Nina Berman/Noor/Redux; p. XVI (top to bottom) Marnie Branfireun; Stephen R. Karr; Courtesy of Andrea Gawrylewski; William

Prouty; p. XVIII Paul Alan Cox; p. XIX Gyro Photo/Amana Images/Getty Images; p. XXI (top) Steve Northup/Time Life Pictures/ Getty Images; (bottom) John E. Marriott/ All Canada Photos/Getty Images; p. XXV Mauricio Handler/National Geographic Stock; p. XXIX Andrew Henderson/National Geographic Stock.

CHAPTER 1:
pp. 0–1 Paul Souders/WorldFoto/Aurora Photos; p. 3 Pete Ryan/National Geographic Stock; p. 4 (left) Ashley Cooper/GHG/ Aurora Photos; (right) Rafn Sigurbjornsson/ Verkis Consulting Engineers; p. 6 NSIDC Courtesy Ted Scambos and Rob Bauer; p. 8 (left to right) Mads Nissen/Panos Pictures; Nurcholis/Rex Features/Associated Press; Martin Roemers/Panos Pictures; p. 11 Danita Delimont/Gallo Images/Getty Images; p. 15 Andreas Strauss/Getty Images; p. 16–17 (left to right) Terry Smith/Time & Life Pictures/Getty Images; Don Mackinnon/ AFP/Newscom; Rachel Carson Council/ Zuma Press/Newscom; Kyodo/Newscom.

CHAPTER 2:
pp. 20–21 Courtesy Linnea Avallone/ Concordiasi Team; p. 23 (top) George Steinmetz/National Geographic Stock; (bottom) Courtesy Dr. Susan Solomon; p. 29 David Hay Jones/Science Source; p. 32 Bettmann/Corbis; p. 34 NASA (x3); p. 35 Bart Coenders/iStockphoto.

CHAPTER 3:
pp. 38–39 ULTRA.F/Digital Vision/Getty Images; p. 41 Bettmann/Corbis; p. 42 Chris Sattlberger/Cultura/Aurora Photos; p. 44 Peter Essick/Aurora Photos; p. 48 camilla$$/ Shutterstock; p. 53 iStockphoto/Thinkstock.

CHAPTER 4:
pp. 56–57 Xinhua/eyevine/Redux; p. 61 Imaginechina/Zuma Press; p. 63 Mufty Munir/EPA/Newscom; p. 68 Bettmann/Corbis.

CHAPTER 5:
p. 76 Jorgen Caris/Hollandse Hoogte/Redux; p. 77 Fabrice Gaëtan/Mountain Equipment Co-op/Newscom; p. 88 Andy Ryan/Corbis Outline; p. 89 (top) Andrew Hetherington/ Redux; (bottom) Tara Walton/The Toronto Star/Zuma Press/Newscom.

CHAPTER 6:
pp. 92–93 UIG/Getty Images; p. 95 (top) Tim Roberts/AFP/Getty Images; (bottom) Judy Natal; pp. 98–99 (left to right) George Burba/Shutterstock; David Noton/ Alamy; IDAK/Shutterstock; iStockphoto/ Thinkstock; Norma Jean Gargasz/ age fotostock; Vladimir Kondrachov/ Shutterstock; p. 102 Judy Natal; p. 108 Judy Natal; p. 109 Peter Essick/Aurora Photos.

CHAPTER 7:
pp. 112–113 Ken Canning/E+/Getty Images; p. 115 National Park Service; p. 117 (top to bottom) Tim Graham/The Image Bank/ Getty Images; John Elk/Lonely Planet Images/ Getty Images; Nature's Images/ Science Source; p. 120 (clockwise from top left) James Balog/Aurora Photos/Getty Images; David R. Frazier/Danita Delimont/ Newscom; Debbie Noda/Modesto Bee/ Zuma Press/Newscom; Lars Thulin/ Johnér Images/Corbis; Martin Harvey/ Getty Images; Karl Gehring/Denver Post/ Getty Images; p. 121 (top row left to right) Norbert Rosing/National Geographic Stock; Wayne Lynch/age fotostock; Juniors Bildarchiv GmbH/Alamy; Michael Quinton/ National Geographic Stock; (bottom row left to right) John W. Bova/Science Source; Taylor S. Kennedy/National Geographic Stock; p. 123 Wild Horizon/Universal Images Group/Getty Images; p. 125 John E. Marriott/All Canada Photos/Getty Images.

CHAPTER 8:
pp. 128–129 Truus van Gog/Hollandse Hoogte/Redux; p. 131 Mike Grandmaison/ Corbis; p. 133 Barry Mansell/Nature Picture Library; p. 137 Marie Read/Nature Picture Library; p. 141 Rolf Nussbaumer/Nature Picture Library; p. 144 Valerie Giles/Science Source/Getty Images; p. 145 Andrew Howe/ Getty Images.

CHAPTER 9:
p. 148 (clockwise from top left) iStockphoto/Thinkstock; Thinkstock; iStockphoto/Thinkstock; Medioimages/ Photodisc/Thinkstock; Thinkstock; Jupiterimages/Thinkstock; iStockphoto/ Thinkstock; p. 149 (counterclockwise from top right) Photos.com/Thinkstock; Siri Stafford/Thinkstock; iStockphoto/ Thinkstock; iStockphoto/Thinkstock; Ingram Publishing/Thinkstock; Tom Brakefield/Thinkstock; iStockphoto/ Thinkstock; p. 150 Randy Olson/National Geographic Stock; p. 151 Paul Alan Cox; p. 154 Paul Alan Cox; p. 155 Danita Delimont/Gallo Images/Getty Images; p. 157 (illustrations, top to bottom) J. Bedmar/Iberfoto/The Image Works; Garo/ Phanie/The Image Works; The Natural History Museum/The Image Works; Nicolle Rager Fuller; The Natural History Museum/ The Image Works; Mary Evans Picture Library/The Image Works; Nicolle Rager Fuller; Jacques Boyer/Roger-Viollet/The Image Works; Custom Medical Stock Photo; p. 157 (photographs) Paul Alan Cox (x3); p. 158 Chris Johns/National Geographic Stock; p. 159 (clockwise from top left) Henk Meijer/Alamy; Tim Laman/Nature Picture Library; Paul Nicklen/National Geographic Stock; Manfred Gottschalk/ Lonely Planet Images/Getty Images; p. 161 (clockwise from top left) Nicolle Rager Fuller; Tavita Togia/National Park Service; Patrick Sahar/Alamy; Georgette Douwma/ Science Source; Nicolle Rager Fuller; Nicolle Rager Fuller; p. 163 Pete McBride/National Geographic Stock.

CHAPTER 10:
pp. 166–167 James Balog/Stone/Getty Images; p. 168 (top) Paul Nicklen/National Geographic Stock; (bottom) Courtesy Dr. Julie Savidge; p. 173 (photograph, left) James Balog/Stone/Getty Images; (illustrations, right) H. Douglas Pratt/ North Carolina State Museum of Natural Sciences (x6); p. 176 Joel Sartore/National Geographic Stock; p. 177 (illustrations) Nicolle Rager Fuller (x5); p. 180 (clockwise from top) DaddyBit/iStockphoto; Jimmy

James Bond/iStockphoto; Eric Isselée/iStockphoto (x2); p. 181 Joel Sartore/National Geographic Stock.

CHAPTER 11:
pp. 184–185 Benjamin Rusnak/Food For The Poor/zReportage.com/Zuma Press; p. 187 James P. Blair/National Geographic Stock; p. 188 (left to right) IDAK/Shutterstock; Chad Ehlers/Alamy; Frans Lan/Mint Images/age fotostock; p. 189 Benjamin Rusnak/Food For The Poor/Zuma Press; p. 191 Abbie Trayler-Smith/Panos Pictures; p. 192 Sergio Ramzzotti/Parallelozero/Aurora Photos; p. 198 Hataigan Doungbal/age fotostock; p. 199 Micheline Pelletier/Corbis; p. 200 Frans Lanting/National Geographic Stock; p. 201 Gary Gardiner.

CHAPTER 12:
pp. 204–205 Barrett Hedges/National Geographic Society/Corbis; p. 207 Randy Olson/National Geographic Stock; p. 208 (clockwise from top left) Danny Warren/Thinkstock; Irina Bazhanova/Thinkstock; Zbynk Buival/age fotostock; Daniel Haller/Thinkstock; Photos.com/Thinkstock; p. 211 James Richardson/National Geographic Stock; p. 212 Steve Northup/Time Life Pictures/Getty Images; p. 214 Mitchell Kanashkevich/The Image Bank/Getty Images; p. 218 Peter McBride/National Geographic Stock.

CHAPTER 13:
pp. 222–223 Courtesy NOAA's Aquarius Reef Base at Florida International University; p. 224 (top) Courtesy NOAA's Aquarius Reef Base at Florida International University; (bottom) Courtesy of the Aquarius Reef Base/NOAA; p. 227 David Littschwager/National Geographic Stock (x2); p. 228 D.P. Wilson/FLPA/ Science Source; p. 229 David Littschwager/National Geographic Stock (x5); p. 231 Courtesy NOAA's Aquarius Reef Base at Florida International University; pp. 232–233 (clockwise from top right) Mauricio Handler/National Geographic Stock (coral reef); Mauricio Handler/National Geographic Stock (open ocean); Doug Perrine/Peter Arnold/Getty Images (meso zones); Jeff Rotman/ The Image Bank/Getty Images (hadal zone); George Grall/National Geographic Stock (intertidal zone); Michael Fay/National Geographic Stock (estuary); p. 235 Tim Laman/National Geographic Stock; p. 236 Erik Olsen/The New York Times/Redux; p. 237 (clockwise from top left) Carlos Villoch/age fotostock (dynamite fishing); Mike Theiss/National Geographic Stock (pollution); Mike Nelson/EPA/Corbis (fish near plastic bag); Dave Martin/Associated Press (oil on wave); Bill Curtsinger/National Geographic Stock (overfishing); Jurgen Freund/Nature Picture Library (cyanide fishing); p. 239 (top) Brian J. Skerry/National Geographic Stock; (bottom) Natureworld/Alamy.

CHAPTER 14:
pp. 242–243 Johannes Arlt/laif/Redux; p. 244 Courtesy of The Rooms, VA21-18, Provincial Archives Division of Newfoundland and Labrador; p. 245 Rare Books Division, The New York Public Library, Astor, Lenox and Tilden Foundations; p. 248 Mark Edwards/stillpictures/Aurora Photos; p. 252 Edwin Remsberg; p. 253 Gyro Photo/Amana Images/Getty Images; p. 254 Courtesy Ocean Farm Technologies; p. 256 Suzanne DeChillo/The New York Times/Redux; p. 257 (left to right) Marine Stewardship Council; Liang Sen/Xinhua/Zuma Press.

CHAPTER 15:
pp. 260–261 Stephanie Roland; p. 263 Courtesy of Mark Greening/Orange County Water District; p. 266 Zuma Press; p. 269 Stephanie Roland; p. 271 David Howell; p. 273 North Light Images/age fotostock; p. 275 iStockphoto/Thinkstock.

CHAPTER 16:
pp. 278–279 MODIS Rapid Response/NASA Earth Observatory; p. 281 (top) Deborah Baic/The Globe and Mail/Canada Press Images; (bottom) Lynn Betts/NRSC/USDA; p. 283 Don Johnston/age fotostock/Getty Images; p. 284 Bettmann/Corbis; p. 285 NASA; p. 288 (top) Jochen Schlenker/Robert Harding World Imagery/Getty Images; (bottom) Kilian Fichou/AFP/Newscom; p. 292 Peter Wynn Thompson/New York Times/Redux; p. 293 Robert Brook/Science Source.

CHAPTER 17:
pp. 296–297 iStockphoto/Thinkstock; p. 299 Courtesy Giora Proskurowski, Sea Education Association (x2); p. 301 Courtesy Giora Proskurowski, Sea Education Association; p. 302 Jens Grossmann/laif/Redux; p. 305 Liang Xu/Xinhua/Zuma Press; p. 306 Ingrid N. Visser/Auscape/The Image Works; p. 307 (top) Alan Marsh/Wave/Corbis; (middle) Chris Price/iStockphoto; (bottom) Mode/MW/Aurora Photos; p. 309 Sarah Leen/National Geographic Stock.

CHAPTER 18:
pp. 314–315 Béatrice Jaud, from the film *Severn the voice of our children* – J+B Séquences; p. 317 (clockwise from top left) Frank Lukasseck/Corbis; Dr. Keith Wheeler/Science Photo Library/Science Source; Hans-Juergen Burkard/eyevine/Redux; Eyecandy Images/Corbis; p. 319 Courtesy Greg Massa; p. 321 (top) Reed Kaestner/Corbis; (bottom) AgStock Images, Inc./Alamy; p. 323 Anjuli Ayer; p. 324 (clockwise from top left) J. Irwin/Robertstock/Aurora Photos (contour farming); Bill Barksdale/AGStockUSA; (reduced tillage); Matt Meadows/Peter Arnold/Getty Images (crop rotation); Brian Gordon/Green/National Geographic Stock (cover crops); Andrew Holt/Photographer's Choice/Getty Images (strip cropping); Frans Lanting/National Geographic Stock (terrace farming); p. 326 Courtesy Massa Organics; p. 328 Lyroky/Alamy.

CHAPTER 19:
pp. 332–333 George Steinmetz/Corbis; p. 334 Antrim Caskey; p. 335 Les Stone; p. 337 Les Stone; p. 338 Antrim Caskey; p. 341 (clockwise from top left) Les Stone; Roger L. Wollenberg/UPI/Newscom; Ron Sachs/CNP/Newscom; Paula Smith/

Alamy; p. 342 Rick Eglinton/Zuma Press; p. 343 Antrim Caskey; p. 347 Antrim Caskey.

CHAPTER 20:
pp. 352–353 The Washington Post/Getty Images; p. 355 (left) Emory Kristof/National Geographic Stock; (right) Todd Korol/Aurora Photos; p. 360 Veronique de Viguerie/Getty Images; p. 363 (clockwise from top left) Greg Wood/AFP/Getty Images; Peter Essick/Aurora Photos; Stephanie Maze/National Geographic Stock; Jon Lowenstein/NOOR/Redux; Joel Sartore/National Geographic Stock; Gerald Herbert/Associated Press; p. 365 Mark Ralston/AFP/Getty Images; p. 366 Barry Williams/Getty Images; p. 367 Jan Sonnenmair/Aurora Photos/Alamy.

CHAPTER 21:
pp. 370–371 Danny Wilcox Frazier/Redux; p. 373 Justin Lambert/Getty Images; p. 375 Janet Foster/Masterfile; p. 379 Chang W. Lee/The New York Times/Redux; p. 384 (left) David Hay Jones/Science Source; (right) Tony Craddock/Science Photo Library/Science Source; p. 385 Ernie Janes/Balance/Photoshot/Zuma Press; p. 387 (top) Pierre Gleizes/REA/Redux; (bottom) Goodshoot/Thinkstock.

CHAPTER 22:
pp. 390–391 Layne Kennedy; p. 393 Layne Kennedy; p. 395 James Balog/Aurora Photos; p. 400 Martin Shields/Science Photo Library/Science Source; p. 403 British Antarctic Survey/Science Photo Library/Science Source; p. 404 NASA/Goddard Scientific Visualization Studio; pp. 406–407 (left to right) Jim Toomey/age fotostock (white fir); Altrendo Nature/Thinkstock (aspen tree); Patrick Palmen/Hollandse Hoogte/Redux (Pied Flycatcher); Bill Draker/ImageBroker/Getty Images (Blue-gray Gnatcatcher); Lars Howlett/Aurora Photos (red maple); Jack Thomas/Alamy (pine beetle).

CHAPTER 23:
pp. 414–415 David Guttenfelder/Associated Press; p. 416 (top) David Guttenfelder/Associated Press; (bottom) Digital Globe/Abaca Press/Newscom; p. 423 David Guttenfelder/Associated Press; p. 427 (top to bottom) Justin Jin/Panos Pictures; SuperStock/Alamy; David Hancock-Skyscans/Auscape/The Image Works; p. 428 Christoph Bangert/Stern/laif/Redux; p. 429 Mark Henley/Panos Pictures; p. 431 Toby Talbot/Associated Press.

CHAPTER 24:
pp. 434–435 Alessandro Grassani/Invision/Aurora Photos; p. 437 Andrew Henderson/National Geographic Stock; p. 439 (top) Dang Ngo/ZUMAPRESS.com; (bottom) Stefan Falke/laif/Redux; p. 440 Alessandro Grassani/Invision/Aurora Photos; p. 441 Alessandro Grassani/Invision/Aurora Photos; p. 442 (left) Gilles Rolle/REA/Redux; (right) Santa Rosa Press Democrat/ZUMAPRESS.com; p. 445 (left) Pep Bonet/Noor/Redux; (right) Oliver Perrot/SIPA; p. 446 Stephanie Roland; p. 447 Gary Weathers/Tetra Images/Getty Images; p. 449 iStockphoto/Thinkstock.

CHAPTER 25:
pp. 452–453 Ben Lowy/Getty Images; p. 454 (top) Courtesy David Tilman; (bottom) Tim Rummelhoff; p. 456 Courtesy Robert Williams; p. 458 Everett Kennedy Brown/EPA/Newscom; p. 461 Ben Lowy/Getty Images; p. 464 David Maung/Bloomberg/Getty Images; p. 465 (left) Sherri Barber/The Coloradoan/Associated Press; (right) Antonio Luis Campos/4SEE/Redux; p. 467 Tracy A. Woodward/The Washington Post/Getty Images; p. 468 (left) Yuriko Nakao/Reuters/Landov; (right) Kfir Sivan/Israel Sun/Landov; p. 469 Ashley Cooper/Corbis.

CHAPTER 26:
pp. 472–473 Nina Berman/Noor/Redux; p. 475 Nina Berman/Noor/Redux; p. 479 Nina Berman/Noor/Redux; p. 482 Nina Berman/Noor/Redux; p. 483 Ken Gillespie Photography/Alamy; p. 487 Monty Rakusen/Cultura RF/Getty Images.

INFORMATION GRAPHICS
MGMT. design

RENDERED ILLUSTRATIONS
Nicolle Rager Fuller, Sayo-Art LLC

INDEX

B

D

E

G

gamma radiation, 425, 425f

H

L

M

N

O

P

Q

R

S

T